Aircraft Dynamics: From Modeling to Simulation

Marcello R. Napolitano, Professor
Department of Mechanical and Aerospace Engineering
Flight Control Research Laboratory, Director
West Virginia University

WILEY

John Wiley & Sons, Inc.

VP AND PUBLISHER	Don Fowley
EXECUTIVE EDITOR	Linda Ratts
EDITORIAL ASSISTANT	Christopher Teja
SENIOR MARKETING MANAGER	Christopher Ruel
PRODUCTION MANAGER	Janis Soo
SENIOR PRODUCTION EDITOR	Joyce Poh

This book was set in 10/12 Times Roman by MPS Limited, a Macmillan Company, and printed and bound by Courier Westford. The cover was printed by Courier Westford.

This book is printed on acid free paper.

Founded in 1807, John Wiley & Sons, Inc. has been a valued source of knowledge and understanding for more than 200 years, helping people around the world meet their needs and fulfill their aspirations. Our company is built on a foundation of principles that include responsibility to the communities we serve and where we live and work. In 2008, we launched a Corporate Citizenship Initiative, a global effort to address the environmental, social, economic, and ethical challenges we face in our business. Among the issues we are addressing are carbon impact, paper specifications and procurement, ethical conduct within our business and among our vendors, and community and charitable support. For more information, please visit our website: www.wiley.com/go/citizenship.

Copyright © 2012 John Wiley & Sons, Inc. All rights reserved. No part of this publication may be reproduced, stored in a retrieval system or transmitted in any form or by any means, electronic, mechanical, photocopying, recording, scanning or otherwise, except as permitted under Sections 107 or 108 of the 1976 United States Copyright Act, without either the prior written permission of the Publisher, or authorization through payment of the appropriate per-copy fee to the Copyright Clearance Center, Inc. 222 Rosewood Drive, Danvers, MA 01923, website www.copyright.com. Requests to the Publisher for permission should be addressed to the Permissions Department, John Wiley & Sons, Inc., 111 River Street, Hoboken, NJ 07030-5774, (201)748-6011, fax (201)748-6008, website http://www.wiley.com/go/permissions.

Evaluation copies are provided to qualified academics and professionals for review purposes only, for use in their courses during the next academic year. These copies are licensed and may not be sold or transferred to a third party. Upon completion of the review period, please return the evaluation copy to Wiley. Return instructions and a free of charge return shipping label are available at www.wiley.com/go/returnlabel. Outside of the United States, please contact your local representative.

Library of Congress Cataloging-in-Publication Data

Napolitano, Marcello R.
    Aircraft dynamics : from modeling to simulation / Marcello R. Napolitano.
      p. cm.
    Includes bibliographical references and index.
    ISBN 978-0-470-62667-2 (hardback : acid-free paper)
      1. Aerodynamics—Mathematics. 2. Airplanes—Simulation methods. 3. Structural dynamics. I. Title.
    TL570.N36 2012
    629.132′30015118—dc23

                                                                                             2011029934

Printed in the United States of America

10 9 8 7 6 5 4 3 2 1

. . . . Dedicato a Zio Piero !

Table of Contents

1. Aircraft Equations of Motion   1

    1.1   Introduction   1
    1.2   Reference Frames and Assumptions   2
    1.3   Conservation of the Linear Momentum Equations (CLMEs)   3
    1.4   Conservation of the Angular Momentum Equations (CAMEs)   6
    1.5   Conservation of the Angular Momentum Equations (CAMEs) with Rotor Effects   10
    1.6   Euler Angles   11
    1.7   Flight Path Equations (FPEs)   12
    1.8   Kinematic Equations (KEs)   14
    1.9   Gravity Equations (GEs)   16
    1.10   Summary of the Aircraft Equations of Motion   16
    1.11   Definition of Steady-State and Perturbation Conditions   17
    1.12   Aircraft Equations of Motion at Steady-State Conditions   18
    1.13   Aircraft Equations of Motion at Perturbed Conditions   19
    1.14   Small Perturbation Equations from a Steady-State Level Flight   22
    1.15   Summary   23
    References   26
    Student Sample Problems   26
    Problems   32

2. Review of Basic Concepts of Aerodynamic Modeling   37

    2.1   Introduction   37
    2.2   Review of Key Aerodynamic Characteristics for Wing Sections   37
    2.3   Wing Planforms and Wing Lift Curve Slope   42
    2.4   Review of the Downwash Effect and Effectiveness of Control Surfaces   48
    2.5   Determination of the Aerodynamic Center for Wing and Wing + Fuselage   53
    2.6   Approaches to the Modeling of Aerodynamic Forces and Moments   57
        2.6.1   Wind Tunnel Analysis   57
        2.6.2   CFD Analysis   58
        2.6.3   Parameter IDentification from Flight Data   59
        2.6.4   Correlation from Wind Tunnel Data and Empirical "Build-Up" Analysis   60
    2.7   Summary   60
    References   61
    Student Sample Problems   62
    Problems   75

3. Modeling of Longitudinal Aerodynamic Forces and Moments   78

    3.1   Introduction   78
    3.2   Aircraft Stability Axes   79
    3.3   Modeling of the Longitudinal Steady-State Aerodynamic Forces and Moment   79
    3.4   Modeling of $F_{A_{X_1}}$ 80
    3.5   Modeling of $F_{A_{Z_1}}$ 83
    3.6   Modeling of M_{A_1} 87
    3.7   Aircraft Aerodynamic Center   89
    3.8   Summary of the Longitudinal Steady-State Aerodynamic Forces and Moment   91
    3.9   Modeling of the Longitudinal Small Perturbation Aerodynamic Forces and Moments   91
        3.9.1   Modeling of $(c_{D_1}, c_{L_1}, c_{m_1})$ 93
        3.9.2   Modeling of $(c_{D_u}, c_{L_u}, c_{m_u})$ 93
        3.9.3   Modeling of $(c_{D_{\dot{\alpha}}}, c_{L_{\dot{\alpha}}}, c_{m_{\dot{\alpha}}})$ and $(c_{D_q}, c_{L_q}, c_{m_q})$ 94
    3.10   Summary of Longitudinal Stability and Control Derivatives   96

3.11 Summary   100
References   100
Student Sample Problems   101
Case Study   110
Short Problems   127
Problems   128

4. Modeling of Lateral Directional Aerodynamic Forces and Moments   135

4.1 Introduction   135
4.2 Modeling of $F_{A_{Y_1}}$ 137
    4.2.1 Conceptual Modeling of c_{Y_β} 138
    4.2.2 Mathematical Modeling of c_{Y_β} 140
    4.2.3 Modeling of $c_{Y_{\delta_A}}$ 147
    4.2.4 Modeling of $c_{Y_{\delta_R}}$ 147
4.3 Modeling of L_{A_1} 149
    4.3.1 Conceptual Modeling of $c_{l\beta}$ 150
    4.3.2 Mathematical Modeling of $c_{l\beta}$ 155
    4.3.3 Modeling of $c_{l_{\delta_A}}$ 160
    4.3.4 Modeling of $c_{l_{\delta_R}}$ 166
4.4 Modeling of N_{A_1} 168
    4.4.1 Conceptual Modeling of $c_{n\beta}$ 169
    4.4.2 Mathematical Modeling of $c_{n\beta}$ 172
    4.4.3 Modeling of $c_{n_{\delta_A}}$ 174
    4.4.4 Modeling of $c_{n_{\delta_R}}$ 176
4.5 Summary of the Lateral Directional Steady-State Force and Moments   177
4.6 Modeling of the Small Perturbation Lateral Directional Aerodynamic Force and Moments   178
    4.6.1 Modeling of $c_{Y_{\dot\beta}}, c_{l_{\dot\beta}}, c_{n_{\dot\beta}}$ 180
    4.6.2 Modeling of c_{Y_p} 180
    4.6.3 Modeling of c_{l_p} 181
    4.6.4 Modeling of c_{n_p} 183
    4.6.5 Modeling of c_{Y_r} 185
    4.6.6 Modeling of c_{l_r} 185
    4.6.7 Modeling of c_{n_r} 187
4.7 Summary of Longitudinal and Lateral Directional Aerodynamic Stability and Control Derivatives   189
4.8 Final Overview and Ranking of the Importance of the Aerodynamic Coefficients   196
4.9 Summary of the Modeling of the Longitudinal and Lateral-Directional Aerodynamic Forces and Moments   198
References   200
Student Sample Problems   200
Case Study   236
Short Problems   262
Problems   263

5. Review of Basic Aircraft Performance and Modeling of Thrust Forces and Moments   268

5.1 Introduction   268
5.2 Review of Different Aircraft Propulsion Systems   268
    5.2.1 Piston Engine (Propeller) Aircraft Engines   269
    5.2.2 Turboprop Aircraft Engines   270
    5.2.3 Turbojet Aircraft Engines   271
    5.2.4 Turbofan Aircraft Engines   272
    5.2.5 Ramjet Aircraft Engines   273
5.3 Review of Basic Aircraft Performance   273

5.4 Power at Level Flight   274
    5.4.1 Maximum Aerodynamic Efficiency   275
    5.4.2 Minimum Aerodynamic Drag   275
    5.4.3 Minimum Power Required   277
5.5 Determination of Power Required   279
5.6 Determination of Power Available   282
5.7 Modeling of the Thrust Forces and Moments   287
    5.7.1 Modeling of the Steady-State Thrust Forces and Moments   288
    5.7.2 Modeling of the Small Perturbation Thrust Forces and Moments   291
5.8 Summary   294
References   296
Student Sample Problems   296
Problems   304

6. Aircraft Stability and Design for Trim Conditions   305

6.1 Introduction   305
6.2 Concept of Aircraft Stability   305
6.3 Criteria for Aircraft Static Stability   306
    6.3.1 Static Stability Criteria #1 (SSC #1)   307
    6.3.2 Static Stability Criteria #2 (SSC #2)   308
    6.3.3 Static Stability Criteria #3 (SSC #3)   308
    6.3.4 Static Stability Criteria #4 (SSC #4)   309
    6.3.5 Static Stability Criteria #5 (SSC #5)   310
    6.3.6 Static Stability Criteria #5, #6, and #7 (SSC #5, SSC #6, SSC #7)   311
    6.3.7 Static Stability Criteria #9 (SSC #9)   311
    6.3.8 Static Stability Criteria #10 (SSC #10)   312
6.4 Longitudinal Analysis of Steady-State Straight Flight   313
6.5 Lift Chart and Trim Diagram   322
    6.5.1 Lift Chart   322
    6.5.2 Trim Diagram   324
    6.5.3 Trim Diagrams for Different Classes of Aircraft   329
    6.5.4 Trim Diagrams for Thrust Axis Above/Below Center of Gravity   329
6.6 Lateral Directional Analysis of Steady-State Straight Flight   332
6.7 Summary   340
References   340
Student Sample Problems   340
Problems   349

7. Solution of the Aircraft Equations of Motion Based on Laplace Transformations and Transfer Functions   352

7.1 Introduction   352
7.2 Application of Laplace Transformations to the Longitudinal Small Perturbation Equations   353
7.3 Routh–Hurwitz Analysis of the Longitudinal Stability   358
7.4 Longitudinal Dynamic Modes: Short Period and Phugoid   360
7.5 Solution of the Longitudinal Equations   361
7.6 Short Period Approximation   363
7.7 Phugoid Approximation   366
7.8 Summary of the Longitudinal Equations   369
7.9 Application of Laplace Transformations to the Lateral Directional Small Perturbation Equations   371
7.10 Routh–Hurwitz Analysis of the Lateral Directional Stability   376
7.11 Lateral Directional Dynamic Modes: Rolling, Spiral, and Dutch Roll   377
7.12 Solution of the Lateral Directional Equations   379

7.13 Rolling Approximation   382
7.14 Summary of Lateral Directional Equations   385
7.15 Sensitivity Analysis for the Aircraft Dynamics   386
    7.15.1 Short Period Sensitivity Analysis   387
    7.15.2 Phugoid Sensitivity Analysis   394
    7.15.3 Sensitivity Analysis for the Lateral Directional Parameters   398
7.16 Summary   407
References   407
Student Sample Problems   407
Problems   430

8. State Variable Modeling of the Aircraft Dynamics   432

8.1 Introduction   432
8.2 Introduction to State Variables for Nonlinear Systems   433
8.3 Introduction to State Variables for Linear/Linearized Systems   433
8.4 State Variable Modeling of the Longitudinal Dynamics   435
8.5 State Variable Modeling of the Lateral Directional Dynamics   440
8.6 Augmentation of the Aircraft State Variable Modeling   445
    8.6.1 Modeling of the Altitude (h)   445
    8.6.2 Modeling of the Flight Path Angle (γ)   446
    8.6.3 Modeling of the Engine Dynamics   446
    8.6.4 Modeling of the Actuator Dynamics   446
    8.6.5 Modeling of the Atmospheric Turbulence   447
8.7 Summary of State Variable Modeling of the Aircraft Dynamics   447
8.8 Summary   450
References   450
Student Sample Problems   450
Problems   470

9. Introduction to Modern Flight Simulation Codes   476

9.1 Introduction   476
9.2 Introduction to the Flight Dynamics & Control (FDC) Toolbox   479
    9.2.1 Equations of Motion within the FDC Simulation Environment   479
    9.2.2 FDC Modeling of Beaver Aerodynamic Forces and Moments   483
    9.2.3 Alternative Approach for FDC Modeling of Aerodynamic Forces and Moments   485
    9.2.4 Case Study #1: FDC Modeling of Look-Up Tables Based Aerodynamic Coefficients   486
    9.2.5 FDC Modeling of the Gravity Force   493
    9.2.6 FDC Modeling of the Atmospheric Turbulence Force   493
    9.2.7 FDC Modeling of the Beaver Propulsive Forces and Moments   494
    9.2.8 Case Study #2: FDC Modeling of Propulsive Forces and Moments   496
    9.2.9 Auxiliary FDC Blocks   498
    9.2.10 Additional FDC Blocks   503
9.3 Introduction to the Aerospace Blockset by Mathworks   503
    9.3.1 General Organization of the Aerospace Blockset   503
    9.3.2 Introduction to the Environment Library   504
    9.3.3 Introduction to the Flight Parameters Library   506
    9.3.4 Introduction to the Equations of Motion Library   506
    9.3.5 Introduction to the Aerodynamics Library   508
    9.3.6 Introduction to the Propulsion Library   508
    9.3.7 Introduction to the Utilities Library   509
    9.3.8 Introduction to the Mass Properties Library   510
    9.3.9 Introduction to the Actuators Library   511
    9.3.10 Introduction to the GNC and Animation Libraries   511

9.4 Introduction to AIRLIB 512
    9.4.1 AIRLIB's Strucure 512
    9.4.2 Generic Aircraft Model: Continuous-time Block 512
    9.4.3 Generic Aircraft Model: Discrete-time Block 515
    9.4.4 Collection of Aircraft Models 516
    9.4.5 Alternative Model Implementation 517
    9.4.6 Additional Tools within AIRLIB: The Function '*air3m*' 517
    9.4.7 Additional Tools within AIRLIB: The Function '*ab2dv*' 518
9.5 Summary 518
References 518
Student Sample Problems 519

10. Pilot Ratings and Aircraft Handling Qualities 523

10.1 Introduction 523
10.2 Aircraft Flight Envelope 524
10.3 Levels of Aircraft Flying Qualities: Cooper-Harper Pilot Rating 526
    10.3.1 Aircraft Control Authority 526
    10.3.2 Pilot Workload 526
    10.3.3 Pilot Compensation 529
    10.3.4 Levels of Flying Qualities 529
10.4 Classes of Aircraft 531
10.5 Classification of Aircraft Maneuvers and Mission Profile 531
10.6 Flying Quality Requirements for the Longitudinal Dynamics 532
    10.6.1 Longitudinal Control Forces 533
    10.6.2 Requirements for the Damping for the Phugoid Mode 535
    10.6.3 Requirements for the Short Period Mode 536
10.7 Flying Quality Requirements for the Lateral Directional Dynamics 536
    10.7.1 Lateral Directional Control Forces 536
    10.7.2 Requirements for the Dutch Roll Mode 538
    10.7.3 Requirements for the Spiral Mode 539
    10.7.4 Requirements for the Rolling Mode 539
    10.7.5 Requirements for the Roll Control Effectiveness 539
    10.7.6 Additional Requirements for Steady Sideslips 541
10.8 Summary 541
References 541

Appendix A Review of Useful Topics 543

Appendix A.1 Review of Vector Operations 544
Appendix A.2 Review of Matrix Operations 548
Appendix A.3 Review of Center of Gravity and Inertial Properties 558
Appendix A.4 Review of Application of Laplace Transform to Linear Constant Coefficients Differential Equations 564
Appendix A.5 Review of First and Second Order Systems 575
Appendix A.6 Review of Standard Atmospheric Model 581

Appendix B Data for Different Aircraft 584

Appendix B.1 Introduction 584
Appendix B.2 Aircraft 1—Cessna 182 586
Appendix B.3 Aircraft 2—Cessna 310 589
Appendix B.4 Aircraft 3—Beech 99 592
Appendix B.5 Aircraft 4—Cessna T37-A 595
Appendix B.6 Aircraft 5—Cessna 620 598
Appendix B.7 Aircraft 6—Learjet 24 601

Appendix B.8 Aircraft 7—Boeing 747-200 604
Appendix B.9 Aircraft 8—SIAI Marchetti S-211 607
Appendix B.10 Aircraft 9—Lockheed F-104 610
Appendix B.11 Aircraft 10—McDonnell Douglas F-4 613
Reference 615

Appendix C Detailed Drawings for Different Aircraft 616

Appendix C.1 Introduction 617
Appendix C.2 Aircraft 1—Aeritalia Fiat G-91 618
Appendix C.3 Aircraft 2—Beech 99 621
Appendix C.4 Aircraft 3—Boeing B52 624
Appendix C.5 Aircraft 4—Boeing B727-200 627
Appendix C.6 Aircraft 5—Boeing B737-600 630
Appendix C.7 Aircraft 6—Boeing B747-200 633
Appendix C.8 Aircraft 7—Boeing B757-200 637
Appendix C.9 Aircraft 8—Boeing B767-200 640
Appendix C.10 Aircraft 9—Cessna Citation CJ3 643
Appendix C.11 Aircraft 10—Cessna T37 645
Appendix C.12 Aircraft 11—General Dynamics F-16 649
Appendix C.13 Aircraft 12—Grumman F-14 652
Appendix C.14 Aircraft 13—Learjet 24 655
Appendix C.15 Aircraft 14—Lockheed F-104 658
Appendix C.16 Aircraft 15—Lockheed F-22 661
Appendix C.17 Aircraft 16—Lockheed L-1011 664
Appendix C.18 Aircraft 17—McDonnell Douglas C-17 667
Appendix C.19 Aircraft 18—McDonnell Douglas DC-8 670
Appendix C.20 Aircraft 19.1—McDonnell Douglas DC-9 Series 10 673
Appendix C.21 Aircraft 19.2—McDonnell Douglas DC-9 Series 30 677
Appendix C.22 Aircraft 19.3—McDonnell Douglas DC-9 Series 40 679
Appendix C.23 Aircraft 19.4—McDonnell Douglas DC-9 Series 50 681
Appendix C.24 Aircraft 20—McDonnell Douglas DC-10 683
Appendix C.25 Aircraft 21—McDonnell Douglas F-4 686
Appendix C.26 Aircraft 22—McDonnell Douglas F-15 689
Appendix C.27 Aircraft 23—Rockwell B-1 692
Appendix C.28 Aircraft 24—SIAI Marchetti S211 695
Appendix C.29 Aircraft 25—Supermarine Spitfire 699

Index 703

Acknowledgments

First, I would like to thank my dear Zio Piero for all his wisdom and directions in my life. I know you are smiling up there with your "*sorriso da toscanaccio.*"

I would like to thank my family for tolerating and understanding my many evenings, nights, and weekends of seclusion in my office while working on this major editorial effort.

I also would like to thank all my colleagues all over the world for encouraging me in this project. Special thanks to my former student Dr. David Doman, who took time from a very busy schedule to provide me with a very useful review.

Next, I would like to thank all my past, current, and future students in the Aerospace Engineering curriculum at West Virginia University. I will never forget the following comment from a Student Evaluation of Instructor form "*Napy: I know you are trying your best but it is tough to learn when you need four or five level of subscripts to describe an aerodynamic coefficient.*" This book has been conceived with the goal of making the learning of a tough subject as smooth and as friendly as possible.

Finally, I would like to thank my graduate student Andres Felipe Velasquez for volunteering his very precious editorial assistance.

Introduction

This textbook was conceived and written for undergraduate aeronautical and aerospace engineering students for a course at junior or senior undergraduate level. This course can be later followed by senior or graduate courses in design of flight control systems as well as flight simulation.

As for most of the authors, my main motivation for writing a text book originates from the experience and the teaching challenges encountered in more than twenty years as a faculty member of the department of Mechanical and Aerospace Engineering at West Virginia University.

There are some disciplines which are continuously evolving; flight dynamics for a rigid body aircraft with conventional design at subsonic airspeeds is *not* one of them. In fact, the aircraft dynamic characteristics have been well understood and have virtually remained unchanged since the introduction of jet propulsion. However, the tools available today to aerospace engineering students and professionals for modeling and simulation of the aircraft dynamics are much more advanced with respect to only a few years ago. Nevertheless, most of the available textbooks do not challenge our students to take advantage of these new classes of capabilities.

The first motivation is to introduce a textbook 'designed' to take advantage of the extensive computational resources commonly available to today's students. MATLAB® and its several toolboxes, including Simulink®, have totally changed the landscape in the education of flight dynamics/simulation. Today's students are now able to develop their own simple but accurate simulation codes, with or without graphics, on their own laptops. These capabilities were confined to a few high-performance work stations only a few years ago. However, most of the available textbooks were introduced *before* this computational revolution, which started in the late 1980s/early 1990s. This textbook has been developed with the specific objective of capitalizing on the widespread availability of these capabilities. Over the years I have found that the teaching effectiveness greatly improves if the instructor clearly sets as a goal for the student the development of these simulation capabilities; this ultimately provides the students with a feeling of empowerment, which greatly benefits their learning performance.

The second motivation is to introduce a textbook to help students to be able to extrapolate from low-level formulas, equations, and details to high-level comprehensive views of the main concepts. Throughout my twenty+-years long experience I realized that a large percentage of the students feel overwhelmed by the level of complexity in most textbooks on this topic. Thus, it is common for them to lose the "big picture." The task of providing direction to our students in a jungle of symbols is typically left to the discretion and mastery of the instructors. A trademark of this textbook is a number of charts, block diagrams, directional prospects, and other tools with the clear objective of helping them to understand the big picture. The goal is to assist them in a process where their minds have to be able to go from the small detail to the highest level view, from a view of the leaf to a view of the forest.

The third motivation is to introduce a textbook to help students in the fundamental skills of learning the basic modeling of aircraft aerodynamics and dynamics. A specific objective is to provide basic understanding of *who does what* in aircraft aerodynamic modeling with references to effects on the aircraft dynamic stability. This process is quite delicate, since this modeling requires an extensive amount of symbols with several levels of subscripts; furthermore, at the end of this modeling, it is also fundamental to provide the students with a ranking of the significance of the individual coefficients in the overall aerodynamic modeling. The importance of this process is fundamental for the analysis of the simulated flight data; in fact, based on the dynamic responses, the students need to have the skills of drawing a direct correlation between aerodynamic modeling and aircraft handling qualities through a number of sensitivity analysis tools.

A fourth motivation is to introduce a textbook specifically designed for a junior-level student. This issue might not be immediately clear but it has major implications for enhancing the learning outcomes by the students. Interactions and discussions with several colleagues from several universities have confirmed this point. Approximately twenty years ago a "Flight Dynamics" course was typically taught at senior level in most academic institutions. However, in the last twenty years a number of courses have been introduced in the curriculum of most aeronautical programs. Courses such as "Design of Flight Control Systems," "UAV Design/Build/Fly," "Flight Testing," and "Flight Simulation" are becoming more and more popular in the aeronautical/aerospace curriculums across U.S. academic institutions and throughout the world to address specific needs by the aerospace industry to hire graduates with capabilities in those areas. Since each of these courses requires a detailed understanding of the open-loop aircraft dynamics, the trend has been to anticipate the "Flight Dynamics" course at junior level.

However, the majority of the available textbooks in "Flight Dynamics" are typically directed toward a more mature audience of senior-level and even graduate-level students. Therefore, a discrepancy exists between the level of students taking "Flight Dynamics" courses and the level for which most of the available textbooks are written.

A fifth motivation is to introduce an "instructor friendly" textbook.

This textbook has been designed with a variety of features to assist the instructors. A key feature of this textbook is a variety of problems introduced for almost all the chapters. Each of the key chapters has a number of Student Sample Problems at the end of the chapter followed by a number of problems to be used for homework assignment. Special emphasis has been placed on the problems for Chapters 3 and 4. Given the level of complexity of the aerodynamic modeling in those chapters, a very detailed Case Study and a number of Short Problems have been introduced before the Problems. The Problems can involve a substantial level of effort and could be used for large or group assignments while the Short Problems can be used for shorter assignments. Instructors are suggested to select a specific aircraft and assign the sequential problems relative to the specific aircraft such that a number of specific parameters can be used from previous assignments into the future assignments. The number of problems is such that an instructor has a variety of similar problems for different aircraft. Another useful feature is a number of very useful MATLAB® codes in Chapters 7 and 8. Finally, a large and complete set of very detailed PowerPoint-based class notes are available to all the instructors adopting the textbook.

In terms of content, approach, and organization, the main objective is to organize the topics into modular blocks, each of them leading to the understanding of the inner mechanisms of the aircraft aerodynamics and dynamics, eventually leading to the development of simple flight simulation schemes. The title of the textbook, *Aircraft Dynamics: From Modeling to Simulation*, describes this general philosophy.

The book is organized into ten chapters and three Appendixes.

Chapter 1 provides a comprehensive analysis of the process leading to the derivation of the fifteen scalar equations needed to provide a simple simulation of aircraft dynamics, both with respect to a body frame and with respect to an Earth-fixed frame. The chapter starts from Newton's second law and gradually builds up to the full set of equations. Several figures are provided to explain all the different forces and moments involved in the equations of motion. Additionally, a number of block diagrams are provided to show the integrations of all the different sets of equations.

Chapter 2 provides a review of the basic concepts from aerodynamics. The chapter starts with a review of aerodynamic concepts for a wing section, followed by their extension to the aircraft wing up to the extension to the full-size aircraft. An important goal of this chapter is to provide students with a general overview of the different approaches available for the modeling of the aerodynamic forces and moments.

Chapter 3 and **Chapter 4** provide a very detailed modeling of the longitudinal and lateral directional aerodynamic forces and moments acting on the aircraft using the DATCOM-based build-up approach. The individual contributions leading to the modeling for each of the aerodynamic forces and moments are introduced with details. An extensive number of figures are used for the purpose of visualizing these contributions. Major emphasis is placed on clearly introducing these coefficients to avoid confusion, given the overwhelming number of symbols with different levels of subscripts. Several detailed summary charts are provided in both chapters.

Chapter 5 provides a brief review of aircraft propulsion systems followed by a review of basic aircraft general and propulsive performance leading to the modeling of the thrust forces and moments.

Chapter 6 provides a description of the criteria to be satisfied for aircraft static stability followed by a detailed discussion of design for trim conditions and trim diagrams. Emphasis is placed on a detailed explanation of the trim diagram as an important tool to understand a number of issues relative to the general stability of the aircraft as well as a clear understanding of the balancing envelope for a given design. The chapter is concluded by a detailed analysis of the lateral directional dynamics in the case of engine(s)-out conditions.

Chapter 7 is a critical chapter where the students are introduced to the actual solution of the equations of motion of the aircraft. First, both sets of longitudinal and lateral directional equations, reduced to linear constant coefficients differential equations using small perturbation assumptions, are introduced and solved using the Laplace transformation based on the concept of transfer functions. The modes for both the longitudinal and the lateral directional dynamics are introduced and discussed along with an extension of the transfer function approach to the case of multiple control surfaces. This section is followed by a detailed and complete discussion of the important sensitivity analysis, allowing the student to draw the important correlation, often missing in other textbooks, between the aircraft aerodynamic modeling and the aircraft dynamic response. A few detailed summary charts are provided in this important chapter.

Chapter 8 provides a detailed description of state variable modeling, an important alternative to the simple transfer function approach introduced in Chapter 7. Students are introduced to a more comprehensive modeling of the aircraft dynamics through the state variable model, specifically suitable for the modeling of multi-input, multi-output (MIMO) systems.

Chapter 9 reviews a variety of options for simulating the aircraft dynamics. The objective is to introduce the students to modern flight dynamic simulation tools, such as the freely available FDC (Flight Dynamics and Control) toolbox, the Mathwork Aerospace Blockset toolbox, and the freely available AIRLIB, a library of mathematical models. Students are therefore shown different approaches toward automating the solution of the aircraft equations of motion using Simulink-based models.

Chapter 10 concludes the textbook with a discussion on the concept of flight envelope and handling qualities requirements, specified for classes of aircraft, categories of maneuvers, levels of handling qualities, for both military and civilian aircraft.

Appendix A contains a number of math topics to review analytical tools necessary for a detailed understanding of specific topics in the course.

Appendix B contains a list of geometric characteristics, mass and inertial data, flight conditions, and aerodynamic coefficients for a total of ten different aircraft at specific flight conditions along with their transfer function-based mathematical models.

Appendix C can be considered a unique feature of this textbook. It contains very detailed drawings of twenty-five different aircraft, along with tables showing *all* the different parameters used for the aerodynamic modeling discussed in Chapters 2, 3, and 4. This material provides an ideal data bank for an instructor willing to engage his/her students in substantial efforts in the modeling of the different aerodynamic coefficients.

List of Symbols

ENGLISH

AR	Aspect ratio
A	System matrix (within the state variable model)
b	Wing span
B	Input matrix (within the state variable model)
c	Aerodynamic coefficient
$\bar{c}$	Mean aerodynamic chord
C	Observation matrix (within the state variable model)
D	Observation input matrix (within the state variable model)
D	Aerodynamic drag force
e	Oswald efficiency factor
F	Generic force
g	Gravity acceleration
i	Incidence angle
l	Generic distance
l	Relative to rolling moment
L	Aerodynamic lift force
L	Aerodynamic rolling moment
m	Dimensionless geometric coefficient for the downwash
m	Relative to pitching moment
M	Aerodynamic pitching moment
$Mach$	Mach number
n	Relative to yawing moment
N	Aerodynamic yawing moment
p	Roll angular rate (around the X axis)
q	Pitch angular rate (around the Y axis)
$\bar{q}$	Dynamic pressure
r	Yaw angular rate (around the Z axis)
S	Wing Surface
T	Time
u	Column vector of system inputs
U	Linear longitudinal velocity (along the X axis)
V	Linear lateral velocity (along the Y axis)
V_P	Aircraft airspeed
W	Linear vertical velocity (along the Z axis)
x	Column vector of state variables
x	Generic station on the longitudinal axis X
$\bar{x}$	Generic station on the longitudinal axis X as fraction of $\bar{c}$
X	Longitudinal axis
y	Column vector of measurable outputs
Y	Lateral axis
Z	Directional axis

GREEK

α	Longitudinal angle of attack
β	Lateral angle of attack (a.k.a. sideslip angle)
γ	Climb or descent angle
Λ	Sweep angle
δ	Angular deflection of control surfaces
Δ	Generic difference
ε	Downwash angle
ε	Wing twist angle
ϕ	Euler roll angle
η	Ratio of dynamic pressure
η	Efficiency
Γ	Geometric dihedral angle
ζ	Damping coefficient
λ	Tip ratio
θ	Euler pitch angle
σ	Sidewash angle
τ	Effectiveness of control surface
ψ	Euler yaw angle
ω	Angular velocity
ω	Frequency

SUBSCRIPT

A	Aileron
B	Fuselage (a.k.a body)
E	Elevator
H	Horizontal tail
Lat_Dir	Relative to the lateral directional dynamics
$Long$	Relative to the longitudinal dynamics
R	Root chord
R	Rudder
T	Tip chord
V	Vertical tail
W	Wing
WB	Wing + Body

ACRONYMS

AC	Aerodynamic center
CG	Center of gravity
LE	Leading edge
MAC	Mean aerodynamic chord
TE	Trailing edge

Chapter 1

Aircraft Equations of Motion

TABLE OF CONTENTS

1.1 Introduction
1.2 Reference Frames and Assumptions
1.3 Conservation of the Linear Momentum Equations (CLMEs)
1.4 Conservation of the Angular Momentum Equations (CAMEs)
1.5 Conservation of the Angular Momentum Equations (CAMEs) with Rotor Effects
1.6 Euler Angles
1.7 Flight Path Equations (FPEs)
1.8 Kinematic Equations (KEs)
1.9 Gravity Equations (GEs)
1.10 Summary of the Aircraft Equations of Motion
1.11 Definition of Steady-State and Perturbation Conditions
1.12 Aircraft Equations of Motion at Steady-State Conditions
1.13 Aircraft Equations of Motion at Perturbed Conditions
1.14 Small Perturbation Equations from a Steady-State Level Flight
1.15 Summary
     References
     Student Sample Problems
     Problems

1.1 INTRODUCTION

The study of aircraft dynamics leading to the development of the aircraft equations of motion started after the first flight by the Wright brothers in 1903. The initial theoretical work on this subject by Lanchester and Bryan dates back to 1908 and 1911, respectively.[1,2] Remarkably, the work by Bryan[2] led to the development of the aircraft equations of motion essentially in the same form as they are known today.

The purpose of this chapter is to provide students with a detailed understanding of the different sets of equations used for describing the dynamics of the aircraft system. The equations will be derived with respect to an inertial Earth-based frame as well as a body reference frame.

Starting from the simplest form of Newton's second law, a detailed analysis will be conducted leading to six core scalar equations (from the conservation of the linear and angular momentum), followed by the flight path equations (used for navigation purposes for tracking the aircraft flight with respect to an Earth-based frame), and the kinematic equations (providing a relationship for the Euler angles used for expressing the orientation of the body axes with respect to the inertial ground frame). A final set of equations provides the modeling of the gravity force along the body axes.

Following a detailed derivation of all of these equations, a functional block diagram is introduced with the goal of showing how these equations are integrated within a generic flight simulation environment. Next, the concept of steady-state and perturbed flight conditions are introduced, leading to the assumption of small perturbation, which is critical for a linearization of the aircraft equations of motion.[3] Finally, the conservation of linear and angular momentum equations are re-derived under these conditions, leading to the final sets of equations that will be solved following detailed modeling of the aerodynamic and thrust forces and moments acting on the aircraft.

1.2 REFERENCE FRAMES AND ASSUMPTIONS

The derivation of the equations describing the aircraft dynamics starts from a clear identification of the ground and body reference frames. It is critical to recall that Newton's second law is expressed with respect to an inertial frame, which is a reference frame that is not accelerating or rotating.[4] By standard definition, an *inertial frame is a frame aligned with the fixed stars*. The selection of such a reference frame is realistically unnecessary for describing the motion of a system in the lower regions of the Earth's atmosphere. Therefore, an Earth-based reference system is selected and assumed to be *"inertial enough"* to satisfy the conditions under which Newton's second law is valid. The selection of this frame implies that the effects of the rotational velocity of the earth can be neglected within this context. This approximation is acceptable for subsonic as well as supersonic aircraft; however, this assumption cannot be valid for hypersonic aircraft flying in the upper levels of the atmosphere.

According to the previous discussion, the following reference frames are introduced and shown in Figure 1.1:

Aircraft-based body frame: X, Y, Z
Earth-based inertial frame: X', Y', Z'

The aircraft-based body frame X, Y, Z is located at the center of gravity of the aircraft. For the purpose of the following discussion, the aircraft is considered a *continuous system* with mass, density, volume, and surface indicated, respectively, by m, ρ_A, V, S. Elementary mass, volume, and surface are indicated, respectively, by dm, dV, dS with $dm = \rho_A \, dV$. Therefore, the total mass of the aircraft system can be expressed as follows:

$$m = \int_V \rho_A \, dV$$

The aircraft mass is subjected to the gravity acceleration $\bar{g}$. According to the "flat-Earth" assumption, the $\bar{g}$ vector is aligned with the Z' axis of the Earth-based reference frame. In addition, the following assumptions are in place for the derivation of the aircraft equations of motion:

Assumption 1: The aircraft is assumed to be a rigid body.[2] This implies that the distance between two generic points of the aircraft is time-invariant when measured with respect to the aircraft body frame X, Y, Z. This assumption is realistic for small-size/low-weight aircraft and aircraft with relatively low values of the wing aspect ratio (see Chapter 2). General aviation and fighter aircraft typically satisfy this assumption. However, this assumption is somewhat violated for aircraft experiencing substantial structural elastic deformation during flight, such as large commercial jetliners, military cargo, and bomber aircraft with fairly high values of the wing aspect ratio.

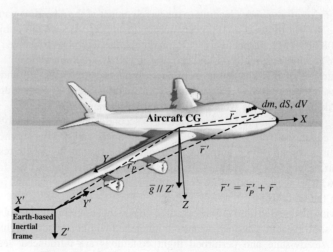

Figure 1.1 Body Axes and Earth-Based Inertial Frame

Assumption 2: The aircraft mass is assumed to be constant, that is $dm/dt = 0$. This assumption might seem unrealistic considering the fuel consumption alone; however, it can be considered acceptable over a limited amount of time. This assumption is clearly not valid for rockets, where a substantial reduction of the mass is experienced in a short amount of time during the launch phase.

Assumption 3: The mass distribution is assumed to be time constant with time. This implies that the inertial characteristics of the aircraft (that is, the moments and the products of inertia) can be assumed constant over a limited amount of time. In reality, these inertial characteristics change with fuel consumption; however, the rate of change for these parameters is low because, by design, the center of gravity of the fuel tanks is located close to the aircraft center of gravity. Nevertheless, rapid changes in the moments of inertia can be experienced due to different reasons, such as fuel slosh, dropping of wing stores for military aircraft, or non-nominal events, such as sudden shifts of cargos within the fuselage. At least one catastrophic accident in recent aviation history has been attributed to the failure of securing cargos in the fuselage, leading to a major shift in the balance of the aircraft, a change of its inertial characteristics, a loss of dynamic stability, and eventually to unrecoverable flight conditions.[5]

1.3 CONSERVATION OF THE LINEAR MOMENTUM EQUATIONS (CLMEs)

The derivation of the conservation of the linear momentum equations starts from the well-known expression for Newton's second law

$$\overline{F} = m\overline{a}$$

where $\overline{a}$ is evaluated with respect to an inertial frame. Considering that the linear momentum is defined as $(m\overline{v})$, a general expression for the conservation of the linear momentum equation is given by

$$\overline{F} = \frac{d}{dt}(m\overline{v}) = m\frac{d\overline{v}}{dt} + \overline{v}\cancel{\frac{dm}{dt}} = m\frac{d\overline{v}}{dt}$$

given the $\frac{dm}{dt} = 0$ assumption

The forces acting on the aircraft are the gravity force, the aerodynamic forces, and the thrust forces. Thus, the previous vectorial equation takes on the form[6]

$$\frac{d}{dt}\int_V \rho_A \frac{d\overline{r}'}{dt} dV = \int_V \rho_A \overline{g}\, dV + \int_S (\overline{F}_A + \overline{F}_T)\, dS$$

The relationship is referred to as *Conservation of the Linear Momentum Equation* (CLME) in the vectorial form with respect to the X', Y', Z' frame. Note that the aerodynamic and thrust forces are here considered as surface forces, whereas gravity is considered as a volumetric force. The position vectors $\overline{r}'$, $\overline{r}'_P$, $\overline{r}$ are shown in Figure 1.1. The relationship among those vectors is given by

$$\overline{r}' = \overline{r}'_P + \overline{r} \quad \Rightarrow \quad \overline{r} = \overline{r}' - \overline{r}'_P$$

Recall from basic concepts of statics that the center of gravity of a system is considered as the point with respect to which the integral of the moments is zero. Therefore, since the aircraft body frame X, Y, Z is located at the center of gravity, the following relationship applies:

$$\int_V \overline{r} \rho_A dV = 0$$

Using $\overline{r} = \overline{r}' - \overline{r}'_P$ leads to

$$\int_V \overline{r} \rho_A dV = \int_V (\overline{r}' - \overline{r}'_P) \rho_A dV = 0$$

Thus, we have

Chapter 1 Aircraft Equations of Motion

$$\int_V \bar{r}'\rho_A \, dV = \int_V \bar{r}'_P \rho_A \, dV = \bar{r}'_P \int_V \rho_A \, dV = m\,\bar{r}'_P$$

$$\bar{r}'_P = \frac{1}{m}\int_V \rho_A \bar{r}' \, dV$$

Using the previous relationship for $\bar{r}'_P$ we have

$$\frac{d}{dt}\int_V \rho_A \frac{d\bar{r}'}{dt} dV = \frac{d}{dt}\frac{d}{dt}\int_V \rho_A \bar{r}' \, dV = \frac{d}{dt}\frac{d}{dt}\int_V \rho_A (\bar{r}'_P + \bar{r}) \, dV$$

By taking advantage of the property of the center of gravity $\left(\int_V \bar{r} \rho_A \, dV = 0\right)$ we have

$$\frac{d}{dt}\frac{d}{dt}\int_V \rho_A \bar{r}'_P \, dV = \frac{d}{dt}\frac{d}{dt}(m\,\bar{r}'_P) = m\frac{d\overline{V}_P}{dt}$$

From the right side of the CLME we have

$$\int_V \rho_A \bar{g} \, dV + \int_S \overline{F} \, dS = m\bar{g} + (\overline{F}_A + \overline{F}_T)$$

leading to a first intermediate vectorial relationship for the CLME:

$$\boxed{m\frac{d\overline{V}_P}{dt} = m\bar{g} + (\overline{F}_A + \overline{F}_T)}$$

Once again, the previous relationship has been derived with respect to the Earth-based *inertial enough* reference frame X', Y', Z'. For the purpose of describing the dynamics with respect to the body frame X, Y, Z located at the aircraft center of gravity, the relative motion of the X, Y, Z frame with respect to the X', Y', Z' needs to be expressed. Consider a generic vector $\overline{C}$ initially defined with respect to the X', Y', Z' frame. To express that vector with respect to the X, Y, Z frame, the angular velocity $\bar{\omega}$ of X, Y, Z with respect to X', Y', Z' must be introduced. Thus, we have

$$\frac{d\overline{C}}{dt} = \frac{\partial \overline{C}}{\partial t} + \bar{\omega} \times \overline{C}$$

Using the previous relationship for $\overline{V}_P$ in the CLME we have

$$m\frac{d\overline{V}_P}{dt} = m\left(\frac{\partial \overline{V}_P}{\partial t} + \bar{\omega} \times \overline{V}_P\right) = m\left(\dot{\overline{V}}_P + \bar{\omega} \times \overline{V}_P\right) = m\bar{g} + (\overline{F}_A + \overline{F}_T)$$

Next, each of the vectors is to be expressed with respect to the X, Y, Z frame using these relationships:

$$\overline{V}_P = U\bar{i} + V\bar{j} + W\bar{k}$$
$$\dot{\overline{V}}_P = \dot{U}\bar{i} + \dot{V}\bar{j} + \dot{W}\bar{k}$$
$$\bar{\omega} = P\bar{i} + Q\bar{j} + R\bar{k}$$
$$\overline{F}_A = F_{A_X}\bar{i} + F_{A_Y}\bar{j} + F_{A_Z}\bar{k}$$
$$\overline{F}_T = F_{T_X}\bar{i} + F_{T_Y}\bar{j} + F_{T_Z}\bar{k}$$
$$\bar{g} = g_X\bar{i} + g_Y\bar{j} + g_Z\bar{k}$$

The previous components are shown in Figure 1.2.

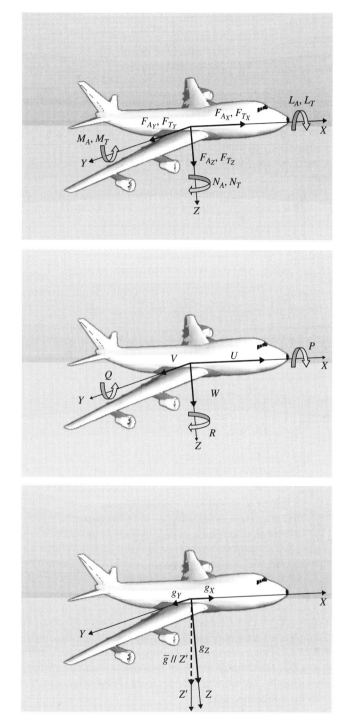

Figure 1.2 Forces, Moments, Linear Velocity, Angular Velocity, and Gravity Components (Positive Direction)

The expansion of the cross product $\left(\overline{\omega} \times \overline{V}_P\right)$ leads to

$$\overline{\omega} \times \overline{V}_P = \begin{vmatrix} \overline{i} & \overline{j} & \overline{k} \\ P & Q & R \\ U & V & W \end{vmatrix} = \overline{i}(QW - RV) + \overline{j}(UR - PW) + \overline{k}(PV - QU)$$

Thus, the conservation of linear momentum equations (CLMEs) with respect to the body axes X, Y, Z are given by

$$\boxed{\begin{aligned} m(\dot{U} + QW - RV) &= mg_X + (F_{A_X} + F_{T_X}) \\ m(\dot{V} + UR - PW) &= mg_Y + (F_{A_Y} + F_{T_Y}) \\ m(\dot{W} + PV - QU) &= mg_Z + (F_{A_Z} + F_{T_Z}) \end{aligned}}$$

Conservation of Angular Momentum Equations

The previous CLMEs form a system of nonlinear differential equations in the unknown variables U, V, W, which are the body-axes components of the linear velocity vector. The known terms of these equations (that is, the inputs to the system) are the forces on the right-hand side. The modeling of the gravity components g_X, g_Y, g_Z will be provided later in this chapter. The modeling of the aerodynamic and thrust forces acting on the aircraft will be the main focus of Chapters 3, 4, and 5. The solution of this system of equations also requires the knowledge of the body-axes components of the angular velocity vector P, Q, R. The variables P, Q, R will be the outputs of the Conservation of the Angular Momentum Equations (CAMEs), which will be discussed next.

1.4 CONSERVATION OF THE ANGULAR MOMENTUM EQUATIONS (CAMEs)

The development of the Conservation of the Angular Momentum Equations (CAMEs) originates from the initial expression for the previous CLMEs:[6]

$$\frac{d}{dt}\int_V \rho_A \frac{d\bar{r}'}{dt} dV = \int_V \rho_A \bar{g}\, dV + \int_S (\bar{F}_A + \bar{F}_T)\, dS$$

Considering the introduction of the moment arm $\bar{r}' = \bar{r}_p + \bar{r}$, where $\bar{r}'$ is the distance between the Earth-based inertial reference frame X', Y', Z' and a generic point on the aircraft, an initial expression of the vectorial CAME is given by

$$\frac{d}{dt}\int_V \bar{r}' \times \rho_A \frac{d\bar{r}'}{dt} dV = \int_V \bar{r}' \times \rho_A \bar{g}\, dV + \int_S \bar{r}' \times (\bar{F}_A + \bar{F}_T)\, dS$$

Using the relationship $\bar{r}' = \bar{r}_p + \bar{r}$ and taking advantage again of the properties of the center of gravity $\left(\int_V \bar{r}\rho_A dV = 0\right)$ it can be demonstrated (see Student Sample Problem 1.1) that the expression reduces itself to

$$\frac{d}{dt}\int_V \bar{r} \times \rho_A \frac{d\bar{r}}{dt} dV = \bar{M}_A + \bar{M}_T$$

where $\bar{M}_A + \bar{M}_T = \int_S \bar{r} \times (\bar{F}_A + \bar{F}_T)\, dS$.

The previous generic vectorial relationship is penalized by a challenging time-dependency of the argument of the integral. Additionally, this relationship is formulated with respect to the Earth-based inertial reference frame X', Y', Z'. Thus, a new relationship has to be introduced with respect to the body frame X, Y, Z to express the rotational dynamics of the aircraft with respect to its center of gravity. First, the "time" of the differentiation is reformulated into the volumetric integral:

$$\frac{d}{dt}\int_V \bar{r} \times \rho_A \frac{d\bar{r}}{dt} dV = \int_V \frac{d\bar{r}}{dt} \times \rho_A \frac{d\bar{r}}{dt} dV + \int_V \bar{r} \times \rho_A \frac{d}{dt}\frac{d\bar{r}}{dt} dV = \int_V \bar{r} \times \rho_A \frac{d}{dt}\frac{d\bar{r}}{dt} dV$$

Next, the challenge is to differentiate the vector $\bar{r}$ twice with respect to the X, Y, Z frame using the relative motion relationship introduced in the previous section.

The first differentiation provides $\frac{d\bar{r}}{dt} = \frac{\partial \bar{r}}{\partial t} + \bar{\omega} \times \bar{r} = \dot{\bar{r}} + \bar{\omega} \times \bar{r}$ leading to

$$\int_V \bar{r} \times \rho_A \frac{d}{dt}\frac{d\bar{r}}{dt} dV = \int_V \bar{r} \times \rho_A \frac{d}{dt}(\dot{\bar{r}} + \bar{\omega} \times \bar{r}) dV$$

Next, introducing a generic vector $\bar{c} = (\dot{\bar{r}} + \bar{\omega} \times \bar{r})$ such that

$$\frac{d\bar{c}}{dt} = \frac{\partial(\dot{\bar{r}} + \bar{\omega} \times \bar{r})}{\partial t} + \bar{\omega} \times (\dot{\bar{r}} + \bar{\omega} \times \bar{r}) = \left[(\ddot{\bar{r}} + \dot{\bar{\omega}} \times \bar{r} + \bar{\omega} \times \dot{\bar{r}}) + \bar{\omega} \times (\dot{\bar{r}} + \bar{\omega} \times \bar{r}) \right]$$

Thus, the CAME becomes

$$\int_V \bar{r} \times \rho_A \frac{d}{dt}(\dot{\bar{r}} + \bar{\omega} \times \bar{r}) dV = \int_V \bar{r} \times \left[(\ddot{\bar{r}} + \dot{\bar{\omega}} \times \bar{r} + \bar{\omega} \times \dot{\bar{r}}) + \bar{\omega} \times (\dot{\bar{r}} + \bar{\omega} \times \bar{r}) \right] \rho_A dV$$

$$= \int_V \bar{r} \times \left[\ddot{\bar{r}} + \dot{\bar{\omega}} \times \bar{r} + \bar{\omega} \times \dot{\bar{r}} + 2\bar{\omega} \times \dot{\bar{r}} + \bar{\omega} \times (\bar{\omega} \times \bar{r}) \right] \rho_A dV = \overline{M}_A + \overline{M}_T$$

At this point, the rigid body assumption described in Section 1.2 is introduced. Because the distance between two points within a rigid body is constant, by definition we have $\dot{\bar{r}} = \ddot{\bar{r}} = 0$. This allows a substantial simplification of the CAME as shown

$$\int_V \bar{r} \times \left[\cancel{\ddot{\bar{r}}} + \dot{\bar{\omega}} \times \bar{r} + \cancel{\bar{\omega} \times \dot{\bar{r}}} + \cancel{2\bar{\omega} \times \dot{\bar{r}}} + \bar{\omega} \times (\bar{\omega} \times \bar{r}) \right] \rho_A dV = \overline{M}_A + \overline{M}_T$$

leading finally to

$$\int_V \bar{r} \times \left[\dot{\bar{\omega}} \times \bar{r} + \bar{\omega} \times (\bar{\omega} \times \bar{r}) \right] \rho_A dV = \overline{M}_A + \overline{M}_T$$

The new expression for the CAME presents one additional computational challenge for the calculation of the double cross product. Taking advantage of the so-called *BAC/CAB* rule (see Appendix A.1), a generic double cross product can be expressed using

$$\overline{A} \times (\overline{B} \times \overline{C}) = \overline{B}(\overline{A} \cdot \overline{C}) - \overline{C}(\overline{A} \cdot \overline{B})$$

In the previous CAME we have

$$\bar{\omega} \times (\bar{\omega} \times \bar{r}) = \bar{\omega}(\bar{\omega} \cdot \bar{r}) - \bar{r}(\bar{\omega} \cdot \bar{\omega})$$

leading to

$$\int_V \bar{r} \times \left[\dot{\bar{\omega}} \times \bar{r} + \bar{\omega}(\bar{\omega} \cdot \bar{r}) - \bar{r}(\bar{\omega} \cdot \bar{\omega}) \right] \rho_A dV = \overline{M}_A + \overline{M}_T$$

Thus, the solution of the first double cross product leads to a second double cross product:

$$\bar{r} \times (\dot{\bar{\omega}} \times \bar{r}) = \dot{\bar{\omega}}(\bar{r} \cdot \bar{r}) - \bar{r}(\bar{r} \cdot \dot{\bar{\omega}})$$

Therefore, we have

$$\int_V \bar{r} \times \left[\dot{\bar{\omega}} \times \bar{r} + \bar{\omega}(\bar{\omega} \cdot \bar{r}) - \bar{r}(\bar{\omega} \cdot \bar{\omega}) \right] \rho_A dV = \int_V \left[\dot{\bar{\omega}}(\bar{r} \cdot \bar{r}) - \bar{r}(\bar{r} \cdot \dot{\bar{\omega}}) + \bar{r} \times \bar{\omega}(\bar{\omega} \cdot \bar{r}) - \bar{r} \times \bar{r}(\bar{\omega} \cdot \bar{\omega}) \right] \rho_A dV$$

$$= \int_V \left[\dot{\bar{\omega}}(\bar{r} \cdot \bar{r}) - \bar{r}(\bar{r} \cdot \dot{\bar{\omega}}) + \bar{r} \times \bar{\omega}(\bar{\omega} \cdot \bar{r}) \right] \rho_A dV = \overline{M}_A + \overline{M}_T$$

The complexity of these relationships suggests a breakdown into the following components:

$$\int_V \left[\dot{\bar{\omega}}(\bar{r} \cdot \bar{r}) \right] \rho_A dV = (I)$$

$$\int_V \left[-\bar{r}(\bar{r} \cdot \dot{\bar{\omega}}) \right] \rho_A dV = (II)$$

$$\int_V [\bar{r} \times \bar{\omega}(\bar{\omega} \cdot \bar{r})] \rho_A dV = (III)$$

$$(I) + (II) + (III) = \overline{M}_A + \overline{M}_T$$

Next, the components can be converted to a scalar form using the appropriate expressions for the vectors $\bar{r}, \bar{\omega}, \dot{\bar{\omega}}$ and the vectorial operations $\bar{r} \cdot \bar{r}, \bar{r} \cdot \dot{\bar{\omega}}, \bar{\omega} \cdot \bar{r}$. Using the scalar components shown in Figure 1.2

$$\bar{r} = X\bar{i} + Y\bar{j} + Z\bar{k}$$
$$\bar{\omega} = P\bar{i} + Q\bar{j} + R\bar{k}$$
$$\dot{\bar{\omega}} = \dot{P}\bar{i} + \dot{Q}\bar{j} + \dot{R}\bar{k}$$

and these scalar vectorial products

$$\bar{r} \cdot \bar{r} = X^2 + Y^2 + Z^2$$
$$\bar{r} \cdot \dot{\bar{\omega}} = X\dot{P} + Y\dot{Q} + Z\dot{R}$$
$$\bar{\omega} \cdot \bar{r} = XP + YQ + ZR$$

the following expressions are derived for $(I), (II)$:

$$(I) = \int_V [\dot{\bar{\omega}}(\bar{r} \cdot \bar{r})] \rho_A dV = \dot{\bar{\omega}} \int_V (\bar{r} \cdot \bar{r}) \rho_A dV = (\dot{P}\bar{i} + \dot{Y}\bar{j} + \dot{Z}\bar{k}) \int_V (X^2 + Y^2 + Z^2) \rho_A dV$$

$$(II) = \int_V [-\bar{r}(\bar{r} \cdot \dot{\bar{\omega}})] \rho_A dV = -\int_V [\bar{r}(\bar{r} \cdot \dot{\bar{\omega}})] \rho_A dV = -\int_V (X\bar{i} + Y\bar{j} + Z\bar{k})(X\dot{P} + Y\dot{Q} + Z\dot{R}) \rho_A dV$$

Next, grouping all the scalar components of these equations, we have

$$(I) + (II) = \bar{i}\left\{\dot{P}\int_V (Y^2+Z^2)\rho_A dV + \dot{P}\int_V \cancel{X^2 \rho_A dV} - \dot{P}\int_V \cancel{X^2 \rho_A dV} - \dot{Q}\int_V XY\rho_A dV - \dot{R}\int_V XZ\rho_A dV\right\}$$

$$+\bar{j}\left\{\dot{Q}\int_V (X^2+Z^2)\rho_A dV + \dot{Q}\int_V \cancel{Y^2 \rho_A dV} - \dot{Q}\int_V \cancel{Y^2 \rho_A dV} - \dot{P}\int_V XY\rho_A dV - \dot{R}\int_V YZ\rho_A dV\right\}$$

$$+\bar{k}\left\{\dot{R}\int_V (X^2+Y^2)\rho_A dV + \dot{R}\int_V \cancel{Z^2 \rho_A dV} - \dot{R}\int_V \cancel{Z^2 \rho_A dV} - \dot{P}\int_V XZ\rho_A dV - \dot{Q}\int_V YZ\rho_A dV\right\}$$

Note that the integrals no longer contain any time dependency. They are all reduced to the following well-known expressions for the moments and the products of inertia of a continuous system (see also Appendix A):

$$I_{XX} = \int_V (Y^2+Z^2)\rho_A dV, \quad I_{YY} = \int_V (X^2+Z^2)\rho_A dV, \quad I_{ZZ} = \int_V (X^2+Y^2)\rho_A dV$$

$$I_{XY} = I_{YX} = \int_V XY\rho_A dV = \int_V YX\rho_A dV,$$

$$I_{XZ} = I_{ZX} = \int_V XZ\rho_A dV = \int_V ZX\rho_A dV,$$

$$I_{YZ} = I_{ZY} = \int_V YZ\rho_A dV = \int_V ZY\rho_A dV$$

Therefore, we have

$$(I) + (II) = \bar{i}\{\dot{P} I_{XX} - \dot{Q} I_{XY} - \dot{R} I_{XZ}\} + \bar{j}\{\dot{Q} I_{YY} - \dot{P} I_{XY} - \dot{R} I_{YZ}\}$$
$$+\bar{k}\{\dot{R} I_{ZZ} - \dot{P} I_{XZ} - \dot{Q} I_{ZY}\}$$

Using a similar process, it can be shown (see Student Sample Problem 1.2) that the following expression is derived for (III):

$$(III) = \int_V [\bar{r} \times \bar{\omega}(\bar{\omega} \cdot \bar{r})] \rho_A dV$$

$$= \int_V [(X\bar{i} + Y\bar{j} + Z\bar{k}) \times (P\bar{i} + Q\bar{j} + R\bar{k})(XP + YQ + ZR)] \rho_A dV$$

$$= \bar{i}\{PR\, I_{XY} + (R^2 - Q^2)\, I_{YZ} - PQ\, I_{XZ} + (I_{ZZ} - I_{YY})RQ\}$$
$$+ \bar{j}\{PR(I_{XX} - I_{ZZ}) + (P^2 - R^2)\, I_{XZ} - QR\, I_{XY} + PQI_{YZ}\}$$
$$+ \bar{k}\{PQ(I_{YY} - I_{XX}) + (Q^2 - P^2)\, I_{XY} + QR\, I_{XZ} - PR\, I_{YZ}\}$$

This finally leads to

$$(I) + (II) + (III) = \bar{i}\{\dot{P}\, I_{XX} - \dot{Q}\, I_{XY} - \dot{R}\, I_{XZ}\} + \bar{j}\{\dot{Q}\, I_{YY} - \dot{P}\, I_{XY} - \dot{R}\, I_{YZ}\} + \bar{k}\{\dot{R}\, I_{ZZ} - \dot{P}\, I_{XZ} - \dot{Q}\, I_{ZY}\}$$
$$+ \bar{i}\{PR\, I_{XY} + (R^2 - Q^2)\, I_{YZ} - PQ\, I_{XZ} + (I_{ZZ} - I_{YY})RQ\}$$
$$+ \bar{j}\{PR(I_{XX} - I_{ZZ}) + (P^2 - R^2)\, I_{XZ} - QR\, I_{XY} + PQI_{YZ}\}$$
$$+ \bar{k}\{PQ(I_{YY} - I_{XX}) + (Q^2 - P^2)\, I_{XY} + QR\, I_{XZ} - PR\, I_{YZ}\}$$

These relationships can be substantially simplified by taking advantage of the fact that in the aircraft system, the XZ plane is a plane of symmetry, as shown in Figure 1.3.

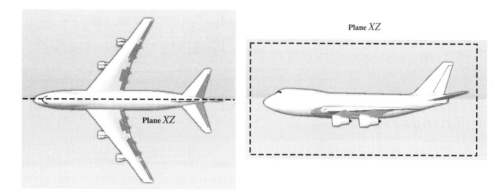

Figure 1.3 Aircraft XZ Plane of Symmetry

Because of the aircraft symmetry, the only nonzero product of inertia is the one associated with the plane of symmetry. Thus, we have $I_{XY} = I_{YZ} = 0$, $I_{XZ} \neq 0$. Therefore,

$$(I) + (II) + (III) = \bar{i}\{\dot{P}\, I_{XX} - \cancel{\dot{Q}\, I_{XY}} - \dot{R}\, I_{XZ}\} + \bar{j}\{\dot{Q}\, I_{YY} - \cancel{\dot{P}\, I_{XY}} - \cancel{\dot{R}\, I_{YZ}}\} + \bar{k}\{\dot{R}\, I_{ZZ} - \dot{P}\, I_{XZ} - \cancel{\dot{Q}\, I_{ZY}}\}$$
$$+ \bar{i}\{\cancel{PR\, I_{XY}} + \cancel{(R^2 - Q^2)\, I_{YZ}} - PQ\, I_{XZ} + (I_{ZZ} - I_{YY})RQ\}$$
$$+ \bar{j}\{PR(I_{XX} - I_{ZZ}) + (P^2 - R^2)\, I_{XZ} - \cancel{QR\, I_{XY}} + \cancel{PQI_{YZ}}\}$$
$$+ \bar{k}\{PQ(I_{YY} - I_{XX}) + \cancel{(Q^2 - P^2)\, I_{XY}} + QR\, I_{XZ} - \cancel{PR\, I_{YZ}}\}$$

Next, using the scalar components of the aerodynamic and thrust moments from Figure 1.2, we have

$$\overline{M}_A = L_A\, \bar{i} + M_A\, \bar{j} + N_A\, \bar{k}$$
$$\overline{M}_T = L_T\, \bar{i} + M_T\, \bar{j} + N_T\, \bar{k}$$

Finally, by grouping the terms along X, Y, Z in $(I) + (II) + (III) = \overline{M}_A + \overline{M}_T$ we have the final expression for the CAMEs along the X, Y, Z body frame:

$$\boxed{\begin{aligned}
\dot{P}\,I_{XX} - \dot{R}\,I_{XZ} - PQ\,I_{XZ} + RQ(I_{ZZ} - I_{YY}) &= L_A + L_T \\
\dot{Q}\,I_{YY} + PR(I_{XX} - I_{ZZ}) + (P^2 - R^2)\,I_{XZ} &= M_A + M_T \\
\dot{R}\,I_{ZZ} - \dot{P}\,I_{XZ} + PQ(I_{YY} - I_{XX}) + QR\,I_{XZ} &= N_A + N_T
\end{aligned}}$$

Conservation of Angular Momentum Equations

In these equations it is important to underline how rotations in two axes (for example, pitch and roll) generate a rotation in the remaining axis whose magnitude is directly correlated with the differences in the moments of the inertia and the product of inertia. These effects, known as gyroscopic effects, are typically present in nonlinear dynamic and aerodynamic flight conditions, such as those associated with high angular velocities and post-stall angles of attack.

The outputs of the CAMEs are P, Q, R, which are the body-axes components of the angular velocity vector. The inputs to the CAMEs are the body-axes components of the aerodynamic and thrust moments acting on the aircraft, which will be extensively discussed in Chapters 3, 4, and 5. The derivations of the previous CLMEs and the CAMEs are conceptually separated. However, recall that P, Q, R are also required for the solution of the previous CLMEs. Therefore, given this coupling, the CLMEs and the CAMEs must be considered as a single system of equations:

$$\boxed{\begin{aligned}
m(\dot{U} + QW - RV) &= mg_X + (F_{A_X} + F_{T_X}) \\
m(\dot{V} + UR - PW) &= mg_Y + (F_{A_Y} + F_{T_Y}) \\
m(\dot{W} + PV - QU) &= mg_Z + (F_{A_Z} + F_{T_Z}) \\
\dot{P}\,I_{XX} - \dot{R}\,I_{XZ} - PQ\,I_{XZ} + RQ(I_{ZZ} - I_{YY}) &= L_A + L_T \\
\dot{Q}\,I_{YY} + PR(I_{XX} - I_{ZZ}) + (P^2 - R^2)\,I_{XZ} &= M_A + M_T \\
\dot{R}\,I_{ZZ} - \dot{P}\,I_{XZ} + PQ(I_{YY} - I_{XX}) + QR\,I_{XZ} &= N_A + N_T
\end{aligned}}$$

Conservation of Linear and Angular Momentum Equations

1.5 CONSERVATION OF THE ANGULAR MOMENTUM EQUATIONS (CAMEs) WITH ROTOR EFFECTS

An "ad hoc" modification to the CAME is required for modeling the gyroscopic effects associated with the rotating components of the propulsion system.[6] These gyroscopic effects can potentially play a substantial role in generating angular moments. For aircraft with an even number of engines a trivial solution for canceling these effects is provided by using opposite signs for the angular rotations of the rotating components of the propulsion system. The problem can be significant for aircraft with an odd number of engines. This problem was first encountered with the first generation of jet-powered fighter aircraft in the early 1950s (the U.S. F-100 and the Russian MiG 15). The problem originated from the massive weight of the propulsion system versus the entire weight of the aircraft due to the use of heavy steel (compared with today's lightweight metal alloys) for the production of the compressor blades and all the rotating components of the propulsion system.

Considering an aircraft with N engines (with N being an odd number), the contribution to the angular momentum from the rotating components of the propulsion system is given by

$$\overline{h} = \sum_{i=1}^{N} \overline{h}_i = \sum_{i=1}^{N} I_{RR}\,\overline{\omega}_{RR_i} = h_X \overline{i} + h_Y \overline{j} + h_Z \overline{k}$$

Thus, the vectorial relationship for the CAME with respect to the X', Y', Z' frame will be modified as

$$\frac{d}{dt}\int_V \overline{r} \times \rho_A \frac{d\overline{r}}{dt} dV + \frac{d\overline{h}}{dt} = \overline{M}_A + \overline{M}_T$$

Considering the relative motion to express $\overline{h}$ with respect to the X, Y, Z frame starting from the X', Y', Z' frame, we have

$$\frac{d\bar{h}}{dt} = \dot{\bar{h}} + \bar{\omega} \times \bar{h}$$

Assuming constant angular velocity for the rotating components of the propulsion system, we have

$$\bar{\dot{\omega}}_{RR_i} = 0 \quad \Rightarrow \quad \dot{\bar{h}}_i = 0$$

Therefore,

$$\frac{d\bar{h}}{dt} = \bar{\omega} \times \bar{h} = \begin{vmatrix} \bar{i} & \bar{j} & \bar{k} \\ P & Q & R \\ h_X & h_Y & h_Z \end{vmatrix} = \bar{i}(Qh_Z - Rh_Y) + \bar{j}(Rh_X - Ph_Z) + \bar{k}(Ph_Y - Qh_X)$$

Thus, the CAMEs including the propulsive gyroscopic effects are given by

$$\boxed{\begin{aligned} \dot{P}I_{XX} - \dot{R}I_{XZ} - PQ\,I_{XZ} + RQ(I_{ZZ} - I_{YY}) + Qh_Z - Rh_Y &= L_A + L_T \\ \dot{Q}I_{YY} + PR(I_{XX} - I_{ZZ}) + (P^2 - R^2)\,I_{XZ} + Rh_X - Ph_Z &= M_A + M_T \\ \dot{R}I_{ZZ} - \dot{P}I_{XZ} + PQ(I_{YY} - I_{XX}) + QR\,I_{XZ} + Ph_Y - Qh_X &= N_A + N_T \end{aligned}}$$

Conservation of Angular Momentum Equations with Gyroscopic Effects

1.6 EULER ANGLES

The solution of the previously introduced system of CLMEs and CAMEs provides the linear and angular velocities relative to the dynamics of the aircraft with respect to the reference frame X, Y, Z at the center of gravity. However, these equations are derived from the inertial reference frame X', Y', Z'. Therefore, a complete analysis of the aircraft dynamics requires a detailed understanding of the relative motion of the body frame X, Y, Z with respect to the Earth-based inertial frame X', Y', Z'. This is critical from a navigational point of view. There are several methods for tracking the orientation of the X, Y, Z frame with respect to the Earth-based inertial frame X', Y', Z'.

The most common approach features the use of the so-called Euler angles.[3,6,7,8,9] An alternative but more complex approach is given by the use of quaternion.[10,11] The introduction of the Euler angles is based on a rigorous sequence that involves the introduction of a number of reference frames based on successive rotations. The associated four-step process is outlined here:

Step 1: Introduce a reference frame X_1, Y_1, Z_1 that moves with the aircraft center of gravity while being parallel to the Earth-based frame X', Y', Z', as shown in Figure 1.4. This frame is sometimes called a "local horizon" frame or a North-East-Down (NED) frame.

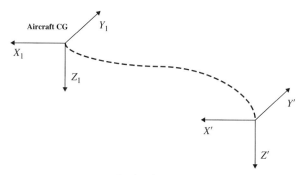

Figure 1.4 From the X', Y', Z' Frame to the X_1, Y_1, Z_1 Frame

Step 2: Rotation around Z_1 of an angle Ψ from the frame X_1, Y_1, Z_1 to a new frame X_2, Y_2, Z_2 with $Z_1 = Z_2$, as shown in Figure 1.5.

Step 3: Rotation around Y_2 of an angle Θ from the frame X_2, Y_2, Z_2 to a new frame X_3, Y_3, Z_3 with $Y_2 = Y_3$, as shown in Figure 1.5.

Step 4: Rotation around X_3 of an angle Φ from the frame X_3, Y_3, Z_3 to the aircraft body frame X, Y, Z with $X_3 = X$, as shown in Figure 1.5.

It is critical to observe the specific order of the sequence since the above rotations do not behave like vectors and individual rotations are not commutative.[7] The sequence introduced next summarizes the widely adopted practice of describing the aircraft orientation following the Z, Y, X right-handed rotation sequence (also known as "$3,2,1$" sequence) from the inertial Earth-based frame to the body frame.

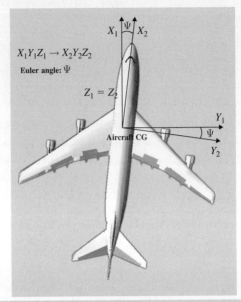

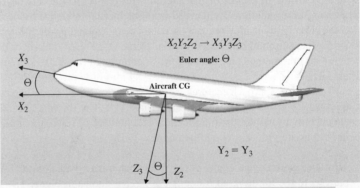

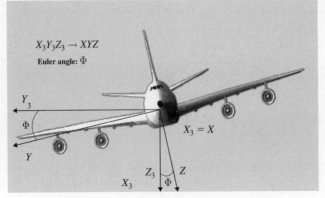

Figure 1.5 Introduction of the Euler Angles Ψ, Θ, Φ

1.7 FLIGHT PATH EQUATIONS (FPEs)

The use of the Euler angles Ψ, Θ, Φ for expressing the orientation of the body frame X, Y, Z with respect to the Earth-based inertial frame X', Y', Z' is critical for navigational purposes. The solution of the system of the CLMEs and the CAMEs describes the motion of the aircraft with respect to the body frame. However, a different set of equations is needed for expressing the trajectory of the aircraft with respect to the Earth-based inertial frame X', Y', Z'.

The Flight Path Equations (FPEs) are specifically used for this purpose.[6] The outputs of this system of equations are the displacements of the aircraft with respect to the Earth-based frame X', Y', Z'. The reduced 2D version of the FPEs provides the so-called ground tracks relative to the aircraft flight.

The development of the FPEs requires the relationships of the components of the linear velocities between the different frames previously used for the introduction of the Euler angles. Since the reference frames X', Y', Z' and X_1, Y_1, Z_1 are parallel to each other, we have

$$U_1 = \dot{X}_1 = \dot{X}'$$
$$V_1 = \dot{Y}_1 = \dot{Y}'$$
$$W_1 = \dot{Z}_1 = \dot{Z}'$$

Next, the following relationship applies to the components of the linear velocities during the transformation from the X_1, Y_1, Z_1 frame to the X_2, Y_2, Z_2 frame:

$$\begin{Bmatrix} U_1 \\ V_1 \\ W_1 \end{Bmatrix} = \begin{bmatrix} \cos\Psi & -\sin\Psi & 0 \\ \sin\Psi & \cos\Psi & 0 \\ 0 & 0 & 1 \end{bmatrix} \begin{Bmatrix} U_2 \\ V_2 \\ W_2 \end{Bmatrix} = R^{X_1Y_1Z_1 \to X_2Y_2Z_2} \begin{Bmatrix} U_2 \\ V_2 \\ W_2 \end{Bmatrix}$$

Similarly, the following relationship applies to the components of the linear velocities during the transformation from the frame X_2, Y_2, Z_2 to the frame X_3, Y_3, Z_3:

$$\begin{Bmatrix} U_2 \\ V_2 \\ W_2 \end{Bmatrix} = \begin{bmatrix} \cos\Theta & 0 & \sin\Theta \\ 0 & 1 & 0 \\ -\sin\Theta & 0 & \cos\Theta \end{bmatrix} \begin{Bmatrix} U_3 \\ V_3 \\ W_3 \end{Bmatrix} = R^{X_2Y_2Z_2 \to X_3Y_3Z_3} \begin{Bmatrix} U_3 \\ V_3 \\ W_3 \end{Bmatrix}$$

Finally, the following relationship applies to the components of the linear velocities during the transformation from the X_3, Y_3, Z_3 frame to the aircraft body frame X, Y, Z:

$$\begin{Bmatrix} U_3 \\ V_3 \\ W_3 \end{Bmatrix} = \begin{bmatrix} 1 & 0 & 0 \\ 0 & \cos\Phi & -\sin\Phi \\ 0 & \sin\Phi & \cos\Phi \end{bmatrix} \begin{Bmatrix} U \\ V \\ W \end{Bmatrix} = R^{X_3Y_3Z_3 \to XYZ} \begin{Bmatrix} U \\ V \\ W \end{Bmatrix}$$

Thus, the following general expression relates the components of the linear velocities along the inertial frame X', Y', Z' with the components of the linear velocities along the body frame X, Y, Z:

$$\begin{Bmatrix} \dot{X}' \\ \dot{Y}' \\ \dot{Z}' \end{Bmatrix} = \begin{Bmatrix} U_1 \\ V_1 \\ W_1 \end{Bmatrix} = R^{X_1Y_1Z_1 \to X_2Y_2Z_2} \cdot R^{X_2Y_2Z_2 \to X_3Y_3Z_3} \cdot R^{X_3Y_3Z_3 \to XYZ} \begin{Bmatrix} U \\ V \\ W \end{Bmatrix}$$

where $\begin{Bmatrix} U \\ V \\ W \end{Bmatrix} = \begin{Bmatrix} \dot{X} \\ \dot{Y} \\ \dot{Z} \end{Bmatrix}$

Therefore, we have

$$\begin{Bmatrix} \dot{X}' \\ \dot{Y}' \\ \dot{Z}' \end{Bmatrix} = \begin{Bmatrix} U_1 \\ V_1 \\ W_1 \end{Bmatrix} = \begin{bmatrix} \cos\Psi & -\sin\Psi & 0 \\ \sin\Psi & \cos\Psi & 0 \\ 0 & 0 & 1 \end{bmatrix} \begin{bmatrix} \cos\Theta & 0 & \sin\Theta \\ 0 & 1 & 0 \\ -\sin\Theta & 0 & \cos\Theta \end{bmatrix} \begin{bmatrix} 1 & 0 & 0 \\ 0 & \cos\Phi & -\sin\Phi \\ 0 & \sin\Phi & \cos\Phi \end{bmatrix} \begin{Bmatrix} U \\ V \\ W \end{Bmatrix}$$

leading to the final expression for the FPEs:

$$\begin{Bmatrix} \dot{X}' \\ \dot{Y}' \\ \dot{Z}' \end{Bmatrix} = \begin{bmatrix} \cos\Psi\cos\Theta & -\sin\Psi\cos\Phi + \cos\Psi\sin\Theta\sin\Phi & \sin\Psi\sin\Phi + \cos\Psi\sin\Theta\cos\Phi \\ \sin\Psi\cos\Theta & \cos\Psi\cos\Phi + \sin\Psi\sin\Theta\sin\Phi & -\sin\Phi\cos\Psi + \sin\Psi\sin\Theta\cos\Phi \\ -\sin\Theta & \cos\Theta\sin\Phi & \cos\Theta\cos\Phi \end{bmatrix} \begin{Bmatrix} U \\ V \\ W \end{Bmatrix}$$

Flight Path Equations

14 Chapter 1 Aircraft Equations of Motion

The inputs for the FPEs are the Euler angles along with the components of the linear velocity (U, V, W) with respect to the body frame X, Y, Z. The solution of the FPEs provides the complete 3D trajectory of the motion of the aircraft with respect to the Earth-based frame X', Y', Z'. Recall that (U, V, W) are provided by the solution of the combined (CLMEs + CAMEs) system. The Euler angles instead will be derived from the components of the angular velocity (P, Q, R) through the solution of another set of equations, the so-called Kinematic Equations (KEs), to be discussed next.[6,7]

1.8 KINEMATIC EQUATIONS (KEs)

The Euler angles allow expressing the angular velocity along the body frame X, Y, Z in terms of rates of changes of the Euler angles. In fact, according to the four-step process introduced in Section 1.6, an expression for the angular velocity vector is given by $\bar{\omega} = \dot{\bar{\Psi}} + \dot{\bar{\Theta}} + \dot{\bar{\Phi}}$. However, by definition, we also have $\bar{\omega} = P\bar{i} + Q\bar{j} + R\bar{k}$.

Therefore, $\bar{\omega} = P\bar{i} + Q\bar{j} + R\bar{k} = \dot{\bar{\Psi}} + \dot{\bar{\Theta}} + \dot{\bar{\Phi}}$.

The purpose of the Kinematic Equations (KEs) is to derive an expression for the Euler angles in terms of the components of the aircraft angular velocity P, Q, R. The derivation of the KEs will be based on the transformations introduced in Section 1.6. Starting from the transformation $X_1, Y_1, Z_1 \to X_2, Y_2, Z_2$ (see Figure 1.5), we have $Z_1 = Z_2$. In terms of unity vectors this implies that $\bar{k}_1 = \bar{k}_2$. Therefore, we have

$$\dot{\bar{\Psi}} = \dot{\Psi}\bar{k}_1 = \dot{\Psi}\bar{k}_2$$

Similarly, in the transformation $X_2, Y_2, Z_2 \to X_3, Y_3, Z_3$ (see Figure 1.5), we have. $Y_2 = Y_3$.
In terms of unity vectors this implies that $\bar{j}_2 = \bar{j}_3$. Therefore, we have

$$\dot{\bar{\Theta}} = \dot{\Theta}\bar{j}_2 = \dot{\Theta}\bar{j}_3$$

Similarly, in the transformation $X_3, Y_3, Z_3 \to X, Y, Z$ (see Figure 1.5), we have: $X_3 = X$. In terms of unity vectors this implies $\bar{i}_3 = \bar{i}$. Therefore, we have

$$\dot{\bar{\Phi}} = \dot{\Phi}\bar{i}_3 = \dot{\Phi}\bar{i}$$

Summarizing, we have

$$\bar{\omega} = \dot{\bar{\Psi}} + \dot{\bar{\Theta}} + \dot{\bar{\Phi}} = P\bar{i} + Q\bar{j} + R\bar{k} = \dot{\Psi}\bar{k}_2 + \dot{\Theta}\bar{j}_3 + \dot{\Phi}\bar{i}$$

Next, the objective is to derive expressions for $\bar{k}_2, \bar{j}_3$ in terms of $\bar{i}, \bar{j}, \bar{k}$. In the $X_2, Y_2, Z_2 \to X_3, Y_3, Z_3$ transformation outlined in Section 1.7, we have

$$\begin{Bmatrix} U_2 \\ V_2 \\ W_2 \end{Bmatrix} = \begin{bmatrix} \cos\Theta & 0 & \sin\Theta \\ 0 & 1 & 0 \\ -\sin\Theta & 0 & \cos\Theta \end{bmatrix} \begin{Bmatrix} U_3 \\ V_3 \\ W_3 \end{Bmatrix}$$

The same relationship also applies in terms of the unity vectors. In fact, we have

$$\begin{Bmatrix} \bar{i}_2 \\ \bar{j}_2 \\ \bar{k}_2 \end{Bmatrix} = \begin{bmatrix} \cos\Theta & 0 & \sin\Theta \\ 0 & 1 & 0 \\ -\sin\Theta & 0 & \cos\Theta \end{bmatrix} \begin{Bmatrix} \bar{i}_3 \\ \bar{j}_3 \\ \bar{k}_3 \end{Bmatrix}$$

Similarly, in the $X_3, Y_3, Z_3 \to X, Y, Z$ transformation outlined in Section 1.7 we have

$$\begin{Bmatrix} U_3 \\ V_3 \\ W_3 \end{Bmatrix} = \begin{bmatrix} 1 & 0 & 0 \\ 0 & \cos\Phi & -\sin\Phi \\ 0 & \sin\Phi & \cos\Phi \end{bmatrix} \begin{Bmatrix} U \\ V \\ W \end{Bmatrix}$$

The same relationship also applies in terms of the unity vectors, leading to

$$\left\{\begin{array}{c} \bar{i}_3 \\ \bar{j}_3 \\ \bar{k}_3 \end{array}\right\} = \begin{bmatrix} 1 & 0 & 0 \\ 0 & \cos\Phi & -\sin\Phi \\ 0 & \sin\Phi & \cos\Phi \end{bmatrix} \left\{\begin{array}{c} \bar{i} \\ \bar{j} \\ \bar{k} \end{array}\right\}$$

First, a relationship for $\bar{k}_2$ can be found using

$$\bar{k}_2 = -\sin\Theta\, \bar{i}_3 + \cos\Theta\, \bar{k}_3 = -\sin\Theta\, \bar{i} + \cos\Theta\, \bar{k}_3$$

where $\bar{k}_3 = \sin\Phi\, \bar{j} + \cos\Phi\, \bar{k}$, leading to

$$\bar{k}_2 = -\sin\Theta\, \bar{i} + \cos\Theta\, (\sin\Phi\, \bar{j} + \cos\Phi\, \bar{k}) = -\sin\Theta\, \bar{i} + \cos\Theta\sin\Phi\, \bar{j} + \cos\Theta\cos\Phi\, \bar{k}$$

Similarly, a simpler relationship for $\bar{j}_3$ is found using

$$\bar{j}_3 = \cos\Phi\, \bar{j} - \sin\Phi\, \bar{k}$$

The availability of $\bar{k}_2, \bar{j}_3$ in terms of $\bar{i}, \bar{j}, \bar{k}$ allows deriving

$$\begin{aligned}\bar{\omega} &= P\bar{i} + Q\bar{j} + R\bar{k} = \dot{\bar{\Psi}} + \dot{\bar{\Theta}} + \dot{\bar{\Phi}} = \dot{\Psi}\bar{k}_2 + \dot{\Theta}\bar{j}_3 + \dot{\Phi}\bar{i} \\ &= \dot{\Psi}(-\sin\Theta\, \bar{i} + \cos\Theta\sin\Phi\, \bar{j} + \cos\Theta\cos\Phi\, \bar{k}) + \dot{\Theta}(\cos\Phi\, \bar{j} - \sin\Phi\, \bar{k}) + \dot{\Phi}\bar{i}\end{aligned}$$

By grouping the terms along $\bar{i}, \bar{j}, \bar{k}$ we have

$$\begin{aligned} P &= \dot{\Phi} - \sin\Theta\, \dot{\Psi} \\ Q &= \cos\Phi\, \dot{\Theta} + \cos\Theta\sin\Phi\, \dot{\Psi} \\ R &= \cos\Theta\cos\Phi\, \dot{\Psi} - \sin\Phi\, \dot{\Theta} \end{aligned}$$

Rearranging in a matrix format leads to

$$\left\{\begin{array}{c} P \\ Q \\ R \end{array}\right\} = \begin{bmatrix} 1 & 0 & -\sin\Theta \\ 0 & \cos\Phi & \cos\Theta\sin\Phi \\ 0 & -\sin\Phi & \cos\Theta\cos\Phi \end{bmatrix} \left\{\begin{array}{c} \dot{\Phi} \\ \dot{\Theta} \\ \dot{\Psi} \end{array}\right\}$$

The above relationship is also known as the Inverse Kinematic Equations (IKEs). Finally, the Kinematic Equations (KEs) are found by inverting the previous relationship:

$$\left\{\begin{array}{c} \dot{\Phi} \\ \dot{\Theta} \\ \dot{\Psi} \end{array}\right\} = \begin{bmatrix} 1 & 0 & -\sin\Theta \\ 0 & \cos\Phi & \cos\Theta\sin\Phi \\ 0 & -\sin\Phi & \cos\Theta\cos\Phi \end{bmatrix}^{-1} \left\{\begin{array}{c} P \\ Q \\ R \end{array}\right\}$$

The final expression for the Kinematic Equations (KEs) in a matrix format (see Student Sample Problem 1.3) is given by

$$\boxed{\left\{\begin{array}{c} \dot{\Phi} \\ \dot{\Theta} \\ \dot{\Psi} \end{array}\right\} = \begin{bmatrix} 1 & \sin\Phi\tan\Theta & \cos\Phi\tan\Theta \\ 0 & \cos\Phi & -\sin\Phi \\ 0 & \sin\Phi\sec\Theta & \cos\Phi\sec\Theta \end{bmatrix} \left\{\begin{array}{c} P \\ Q \\ R \end{array}\right\}}$$

Kinematic Equations

It should be pointed out that the previous equations have a singularity associated with the condition $\Theta = 90°$. This is one of the reasons why an alternative approach based on the use of quaternion[10,11] is used for large-scale simulations.

In summary, the inputs to the Kinematic Equations are the components of the angular velocity (P, Q, R). The solution of the KEs provides the numerical values of the Euler angles (Ψ, Θ, Φ).

1.9 GRAVITY EQUATIONS (GEs)

The last set of equations to complete the description of the aircraft dynamics is represented by the Gravity Equations (GEs). The GEs are necessary for expressing the components of the gravity vector—which is parallel to the axis Z' in the Earth-based inertial reference frame X', Y', Z'—in the body frame X, Y, Z. Once again, the transformations introduced in Section 1.6 are used for expressing the relationships between the unity vectors in the different frames. Starting from the basic definition:

$$\bar{g} = \bar{k}'g = \bar{k}_1 g = \bar{k}_2 g = g_X \bar{i} + g_Y \bar{j} + g_Z \bar{k}$$

The objective is to express $\bar{k}_2$ in terms of $\bar{i}, \bar{j}, \bar{k}$. However, this relationship was already derived in Section 1.8.

$$\bar{k}_2 = -\sin\Theta\,\bar{i} + \cos\Theta\left(\sin\Phi\,\bar{j} + \cos\Phi\,\bar{k}\right)$$
$$= -\sin\Theta\,\bar{i} + \cos\Theta\sin\Phi\,\bar{j} + \cos\Theta\cos\Phi\,\bar{k}$$

Therefore, we have

$$g\,\bar{k}_2 = g\left(-\sin\Theta\,\bar{i} + \cos\Theta\sin\Phi\,\bar{j} + \cos\Theta\cos\Phi\,\bar{k}\right)$$
$$= g_X \bar{i} + g_Y \bar{j} + g_Z \bar{k}$$

By grouping the terms along $\bar{i}, \bar{j}, \bar{k}$ the Gravity Equations (GEs) are finally derived:

$$\boxed{\begin{aligned} g_X &= -g\sin\Theta \\ g_Y &= g\cos\Theta\sin\Phi \\ g_Z &= g\cos\Theta\cos\Phi \end{aligned}}$$

Gravity Equations

In summary, the inputs to the GEs are the numerical values of the Euler angles (Ψ, Θ, Φ). The outputs of the GEs are the components of the gravity vector (g_X, g_Y, g_Z) along the body axes X, Y, Z. The ultimate purpose of the GEs is to provide one of the three sets of inputs to the CLMEs previously introduced, as recalled here:

$$m(\dot{U} + QW - RV) = \underline{-mg\sin\Theta} + (F_{A_X} + F_{T_X})$$
$$m(\dot{V} + UR - PW) = \underline{mg\cos\Theta\sin\Phi} + (F_{A_Y} + F_{T_Y})$$
$$m(\dot{W} + PV - QU) = \underline{mg\cos\Theta\cos\Phi} + (F_{A_Z} + F_{T_Z})$$

1.10 SUMMARY OF THE AIRCRAFT EQUATIONS OF MOTION

Based on the previous discussion, the complete set of equations describing the dynamics of the aircraft is given by the following:

CLMEs and CAMEs (six scalar equations)

$$m(\dot{U} + QW - RV) = -mg\sin\Theta + (F_{A_X} + F_{T_X})$$
$$m(\dot{V} + UR - PW) = mg\cos\Theta\sin\Phi + (F_{A_Y} + F_{T_Y})$$
$$m(\dot{W} + PV - QU) = mg\cos\Theta\cos\Phi + (F_{A_Z} + F_{T_Z})$$
$$\dot{P} I_{XX} - \dot{R} I_{XZ} - PQ\,I_{XZ} + RQ(I_{ZZ} - I_{YY}) = L_A + L_T$$
$$\dot{Q} I_{YY} + PR(I_{XX} - I_{ZZ}) + (P^2 - R^2) I_{XZ} = M_A + M_T$$
$$\dot{R} I_{ZZ} - \dot{P} I_{XZ} + PQ(I_{YY} - I_{XX}) + QR\,I_{XZ} = N_A + N_T$$

FPEs (three scalar equations)

$$\begin{Bmatrix} \dot{X}' \\ \dot{Y}' \\ \dot{Z}' \end{Bmatrix} = \begin{Bmatrix} U_1 \\ V_1 \\ W_1 \end{Bmatrix} = \begin{bmatrix} \cos\Psi\cos\Theta & -\sin\Psi\cos\Phi + \cos\Psi\sin\Theta\sin\Phi & \sin\Psi\sin\Phi + \cos\Psi\sin\Theta\cos\Phi \\ \sin\Psi\cos\Theta & \cos\Psi\cos\Phi + \sin\Psi\sin\Theta\sin\Phi & -\sin\Phi\cos\Psi + \sin\Psi\sin\Theta\cos\Phi \\ -\sin\Theta & \cos\Theta\sin\Phi & \cos\Theta\cos\Phi \end{bmatrix} \begin{Bmatrix} U \\ V \\ W \end{Bmatrix}$$

KEs (three scalar equations)

$$\begin{Bmatrix} \dot{\Phi} \\ \dot{\Theta} \\ \dot{\Psi} \end{Bmatrix} = \begin{bmatrix} 1 & \sin\Phi\tan\Theta & \cos\Phi\tan\Theta \\ 0 & \cos\Phi & -\sin\Phi \\ 0 & \sin\Phi\sec\Theta & \cos\Phi\sec\Theta \end{bmatrix} \begin{Bmatrix} P \\ Q \\ R \end{Bmatrix}$$

GEs embedded into the CLMEs (three scalar equations)

$$g_X = -g\sin\Theta$$
$$g_Y = g\cos\Theta\sin\Phi$$
$$g_Z = g\cos\Theta\cos\Phi$$

A block diagram showing the integration of the aircraft equations of motion is given in Figure 1.6.

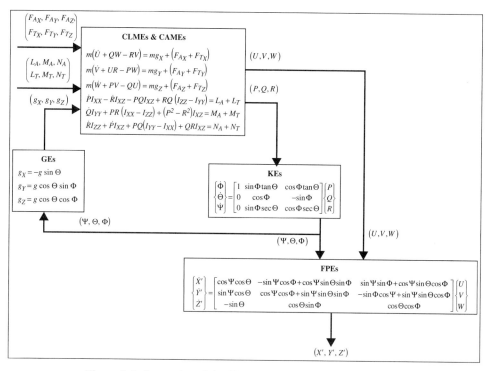

Figure 1.6 Integration of the CLMEs, CAMEs, KEs, FPEs, and GEs

1.11 DEFINITION OF STEADY-STATE AND PERTURBATION CONDITIONS

The equations previously introduced are relative to a generic flight condition. However, for practical applications, other than the terminal phases (that is, take off, climbing, descent, approach, and landing) most aircraft fly at a few specific flight conditions for most of their mission and perform a limited number of maneuvers, starting from those nominal flight conditions. Therefore, it is customary to study and to solve the aircraft equations of motion under the two following conditions:[3,6,9]

- steady-state flight;
- perturbed flight.

This is essentially the approach introduced by Bryan[2] in 1911.

Steady-state flight conditions are defined as those conditions where the linear and the angular accelerations with respect to the aircraft body frame X, Y, Z are zero.

Therefore, by definition, we have

$$\dot{\vec{V}}_P = 0 \quad \rightarrow \quad \vec{V}_P = const.$$
$$\dot{\vec{\omega}} = 0 \quad \rightarrow \quad \vec{\omega} = const.$$

The steady-state conditions for the motion variables will be here indicated with the subscript "1".

Perturbed flight conditions are defined as those conditions in which all the motion variables of the aircraft dynamics deviate from a set of original steady-state values.

Based on these definitions, the following notations characterize the aircraft motion variables at steady-state and perturbed flight conditions:

$$U = U_1 + u, \quad V = V_1 + v, \quad W = W_1 + w$$
$$P = P_1 + p, \quad Q = Q_1 + q, \quad R = R_1 + r$$
$$\Phi = \Phi_1 + \phi, \quad \Theta = \Theta_1 + \theta, \quad \Psi = \Psi_1 + \psi$$
$$F_{A_X} = F_{A_{X_1}} + f_{A_X}, \quad F_{A_Y} = F_{A_{Y_1}} + f_{A_Y}, \quad F_{A_Z} = F_{A_{Z_1}} + f_{A_Z}$$
$$L_A = L_{A_1} + l_A, \quad M_A = M_{A_1} + m_A, \quad N_A = N_{A_1} + n_A$$
$$F_{T_X} = F_{T_{X_1}} + f_{T_X}, \quad F_{T_Y} = F_{T_{Y_1}} + f_{T_Y}, \quad F_{T_Z} = F_{T_{Z_1}} + f_{T_Z}$$
$$L_T = L_{T_1} + l_T, \quad M_T = M_{T_1} + m_T, \quad N_T = N_{T_1} + n_T$$

The linear and angular accelerations with respect to X, Y, Z are zero by definition; thus we have

$$\dot{U}_1 = \dot{V}_1 = \dot{W}_1 = 0$$
$$\dot{P}_1 = \dot{Q}_1 = \dot{R}_1 = 0$$
$$\dot{U} = \dot{u}, \quad \dot{V} = \dot{v}, \quad \dot{W} = \dot{w}$$
$$\dot{P} = \dot{p}, \quad \dot{Q} = \dot{q}, \quad \dot{R} = \dot{r}$$
$$\dot{\Phi} = \dot{\phi}, \quad \dot{\Theta} = \dot{\theta}, \quad \dot{\Psi} = \dot{\psi}$$

In the next sections, the aircraft equations of motion will be characterized for both steady-state and perturbed flight conditions.

1.12 AIRCRAFT EQUATIONS OF MOTION AT STEADY-STATE CONDITIONS

Starting from the previous general expressions for the CLMEs, CAMEs, and KEs

$$m(\dot{U} + QW - RV) = -mg \sin\Theta + (F_{A_X} + F_{T_X})$$
$$m(\dot{V} + UR - PW) = mg \cos\Theta \sin\Phi + (F_{A_Y} + F_{T_Y})$$
$$m(\dot{W} + PV - QU) = mg \cos\Theta \cos\Phi + (F_{A_Z} + F_{T_Z})$$
$$\dot{P} I_{XX} - \dot{R} I_{XZ} - PQ I_{XZ} + RQ(I_{ZZ} - I_{YY}) = L_A + L_T$$
$$\dot{Q} I_{YY} + PR(I_{XX} - I_{ZZ}) + (P^2 - R^2) I_{XZ} = M_A + M_T$$
$$\dot{R} I_{ZZ} - \dot{P} I_{XZ} + PQ(I_{YY} - I_{XX}) + QR I_{XZ} = N_A + N_T$$

$$\begin{Bmatrix} \dot{\Phi} \\ \dot{\Theta} \\ \dot{\Psi} \end{Bmatrix} = \begin{bmatrix} 1 & \sin\Phi \tan\Theta & \cos\Phi \tan\Theta \\ 0 & \cos\Phi & -\sin\Phi \\ 0 & \sin\Phi \sec\Theta & \cos\Phi \sec\Theta \end{bmatrix} \begin{Bmatrix} P \\ Q \\ R \end{Bmatrix}$$

The introduction of the steady-state conditions

$$\dot{U}_1 = \dot{V}_1 = \dot{W}_1 = 0$$
$$\dot{P}_1 = \dot{Q}_1 = \dot{R}_1 = 0$$

leads to

$$\boxed{\begin{aligned} m(Q_1 W_1 - R_1 V_1) &= -mg \sin\Theta_1 + (F_{A_{X_1}} + F_{T_{X_1}}) \\ m(U_1 R_1 - P_1 W_1) &= mg \cos\Theta_1 \sin\Phi_1 + (F_{A_{Y_1}} + F_{T_{Y_1}}) \\ m(P_1 V_1 - Q_1 U_1) &= mg \cos\Theta_1 \cos\Phi_1 + (F_{A_{Z_1}} + F_{T_{Z_1}}) \\ P_1 Q_1 I_{XZ} + R_1 Q_1 (I_{ZZ} - I_{YY}) &= L_{A_1} + L_{T_1} \\ P_1 R_1 (I_{XX} - I_{ZZ}) + (P_1^2 - R_1^2) I_{XZ} &= M_{A_1} + M_{T_1} \\ P_1 Q_1 (I_{YY} - I_{XX}) + Q_1 R_1 I_{XZ} &= N_{A_1} + N_{T_1} \end{aligned}}$$

CLMEs and CAMEs at Steady-State Conditions

$$\left\{\begin{array}{c}\dot{\Phi}_1\\ \dot{\Theta}_1\\ \dot{\Psi}_1\end{array}\right\} = \begin{bmatrix} 1 & \sin\Phi_1\tan\Theta_1 & \cos\Phi_1\tan\Theta_1\\ 0 & \cos\Phi_1 & -\sin\Phi_1\\ 0 & \sin\Phi_1\sec\Theta_1 & \cos\Phi_1\sec\Theta_1\end{bmatrix}\left\{\begin{array}{c}P_1\\ Q_1\\ R_1\end{array}\right\}$$

KEs at Steady-State Conditions

A detailed analysis of the aircraft equations of motion at steady-state conditions reveals that under the conditions of this definition, the following conditions can be considered:[6]

- steady-state rectilinear flight;
- steady-state turning flight;
- steady-state symmetrical pull-up.

Although mathematically all of these flight conditions are possible, in reality only one condition (steady-state rectilinear flight, also known as straight line flight) applies to actual flight conditions. Particularly, steady-state rectilinear flight implies that in addition to $\dot{U}_1 = \dot{V}_1 = \dot{W}_1 = 0, \dot{P}_1 = \dot{Q}_1 = \dot{R}_1 = 0$ we have

$$P_1 = Q_1 = R_1 = 0$$

Under these conditions, the previously introduced KEs are trivially satisfied, and the CLMEs and CAMEs steady-state equations reduce themselves to the following:

$$0 = -mg\sin\Theta_1 + \left(F_{A_{X_1}} + F_{T_{X_1}}\right)$$
$$0 = mg\cos\Theta_1\sin\Phi_1 + \left(F_{A_{Y_1}} + F_{T_{Y_1}}\right)$$
$$0 = mg\cos\Theta_1\cos\Phi_1 + \left(F_{A_{Z_1}} + F_{T_{Z_1}}\right)$$
$$0 = L_{A_1} + L_{T_1}$$
$$0 = M_{A_1} + M_{T_1}$$
$$0 = N_{A_1} + N_{T_1}$$

The steady-state turning flight (also known as steady-state turn) characterizes a very specific flight condition in which the aircraft is in a constant altitude circular pattern. Therefore, at this condition the aircraft angular velocity along the body axes X, Y, Z is perfectly aligned with the Z' axis of the Earth-based inertial frame $X'\, Y'\, Z'$. Thus,

$$\overline{\omega} = \overline{k}'\dot{\Psi} = \overline{k}_1\,\dot{\Psi} = \overline{k}_2\,\dot{\Psi}$$

This flight condition also implies constant values for the pitch and roll angles Θ, Φ. While theoretically possible, this approximation is rarely realistic because of the strict requirements on the pilot and the flight control laws to hold constant altitude while also holding Θ, Φ constant.

The steady-state symmetrical pull-up (also known as the infinite loop) defines a theoretical but practically impossible flight condition in which the aircraft is in a 360° vertical loop with a pitch-only angular velocity $\overline{\omega} = \dot{\overline{\Theta}}$. While mathematically feasible, this steady-state flight condition is virtually unachievable in reality because the basic assumptions of zero angular and linear accelerations cannot be preserved in both climbing phase and descent phase.

1.13 AIRCRAFT EQUATIONS OF MOTION AT PERTURBED CONDITIONS

While the aircraft equations of motion at steady-state conditions describe realistic conditions only for rectilinear flight, an analysis of the aircraft equations of motion at perturbed conditions provides an important frame for analyzing the aircraft dynamics under typical maneuvers or when the aircraft is subjected to atmospheric turbulence.[3,6,9] The most important benefit is that the development of the aircraft equations of motion at perturbed conditions will lead to a fairly simple mathematical approach for the solution of the equations following their linearization. Starting with the basic definitions of steady-state and perturbed flight

$$U = U_1 + u, \quad V = V_1 + v, \quad W = W_1 + w$$
$$P = P_1 + p, \quad Q = Q_1 + q, \quad R = R_1 + r$$
$$\Phi = \Phi_1 + \phi, \quad \Theta = \Theta_1 + \theta, \quad \Psi = \Psi_1 + \psi$$
$$F_{A_X} = F_{A_{X_1}} + f_{A_X}, \quad F_{A_Y} = F_{A_{Y_1}} + f_{A_Y}, \quad F_{A_Z} = F_{A_{Z_1}} + f_{A_Z}$$
$$L_A = L_{A_1} + l_A, \quad M_A = M_{A_1} + m_A, \quad N_A = N_{A_1} + n_A$$
$$F_{T_X} = F_{T_{X_1}} + f_{T_X}, \quad F_{T_Y} = F_{T_{Y_1}} + f_{T_Y}, \quad F_{T_Z} = F_{T_{Z_1}} + f_{T_Z}$$
$$L_T = L_{T_1} + l_T, \quad M_T = M_{T_1} + m_T, \quad N_T = N_{T_1} + n_T$$

and imposing the conditions associated with steady-state flight conditions:

$$\dot{U} = \dot{U}_1 + \dot{u} = \dot{u}, \quad \dot{V} = \dot{V}_1 + \dot{v} = \dot{v}, \quad \dot{W} = \dot{W}_1 + \dot{w} = \dot{w}$$
$$\dot{P} = \dot{P}_1 + \dot{p} = \dot{p}, \quad \dot{Q} = \dot{Q}_1 + \dot{q} = \dot{q}, \quad \dot{R} = \dot{R}_1 + \dot{r} = \dot{r}$$

The substitution of these relationships in the CLMEs and CAMEs leads to

$$m[\dot{u} + (Q_1 + q)(W_1 + w) - (R_1 + r)(V_1 + v)] = -mg\sin(\Theta_1 + \theta) + \left(F_{A_{X_1}} + f_{A_X}\right) + \left(F_{T_{X_1}} + f_{T_X}\right)$$
$$m(\dot{v} + (U_1 + u)(R_1 + r) - (P_1 + p)(W_1 + w)) = mg\cos(\Theta_1 + \theta)\sin(\Phi_1 + \phi) + \left(F_{A_{Y_1}} + f_{A_Y}\right) + \left(F_{T_{Y_1}} + f_{T_Y}\right)$$
$$m(\dot{w} + (P_1 + p)(V_1 + v) - (Q_1 + q)(U_1 + u)) = mg\cos(\Theta_1 + \theta)\cos(\Phi_1 + \phi) + \left(F_{A_{Z_1}} + f_{A_Z}\right) + \left(F_{T_{Z_1}} + f_{T_Z}\right)$$
$$\dot{p}I_{XX} - \dot{r}I_{XZ} - (P_1 + p)(Q_1 + q)I_{XZ} + (R_1 + r)(Q_1 + q)(I_{ZZ} - I_{YY}) = (L_{A_1} + l_A) + (L_{T_1} + l_T)$$
$$\dot{q}I_{YY} + (P_1 + p)(R_1 + r)(I_{XX} - I_{ZZ}) + \left((P_1 + p)^2 - (R_1 + r)^2\right)I_{XZ} = (M_{A_1} + m_A) + (M_{T_1} + m_T)$$
$$\dot{r}I_{ZZ} - \dot{p}I_{XZ} + (P_1 + p)(Q_1 + q)(I_{YY} - I_{XX}) + (Q_1 + q)(R_1 + r)I_{XZ} = (N_{A_1} + n_A) + (N_{T_1} + n_T)$$

These equations provide the framework for additional analysis. The condition of perturbed state for a motion variable identifies a deviation from a previous steady-state condition. This deviation (or perturbation) can be further classified as

- large perturbation;
- medium perturbation;
- small perturbation.

Each of these classifications carries a number of specific assumptions of the magnitude of the perturbation terms. Calcara[9] provides a systematic approach leading to the formulation of specific sets of equations for each of the above classes of perturbation. However, by far the most extensively used formulation is relative to the small perturbation conditions.[3,6,9] Thus, the aircraft equations of motion at perturbed conditions under the assumption of small perturbations are analyzed next.

The small perturbations condition is based on the following assumptions:

Assumption 1: For the perturbation terms of the linear and angular velocities (u, v, w, p, q, r) the products of two "small perturbation" terms are always negligible.

For example: $up = 0, wr = 0, pq = 0, p^2 = 0, \ldots$

Assumption 2: For the perturbation terms of the Euler angles, the nonlinear products are also negligible; additionally, the following trigonometric conditions apply:

$$\sin(x) \approx x, \quad \cos(x) \approx 1, \quad \tan(x) = \frac{\sin(x)}{\cos(x)} \approx x$$

Therefore, we have

$$\psi\theta = \psi\phi = \theta\phi \approx 0$$
$$\sin\psi \approx \psi, \quad \sin\theta \approx \theta, \quad \sin\phi \approx \phi$$
$$\cos\psi \approx 1, \quad \cos\theta \approx 1, \quad \cos\phi \approx 1$$

This assumption yields accurate results for values of the Euler angles up to $[16°-18°]$.

Additionally, under this assumption the following relationships can be used for the GEs terms $(g\sin(\Theta_1+\theta), g\cos(\Theta_1+\theta)\sin(\Phi_1+\phi), g\cos(\Theta_1+\theta)\cos(\Phi_1+\phi))$ embedded in the CLMEs:

$$\sin(\Theta_1+\theta) \approx \sin\Theta_1 + \theta\cos\Theta_1$$
$$\cos(\Theta_1+\theta)\sin(\Phi_1+\phi) \approx \sin\Phi_1\cos\Theta_1 - \theta\sin\Phi_1\sin\Theta_1 + \phi\cos\Phi_1\cos\Theta_1 - \theta\phi\cos\Phi_1\sin\Theta_1$$
$$\cos(\Theta_1+\theta)\cos(\Phi_1+\phi) \approx \cos\Phi_1\cos\Theta_1 - \theta\cos\Phi_1\sin\Theta_1 - \phi\sin\Phi_1\cos\Theta_1 + \theta\phi\sin\Phi_1\sin\Theta_1$$

After inserting the previous trigonometric relationships in the CLMEs and CAMEs and after working out all the different products we have

$$m[\dot{u} + (Q_1W_1 + Q_1w + qW_1 + \underline{\underline{qw}}) - (R_1V_1 + R_1v + rV_1 + \underline{\underline{vr}})]$$
$$= -mg\sin\Theta_1 - mg\theta\cos\Theta_1 + (F_{A_{X_1}} + f_{A_X}) + (F_{T_{X_1}} + f_{T_X})$$
$$m[\dot{v} + (U_1R_1 + U_1r + uR_1 + \underline{\underline{ur}}) - (P_1W_1 + P_1w + pW_1 + \underline{\underline{pw}})]$$
$$= mg\sin\Phi_1\cos\Theta_1 - mg\theta\sin\Phi_1\sin\Theta_1 + mg\phi\cos\Phi_1\cos\Theta_1 - \underline{\underline{mg\theta\phi\cos\Phi_1\sin\Theta_1}}$$
$$+ (F_{A_{Y_1}} + f_{A_Y}) + (F_{T_{Y_1}} + f_{T_Y})$$
$$m[\dot{w} + (P_1V_1 + P_1v + pV_1 + \underline{\underline{pv}}) - (Q_1U_1 + Q_1u + U_1q + \underline{\underline{uq}})]$$
$$= mg\cos\Phi_1\cos\Theta_1 - mg\theta\cos\Phi_1\sin\Theta_1 - mg\phi\sin\Phi_1\cos\Theta_1 + \underline{\underline{mg\theta\phi\sin\Phi_1\sin\Theta_1}}$$
$$+ (F_{A_{Z_1}} + f_{A_Z}) + (F_{T_{Z_1}} + f_{T_Z})$$
$$\dot{p}I_{XX} - \dot{r}I_{XZ} - (P_1Q_1 + \overline{P_1q} + Q_1p + \underline{\underline{pq}})I_{XZ} + (R_1Q_1 + R_1q + Q_1r + \underline{\underline{rq}})(I_{ZZ} - I_{YY})$$
$$= (L_{A_1} + l_A) + (L_{T_1} + l_T)$$
$$\dot{q}I_{YY} + (P_1R_1 + P_1r + pR_1 + \underline{\underline{pr}})(I_{XX} - I_{ZZ}) + (P_1^2 + 2P_1p + \underline{\underline{p^2}} - R_1^2 - 2R_1r \underline{\underline{- r^2}})I_{XZ}$$
$$= (M_{A_1} + m_A) + (M_{T_1} + m_T)$$
$$\dot{r}I_{ZZ} - \dot{p}I_{XZ} + (P_1Q_1 + P_1q + pQ_1 + \underline{\underline{pq}})(I_{YY} - I_{XX}) + (Q_1R_1 + Q_1r + R_1q + \underline{\underline{rq}})I_{XZ}$$
$$= (N_{A_1} + n_A) + (N_{T_1} + n_T)$$

Next, an analysis is required for the <u>single</u> and <u>double</u> underlined terms in the preceding equations. The <u>single</u> underlined terms identify all the steady-state terms. All these terms cancel out, since they are essentially the steady-state CLMEs and CAMEs from Section 1.12 that are embedded in the small perturbation CLMEs and CAMEs. The <u>double</u> underlined terms instead cancel out because they all contain products of small perturbation terms, which are negligible according to the previous Assumptions 1 and 2. Therefore, the final small perturbation CLME and CAME are given by the following:

$$\boxed{\begin{aligned}
&m[\dot{u} + Q_1w + qW_1 - R_1v - rV_1] = -mg\theta\cos\Theta_1 + (f_{A_X} + f_{T_X})\\
&m[\dot{v} + U_1r + uR_1 - P_1w - pW_1] = -mg\theta\sin\Phi_1\sin\Theta_1 + mg\phi\cos\Phi_1\cos\Theta_1 + (f_{A_Y} + f_{T_Y})\\
&m[\dot{w} + P_1v + pV_1 - Q_1u - U_1q] = -mg\theta\cos\Phi_1\sin\Theta_1 - mg\phi\sin\Phi_1\cos\Theta_1 + (f_{A_Z} + f_{T_Z})\\
&\dot{p}I_{XX} - \dot{r}I_{XZ} - (P_1q + Q_1p)I_{XZ} + (R_1q + Q_1r)(I_{ZZ} - I_{YY}) = (l_A + l_T)\\
&\dot{q}I_{YY} + (P_1r + pR_1)(I_{XX} - I_{ZZ}) + (2P_1p - 2R_1r)I_{XZ} = (m_A + m_T)\\
&\dot{r}I_{ZZ} - \dot{p}I_{XZ} + (P_1q + pQ_1)(I_{YY} - I_{XX}) + (Q_1r + R_1q)I_{XZ} = (n_A + n_T)
\end{aligned}}$$

Small Perturbation CLMEs and CAMEs

Following the analysis of the CLMEs and the CAMEs, the attention now shifts to the IKEs and the KEs. Starting from the Inverted Kinematic Equations (IKEs)

$$P = \dot{\Phi} - \sin\Theta\,\dot{\Psi}$$
$$Q = \cos\Phi\,\dot{\Theta} + \cos\Theta\sin\Phi\,\dot{\Psi}$$
$$R = \cos\Theta\cos\Phi\,\dot{\Psi} - \sin\Phi\,\dot{\Theta}$$

after inserting

$$P = P_1 + p, Q = Q_1 + q, R = R_1 + r, \quad \dot{P} = \dot{p}, \dot{Q} = \dot{q}, \dot{R} = \dot{r}$$
$$\Phi = \Phi_1 + \phi, \Theta = \Theta_1 + \theta, \Psi = \Psi_1 + \psi, \quad \dot{\Phi} = \dot{\phi}, \dot{\Theta} = \dot{\theta}, \dot{\Psi} = \dot{\psi}$$

we have

$$P_1 + p = (\dot{\Phi}_1 + \dot{\phi}) - (\dot{\Psi}_1 + \dot{\psi})\sin(\Theta_1 + \theta)$$
$$Q_1 + q = (\dot{\Theta}_1 + \dot{\theta})\cos(\Phi_1 + \phi) + (\dot{\Psi}_1 + \dot{\psi})\cos(\Theta_1 + \theta)\sin(\Phi_1 + \phi)$$
$$R_1 + r = (\dot{\Psi}_1 + \dot{\psi})\cos(\Theta_1 + \theta)\cos(\Phi_1 + \phi) - (\dot{\Theta}_1 + \dot{\theta})\sin(\Phi_1 + \phi)$$

Next, inserting the following previous trigonometric relationships for small angles

$$\sin(\Theta_1 + \theta) \approx \sin\Theta_1 + \theta\cos\Theta_1$$
$$\cos(\Phi_1 + \phi) \approx \cos\Phi_1 - \phi\sin\Phi_1$$
$$\cos(\Theta_1 + \theta)\sin(\Phi_1 + \phi) \approx \sin\Phi_1\cos\Theta_1 - \theta\sin\Phi_1\sin\Theta_1 + \phi\cos\Phi_1\cos\Theta_1 - \theta\phi\cos\Phi_1\sin\Theta_1$$
$$\cos(\Theta_1 + \theta)\cos(\Phi_1 + \phi) \approx \cos\Phi_1\cos\Theta_1 - \theta\cos\Phi_1\sin\Theta_1 - \phi\sin\Phi_1\cos\Theta_1 + \theta\phi\sin\Phi_1\sin\Theta_1$$

we have

$$P_1 + p = (\dot{\Phi}_1 + \dot{\phi}) - (\dot{\Psi}_1 + \dot{\psi})(\sin\Theta_1 + \theta\cos\Theta_1)$$
$$Q_1 + q = (\dot{\Theta}_1 + \dot{\theta})(\cos\Phi_1 - \phi\sin\Phi_1) + (\dot{\Psi}_1 + \dot{\psi})(\sin\Phi_1\cos\Theta_1 - \theta\sin\Phi_1\sin\Theta_1$$
$$+ \phi\cos\Phi_1\cos\Theta_1 - \theta\phi\cos\Phi_1\sin\Theta_1)$$
$$R_1 + r = (\dot{\Psi}_1 + \dot{\psi})(\cos\Phi_1\cos\Theta_1 - \theta\cos\Phi_1\sin\Theta_1 - \phi\sin\Phi_1\cos\Theta_1 + \theta\phi\sin\Phi_1\sin\Theta_1)$$
$$- (\dot{\Theta}_1 + \dot{\theta})(\sin\Phi_1 + \phi\cos\Phi_1)$$

After multiplying the internal products we have

$$\underline{P_1} + p = \underline{\dot{\Phi}_1} + \dot{\phi} \underline{- \dot{\Psi}_1\sin\Theta_1} - \underline{\underline{\dot{\Psi}_1\theta\cos\Theta_1}} - \dot{\psi}\sin\Theta_1 - \underline{\underline{\dot{\psi}\theta\cos\Theta_1}}$$

$$\underline{Q_1} + q = \underline{\dot{\Theta}_1\cos\Phi_1} - \underline{\underline{\dot{\Theta}_1\phi\sin\Phi_1}} + \dot{\theta}\cos\Phi_1 - \underline{\underline{\dot{\theta}\phi\sin\Phi_1}} + \underline{\dot{\Psi}_1\sin\Phi_1\cos\Theta_1}$$
$$- \underline{\underline{\dot{\Psi}_1\theta\sin\Phi_1\sin\Theta_1}} + \underline{\underline{\dot{\Psi}_1\phi\cos\Phi_1\cos\Theta_1}} - \underline{\underline{\dot{\Psi}_1\theta\phi\cos\Phi_1\sin\Theta_1}}$$
$$+ \dot{\psi}\sin\Phi_1\cos\Theta_1 - \underline{\underline{\dot{\psi}\theta\sin\Phi_1\sin\Theta_1}} + \underline{\underline{\dot{\psi}\phi\cos\Phi_1\cos\Theta_1}} - \underline{\underline{\dot{\psi}\theta\phi\cos\Phi_1\sin\Theta_1}}$$

$$\underline{R_1} + r = \underline{\dot{\Psi}_1\cos\Phi_1\cos\Theta_1} - \underline{\underline{\dot{\Psi}_1\theta\cos\Phi_1\sin\Theta_1}} - \underline{\underline{\dot{\Psi}_1\phi\sin\Phi_1\cos\Theta_1}} + \underline{\underline{\dot{\Psi}_1\theta\phi\sin\Phi_1\sin\Theta_1}}$$
$$+ \dot{\psi}\cos\Phi_1\cos\Theta_1 - \underline{\underline{\dot{\psi}\theta\cos\Phi_1\sin\Theta_1}} - \underline{\underline{\dot{\psi}\phi\sin\Phi_1\cos\Theta_1}} + \underline{\underline{\dot{\psi}\theta\phi\sin\Phi_1\sin\Theta_1}}$$
$$- \underline{\dot{\Theta}_1\sin\Phi_1} - \underline{\underline{\dot{\Theta}_1\phi\cos\Phi_1}} - \dot{\theta}\sin\Phi_1 - \underline{\underline{\dot{\theta}\phi\cos\Phi_1}}$$

Next, as for the previous CLMEs and CAMEs, the single and the double underlined terms are considered in the above equations. The single underlined terms identify all the steady-state terms. Once again, these terms cancel out because they are essentially the steady-state IKEs embedded in the small perturbation IKEs. The double underlined terms instead cancel out because they all contain nonlinear products of small perturbation terms, which are negligible according to the previously described assumptions. Thus, the final small perturbation IKEs are given by

$$\boxed{\begin{aligned} p &= \dot{\phi} - \dot{\Psi}_1\theta\cos\Theta_1 - \dot{\psi}\sin\Theta_1 \\ q &= -\dot{\Theta}_1\phi\sin\Phi_1 + \dot{\theta}\cos\Phi_1 - \dot{\Psi}_1\theta\sin\Phi_1\sin\Theta_1 + \dot{\Psi}_1\phi\cos\Phi_1\cos\Theta_1 + \dot{\psi}\sin\Phi_1\cos\Theta_1 \\ r &= -\dot{\Psi}_1\theta\cos\Phi_1\sin\Theta_1 - \dot{\Psi}_1\phi\sin\Phi_1\cos\Theta_1 + \dot{\psi}\cos\Phi_1\cos\Theta_1 - \dot{\Theta}_1\phi\cos\Phi_1 - \dot{\theta}\sin\Phi_1 \end{aligned}}$$

Small Perturbation IKEs

1.14 SMALL PERTURBATION EQUATIONS FROM A STEADY-STATE LEVEL FLIGHT

The previously introduced small perturbation CLMEs, CAMEs, and IKEs are relative to general perturbed conditions. For pedagogical purposes, the solution of these equations starting from the most common initial condition, which is steady-state rectilinear wing-level flight, is of specific interest. At these conditions the following applies:

- angular velocities $P_1 = Q_1 = R_1 = 0$;
- Euler angles $\Phi_1 = const.$, $\Theta_1 = const.$, $\Psi_1 = const.$;
- lateral velocity $V_1 = 0$;
- roll angle $\Phi_1 = 0, \sin \Phi_1 = 0, \cos \Phi_1 = 1$ (wing level).

At these conditions the associated CLMEs, CAMEs, and IKEs equations take on the following simplified form:

$$\begin{aligned}
m[\dot{u} + qW_1] &= -mg\,\theta \cos \Theta_1 + \left(f_{A_X} + f_{T_X}\right) \\
m[\dot{v} + U_1 r - pW_1] &= mg\,\phi \cos \Theta_1 + \left(f_{A_Y} + f_{T_Y}\right) \\
m[\dot{w} - U_1 q] &= -mg\,\theta \sin \Theta_1 + \left(f_{A_Z} + f_{T_Z}\right) \\
\dot{p}I_{XX} - \dot{r}I_{XZ} &= (l_A + l_T) \\
\dot{q}I_{YY} &= (m_A + m_T) \\
\dot{r}I_{ZZ} - \dot{p}I_{XZ} &= (n_A + n_T) \\
p &= \dot{\phi} - \dot{\psi} \sin \Theta_1 \\
q &= \dot{\theta} \\
r &= \dot{\psi} \cos \Theta_1
\end{aligned}$$

Small Perturbations CLMEs, CAMEs, IKEs at Steady-State Rectilinear Flight

1.15 SUMMARY

This chapter introduced and discussed different sets of equations describing aircraft dynamics. Starting from the basic expression of Newton's second law, the CLMEs and CAMEs were derived, first with respect to an inertial Earth-based $X'Y'Z'$ frame and later with respect to the body frame XYZ. Next, the Euler angles were introduced; these angles are used to express the orientation of the XYZ frame with respect to the $X'Y'Z'$ frame. The Euler angles can be evaluated from the components of the body axis angular velocity using the Kinematic Equations (KEs). The information of the Euler angles along with the components of the body axis linear velocity allow the derivation of the trajectory of the aircraft with respect to the Earth-based frame $X'Y'Z'$ through the use of the Flight Path Equations (FPEs). The final use of the Euler angles is for the evaluation of the components of the gravity vector with respect to the body frame XYZ. Next, the equations of motions have been characterized for two specific flight conditions: steady-state flight conditions and perturbed flight conditions under the small perturbation assumptions. In particular, the small perturbation equations of motion are important for setting up the mathematical framework for a simplified solution because of the assumption of neglecting the nonlinear terms associated with the products of small perturbation terms.

The driving inputs to the CLMEs and the CAMEs are the aerodynamic and thrust forces and moments. The modeling of these important quantities will be the focus of Chapters 2, 3, and 4. A summary of the equations introduced in this chapter is provided here. A detailed flow chart of the chapter is shown in Figure 1.7.

General CLMEs and CAMEs (six scalar equations)

$$\begin{aligned}
m(\dot{U} + QW - RV) &= -mg \sin \Theta + \left(F_{A_X} + F_{T_X}\right) \\
m(\dot{V} + UR - PW) &= mg \cos \Theta \sin \Phi + \left(F_{A_Y} + F_{T_Y}\right) \\
m(\dot{W} + PV - QU) &= mg \cos \Theta \cos \Phi + \left(F_{A_Z} + F_{T_Z}\right) \\
\dot{P}I_{XX} - \dot{R}I_{XZ} - PQI_{XZ} + RQ(I_{ZZ} - I_{YY}) &= L_A + L_T \\
\dot{Q}I_{YY} + PR(I_{XX} - I_{ZZ}) + (P^2 - R^2)I_{XZ} &= M_A + M_T \\
\dot{R}I_{ZZ} - \dot{P}I_{XZ} + PQ(I_{YY} - I_{XX}) + QRI_{XZ} &= N_A + N_T
\end{aligned}$$

General FPEs (three scalar equations)

$$\begin{Bmatrix} \dot{X}' \\ \dot{Y}' \\ \dot{Z}' \end{Bmatrix} = \begin{bmatrix} \cos \Psi \cos \Theta & -\sin \Psi \cos \Phi + \cos \Psi \sin \Theta \sin \Phi & \sin \Psi \sin \Phi + \cos \Psi \sin \Theta \cos \Phi \\ \sin \Psi \cos \Theta & \cos \Psi \cos \Phi + \sin \Psi \sin \Theta \sin \Phi & -\sin \Phi \cos \Psi + \sin \Psi \sin \Theta \cos \Phi \\ -\sin \Theta & \cos \Theta \sin \Phi & \cos \Theta \cos \Phi \end{bmatrix} \begin{Bmatrix} U \\ V \\ W \end{Bmatrix}$$

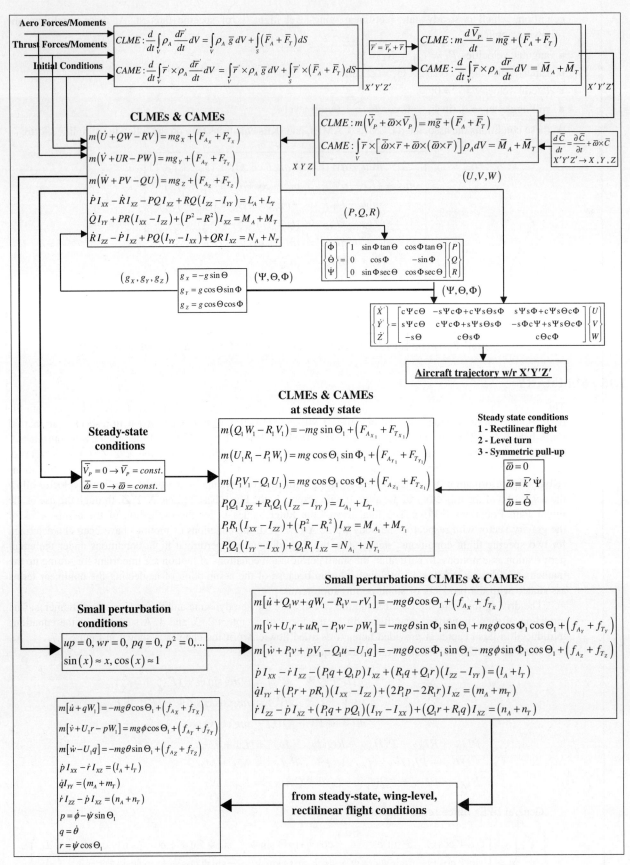

Figure 1.7 Development of the Aircraft Equations of Motion

General KEs (three scalar equations)

$$\left\{\begin{array}{c}\dot{\Phi}\\ \dot{\Theta}\\ \dot{\Psi}\end{array}\right\} = \begin{bmatrix} 1 & \sin\Phi\tan\Theta & \cos\Phi\tan\Theta \\ 0 & \cos\Phi & -\sin\Phi \\ 0 & \sin\Phi\sec\Theta & \cos\Phi\sec\Theta \end{bmatrix} \left\{\begin{array}{c}P\\ Q\\ R\end{array}\right\}$$

General GEs—embedded into the CLMEs (three scalar equations)

$$g_X = -g\sin\Theta$$
$$g_Y = g\cos\Theta\sin\Phi$$
$$g_Z = g\cos\Theta\cos\Phi$$

Steady-state CLMEs and CAMEs (six scalar equations)

$$m(Q_1 W_1 - R_1 V_1) = -mg\sin\Theta_1 + \left(F_{A_{X_1}} + F_{T_{X_1}}\right)$$
$$m(U_1 R_1 - P_1 W_1) = mg\cos\Theta_1 \sin\Phi_1 + \left(F_{A_{Y_1}} + F_{T_{Y_1}}\right)$$
$$m(P_1 V_1 - Q_1 U_1) = mg\cos\Theta_1 \cos\Phi_1 + \left(F_{A_{Z_1}} + F_{T_{Z_1}}\right)$$
$$P_1 Q_1 I_{XZ} + R_1 Q_1 (I_{ZZ} - I_{YY}) = L_{A_1} + L_{T_1}$$
$$P_1 R_1 (I_{XX} - I_{ZZ}) + (P_1^2 - R_1^2) I_{XZ} = M_{A_1} + M_{T_1}$$
$$P_1 Q_1 (I_{YY} - I_{XX}) + Q_1 R_1 I_{XZ} = N_{A_1} + N_{T_1}$$

Steady-state KEs (three scalar equations)

$$\left\{\begin{array}{c}\dot{\Phi}_1\\ \dot{\Theta}_1\\ \dot{\Psi}_1\end{array}\right\} = \begin{bmatrix} 1 & \sin\Phi_1\tan\Theta_1 & \cos\Phi_1\tan\Theta_1 \\ 0 & \cos\Phi_1 & -\sin\Phi_1 \\ 0 & \sin\Phi_1\sec\Theta_1 & \cos\Phi_1\sec\Theta_1 \end{bmatrix} \left\{\begin{array}{c}P_1\\ Q_1\\ R_1\end{array}\right\}$$

Small perturbation CLMEs and CAMEs (six scalar equations)

$$m[\dot{u} + Q_1 w + qW_1 - R_1 v - rV_1] = -mg\theta\cos\Theta_1 + \left(f_{A_X} + f_{T_X}\right)$$
$$m[\dot{v} + U_1 r + uR_1 - P_1 w - pW_1] = -mg\theta\sin\Phi_1\sin\Theta_1 + mg\phi\cos\Phi_1\cos\Theta_1 + \left(f_{A_Y} + f_{T_Y}\right)$$
$$m[\dot{w} + P_1 v + pV_1 - Q_1 u - U_1 q] = -mg\theta\cos\Phi_1\sin\Theta_1 - mg\phi\sin\Phi_1\cos\Theta_1 + \left(f_{A_Z} + f_{T_Z}\right)$$
$$\dot{p} I_{XX} - \dot{r} I_{XZ} - (P_1 q + Q_1 p) I_{XZ} + (R_1 q + Q_1 r)(I_{ZZ} - I_{YY}) = (l_A + l_T)$$
$$\dot{q} I_{YY} + (P_1 r + pR_1)(I_{XX} - I_{ZZ}) + (2P_1 p - 2R_1 r) I_{XZ} = (m_A + m_T)$$
$$\dot{r} I_{ZZ} - \dot{p} I_{XZ} + (P_1 q + pQ_1)(I_{YY} - I_{XX}) + (Q_1 r + R_1 q) I_{XZ} = (n_A + n_T)$$

Small perturbation IKEs (three scalar equations)

$$p = \dot{\phi} - \dot{\Psi}_1 \theta\cos\Theta_1 - \dot{\psi}\sin\Theta_1$$
$$q = -\dot{\Theta}_1 \phi\sin\Phi_1 + \dot{\theta}\cos\Phi_1 - \dot{\Psi}_1 \theta\sin\Phi_1\sin\Theta_1$$
$$\quad + \dot{\Psi}_1 \phi\cos\Phi_1\cos\Theta_1 + \dot{\psi}\sin\Phi_1\cos\Theta_1$$
$$r = -\dot{\Psi}_1 \theta\cos\Phi_1\sin\Theta_1 - \dot{\Psi}_1 \phi\sin\Phi_1\cos\Theta_1$$
$$\quad + \dot{\psi}\cos\Phi_1\cos\Theta_1 - \dot{\Theta}_1 \phi\cos\Phi_1 - \dot{\theta}\sin\Phi_1$$

Small perturbation CLMEs, CAMEs, and IKEs at steady-state level flight (nine scalar equations)

$$m[\dot{u} + qW_1] = -mg\theta\cos\Theta_1 + \left(f_{A_X} + f_{T_X}\right)$$
$$m[\dot{v} + U_1 r - pW_1] = mg\phi\cos\Theta_1 + \left(f_{A_Y} + f_{T_Y}\right)$$
$$m[\dot{w} - U_1 q] = -mg\theta\sin\Theta_1 + \left(f_{A_Z} + f_{T_Z}\right)$$
$$\dot{p} I_{XX} - \dot{r} I_{XZ} = (l_A + l_T)$$

$$\dot{q}I_{YY} = (m_A + m_T)$$
$$\dot{r}I_{ZZ} - \dot{p}I_{XZ} = (n_A + n_T)$$
$$p = \dot{\phi} - \dot{\psi}\sin\Theta_1$$
$$q = \dot{\theta}$$
$$r = \dot{\psi}\cos\Theta_1$$

REFERENCES

1. Lanchester, F. W. *Aerodonetics*. Constable and Co. Ltd., London, 1908.
2. Bryan, G. H. *Stability in Aviation*. Macmillan and Co., London, 1911.
3. Cook, M. V. *Flight Dynamic Principles* (2nd ed.), Elsevier Aerospace Engineering Series, 2007.
4. Hull, D. G. *Fundamentals of Airplane Flight Mechanics*, Springer, 2007.
5. Multiple sources from Google search, "DC-8 Miami Crash."
6. Roskam, J. *Airplane Flight Dynamics and Automatic Flight Controls – Part I*. Design, Analysis, and Research Corporation, Lawrence, Kansas, 1995.
7. Stevens, B. L., and Lewis, F. L. *Aircraft Control and Simulation*. John Wiley & Sons, 2003.
8. Schmidt, L. V. *Introduction to Aircraft Flight Dynamics*. AIAA Education Series, 1998.
9. Calcara, M. *Elementi di Dinamica del Velivolo*. Collana di Aeronautica, CUEN, Napoli, 1981.
10. Kuipers, J. B. *Quaternion and Rotation Sequences: A Primer with Applications to Orbits, Aerospace, and Virtual Reality*. Princeton University Press, 1999.
11. Altmann, S. L. *Rotations, Quaternions, and Double Groups*. Dover Publications, 1986.

STUDENT SAMPLE PROBLEMS

Student Sample Problem 1.1

Consider again the aircraft system and its body and Earth reference frames, as shown in the figure.

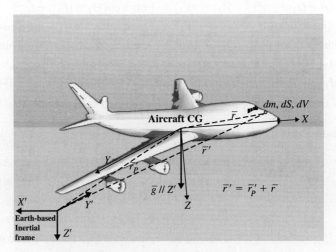

Figure SSP1.1.1 Earth and Body Reference Frames

Show that the conservation of the angular momentum equation (CAME) with respect to X', Y', Z'

$$\frac{d}{dt}\int_V \bar{r}' \times \rho_A \frac{d\bar{r}'}{dt}dV = \int_V \bar{r}' \times \rho_A \bar{g}dV + \int_s \bar{r}' \times \overline{F}ds \text{ with } \overline{F} = \overline{F}_A + \overline{F}_T$$

can be redefined as

$$\boxed{\frac{d}{dt}\int_V \bar{r} \times \frac{d\bar{r}}{dt} \rho_A dV = \overline{M}}$$

where $\overline{M} = \overline{M}_A + \overline{M}_T = \int_S \bar{r} \times \overline{F} ds$.

First, introduce $\bar{r}' = \bar{r}'_P + \bar{r}$ on both sides of the CAME.

$$\frac{d}{dt}\int_V \left[(\bar{r}'_P + \bar{r}) x \frac{d(\bar{r}'_P + \bar{r})}{dt}\right] \rho_A dV = \int_V [(\bar{r}'_P + \bar{r}) \times \bar{g}] \rho_A dV + \int_S [(\bar{r}'_P + \bar{r}) \times \overline{F}] ds$$

The left-hand side (LHS) has the following terms:

$$\frac{d}{dt}\int_V \left[(\bar{r}'_P + \bar{r}) x \frac{d(\bar{r}'_P + \bar{r})}{dt}\right] \rho_A dV = \frac{d}{dt}\int_V \left[\bar{r}'_P x \frac{d\bar{r}'_P}{dt}\right] \rho_A dV + \frac{d}{dt}\int_V \left[\bar{r}'_P x \frac{d\bar{r}}{dt}\right] \rho_A dV + \frac{d}{dt}\int_V \left[\bar{r} x \frac{d\bar{r}'_P}{dt}\right] \rho_A dV$$

$$+ \frac{d}{dt}\int_V \left[\bar{r} x \frac{d\bar{r}}{dt}\right] \rho_A dV = (I) + (II) + (III) + (IV)$$

The right-hand side (RHS) has the following terms:

$$\int_V [(\bar{r}'_P + \bar{r}) \times \bar{g}] \rho_A dV + \int_S [(\bar{r}'_P + \bar{r}) \times \overline{F}] ds = \int_V [\bar{r}'_P \times \bar{g}] \rho_A dV + \int_V [\bar{r} \times \bar{g}] \rho_A dV + \int_S [\bar{r}'_P \times \overline{F}] ds + \int_S [\bar{r} \times \overline{F}] ds$$

$$= (V) + (VI) + (VII) + (VIII)$$

Consider (I) in the LHS.

$$(I) = \frac{d}{dt}\int_V \left[\bar{r}'_P x \frac{d\bar{r}'_P}{dt}\right] \rho_A dV = \int_V \left[\cancel{\frac{d\bar{r}'_P}{dt} x \frac{d\bar{r}'_P}{dt}}\right] \rho_A dV + \int_V \left[\bar{r}'_P x \frac{d}{dt}\frac{d\bar{r}'_P}{dt}\right] \rho_A dV$$

$$= \int_V \left[\bar{r}'_P x \frac{d\overline{V}_P}{dt}\right] \rho_A dV$$

Since $\bar{r}'_P$ does not depend on the volume (and mass) of the aircraft, it can be taken out of the integral. Additionally, recall that the first vectorial expression of the CLME with respect to the frame X', Y', Z' is given by

$$m\frac{d\overline{V}'_P}{dt} = m\bar{g} + \overline{F}$$

Therefore, (I) becomes

$$(I) = \int_V \left[\bar{r}'_P x \frac{d\overline{V}'_P}{dt}\right] \rho_A dV = \bar{r}'_P x \int_V \rho_A \frac{d\overline{V}'_P}{dt} dV = \bar{r}'_P x m \frac{d\overline{V}'_P}{dt}$$

Thus, (I) will cancel out with (V) and (VII), since

$$\bar{r}'_P x m \frac{d\overline{V}'_P}{dt} = \int_V [\bar{r}'_P \times \bar{g}] \rho_A dV + \int_S [\bar{r}'_P \times \overline{F}] ds = \bar{r}'_P x \int_V \rho_A \bar{g} dV + \bar{r}'_P x \int_S \overline{F} ds$$

$$\Rightarrow \cancel{(I)} = \cancel{(V)} + \cancel{(VII)}$$

Next, consider (II) in the LHS.

28 Chapter 1 Aircraft Equations of Motion

$$(II) = \frac{d}{dt}\int_V \left[\bar{r}'_P x \frac{d\bar{r}}{dt}\right]\rho_A dV = \int_V \left[\frac{d\bar{r}'_P}{dt} x \frac{d\bar{r}}{dt}\right]\rho_A dV + \int_V \left[\bar{r}'_P x \frac{d}{dt}\frac{d\bar{r}}{dt}\right]\rho_A dV$$

Again, $\bar{r}'_P$ does not depend on the volume (and mass) of the aircraft. Therefore, it can be taken out of the integral. Similarly, the aircraft density and volume are not functions of time. Therefore, (II) will become

$$(II) = \frac{d\bar{r}'_P}{dt} x \int_V \frac{d\bar{r}}{dt}\rho_A dV + \bar{r}'_P x \int_V \frac{d}{dt}\frac{d\bar{r}}{dt}\rho_A dV = \frac{d\bar{r}'_P}{dt} x \frac{d(\cancel{\int_V \bar{r}\rho_A dV})}{dt} + \bar{r}'_P x \frac{d}{dt}\frac{(\cancel{\int_V \bar{r}\rho_A dV})}{dt}$$

However, because of the property of the center of gravity, since $\bar{r}$ is the distance of a generic point of the aircraft from the center of gravity, taking the integral with respect the entire aircraft, we have $\left(\int_V \bar{r}\rho_A dV = 0\right)$.
Therefore:

$$(II) = \frac{d\bar{r}_P}{dt} x \frac{\overline{d(\cancel{\int_V \bar{r}\rho_A dV})}}{dt} + \bar{r}_P x \frac{d}{dt}\frac{\overline{(\cancel{\int_V \bar{r}\rho_A dV})}}{dt} = 0$$

Thus, (II) drops out. Next, consider (III) in the LHS.

$$(III) = \frac{d}{dt}\int_V \left[\bar{r} x \frac{d\bar{r}'_P}{dt}\right]\rho_A dV = \int_V \left[\frac{d\bar{r}}{dt} x \frac{d\bar{r}'_P}{dt}\right]\rho_A dV + \int_V \left[\bar{r} x \frac{d}{dt}\frac{d\bar{r}'_P}{dt}\right]\rho_A dV$$

At this point, recall a basic property of the cross product between two generic vectors:

$$\overline{A} \times \overline{B} = -\left(\overline{B} \times \overline{A}\right)$$

Therefore, taking $\bar{r}'_P$ out of the integral and taking advantage of the property of the center of gravity:

$$(III) = -\int_V \left[\frac{d\bar{r}'_P}{dt} \times \frac{d\bar{r}}{dt}\right]\rho_A dV - \int_V \left[\frac{d}{dt}\frac{d\bar{r}'_P}{dt} x \bar{r}\right]\rho_A dV$$

$$= -\frac{d\bar{r}'_P}{dt} \times \int_V \frac{d\bar{r}}{dt}\rho_A dV - \frac{d}{dt}\frac{d\bar{r}'_P}{dt} x \int_V \bar{r}\rho_A dV$$

$$= -\frac{d\bar{r}'_P}{dt} \times \frac{d}{dt}\cancel{\int_V \bar{r}\rho_A dV} - \frac{d}{dt}\frac{d\bar{r}'_P}{dt} x \cancel{\int_V \bar{r}\rho_A dV} = 0$$

In the LHS the only residual term is (IV), which is the desired term. On the RHS, the residual terms are (VI) and (VIII). Therefore

$$\int_V [\bar{r} \times \bar{g}]\rho_A\, dV + \int_S [\bar{r} \times \bar{F}]\, ds = (VI) + (VIII)$$

Next, consider (VI) in the RHS:

$$\int_V [\bar{r} \times \bar{g}]\rho_A\, dV + \int_S [\bar{r} \times \bar{F}]\, ds = (VI) + (VIII)$$

$$(VI) = \int_V [\bar{r} \times \bar{g}]\rho_A\, dV = \int_V -[\bar{g} \times \bar{r}]\rho_A\, dV = -\bar{g} \times \cancel{\int_V \bar{r}\rho_A dV}$$

Following these calculations, the residual terms are (IV) and (VIII).

$$(IV) = \frac{d}{dt}\int_V \left[\bar{r} x \frac{d\bar{r}}{dt}\right]\rho_A dV$$

$$(VIII) = \int_S [\bar{r} \times \bar{F}]\, ds$$

Therefore, it has been shown that:

$$\boxed{\frac{d}{dt}\int_V \bar{r} \times \frac{d\bar{r}}{dt}\rho_A dV = \overline{M}}$$

where

$$\overline{M} = \overline{M}_A + \overline{M}_T = \int_S [\bar{r} \times \overline{F}]ds$$

Student Sample Problem 1.2

Show that

$$\int_V [\bar{r} \times \overline{\omega}(\overline{\omega} \cdot \bar{r})]\rho_A dV$$

can be reduced to

$$\bar{i}\{PR\,I_{XY} + (R^2 - Q^2)\,I_{YZ} - PQ\,I_{XZ} + RQ(I_{ZZ} - I_{YY})\}$$
$$+ \bar{j}\{PR(I_{XX} - I_{ZZ}) + (P^2 - R^2)\,I_{XZ} - QR\,I_{XY} + PQI_{YZ}\}$$
$$+ \bar{k}\{PQ(I_{YY} - I_{XX}) + (Q^2 - P^2)\,I_{XY} + QR\,I_{XZ} - PR\,I_{YZ}\}$$

First, consider the scalar product: $(\overline{\omega}\bar{r}) = (XP + YQ + ZR)$.

Next, consider the cross product:

$$\bar{r} \times \overline{\omega} = \begin{vmatrix} \bar{i} & \bar{j} & \bar{k} \\ X & Y & Z \\ P & Q & R \end{vmatrix} = \bar{i}(YR - ZQ) + \bar{j}(ZP - XR) + \bar{k}(XQ - PY)$$

Using the above scalar and cross products, we will have

$$\int_V [\bar{r} \times \overline{\omega}(\overline{\omega} \cdot \bar{r})]\rho_A dV = \int_V [(\bar{i}(YR-ZQ) + \bar{j}(ZP-XR) + \bar{k}(XQ-PY))(XP+YQ+ZR)]\rho_A dV$$
$$= \bar{i}\int_V [(YR-ZQ)(XP+YQ+ZR)]\rho_A dV + \bar{j}\int_V [(ZP-XR)(XP+YQ+ZR)]\rho_A dV$$
$$+ \bar{k}\int_V [(XQ-PY)(XP+YQ+ZR)]\rho_A dV$$

Next, recall the basic definitions of products and moments of inertia:

$$I_{XX} = \int_V (Y^2 + Z^2)\rho_A dV, \quad I_{YY} = \int_V (X^2 + Z^2)\rho_A dV, \quad I_{ZZ} = \int_V (X^2 + Y^2)\rho_A dV$$

$$I_{XY} = I_{YX} = \int_V XY\rho_A dV = \int_V YX\rho_A dV,$$

$$I_{XZ} = I_{ZX} = \int_V XZ\rho_A dV = \int_V ZX\rho_A dV,$$

$$I_{YZ} = I_{ZY} = \int_V YZ\rho_A dV = \int_V ZY\rho_A dV$$

Therefore, for the terms associated with $\bar{i}$ we have

$$\int_V [(YR - ZQ)(XP + YQ + ZR)]\rho_A dV$$

$$= \int_V [(YRXP + YRYQ + YRZR - ZQXP - ZQYQ - ZQZR)]\rho_A dV$$

$$= \left(PR \int_V YX\rho_A dV + RQ \int_V Y^2\rho_A dV + R^2 \int_V YZ\rho_A dV - PQ \int_V XZ\rho_A dV - Q^2 \int_V YZ\rho_A dV - RQ \int_V Z^2\rho_A dV \right)$$

Next, consider the following:

$$\left(RQ \int_V Y^2\rho_A dV - RQ \int_V Z^2\rho_A dV \right) = \left(RQ \int_V (X^2 + Y^2)\rho_A dV - RQ \int_V (X^2 + Z^2)\rho_A dV \right)$$

$$= (RQ\, I_{ZZ} - RQ\, I_{YY}) = RQ(I_{ZZ} - I_{YY})$$

Therefore, combining all the terms associated with $\bar{i}$, we have

$$(PR\, I_{XY} + RQ(I_{ZZ} - I_{YY}) + (R^2 - Q^2)I_{YZ} - PQ\, I_{XZ})$$

Similarly, for the terms associated with $\bar{j}$ we have

$$\int_V [(ZP - XR)(XP + YQ + ZR)]\rho_A dV$$

$$= \int_V [(ZPXP + ZPYQ + ZPZR - XRXP - XRYQ - XRZR)]\rho_A dV$$

$$= \left(P^2 \int_V XZ\rho_A dV + PQ \int_V YZ\rho_A dV + PR \int_V Z^2\rho_A dV - PR \int_V X^2\rho_A dV - RQ \int_V XY\rho_A dV - R^2 \int_V XZ\rho_A dV \right)$$

Next, consider the following:

$$\left(PR \int_V Z^2\rho_A dV - PR \int_V X^2\rho_A dV \right) = \left(PR \int_V (Y^2 + Z^2)\rho_A dV - PR \int_V (Y^2 + X^2)\rho_A dV \right)$$

$$= (PR\, I_{XX} - PR\, I_{ZZ}) = PR(I_{XX} - I_{ZZ})$$

Therefore, combining all the terms associated with $\bar{j}$, we have

$$((P^2 - R^2)I_{XZ} + PQ\, I_{YZ} + PR(I_{XX} - I_{ZZ}) - RQ\, I_{XY})$$

Finally, for the terms associated with $\bar{k}$ we have

$$\int_V [(XQ - PY)(XP + YQ + ZR)]\rho_A dV$$

$$= \int_V \left[\left(XQXP + XQYQ + XQZR - PYXP - PYYQ - PYZR \right) \right] \rho_A dV$$

$$= \left(PQ \int_V X^2\rho_A dV + Q^2 \int_V XY\rho_A dV + QR \int_V XZ\rho_A dV - P^2 \int_V XY\rho_A dV - PQ \int_V Y^2\rho_A dV - PR \int_V YZ\rho_A dV \right)$$

Next, consider the following:

$$\left(PQ \int_V X^2\rho_A dV - PQ \int_V Y^2\rho_A dV \right) = \left(PQ \int_V (X^2+Z^2)\rho_A dV - PQ \int_V (Y^2+Z^2)\rho_A dV \right)$$

$$= (PQ\, I_{YY} - PQ\, I_{XX}) = PQ(I_{YY} - I_{XX})$$

Therefore, combining all the terms associated with $\bar{k}$, we have

$$(QR\, I_{XZ} + PQ(I_{YY} - I_{XX}) + (Q^2 - P^2)\, I_{XY} - PR\, I_{YZ})$$

Combining all the terms, we finally have:

$$\int_V [\bar{r} \times \bar{\omega}(\bar{\omega} \cdot \bar{r})]\rho_A dV = \bar{i}(PRI_{XY} + RQ(I_{ZZ} - I_{YY}) \\
+ (R^2 - Q^2)I_{YZ} - PQI_{XZ}) + \bar{j}(PR(I_{XX} - I_{ZZ}) \\
+ (P^2 - R^2)I_{XZ} - QRI_{XY} + PQI_{YZ}) \\
+ \bar{k}(PQ(I_{YY} - I_{XX}) + (Q^2 - P^2)I_{XY} + QRI_{XZ} - PRI_{YZ})$$

Student Sample Problem 1.3

Starting from the Inverse Kinematic Equation (IKE)

$$\begin{Bmatrix} P \\ Q \\ R \end{Bmatrix} = \begin{bmatrix} 1 & 0 & -\sin\Theta \\ 0 & \cos\Phi & \cos\Theta \sin\Phi \\ 0 & -\sin\Phi & \cos\Theta \cos\Phi \end{bmatrix} \begin{Bmatrix} \dot{\Phi} \\ \dot{\Theta} \\ \dot{\Psi} \end{Bmatrix} = A \begin{Bmatrix} \dot{\Phi} \\ \dot{\Theta} \\ \dot{\Psi} \end{Bmatrix}$$

the objective is to determine A^{-1} such that

$$\begin{Bmatrix} \dot{\Phi} \\ \dot{\Theta} \\ \dot{\Psi} \end{Bmatrix} = A^{-1} \begin{Bmatrix} P \\ Q \\ R \end{Bmatrix}$$

From Appendix A.2, given a generic $A_{(3 \times 3)}$ matrix, A^{-1} can be found using

$$A^{-1} = inv(A) = \frac{adj(A)}{\det(A)} \text{ where } adj(A) = cof(A)^T$$

Therefore, starting from

$$A = \begin{bmatrix} 1 & 0 & -\sin\Theta \\ 0 & \cos\Phi & \cos\Theta \sin\Phi \\ 0 & -\sin\Phi & \cos\Theta \cos\Phi \end{bmatrix} = \begin{bmatrix} a_{11} & a_{12} & a_{13} \\ a_{21} & a_{22} & a_{23} \\ a_{31} & a_{32} & a_{33} \end{bmatrix}$$

the matrix $cof(A)$ is here defined as

$$cof(A) = \begin{bmatrix} c_{11} & c_{12} & c_{13} \\ c_{21} & c_{22} & c_{23} \\ c_{31} & c_{32} & c_{33} \end{bmatrix}$$

where

$$c_{11} = (a_{22}a_{33} - a_{32}a_{23}) = \cos(\Theta)\cos^2(\Phi) + \cos(\Theta)\sin^2(\Phi)$$
$$= \cos(\Theta)(\cos^2(\Phi) + \sin^2(\Phi)) = \cos(\Theta)$$
$$c_{12} = (a_{31}a_{23} - a_{21}a_{33}) = 0$$
$$c_{13} = (a_{21}a_{32} - a_{31}a_{22}) = 0$$
$$c_{21} = (a_{32}a_{13} - a_{12}a_{33}) = \sin(\Phi)\sin(\Theta)$$
$$c_{22} = (a_{11}a_{33} - a_{31}a_{13}) = \cos(\Phi)\cos(\Theta)$$
$$c_{23} = (a_{31}a_{12} - a_{11}a_{32}) = \sin(\Phi)$$
$$c_{31} = (a_{12}a_{23} - a_{22}a_{13}) = \cos(\Phi)\sin(\Theta)$$
$$c_{32} = (a_{21}a_{13} - a_{11}a_{23}) = -\sin(\Phi)\cos(\Theta)$$
$$c_{33} = (a_{11}a_{22} - a_{21}a_{12}) = \cos(\Phi)$$

Therefore, we will have

$$cof(A) = \begin{bmatrix} \cos(\Theta) & 0 & 0 \\ \sin(\Phi)\sin(\Theta) & \cos(\Phi)\cos(\Theta) & \sin(\Phi) \\ \cos(\Phi)\sin(\Theta) & -\sin(\Phi)\cos(\Theta) & \cos(\Phi) \end{bmatrix}$$

leading to

$$adj(A) = cof(A)^T = \begin{bmatrix} \cos(\Theta) & \sin(\Phi)\sin(\Theta) & \cos(\Phi)\sin(\Theta) \\ 0 & \cos(\Phi)\cos(\Theta) & -\sin(\Phi)\cos(\Theta) \\ 0 & \sin(\Phi) & \cos(\Phi) \end{bmatrix}$$

Next,

$$\det(A) = 1\big(\cos(\Theta)\cos^2(\Phi) + \cos(\Theta)\sin^2(\Phi)\big)$$
$$= \cos(\Theta)\big(\cos^2(\Phi) + \sin^2(\Phi)\big) = \cos(\Theta)(1) = \cos(\Theta)$$

leading to

$$A^{-1} = inv(A) = \frac{adj(A)}{\det(A)} = \frac{1}{\cos(\Theta)}\begin{bmatrix} \cos(\Theta) & \sin(\Phi)\sin(\Theta) & \cos(\Phi)\sin(\Theta) \\ 0 & \cos(\Phi)\cos(\Theta) & -\sin(\Phi)\cos(\Theta) \\ 0 & \sin(\Phi) & \cos(\Phi) \end{bmatrix}$$
$$= \begin{bmatrix} 1 & \sin(\Phi)\tan(\Theta) & \cos(\Phi)\tan(\Theta) \\ 0 & \cos(\Phi) & -\sin(\Phi) \\ 0 & \sin(\Phi)/\cos(\Theta) & \cos(\Phi)/\cos(\Theta) \end{bmatrix} = \begin{bmatrix} 1 & \sin(\Phi)\tan(\Theta) & \cos(\Phi)\tan(\Theta) \\ 0 & \cos(\Phi) & -\sin(\Phi) \\ 0 & \sin(\Phi)\sec(\Theta) & \cos(\Phi)\sec(\Theta) \end{bmatrix}$$

where $\sec(\Theta) = \dfrac{1}{\cos(\Theta)}$.

Therefore, the final expression for the Kinematic Equations (KEs) is given by

$$\begin{Bmatrix} \dot{\Phi} \\ \dot{\Theta} \\ \dot{\Psi} \end{Bmatrix} = \begin{bmatrix} 1 & \sin(\Phi)\tan(\Theta) & \cos(\Phi)\tan(\Theta) \\ 0 & \cos(\Phi) & -\sin(\Phi) \\ 0 & \sin(\Phi)\sec(\Theta) & \cos(\Phi)\sec(\Theta) \end{bmatrix} \begin{Bmatrix} P \\ Q \\ R \end{Bmatrix}$$

PROBLEMS

Problem 1.1

Consider the Conservation of Linear Momentum Equations (CLMEs) with respect to the body axes X, Y, Z:

$$m(\dot{U} + QW - RV) = m\,g_X + (F_{A_X} + F_{T_X})$$
$$m(\dot{V} + UR - PW) = m\,g_Y + (F_{A_Y} + F_{T_Y})$$
$$m(\dot{W} + PV - QU) = m\,g_Z + (F_{A_Z} + F_{T_Z})$$

Rewrite the above equations under the following conditions:

- constant pitching maneuver (loop in the XZ plane);
- constant rolling around the X axis;
- steady turn at constant altitude (with ailerons and rudder maneuver).

Problem 1.2

Consider a Vertical Take Off Landing (VTOL) aircraft, such as the AV8 Harrier aircraft. Assume that the aircraft has a jet engine with a constant angular momentum $\bar{h} = I_R \bar{\omega}_R$. Derive expressions for h_X, h_Y, h_Z at the following configurations:

- rotor axis at 90° tilted upward with respect to the X axis;
- rotor axis at 60° tilted upward with respect to the X axis;
- rotor axis at 30° tilted upward with respect to the X axis;
- rotor axis aligned with the X axis.

Problem 1.3

Consider a Vertical Take Off Landing (VTOL) aircraft with a thrust vectoring system. Assume that the aircraft has a jet engine with a constant angular momentum $\bar{h} = I_R \bar{\omega}_R$. Derive expressions for h_X, h_Y, h_Z at the following configuration:

- rotor axis at 30° tilted upward with respect to the X axis and at 20° tilted to the right with respect to the X axis.

Problem 1.4

Demonstrate that the relationship between the components of the aircraft velocity in the body axes and the Earth inertial frame (also known as Flight Path Equations) is given by the following:

$$\begin{Bmatrix} \dot{X}' \\ \dot{Y}' \\ \dot{Z}' \end{Bmatrix} = \begin{bmatrix} \cos\Psi\cos\Theta & -\sin\Psi\cos\Phi + \cos\Psi\sin\Theta\sin\Phi & \sin\Psi\sin\Phi + \cos\Psi\sin\Theta\cos\Phi \\ \sin\Psi\cos\Theta & \cos\Psi\cos\Phi + \sin\Psi\sin\Theta\sin\Phi & -\sin\Phi\cos\Psi + \sin\Psi\sin\Theta\cos\Phi \\ -\sin\Theta & \cos\Theta\sin\Phi & \cos\Theta\cos\Phi \end{bmatrix} \begin{Bmatrix} U \\ V \\ Z \end{Bmatrix}$$

Problem 1.5

Consider the general expression of the Conservation of Linear Momentum Equations (CLMEs) with respect to the body axes X, Y, Z.

$$m(\dot{U} + QW - RV) = -mg\sin\Theta + (F_{A_X} + F_{T_X})$$
$$m(\dot{V} + UR - PW) = mg\cos\Theta\sin\Phi + (F_{A_Y} + F_{T_Y})$$
$$m(\dot{W} + PV - QU) = mg\cos\Theta\cos\Phi + (F_{A_Z} + F_{T_Z})$$

Demonstrate step-by-step how the introduction of the steady-state and perturbed flight conditions—along with the introduction of the small perturbation assumptions—leads to the following small perturbations expression for the CLMEs with respect to the body axes X, Y, Z.

$$m[\dot{u} + Q_1 w + qW_1 - R_1 v - rV_1] = -mg\theta\cos\Theta_1 + (f_{A_X} + f_{T_X})$$
$$m[\dot{v} + U_1 r + uR_1 - P_1 w - pW_1] = -mg\theta\sin\Phi_1\sin\Theta_1 + mg\phi\cos\Phi_1\cos\Theta_1 + (f_{A_Y} + f_{T_Y})$$
$$m[\dot{w} + P_1 v + pV_1 - Q_1 u - U_1 q] = -mg\theta\cos\Phi_1\sin\Theta_1 - mg\phi\sin\Phi_1\cos\Theta_1 + (f_{A_Z} + f_{T_Z})$$

Specifically, explain why and how different terms cancel out, leading to the previous expressions.

Problem 1.6

Consider the general expression of the Conservation of Angular Momentum Equations (CAMEs) with respect to the body axes X, Y, Z.

$$\dot{P} I_{XX} - \dot{R} I_{XZ} - PQ I_{XZ} + RQ(I_{ZZ} - I_{YY}) = L_A + L_T$$
$$\dot{Q} I_{YY} + PR(I_{XX} - I_{ZZ}) + (P^2 - R^2) I_{XZ} = M_A + M_T$$
$$\dot{R} I_{ZZ} - \dot{P} I_{XZ} + PQ(I_{YY} - I_{XX}) + QR I_{XZ} = N_A + N_T$$

Demonstrate step-by-step how the introduction of the steady-state and perturbed flight conditions—along with the introduction of the small perturbation assumptions leads—to the following small perturbations expression for the CAMEs with respect to the body axes X, Y, Z.

$$\dot{p} I_{XX} - \dot{r} I_{XZ} - (P_1 q + Q_1 p) I_{XZ} + (R_1 q + Q_1 r)(I_{ZZ} - I_{YY}) = (l_A + l_T)$$
$$\dot{q} I_{YY} + (P_1 r + p R_1)(I_{XX} - I_{ZZ}) + (2 P_1 p - 2 R_1 r) I_{XZ} = (m_A + m_T)$$
$$\dot{r} I_{ZZ} - \dot{p} I_{XZ} + (P_1 q + p Q_1)(I_{YY} - I_{XX}) + (Q_1 r + R_1 q) I_{XZ} = (n_A + n_T)$$

Specifically, explain why and how different terms out cancel out, leading to the previous expressions.

Problem 1.7

Consider the Inverted Kinematic Equations (IKEs) under the assumption of small perturbations:

$$p = \dot{\phi} - \dot{\Psi}_1 \theta \cos \Theta_1 - \dot{\psi} \sin \Theta_1$$
$$q = -\dot{\Theta}_1 \phi \sin \Phi_1 + \dot{\theta} \cos \Phi_1 - \dot{\Psi}_1 \theta \sin \Phi_1 \sin \Theta_1 + \dot{\Psi}_1 \phi \cos \Phi_1 \cos \Theta_1 + \dot{\psi} \sin \Phi_1 \cos \Theta_1$$
$$r = -\dot{\Psi}_1 \theta \cos \Phi_1 \sin \Theta_1 - \dot{\Psi}_1 \phi \sin \Phi_1 \cos \Theta_1 + \dot{\psi} \cos \Phi_1 \cos \Theta_1 - \dot{\Theta}_1 \phi \cos \Phi_1 - \dot{\theta} \sin \Phi_1$$

Explain under which conditions these equations can be reduced to the following expressions:

$$p = \dot{\phi}$$
$$q = \dot{\theta}$$
$$r = \dot{\psi}$$

Problem 1.8

Consider the drawings here with the dimensions of four extra large aircraft.

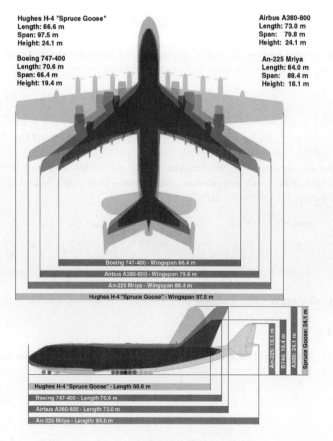

Figure P1.8.1 (Source: http://upload.wikimedia.org/wikipedia/commons/9/96/Giant_Plane_Comparison.jpg)

Using your best technical judgment, rank the values of all the moments of inertia for the different aircraft. Document your response with simple calculations using approximate distances.

Problem 1.9

Consider the F111 aircraft shown in this drawing.

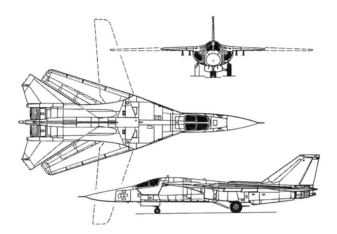

Figure P1.9.1 3D View of the General Dynamic F-111 Aircraft
(Source: http://www.aviastar.org/index2.html)

Using your best technical judgment, explain how the values of all the moments of inertia change when the configuration of the aircraft varies from the low-subsonic forward wing position (low sweep angle) to the high-subsonic, high-sweep angle position.

Problem 1.10

Consider the Lockheed L-1011 Tristar aircraft shown in the drawing.

Assuming that at steady-state conditions the throttle setting is the same for each of the three aircraft engines, first introduce generic distances along the body axes X, Y, Z of the distances of each of the engines with respect to the aircraft center of gravity. Next, qualitatively describe how the terms $F_{T_X}, F_{T_Y}, F_{T_Z}, L_T, M_T, N_T$ in the CLMEs and CAMEs are modified under the following engine failure conditions:

- failure of the tail engine;
- failure of the wing engine on the right of the pilot.

Assume that in both cases the failures occur while the aircraft is at steady-state rectilinear flight conditions.

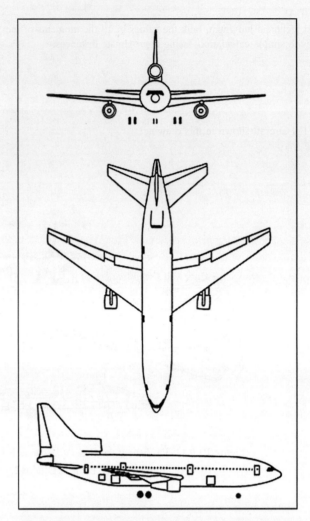

Figure P1.10.1 3D View of the Lockheed L-1011 Aircraft
(Source: http://www.aviastar.org/index2.html)

Chapter 2

Review of Basic Concepts of Aerodynamic Modeling

TABLE OF CONTENTS

2.1 Introduction
2.2 Review of Key Aerodynamic Characteristics for Wing Sections
2.3 Wing Planforms and Wing Lift Curve Slope
2.4 Review of the Downwash Effect and Effectiveness of Control Surfaces
2.5 Determination of the Aerodynamic Center for Wing and Wing + Fuselage
2.6 Approaches to the Modeling of Aerodynamic Forces and Moments
    2.6.1 Wind Tunnel Analysis
    2.6.2 CFD Analysis
    2.6.3 Parameter IDentification from Flight Data
    2.6.4 Correlation from Wind Tunnel Data and Empirical "Build-Up" Analysis
2.7 Summary
References
Student Sample Problems
Problems

2.1 INTRODUCTION

The purpose of this section is to provide a brief review of key aerodynamic concepts starting from the level of wing sections followed by an extension to the wing planform and to the wing+fuselage system. The important concepts of the wing lift curve slope, the downwash effect on the horizontal tail, and aerodynamic center for wing and wing + body are also reviewed. It is assumed that the students have been exposed to an undergraduate level course in aerodynamics. A review of the basic aerodynamic concepts and tools in the well-known textbook by Anderson[1] is recommended to students approaching the modeling of the aerodynamic forces and moments acting on the aircraft. A brief description of all the available approaches to the aerodynamic modeling is also provided.

The goal is to introduce basic tools which will later be used for the aerodynamic modeling outlined in Chapter 3 and in Chapter 4 for a detailed understanding of the longitudinal and lateral directional aerodynamic forces.

2.2 REVIEW OF KEY AERODYNAMIC CHARACTERISTICS FOR WING SECTIONS

The purpose of this section is to provide a brief review of key aerodynamic concepts starting from the level of wing sections. The study of wing sections (also known in the technical literature as airfoil sections) has been the focus of a comprehensive effort in the 1940s and 1950s by NASA (at that time called NACA) researchers. That effort led to a massive amount of experimental data and results summarized in the well-known "Theory of Wing Sections" handbook.[2]

A generic 3D wing can be considered as a sequence of wing sections along the wing span,[3] as shown in Figure 2.1.

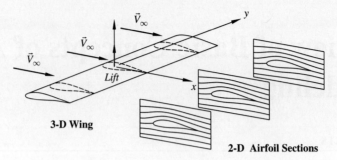

Figure 2.1 3D Wing as a Sequence of 2D Wing Sections

The aerodynamic behavior of a 2D wing section is a function of the following geometric parameters, which are shown in Figure 2.2:

- Camber, also referred as the shape of the mean line. A specific subset of wing sections is given by the so-called symmetric profiles, where the mean line is a straight line;
- Maximum thickness to chord ratio, expressed as a percentage and indicated by $(t/c)|_{MAX}$;
- Shape of the leading edge, indicated by the radius of the leading edge r_{LE};
- Shape of the trailing edge, indicated by the angle Φ_{TE}. Note that the value of Φ_{TE} is dependent on the value of the previous parameters.

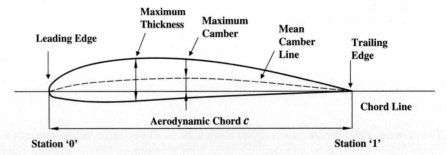

Figure 2.2 Geometric Parameters of a Wing Section

The following parameters are required for uniquely identifying a wing section:

- A relationship for the mean line;
- A relationship for the thickness;
- A relationship for the leading-edge radius.

Abbott and Von Doenhoff[2] provide an interesting historical review of the extensive research on wing sections that started in the 1920s and 1930s in Europe and in the United States. The current classification of wing sections is based on a number of "families"; the differences among these families are associated with different methods for setting the equations for the mean line, the thickness ratio, and the leading-edge radius.

The most important families of wing sections are

NACA 4-digit wing sections
NACA 5-digit wing sections
NACA 6-series wing sections.

Interested students are strongly encouraged to consult[2] to get familiar with the different families of wing sections and the trends of their aerodynamic characteristics as function of their geometric parameters.

A common approach for drawing the wing sections is the so-called canonical form, in which all the distances are dimensionless through the ratio with respect to the chord. Figure 2.3 shows the canonical form for the NACA 0012 wing sections.

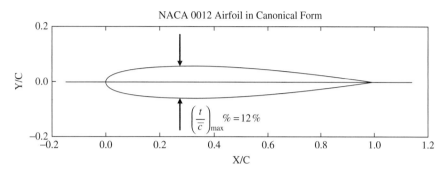

Figure 2.3 NACA 0012 Wing Section[4]

Specifically, the wing section in Figure 2.3 belongs to the subset of symmetric profiles. These profiles are typically used for control surfaces, such as horizontal tail and vertical tail, because they provide equal aerodynamic forces regardless of the direction of the deflections.

The aerodynamic forces and the pitching moment acting on a wing section are shown in Figure 2.4. The aerodynamic drag is aligned with the direction of the free-stream velocity vector while the aerodynamic lift is perpendicular to the drag.

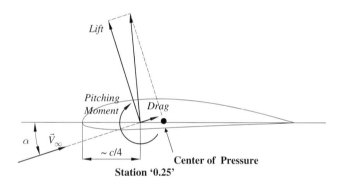

Figure 2.4 Aerodynamic Forces on a Wing Section

Note that the described modeling assumes the forces acting on the wing section aerodynamic center—in lieu of the wing section center of pressure (CP)—through the introduction of the wing section pitching moment M acting at the aerodynamic center. The wing section CP is essentially the point of application of the resultant of the aerodynamic forces L and D. The wing section aerodynamic center x_{AC} is instead defined as the point with respect to which the pitching moment coefficient does not change with the aerodynamic angle of attack α, also referred to as the longitudinal flow angle. Therefore, at the aerodynamic center x_{AC} we have the condition $c_{m_\alpha} = 0$. Based on this description, considering the pitching moment as negative in clockwise direction, the pitching moment coefficient with respect to x_{AC} is given by

$$c_{m_{AC}} = -c_l \frac{(x_{CP} - x_{AC})}{c}$$

The x_{AC} is mainly referred to in terms of percentage of the chord using $\bar{x}_{AC} = x_{AC}/c$. The evaluation of this important parameter starts at the level of wing section, and it evolves to the level of wing planform, wing + fuselage system, up to the entire aircraft system. At subsonic conditions, the aerodynamic center is located at 25 percent of the chord for a flat plate.[5] For a wing section with moderate camber and average thickness ratio this value slightly increases.

One of the key aerodynamic parameters of a wing section is the c_l vs. α curve; a sample of this curve is shown in Figure 2.5.

Chapter 2   Review of Basic Concepts of Aerodynamic Modeling

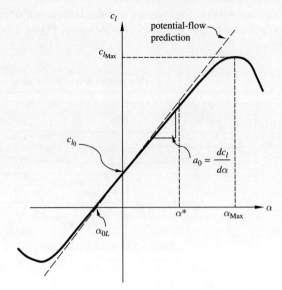

Figure 2.5 Typical c_l vs. α for a Wing Section[3]

The following parameters have critical aerodynamic importance in the above curve:

- α_{0L}: The value of α associated with the condition $c_l = 0$;
- $\alpha_{\max}$: The value of α associated with the condition $c_l = c_{l_{\max}}$;
- α^*: The value of α associated with the end of the linear c_l behavior;
- c_{l_0}: The value of c_l associated with the condition $\alpha = 0$;
- $c_{l_{\text{Max}}}$: The maximum value of c_l.

Particularly, the values of $\alpha_{\max}$ and $c_{l_{\text{Max}}}$ are critical for the analysis of the stall performance of the wing section and, at a larger scale, of the entire wing.

Another key aerodynamic parameter of a wing section is the c_l vs. c_d curve, also known as the drag polar curve. A sample of this curve is shown in Figure 2.6.

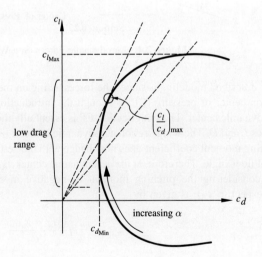

Figure 2.6 Typical c_l vs. c_d (Polar Curve) for a Wing Section[3]

A typical design objective is to operate in the linear α range, providing "enough" c_l with a "low enough" c_d. Toward this goal, the determination of the following aerodynamic parameters is critical:

- $c_{d_{\min}}$, which is typically associated with excessively low values of c_l;
- $c_d\big|_{\left(\frac{c_l}{c_d}\right)_{\max}}$, which is associated with the important condition of $(c_l/c_d)_{\max}$.

Other important aerodynamic parameters of a wing section are the c_d vs. α and the c_m vs. α curves; a sample of these curves is shown in Figure 2.7.

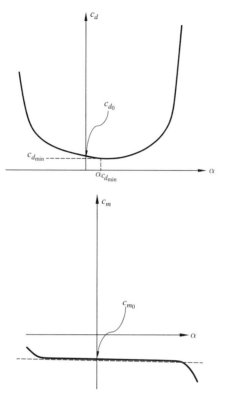

Figure 2.7 Typical c_d vs. α and c_m vs. α Curves for a Wing Section[3]

An experimental example of these curves from wind tunnel analysis is shown in Figure 2.8, relative to two values of the Reynolds number.

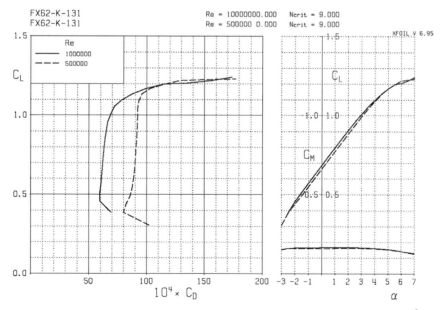

Figure 2.8 Experimental c_l vs. c_d, c_l vs. α, and c_m vs. α Curves for a Wing Section[3]

Following a review of the most important parameter describing the aerodynamic behavior of a wing section, a fundamental task of the designer is to properly consider the following functional relationships between the wing section geometric and aerodynamic parameters:

- variation of the camber (shape of the mean line) ⇒ variations of $c_{l_{\max}}, \alpha_0, c_{m_0}$;
- variation of the maximum thickness ⇒ variations of $c_{l_{\max}}, x_{AC}$;
- variation of the leading edge ⇒ variations of $\alpha^*, c_{l_{\max}}$.

A critical aerodynamic concept for a wing section is the lift distribution. For a generic nonsymmetric wing section, the lift distribution includes the following components:

- "Basic" lift distribution, which depends on the curvature of the mean line. This lift distribution is associated with the negative pitching moment $c_{m_{AC}}$;
- "Additional" lift distribution, which is associated with the angle of attack α.

By definition, a symmetric wing section will only feature the "additional" lift distribution; therefore, this type of wing sections will not provide lift when $\alpha = 0$.

A key aerodynamic characteristic of a wing section is the so-called lift-curve slope gradient c_{l_α} indicated in the c_l vs. α curve previously described. The c_{l_α} relationship with Mach number is described by the Prandtl-Glauert transformation:[1,6]

$$c_{l\alpha}|_{Mach} = \frac{c_{l\alpha}|_{Mach=0}}{\sqrt{1 - Mach^2}}$$

This relationship provides realistic c_{l_α} values at subsonic conditions; it is not valid around transonic conditions, approximately defined as $Mach \in [0.85 - 1.15]$.

A final consideration about the aerodynamic characteristics of the wing section is on the dependency of the location of the aerodynamic center with Mach. At subsonic conditions, for moderately thick and average camber wing sections, $\bar{x}_{AC} \in [0.25 - 0.27]$; however, the wing section $\bar{x}_{AC}$ shifts suddenly to 0.5 during transonic flight conditions. This aerodynamic phenomenon is known as tuck. When extrapolated for the aerodynamics of the entire aircraft system, this drastic shift has major implications on the overall dynamic stability of the aircraft. Therefore, the aerodynamic robustness to the transonic crisis due to the sudden shift of $\bar{x}_{AC}$ is a major design factor for the wings of supersonic aircraft.

2.3 WING PLANFORMS AND WING LIFT CURVE SLOPE

The integration of a continuous string of wing sections along the direction of the wing span leads to a wing planform, commonly known as the wing. The wing planform is clearly shown in the top view of a Boeing 747 aircraft in Figure 2.9. From a geometric construction point, the wing area might be considered to be extended to the aircraft longitudinal axis (as shown in the figure), also known as reference wing surface, or it might be considered to end at the intersection with the aircraft fuselage, also known as exposed wing surface.

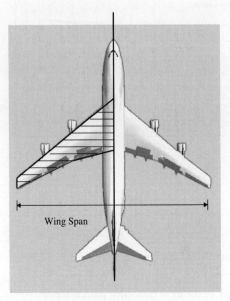

Figure 2.9 B747 Wing Planform

For the purpose of outlining clearly the geometric parameters of a planform, only the left semiwing is shown in Figure 2.10.

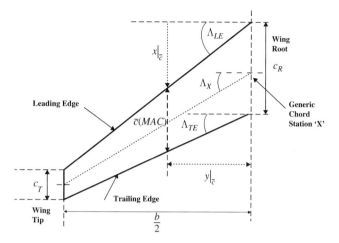

Figure 2.10 Generic Straight Wing Planform

The planform shown is relative to a so-called straight wing. This implies that both leading-edge and trailing-edge angles do not change in the direction of the wing span. An example of a nonstraight wing is shown in Figure 2.11, relative to the specific case of varying sweep angles on the trailing edge.

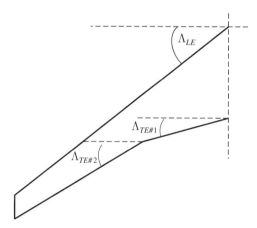

Figure 2.11 Nonstraight Wing Planform

The following geometric parameters are identified for a generic straight wing planform:[6]

$\lambda = \frac{c_T}{c_R} \in [0 - 1]$: taper ratio

$AR = \frac{b^2}{S}$: aspect ratio

$\Lambda_{LE}, \Lambda_{TE}$: leading-edge and trailing-edge sweep angles

Special cases of wing planforms associated with specific values of the above parameters are

$\lambda = 1, \quad \Lambda_{LE} = 0°$: rectangular wing
$\lambda = 0, \quad \Lambda_{LE} > 0°, \Lambda_{TE} = 0°$: delta wing

The fundamental importance of the sweep angle for aircraft operating at high subsonic conditions is shown in Figure 2.12.

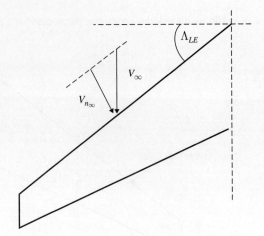

Figure 2.12 Airspeed on the Wing Planform

The airspeed component associated with the generation of the lift on a wing planform is the normal component $V_{n\infty}$ given by

$$V_{n\infty} = V_\infty \cos(\Lambda_{LE})$$

Clearly, we have $V_{n\infty} < V_\infty$; therefore, one effect of the sweep angle (Λ_{LE}) is a decrease of the speed of the airflow around the wing section. This, in turn, increases the value of the parameter M_{CRIT}, which is defined as the upstream value of M at which somewhere on the wing section the transonic condition $M = 1$ is reached. The critical importance of the sweep angle is therefore to allow wings with moderate thickness ratio to fly at fairly high subsonic conditions without experiencing the problems associated with the transonic crisis. Virtually all the commercial jetliners take advantage of the so-called sweep angle cosine bonus. allowing cruise speed around $M \in [0.85 - 0.9]$ with $\Lambda_{LE} \in [25 - 35]°$.

Using the described geometric parameters. the following geometric characteristics can be introduced for straight wing planforms:[6]

- Wing surface: $S = \frac{b}{2} c_R (1 + \lambda)$;
- Mean aerodynamic chord (MAC): $MAC = \bar{c} = \frac{2}{3} c_R \frac{(1 + \lambda + \lambda^2)}{(1 + \lambda)}$;
- Sweep angle at a generic chord station x: $\tan \Lambda_x = \tan \Lambda_{LE} - \frac{4x(1 - \lambda)}{AR(1 + \lambda)}$ where $x \in [0 - 1]$ ($x = 0$ at the leading edge, $x = 1$ at the trailing edge);
- Location of MAC on the lateral axis (w.r.t. wing span): $y_{MAC} = \frac{b}{6} \frac{(1 + 2\lambda)}{(1 + \lambda)}$;
- Location of MAC on the longitudinal axis (w.r.t. c_R): $x_{MAC} = \frac{b}{6} \frac{(1 + 2\lambda)}{(1 + \lambda)} \tan(\Lambda_{LE})$.

The above geometric parameters can be all extended to the general case of a nonstraight wing planform. Consider for example the two-section nonstraight wing planform shown in Figure 2.13.

The previously introduced geometric parameters can be extended for a two-section nonstraight wing planform as in the following:

- Wing surface;

$$S_1 = \frac{b_1}{2} c_{R_1}(1 + \lambda_1), \quad S_2 = \frac{b_2}{2} c_{R_2}(1 + \lambda_2), \quad S = (S_1 + S_2)$$

- Mean aerodynamic chord (MAC);

$$MAC_1 = \bar{c}_1 = \frac{2}{3} c_{R_1} \frac{(1 + \lambda_1 + \lambda_1^2)}{(1 + \lambda_1)}, \quad MAC_2 = \bar{c}_2 = \frac{2}{3} c_{R_2} \frac{(1 + \lambda_2 + \lambda_2^2)}{(1 + \lambda_2)}$$

$$MAC = \bar{c} = \frac{(\bar{c}_1 \cdot S_1) + (\bar{c}_2 \cdot S_2)}{S}$$

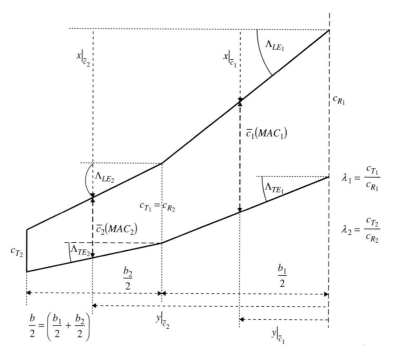

Figure 2.13 Two-Section Nonstraight Wing Planform

- Sweep angle at a generic station x along the first or the second section;

$$\tan \Lambda_{x_1} = \tan \Lambda_{LE_1} - \frac{4x_1(1-\lambda_1)}{AR_1(1+\lambda_1)}$$

$$\tan \Lambda_{x_2} = \tan \Lambda_{LE_2} - \frac{4x_2(1-\lambda_2)}{AR_2\ (1+\lambda_2)}$$

where $x \in [0-1]$ ($x = 0$ at the leading edge, $x = 1$ at the trailing edge)
- Location of MAC on the lateral axis;

$$y_{MAC_1} = \frac{b_1}{6}\frac{(1+2\lambda_1)}{(1+\lambda_1)}, \quad y_{MAC_2} = \frac{b_2}{6}\frac{(1+2\lambda_2)}{(1+\lambda_2)}$$

$$y_{MAC} = \frac{(y_{MAC_1}\cdot S_1) + (y_{MAC_2}\cdot S_2)}{S}$$

- Location of MAC on the longitudinal axis.

$$x_{MAC_1} = \frac{b_1}{6}\frac{(1+2\lambda_1)}{(1+\lambda_1)}\tan(\Lambda_{LE_1}), \quad x_{MAC_2} = \frac{b_2}{6}\frac{(1+2\lambda_2)}{(1+\lambda_2)}\tan(\Lambda_{LE_2})$$

$$x_{MAC} = \frac{(x_{MAC_1}\cdot S_1) + (x_{MAC\ 2}\cdot S_2)}{S}$$

These parameters are summarized in Figure 2.14.

Note that in the given example the first section of the wing is larger than the second section; therefore, the longitudinal and lateral locations of the mean aerodynamic chord are within the first section.

Without any loss of generality the described relationships can be extended to the case of n-th section nonstraight wing planforms. In general terms, an n-th section nonstraight wing can be approximated to a straight wing with the following geometric parameters:

46 Chapter 2 Review of Basic Concepts of Aerodynamic Modeling

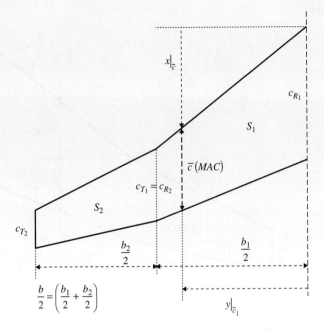

Figure 2.14 Two-Section Nonstraight Wing Planform (Summary)

$$S = \sum_{i=1}^{n} S_i = \sum_{i=1}^{n} \frac{b_i}{2} c_{R_1}(1 + \lambda_i)$$

$$\lambda = \frac{\sum_{i=1}^{n} \lambda_i S_i}{S}$$

$$AR = \frac{\sum_{i=1}^{n} AR_i S_i}{S}$$

The most important aerodynamic parameter associated with the wing planform is the lift-curve slope c_{L_α}. For a given selection of wing section, at a given Reynolds number, a functional relationship for c_{L_α} is given by

$$c_{L_\alpha} = c_{L_\alpha}(\text{Mach}, AR, \lambda, \Lambda_{LE})$$

Wind tunnel analysis is typically required to estimate this critical aerodynamic coefficient at transonic and supersonic conditions. However, at subsonic conditions with $M < M_{CRIT}$, the following empirical relationships can be used for wing planforms with moderate sweep angles $\Lambda_{LE} < [30 - 32]°$, moderate aspect ratio $AR \in [3 - 8]$, and moderately high taper ratio $\lambda \in [0.4 - 1.0]$:

$$c_{L_\alpha} = \frac{2\pi\, AR}{2 + \sqrt{\left\{\left[\frac{AR^2(1 - \text{Mach}^2)}{k^2}\left(1 + \frac{\tan^2(\Lambda_{0.5})}{(1 - \text{Mach}^2)}\right)\right] + 4\right\}}}$$

where $\Lambda_{0.5}$ is the sweep angle calculated at a station at 50 percent of the chord. The constant k is a function of the wing geometry and takes on the following values:

$$\text{for } AR < 4 \quad k = 1 + \frac{AR(1.87 - 0.000233\, \Lambda_{LE})}{100}, \text{ with } \Lambda_{LE} \text{ in rad}$$

$$\text{for } AR \geq 4 \quad k = 1 + \frac{[(8.2 - 2.3\, \Lambda_{LE}) - AR(0.22 - 0.153\Lambda_{LE})]}{100}, \text{ with } \Lambda_{LE} \text{ in rad}$$

This relationship, known as the Polhamus formula,[7] was derived from correlation studies using a large amount of experimental data from wind tunnel analysis of several classes of wing planforms. Caution: Properly use the Polhamus formula within the validity ranges provided above; the use of the Polhamus formula using the wing geometric parameters outside the above range will lead to approximated results. Note that the maximum theoretical value for c_{L_α} is 2π. Figures 2.15, 2.16, and 2.17 provide typical trends for c_{L_α} for classes of wing planforms at subsonic conditions.

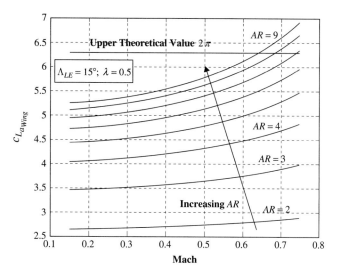

Figure 2.15 c_{L_α} vs. Mach with Varying AR

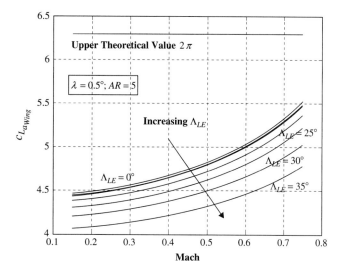

Figure 2.16 c_{L_α} vs. Mach with Varying Λ_{LE}

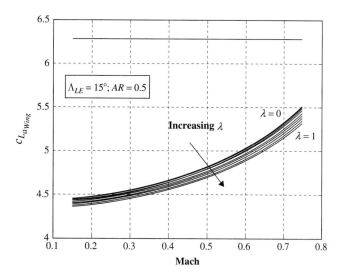

Figure 2.17 c_{L_α} vs. Mach with Varying λ

The trend in the previous figures shows that c_{L_α} increases with increases in aspect ratio (AR), while it substantially decreases with increases in the sweep angle (Λ). Variations of the taper ratio (λ) do not lead to significant effects in the values of c_{L_α}. An interesting observation is that c_{L_α} exceeds the maximum theoretical value (2π) as soon as the aspect ratio exceeds the recommended range $(AR \in [3-8])$. Students and other interested readers are encouraged to make use of this empirical but fairly accurate relationship for estimating c_{L_α} as long as the wing geometric parameters are within the provided ranges.

2.4 REVIEW OF THE DOWNWASH EFFECT AND EFFECTIVENESS OF CONTROL SURFACES

An important longitudinal aerodynamic effect is the so-called downwash effect. In general, this effect can be considered to be an aerodynamic "interference" generated by the wing on the horizontal tail due to the system of vortices created by the wing itself. A view of the system of vortices creating the downwash effect is evident in Figure 2.18.

Figure 2.18 Downwash Effect from a Cessna Citation above a Cloud[8]

A detailed aerodynamic analysis of the system of vortices generated by the lifting wing is available through the vortex laws introduced by Helmotz.[9] In general, it can be said that the magnitude of the downwash effect is strongly dependent on the distribution of the lift along the wingspan.

Classic aerodynamic theory suggests that the optimal elliptical lift distribution is provided by an elliptically shaped wing from wing tip to wing tip. Only a few aircraft have approximated this ideal lift distribution, with the English World War II fighter Supermarine Spitfire aircraft (shown in Figure 2.19) being the most famous example.

Figure 2.19 Supermarine Spitfire Aircraft

Any deviation from this elliptical distribution is reflected through the Oswald efficiency factor $e \in [0 - 1]$ with e taking on lower values as the lift distribution deviates significantly from the elliptical shape.

Most of the aircraft designed prior to World War II featured a somewhat elliptical lift distribution over the wing span. However, with the increase in speed associated with the introduction of jet propulsion, followed later by the introduction of the sweep angle to increase M_{CRIT} (necessary for allowing flying at higher subsonic speed without experiencing the transonic crisis), the lift distribution along the wingspan deviated substantially from the elliptical distribution. Particularly, the simultaneous increase in sweep angles Λ_{LE} along with decreases in AR and λ have all contributed to shift the lift distribution toward the tip of the wing, as shown in Figure 2.20.

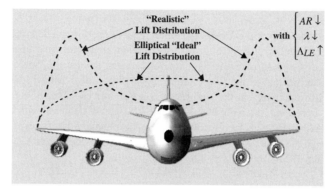

Figure 2.20 Lift Distribution along the Wingspan of a B747

Note that this description is related to steady-state flight conditions and is not associated with the conditions when flaps are deflected. Obviously, the deflections of the flaps at lower speeds has the effect of moving the lift distribution toward the center of the aircraft.

An effect of shifting the lift distribution toward the wing tips is an increase in the strength of the system of vortices at the wing tip, as shown in Figure 2.21.

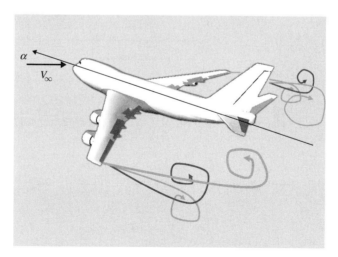

Figure 2.21 Systems of Vortices at the Wing Tips of a B747

From a modeling point of view, the downwash effect can be interpreted as an effective decrease of the longitudinal incidence angle at the horizontal tail, denoted as α_H, compared to the same angle at free stream conditions α. Thus, the modeling of the downwash effect can be introduced as

$$\alpha_H = \alpha - \varepsilon$$

where ε is the downwash angle, which is a function of α. Therefore, using a Taylor expansion around the condition $(\alpha_0 = 0°)$, we have

$$\varepsilon = \varepsilon|_{\alpha=0°} + \frac{d\varepsilon}{d\alpha}\alpha = \varepsilon_0 + \frac{d\varepsilon}{d\alpha}\alpha$$

Next, assuming $\varepsilon_0 \approx 0°$, we have

$$\alpha_H = \alpha - \varepsilon = \alpha - \frac{d\varepsilon}{d\alpha}\alpha = \alpha\left(1 - \frac{d\varepsilon}{d\alpha}\right)$$

Therefore, the modeling of the downwash effect reduces itself to the determination of the gradient $(d\varepsilon/d\alpha)$. From the previous discussion relative to the lift distribution along the wing span, the downwash effect will increase when the lift distribution over the wing span is shifted toward the tips of the wing. Therefore, $(d\varepsilon/d\alpha)$ will increase with an increase in the sweep angle Λ_{LE} and will decrease with an increase in aspect ratio AR and tip ratio λ. It is also intuitive that the strength of the tip vortices—and, therefore, the magnitude of $(d\varepsilon/d\alpha)$—will increase with increasing Mach number. Finally, $(d\varepsilon/d\alpha)$ will decrease with increases in the geometric distances—in both longitudinal and vertical directions—of the horizontal tail with respect to the wing. These distances can be modeled through the dimensionless coefficients m and r shown in Figure 2.22. Typical ranges for the coefficients r and m are $[0.5 - 1.5]$ and $[0.15 - 0.5]$, respectively.

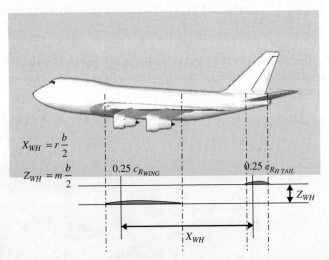

Figure 2.22 Wing-Tail Distances in the Longitudinal and Vertical Directions

From this discussion, a functional relationship for $(d\varepsilon/d\alpha)$ is given by

$$\left(\frac{d\varepsilon}{d\alpha}\right) = f(Mach, m, r, \Lambda_{LE}, \lambda, AR)$$

A closed-form expression for $(d\varepsilon/d\alpha)$ is given by[7,10,11]

$$\left(\frac{d\varepsilon}{d\alpha}\right)\bigg|_{Mach} = \left(\frac{d\varepsilon}{d\alpha}\right)\bigg|_{Mach=0} \sqrt{1 - Mach^2}$$

where

2.4 Review of the Downwash Effect and Effectiveness of Control Surfaces

$$\left(\frac{d\varepsilon}{d\alpha}\right)\bigg|_{Mach=0} = 4.44 \ \left(K_{AR} \ K_\lambda \ K_{mr}\sqrt{\cos\left(\Lambda_{0.25}\right)}\right)^{1.19}$$

with

$$K_{AR} = \frac{1}{AR} - \frac{1}{1+(AR)^{1.7}}, \quad K_\lambda = \frac{10-3\lambda}{7}, \quad K_{mr} = \frac{1-\frac{m}{2}}{(r)^{0.333}}$$

The coefficients are also shown in Figure 2.23.

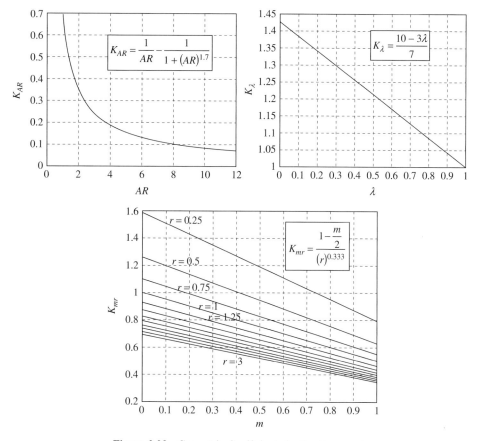

Figure 2.23 Geometric Coefficients for the Downwash

Note that $\Lambda_{0.25}$ is the sweep angle calculated at a station at 25 percent of the chord. This angle can be evaluated using the previously introduced relationship for wing planforms:

$$\tan \Lambda_x = \tan \Lambda_{LE} - \frac{4x(1-\lambda)}{AR \ (1+\lambda)} \quad \rightarrow \quad \tan \Lambda_{0.25} = \tan \Lambda_{LE} - \frac{4 \cdot 0.25(1-\lambda)}{AR \ (1+\lambda)}$$

The empirical relationship has the same range of validity of the previously introduced Polhamus formula for estimating c_{L_α}.

The modeling of the aerodynamic forces associated with the deflection of the conventional subsonic control surfaces (horizontal tail and vertical tail) requires the introduction of additional parameters.

The first set of parameters is associated with the variation in the values of the dynamic pressure $\bar{q} = \frac{1}{2}\rho V^2$ when the flow reaches the tail region. Due to the aerodynamic effects of the wing, a mild decrease of the airspeed values is typically experienced, especially for aircraft with lower values of the X_{WH}, Z_{WH} geometric parameters. This can be modeled through the following dimensionless coefficients, known as dynamic pressure ratios (η):

$$\eta_H = \frac{\frac{1}{2}\rho V_H^2}{\frac{1}{2}\rho V^2} = \frac{\bar{q}_H}{\bar{q}}, \qquad \eta_V = \frac{\frac{1}{2}\rho V_V^2}{\frac{1}{2}\rho V^2} = \frac{\bar{q}_V}{\bar{q}}$$

where the subscripts H and V indicate the horizontal tail and the vertical tail, respectively. A typical range for both coefficients is $[0.85 - 0.95]$ with the higher values associated with aircraft with large values of the geometric distances X_{WH}, Z_{WH}.

A second set of parameters is associated with the specific geometry of the control surfaces located at the horizontal and vertical tails. The design of both horizontal and vertical tails is based on a symmetric profile due to the obvious need of generating equal aerodynamic forces and moments with the same deflections in the opposite directions. A control surface is essentially a portion of the entire tail capable of deflecting around a hinge axis, as shown in Figure 2.24. By convention, the deflections of the control surfaces are assumed positive for trailing-edge down deflections for the horizontal tail and trailing-edge left side with respect to the pilot for the vertical tail.

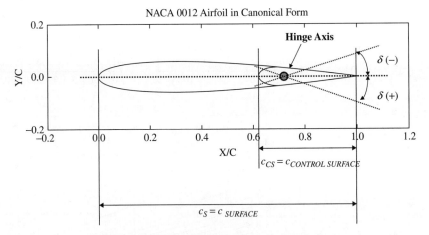

Figure 2.24 Typical Geometry of a Control Surface

From Figure 2.24 visualize that the aerodynamic force associated with the deflection of the control surface is a fraction of the force associated with the deflection of the entire surface. This can be modeled through a generic control surface effectiveness factor τ, which is a function of the ratio c_{CS}/c_S. Typically the value for this factor is around 0.5 for values of c_{CS}/c_S in range of $[0.4 - 0.6]$, as shown in Figure 2.25.

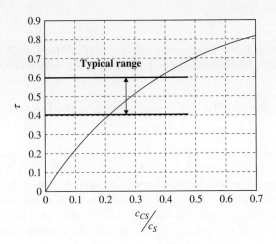

Figure 2.25 Typical Range for Control Surface Effectiveness τ

The control surface effectiveness factors τ_E, τ_A, τ_R will be introduced later for the elevators, ailerons, and rudder, respectively, for modeling of the aerodynamic forces and moments associated with the deflections of those control surfaces.

2.5 DETERMINATION OF THE AERODYNAMIC CENTER FOR WING AND WING + FUSELAGE

In addition to c_{L_α}, another critical aerodynamic parameter for a wing planform is the wing aerodynamic center (AC), indicated as $\bar{x}_{AC_{Wing}}$ or $\bar{x}_{AC_W}$ and expressed as a percentage of $\bar{c}$ (MAC). Once again, the wing aerodynamic center is defined as the point on the $\bar{c}$ where the pitching moment does not change with the angle of attack α.

The same concept also applies to the planform of the horizontal tail. Furthermore, the concept is extended to the aerodynamic center of the wing + fuselage system, indicated as $\bar{x}_{AC_{WB}}$. An approach for the estimation of $\bar{x}_{AC_{WB}}$ is to evaluate first $\bar{x}_{AC_W}$; next, the contribution to the AC from the fuselage is calculated separately;[7,11] therefore, the resulting expression is given by

$$\bar{x}_{AC_{WB}} = \bar{x}_{AC_W} + \Delta \bar{x}_{AC_B}$$

The calculation of $\bar{x}_{AC_W}$ is based on the calculation of the geometric parameter $\left(\frac{x'_{AC}}{c_R}\right)$, shown in Figure 2.26.

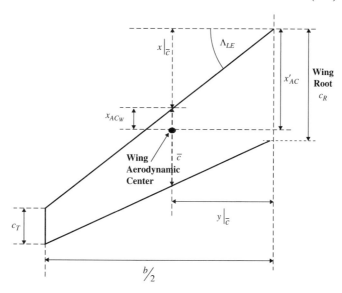

Figure 2.26 Location of the Wing Aerodynamic Center

For subsonic wings with wing sections with moderate thickness and curvature and planform with moderate AR and Λ_{LE} the value of $\left(\frac{x'_{AC}}{c_R}\right)$ is provided through the interpolation from the different curves in Figure 2.27. Next, $\bar{x}_{AC_W}$ is calculated using

$$\bar{x}_{AC_W} = K_1 \left(\frac{x'_{AC}}{c_R} - K_2\right)$$

where the coefficients K_1, K_2 associated with the geometry of the wing are found through the use of Figure 2.28 and Figure 2.29, respectively.

Next, the attention shifts to the calculation of $\Delta \bar{x}_{AC_{WB}}$. Using Munk's theory,[6,7,11] the aerodynamic center shift due to the presence of the fuselage can be calculated using the following relationship:

$$\Delta \bar{x}_{AC_B} = \frac{-\left(\frac{dM}{d\alpha}\right)}{\bar{q} \, S \, \bar{c} \, c_{L_{\alpha W}}} = -\frac{\left(\frac{\bar{q}}{36.5} \frac{c_{L_{\alpha W}}}{0.08}\right)}{\bar{q} \, S \, \bar{c} \, c_{L_{\alpha W}}} \sum_{i=1}^{N} w_{B_i}^2 \left(\frac{d\bar{\varepsilon}}{d\alpha}\right)_i \Delta x_i$$

$$= -\frac{1}{2.92 \cdot S \, \bar{c}} \sum_{i=1}^{N} w_{B_i}^2 \left(\frac{d\bar{\varepsilon}}{d\alpha}\right)_i \Delta x_i$$

54 Chapter 2 Review of Basic Concepts of Aerodynamic Modeling

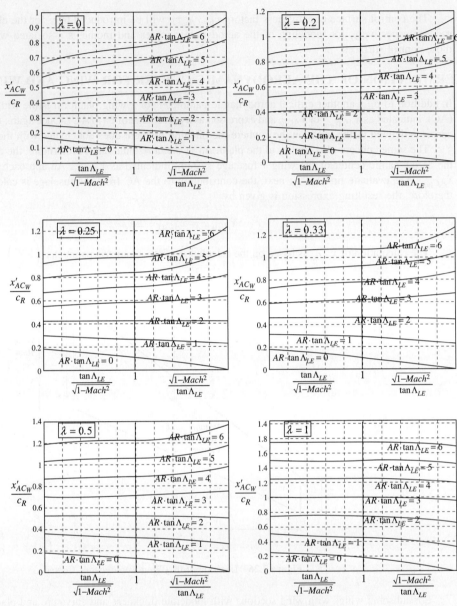

Figure 2.27 $\left(x'_{AC}/c_R\right)$ for Different Airspeed[11]

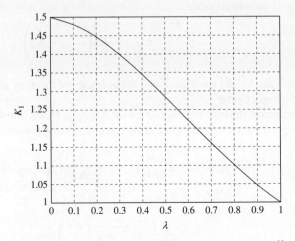

Figure 2.28 K_1 Coefficient for the Calculation of $\bar{x}_{AC_W}$[11]

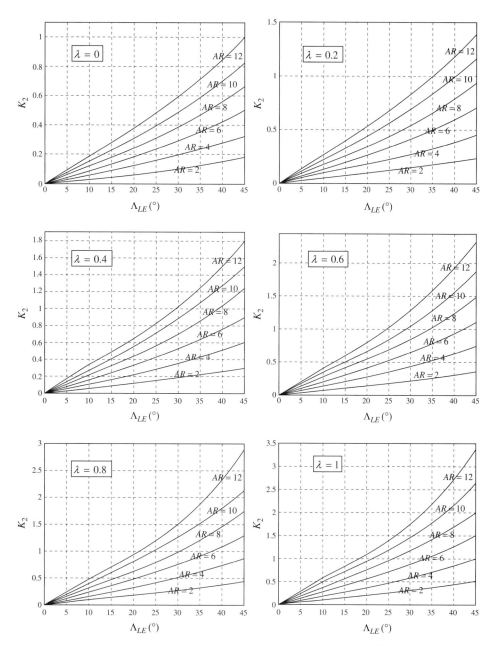

Figure 2.29 K_2 Coefficient for the Calculation of $\bar{x}_{AC_W}$[11]

where w_{B_i}, Δx_i are geometric parameters for N discretized aircraft sections. Particularly, w_{B_i} and Δx_i represent the width and the length of a generic i-th section of the fuselage. A numerical example is shown in Figure 2.30; note that $N = 13$ has been selected without loss of generality.

The parameter x_i represents the longitudinal distance between the midpoint of the i-th section and the leading edge of the root chord (c_{Root}) for the stations of the fuselage ahead of the leading edge of the root chord. Instead, for the stations of the fuselage behind the leading edge of the root chord, x_i represents the distance between the midpoint of the i-th section and the trailing edge of the root chord. Figure 2.30 shows all these parameters for one station in front and one station behind the leading edge of the root chord (the 4th and the 9th stations, respectively). Finally, the parameter x_H represents the longitudinal distance between the aerodynamic center of the horizontal tail (typically located around $0.25 \, \bar{c}_H$) and the trailing edge of the root chord.

The evaluation of the parameter $(d\bar{\varepsilon}/d\alpha)_i$ requires the two curves in Figure 2.31.

First, consider that the fuselage stations ahead of the wing are subjected to an upwash angle effect (indicated as $\bar{\varepsilon}$) while the fuselage stations behind the wing are exposed to a downwash effect (indicated as ε). Therefore,

56 Chapter 2 Review of Basic Concepts of Aerodynamic Modeling

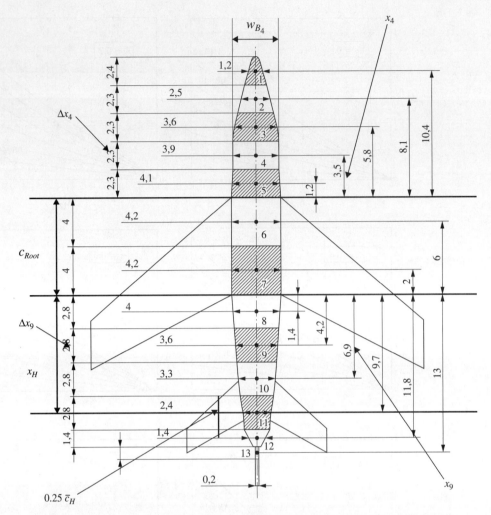

Figure 2.30 Example of Aircraft Sections for Calculations of $\Delta \bar{x}_{AC_{WB}}$ (dimensions in ft.)

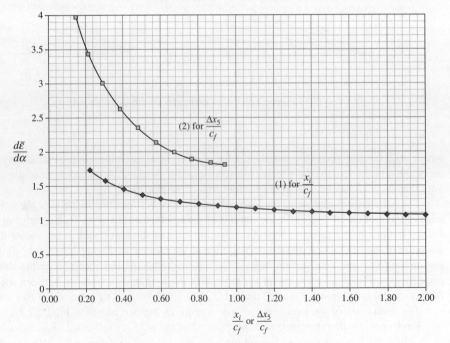

Figure 2.31 Upwash Calculations for $\Delta \bar{x}_{AC_{WB}}$[11]

consider the integers N, M as the number of stations in front and behind the leading edge of the root chord, exposed, respectively, to the upwash and the downwash effects.

Next, $(d\bar{\varepsilon}/d\alpha)_i$ is calculated as in the following: $(d\bar{\varepsilon}/d\alpha)_i$ for $i = 1,..,N-1$ is obtained using curve (1) using the required values for x_i with $c_f = c_{Root}$. Special attention is placed on the Nth section of the fuselage, that is, the section located immediately in the front of the leading edge of the root chord. In fact, for $i = N$, $(d\bar{\varepsilon}/d\alpha)_i$ is obtained using curve (2) in Figure 2.31 (in this specific case using $\Delta x_5 = 2.3$ft).

Finally, for $i = 1,..,M$ (that is, for the stations located behind the leading edge of the root chord), the upwash effect becomes a downwash effect; therefore, we have

$$\left(\frac{d\bar{\varepsilon}}{d\alpha}\right)_i = \left(\frac{x_i}{x_H}\right)\left(1 - \frac{d\varepsilon}{d\alpha}\bigg|_{m=0}\right)$$

where the value of $d\varepsilon/d\alpha\big|_{m=0}$ is the downwash effect evaluated with $m = 0$.

The partitioning of the aircraft fuselage into a number of discrete sections is known in the technical literature as the Multhopp strip-integration method.[7,11]

2.6 APPROACHES TO THE MODELING OF AERODYNAMIC FORCES AND MOMENTS

Starting from the review of the key aerodynamic concepts in the previous sections, the objectives of Chapters 3 and 4 are to provide basic tools for the modeling of the aerodynamic forces and moments acting on the aircraft. According to the small perturbation theory in Chapter 1, the aerodynamic forces and moments are given by

Aerodynamic Longitudinal Forces and Moment

Along X-axis $\quad F_{A_X} = F_{A_{X_1}} + f_{A_X}$

Along Z-axis $\quad F_{A_Z} = F_{A_{Z_1}} + f_{A_Z}$

Around Y-axis $\quad M_A = M_{A_1} + m_A$

Aerodynamic Lateral Directional Force and Moments

Around X-axis $\quad L_A = L_{A_1} + l_A$

Around Z-axis $\quad N_A = N_{A_1} + n_A$

Along Y-axis $\quad F_{A_Y} = F_{A_{Y_1}} + f_{A_Y}$

There are a number of different approaches available today in the aerospace industry for estimating aerodynamic forces and moments acting on the aircraft. The selection of the specific general approach is a complex function of several factors, including but not limited to budget and time constraints, conventional versus unconventional aircraft configurations, and desired level of accuracy. A list of these general approaches is provided here:

- Wind tunnel analysis;
- Computation Fluid Dynamics (CFD) analysis;
- Parameter identification (PID) from flight data;
- Correlation from wind tunnel data and empirical "*Build-Up*" analysis.

Brief descriptions of the approaches are provided next.

2.6.1 Wind Tunnel Analysis

The highest fidelity and lowest risk methodology for an accurate estimate of all the aircraft aerodynamic characteristics is through a well-planned and well-executed wind tunnel test program. Unfortunately, this method of obtaining aerodynamic data can also be very expensive and time-consuming.

Wind tunnels are essentially machines that allow the "flight" of an aircraft on the ground. As a general description, they are tubelike structures or passages in which wind is produced, using a large fan, allowing airflows over objects such as full-size aircraft, engines, wings, or models of these objects. A stationary scale model of the aircraft or the aircraft itself is placed in the test section of a tunnel and connected to instruments measuring and recording airflow around the object and the aerodynamic forces and moments acting upon it. From all the data gathered in these observations, wind tunnel engineers can then determine the behavior of an aircraft or its components at takeoff, cruise, descent, landing, and other flight conditions of interest. Wind tunnels are also extensively used for the purpose of testing specific modifications for an existing aircraft.

NASA has a long history in the development of wind tunnels, starting from the early 1920s. Currently NASA operates more than thirty wind tunnels located at NASA Langley (Virginia), NASA Ames (California), NASA Glenn (Ohio), and NASA Marshall (Alabama), with test sections ranging from a few inches square up to sections large enough to test a full-size aircraft.[12] A number of wind tunnels are also operated at major research and academic institutions all over the world.

Some wind tunnels are specifically designed for the study of the shapes of wings and fuselage. Other wind tunnels are used exclusively for the testing of propulsion systems. In general, wind tunnels are classified according to the speed of the air flowing through the tunnel. Therefore, we have subsonic, transonic, supersonic, and hypersonic wind tunnels. Wind tunnels of the latest generation are equipped with lasers for a technique called Laser Doppler Velocimetry (LDV). This is a nonintrusive technique allowing accurate determination of velocities with light beams. The light beams do not-interfere with the airflow, which is a known problem with the instrumentation requiring a physical presence in the test section.

2.6.2 CFD Analysis

Computational Fluid Dynamics (CFD) is a specific branch of fluid mechanics that features numerical methods and algorithms to solve and analyze problems involving fluid flows. Computers are used to perform the millions of calculations required to simulate the interaction of fluids with the complex aircraft shape. Until a few years ago, even with simplified equations and high-speed supercomputers, only approximate solutions could be achieved for many practical applications. Today ongoing research and recent development of sophisticated algorithms have improved the accuracy and speed of complex simulation scenarios, especially for transonic or turbulent flows. Validation of such software is often performed using a wind tunnel.

The fundamental basis of any CFD problem is the Navier-Stokes equations, which define any single-phase fluid flow. Toward reaching approximate solutions, these equations can be simplified by removing terms modeling viscosity to yield the Euler equations. Further simplifications by removing terms describing vorticity lead to the full potential equations. These equations can be linearized, yielding the linearized potential equations. Historically, methods were first developed to solve the linearized potential equations. Two-dimensional methods, using conformal transformations of the flow about a cylinder to the flow about an airfoil, were developed in the 1930s. The computational power associated with the introduction of analog computers and later digital computers led to the development of 3D methods starting in the 1960s.

The solution of the Navier-Stokes equations around the geometry of the aircraft is the ultimate target for CFD purposes. 2D codes first emerged in the 1980s and early 1990s. A number of 3D codes were more recently introduced, leading to numerous commercial packages running on a number of systems ranging from conventional desktops to high-performance workstations, parallel computers, and supercomputers.

In all of the detailed CFD approaches, the same basic procedure applies:

Preprocessing

- The geometry of the aircraft system is defined;
- The volume occupied by the fluid is divided into discrete cells (mesh). The mesh may be uniform or nonuniform;
- The physical modeling is defined, for example, the equations of motions + enthalpy + radiation + species conservation;
- Boundary conditions are defined. This involves specifying the fluid behavior and properties at the boundaries of the problem. For transient problems, the initial conditions are also defined.

Simulation

- The equations are solved iteratively as a steady-state or transient.

Postprocessing

- For the analysis and visualization of the resulting solution.

In recent years CFD codes have been extensively used for estimating the aerodynamic characteristics of several aircraft. At times, a hybrid approach is used in which the aerodynamic behavior is predicted with a combination of CFD codes and wind tunnel analysis. An example of this approach was given by the aerodynamic design of the NASA Space Shuttle. A measure of the effectiveness of such methods can be found in.[13] Particularly, interested students are referred to[13] for an appreciation of the level of uncertainty one can expect from empirically derived and CFD-based aero models.

The advantages of the application of CFD-based approaches for accurate evaluation of the aircraft aerodynamic behavior are lower costs and lower time compared with a detailed wind tunnel analysis. This trend has been accelerated in recent years by the widespread availability of commercial CFD packages, along with the availability of massive computational power at a reasonable cost.

The main drawback of this technology is that for nonconventional configurations or high-speed aircraft, CFD-based results might still have high levels of uncertainty for some of the aerodynamic coefficients.

2.6.3 Parameter IDentification from Flight Data

The determination of the parameters describing the aerodynamic behavior of the aircraft from flight data is the most recent approach. The first and intuitive drawback of this approach is that it cannot be used for the actual design of the aircraft, since it requires the availability of flight data from the aircraft itself. However, the application of this technology provides a number of advantages, such as the validation of the methods originally used for the estimates of the aerodynamic coefficients, the assessment of the performance of flight control laws for an analysis of the closed-loop characteristics, as well as the possibility of evaluating the new aerodynamic coefficients following modifications to an existing aircraft (without having to use expensive wind tunnel analysis).

The general idea associated with this approach is to develop a mathematical model following the processing of the input and output data of the aircraft. In the technical literature, this process is known as PID, **P**arameter **ID**entification. A conceptual block diagram of the PID process is shown in Figure 2.32.

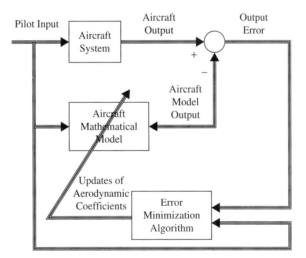

Figure 2.32 Conceptual Block Diagram of the Aircraft PID Process

For a given aircraft fully instrumented for recording pilot input data along with conventional flight data and for a given general mathematical model of the aircraft with initial estimates of the aerodynamic coefficients, the objective is to use an error minimization scheme minimizing iteratively the (quadratic) difference between the actual aircraft output and the output of the mathematical model. Following a number of iterations, the error minimization algorithm is expected to provide corrections to the values of the aerodynamic coefficients, which will eventually lead to a matching between the output of the model and the actual aircraft output. An example of this matching is shown in Figure 2.33, relative to the application of the PID process to the WVU jet-powered YF-22 research aircraft.[14]

Following the introduction of some empirical matching techniques in the 1950s,[15,16] researchers at NASA Dryden in the 1960s and 1970s pioneered the use of PID techniques for extracting all the aerodynamic parameters from flight data. The first formal PID method was the Maximum Likelihood (ML) method,[17,18,19] which was extensively used by NASA researchers for estimating the aerodynamic coefficients from a number of military and research aircraft, including the space shuttle. Within the ML method the error minimization algorithm in Figure 2.32 was based on the Newton-Raphson algorithm, which is essentially a minimization algorithm based on the use of the Hessian matrix.[17,18,19] In the 1980s the availability of digital computers allowed the widespread application of this technology. Other suitable techniques were all based on modification of the wellknow least squares (LS) algorithms.[20,21] All the above techniques—based on analysis in the time domain— were suitable for offline application of the PID process.

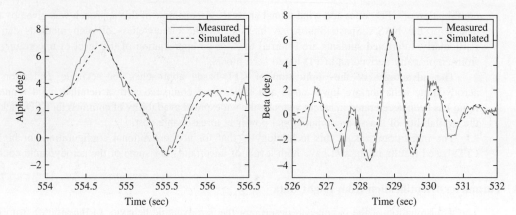

Figure 2.33 Comparison between WVU YF-22 Flight Data and Mathematical Model[14]

In recent years, the introduction of frequency domain-based techniques along with the availability of substantial on-board computational power has given the potential of allowing estimates of the aerodynamic coefficients in real time.[22] Frequency-domain techniques have the important advantage of filtering the effects of the noise on the flight data along with a lower computational complexity compared to the time domain-based PID techniques. The application of real-time PID techniques can potentially allow the design of advanced flight control laws using real-time estimates of the aerodynamic coefficients for assisting the pilot in the difficult tasks of regaining control of an aircraft following adverse flight conditions, such as those associated to failures of primary control surfaces.

In addition to a large number of research publications in this fascinating topic, interested readers are referred to[23] and,[24] which are two outstanding textbooks providing an excellent review of all the currently available techniques for estimating aerodynamic coefficients from flight data.

2.6.4 Correlation from Wind Tunnel Data and Empirical "Build-Up" Analysis

While wind tunnel analysis and, more recently, CFD-based analysis are the main approaches for a detailed evaluation of the aerodynamic coefficients leading to an estimate of the aerodynamic forces and moments acting on the aircraft, for typical subsonic configurations featuring a conventional wing + fuselage + horizontal tail + vertical tail architecture, a suitable low-budget approach consists of the use of empirical methods using data and relationships derived from correlation studies from years of wind tunnel analysis performed by NASA researchers in the 1940s and 1950s.

Specifically, correlation studies on this massive quantity of experimental data have led to a number of closed-form relationships for all the aerodynamic coefficients contributing to the aerodynamic forces and moments. The well-known DATCOM[11] is the single most important reference for an aircraft designer for tackling this task;[7] also provides a summary of these important tools.

In more recent years some of these aerodynamic tools have been packaged within user-friendly commercial codes which calculate the aircraft aerodynamic coefficients—and therefore the aerodynamic forces and moments—for a given specified aircraft geometry. The most relevant examples of these sophisticated yet user-friendly computer codes are the Advanced Aircraft Analysis (AAA), developed by DARCorporation.[25] and SimGen, developed by Bihrle Applied Research.[26] Both codes provide essentially all the tools necessary for the complete aircraft design.

In the following sections of this chapter, the students will be introduced to a detailed explanation of the modeling for the aerodynamic forces and moments acting on an aircraft. Additionally, simple but effective modeling techniques suitable for conventional subsonic configurations allowing detailed estimates of the aerodynamic coefficients will be introduced. The framework for this modeling is valid even for supersonic and more sophisticated aircraft configurations. However, for these configurations, the use of wind tunnel analysis or CFD-based methods is also encouraged.

2.7 SUMMARY

This chapter has provided a brief review of key aerodynamic concepts needed for modeling aerodynamic forces and moments in Chapters 3 and 4. Additionally, brief descriptions of the available approaches to the modeling of the aerodynamic characteristics have been provided. Chapters 3 and 4 will provide a detailed description of the DATCOM-based modeling of the longitudinal and lateral-directional aerodynamic forces and moments.

REFERENCES

1. Anderson, J. D. *Fundamentals of Aerodynamics*. McGraw Hill, Inc., 1984.
2. Abbott, I. H., Von Doenhoff, A. E. *Theory of Wing Sections*. Dover Publications Inc., 1959.
3. MIT OpenCourseWare (OCW). Fall 2005–Spring 2006 course notes.
4. Chang I.-C., Torres F. J., Tung C. "Geometric Analysis of Wing Sections." NASA Technical Memorandum 110346, NASA, April 1995.
5. http://wright.nasa.gov/airplane/ac.html
6. Roskam, J. *Airplane Flight Dynamics and Automatic Flight Controls–Part I*. Design, Analysis, and Research Corporation, Lawrence, KS, 1995
7. Roskam, J. *Airplane Design. Part I – Part VIII*. Roskam Aviation and Engineering Corporation, Lawrence, KS, 1990.
8. http://www.grc.nasa.gov/WWW/K-12/airplane/downwash.html (Picture sent to NASA by Jan-Olov Newborg and originally taken by Paul Bowens).
9. Kuethe, A. M., Schetzer. J. D. *Foundation of Aerodynamics*. John Wiley & Sons, New York, 1959.
10. Pamadi, B. N. "Performance, Stability, Dynamics, and Control of Airplanes." AIAA Education Series, 2nd ed., 2004.
11. Hoak, D. E., et al. "The USAF Stability and Control DATCOM." Air Force Wright Aeronautical Laboratories, TR-83-2048, October 1960 (revised in 1978).
12. Google search, "NASA wind tunnels."
13. Cobleigh, B. R. *Development of the X-33 Aerodynamic Uncertainty Model*, NASA Dryden Technical Report, 1998.
14. Napolitano, M. R. *Development of Formation Flight Control Algorithms Using 3 YF-22 Flying Models*. Final Report, AFOSR Grant F49620-01-1-0373, 2005.
15. Greenberg, H. *A Survey of Methods for Determining Stability Parameters of an Airplane from Dynamic Flight Measurements*. NACA TN2340, 1951.
16. Shinbrot, M. *A Least Square Curve Fitting Method with Application of the Calculation of Stability Coefficients from Transient-Response Data*. NACA TN 2341, 1951.
17. Iliff, K. W., Taylor, L. W. *Determination of Stability Derivatives from Flight Data Using a Newton-Raphson Minimization Technique*. NASA TN-D6579, 1971.
18. Taylor, L. W., Iliff K. W. "*System Identification Using a Modified Newton-Raphson Method: a FORTRAN Program*", NASA TN D-6734, 1972.
19. Klein, V. *Maximum Likelihood Method for Estimating Airplane Stability and Control Parameters from Flight Data in Frequency Domain*. NASA TP-1637, 1980.
20. Maine, R. E., Iliff K. W. *The Theory and Practice of Estimating the Accuracy of Dynamic Flight-Determined Coefficients*. NASA RP 1077, 1981.
21. Maine, R. E., Iliff K. W. *Application of Parameter Estimation to Aircraft Stability and Control. The Output Error Approach*. NASA RP 1168, 1986.
22. Morelli, E. A. "Real-Time Parameter Estimation in the Frequency Domain." Journal of Guidance, Control, and Dynamics, Vol. 23, No. 5, 2000, pp. 812-818.
23. Klein, V., Morelli, E. A. *Aircraft System Identification: Theory and Practice*. American Institute of Aeronautics and Astronautics (AIAA) Education Series, 2006.
24. Jategaonkar, R. *Flight Vehicle System Identification: A Time Domain Methodology*. American Institute of Aeronautics and Astronautics (AIAA), 2006.
25. "Advanced Aircraft Analysis." Design, Analysis, and Research Corporation, www.darcorp.com.
26. "SimGen." Bihrle Applied Research Inc. www.bihrle.com.
27. Napolitano, M. R. *WVU MAE 242 Flight Testing—Class Notes*. Morgantown, WV, 2001.
28. Philips, W. *Mechanics of Flight*. John Wiley and Sons, 2004.
29. Multhopp, H. *Aerodynamics of the Fuselage*. NASA Technical Memorandum No. 1036, December 1942.
30. "DC-9 Airplane Characteristics for Airport Planning." June 1984, Douglas Aircraft Company. *http://www.boeing.com/commercial/airports/dc9.htm* (PDF format).

62 Chapter 2 Review of Basic Concepts of Aerodynamic Modeling

STUDENT SAMPLE PROBLEMS

Student Sample Problem 2.1

Consider the data relative to the McDonnell Douglas DC-9 aircraft in Appendix C. Use the provided drawings to extract all the relevant geometric characteristics of the aircraft. Provide estimates for the following parameters:

$$c_{L_{\alpha_W}}, \frac{d\varepsilon}{d\alpha}, \bar{x}_{AC_W}, \Delta\bar{x}_{AC_B}$$

Solution of Student Sample Problem 2.1

All the relevant geometric parameters were extracted from the images in Appendix C, as shown in the figure.

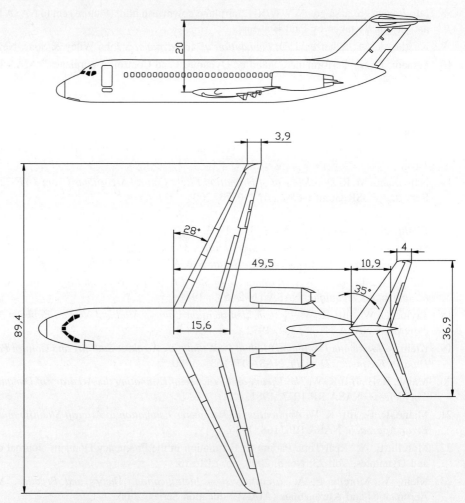

Figure SSP2.1.1 MODIFIED 3D View of McDonnell Douglas DC9 Series 10 Aircraft.
(Source: http://www.aviastar.org/index2.html)

Specifically, the following geometric parameters were determined:

$$c_T = 3.9\,\text{ft}, \quad c_R = 15.6\,\text{ft}, \quad \Lambda_{LE} = 28° = 0.489\,\text{rad}, \quad X_{WH_R} = 49.5\,\text{ft}, \quad Z_{WH} = 20\,\text{ft}$$
$$b_H = 38.5\,\text{ft}, \quad c_{T_H} = 5.6\,\text{ft}, \quad c_{R_H} = 10.9\,\text{ft}, \quad \Lambda_{LE_H} = 34° = 0.593\,\text{rad}$$

Next, using the assumption of straight wings, the following additional wing and tail geometric parameters were derived using the above values:

Wing Geometric Parameters

$$\lambda = \frac{c_T}{c_R} = \frac{3.9}{15.6} = 0.25, \ AR = \frac{b^2}{S} = \frac{89.4^2}{928} = 8.612$$

$$\bar{c} = MAC = \frac{2}{3}c_R\frac{(1+\lambda+\lambda^2)}{(1+\lambda)} = \frac{2}{3} \cdot 15.6 \cdot \frac{(1+0.25+0.25^2)}{(1+0.25)} = 10.92\,\text{ft}$$

$$x_{MAC} = \frac{b}{6}\frac{(1+2\lambda)}{(1+\lambda)}\tan(\Lambda_{LE}) = \frac{89.4}{6} \cdot \frac{(1+2\cdot 0.25)}{(1+0.25)}\tan(0.489) = 9.507\,\text{ft}$$

$$\tan\Lambda_{0.5} = \tan\Lambda_{LE} - \frac{4\cdot 0.5\cdot(1-\lambda)}{AR(1+\lambda)} = \tan(0.48) - \frac{4\cdot 0.5\cdot(1-0.25)}{8.612\cdot(1+0.25)} \Rightarrow \Lambda_{0.5} = 0.374\,\text{rad}$$

$$\tan\Lambda_{0.25} = \tan\Lambda_{LE} - \frac{4\cdot 0.25\cdot(1-\lambda)}{AR(1+\lambda)} = \tan(0.489) - \frac{4\cdot 0.25\cdot(1-0.25)}{8.612\cdot(1+0.25)} \Rightarrow \Lambda_{0.25} = 0.433\,\text{rad}$$

Horizontal Tail Geometric Parameters

$$\lambda_H = \frac{c_{T_H}}{c_{R_H}} = \frac{4}{10.9} = 0.367$$

$$\bar{c}_H = \frac{2}{3}c_{R_H}\frac{(1+\lambda_H+\lambda_H^2)}{(1+\lambda_H)} = \frac{2}{3} \cdot 10.9 \cdot \frac{(1+0.367+0.367^2)}{(1+0.367)} = 7.98\,\text{ft}$$

$$x_{MAC_H} = \frac{b_H}{6}\frac{(1+2\lambda_H)}{(1+\lambda_H)}\tan(\Lambda_{LE_H}) = \frac{36.9}{6} \cdot \frac{(1+2\cdot 0.367)}{(1+0.367)}\tan(0.611) = 5.46\,\text{ft}$$

Wing-Tail Geometric Parameters

The sketch in Figure SSP2.1.2 was derived to illustrate the wing-tail geometric distances.

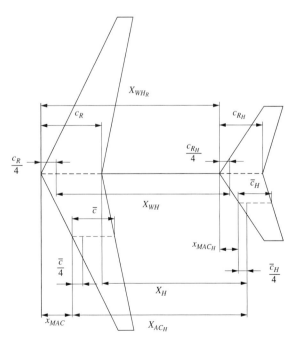

Figure SSP2.1.2 Wing-Horizontal Tail Geometric Distances for the DC-9 Aircraft

From the above figure:

$$X_{WH_R} + \frac{c_{R_H}}{4} = \frac{c_R}{4} + X_{WH}$$

Knowing that $X_{WH_R} = 49.5$ ft:

$$X_{WH} = X_{WH_R} + \frac{c_{R_H}}{4} - \frac{c_R}{4} = 49.5 + \frac{10.9}{4} - \frac{15.6}{4} = 48.325 \text{ ft}$$

Also, knowing that $Z_{WH} = 20$ ft, the wing-tail geometric parameters needed for the analysis of the downwash effect are given by

$$r = \frac{2X_{WH}}{b} = \frac{2 \cdot 48.325}{89.4} = 1.081$$

$$m = \frac{2Z_{WH}}{b} = \frac{2 \cdot 20}{89.4} = 0.447$$

Recall that the meaning of the geometric parameters m and r (shown in Figure SSP2.1.3) was introduced in the analysis of the downwash effect (Section 2.4).

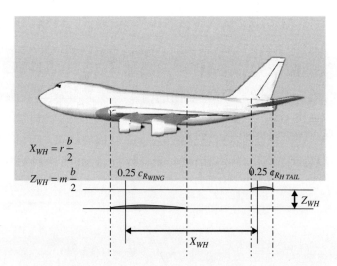

Figure SSP2.1.3 Geometric Parameters for the Downwash Effect

From Figure SSP2.1.2, the coordinate $\bar{x}_{AC_H}$ can be found using

$$X_{WH_R} + x_{MAC_H} + \frac{\bar{c}_H}{4} = x_{MAC} + X_{AC_H}$$

Solving for X_{AC_H}

$$X_{AC_H} = X_{WH_R} + x_{MAC_H} + \frac{\bar{c}_H}{4} - x_{MAC} = 49.5 + 5.462 + \frac{7.98}{4} - 9.51 = 47.45 \text{ ft}$$

Wing Lift-Slope Coefficient

Since $AR = 8.612 \geq 4$

$$k = 1 + \frac{[(8.2 - 2.3\Lambda_{LE}) - AR(0.22 - 0.153\Lambda_{LE})]}{100}$$

$$= 1 + \frac{[(8.2 - 2.3 \cdot 0.489) - 8.612 \cdot (0.22 - 0.153 \cdot 0.489)]}{100} = 1.058$$

$$c_{L_{\alpha W}} = \frac{2\pi AR}{2 + \sqrt{\left\{\left[\frac{AR^2(1 - Mach^2)}{k^2}\left(1 + \frac{\tan^2(\Lambda_{0.5})}{(1 - Mach^2)}\right)\right] + 4\right\}}}$$

$$= \frac{2\pi \cdot 8.612}{2 + \sqrt{\left\{\left[\frac{8.612^2(1 - 0.7^2)}{1.058^2}\left(1 + \frac{\tan^2(0.374)}{(1 - 0.7^2)}\right)\right] + 4\right\}}}$$

$$= 6.062$$

It is expected that the shown value is higher than the actual $c_{L_{\alpha W}}$ value. This is due to the fact that $AR = 8.612$ is slightly above the suitable ranges for the Polhamus formula. Although less critical, $\lambda = \frac{c_T}{c_R} = 0.25$ is also slightly below the acceptable range.

Modeling of the Downwash Effect

$$K_{AR} = \frac{1}{AR} - \frac{1}{1 + (AR)^{1.7}} = \frac{1}{8.612} - \frac{1}{1 + (8.612)^{1.7}} = 0.091$$

$$K_\lambda = \frac{10 - 3\lambda}{7} = \frac{10 - 3 \cdot 0.25}{7} = 1.321$$

$$K_{mr} = \frac{1 - \frac{m}{2}}{(r)^{0.33}} = \frac{1 - \frac{0.447}{2}}{(1.081)^{0.33}} = 0.756$$

$$\left(\frac{d\varepsilon}{d\alpha}\right)\bigg|_{Mach=0} = 4.44 \left(K_{AR} K_\lambda K_{mr} \sqrt{\cos(\Lambda_{0.25})}\right)^{1.19}$$

$$= 4.44 \left(0.091 \cdot 1.321 \cdot 0.756 \sqrt{\cos(0.433)}\right)^{1.19} = 0.242$$

$$c_{L_{\alpha W}}\bigg|_{Mach=0} = \frac{2\pi AR}{2 + \sqrt{\left\{\left[\frac{AR^2}{k^2}\left(1 + \tan^2(\Lambda_{0.5})\right)\right] + 4\right\}}}$$

$$= \frac{2\pi \cdot 8.612}{2 + \sqrt{\left\{\left[\frac{8.612^2}{1.058^2}\left(1 + \tan^2(0.374)\right)\right] + 4\right\}}} = 4.934$$

$$\left(\frac{d\varepsilon}{d\alpha}\right)\bigg|_{Mach} = \left(\frac{d\varepsilon}{d\alpha}\right)\bigg|_{Mach=0} \frac{c_{L_{\alpha W}}|_{Mach}}{c_{L_{\alpha W}}|_{Mach=0}} = 0.242 \frac{6.062}{4.934} = 0.297$$

Wing Aerodynamic Center

This parameter can be obtained using the following relationship (Section 2.5):

$$\bar{x}_{AC_W} = K_1 \left(\frac{x'_{AC}}{c_R} - K_2\right)$$

where $\frac{x'_{AC}}{c_R}$, K_1 and K_2 are obtained from Figures 2.27, 2.28, and 2.29, respectively. The following parameters are required in Figure 2.27:

$$\frac{\tan \Lambda_{LE}}{\sqrt{1 - M^2}} = \frac{\tan 0.489}{\sqrt{1 - 0.7^2}} = 0.745$$

$$AR \cdot \tan \Lambda_{LE} = 8.612 \cdot \tan 0.489 = 4.58$$

66 Chapter 2 Review of Basic Concepts of Aerodynamic Modeling

Interpolating between the curves of the plots of Figure 2.27 using

$$\lambda = 0.25 \text{ and } AR \cdot \tan \Lambda_{LE} = 4.58 \text{ we have } \frac{x'_{AC}}{c_R} = 0.78$$

From interpolation of Figure 2.28, using $\lambda = 0.25 \Rightarrow K_1 = 1.425$.
From interpolation of Figure 2.29, using

$$\Lambda_{LE} = 28°, \quad \lambda = 0.25, \quad AR = 8.612 \Rightarrow K_2 = 0.58$$

Thus, the location of the wing aerodynamic center is estimated at

$$\boxed{\bar{x}_{AC_W} = K_1 \left(\frac{x'_{AC}}{c_R} - K_2 \right) = 1.425 \cdot (0.78 - 0.58) = 0.285}$$

Modeling of $\Delta \bar{x}_{AC_B}$

The aerodynamic center shift due to the body is calculated using the Munk theory (Section 2.5):

$$\Delta \bar{x}_{AC_B} = \frac{-\left(\frac{dM}{d\alpha}\right)}{\bar{q} \, S \, \bar{c} \, c_{L_{\alpha W}}} = -\frac{\left(\frac{\bar{q}}{36.5} \frac{c_{L_{\alpha W}}}{0.08}\right)}{\bar{q} \, S \, \bar{c} \, c_{L_{\alpha W}}} \sum_{i=1}^{13} w_{B_i}^2 \left(\frac{d\bar{\varepsilon}}{d\alpha}\right)_i \Delta x_i$$

$$= -\frac{1}{2.92 \cdot S\bar{c}} \sum_{i=1}^{13} w_{B_i}^2 \left(\frac{d\bar{\varepsilon}}{d\alpha}\right)_i \Delta x_i$$

$w_{b_i}, \Delta x_i$ are geometric parameters for discretized aircraft sections, as shown in Figure SSP2.1.4.

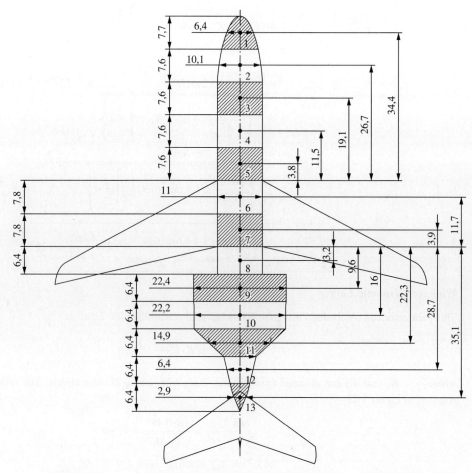

Figure SSP2.1.4 Aircraft Sections for Calculations of $\Delta \bar{x}_{AC_{WB}}$ (dimensions in ft)

The parameter $(d\bar{\varepsilon}/d\alpha)_i$ is calculated through the two curves in Figure SSP2.1.5.

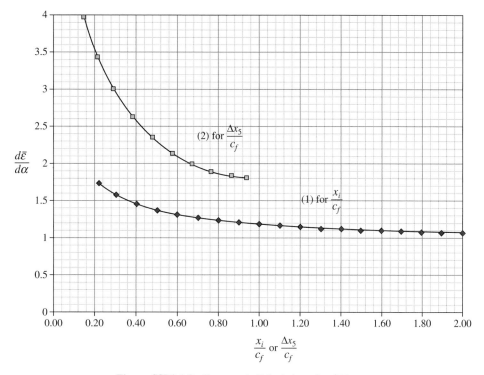

Figure SSP2.1.5 Downwash Calculations for $\Delta \bar{x}_{AC_{WB}}$

Specifically, $(d\bar{\varepsilon}/d\alpha)_i$ for $i = 1, 2, 3, 4, 5$ is obtained using curve (1) using the required values for x_i with $c_f = c_{Root} = 15.6$ ft. For $i = 5$, $(d\bar{\varepsilon}/d\alpha)_i$ is obtained using curve (2) and taking into account that $\Delta x_5 = 7.6$ ft. Finally, for $i = 6...13$,

$$\left(\frac{d\bar{\varepsilon}}{d\alpha}\right)_i = \left(\frac{x_i}{x_H}\right)\left(1 - \frac{d\varepsilon}{d\alpha}\bigg|_{m=0}\right)$$

The value of $d\varepsilon/d\alpha\big|_{m=0}$ is the downwash effect with $m = 0$, thus

$$K_{mr} = \frac{1 - \dfrac{m}{2}}{(r)^{0.33}} = \frac{1 - 0}{(1.081)^{0.33}} = 0.974$$

$$\left(\frac{d\varepsilon}{d\alpha}\right)\bigg|_{Mach=0, m=0} = 4.44\left(K_{AR}K_\lambda K_{mr}\sqrt{\cos(\Lambda_{0.25})}\right)^{1.19}$$

$$= 4.44(0.091 \cdot 1.321 \cdot 0.974\sqrt{\cos(0.433)})^{1.19} = 0.327$$

$$\left(\frac{d\varepsilon}{d\alpha}\right)\bigg|_{Mach, m=0} = \left(\frac{d\varepsilon}{d\alpha}\right)\bigg|_{Mach=0, m=0} \frac{c_{L_{\alpha w}}|_{Mach}}{c_{L_{\alpha w}}|_{Mach=0}} = 0.327\frac{6.062}{4.934} = 0.402$$

From Figure SSP2.1.2:

$$c_R + x_H = x_{MAC} + X_{AC_H}$$

Therefore

$$x_H = x_{MAC} + X_{AC_H} - c_R = 9.507 + 47.451 - 15.6 = 41.358 \, \text{ft}$$

68 Chapter 2 Review of Basic Concepts of Aerodynamic Modeling

The values for $(d\bar{\varepsilon}/d\alpha)_i$ and the final results are summarized in Table SSP2.1.1.

Table SSP2.1.1 Calculations for $\Delta \bar{x}_{AC_B}$

i	x_i[ft]	Δx_i[ft]	$\dfrac{x_i}{c_f}$	w_i[ft]	$\left(\dfrac{d\bar{\varepsilon}}{d\alpha}\right)_i$	$w_{b i}^2 \left(\dfrac{d\bar{\varepsilon}}{d\alpha}\right)_i \Delta x_i$
1	34.4	7.7	2.21	6.4	1.07	337.469
2	26.7	7.6	1.71	10.1	1.09	843.441
3	19.1	7.6	1.22	10.1	1.14	886.581
4	11.5	7.6	0.74	10.1	1.26	978.503
5	3.8	7.6	0.24	10.1	2.34	1813.249
6	11.7	7.8	0.75	10.1	0.17	134.669
7	3.9	7.8	0.25	10.1	0.06	44.890
8	3.2	6.4	0.21	10.1	0.05	30.222
9	9.6	6.4	0.62	22.4	0.14	445.957
10	16	6.4	1.03	22.2	0.23	730.048
11	22.3	6.4	1.43	14.9	0.32	458.356
12	28.7	6.4	1.84	6.4	0.42	108.835
13	35.1	6.4	2.25	2.9	0.51	27.329

$$\Delta x_{AC_B} = -\frac{1}{2.92 \cdot S\bar{c}} \sum_{i=1}^{i=13} w_{b_i}^2 \left(\frac{d\bar{\varepsilon}}{d\alpha}\right)_i \Delta x_i = -0.231$$

$$\boxed{\Delta \bar{x}_{AC_B} = -\frac{1}{2.92 \cdot S\bar{c}} \sum_{i=1}^{13} w_{B_i}^2 \left(\frac{d\bar{\varepsilon}}{d\alpha}\right)_i \Delta x_i = -0.231}$$

Student Sample Problem 2.2

Consider the data relative to the SIAI Marchetti S211 aircraft in Appendixes B and C. Use the provided drawings to extract all the relevant geometric characteristics of the aircraft.

Provide estimates for the following parameters:

$$c_{L_{\alpha_W}}, \frac{d\varepsilon}{d\alpha}, \bar{x}_{AC_W}, \Delta \bar{x}_{AC_B}$$

Solution of Student Sample Problem 2.2

All the relevant geometric parameters were extracted from the images in Appendix C, as shown in Figure SSP2.2.1.
Specifically, the following geometric parameters were identified:

$$c_T = 3.1\,\text{ft}, \; c_R = 6\,\text{ft}, \; \Lambda_{LE} = 19.5° = 0.34\,\text{rad}, \; X_{WH_R} = 13.1\,\text{ft}, \; Z_{WH} = 1.3\,\text{ft}$$
$$b_H = 13.3\,\text{ft}, \; c_{T_H} = 1.55\,\text{ft}, \; c_{R_H} = 3.5\,\text{ft}, \; \Lambda_{LE_H} = 18.5° = 0.323\,\text{rad}$$

Next, using the assumption of straight wings, the following additional wing and tail geometric parameters were derived using the above values:

Wing Geometric Parameters

$$\lambda = \frac{c_T}{c_R} = \frac{3.1}{6} = 0.517, \; AR = \frac{b^2}{S} = \frac{26.3^2}{136} = 5.086$$

$$\bar{c} = MAC = \frac{2}{3} c_R \frac{(1+\lambda+\lambda^2)}{(1+\lambda)} = \frac{2}{3} \cdot 6 \cdot \frac{(1+0.517+0.517^2)}{(1+0.517)} = 4.704\,\text{ft}$$

$$x_{MAC} = \frac{b}{6} \frac{(1+2\lambda)}{(1+\lambda)} \tan(\Lambda_{LE}) = \frac{26.3}{6} \cdot \frac{(1+2 \cdot 0.517)}{(1+0.517)} \tan(0.34) = 2.081\,\text{ft}$$

$$\tan \Lambda_{0.5} = \tan \Lambda_{LE} - \frac{4 \cdot 0.5 \cdot (1-\lambda)}{AR(1+\lambda)} = \tan(0.34) - \frac{4 \cdot 0.5 \cdot (1-0.517)}{5.086 \cdot (1+0.517)} \Rightarrow \Lambda_{0.5} = 0.225\,\text{rad}$$

$$\tan \Lambda_{0.25} = \tan \Lambda_{LE} - \frac{4 \cdot 0.25 \cdot (1-\lambda)}{AR(1+\lambda)} = \tan(0.34) - \frac{4 \cdot 0.25 \cdot (1-0.517)}{5.086 \cdot (1+0.517)} \Rightarrow \Lambda_{0.25} = 0.284\,\text{rad}$$

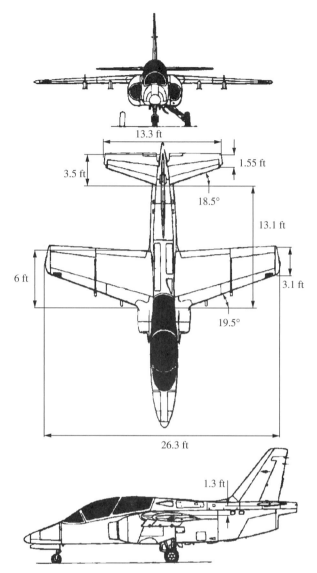

Figure SSP2.2.1 MODIFIED 3D View of SIAI Marchetti S211 Aircraft
(Source: http://www.aviastar.org/index2.html)

Horizontal Tail Geometric Parameters

$$\lambda_H = \frac{c_{T_H}}{c_{R_H}} = \frac{1.55}{3.5} = 0.443$$

$$\bar{c}_H = \frac{2}{3} c_{R_H} \frac{(1 + \lambda_H + \lambda_H^2)}{(1 + \lambda_H)} = \frac{2}{3} \cdot 3.5 \cdot \frac{(1 + 0.443 + 0.443^2)}{(1 + 0.443)} = 2.65 \, \text{ft}$$

$$x_{MAC_H} = \frac{b_H}{6} \frac{(1 + 2\lambda_H)}{(1 + \lambda_H)} \tan(\Lambda_{LE_H}) = \frac{13.3}{6} \cdot \frac{(1 + 2 \cdot 0.443)}{(1 + 0.443)} \tan(0.323) = 0.969 \, \text{ft}$$

Wing-Tail Geometric Parameters

The sketch in Figure SSP2.2.2 shows the wing-tail geometric distances.
From Figure SSP2.2.2

$$X_{WH_R} + \frac{c_{R_H}}{4} = \frac{c_R}{4} + X_{WH}$$

Knowing that $X_{WH_R} = 13.1 \, \text{ft}$

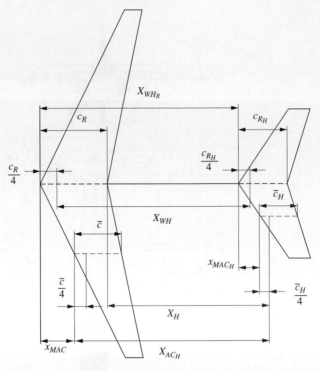

Figure SSP2.2.2 Wing-Horizontal Tail Geometric Distances for the S211 Aircraft

$$X_{WH} = X_{WH_R} + \frac{c_{R_H}}{4} - \frac{c_R}{4} = 13.1 + \frac{3.5}{4} - \frac{6}{4} = 12.475 \,\text{ft}$$

Also, knowing that $Z_{WH} = 1.3$ ft, the wing-tail geometric parameters needed for the analysis of the downwash effect are given by

$$r = \frac{2X_{WH}}{b} = \frac{2 \cdot 12.48}{26.3} = 0.949$$

$$m = \frac{2Z_{WH}}{b} = \frac{2 \cdot 1.3}{26.3} = 0.099$$

Recall that the meaning of the geometric parameters m and r (shown in Figure SSP2.2.3) was introduced in the analysis of the downwash effect (Section 2.4).

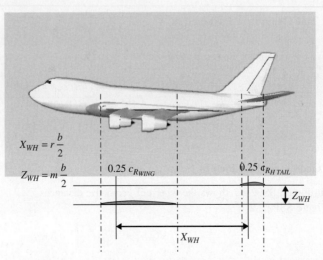

Figure SSP2.2.3 Geometric Parameters for the Downwash Effect

From Figure SSP2.2.2 the coordinate $\bar{x}_{AC_H}$ can be found using

$$X_{WH_R} + x_{MAC_H} + \frac{\bar{c}_H}{4} = x_{MAC} + X_{AC_H}$$

Solving for X_{AC_H}:

$$X_{AC_H} = X_{WH_R} + x_{MAC_H} + \frac{\bar{c}_H}{4} - x_{MAC} = 13.1 + 0.969 + \frac{2.65}{4} - 2.081 = 12.651\,\text{ft}$$

Wing Lift-Slope Coefficient

Since $AR = 5.09 \geq 4$

$$k = 1 + \frac{[(8.2 - 2.3\Lambda_{LE}) - AR(0.22 - 0.153\Lambda_{LE})]}{100}$$

$$= 1 + \frac{[(8.2 - 2.3 \cdot 0.34) - 5.09 \cdot (0.22 - 0.153 \cdot 0.34)]}{100} = 1.07$$

$$\boxed{c_{L_{\alpha_W}} = \frac{2\pi AR}{2 + \sqrt{\left\{\left[\frac{AR^2(1 - Mach^2)}{k^2}\left(1 + \frac{\tan^2(\Lambda_{0.5})}{(1 - Mach^2)}\right)\right] + 4\right\}}}}$$

$$= \frac{2\pi \cdot 5.09}{2 + \sqrt{\left\{\left[\frac{5.09^2(1 - 0.6^2)}{1.07^2}\left(1 + \frac{\tan^2(0.225)}{(1 - 0.6^2)}\right)\right] + 4\right\}}} = 4.96$$

Given that the wing geometric parameters are within the suitable ranges for the Polhamus formula, it is expected that the above value of $c_{L_{\alpha_W}}$ is fairly accurate.

Modeling of the Downwash Effect

$$K_{AR} = \frac{1}{AR} - \frac{1}{1 + (AR)^{1.7}} = \frac{1}{5.27} - \frac{1}{1 + (5.27)^{1.7}} = 0.1374$$

$$K_\lambda = \frac{10 - 3\lambda}{7} = \frac{10 - 3 \cdot 0.52}{7} = 1.2071$$

$$K_{mr} = \frac{1 - \frac{m}{2}}{(r)^{0.33}} = \frac{1 - \frac{0.10}{2}}{(0.95)^{0.33}} = 0.9674$$

$$\left(\frac{d\varepsilon}{d\alpha}\right)\bigg|_{Mach=0} = 4.44\left(K_{AR}K_\lambda K_{mr}\sqrt{\cos(\Lambda_{0.25})}\right)^{1.19}$$

$$= 4.44\left(0.1374 \cdot 1.2071 \cdot 0.9674\sqrt{\cos(0.284)}\right)^{1.19} = 0.4910$$

$$c_{L_{\alpha_W}}\bigg|_{Mach=0} = \frac{2\pi AR}{2 + \sqrt{\left\{\left[\frac{AR^2}{k^2}\left(1 + \tan^2(\Lambda_{0.5})\right)\right] + 4\right\}}}$$

$$= \frac{2\pi \cdot 5.09}{2 + \sqrt{\left\{\left[\frac{5.09^2}{1.07^2}\left(1 + \tan^2(0.225)\right)\right] + 4\right\}}} = 4.38$$

$$\boxed{\left(\frac{d\varepsilon}{d\alpha}\right)\bigg|_{Mach} = \left(\frac{d\varepsilon}{d\alpha}\right)\bigg|_{Mach=0} \frac{c_{L_{\alpha w}}|_{Mach}}{c_{L_{\alpha w}}|_{Mach=0}} = 0.491 \frac{4.96}{4.38} = 0.555}$$

Wing Aerodynamic Center

This parameter can be obtained using the following relationship,

$$\bar{x}_{AC_W} = K_1\left(\frac{x'_{AC}}{c_R} - K_2\right)$$

where $\frac{x'_{AC}}{c_R}$, K_1 and K_2 are obtained from Figures 2.27, 2.28, and 2.29, respectively.

The following parameters are required in Figure 2.27:

$$\frac{\tan \Lambda_{LE}}{\sqrt{1-M^2}} = \frac{\tan 0.34}{\sqrt{1-0.6^2}} = 0.44$$

$$AR \cdot \tan \Lambda_{LE} = 5.086 \cdot \tan 0.34 = 1.80$$

Interpolating between the curves of the plots of Figure 2.27 using $\lambda = 0.517$ and $AR \cdot \tan \Lambda_{LE} = 1.80$ we have $\frac{x'_{AC}}{c_R} = 0.491$.

From interpolation of Figure 2.28, using $\lambda = 0.517 \Rightarrow K_1 = 1.275$.
From interpolation of Figure 2.29, using $\Lambda_{LE} = 19.5°$, $\lambda = 0.517$, $AR = 5.086 \Rightarrow K_2 = 0.309$.
The location of the wing aerodynamic center is

$$\boxed{\bar{x}_{AC_W} = K_1\left(\frac{x'_{AC}}{c_R} - K_2\right) = 1.275 \cdot (0.491 - 0.309) = 0.232}$$

Modeling of $\Delta \bar{x}_{AC_B}$

The aerodynamic center shift due to the body is calculated using the Munk theory (Section 2.5):

$$\Delta \bar{x}_{AC_B} = \frac{-\left(\frac{dM}{d\alpha}\right)}{\bar{q}\, S\, \bar{c}\, c_{L_{\alpha W}}} = -\frac{\left(\frac{\bar{q}}{36.5} \frac{c_{L_\alpha\, W}}{0.08}\right)}{\bar{q}\, S\, \bar{c}\, c_{L_{\alpha W}}} \sum_{i=1}^{13} w_{B_i}^2 \left(\frac{d\bar{\varepsilon}}{d\alpha}\right)_i \Delta x_i$$

$$= -\frac{1}{2.92 \cdot S\bar{c}} \sum_{i=1}^{13} w_{B_i}^2 \left(\frac{d\bar{\varepsilon}}{d\alpha}\right)_i \Delta x_i$$

w_{b_i}, Δx_i are geometric parameters for discretized aircraft sections, as shown in Figure SSP2.2.4.

The parameter $(d\bar{\varepsilon}/d\alpha)_i$ is calculated through the two curves in Figure SSP2.2.5.

Specifically, $(d\bar{\varepsilon}/d\alpha)_i$ for $i = 1, 2, 3, 4, 5$ is obtained using curve (1) using the required values for x_i with $c_f = c_{Root} = 6$ ft. For $i = 5$, $(d\bar{\varepsilon}/d\alpha)_i$ is obtained using curve (2) and taking into account that $\Delta x_5 = 3$ ft. Finally, for $i = 6...13$,

$$\left(\frac{d\bar{\varepsilon}}{d\alpha}\right)_i = \left(\frac{x_i}{x_H}\right)\left(1 - \frac{d\varepsilon}{d\alpha}\right)$$

The value of $d\varepsilon/d\alpha\big|_{m=0}$ is the downwash effect evaluated with $m = 0$, thus

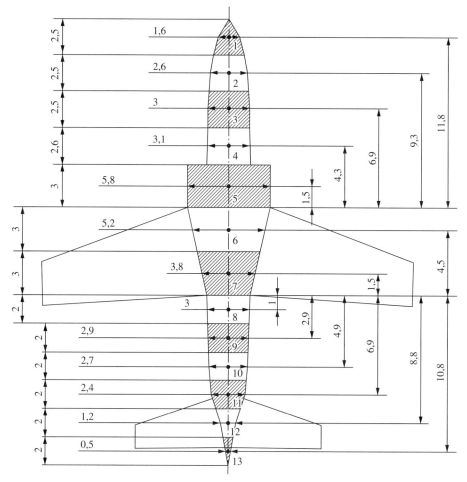

Figure SSP2.2.4 Aircraft Sections for Calculations of $\Delta \bar{x}_{AC_{WB}}$ (dimensions in ft)

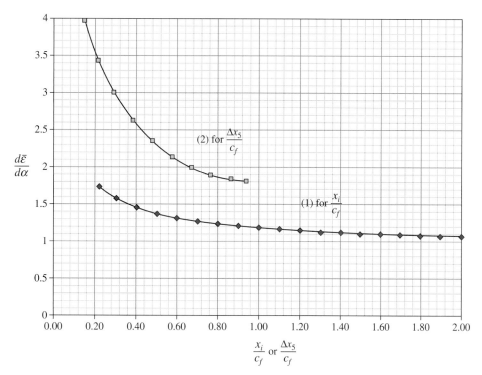

Figure SSP2.2.5 Downwash Calculations for $\Delta \bar{x}_{AC_{WB}}$

$$K_{mr} = \frac{1 - \frac{m}{2}}{(r)^{0.33}} = \frac{1 - 0}{(0.949)^{0.33}} = 1.018$$

$$\left(\frac{d\varepsilon}{d\alpha}\right)\bigg|_{Mach=0,\, m=0} = 4.44\left(K_{AR}K_\lambda K_{mr}\sqrt{\cos(\Lambda_{0.25})}\right)^{1.19}$$

$$= 4.44\left(0.137 \cdot 1.207 \cdot 1.018\sqrt{\cos(0.284)}\right)^{1.19} = 0.522$$

$$\left(\frac{d\varepsilon}{d\alpha}\right)\bigg|_{Mach,\, m=0} = \left(\frac{d\varepsilon}{d\alpha}\right)\bigg|_{Mach=0,\, m=0} \frac{c_{L_{\alpha w}}|_{Mach}}{c_{L_{\alpha w}}|_{Mach=0}} = 0.522\frac{6.009}{4.901} = 0.59$$

From Figure SSP2.2.2:

$$c_R + x_H = x_{MAC} + X_{AC_H}$$

Therefore:

$$x_H = x_{MAC} + X_{AC_H} - c_R = 2.08 + 12.65 - 6 = 8.73\,\text{ft}$$

The values for $(d\bar{\varepsilon}/d\alpha)_i$ and the final results are summarized in Table SSP2.2.1.

Table SSP2.2.1 Calculations for $\Delta\bar{x}_{AC_B}$

i	x_i[ft]	Δx_i[ft]	$\dfrac{x_i}{c_f}$	w_i[ft]	$\left(\dfrac{d\bar{\varepsilon}}{d\alpha}\right)_i$	$w_{bi}^2\left(\dfrac{d\bar{\varepsilon}}{d\alpha}\right)_i \Delta x_i$
1	11.8	2.5	1.97	1.6	1.07	6.849
2	9.3	2.5	1.55	2.6	1.10	18.641
3	6.9	2.5	1.15	3	1.16	26.035
4	4.3	2.6	0.72	3.1	1.27	31.705
5	1.5	3	0.25	5.8	2.31	232.872
6	4.5	3	0.75	5.2	0.21	17.151
7	1.5	3	0.25	3.8	0.07	3.053
8	1	2	0.17	3	0.05	0.846
9	2.9	2	0.48	2.9	0.14	2.292
10	4.9	2	0.82	2.7	0.23	3.357
11	6.9	2	1.15	2.4	0.32	3.735
12	8.8	2	1.47	1.2	0.41	1.191
13	10.8	2	1.80	0.5	0.51	0.254

$$\Delta x_{AC_B} = -\frac{1}{2.92 \cdot S\,\bar{c}} \sum_{i=1}^{i=13} w_{bi}^2\left(\frac{d\bar{\varepsilon}}{d\alpha}\right)_i \Delta x_i = -0.186$$

$$\boxed{\Delta\bar{x}_{AC_B} = -\frac{1}{2.92 \cdot S\,\bar{c}} \sum_{i=1}^{13} w_{B_i}^2\left(\frac{d\bar{\varepsilon}}{d\alpha}\right)_i \Delta x_i = -0.186}$$

PROBLEMS

Problem 2.1

Consider the data relative to the Aeritalia G91 aircraft in Appendix C. Use the provided drawings to extract all the relevant geometric characteristics of the aircraft. Provide estimates for the following parameters:

$$c_{L_{\alpha_W}}, \frac{d\varepsilon}{d\alpha}, \bar{x}_{AC_W}, \Delta\bar{x}_{AC_B}$$

Problem 2.2

Consider the data relative to the Beech 99 aircraft in Appendixes B and C. Use the provided drawings to extract all the relevant geometric characteristics of the aircraft.

Provide estimates for the following parameters:

$$c_{L_{\alpha_W}}, \frac{d\varepsilon}{d\alpha}, \bar{x}_{AC_W}, \Delta\bar{x}_{AC_B}$$

Problem 2.3

Consider the data relative to the Cessna T37 aircraft in Appendixes B and C. Use the provided drawings to extract all the relevant geometric characteristics of the aircraft.

Provide estimates for the following parameters:

$$c_{L_{\alpha_W}}, \frac{d\varepsilon}{d\alpha}, \bar{x}_{AC_W}, \Delta\bar{x}_{AC_B}$$

Problem 2.4

Consider the data relative to the McDonnell Douglas DC-9 aircraft in Appendix C. Next, consider the later versions of the DC-9, which were introduced later in the operational life of the aircraft.

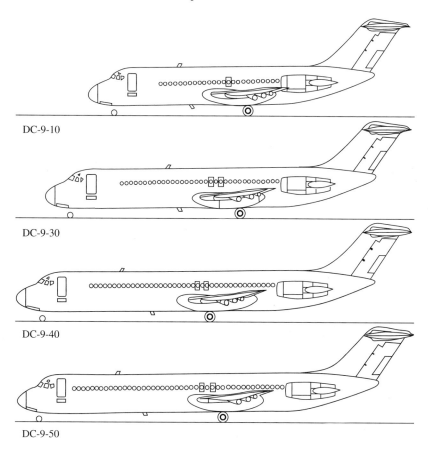

Figure P2.4.1 Lateral Views McDonnell Douglas DC 9 Series 10/30/40/50
(Source: http://www.aviastar.org/index2.html)

Ref.[30] provides detailed information about the differences in the geometric parameters. The geometric parameters for the DC-9 Series 10/30/40/50 aircraft are summarized in the following Tables P2.4.1–P2.4.3 and Figure P.2.4.2.

Wing Geometric Parameters (as shown in Student Sample Problem #1)

Table P2.4.1 Wing Geometric Parameters

	DC9-10	DC9-30	DC9-40	DC9-50
b [ft]	89.4	93.3	93.3	93.32
c_T [ft]	3.9	3.9	3.9	3.9
c_R [ft]	15.6	16.1	16.1	16.1
Λ_{LE} [rad]	0.489	0.489	0.489	0.489
Λ_{LE} [deg]	28	28	28	28
λ	0.250	0.242	0.242	0.242
S [ft^2]	928.00	996.00	996.00	996.00
AR	8.612	8.740	8.740	8.744
$\bar{c}$ [ft]	10.92	11.240	11.240	11.240
x_{MAC} [ft]	9.51	9.880	9.880	9.882

Horizontal Tail Geometric Parameters (as shown in Student Sample Problem #1)

Table P2.4.2 Horizontal Tail Geometric Parameters

	DC9-10	DC9-30	DC9-40	DC9-50
b_H [ft]	36.900	36.800	36.800	36.850
c_{T_H} [ft]	4.000	4.000	4.000	4.000
c_{R_H} [ft]	10.900	10.900	10.900	10.900
Λ_{LE_H} [rad]	0.611	0.611	0.611	0.611
Λ_{LE_H} [deg]	35	35	35	35
λ_H	0.367	0.367	0.367	0.367
S_H [ft^2]	274.91	274.16	274.16	274.53
AR_H	4.95	4.94	4.94	4.95
$\bar{c}_H$ [ft]	7.98	7.98	7.98	7.98
x_{MAC_H} [ft]	5.46	5.45	5.45	5.45

Wing-Tail Geometric Parameters (as shown in Student Sample Problem #1)

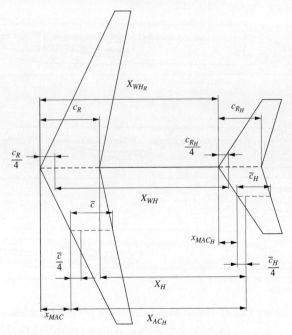

Figure P2.4.2 Wing-Horizontal Tail Geometric Distances

Table P2.4.3 Wing Geometric Parameters

	DC9-10	DC9-30	DC9-40	DC9-50
X_{WH_R} [ft]	49.500	54.900	58.300	61.500
Z_{WH} [ft]	20	20	20	20
X_{WH} [ft]	48.33	53.60	57.00	60.20
r	1.081	1.149	1.222	1.290
m	0.447	0.429	0.429	0.429
X_{AC_H} [ft]	47.451	52.463	55.863	59.068
$\bar{x}_{AC_H}$	4.345	4.667	4.970	5.255
x_H [ft]	41.358	46.243	49.643	52.851

Using the geometric information derived from Ref.,[30] provide estimates for the following parameters for the DC-9 30, DC-9 40, and DC-9 50: $c_{L_{\alpha_W}}, \dfrac{d\varepsilon}{d\alpha}, \bar{x}_{AC_W}, \Delta \bar{x}_{AC_B}$

Problem 2.5

Consider the data relative to the McDonnell Douglas F-4 aircraft in Appendix C. Use the provided drawings to extract all the relevant geometric characteristics of the aircraft. Provide estimates for the following parameters:

$$c_{L_{\alpha_W}}, \frac{d\varepsilon}{d\alpha}, \bar{x}_{AC_W}, \Delta \bar{x}_{AC_B}$$

Problem 2.6

Consider the data relative to the McDonnell Douglas DC-8 aircraft in Appendix C. Use the provided drawings to extract all the relevant geometric characteristics of the aircraft. Provide estimates for the following parameters:

$$c_{L_{\alpha_W}}, \frac{d\varepsilon}{d\alpha}, \bar{x}_{AC_W}, \Delta \bar{x}_{AC_B}$$

Chapter 3

Modeling of Longitudinal Aerodynamic Forces and Moments

TABLE OF CONTENTS

- 3.1 Introduction
- 3.2 Aircraft Stability Axes
- 3.3 Modeling of the Longitudinal Steady-State Aerodynamic Forces and Moment
- 3.4 Modeling of $F_{A_{X_1}}$
- 3.5 Modeling of $F_{A_{Z_1}}$
- 3.6 Modeling of M_{A_1}
- 3.7 Aircraft Aerodynamic Center
- 3.8 Summary of the Longitudinal Steady-State Aerodynamic Forces and Moment
- 3.9 Modeling of the Longitudinal Small Perturbation Aerodynamic Forces and Moments
  - 3.9.1 Modeling of $(c_{D_1}, c_{L_1}, c_{m_1})$
  - 3.9.2 Modeling of $(c_{D_u}, c_{L_u}, c_{m_u})$
  - 3.9.3 Modeling of $(c_{D_{\dot\alpha}}, c_{L_{\dot\alpha}}, c_{m_{\dot\alpha}})$ and $(c_{D_q}, c_{L_q}, c_{m_q})$
- 3.10 Summary of Longitudinal Stability and Control Derivatives
- 3.11 Summary
- References
- Student Sample Problems
- Case Study
- Short Problems
- Problems

3.1 INTRODUCTION

The main purpose of this chapter is to introduce the tools for the challenging task of modeling the aerodynamic forces and moments acting on the aircraft. Students are assumed to be familiar with the main concepts in subsonic aerodynamics from basic aerodynamic courses and partially reviewed in Chapter 2.[1,2,3]

The specific goal of this chapter is to introduce a simple empirical modeling approach based on a detailed understanding of the different aerodynamic contributions from different components of an aircraft at subsonic conditions.[4] The starting point is given by the following breakdown of a generic aerodynamic force and moment:

$$F_{Aerodynamic} = F_{Aero_{WING}} + F_{Aero_{BODY}} + F_{Aero\ CONTROL\ SURFACES}$$
$$M_{Aerodynamic} = M_{Aero_{WING}} + M_{Aero_{BODY}} + M_{Aero_{CONTROL\ SURFACES}}$$

For a conventional subsonic design, the aircraft features a horizontal tail and a vertical tail. Additionally, for the modeling of specific forces or moments, it might be convenient to combine the aerodynamic contributions from the wing and the body (fuselage). Under the above conditions, we have

$$F_{Aerodynamic} \approx F_{Aero\ WING+BODY} + F_{Aero\ HORIZ.TAIL} + F_{Aero\ VERTICAL\ TAIL}$$
$$M_{Aerodynamic} \approx M_{Aero\ WING+BODY} + M_{Aero\ HORIZ.TAIL} + M_{Aero\ VERTICAL\ TAIL}$$

The implementation of this build-up approach to a wider scale for all possible aircraft configurations for all the ranges of speed (subsonic, transonic, supersonic, and hypersonic) led to the development of DATCOM.[5] To date,[5] and[6] (which is derived from)[5] are still the most comprehensive tools for the empirical modeling of the aircraft aerodynamic behavior.

Although this approach does not have the accuracy of a complete wind tunnel analysis, it provides fairly accurate results for most of the aerodynamic coefficients at subsonic conditions. Because the method is based on breaking down the configurations from different components of the aircraft, it also has the specific advantage of allowing students to gain a detailed localized understanding of the generation of aerodynamic forces and moments. The modeling of the aerodynamic forces and moments is presented in terms of the steady-state components and later in terms of the small perturbation components.[4]

This chapter focuses on the modeling of the longitudinal aerodynamic forces and moment whereas Chapter 4 will describe the modeling of the lateral directional force and moments. The main outcomes of Chapters 3 and 4 are detailed descriptions of the small perturbation aerodynamic forces and moments, that is, the single underlined terms in the small perturbation equations of motion (CLMEs and CAMEs) from Chapter 1, which are reported here:

$$m[\dot{u} + Q_1 w + qW_1 - R_1 v - rV_1] = -mg\theta\cos\Theta_1 + \left(\underline{f_{A_X}} + \underline{\underline{f_{T_X}}}\right)$$

$$m[\dot{v} + U_1 r + uR_1 - P_1 w - pW_1] = -mg\theta\sin\Phi_1\sin\Theta_1 + mg\phi\cos\Phi_1\cos\Theta_1 + \left(\underline{f_{A_Y}} + \underline{\underline{f_{T_Y}}}\right)$$

$$m[\dot{w} + P_1 v + pV_1 - Q_1 u - U_1 q] = -mg\cos\Phi_1\sin\Theta_1 - mg\phi\sin\Phi_1\cos\Theta_1 + \left(\underline{f_{A_Z}} + \underline{\underline{f_{T_Z}}}\right)$$

$$\dot{p}I_{XX} - \dot{r}I_{XZ} - (P_1 q + Q_1 p)I_{XZ} + (R_1 q + Q_1 r)(I_{ZZ} - I_{YY}) = \left(\underline{l_A} + \underline{\underline{l_T}}\right)$$

$$\dot{q}I_{YY} + (P_1 r + pR_1)(I_{XX} - I_{ZZ}) + (2P_1 p - 2R_1 r)I_{XZ} = \left(\underline{m_A} + \underline{\underline{m_T}}\right)$$

$$\dot{r}I_{ZZ} - \dot{p}I_{XZ} + (P_1 q + pQ_1)(I_{YY} - I_{XX}) + (Q_1 r + R_1 q)I_{XZ} = \left(\underline{n_A} + \underline{\underline{n_T}}\right)$$

The modeling of the small perturbation thrust forces and moments (the double underlined terms in the previous equations) will be discussed in Chapter 5. Although the steady-state aerodynamic forces and moments are missing in the relationships, a detailed analysis of the steady-state aerodynamic forces and moments is required because the small perturbation terms are calculated from the steady-state terms.

3.2 AIRCRAFT STABILITY AXES

The modeling of the aircraft dynamics in Chapter 1 is based on the use of the aircraft body axes X, Y, Z. This reference frame is typically used for the calculation of the aircraft mass properties. However, a different reference frame is needed for modeling the aerodynamic forces and moments acting on the aircraft. This reference frame is known as Stability Axes X_S, Y_S, Z_S, also referred to as Wind Axes in the technical literature. The Stability Axes are defined with the property of the X_S axis being aligned with the direction of the steady-state airspeed, that is, $V_{P_1} = U_{1_S}$. Therefore the vertical component of the airspeed along Z_S is zero, $W_{1_S} = 0$. The angle between X_S and X is the steady-state angle of attack α_1. Both body and stability axes are shown in Figure 3.1 along with the steady-state climb (or descent) angle γ_1 and the pitch angle Θ_1.

3.3 MODELING OF THE LONGITUDINAL STEADY-STATE AERODYNAMIC FORCES AND MOMENT

For the purpose of modeling the longitudinal steady-state forces and moments for a given aircraft, it is necessary to specify the configuration of the horizontal tail. For conventional (horizontal + vertical tail) architectures two longitudinal control surfaces—the elevator and the stabilator—can be used; a third surface, known as trim tab, might also be available on the trailing edge of the horizontal tail. Therefore, the following configurations can be commonly found:

80   Chapter 3   Modeling of Longitudinal Aerodynamic Forces and Moments

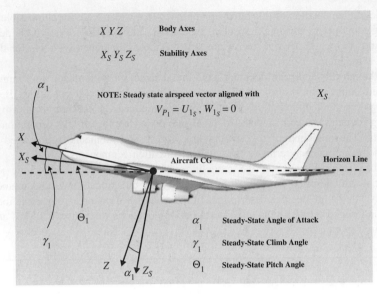

Figure 3.1   Aircraft Body and Stability Axes

Configuration 1:   Fixed stabilators; symmetric elevators with angular deflection δ_E used for both trimming and maneuvering, with an optional trim tab acting as servo surface.

Configuration 2:   No elevators; symmetric stabilator with angular deflection i_H used for both trimming and maneuvering (also known as all tail longitudinal control surface).

Configuration 3:   Stabilators with deflection i_H used for trimming and elevators with deflection δ_E used for maneuvering, with an optional trim tab acting as servo surface.

The last configuration, shown in Figure 3.2, is the most generic configuration, and it will be used here for modeling purposes.

The longitudinal steady state aerodynamic forces and moment are shown in Figure 3.3.

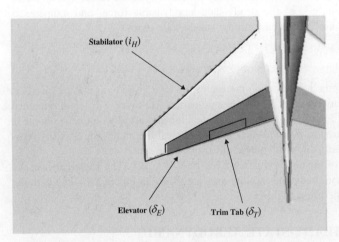

Figure 3.2   (Stabilator + Elevator) Longitudinal Control Surfaces

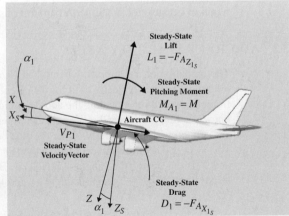

Figure 3.3   Longitudinal Steady-State Forces and Moments

3.4   MODELING OF $F_{A_{X_1}}$

From Figure 3.3 the steady-state aerodynamic force along X_S is given by

$$F_{A_{X_1}} = F_{A_{X_{1S}}} = -D_1 \quad D_1 = c_{D_1} \bar{q} S$$

where

$$D_1 = c_{D_1} \bar{q} S$$

c_{D_1} is the aircraft dimensionless drag coefficient at steady-state conditions
$\bar{q}$ is the dynamic pressure acting on the aircraft
S is the reference wing surface

Thus, the modeling of the aerodynamic drag at steady-state conditions reduces itself to the modeling of the different effects within the coefficient c_{D_1}. For a given Mach number and for the Reynolds number associated with the aerodynamic design, the drag coefficient can be expressed as function of the following variables:

$$c_{D_1} = f(\alpha, \delta_E, i_H)$$

The introduction of a Taylor expansion with a first-order approximation around the condition $\left[\alpha_0 = \delta_{E_0} = i_{H_0} = 0°\right]$ leads to

$$c_{D_1} = c_{D_0} + \frac{\partial c_D}{\partial \alpha}(\alpha - \alpha_0) + \frac{\partial c_D}{\partial \delta_E}(\delta_E - \delta_{E_0}) + \frac{\partial c_D}{\partial i_H}(i_H - i_{H_0})$$
$$= c_{D_0} + c_{D_\alpha}\alpha + c_{D_{\delta_E}}\delta_E + c_{D_{i_H}}i_H$$

where
$c_{D_0} = c_D|_{\alpha=\delta_E=i_H=0°}$ is the drag coefficient evaluated at the initial condition
c_{D_α} is the drag stability derivative with respect to the angle of attack
$c_{D_{\delta_E}}$ is the drag stability derivative with respect to the elevator deflection
$c_{D_{i_H}}$ is the drag stability derivative with respect to the stabilator deflection

As shown in Figure 3.4, the longitudinal control surfaces (stabilators and elevator) typically provide limited cross sections exposed to the airflow, also known as wetted area. Therefore, for most aircraft the derivatives $c_{D_{\delta_E}}$, $c_{D_{i_H}}$ can be approximated to be negligible; a special case is represented by the McDonnell Douglas F4 fighter aircraft, which features a sizable horizontal tail with a large anhedral angle, resulting in a non-negligible wetted area.

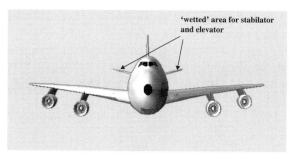

Figure 3.4 Wetted Area for Longitudinal Control Surfaces

Therefore, the following approximation can be made:

$$c_D \approx c_{D_0} + c_{D_\alpha}\alpha$$

It turns out that the drag coefficient c_D is the only longitudinal aerodynamic coefficient for which the implementation of the previously described build-up approach does not provide accurate and reliable results. This is due to the difficulty in the modeling of the wing + body integration in addition to the complexity in the modeling of the aerodynamic drag associated with the propulsion system, for both propeller-type and jet-based propulsion, as well as the drag associated with the aerodynamic shape of different objects under the wing. Hoak and Roskam[5,6] provide a number of empirical tools for the modeling of the coefficients c_{D_0}, c_{D_α}. However, wind tunnel analysis is still the most accurate method when a detailed estimate of the drag coefficients is required.

At a general level, an alternative conceptual approach for the modeling of the overall c_D coefficient is provided by the well-known Prandtl relationship[1] given by

$$c_{D_1} = \bar{c}_{D_0} + \frac{c_L^2}{\pi\, AR\, e}$$

The term $\bar{c}_{D_0}$ is known as the parasite drag coefficient; $\bar{c}_{D_0}$ is essentially the value of c_D associated with the condition of $c_L = 0$, as shown in Figure 3.5.

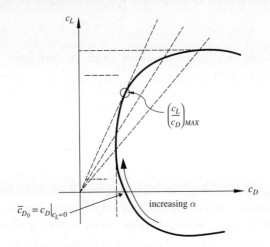

Figure 3.5 Polar Curve for the Entire Aircraft

It should be clear that $\bar{c}_{D_0}$ and the previously introduced c_{D_0} are conceptually different since $\bar{c}_{D_0} = c_D|_{c_L=0}$ while $c_{D_0} = c_D|_{\alpha=\delta_E=i_H=0°}$. The term $\left(\frac{c_L^2}{\pi\, AR\, e}\right)$ is known as the induced drag coefficient—also referred to as c_{D_i} in the technical literature—where e is the Oswald Efficiency factor introduced in Chapter 2, Section 2.4. In practical terms c_{D_i} can be considered as the aerodynamic price to be paid in terms of drag for producing lift. From Prandtl's relationship, it can be seen that c_{D_i} is lower for high-aspect ratio wings as well as for wings with lift distribution approximating an elliptical distribution along the wing span $(e \to 1)$.

In the special case of uncambered airplanes (when the drag polar curve can be approximated to be parabolic) is the condition $\bar{c}_{D_0} \approx c_{D_0}$. Therefore, the following approximate relationship can be derived:

$$\cancel{c_{D_0}} + c_{D_\alpha}\alpha = \cancel{\bar{c}_{D_0}} + \frac{c_L}{\pi\, AR\, e}$$

leading to

$$c_{D_\alpha} \approx \frac{\partial \frac{c_{L_1}^2}{\pi\, AR\, e}}{\partial \alpha} = \left(\frac{2c_{L_1}}{\pi\, AR\, e}\right) c_{L_\alpha}$$

Hoak and Roskam[5,6] describe empirical approaches for the calculation of the drag coefficients using the build-up approach. However, in general, wind tunnel analysis is the best approach for accurate estimates of both $\bar{c}_{D_0}$ and c_{D_i}. A specific technique has been developed for the determination of $\bar{c}_{D_0}$ and e from reduction of flight data; particularly, these coefficients are found within the general framework of evaluating the power required for level flight,[2,7] as shown in Chapter 5.

3.5 MODELING OF $F_{A_{Z_1}}$

According to the forces shown in Figure 3.3 the steady-state aerodynamic force acting along the Z_S axis is given by

$$F_{A_{Z_1}} = F_{A_{Z_{1S}}} = -L_1$$

From conventional aerodynamic modeling an expression for the overall aircraft lift is given by

$$L_1 = c_{L_1} \bar{q} S$$

where c_{L_1} is the dimensionless lift coefficient for the entire aircraft.

For a given Mach number and a given Reynolds number, the lift coefficient can be expressed as function of the following variables:

$$c_{L_1} = f(\alpha, \delta_E, i_H)$$

The introduction of a Taylor expansion with a first order approximation around the condition $\left[\alpha_0 = \delta_{E_0} = i_{H_0} = 0°\right]$ leads to

$$c_{L_1} = c_{L_0} + \frac{\partial c_L}{\partial \alpha}(\alpha - \alpha_0) + \frac{\partial c_L}{\partial \delta_E}(\delta_E - \delta_{E_0}) + \frac{\partial c_L}{\partial i_H}(i_H - i_{H_0})$$

$$= c_{L_0} + c_{L_\alpha} \alpha + c_{L_{\delta_E}} \delta_E + c_{L_{i_H}} i_H$$

where

$c_{L_0} = c_L|_{\alpha = \delta_E = i_H = 0°}$ is the lift coefficient evaluated at the initial condition

c_{L_α} is the lift stability derivative with respect to the angle of attack

$c_{L_{\delta_E}}$ is the lift stability derivative with respect to the elevator deflection

$c_{L_{i_H}}$ is the lift stability derivative with respect to the stabilator deflection

Next, as outlined previously, the general modeling approach is to isolate the contributions from the different components of the aircraft. For a conventional subsonic aircraft, the lift of the entire aircraft at steady-state conditions is given by

$$L_1 = L_W + L_B + L_H$$

Merging the contribution of wing and fuselage leads to

$$L_1 = L_{WB} + L_H$$

In terms of coefficients we have

$$L_1 = c_{L_1} \bar{q} S = c_{L_{WB}} \bar{q} S + c_{L_H} \bar{q}_H S_H$$

Therefore

$$c_{L_1} = c_{L_{WB}} + c_{L_H} \frac{\bar{q}_H S_H}{\bar{q} S} = c_{L_{WB}} + c_{L_H} \eta_H \frac{S_H}{S}$$

Next, the goal is to outline the specific contributions of the variables (α, δ_E, i_H) for the coefficients $(c_{L_{WB}}, c_{L_H})$. Because the wing + body subsystem of the aircraft system does not include the horizontal tail, a Taylor expansion of $c_{L_{WB}}$ with a first order approximation around the condition $(\alpha_0 = 0°)$ is given by

$$c_{L_{WB}} = c_{L_{WB}}(\alpha) = c_{L_{0_{WB}}} + c_{L_{\alpha_{WB}}} \alpha$$

where

$c_{L_{0_{WB}}} = c_{L_{WB}}|_{\alpha = 0°}$ is the (wing + body) lift coefficient evaluated at the initial condition $(\alpha_0 = 0°)$

$c_{L_{\alpha_{WB}}}$ is the lift stability derivative of the (wing + body) with respect to the angle of attack

Considering the aircraft geometry in Figure 3.6, assuming a ratio $(b/d) > 4$, the following approximation can be used:[5]

$$c_{L_{\alpha_{WB}}} \approx K_{WB}\, c_{L_{\alpha_W}}$$

where $K_{WB} = 1 + 0.025(d/b) - 0.25(d/b)^2$. For most subsonic configurations the approximation $K_{WB} \approx 1$ can be used.

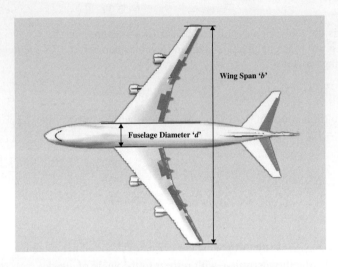

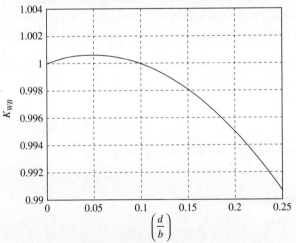

Figure 3.6 Wing Span vs. Fuselage Diameter[5]

Next, consider the coefficient $c_{L_H}(\alpha_H, \delta_E, i_H)$. Using a first order Taylor expansion around the initial condition $(\alpha_0 = \delta_{E_0} = i_{H_0} = 0°)$

$$c_{L_H} = c_{L_{0_H}} + c_{L_{\alpha_H}} \alpha_H + c_{L_{\alpha_H}} \tau_E \delta_E + c_{L_{\alpha_H}} i_H$$

where

$$c_{L_{0_H}} = c_{L_H}\big|_{\alpha = \delta_E = i_H = 0°}$$

is the horizontal tail lift coefficient evaluated at the initial condition $(\alpha_{H_0} = \delta_{E_0} = i_{H_0} = 0°)$. Symmetric profiles are used for the design of the horizontal tail; therefore, we have $c_{L_{0_H}} = 0$.

$c_{L_{\alpha_H}}$ is the lift stability derivative of the horizontal tail with respect to the angle of attack.

τ_E is the effectiveness of the elevator (discussed in Chapter 2, Section 2.4), shown in Figure 3.7.

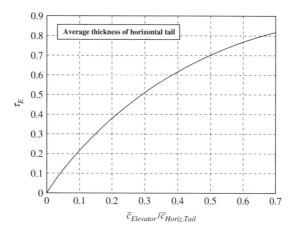

Figure 3.7 Effectiveness of the Elevator τ_E as Function of $\bar{c}_{Elevator}/\bar{c}_{Horiz.Tail}$

For modeling purposes the horizontal tail is considered to be exposed to three different angles: the aerodynamic angle of attack evaluated at the horizontal tail (α_H), the deflection of the elevator (δ_E), and the deflection of the stabilator (i_H), as shown in Figure 3.8.

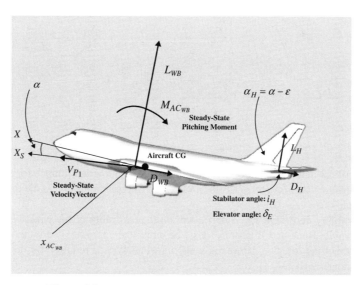

Figure 3.8 Longitudinal Forces and Aerodynamic Angles

Accounting for the modeling of the downwash effect, the angle of attack at the horizontal tail is given by

$$\alpha_H = \alpha - \varepsilon(\alpha)$$

The use of a first order approximation Taylor expansion of the downwash effect around the condition $(\alpha_0 = 0°)$ leads to

$$\alpha_H = \alpha - \left(\varepsilon_0 + \frac{d\varepsilon}{d\alpha}\alpha\right)$$

As discussed in Chapter 2, the downwash effect is approximated to be negligible when $\alpha = 0°$. Therefore

$$\varepsilon_0 = \varepsilon|_{\alpha=0°} \approx 0$$

leading to

$$\alpha_H = \alpha - \frac{d\varepsilon}{d\alpha}\alpha = \alpha\left[1 - \frac{d\varepsilon}{d\alpha}\right]$$

finally yielding to

$$c_{L_H} = c_{L_{\alpha_H}}\left[\left(1 - \frac{d\varepsilon}{d\alpha}\right)\alpha + \tau_E \delta_E + i_H\right]$$

In summary, the described relationship leads to

$$c_{L_1} = c_{L_{WB}} + c_{L_H}\eta_H \frac{S_H}{S} = (c_{L_{0WB}} + c_{L_{\alpha W}}\alpha) + \eta_H \frac{S_H}{S} c_{L_{\alpha_H}}\left[\left(1 - \frac{d\varepsilon}{d\alpha}\right)\alpha + \tau_E \delta_E + i_H\right]$$

Next, equating the given expression to the previous general expression

$$c_{L_1} = c_{L_0} + c_{L_\alpha}\alpha + c_{L_{\delta_E}}\delta_E + c_{L_{i_H}}i_H$$

leads to the following general expressions for the lift stability and control derivatives:[4]

$$\begin{aligned}
c_{L_0} &= c_{L_{0WB}} \\
c_{L_\alpha} &= c_{L_{\alpha W}} + c_{L_{\alpha_H}}\eta_H \frac{S_H}{S}\left(1 - \frac{d\varepsilon}{d\alpha}\right) \\
c_{L_{i_H}} &= \eta_H \frac{S_H}{S} c_{L_{\alpha_H}} \\
c_{L_{\delta_E}} &= \eta_H \frac{S_H}{S} c_{L_{\alpha_H}} \tau_E = c_{L_{i_H}} \tau_E
\end{aligned}$$

It is critical that students become familiar with the numerical ranges for the above coefficients; a review of the aerodynamic data in Appendix B for a variety of different aircraft and the Student Sample Problems is strongly recommended.

The stability derivative c_{L_0} is dependent on the capability of the wing + body system to provide lift at zero incidence angle. Hoak and Roskam[5,6] describe empirical approximated methods for evaluating c_{L_0} based on the cross section of the fuselage. However, wind tunnel analysis is required for detailed estimates for this coefficient. A typical range is [0.1–0.4].

The stability derivative c_{L_α} is one of the most critical aerodynamic coefficients. Its value for the entire aircraft is mainly a function of the corresponding values for the wing $(c_{L_{\alpha W}})$ and to a lower extent the tail $(c_{L_{\alpha_H}})$. The Polhamus formula can be used for an estimate of these values for conventional subsonic configurations with $M \leq M_{CRIT}$; a typical range for both $c_{L_{\alpha W}}$ and $c_{L_{\alpha_H}}$ is [4–5.5]. A detailed description of the modeling of the downwash effect $(d\varepsilon/d\alpha)$ has been provided in Chapter 2. Typical ranges for the coefficients η_H and S_H/S are [0.85–0.95] and [0.15–0.3], respectively. For example, the use of the following midrange values:

$$c_{L_{\alpha W}} = 5, \quad c_{L_{\alpha_H}} = 5, \quad \eta_H = 0.9, \quad \frac{S_H}{S} = 0.225, \quad \left(1 - \frac{d\varepsilon}{d\alpha}\right) = 0.65$$

leads to

$$c_{L_\alpha} = c_{L_{\alpha W}} + c_{L_{\alpha_H}}\eta_H \frac{S_H}{S}\left(1 - \frac{d\varepsilon}{d\alpha}\right) = 5 + 5 \cdot 0.9 \cdot 0.225 \cdot 0.65 \approx 5.66$$

Therefore, for typical subsonic configurations, the contribution of the horizontal tail to the value of c_{L_α} is approximately [10–15] percent of the entire value. Finally, the numerical values of the control derivatives $c_{L_{i_H}}, c_{L_{\delta_E}}$ are approximately [10–15] percent and [5–10] percent of the value for c_{L_α}.

3.6 MODELING OF M_{A_1}

According to the modeling of the longitudinal forces in Figure 3.8, the steady-state aerodynamic moment acting around Y_S is given by

$$M_{A_1} = M_1$$

An expression for the overall aircraft pitching moment is given by

$$M_1 = c_{m_1} \bar{q} S \bar{c}$$

As for the lift coefficient, for a given Mach number and a given Reynolds number, the pitching moment coefficient can be expressed as a function of the following variables:

$$c_{m_1} = f(\alpha, \delta_E, i_H)$$

The use of first-order approximation for the Taylor expansion around the condition $\left[\alpha_0 = \delta_{E_0} = i_{H_0} = 0°\right]$ leads to

$$c_{m_1} = c_{m_0} + \frac{\partial c_m}{\partial \alpha}(\alpha - \alpha_0) + \frac{\partial c_m}{\partial \delta_E}(\delta_E - \delta_{E_0}) + \frac{\partial c_m}{\partial i_H}(i_H - i_{H_0})$$

$$= c_{m_0} + c_{m_\alpha}\alpha + c_{m_{\delta_E}}\delta_E + c_{m_{i_H}}i_H$$

where

$c_{m_0} = c_m|_{\alpha=\delta_E=i_H=0°}$ is the pitching moment coefficient evaluated at the condition $(\alpha_0 = \delta_{E_0} = i_{H_0} = 0°)$

c_{m_α} is the pitching moment stability derivative with respect to the angle of attack

$c_{m_{\delta_E}}$ is the pitching moment stability derivative with respect to the elevator deflection

$c_{m_{i_H}}$ is pitching moment stability derivative with respect to the stabilator deflection

Figure 3.9 shows the geometric parameters necessary for modeling the different contributions to the pitching moment from the aircraft components.

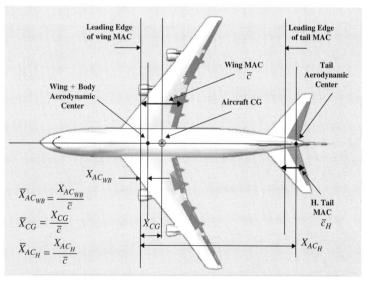

Figure 3.9 Geometric Parameters for the Modeling of the Pitching Moment

For a conventional subsonic aircraft the pitching moment coefficient for the entire aircraft[4] is given by

$$M_1 = M_{AC_{WB}} + L_{WB}(x_{CG} - x_{AC_{WB}})\cos(\alpha) + D_{WB}(x_{CG} - x_{AC_{WB}})\sin(\alpha)$$
$$- L_H(x_{AC_H} - x_{CG})\cos(\alpha_H) + D_H(x_{AC_H} - x_{CG})\sin(\alpha_H)$$

with $\alpha_H = \alpha - \varepsilon = \alpha\left(1 - \dfrac{d\varepsilon}{d\alpha}\right)$.

Chapter 3 Modeling of Longitudinal Aerodynamic Forces and Moments

At small values for the angles of attack we can approximate

$$\cos(\alpha) \approx 1 \quad >> \quad \sin(\alpha) \approx \alpha$$
$$\cos(\alpha_H) \approx 1 \quad >> \quad \sin(\alpha_H) \approx \alpha_H$$

Also, it can be approximated that

$$L_{WB} >> D_{WB}, \quad D_H \approx 0$$

The above assumptions lead to

$$D_{WB}(x_{CG} - x_{AC_{WB}}) \sin(\alpha) \approx 0$$
$$D_H(x_{AC_H} - x_{CG}) \sin(\alpha_H) \approx 0$$

Therefore

$$M_1 = M_{AC_{WB}} + L_{WB}(x_{CG} - x_{AC_{WB}})\cos(\alpha) + \cancel{D_{WB}(x_{CG} - x_{AC_{WB}})\sin(\alpha)}$$
$$- L_H(x_{AC_H} - x_{CG})\cos(\alpha_H) + \cancel{D_H(x_{AC_H} - x_{CG})\sin(\alpha_H)}$$

Next, expressing (L_{WB}, L_H) using the relationships previously introduced

$$M_1 = c_{m_{AC_{WB}}} \bar{q} S \bar{c} + c_{L_{WB}} \bar{q} S (x_{CG} - x_{AC_{WB}}) - c_{L_H} \bar{q}_H S_H (x_{AC_H} - x_{CG})$$

Therefore

$$M_1 = c_{m_1} \bar{q} S \bar{c} = c_{m_{AC_{WB}}} \bar{q} S \bar{c} + (c_{L_{0_{WB}}} + c_{L_{\alpha_{WB}}} \alpha) \bar{q} S (x_{CG} - x_{AC_{WB}})$$
$$- c_{L_{\alpha_H}} \left[\left(1 - \frac{d\varepsilon}{d\alpha}\right)\alpha + \tau_E \delta_E + i_H \right] \bar{q}_H S_H (x_{AC_H} - x_{CG})$$

with $c_{L_{\alpha_{WB}}} \approx c_{L_{\alpha_W}}$ as discussed in the previous section. In terms of the pitching moment coefficient

$$c_{m_1} = c_{m_{AC_{WB}}} \frac{\cancel{\bar{q} S \bar{c}}}{\cancel{\bar{q} S \bar{c}}} + (c_{L_{0_{WB}}} + c_{L_{\alpha_W}} \alpha) \frac{\cancel{\bar{q} S}(x_{CG} - x_{AC_{WB}})}{\cancel{\bar{q} S} \bar{c}}$$
$$- c_{L_{\alpha_H}} \left[\left(1 - \frac{d\varepsilon}{d\alpha}\right)\alpha + \tau_E \delta_E + i_H \right] \frac{\bar{q}_H S_H (x_{AC_H} - x_{CG})}{\bar{q} S \bar{c}}$$

leading to

$$c_{m_1} = c_{m_{AC_{WB}}} + (c_{L_{0_{WB}}} + c_{L_{\alpha_W}} \alpha) \frac{(x_{CG} - x_{AC_{WB}})}{\bar{c}}$$
$$- c_{L_{\alpha_H}} \left[\left(1 - \frac{d\varepsilon}{d\alpha}\right)\alpha + \tau_E \delta_E + i_H \right] \eta_H \frac{S_H (x_{AC_H} - x_{CG})}{S \bar{c}}$$

Introducing the dimensionless longitudinal distances shown in Figure 3.9

$$c_{m_1} = c_{m_{AC_{WB}}} + (c_{L_{0_{WB}}} + c_{L_{\alpha_W}} \alpha)(\bar{x}_{CG} - \bar{x}_{AC_{WB}})$$
$$- c_{L_{\alpha_H}} \left[\left(1 - \frac{d\varepsilon}{d\alpha}\right)\alpha + \tau_E \delta_E + i_H \right] \eta_H \frac{S_H}{S} (\bar{x}_{AC_H} - \bar{x}_{CG})$$

Next, the objective is to equate the previous expression to the generic relationship:

$$c_{m_1} = c_{m_0} + c_{m_\alpha} \alpha + c_{m_{\delta_E}} \delta_E + c_{m_{i_H}} i_H$$

Isolating the terms which are functions of (α, δ_E, i_H) respectively leads to

$$c_{m_0} = c_{m_{AC_{WB}}} + c_{L_{0_{WB}}}(\bar{x}_{CG} - \bar{x}_{AC_{WB}})$$

$$c_{m_\alpha} = c_{L_{\alpha_W}}(\bar{x}_{CG} - \bar{x}_{AC_{WB}}) - c_{L_{\alpha_H}}\eta_H \frac{S_H}{S}\left(1 - \frac{d\varepsilon}{d\alpha}\right)(\bar{x}_{AC_H} - \bar{x}_{CG})$$

$$c_{m_{i_H}} = -c_{L_{\alpha_H}}\eta_H \frac{S_H}{S}(\bar{x}_{AC_H} - \bar{x}_{CG})$$

$$c_{m_{\delta_E}} = -c_{L_{\alpha_H}}\eta_H \frac{S_H}{S}(\bar{x}_{AC_H} - \bar{x}_{CG})\tau_E = c_{m_{i_H}}\tau_E$$

Once again, it is critical for the students to become familiar with the numerical ranges for the above coefficients; a review of the aerodynamic data in Appendix B and the Student Sample Problems is strongly recommended.

The stability derivative c_{m_0} is dependent on the capability of the (wing + body) system to provide a pitching moment at zero incidence angle. A typical range is [0.01–0.1] for a natural tendency of the aircraft to nose up. The stability derivative c_{m_α} is another very critical aerodynamic coefficient. Its value is a function of $(c_{L_{\alpha_{WB}}})$ and $(c_{L_{\alpha_H}})$. As described in Chapter 2, the Polhamus formula can be used for an estimate of these values for conventional subsonic configurations with $M \leq M_{CRIT}$; a typical range for both $c_{L_{\alpha_W}}$ and $c_{L_{\alpha_H}}$ is [4–5.5]. A detailed description of the modeling of the downwash effect $(d\varepsilon/d\alpha)$ has also been provided in Chapter 2, Section 2.4. Typical ranges for the coefficients η_H, S_H/S are [0.85–0.95] and [0.15–0.3], respectively. The geometric parameter $(\bar{x}_{CG} - \bar{x}_{AC_{WB}})$ is typically small since the aerodynamic center of the wing + body system is located close to the center of gravity on the longitudinal axis. The most important geometric parameter is $(\bar{x}_{AC_H} - \bar{x}_{CG})$, which is the dimensionless moment arm of the horizontal tail with respect to the aircraft center of gravity. Typical ranges for $\bar{x}_{AC_H}$ of values vary depending on the aircraft classes; suitable ranges are

[3–4] for general aviation
[3–6] for commercial aviation and military transport
[2–3] for fighter aircraft

Because of the difference in the value of $(\bar{x}_{AC_H} - \bar{x}_{CG})$ with respect to $(\bar{x}_{CG} - \bar{x}_{AC_{WB}})$, c_{m_α} is negative. The implications of the sign of c_{m_α} on the aircraft dynamic stability will be clarified in the later chapters. Finally, the aerodynamic coefficients $c_{m_{i_H}}$, $c_{m_{\delta_E}}$—which are both negative—model the pitching authority of the horizontal tail. Because the elevator effectiveness factor τ_E is in range [0.4–0.6], we typically have the condition $c_{m_{i_H}} \approx 2\, c_{m_{\delta_E}}$.

3.7 AIRCRAFT AERODYNAMIC CENTER

A critical concept associated with the modeling of the pitching moment is the aircraft Aerodynamic Center (AC). Chapter 2 has reviewed the definitions of the concept of aerodynamic center for wing sections, wing planforms, and the (wing + body) system.[6,8] Within this context, the concept is now extended to the entire aircraft.

The aircraft aerodynamic center is defined as the point with respect to which the pitching moment for the entire aircraft does not change with variations in the longitudinal angle of attack. Mathematically, this implies

$$c_{m_\alpha} = 0$$

Therefore, the aircraft aerodynamic center can be considered to be that specific location for the aircraft center of gravity at which the following relationship applies:

$$\bar{x}_{AC} = \bar{x}_{CG}|_{c_{m_\alpha}=0}$$

Therefore, setting to zero the previously introduced c_{m_α} relationship

$$c_{m_\alpha} = 0 = c_{L_{\alpha_W}}(\bar{x}_{AC} - \bar{x}_{AC_{WB}}) - c_{L_{\alpha_H}}\eta_H \frac{S_H}{S}\left(1 - \frac{d\varepsilon}{d\alpha}\right)(\bar{x}_{AC_H} - \bar{x}_{AC})$$

leads to the following expression:

$$\bar{x}_{AC} = \frac{\bar{x}_{AC_{WB}} + \dfrac{c_{L_{\alpha H}}}{c_{L_{\alpha W}}} \eta_H \dfrac{S_H}{S} \left(1 - \dfrac{d\varepsilon}{d\alpha}\right) \bar{x}_{AC_H}}{1 + \dfrac{c_{L_{\alpha H}}}{c_{L_{\alpha W}}} \eta_H \dfrac{S_H}{S} \left(1 - \dfrac{d\varepsilon}{d\alpha}\right)}$$

Typical ranges for the value of $\bar{x}_{AC}$ are

[0.4–0.6] for general aviation
[0.35–0.45] for commercial aviation and military transport
[0.3–0.35] for fighter aircraft

Once $\bar{x}_{AC}$ is known, the distance between the aerodynamic center and the center of gravity normalized with respect to the wing MAC is known as the longitudinal Static Margin SM:

$$c_{m_\alpha} = c_{L_\alpha}(\bar{x}_{CG} - \bar{x}_{AC}) = c_{L_\alpha} \, SM$$

SM has major implications on the aircraft longitudinal dynamic stability. In general, civilian and military cargo aircraft have larger values for the Static Margin, allowing for loading flexibility and thus wider allowed ranges of the aircraft center of gravity. Fighter aircraft have instead lower values for SM due to lower values for $\bar{x}_{AC}$, ultimately due to small values for the geometric parameter $\bar{x}_{AC_H}$ shown in Figure 3.9. Figure 3.10 shows typical SMs for a jetliner (left) and a fighter aircraft (right). Note that the SM is a negative parameter because the aircraft center of gravity is located in front of the aircraft aerodynamic center.

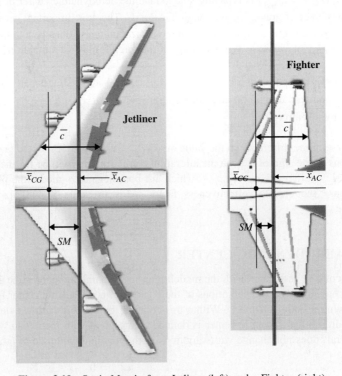

Figure 3.10 Static Margin for a Jetliner (left) and a Fighter (right)

It should be pointed out that the reduced SM values for fighter aircraft (ultimately leading to reduced or relaxed stability, as will be shown in Chapters 6 and 7) bring enormous benefits in terms of maneuverability. This trend led in the 1970s to the design and development of fighter aircraft naturally "unstable" (with a positive SM) but very maneuverable, with the stability provided through the use of closed-loop control laws activated by the flight computer. This implied that these aircraft required the continuous use of flight computers throughout the flight. The General Dynamics F-16 fighter aircraft was the first example of this revolutionary design.

Typical ranges of values for the absolute value of *SM* are

[0.15−0.25] for commercial aviation and military transport

[0.05−0.15] for open loop stable fighter aircraft

Chapter 6 will provide additional definitions for the concept of aircraft aerodynamic center. This concept is very critical from a stability and maneuverability point of view in addition to being a critical aerodynamic parameter.

3.8 SUMMARY OF THE LONGITUDINAL STEADY-STATE AERODYNAMIC FORCES AND MOMENT

A summary of the longitudinal steady-state forces and moment in a matrix format is provided:

$$\begin{Bmatrix} F_{A_{X_{1_S}}} \\ F_{A_{Z_{1_S}}} \\ M_{A_{1_S}} \end{Bmatrix} = \begin{Bmatrix} F_{A_{X_1}} \\ F_{A_{Z_1}} \\ M_{A_1} \end{Bmatrix} = \begin{Bmatrix} -(\bar{q}\, S\, c_{D_1}) \\ -(\bar{q}\, S\, c_{L_1}) \\ \bar{q}\, S\, \bar{c}\, c_{m_1} \end{Bmatrix}$$

where

$$\begin{Bmatrix} c_{D_1} \\ c_{L_1} \\ c_{m_1} \end{Bmatrix} = \begin{bmatrix} c_{D_0} & c_{D_\alpha} & \cancel{c_{D_{\delta_E}}} & \cancel{c_{D_{i_H}}} \\ c_{L_0} & c_{L_\alpha} & c_{L_{\delta_E}} & c_{L_{i_H}} \\ c_{m_0} & c_{m_\alpha} & c_{m_{\delta_E}} & c_{m_{i_H}} \end{bmatrix} \begin{Bmatrix} 1 \\ \alpha \\ \delta_E \\ i_H \end{Bmatrix}$$

3.9 MODELING OF THE LONGITUDINAL SMALL PERTURBATION AERODYNAMIC FORCES AND MOMENTS

Following the modeling of the steady-state forces and moments for the longitudinal dynamics, the next objective is the modeling of the small perturbation components of the lift and drag forces and the pitching moment. The origin of these components is associated with the perturbation terms in all the components of the linear and angular velocities, shown in Figure 3.11.

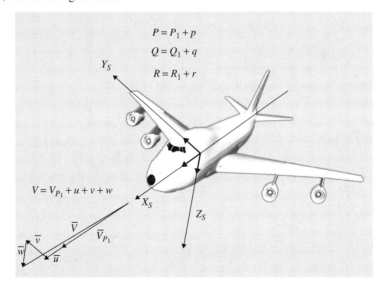

Figure 3.11 Small Perturbations in Linear and Agular Velocities

Note that additional small perturbation aerodynamic forces are associated with the rate of change of the small perturbation term for the vertical velocity, $\dot{w}$. However, the perturbation terms in terms of the vertical linear velocity component can also be interpreted as a perturbation in the angle of attack. Therefore, taking advantage of the assumption of small perturbations, we have

$$\tan\alpha = \frac{\sin\alpha}{\cos\alpha} \approx \frac{w}{V_{P_1}} \rightarrow \alpha \approx \frac{w}{V_{P_1}}, \dot{\alpha} \approx \frac{\dot{w}}{V_{P_1}}$$

In addition, the deflections of the control surfaces for controlling the longitudinal and the lateral directional dynamics can also be considered as small perturbation terms. As for the modeling of the longitudinal steady-state forces and moment, assume that the aircraft features both elevators (δ_E) and stabilators (i_H).

Taking advantage of the decoupling between longitudinal and lateral directional dynamics, the following relationships describe the functionalities of the longitudinal small perturbation forces f_{A_X}, f_{A_Z} and moment m_A:

$$f_{A_X} = f_{A_X}(u, w, \dot{w}, q, \delta_E, i_H)$$
$$f_{A_Z} = f_{A_Z}(u, w, \dot{w}, q, \delta_E, i_H)$$
$$m_A = m_A(u, w, \dot{w}, q, \delta_E, i_H)$$

The relationships are formally correct; however, the aerodynamic stability derivatives are evaluated with respect to the longitudinal angle of attack α. Therefore, in the given relationships $(\alpha, \dot{\alpha})$ will be used in lieu of $(w, \dot{w})$, leading to

$$f_{A_X} = f_{A_X}(u, \alpha, \dot{\alpha}, q, \delta_E, i_H)$$
$$f_{A_Z} = f_{A_Z}(u, \alpha, \dot{\alpha}, q, \delta_E, i_H)$$
$$m_A = m_A(u, \alpha, \dot{\alpha}, q, \delta_E, i_H)$$

Next, the modeling of the forces and moments is set up as a first-order Taylor expansion with respect to the longitudinal set of independent variables $(u, \alpha, \dot{\alpha}, q, \delta_E, i_H)$:

$$f_{A_X} = \left.\frac{\partial F_{A_X}}{\partial u}\right|_{SS} u + \left.\frac{\partial F_{A_X}}{\partial \alpha}\right|_{SS} \alpha + \left.\frac{\partial F_{A_X}}{\partial \dot{\alpha}}\right|_{SS} \dot{\alpha} + \left.\frac{\partial F_{A_X}}{\partial q}\right|_{SS} q + \left.\frac{\partial F_{A_X}}{\partial \delta_E}\right|_{SS} \delta_E + \left.\frac{\partial F_{A_X}}{\partial i_H}\right|_{SS} i_H$$

$$f_{A_Z} = \left.\frac{\partial F_{A_Z}}{\partial u}\right|_{SS} u + \left.\frac{\partial F_{A_Z}}{\partial \alpha}\right|_{SS} \alpha + \left.\frac{\partial F_{A_Z}}{\partial \dot{\alpha}}\right|_{SS} \dot{\alpha} + \left.\frac{\partial F_{A_Z}}{\partial q}\right|_{SS} q + \left.\frac{\partial F_{A_Z}}{\partial \delta_E}\right|_{SS} \delta_E + \left.\frac{\partial F_{A_Z}}{\partial i_H}\right|_{SS} i_H$$

$$m_A = \left.\frac{\partial M_A}{\partial u}\right|_{SS} u + \left.\frac{\partial M_A}{\partial \alpha}\right|_{SS} \alpha + \left.\frac{\partial M_A}{\partial \dot{\alpha}}\right|_{SS} \dot{\alpha} + \left.\frac{\partial M_A}{\partial q}\right|_{SS} q + \left.\frac{\partial M_A}{\partial \delta_E}\right|_{SS} \delta_E + \left.\frac{\partial M_A}{\partial i_H}\right|_{SS} i_H$$

where the angles (α, δ_E, i_H) are expressed in radians and are, therefore, dimensionless. Note that the values of the gradients in the above relationships are evaluated at the steady-state conditions. An analysis of the relationships reveals that the dimensions of the gradients are different. It would be desirable to modify the relationships such that each of the terms on the right-hand side has the same dimensions of the left-hand side. Thus, an ad-hoc non-dimensionalization process needs to be introduced for the $(u, \dot{\alpha}, q)$ coefficients with the objective of making them dimensionless. The goal can be achieved by introducing the following transformations of variables:

$$u\left[\frac{\text{ft}}{\text{sec}}\right] \to \frac{u}{V_{P_1}}, \quad \dot{\alpha}\left[\frac{\text{rad}}{\text{sec}}\right] \to \frac{\dot{\alpha}\bar{c}}{2V_{P_1}}, \quad q\left[\frac{\text{rad}}{\text{sec}}\right] \to \frac{q\bar{c}}{2V_{P_1}}$$

The resulting expressions for the longitudinal small perturbation terms are given by

$$f_{A_X} = \left.\frac{\partial F_{A_X}}{\partial\left(\frac{u}{V_{P_1}}\right)}\right|_{SS}\left(\frac{u}{V_{P_1}}\right) + \left.\frac{\partial F_{A_X}}{\partial \alpha}\right|_{SS}\alpha + \left.\frac{\partial F_{A_X}}{\partial\left(\frac{\dot{\alpha}\bar{c}}{2V_{P_1}}\right)}\right|_{SS}\left(\frac{\dot{\alpha}\bar{c}}{2V_{P_1}}\right) + \left.\frac{\partial F_{A_X}}{\partial\left(\frac{q\bar{c}}{2V_{P_1}}\right)}\right|_{SS}\left(\frac{q\bar{c}}{2V_{P_1}}\right) + \left.\frac{\partial F_{A_X}}{\partial \delta_E}\right|_{SS}\delta_E + \left.\frac{\partial F_{A_X}}{\partial i_H}\right|_{SS}i_H$$

$$f_{A_Z} = \left.\frac{\partial F_{A_Z}}{\partial\left(\frac{u}{V_{P_1}}\right)}\right|_{SS}\left(\frac{u}{V_{P_1}}\right) + \left.\frac{\partial F_{A_Z}}{\partial \alpha}\right|_{SS}\alpha + \left.\frac{\partial F_{A_Z}}{\partial\left(\frac{\dot{\alpha}\bar{c}}{2V_{P_1}}\right)}\right|_{SS}\left(\frac{\dot{\alpha}\bar{c}}{2V_{P_1}}\right) + \left.\frac{\partial F_{A_Z}}{\partial\left(\frac{q\bar{c}}{2V_{P_1}}\right)}\right|_{SS}\left(\frac{q\bar{c}}{2V_{P_1}}\right) + \left.\frac{\partial F_{A_Z}}{\partial \delta_E}\right|_{SS}\delta_E + \left.\frac{\partial F_{A_Z}}{\partial i_H}\right|_{SS}i_H$$

$$m_A = \left.\frac{\partial M_A}{\partial\left(\frac{u}{V_{P_1}}\right)}\right|_{SS}\left(\frac{u}{V_{P_1}}\right) + \left.\frac{\partial M_A}{\partial \alpha}\right|_{SS}\alpha + \left.\frac{\partial M_A}{\partial\left(\frac{\dot{\alpha}\bar{c}}{2V_{P_1}}\right)}\right|_{SS}\left(\frac{\dot{\alpha}\bar{c}}{2V_{P_1}}\right) + \left.\frac{\partial M_A}{\partial\left(\frac{q\bar{c}}{2V_{P_1}}\right)}\right|_{SS}\left(\frac{q\bar{c}}{2V_{P_1}}\right) + \left.\frac{\partial M_A}{\partial \delta_E}\right|_{SS}\delta_E + \left.\frac{\partial M_A}{\partial i_H}\right|_{SS}i_H$$

Roskam[4] provides a detailed analysis of the modeling for each of the terms in the above relationships. It can be shown that

$$f_{A_X} = \bar{q}\,S\left\{-[c_{D_u} + 2c_{D_1}]\left(\frac{u}{V_{P_1}}\right) + [-c_{D_\alpha} + c_{L_1}]\alpha - c_{D_{\dot\alpha}}\left(\frac{\dot\alpha\,\bar{c}}{2V_{P_1}}\right) - c_{D_q}\left(\frac{q\,\bar{c}}{2V_{P_1}}\right) - c_{D_{\delta_E}}\delta_E - c_{D_{i_H}}i_H\right\}$$

$$f_{A_z} = \bar{q}\,S\left\{-[c_{L_u} + 2c_{L_1}]\left(\frac{u}{V_{P_1}}\right) - [c_{L_\alpha} + c_{D_1}]\alpha - c_{L_{\dot\alpha}}\left(\frac{\dot\alpha\,\bar{c}}{2V_{P_1}}\right) - c_{L_q}\left(\frac{q\,\bar{c}}{2V_{P_1}}\right) - c_{L_{\delta_E}}\delta_E - c_{L_{i_H}}i_H\right\}$$

$$m_A = \bar{q}\,S\,\bar{c}\left\{[c_{m_u} + 2c_{m_1}]\left(\frac{u}{V_{P_1}}\right) + c_{m_\alpha}\alpha + c_{m_{\dot\alpha}}\left(\frac{\dot\alpha\,\bar{c}}{2V_{P_1}}\right) + c_{m_q}\left(\frac{q\,\bar{c}}{2V_{P_1}}\right) + c_{m_{\delta_E}}\delta_E + c_{m_{i_H}}i_H\right\}$$

Refer to[4,5,6] for a complete and detailed understanding of the aerodynamic origin and modeling for each of the aerodynamic terms in the relationships. The α − derivatives $(c_{D_\alpha}, c_{L_\alpha}, c_{m_\alpha})$ and the control derivatives $(c_{D_{\delta_E}}, c_{L_{\delta_E}}, c_{m_{\delta_E}})$, $(c_{D_{i_H}}, c_{L_{i_H}}, c_{m_{i_H}})$ have already been discussed and modeled in the previous analysis of the steady-state forces and moments. A conceptual review and the modeling of all the other aerodynamic coefficients will be provided.

3.9.1 Modeling of $(c_{D_1}, c_{L_1}, c_{m_1})$

The terms with the 1 subscript are the steady-state terms just discussed. The values of c_{D_1}, c_{L_1} are depending upon the specific steady-state conditions for the aircraft while $c_{m_1} = 0$ is by definition of the trim conditions. The design for trim conditions is discussed in Chapter 6.

3.9.2 Modeling of $(c_{D_u}, c_{L_u}, c_{m_u})$

The u − derivatives $(c_{D_u}, c_{L_u}, c_{m_u})$ are associated with the (small) perturbation in the forward speed.

The derivative c_{D_u} models the change (increase) in drag for small, positive increments in the forward speed. This derivative is negligible at the low subsonic conditions associated with incompressible aerodynamics; however, it takes on significant positive values as the airspeed approaches transonic conditions. In general, it can be modeled using

$$c_{D_u} = \text{Mach} \cdot \frac{\partial c_D}{\partial \text{Mach}}$$

An empirical modeling of this derivative using the build-up approach is very unpractical. A more suitable approach is to derive the slope $\partial c_D / \partial \text{Mach}$ once experimental drag results are available from wind tunnel analysis.

The derivative c_{L_u} models the change (increase) in lift for small, positive increments in the forward speed. Similar to c_{D_u}, c_{L_u} is negligible at the low subsonic conditions associated with incompressible aerodynamics; however, it takes on significant positive values as the airspeed increases. In general, it can be modeled using

$$c_{L_u} = \left(\frac{\text{Mach}^2}{1 - \text{Mach}^2}\right) \cdot c_{L_1}$$

The derivative c_{m_u} is the most important aerodynamic parameter in this group; it models the change in pitching moment associated with small, positive increments in the forward speed. Similar to both c_{D_u} and c_{L_u}, c_{m_u} is negligible at the low subsonic conditions associated with incompressible aerodynamics; however, it takes on significant positive values as the airspeed increases. Further, its value becomes extremely critical in the transonic region. In general, it can be modeled using

$$c_{m_u} = -c_{L_1} \frac{\partial \bar{x}_{AC_W}}{\partial\, Mach}$$

A wind tunnel analysis is required for the estimate of c_{m_u} at high subsonic, transonic, and supersonic conditions. The relationship between $\bar{x}_{AC_W}$ and $Mach$ was briefly reviewed in Chapter 2, Section 2.5. Note that a numerical approximation for $(\partial \bar{x}_{AC_W}/\partial\, Mach)$, which is considered positive for a shift of $\bar{x}_{AC_W}$ toward the tail, can be found by plotting $\bar{x}_{AC_W}$ versus $Mach$ around the steady-state value of $Mach$ and evaluating the associated slope. The importance of c_{m_u} for the aircraft static and dynamic stability will be clearly shown in later chapters.

3.9.3 Modeling of $(c_{D_{\dot\alpha}}, c_{L_{\dot\alpha}}, c_{m_{\dot\alpha}})$ and $(c_{D_q}, c_{L_q}, c_{m_q})$

These sets of derivatives are described together because they produce similar aerodynamic effects, given the strong correlation between the rate of change in the angle of attack α and the associated pitch rate q.

At subsonic conditions it is customary to assume that $c_{D_{\dot\alpha}} \approx 0$, $c_{D_q} \approx 0$.

Since the contribution from the fuselage and the wing can be neglected, for $c_{L_{\dot\alpha}}$ we have

$$c_{L_{\dot\alpha}} = c_{L_{\dot\alpha_{WB}}} + c_{L_{\dot\alpha_H}} \approx c_{L_{\dot\alpha_H}}$$

A closed-form expression for $c_{L_{\dot\alpha_H}}$ is given by[5]

$$c_{L_{\dot\alpha_H}} \approx 2 c_{L_{\alpha_H}} \eta_H \frac{S_H}{S} (\bar{x}_{AC_H} - \bar{x}_{CG}) \frac{d\varepsilon}{d\alpha}$$

Thus, a final relationship for $c_{L_{\dot\alpha}}$ is given by

$$\boxed{c_{L_{\dot\alpha}} \approx 2 c_{L_{\alpha_H}} \eta_H \frac{S_H}{S} (\bar{x}_{AC_H} - \bar{x}_{CG}) \frac{d\varepsilon}{d\alpha}}$$

For c_{L_q} we have instead

$$c_{L_q} = c_{L_{q_{WB}}} + c_{L_{q_H}} \approx c_{L_{q_W}} + c_{L_{q_H}}$$

In this case, only the contribution from the fuselage can be typically neglected. The following closed-form relationship is available for the evaluation of $c_{L_{q_W}}$:[5]

$$c_{L_{q_W}} = \left[\frac{AR + 2 \cos \Lambda_{c/4}}{AR \cdot B + 2 \cos \Lambda_{c/4}} \right] \cdot c_{L_{q_W}} \bigg|_{Mach=0}$$

where

$$B = \sqrt{1 - Mach^2 (\cos \Lambda_{c/4})^2}$$

$$c_{L_{q_W}} \bigg|_{Mach=0} = \left(\frac{1}{2} + 2 \cdot \left| \bar{x}_{AC_W} - \bar{x}_{CG} \right| \right) c_{L_{\alpha_W}} \bigg|_{Mach=0}$$

The following closed-form relationship is instead available for the evaluation of $c_{L_{q_H}}$:[5]

$$c_{L_{q_H}} \approx 2 c_{L_{\alpha_H}} \eta_H \frac{S_H}{S} (\bar{x}_{AC_H} - \bar{x}_{CG})$$

Therefore, a final relationship for c_{L_q} is given by

$$\boxed{c_{L_q} \approx \left[\frac{AR + 2\cos\Lambda_{c/4}}{AR\cdot\left(\sqrt{1 - Mach^2(\cos\Lambda_{c/4})^2}\right) + 2\cos\Lambda_{c/4}}\right]\cdot\left(\frac{1}{2} + 2\cdot\left|\bar{x}_{AC_W} - \bar{x}_{CG}\right|\right) c_{L_{\alpha_W}}\bigg|_{Mach=0} \\ + 2\, c_{L_{\alpha_H}}\eta_H\frac{S_H}{S}(\bar{x}_{AC_H} - \bar{x}_{CG})}$$

For $c_{m_{\dot\alpha}}$ we have

$$c_{m_{\dot\alpha}} = c_{m_{\dot\alpha_{WB}}} + c_{m_{\dot\alpha_H}}$$

As the contributions from both fuselage and wing can be neglected, we have

$$c_{m_{\dot\alpha}} \approx c_{m_{\dot\alpha_H}}$$

A closed-form relationship for $c_{m_{\dot\alpha}} \approx c_{m_{\dot\alpha_H}}$ is given by[5]

$$\boxed{c_{m_{\dot\alpha}} \approx c_{m_{\dot\alpha_H}} \approx -c_{L_{\dot\alpha_H}}(\bar{x}_{AC_H} - \bar{x}_{CG}) \approx -2 c_{L_{\alpha_H}}\eta_H\frac{S_H}{S}(\bar{x}_{AC_H} - \bar{x}_{CG})^2\frac{d\varepsilon}{d\alpha}}$$

Finally, for c_{m_q} we have

$$c_{m_q} = c_{m_{q_{WB}}} + c_{m_{q_H}} \approx c_{m_{q_W}} + c_{m_{q_H}}$$

with the contribution of the fuselage being typically negligible. A closed-form relationship for $c_{m_{q_W}}$ is given by[5]

$$c_{m_{q_W}} = \left[\frac{\left(\dfrac{AR^3 \tan^2\Lambda_{c/4}}{AR\cdot B + 6\cos\Lambda_{c/4}}\right) + \dfrac{3}{B}}{\left(\dfrac{AR^3 \tan^2\Lambda_{c/4}}{AR + 6\cos\Lambda_{c/4}}\right) + 3}\right]\cdot c_{m_{q_W}}\bigg|_{Mach=0}$$

where again $B = \sqrt{1 - Mach^2(\cos\Lambda_{c/4})^2}$.

Within $c_{m_{q_W}}$, $c_{m_{q_W}}\big|_{Mach=0}$ is found using[5]

$$c_{m_{q_W}}\bigg|_{Mach=0} = -K_q c_{L_{\alpha_W}}\bigg|_{Mach=0}\cos\Lambda_{c/4}\cdot C$$

where the coefficient C is given by

$$C = \left\{\frac{AR\left(0.5\left|\bar{x}_{AC_W} - \bar{x}_{CG}\right| + 2\left|\bar{x}_{AC_W} - \bar{x}_{CG}\right|^2\right)}{AR + 2\cos\Lambda_{c/4}} + \frac{1}{24}\left(\frac{AR^3\tan^2\Lambda_{c/4}}{AR + 6\cos\Lambda_{c/4}}\right) + \frac{1}{8}\right\}$$

while the coefficient K_q is evaluated using Figure 3.12.

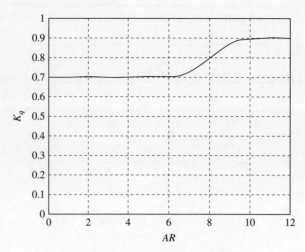

Figure 3.12 K_q Correction Coefficient[5]

Despite its complexity, the value of $c_{m_{q_W}}$ is often negligible.[5] A closed-form relationship for $c_{m_{q_H}}$ is instead given by

$$c_{m_{q_H}} \approx -2c_{L_{\alpha_H}} \eta_H \frac{S_H}{S} (\bar{x}_{AC_H} - \bar{x}_{CG})^2$$

Therefore, a final expression for c_{m_q} is given by

$$\boxed{\begin{aligned}c_{m_q} &= \left[\frac{\left(\dfrac{AR^3 \tan^2 \Lambda_{c/4}}{AR + 6\cos \Lambda_{c/4}}\right) + \dfrac{3}{\sqrt{1 - Mach^2 (\cos \Lambda_{c/4})^2}}}{\left(\dfrac{AR^3 \tan^2 \Lambda_{c/4}}{AR + 6\cos \Lambda_{c/4}}\right) + 3}\right] \cdot \left\{-K_q c_{L_{\alpha_W}}\Big|_{Mach=0} \cos \Lambda_{c/4}\right\} \\ &\cdot \left\{\frac{AR\left(0.5\left|\bar{x}_{AC_W} - \bar{x}_{CG}\right| + 2\left|\bar{x}_{AC_W} - \bar{x}_{CG}\right|^2\right)}{AR + 2\cos \Lambda_{c/4}} + \frac{1}{24}\left(\frac{AR^3 \tan^2 \Lambda_{c/4}}{AR + 6\cos \Lambda_{c/4}}\right) + \frac{1}{8}\right\} \\ &+ 2c_{L_{\alpha_H}} \eta_H \frac{S_H}{S}(\bar{x}_{AC_H} - \bar{x}_{CG})^2 \approx -2c_{L_{\alpha_H}} \eta_H \frac{S_H}{S}(\bar{x}_{AC_H} - \bar{x}_{CG})^2\end{aligned}}$$

In summary, the drag coefficients $c_{D_{\dot{\alpha}}}$, c_{D_q} are always negligible; however, the lift and moment aerodynamic coefficients $c_{L_{\dot{\alpha}}}$, c_{L_q} are significant, whereas $c_{m_{\dot{\alpha}}}$, c_{m_q} are very important for the longitudinal handling qualities and dynamic characteristics of the aircraft. As described in Chapter 7, the coefficients $c_{m_{\dot{\alpha}}}$, c_{m_q} are often referred to in the technical literature as the longitudinal damping coefficients for their critical role in determining the value of the damping coefficient in the longitudinal short period dynamic mode. From a conceptual point of view, higher absolute values of the coefficients $c_{m_{\dot{\alpha}}}$, c_{m_q} imply better capabilities for the aircraft to hold the steady-state values of its longitudinal equilibrium conditions following either atmospheric disturbances or pilot maneuvers.

3.10 SUMMARY OF LONGITUDINAL STABILITY AND CONTROL DERIVATIVES

The following Figures 3.13 to 3.15 will provide a comprehensive graphical summary of all the longitudinal aerodynamic stability and control derivatives introduced in this chapter.

$$\boxed{f_{A_X} = \bar{q}S\left\{-\left[c_{D_u} + 2c_{D_1}\right]\left(\frac{u}{V_{P_1}}\right) + \left[-c_{D_\alpha} + c_{L_1}\right]\alpha - c_{D_{\dot\alpha}}\left(\frac{\dot\alpha \bar c}{2V_{P_1}}\right) - c_{D_q}\left(\frac{q\bar c}{2V_{P_1}}\right) - c_{D_{\delta_E}}\delta_E - c_{D_{i_H}}i_H\right\}}$$

$$c_{D_1} = \frac{T_1}{\bar{q}_1 S} = c_{D_0} + C_{D_\alpha}\alpha_1 + C_{D_{\delta_E}}\delta_{E_1} + c_{D_{i_H}}i_{H_1}$$

$$c_{L_1} = \frac{W}{\bar{q}_1 S} \quad \longleftarrow \quad L_1 = c_{L_1}\bar{q}\,S = W$$

$$c_{D_0} = c_D\big|_{\alpha=\delta_E=i_H=0°} \quad \longleftarrow \quad \text{[5],[6] or wind tunnel analysis}$$

$$c_{D_\alpha} \quad \longleftarrow \quad \text{[5],[6] or wind tunnel analysis}$$

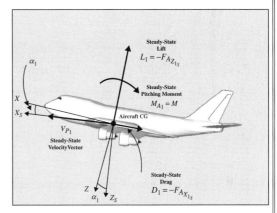

SPECIAL CASE : "uncambered" aircraft (with parabolic "drag" polar curve)

$$\bar{c}_{D_0} = c_{D_0} \quad \longleftarrow$$

$$c_{D_\alpha} \approx \left(\frac{2c_{L_1}}{\pi AR\, e}\right)c_{L_\alpha}$$

$$c_{D_u} = \text{Mach}\cdot\frac{1}{\partial \text{Mach}}\frac{\partial c_D}{\partial \text{Mach}} \approx 0 \quad \text{at low subsonic}$$

$$c_{D_{\dot\alpha}} \approx 0$$

$$c_{D_q} \approx 0$$

$$\left.\begin{array}{l} c_{D_{\delta_E}} \approx 0 \\ c_{D_{i_H}} \approx 0 \end{array}\right\} \text{except for large horizontal tails with large dihedral/anhedral angles}$$

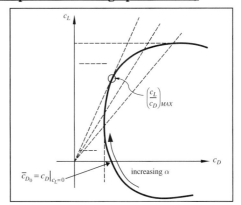

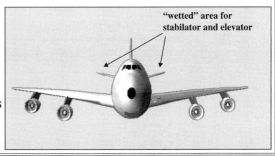

Figure 3.13 Summary of the c_D Stability and Control Derivatives

98 Chapter 3 Modeling of Longitudinal Aerodynamic Forces and Moments

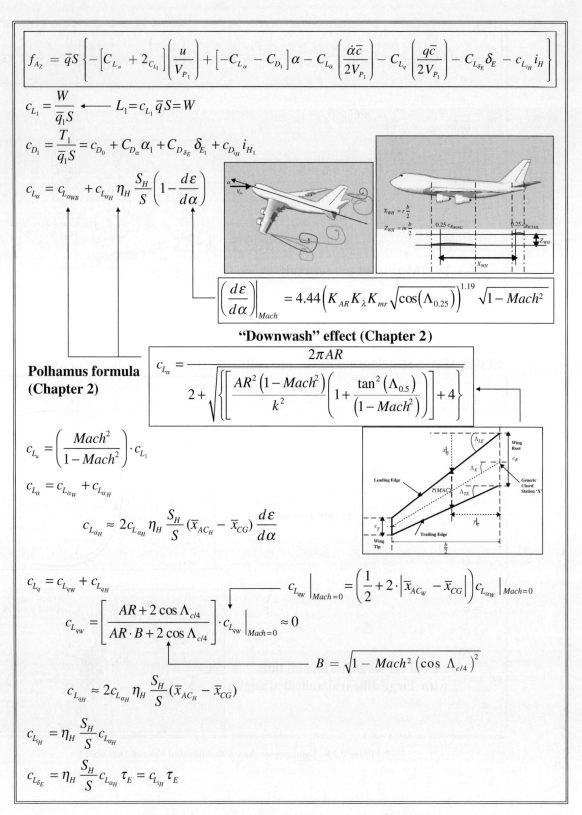

$$f_{A_Z} = \bar{q}S\left\{-\left[C_{L_u} + 2C_{L_1}\right]\left(\frac{u}{V_{P_1}}\right) + \left[-C_{L_\alpha} - C_{D_1}\right]\alpha - C_{L_{\dot\alpha}}\left(\frac{\dot\alpha \bar{c}}{2V_{P_1}}\right) - C_{L_q}\left(\frac{q\bar{c}}{2V_{P_1}}\right) - C_{L_{\delta_E}}\delta_E - c_{L_{i_H}}i_H\right\}$$

$$c_{L_1} = \frac{W}{\bar{q}_1 S} \;\longleftarrow\; L_1 = c_{L_1}\bar{q}S = W$$

$$c_{D_1} = \frac{T_1}{\bar{q}_1 S} = c_{D_0} + C_{D_\alpha}\alpha_1 + C_{D_{\delta_E}}\delta_{E_1} + c_{D_{iH}}i_{H_1}$$

$$c_{L_\alpha} = c_{L_{\alpha_{WB}}} + c_{L_{\alpha_H}}\eta_H \frac{S_H}{S}\left(1 - \frac{d\varepsilon}{d\alpha}\right)$$

$$\left(\frac{d\varepsilon}{d\alpha}\right)_{Mach} = 4.44\left(K_{AR}K_\lambda K_{mr}\sqrt{\cos(\Lambda_{0.25})}\right)^{1.19}\sqrt{1-Mach^2}$$

"Downwash" effect (Chapter 2)

Polhamus formula (Chapter 2)

$$c_{L_\alpha} = \frac{2\pi AR}{2 + \sqrt{\left[\frac{AR^2(1-Mach^2)}{k^2}\left(1 + \frac{\tan^2(\Lambda_{0.5})}{(1-Mach^2)}\right)\right] + 4}}$$

$$c_{L_u} = \left(\frac{Mach^2}{1-Mach^2}\right)\cdot c_{L_1}$$

$$c_{L_{\dot\alpha}} = c_{L_{\dot\alpha_W}} + c_{L_{\dot\alpha_H}}$$

$$c_{L_{\dot\alpha_H}} \approx 2c_{L_{\alpha_H}}\eta_H \frac{S_H}{S}(\bar{x}_{AC_H} - \bar{x}_{CG})\frac{d\varepsilon}{d\alpha}$$

$$c_{L_q} = c_{L_{q_W}} + c_{L_{q_H}}$$

$$c_{L_{q_W}} = \left[\frac{AR + 2\cos\Lambda_{c/4}}{AR\cdot B + 2\cos\Lambda_{c/4}}\right]\cdot c_{L_{q_W}}\Big|_{Mach=0}$$

$$c_{L_{q_W}}\Big|_{Mach=0} = \left(\frac{1}{2} + 2\cdot\left|\bar{x}_{AC_W} - \bar{x}_{CG}\right|\right)c_{L_{\alpha_W}}\Big|_{Mach=0}$$

$$B = \sqrt{1 - Mach^2(\cos\Lambda_{c/4})^2}$$

$$c_{L_{q_H}} \approx 2c_{L_{\alpha_H}}\eta_H \frac{S_H}{S}(\bar{x}_{AC_H} - \bar{x}_{CG})$$

$$c_{L_{i_H}} = \eta_H \frac{S_H}{S}c_{L_{\alpha_H}}$$

$$c_{L_{\delta_E}} = \eta_H \frac{S_H}{S}c_{L_{\alpha_H}}\tau_E = c_{L_{i_H}}\tau_E$$

Figure 3.14 Summary of the c_L Stability and Control Derivatives

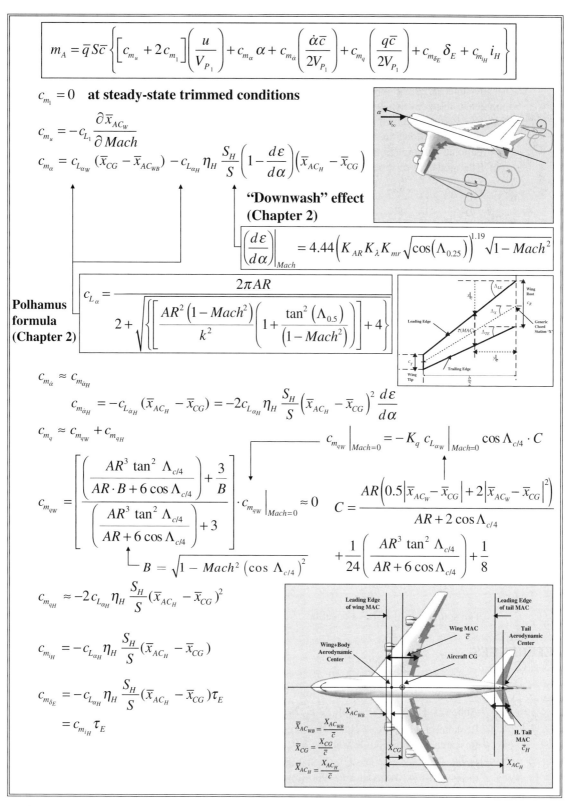

Figure 3.15 Summary of the c_m Stability and Control Derivatives

3.11 SUMMARY

This chapter has introduced basic tools for the modeling of the longitudinal aerodynamic forces and moments acting on the aircraft. A similar set of tools will be introduced for the lateral directional force and moments in Chapter 4. Following the breakdown used in Chapter 1, the modeling has been divided into the modeling of steady-state and small perturbation components, grouped into longitudinal and lateral directional forces and moments, respectively. It should be underlined that this modeling has been developed along the stability axes (X_S, Y_S, Z_S), also known in the technical literature as wind axes, whereas the dynamic modeling in Chapter 1 has been performed along the body axes (X, Y, Z). Without loss of generality, the modeling has been developed for a conventional subsonic aircraft with elevators/stabilators being used for longitudinal control.

The modeling of the longitudinal forces and moment is summarized:

$$\begin{Bmatrix} F_{A_{X_{1_S}}} \\ F_{A_{Z_{1_S}}} \\ M_{A_{1_S}} \end{Bmatrix} = \begin{Bmatrix} F_{A_{X_1}} \\ F_{A_{Z_1}} \\ M_{A_1} \end{Bmatrix} = \begin{Bmatrix} -(\bar{q} S c_{D_1}) \\ -(\bar{q} S c_{L_1}) \\ \bar{q} S \bar{c} c_{m_1} \end{Bmatrix} \text{ with : } \begin{Bmatrix} c_{D_1} \\ c_{L_1} \\ c_{m_1} \end{Bmatrix} = \begin{bmatrix} c_{D_0} & c_{D_\alpha} & \cancel{c_{D_{\delta_E}}} & \cancel{c_{D_{i_H}}} \\ c_{L_0} & c_{L_\alpha} & c_{L_{\delta_E}} & c_{L_{i_H}} \\ c_{m_0} & c_{m_\alpha} & c_{m_{\delta_E}} & c_{m_{i_H}} \end{bmatrix} \begin{Bmatrix} 1 \\ \alpha \\ \delta_E \\ i_H \end{Bmatrix}$$

Steady-State Aerodynamic Forces and Moments

$$\begin{cases} f_{A_X} = \bar{q} S \left\{ -[c_{D_u} + 2c_{D_1}]\left(\dfrac{u}{V_{P_1}}\right) + [-c_{D_\alpha} + c_{L_1}]\alpha - c_{D_{\dot\alpha}}\left(\dfrac{\dot\alpha \bar{c}}{2 V_{P_1}}\right) - c_{D_q}\left(\dfrac{q \bar{c}}{2 V_{P_1}}\right) - c_{D_{\delta_E}}\delta_E - c_{D_{i_H}} i_H \right\} \\ f_{A_z} = \bar{q} S \left\{ -[c_{L_u} + 2c_{L_1}]\left(\dfrac{u}{V_{P_1}}\right) - [c_{L_\alpha} + c_{D_1}]\alpha - c_{L_{\dot\alpha}}\left(\dfrac{\dot\alpha \bar{c}}{2 V_{P_1}}\right) - c_{L_q}\left(\dfrac{q \bar{c}}{2 V_{P_1}}\right) - c_{L_{\delta_E}}\delta_E - c_{L_{i_H}} i_H \right\} \\ m_A = \bar{q} S \bar{c} \left\{ [c_{m_u} + 2c_{m_1}]\left(\dfrac{u}{V_{P_1}}\right) + c_{m_\alpha}\alpha + c_{m_{\dot\alpha}}\left(\dfrac{\dot\alpha \bar{c}}{2 V_{P_1}}\right) + c_{m_q}\left(\dfrac{q \bar{c}}{2 V_{P_1}}\right) + c_{m_{\delta_E}}\delta_E + c_{m_{i_H}} i_H \right\} \end{cases}$$

Small Perturbation Aerodynamic Forces and Moments

Note that the steady-state aerodynamic forces and moments will not actually be used for the solution of the equations of motion developed in Chapter 1. However, they are needed for the derivation of the small perturbation terms. Once available, the small perturbation longitudinal aerodynamic forces and moment (f_{A_X}, f_{A_Z}, m_A) are provided to the small perturbation equations of motion:

$$\begin{cases} m[\dot{u} + Q_1 w + q W_1 - R_1 v - r V_1] = -mg\,\theta \cos\Theta_1 + (f_{A_X} + f_{T_X}) \\ m[\dot{w} + P_1 v + p V_1 - Q_1 u - U_1 q] = -mg\,\theta \cos\Phi_1 \sin\Theta_1 - mg\,\phi \sin\Phi_1 \cos\Theta_1 + (f_{A_Z} + f_{T_Z}) \\ \dot{q} I_{YY} + (P_1 r + p R_1)(I_{XX} - I_{ZZ}) + (2 P_1 p - 2 R_1 r) I_{XZ} = (m_A + m_T) \end{cases}$$

The solution of the above equations will be discussed in Chapter 7.

REFERENCES

1. Anderson, J. D. *Fundamentals of Aerodynamics*. McGraw Hill, Inc., 1984.
2. Philips, W. *Mechanics of Flight*. John Wiley and Sons, 2004.
3. Abbott, I. H., Von Doenhoff, A. E. *Theory of Wing Sections*. Dover Publications Inc., 1959.
4. Roskam, J. *Airplane Flight Dynamics and Automatic Flight Controls – Part I*. Design, Analysis, and Research Corporation, Lawrence, KS, 1995.
5. Hoak, D. E., et al. "The USAF Stability and Control DATCOM." Air Force Wright Aeronautical Laboratories, TR-83-2048, October 1960 (revised in 1978).
6. Roskam, J. *Airplane Design. Part I–Part VIII*. Roskam Aviation and Engineering Corporation, Lawrence, KS, 1990.
7. Napolitano, M. R. *WVU MAE 242 Flight Testing—Class Notes*. West Virginia University, Morgantown, 2001.
8. Multhopp, H. *Aerodynamics of the Fuselage*. NASA Technical Memorandum No. 1036, December 1942.
9. "DC-9 Airplane Characteristics for Airport Planning," June 1984, Douglas Aircraft Company, downloaded from *http://www.boeing.com/commercial/airports/dc9.htm* (PDF format).
10. *http://textron.vo.llnwd.net/o25/CES/cessna_aircraft_docs/citation/cj3/cj3_s&d.pdf*.

STUDENT SAMPLE PROBLEMS

Student Sample Problem 3.1

Consider the data and the drawings for the McDonnell Douglas DC-9 aircraft in Appendix C. Use the provided drawings to extract all the relevant geometric characteristics of the aircraft. Assume that this aircraft features stabilators and elevators for the control of the longitudinal dynamics with $\tau_E = 0.5$, $\eta_H = 0.9$. Using the modeling outlined in this chapter, find a numerical value for the following aerodynamic parameters:

- $\dfrac{d\varepsilon}{d\alpha}, \bar{x}_{AC_W}, \Delta\bar{x}_{AC_B}$;
- $c_{L_\alpha}, c_{L_{\delta_E}}, c_{L_{i_H}}$;
- $\bar{x}_{AC}, c_{m_\alpha}, c_{m_{\delta_E}}, c_{m_{i_H}}$.

Solution of Student Sample Problem 3.1

All the relevant geometric parameters were extracted from the provided images in Appendix C. Specifically, the following geometric parameters were determined:

$$c_T = 3.9 \text{ ft}, \ c_R = 15.6 \text{ ft}, \ \Lambda_{LE} = 28° = 0.489 \text{ rad}, \ X_{WH_R} = 49.5 \text{ ft}, \ Z_{WH} = 20 \text{ ft},$$

$$b_H = 38.5 \text{ ft}, \ c_{T_H} = 5.6 \text{ ft}, \ c_{R_H} = 10.9 \text{ ft}, \ \Lambda_{LE_H} = 34° = 0.593 \text{ rad}$$

Next, using the assumption of straight wings, the following wing and tail geometric parameters were derived using the above values.

Horizontal Tail Geometric Parameters

From Student Sample Problem 2.1:

$$\lambda_H = 0.367, \ S_H = \frac{b_H}{2} c_{R_H}(1 + \lambda_H) = \frac{36.9}{2} \cdot 10.9 \cdot (1 + 0.367) = 274.91 \text{ ft}^2$$

$$AR_H = \frac{b_H^2}{S_H} = \frac{36.9^2}{274.91} = 4.953$$

$$\tan \Lambda_{0.5_H} = \tan \Lambda_{LE_H} - \frac{4 \cdot 0.5 \cdot (1 - \lambda_H)}{AR_H(1 + \lambda_H)} = \tan(0.611) - \frac{4 \cdot 0.5 \cdot (1 - 0.367)}{4.953 \cdot (1 + 0.367)} \Rightarrow \Lambda_{0.5_H} = 0.474 \text{ rad}$$

Wing-Tail Geometric Parameters

From Student Sample Problem 2.1: $\bar{x}_{AC_H} = 4.345$

Wing Lift-Slope Coefficient

From Student Sample Problem 2.1: $c_{L_{\alpha_W}} = 6.062$. Because the geometric parameters of the wing are somewhat outside the nominal range for the Polhamus formula, it is expected that the preceding value is higher than the actual $c_{L_{\alpha_W}}$ value.

Horizontal Tail Lift Slope Coefficient

Since $AR_H = 4.953 \geq 4$:

$$k_H = 1 + \frac{[(8.2 - 2.3\Lambda_{LE_H}) - AR_H(0.22 - 0.153\Lambda_{LE_H})]}{100}$$

$$= 1 + \frac{[(8.2 - 2.3 \cdot 0.611) - 4.953 \cdot (0.22 - 0.153 \cdot 0.611)]}{100} = 1.062$$

$$c_{L_{\alpha_H}} = \frac{2\pi AR_H}{2 + \sqrt{\left\{\left[\dfrac{AR_H^2(1 - \text{Mach}^2)}{k_H^2}\right]\left(1 + \dfrac{\tan^2(\Lambda_{0.5_H})}{(1 - \text{Mach}^2)}\right)\right] + 4\right\}}}$$

$$= \frac{2\pi \cdot 4.953}{2 + \sqrt{\left\{\left[\dfrac{4.953^2(1 - 0.7^2)}{1.062^2}\right]\left(1 + \dfrac{\tan^2(0.474)}{(1 - 0.7^2)}\right)\right] + 4\right\}}} = 4.741$$

As for $c_{L_{\alpha W}}$, it is expected that the preceding value is higher than the actual $c_{L_{\alpha H}}$ value.

Modeling of the Downwash Effect

From Student Sample Problem 2.1: $\left(\dfrac{d\varepsilon}{d\alpha}\right)\bigg|_{Mach} = 0.297$

Modeling of $\bar{x}_{AC_{WB}}$

A relationship for $\bar{x}_{AC_{WB}}$ is given by:

$$\bar{x}_{AC_{WB}} = \bar{x}_{AC_W} + \Delta \bar{x}_{AC_B}$$

From Student Sample Problem 2.1:
$\bar{x}_{AC_W} = 0.285$ and $\Delta \bar{x}_{AC_B} = -0.231$ thus $\bar{x}_{AC_{WB}} = 0.285 - 0.231 = 0.054$

Stability and Control Derivatives

$$c_{L_\alpha} = c_{L_{\alpha W}} + c_{L_{\alpha H}} \eta_H \frac{S_H}{S}\left(1 - \frac{d\varepsilon}{d\alpha}\right) = 6.062 + 4.741 \cdot 0.9 \cdot \frac{274.91}{928}(1 - 0.297) = \boxed{6.950}$$

$$c_{L_{\delta_E}} = \eta_H \frac{S_H}{S} c_{L_{\alpha H}} \tau_E = 0.9 \cdot \frac{274.91}{928} \cdot 4.741 \cdot 0.5 = \boxed{0.632}$$

$$c_{L_{i_H}} = \eta_H \frac{S_H}{S} c_{L_{\alpha H}} = 0.9 \cdot \frac{274.91}{928} \cdot 4.741 = \boxed{1.264}$$

$$c_{m_\alpha} = c_{L_{\alpha W}}(\bar{x}_{CG} - \bar{x}_{AC_{WB}}) - c_{L_{\alpha H}} \eta_H \frac{S_H}{S}\left(1 - \frac{d\varepsilon}{d\alpha}\right)(\bar{x}_{AC_H} - \bar{x}_{CG})$$

$$= 6.062 \cdot (0.3 - 0.054) - 4.741 \cdot 0.9 \cdot \frac{274.91}{928}(1 - 0.297)(4.345 - 0.3) = \boxed{-2.101}$$

$$c_{m_{\delta_E}} = -c_{L_{\alpha H}} \eta_H \frac{S_H}{S}(\bar{x}_{AC_H} - \bar{x}_{CG})\tau_E = -4.741 \cdot 0.9 \cdot \frac{274.91}{928} \cdot (4.345 - 0.3) \cdot 0.5 = \boxed{-2.557}$$

$$c_{m_{i_H}} = -c_{L_{\alpha H}} \eta_H \frac{S_H}{S}(\bar{x}_{AC_H} - \bar{x}_{CG}) = -4.741 \cdot 0.9 \cdot \frac{274.91}{928} \cdot (4.345 - 0.3) = \boxed{-5.113}$$

It should be emphasized that the preceding estimates are affected by the approximation associated with the previous estimates using the Polhamus formula.

Finally, the estimated aircraft aerodynamic center can be determined using

$$\bar{x}_{AC} = \frac{\bar{x}_{AC_{WB}} + \dfrac{c_{L_{\alpha H}}}{c_{L_{\alpha W}}} \eta_H \dfrac{S_H}{S}\left(1 - \dfrac{d\varepsilon}{d\alpha}\right)\bar{x}_{AC_H}}{1 + \dfrac{c_{L_{\alpha H}}}{c_{L_{\alpha W}}} \eta_H \dfrac{S_H}{S}\left(1 - \dfrac{d\varepsilon}{d\alpha}\right)}$$

$$= \frac{0.054 + \dfrac{4.741}{6.062} \cdot 0.9 \cdot \dfrac{274.91}{928}(1 - 0.297) \cdot 4.345}{1 + \dfrac{4.741}{6.062} \cdot 0.9 \cdot \dfrac{274.91}{928}(1 - 0.297)} = \boxed{0.602}$$

Although the "true" values of the aerodynamic and stability control derivatives are unknown, it can be speculated that those estimates are approximate due to the over-estimate of the parameters $c_{L_{\alpha W}} \cdot c_{L_{\alpha H}}$ from the Polhamus formula. Considering a tentative 20 percent overestimate for both $c_{L_{\alpha W}} \cdot c_{L_{\alpha H}}$ and neglecting its effect on the estimate of the downwash effect, more appropriate values can be given by

$$c_{L_{\alpha W}} \approx 4.85, \; c_{L_{\alpha H}} \approx 3.793$$

leading to the following revised estimates.

REVISED stability and control derivatives and $\bar{x}_{AC}$ (using REVISED $c_{L_{\alpha_W}} \cdot c_{L_{\alpha_H}}$)

$$c_{L_\alpha} = c_{L_{\alpha_W}} + c_{L_{\alpha_H}} \eta_H \frac{S_H}{S}\left(1 - \frac{d\varepsilon}{d\alpha}\right) = 4.85 + 3.793 \cdot 0.9 \cdot \frac{274.91}{928}(1 - 0.297) = \boxed{5.56}$$

$$c_{L_{\delta_E}} = \eta_H \frac{S_H}{S} c_{L_{\alpha_H}} \tau_E = 0.9 \cdot \frac{274.91}{928} \cdot 3.793 \cdot 0.5 = \boxed{0.506}$$

$$c_{L_{i_H}} = \eta_H \frac{S_H}{S} c_{L_{\alpha_H}} = 0.9 \cdot \frac{274.91}{928} \cdot 3.793 = \boxed{1.012}$$

$$c_{m_\alpha} = c_{L_{\alpha_W}}\left(\bar{x}_{CG} - \bar{x}_{AC_{WB}}\right) - c_{L_{\alpha_H}} \eta_H \frac{S_H}{S}\left(1 - \frac{d\varepsilon}{d\alpha}\right)\left(\bar{x}_{AC_H} - \bar{x}_{CG}\right)$$

$$= 4.85 \cdot (0.3 - 0.054) - 3.793 \cdot 0.9 \cdot \frac{274.91}{928}(1 - 0.297)(4.345 - 0.3) = \boxed{-1.681}$$

$$c_{m_{\delta_E}} = -c_{L_{\alpha_H}} \eta_H \frac{S_H}{S}\left(\bar{x}_{AC_H} - \bar{x}_{CG}\right)\tau_E = -3.793 \cdot 0.9 \cdot \frac{274.91}{928} \cdot (4.345 - 0.3) \cdot 0.5 = \boxed{-2.045}$$

$$c_{m_{i_H}} = -c_{L_{\alpha_H}} \eta_H \frac{S_H}{S}\left(\bar{x}_{AC_H} - \bar{x}_{CG}\right) = -3.793 \cdot 0.9 \cdot \frac{274.91}{928} \cdot (4.345 - 0.3) = \boxed{-4.091}$$

$$\bar{x}_{AC} = \frac{\bar{x}_{AC_{WB}} + \frac{c_{L_{\alpha_H}}}{c_{L_{\alpha_W}}} \eta_H \frac{S_H}{S}\left(1 - \frac{d\varepsilon}{d\alpha}\right) \bar{x}_{AC_H}}{1 + \frac{c_{L_{\alpha_H}}}{c_{L_{\alpha_W}}} \eta_H \frac{S_H}{S}\left(1 - \frac{d\varepsilon}{d\alpha}\right)}$$

$$= \frac{0.054 + \frac{3.793}{4.85} \cdot 0.9 \cdot \frac{274.91}{928}(1 - 0.297) \cdot 4.345}{1 + \frac{3.793}{4.85} \cdot 0.9 \cdot \frac{274.91}{928}(1 - 0.297)} = \boxed{0.602}$$

The revised estimate for $\bar{x}_{AC}$ is virtually identical to the previous estimate.

Student Sample Problem 3.2

Consider the data relative to the SIAI Marchetti S211 aircraft in Appendix B and Appendix C. Use the provided drawings to extract all the relevant geometric characteristics of the aircraft. This aircraft features both stabilators and elevators for the control of the longitudinal dynamics. Assume $\tau_E = 0.35$, $\eta_H = 0.92$. Using the modeling outlined in this chapter, find a numerical value for the following aerodynamic parameters:

- $\frac{d\varepsilon}{d\alpha}, \bar{x}_{AC_W}, \Delta\bar{x}_{AC_B}$;
- $c_{L_\alpha}, c_{L_{\delta_E}}, c_{L_{i_H}}$;
- $\bar{x}_{AC}, c_{m_\alpha}, c_{m_{\delta_E}}, c_{m_{i_H}}$.

Next, compare the obtained values with the true values listed in Appendix B (Aircraft 8) and evaluate the errors associated with the use of the empirical modeling approach.

Solution of Student Sample Problem 3.2

All the relevant geometric parameters were extracted from the provided images in Appendix C. Specifically, the following geometric parameters were identified:

$$c_T = 3.1 \text{ ft}, \ c_R = 6 \text{ ft}, \ \Lambda_{LE} = 19.5° = 0.34 \text{ rad}, \ X_{WH_R} = 13.1 \text{ ft}, \ Z_{WH} = 1.3 \text{ ft}$$

$$b_H = 13.3 \text{ ft}, \ c_{T_H} = 1.55 \text{ ft}, \ c_{R_H} = 3.5 \text{ ft}, \ \Lambda_{LE_H} = 18.5° = 0.323 \text{ rad}$$

Next, using the assumption of straight wings, the following wing and tail geometric parameters were derived using the preceding values.

Horizontal Tail Geometric Parameters

From Student Sample Problem 2.2:

$$\lambda_H = 0.443, \; S_H = \frac{b_H}{2}c_{R_H}(1+\lambda_H) = \frac{13.3}{2} \cdot 3.5 \cdot (1+0.443) = 33.583 \text{ ft}^2$$

$$AR_H = \frac{b_H^2}{S_H} = \frac{13.3^2}{33.583} = 5.267$$

$$\tan \Lambda_{0.5_H} = \tan \Lambda_{LE_H} - \frac{4 \cdot 0.5 \cdot (1-\lambda_H)}{AR_H(1+\lambda_H)} = \tan(0.323) - \frac{4 \cdot 0.5 \cdot (1-0.443)}{5.267 \cdot (1+0.443)} \Rightarrow \Lambda_{0.5_H} = 0.186 \text{ rad}$$

Wing-Tail Geometric Parameters

From Student Sample Problem 2.2: $\bar{x}_{AC_H} = 2.689$

Wing Lift-Slope Coefficient

From Student Sample Problem 2.2: $c_{L_{\alpha_W}} = 4.96$

Horizontal Tail Lift Slope Coefficient

Because $AR_H = 5.27 \geq 4$:

$$k_H = 1 + \frac{[(8.2 - 2.3\Lambda_{LE_H}) - AR(0.22 - 0.153\Lambda_{LE_H})]}{100}$$

$$= 1 + \frac{[(8.2 - 2.3 \cdot 0.323) - 5.27 \cdot (0.22 - 0.153 \cdot 0.323)]}{100} = 1.07$$

$$c_{L_{\alpha_H}} = \frac{2\pi AR_H}{2 + \sqrt{\left\{\left[\frac{AR_H^2(1-\text{Mach}^2)}{k_H^2}\right]\left(1 + \frac{\tan^2(\Lambda_{0.5_H})}{(1-\text{Mach}^2)}\right)\right] + 4\right\}}}$$

$$= \frac{2\pi \cdot 5.27}{2 + \sqrt{\left\{\left[\frac{5.27^2(1-0.6^2)}{1.07^2}\right]\left(1 + \frac{\tan^2(0.186)}{(1-0.6^2)}\right)\right] + 4\right\}}} = 5.07$$

Given that the tail geometric parameters—with the exception of λ_H—are well within the suitable ranges for the Polhamus formula, the above estimated value of $c_{L_{\alpha_H}}$ can be considered fairly accurate.

Modeling of the Downwash Effect

From Student Sample Problem 2.2: $\left(\frac{d\varepsilon}{d\alpha}\right)\bigg|_{\text{Mach}} = 0.555$

Modeling of $\bar{x}_{AC_{WB}}$

A relationship for $\bar{x}_{AC_{WB}}$ is given by

$$\bar{x}_{AC_{WB}} = \bar{x}_{AC_W} + \Delta\bar{x}_{AC_B}$$

From Student Sample Problem 2.2:
$\bar{x}_{AC_W} = 0.232$ and $\Delta\bar{x}_{AC_B} = -0.186$
thus
$\bar{x}_{AC_{WB}} = 0.232 - 0.186 = 0.046$

Stability and Control Derivatives

$$c_{L_\alpha} = c_{L_{\alpha W}} + c_{L_{\alpha H}} \eta_H \frac{S_H}{S}\left(1 - \frac{d\varepsilon}{d\alpha}\right) = 4.96 + 5.07 \cdot 0.92 \cdot \frac{33.58}{136}(1 - 0.555) = \boxed{5.469}$$

$$c_{L_{\delta_E}} = \eta_H \frac{S_H}{S} c_{L_{\alpha H}} \tau_E = 0.92 \cdot \frac{33.58}{136} \cdot 5.07 \cdot 0.35 = \boxed{0.403}$$

$$c_{L_{i_H}} = \eta_H \frac{S_H}{S} c_{L_{\alpha H}} = 0.92 \cdot \frac{33.58}{136} \cdot 5.07 = \boxed{1.152}$$

$$c_{m_\alpha} = c_{L_{\alpha W}}\left(\bar{x}_{CG} - \bar{x}_{AC_{WB}}\right) - c_{L_{\alpha H}} \eta_H \frac{S_H}{S}\left(1 - \frac{d\varepsilon}{d\alpha}\right)\left(\bar{x}_{AC_H} - \bar{x}_{CG}\right)$$

$$= 4.96 \cdot (0.255 - 0.046) - 5.07 \cdot 0.92 \cdot \frac{33.58}{136}(1 - 0.555)(2.69 - 0.25) = \boxed{-0.237}$$

$$c_{m_{\delta_E}} = -c_{L_{\alpha H}} \eta_H \frac{S_H}{S}\left(\bar{x}_{AC_H} - \bar{x}_{CG}\right)\tau_E = -5.07 \cdot 0.92 \cdot \frac{33.58}{136} \cdot (2.69 - 0.25) \cdot 0.35 = \boxed{-0.983}$$

$$c_{m_{i_H}} = -c_{L_{\alpha H}} \eta_H \frac{S_H}{S}\left(\bar{x}_{AC_H} - \bar{x}_{CG}\right) = -5.07 \cdot 0.92 \cdot \frac{33.58}{136} \cdot (2.69 - 0.25) = \boxed{-2.81}$$

Finally, the aircraft aerodynamic center can be evaluated using:

$$\bar{x}_{AC} = \frac{\bar{x}_{AC_{WB}} + \frac{c_{L_{\alpha H}}}{c_{L_{\alpha W}}} \eta_H \frac{S_H}{S}\left(1 - \frac{d\varepsilon}{d\alpha}\right)\bar{x}_{AC_H}}{1 + \frac{c_{L_{\alpha H}}}{c_{L_{\alpha W}}} \eta_H \frac{S_H}{S}\left(1 - \frac{d\varepsilon}{d\alpha}\right)}$$

$$= \frac{0.046 + \frac{5.07}{4.96} \cdot 0.92 \cdot \frac{33.58}{136}(1 - 0.555) \cdot 2.69}{1 + \frac{5.07}{4.96} \cdot 0.92 \cdot \frac{33.58}{136}(1 - 0.555)} = \boxed{0.293}$$

Comparison between True Values and Empirical Values

A comparison between the true values of the preceding derivatives and their estimates using the empirical aerodynamic modeling approach, along with a calculation of the percentage of the error, is shown in Table SSP3.2.1:

Table SSP3.2.1 Comparison between Empirical and True Values for the SIAI Marchetti S211

	SIAI Marchetti S211		
	Empirical Values	True Values	Percent Error [%]
c_{L_α}	5.469	5.500	−0.6%
$c_{L_{\delta_E}}$	0.403	0.380	6.1%
$c_{L_{i_H}}$	1.152	0.990	16.3%
c_{m_α}	−0.237	−0.240	−1.2%
$c_{m_{\delta_E}}$	−0.983	−0.880	11.7%
$c_{m_{i_H}}$	−2.81	−2.300	22.2%
$\bar{x}_{AC}$	0.293	0.294	−0.1%

These results confirm the effectiveness of the empirical modeling approach for the preceding sets of aerodynamic coefficients. Note that the values of the wing geometry and the considered airspeed are such that using the Polhamus formula—on which the entire empirical modeling is based—provides reliable results.

Student Sample Problem 3.3

Consider the data relative to the McDonnell Douglas DC-9 aircraft in Appendix C. Use the provided drawings to extract all the relevant geometric characteristics of the aircraft. Using the modeling outlined in this chapter, find a numerical value for the following aerodynamic parameters:

- $c_{L_{\dot{\alpha}}}, c_{L_q}$;
- $c_{m_{\dot{\alpha}}}, c_{m_q}$.

Solution of Student Sample Problem 3.3

Wing Parameters

From Student Sample Problems 2.1 and 3.1:

$$AR = 8.372, \; \lambda = 0.309, \; \Lambda_{0.25} = 0.429 \text{ rad}, \; \bar{x}_{AC_W} = 0.289$$
$$c_{L_{\alpha_W}} = 6.062, \; c_{L_{\alpha_W}}\big|_{Mach=0} = 4.934$$

Horizontal Tail Parameters

From Student Sample Problems 2.1 and 3.1:

$$S_H = 274.91 \text{ ft}^2, \; \bar{x}_{AC_H} = 4.345$$
$$c_{L_{\alpha_H}} = 4.741$$

As for $c_{L_{\alpha_W}}$, the above value is expected to be higher than the actual $c_{L_{\alpha_H}}$ value. In fact, $\lambda_H, \Lambda_{LE_H}$ are slightly out of the valid range in the Polhamus formula.

Downwash Effect

From Student Sample Problem 3.1:

$$\left(\frac{d\varepsilon}{d\alpha}\right)\bigg|_{Mach} = 0.297$$

Stability Derivatives

The following derivatives are calculated using the modeling shown in Section 3.9.3.

Calculating $c_{L_{\dot{\alpha}}}$

$$c_{L_{\dot{\alpha}}} \approx 2 c_{L_{\alpha_H}} \eta_H \frac{S_H}{S} (\bar{x}_{AC_H} - \bar{x}_{CG}) \frac{d\varepsilon}{d\alpha} = 2 \cdot 4.741 \cdot 0.9 \cdot \frac{274.91}{928} (4.345 - 0.3) \cdot 0.297 = \boxed{3.039}$$

Calculating c_{L_q}

$$B = \sqrt{1 - Mach^2 (\cos \Lambda_{c/4})^2} = \sqrt{1 - 0.7^2 (\cos 0.433)^2} = 0.772$$

$$c_{L_{q_W}}\big|_{Mach=0} = \left(\frac{1}{2} + 2 \cdot |\bar{x}_{AC_W} - \bar{x}_{CG}|\right) c_{L_{\alpha_W}}\big|_{Mach=0} = \left(\frac{1}{2} + 2 \cdot |0.285 - 0.3|\right) 4.934 = 2.615$$

$$c_{L_{q_W}} = \left[\frac{AR + 2\cos\Lambda_{c/4}}{AR \cdot B + 2\cos\Lambda_{c/4}}\right] \cdot c_{L_{q_W}}\big|_{Mach=0} = \left[\frac{8.612 + 2\cos 0.433}{8.612 \cdot 0.772 + 2\cos 0.433}\right] \cdot 2.615 = 3.221$$

$$c_{L_{q_H}} = 2 c_{L_{\alpha_H}} \eta_H \frac{S_H}{S} (\bar{x}_{AC_H} - \bar{x}_{CG}) = 2 \cdot 4.741 \cdot 0.9 \cdot \frac{274.91}{928} (4.345 - 0.3) = 10.226$$

$$c_{L_q} = c_{L_{q_W}} + c_{L_{q_H}} = 3.221 + 10.226 = \boxed{13.447}$$

Calculating $c_{m_{\dot{\alpha}}}$

$$c_{m_{\dot{\alpha}}} \approx c_{m_{\dot{\alpha}_H}} \approx -2 c_{L_{\alpha_H}} \eta_H \frac{S_H}{S} (\bar{x}_{AC_H} - \bar{x}_{CG})^2 \frac{d\varepsilon}{d\alpha} = -2 \cdot 4.741 \cdot 0.9 \cdot \frac{274.91}{928} (4.345 - 0.3)^2 \cdot 0.297 = \boxed{-12.295}$$

Calculating c_{m_q}

$$C = \left\{ \frac{AR\left(0.5|\bar{x}_{AC_W} - \bar{x}_{CG}| + 2|\bar{x}_{AC_W} - \bar{x}_{CG}|^2\right)}{AR + 2\cos\Lambda_{c/4}} + \frac{1}{24}\left(\frac{AR^3\tan\Lambda_{c/4}}{AR + 6\cos\Lambda_{c/4}}\right) + \frac{1}{8}\right\}$$

$$= \left\{\frac{8.612(0.5|0.285 - 0.3| + 2|0.285 - 0.3|^2)}{8.612 + 2\cos 0.433} + \frac{1}{24}\left(\frac{8.612^3\tan 0.433}{8.612 + 6\cos 0.433}\right) + \frac{1}{8}\right\} = 0.536$$

From Figure 3.12 $K_q = 0.846$. Therefore, we have

$$c_{m_{qW}}\big|_{Mach=0} = -K_q c_{L_{\alpha W}}\big|_{Mach=0} \cos\Lambda_{c/4} \cdot C = -0.846 \cdot 4.934 \cdot \cos 0.433 \cdot 0.536 = -2.030$$

$$B = \sqrt{1 - Mach^2(\cos\Lambda_{c/4})^2} = \sqrt{1 - 0.7^2(\cos 0.433)^2} = 0.772$$

$$c_{m_{qW}} = \left[\frac{\dfrac{AR^3\tan^2\Lambda_{c/4}}{AR \cdot B + 6\cos\Lambda_{c/4}} + \dfrac{3}{B}}{\dfrac{AR^3\tan^2\Lambda_{c/4}}{AR + 6\cos\Lambda_{c/4}} + 3}\right] \cdot c_{m_{qW}}\big|_{Mach=0}$$

$$= \left[\frac{\dfrac{8.612^3\tan^2 0.433}{8.612 \cdot 0.772 + 6\cos 0.433} + \dfrac{3}{0.772}}{\dfrac{8.612^3\tan^2 0.433}{8.612 + 6\cos 0.433} + 3}\right] \cdot (-2.03) = -2.423$$

$$c_{m_{qH}} \approx -2c_{L_{\alpha H}}\eta_H \frac{S_H}{S}(\bar{x}_{AC_H} - \bar{x}_{CG})^2 = -2 \cdot 4.741 \cdot 0.9 \cdot \frac{274.91}{928}(4.345 - 0.3)^2 = -41.369$$

$$c_{m_q} \approx c_{m_{qW}} + c_{m_{qH}} = -2.423 - 41.369 = \boxed{-43.79}$$

Although the "true" values of the preceding stability derivatives are unknown, it can be speculated that those estimates are approximate due to the overestimate of the parameter $c_{L_{\alpha W}} \cdot c_{L_{\alpha H}}$ from the Polhamus formula. Considering a tentative 20 percent overestimate for $c_{L_{\alpha W}}|_{Mach=0}$ and $c_{L_{\alpha H}}$ and neglecting their effect on the estimate of the downwash effect, a more appropriate value can be given by $c_{L_{\alpha W}}|_{Mach=0} = 3.947$, $c_{L_{\alpha H}} = 3.793$ leading to the following revised estimates.

REVISED Stability Derivatives (using REVISED $c_{L_{\alpha W}}|_{Mach=0}$ and $c_{L_{\alpha H}}$)

Calculating $c_{L_{\dot{\alpha}}}$

$$c_{L_{\dot{\alpha}}} \approx 2c_{L_{\alpha H}}\eta_H \frac{S_H}{S}(\bar{x}_{AC_H} - \bar{x}_{CG})\frac{d\varepsilon}{d\alpha} = 2 \cdot 3.793 \cdot 0.9 \cdot \frac{274.91}{928}(4.345 - 0.3) \cdot 0.297 = \boxed{2.431}$$

Calculating c_{L_q}

$$B = 0.772$$

$$c_{L_{qW}}\big|_{Mach=0} = \left(\frac{1}{2} + 2 \cdot |\bar{x}_{AC_W} - \bar{x}_{CG}|\right)c_{L_{\alpha W}}\big|_{Mach=0} = \left(\frac{1}{2} + 2 \cdot |0.285 - 0.3|\right)3.947 = 2.092$$

$$c_{L_{qW}} = \left[\frac{AR + 2\cos\Lambda_{c/4}}{AR \cdot B + 2\cos\Lambda_{c/4}}\right] \cdot c_{L_{qW}}\big|_{Mach=0} = \left[\frac{8.612 + 2\cos 0.433}{8.612 \cdot 0.772 + 2\cos 0.433}\right] \cdot 2.092 = 2.577$$

$$c_{L_{q_H}} = 2c_{L_{\alpha_H}}\eta_H \frac{S_H}{S}\left(\bar{x}_{AC_H} - \bar{x}_{CG}\right) = 2 \cdot 3.793 \cdot 0.9 \cdot \frac{274.91}{928}(4.345 - 0.3) = 8.181$$

$$c_{L_q} = c_{L_{q_W}} + c_{L_{q_H}} = 2.577 + 8.181 = \boxed{10.758}$$

Calculating $c_{m_{\dot{\alpha}}}$

$$c_{m_{\dot{\alpha}}} \approx c_{m_{\dot{\alpha}_H}} \approx -2c_{L_{\alpha_H}}\eta_H \frac{S_H}{S}\left(\bar{x}_{AC_H} - \bar{x}_{CG}\right)^2 \frac{d\varepsilon}{d\alpha} = -2 \cdot 3.793 \cdot 0.9 \cdot \frac{274.91}{928}(4.345 - 0.3)^2 \cdot 0.297 = \boxed{-9.836}$$

Calculating c_{m_q}

As before, $C = 0.536$, $K_q = 0.846$, $B = 0.772$. Therefore, we have:

$$c_{m_{q_W}}\Big|_{Mach=0} = -K_q c_{L_{\alpha_W}}\Big|_{Mach=0} \cos \Lambda_{c/4} \cdot C = -0.846 \cdot 3.947 \cdot \cos 0.433 \cdot 0.536 = -1.624$$

$$c_{m_{q_W}} = \left[\frac{\dfrac{AR^3 \tan^2 \Lambda_{c/4}}{AR \cdot B + 6\cos \Lambda_{c/4}} + \dfrac{3}{B}}{\dfrac{AR^3 \tan^2 \Lambda_{c/4}}{AR + 6\cos \Lambda_{c/4}} + 3}\right] \cdot c_{m_{q_W}}\Big|_{Mach=0}$$

$$= \left[\frac{\dfrac{8.612^3 \tan^2 0.433}{8.612 \cdot 0.772 + 6\cos 0.433} + \dfrac{3}{0.772}}{\dfrac{8.612^3 \tan^2 0.433}{8.612 + 6\cos 0.433} + 3}\right] \cdot (-1.624) = -1.938$$

$$c_{m_{q_H}} \approx -2c_{L_{\alpha_H}}\eta_H \frac{S_H}{S}\left(\bar{x}_{AC_H} - \bar{x}_{CG}\right)^2 = -2 \cdot 3.793 \cdot 0.9 \cdot \frac{274.91}{928}(4.345 - 0.3)^2 = -33.095$$

$$c_{m_q} \approx c_{m_{q_W}} + c_{m_{q_H}} = -1.938 - 33.095 = \boxed{-35.03}$$

Student Sample Problem 3.4

Consider the data relative to the SIAI Marchetti S211 aircraft in Appendixes B and C. Use the provided drawings to extract all the relevant geometric characteristics of the aircraft. Using the modeling outlined in this chapter, find a numerical value for the following aerodynamic parameters:

- $c_{L_{\dot{\alpha}}}, c_{L_q}$;
- $c_{m_{\dot{\alpha}}}, c_{m_q}$.

Next, compare the obtained values with the true values listed in Appendix B (Aircraft 8) and evaluate the error associated with the use of the empirical modeling approach.

Solution of Student Sample Problem 3.4

Wing Parameters

From Student Sample Problems 2.2 and 3.2:

$$AR = 5.086, \ \lambda = 0.517, \ \Lambda_{0.25} = 0.284 \text{ rad}, \ \bar{x}_{AC_W} = 0.232$$

$$c_{L_{\alpha_W}} = 4.957, \ c_{L_{\alpha_W}}\Big|_{Mach=0} = 4.384$$

Horizontal Tail Parameters

From Student Sample Problems 2.2 and 3.2:

$$S_H = 33.583 \text{ ft}^2, \ \bar{x}_{AC_H} = 2.689$$

$$c_{L_{\alpha_H}} = 5.070$$

Downwash Effect

From Student Sample Problem 3.2: $\left.\left(\dfrac{d\varepsilon}{d\alpha}\right)\right|_{Mach} = 0.555$

Stability Derivatives

The following derivatives are calculated using the modeling shown in Section 3.9.3.

Calculating $c_{L_{\dot{\alpha}}}$

$$c_{L_{\dot{\alpha}}} \approx 2c_{L_{\alpha_H}} \eta_H \frac{S_H}{S} \left(\bar{x}_{AC_H} - \bar{x}_{CG}\right) \frac{d\varepsilon}{d\alpha} = 2 \cdot 5.07 \cdot 0.92 \cdot \frac{33.583}{136} (2.689 - 0.25) \cdot 0.555 = \boxed{3.12}$$

Calculating c_{L_q}

$$B = \sqrt{1 - Mach^2 \left(\cos \Lambda_{c/4}\right)^2} = \sqrt{1 - 0.6^2 (\cos 0.284)^2} = 0.817$$

$$\left.c_{L_{qW}}\right|_{Mach=0} = \left(\frac{1}{2} + 2 \cdot \left|\bar{x}_{AC_W} - \bar{x}_{CG}\right|\right) \left.c_{L_{\alpha W}}\right|_{Mach=0} = \left(\frac{1}{2} + 2 \cdot |0.232 - 0.25|\right) 4.384 = 2.350$$

$$c_{L_{qW}} = \left[\frac{AR + 2\cos\Lambda_{c/4}}{AR \cdot B + 2\cos\Lambda_{c/4}}\right] \cdot \left.c_{L_{qW}}\right|_{Mach=0} = \left[\frac{5.086 + 2\cos 0.284}{5.086 \cdot 0.817 + 2\cos 0.284}\right] \cdot 2.350 = 2.709$$

$$c_{L_{qH}} \approx 2c_{L_{\alpha_H}} \eta_H \frac{S_H}{S} \left(\bar{x}_{AC_H} - \bar{x}_{CG}\right) = 2 \cdot 5.07 \cdot 0.92 \cdot \frac{33.583}{136}(2.689 - 0.25) = 5.619$$

$$c_{L_q} = c_{L_{qW}} + c_{L_{qH}} = 2.709 + 5.497 = \boxed{8.328}$$

Calculating $c_{m_{\dot{\alpha}}}$

$$c_{m_{\dot{\alpha}}} \approx -2c_{L_{\alpha_H}} \eta_H \frac{S_H}{S} \left(\bar{x}_{AC_H} - \bar{x}_{CG}\right)^2 \frac{d\varepsilon}{d\alpha} = -2 \cdot 5.07 \cdot 0.92 \cdot \frac{33.583}{136}(2.689 - 0.25)^2 \cdot 0.555 = \boxed{-7.611}$$

Calculating c_{m_q}

$$C = \left\{\frac{AR\left(0.5\left|\bar{x}_{AC_W} - \bar{x}_{CG}\right| + 2\left|\bar{x}_{AC_W} - \bar{x}_{CG}\right|^2\right)}{AR + 2\cos\Lambda_{c/4}} + \frac{1}{24}\left(\frac{AR^3 \tan\Lambda_{c/4}}{AR + 6\cos\Lambda_{c/4}}\right) + \frac{1}{8}\right\}$$

$$= \left\{\frac{5.086(0.5|0.232 - 0.25| + 2|0.232 - 0.25|^2)}{5.086 + 2\cos 0.284} + \frac{1}{24}\left(\frac{5.086^3 \tan 0.284}{5.086 + 6\cos 0.284}\right) + \frac{1}{8}\right\} = 0.175$$

From Figure 3.12 $K_q = 0.7$. Therefore, we have

$$\left.c_{m_{qW}}\right|_{Mach=0} = -K_q \left.c_{L_{\alpha W}}\right|_{Mach=0} \cos\Lambda_{c/4} \cdot C = -0.7 \cdot 4.384 \cdot \cos 0.284 \cdot 0.175 = -0.515$$

$$B = \sqrt{1 - Mach^2\left(\cos\Lambda_{c/4}\right)^2} = \sqrt{1 - 0.6^2(\cos 0.284)^2} = 0.817$$

$$c_{m_{qW}} = \left[\frac{\frac{AR^3 \tan^2 \Lambda_{c/4}}{AR \cdot B + 6\cos\Lambda_{c/4}} + \frac{3}{B}}{\frac{AR^3 \tan^2 \Lambda_{c/4}}{AR + 6\cos\Lambda_{c/4}} + 3} \right] \cdot c_{m_{qW}}\bigg|_{Mach=0}$$

$$= \left[\frac{\frac{5.086^3 \tan^2 0.284}{5.086 \cdot 0.817 + 6\cos 0.284} + \frac{3}{0.817}}{\frac{5.086^3 \tan^2 0.284}{5.086 + 6\cos 0.284} + 3} \right] \cdot (-0.515) = -0.613$$

$$c_{m_{q_H}} \approx -2c_{L_{\alpha_H}} \eta_H \frac{S_H}{S} (\bar{x}_{AC_H} - \bar{x}_{CG})^2 = -2 \cdot 5.07 \cdot 0.92 \cdot \frac{33.583}{136} (2.689 - 0.25)^2 = -13.707$$

$$c_{m_q} \approx c_{m_{qW}} + c_{m_{q_H}} = -0.613 - 13.707 = \boxed{-14.321}$$

Comparison between True Values and Empirical Values

A comparison between the true values of the preceding derivatives from Appendix B and their estimates using the empirical aerodynamic modeling approach, along with a calculation of the percentage of the error, is shown in the Table SSP3.4.1.

Table SSP3.4.1 Comparison between Empirical and True Values

	Empirical Values	True Values	Percent Error [%]
$c_{L_{\dot\alpha}}$	3.12	4.2	−25.7%
c_{L_q}	8.328	10.0	−16.7%
$c_{m_{\dot\alpha}}$	−7.611	−9.6	−20.7%
c_{m_q}	−14.321	−17.7	−19.1%

The relatively large error for the preceding derivatives is typical of the approximation associated with the use of the simple empirical approach for this type of derivatives, as explained earlier in this chapter.

CASE STUDY

Modeling of the Longitudinal Aerodynamics for the Cessna CJ3

The objective of this Case Study is to show a detailed DATCOM-based modeling for the longitudinal aerodynamic coefficients of the Cessna CJ3 aircraft over a 0.2−0.7 Mach range.

1 Modeling of the LIFT aerodynamic coefficients

This section provides the modeling of the following LIFT stability and control derivatives:

$$\left(c_{L_1}, c_{L_\alpha}, c_{L_u}, c_{L_{\dot\alpha}}, c_{L_q}, c_{L_{\delta_E}}, c_{L_{i_H}} \right)$$

1.1 Modeling of (c_{L_1})

c_{L_1} is the aircraft dimensionless lift coefficient at steady-state conditions, relative to the condition:

$$c_{L_1} \bar{q} S = W \quad \rightarrow \quad c_{L_1} = \frac{W}{\bar{q} S}$$

It is considered a design parameter and is related to the design weight at the given flight condition. Ref. [10] provides

Maximum ramp weight: 14,070 lbs
Maximum take-off weight: 13,870 lbs
Maximum landing weight: 12,750 lbs

An average operating reference weight is selected: $W_{Avg.} = 13,000$ lbs.
Considering $Alt. = 40,000$ ft, $Mach = 0.7$, $S = 294.1 \, \text{ft}^2$ we have

$$c_{L_1}\big|_{Mach=0.7, \, Alt.=40K \, \text{ft}} = \frac{W}{\bar{q} \, S} = \frac{13,000}{135 \cdot 294.1} \approx 0.327$$

Including flight conditions with different airspeed (within the given Mach range), different altitudes, and different weight, we have the following approximated range:

$$c_{L_1} \in [0.3 - 1.75]$$

1.2 Modeling of (c_{L_α})

A general expression for c_{L_α} is given by

$$c_{L_\alpha} = c_{L_{\alpha_{WB}}} + c_{L_{\alpha_H}} \eta_H \frac{S_H}{S} \left(1 - \frac{d\varepsilon}{d\alpha}\right)$$

where

$c_{L_{\alpha_{WB}}} \approx K_{WB} \, c_{L_{\alpha_W}}$ with $K_{WB} = 1 + 0.025(d/b) - 0.25(d/b)^2$

(shown in Figure 3.6) where (d, b) are the fuselage diameter and the wing span, respectively. An arbitrary empirical value $\eta_H = \bar{q}_H/\bar{q} = 0.9$ is selected for the dynamic pressure ratio.

As shown in the drawings in Appendix C, the following geometric parameters were identified for the Cessna CJ3 aircraft:

$b = 53.3$ ft, $c_T = 2.5$ ft, $c_R = 8.2$ ft, $\Lambda_{LE} = 4° = 0.07$ rad, $X_{WH_R} = 23.3$ ft, $Z_{WH} = 11.7$ ft
$b_H = 20.83$ ft, $c_{T_H} = 2$ ft, $c_{R_H} = 4.7$ ft, $\Lambda_{LE_H} = 23° = 0.401$ rad

Next, using the approximation of straight wings, the following wing and tail geometric parameters were derived using the preceding values.

Wing Geometric Parameters

$$\lambda = \frac{c_T}{c_R} = \frac{2.5}{8.2} = 0.305, \quad S = 294.1 \, \text{ft}^2, \quad AR = \frac{b^2}{S} = \frac{53.33^2}{294.1} = 9.67$$

$$\bar{c} = MAC = \frac{2}{3} c_R \frac{(1 + \lambda + \lambda^2)}{(1 + \lambda)} = \frac{2}{3} \cdot 8.2 \cdot \frac{(1 + 0.305 + 0.305^2)}{(1 + 0.305)} = 5.856 \, \text{ft}$$

$$x_{MAC} = \frac{b}{6} \frac{(1 + 2\lambda)}{(1 + \lambda)} \tan(\Lambda_{LE}) = \frac{53.33}{6} \cdot \frac{(1 + 2 \cdot 0.305)}{(1 + 0.305)} \tan(0.07) = 0.767 \, \text{ft}$$

$$\tan \Lambda_{0.5} = \tan \Lambda_{LE} - \frac{4 \cdot 0.5 \cdot (1 - \lambda)}{AR(1 + \lambda)} = \tan(0.07) - \frac{4 \cdot 0.5 \cdot (1 - 0.305)}{9.67 \cdot (1 + 0.305)} \Rightarrow \Lambda_{0.5} = -0.040 \, \text{rad}$$

$$\tan \Lambda_{0.25} = \tan \Lambda_{LE} - \frac{4 \cdot 0.25 \cdot (1 - \lambda)}{AR(1 + \lambda)} = \tan(0.07) - \frac{4 \cdot 0.25 \cdot (1 - 0.305)}{9.67 \cdot (1 + 0.305)} \Rightarrow \Lambda_{0.25} = 0.015 \, \text{rad}$$

Horizontal Tail Geometric Parameters

$$\lambda_H = \frac{c_{T_H}}{c_{R_H}} = \frac{2}{4.7} = 0.426, \quad S_H = 70.7 \, \text{ft}^2, \quad AR_H = \frac{b_H^2}{S_H} = \frac{20.83^2}{70.71} = 6.137$$

$$\bar{c}_H = \frac{2}{3} c_{R_H} \frac{(1 + \lambda_H + \lambda_H^2)}{(1 + \lambda_H)} = \frac{2}{3} \cdot 4.7 \cdot \frac{(1 + 0.426 + 0.426^2)}{(1 + 0.426)} = 3.531 \, \text{ft}$$

$$x_{MAC_H} = \frac{b_H}{6}\frac{(1+2\lambda_H)}{(1+\lambda_H)}\tan(\Lambda_{LE_H}) = \frac{20.83}{6} \cdot \frac{(1+2 \cdot 0.426)}{(1+0.426)}\tan(0.401) = 1.914 \text{ ft}$$

$$\tan\Lambda_{0.5_H} = \tan\Lambda_{LE_H} - \frac{4 \cdot 0.5 \cdot (1-\lambda_H)}{AR_H(1+\lambda_H)} = \tan(0.401) - \frac{4 \cdot 0.5 \cdot (1-0.426)}{6.137 \cdot (1+0.426)} \Rightarrow \Lambda_{0.5_H} = 0.285 \text{ rad}$$

Wing-Tail Geometric Parameters

Figure CS.1 shows the wing-tail geometric distances.

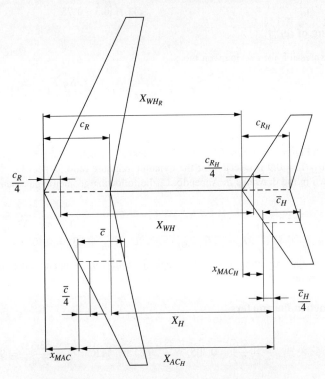

Figure CS.1 Wing-Horizontal Tail Geometric Distances

From Figure CS.1:

$$X_{WH_R} + \frac{c_{R_H}}{4} = \frac{c_R}{4} + X_{WH}$$

Knowing that $X_{WH_R} = 23.3$ ft:

$$X_{WH} = X_{WH_R} + \frac{c_{R_H}}{4} - \frac{c_R}{4} = 23.3 + \frac{4.7}{4} - \frac{8.2}{4} = 22.425 \text{ ft}$$

Also, knowing that $Z_{WH} = 11.7$ ft, the wing-tail geometric parameters needed for the analysis of the downwash effect, shown in Figure CS.2, are given by:

$$r = \frac{2X_{WH}}{b} = \frac{2 \cdot 22.425}{53.33} = 0.841, \quad m = \frac{2Z_{WH}}{b} = \frac{2 \cdot 11.7}{53.33} = 0.439$$

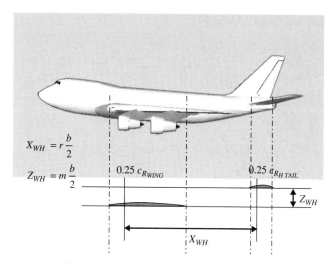

Figure CS.2 Geometric Parameters for the Downwash Effect

Wing Lift-Slope Coefficient

The Polhamus relationship (to be used for both wing and horizontal tail) is given by

$$c_{L_{\alpha_W}} = \frac{2\pi AR}{2 + \sqrt{\left\{\left[\frac{AR^2(1-Mach^2)}{k^2}\left(1+\frac{\tan^2(\Lambda_{0.5})}{(1-Mach^2)}\right)\right]+4\right\}}}$$

where the parameter k takes on the following values, depending on the aspect ratio:

for $AR < 4$ $k = 1 + \frac{AR(1.87 - 0.000233\,\Lambda_{LE})}{100}$, with Λ_{LE} in rad; for $AR \geq 4$ $k = 1 + \frac{[(8.2 - 2.3\,\Lambda_{LE}) - AR(0.22 - 0.153\Lambda_{LE})]}{100}$, with Λ_{LE} in rad.

Since $AR = 9.67 \geq 4$:

$$k = 1 + \frac{[(8.2 - 2.3\Lambda_{LE}) - AR(0.22 - 0.153\Lambda_{LE})]}{100}$$

$$= 1 + \frac{[(8.2 - 2.3 \cdot 0.07) - 9.67 \cdot (0.22 - 0.153 \cdot 0.07)]}{100} = 1.060$$

Note that the value of $AR = 9.67$ is above the allowed range of aspect ratios for the Polhamus formula, leading to a substantial overestimate of $c_{L_{\alpha_W}}$. Therefore, the results from the Polhamus formula have been multiplied by an empirical correction factor <1. Specifically, a 0.8 correction factor was used, leading to the following results shown in Figure CS.3 for the given range of Mach.

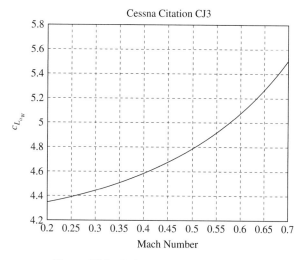

Figure CS.3 Estimates of $c_{L_{\alpha_W}}$ vs. Mach

Horizontal Tail Lift Slope Coefficient

Using the previously calculated geometric parameters

$$\lambda_H = 0.426, \quad AR_H = \frac{b_H^2}{S_H} = \frac{20.83^2}{70.71} = 6.137, \quad \Lambda_{0.5_H} = 0.285 \text{ rad}$$

we have $k_H = 1 + \dfrac{\left[(8.2 - 2.3\Lambda_{LE_H}) - AR_H(0.22 - 0.153\Lambda_{LE_H})\right]}{100} = 1.063$

The Polhamus formula for the horizontal tail can be rewritten as

$$c_{L_{\alpha H}} = \frac{2\pi AR_H}{2 + \sqrt{\left\{\left[\dfrac{AR_H^2(1 - Mach^2)}{k_H^2}\right]\left(1 + \dfrac{\tan^2(\Lambda_{0.5_H})}{(1 - Mach^2)}\right)\right] + 4\right\}}}$$

The preceding relationship provides the following results over the given Mach range:

Mach	0.15	0.20	0.25	0.30	0.35	0.40	0.45	0.50	0.55	0.60	0.65	0.70
k_H							1.063					
$\Lambda_{0.5_H}$ [rad]							0.285					
$c_{L_{\alpha H}}$	4.657	4.683	4.718	4.761	4.814	4.878	4.954	5.044	5.150	5.275	5.423	5.601

Modeling of the Downwash Effect

The closed-form expression for the modeling of the downwash effect is given by

$$\left(\frac{d\varepsilon}{d\alpha}\right)\bigg|_{Mach} = \left(\frac{d\varepsilon}{d\alpha}\right)\bigg|_{Mach=0} \frac{c_{L_{\alpha w}}\big|_{Mach}}{c_{L_{\alpha w}}\big|_{Mach=0}}$$

where $\left(\dfrac{d\varepsilon}{d\alpha}\right)\bigg|_{Mach=0} = 4.44\left(K_{AR}K_\lambda K_{mr}\sqrt{\cos(\Lambda_{0.25})}\right)^{1.19}$

Using the geometric parameters introduced previously:

$$K_{AR} = \frac{1}{AR} - \frac{1}{1 + (AR)^{1.7}} = \frac{1}{9.67} - \frac{1}{1 + (9.67)^{1.7}} = 0.083$$

$$K_\lambda = \frac{10 - 3\lambda}{7} = \frac{10 - 3 \cdot 0.305}{7} = 1.298, \quad K_{mr} = \frac{1 - \dfrac{m}{2}}{(r)^{0.33}} = \frac{1 - \dfrac{0.439}{2}}{(0.841)^{0.33}} = 0.827$$

$$\left(\frac{d\varepsilon}{d\alpha}\right)\bigg|_{Mach=0} = 4.44\left(K_{AR}K_\lambda K_{mr}\sqrt{\cos(\Lambda_{0.25})}\right)^{1.19}$$

$$= 4.44\left(0.083 \cdot 1.298 \cdot 0.827\sqrt{\cos(0.015)}\right)^{1.19} = 0.249$$

$$c_{L_{\alpha w}}\big|_{Mach=0} = \frac{2\pi AR}{2 + \sqrt{\left\{\left[\dfrac{AR^2}{k^2}\left(1 + \tan^2(\Lambda_{0.5})\right)\right] + 4\right\}}}$$

$$= \frac{2\pi \cdot 9.67}{2 + \sqrt{\left\{\left[\dfrac{9.67^2}{1.060^2}\left(1 + \tan^2(-0.040)\right)\right] + 4\right\}}} = 5.355$$

$$\left(\frac{d\varepsilon}{d\alpha}\right)\bigg|_{Mach} = \left(\frac{d\varepsilon}{d\alpha}\right)\bigg|_{Mach=0} \frac{c_{L_{\alpha w}}\big|_{Mach}}{c_{L_{\alpha w}}\big|_{Mach=0}}$$

With $\left(\frac{d\varepsilon}{d\alpha}\right)\bigg|_{Mach=0}$, $c_{L_{\alpha W}}\big|_{Mach=0}$ being constant, the only varying parameter in the preceding relationship is $c_{L_{\alpha W}}\big|_{Mach}$. Over the provided Mach range we have Figure CS.4.

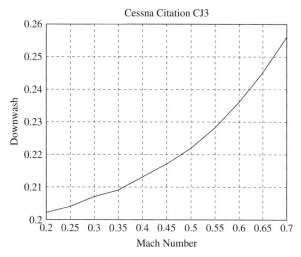

Figure CS.4 Estimates of $\left(\frac{d\varepsilon}{d\alpha}\right)\big|_{Mach}$ with Mach

Aircraft Lift-Slope Coefficient

Using the previously calculated $c_{L_{\alpha W}}$, $c_{L_{\alpha H}}$, $\dfrac{d\varepsilon}{d\alpha}$ along $\eta_H = 0.9$.

Mach	0.15	0.20	0.25	0.30	0.35	0.40	0.45	0.50	0.55	0.60	0.65	0.70
$c_{L_{\alpha W}}$	4.323	4.353	4.394	4.445	4.508	4.584	4.676	4.786	4.918	5.076	5.270	5.509
$c_{L_{\alpha H}}$	4.657	4.683	4.718	4.761	4.814	4.878	4.954	5.044	5.150	5.275	5.423	5.601
$\left(\dfrac{d\varepsilon}{d\alpha}\right)\bigg|_{Mach}$	0.201	0.202	0.204	0.207	0.209	0.213	0.217	0.222	0.228	0.236	0.245	0.256
c_{L_α}	5.128	5.162	5.206	5.262	5.331	5.415	5.515	5.634	5.777	5.949	6.156	6.410

A review of the preceding results for c_{L_α} suggests that this coefficient has been overestimated. While a large value for c_{L_α} is to be expected for the Cessna CJ3 aircraft due to the very high AR, fairly high λ, and very low value of Λ of its wing, under no circumstances can the value of c_{L_α} exceed the theoretical value of $2\pi \approx 6.28$. Therefore, a 0.94 correction factor has been introduced, leading to estimate of c_{L_α} vs. Mach in Figure CS.5.

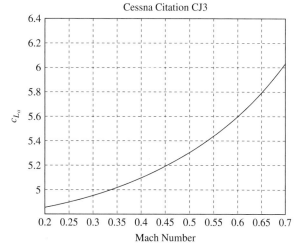

Figure CS.5 Estimates of c_{L_α} vs. Mach

1.3 Modeling of (c_{L_u})

c_{L_u} is negligible at the low subsonic conditions associated with incompressible aerodynamics; however, it takes on significant positive values around transonic conditions. In general, it can be modeled using

$$c_{L_u} = \left(\frac{Mach^2}{1 - Mach^2}\right) \cdot c_{L_1}$$

Because c_{L_1} depends on the flight conditions, there is no practical use for providing a relationship for c_{L_u} as function of Mach.

1.4 Modeling of ($c_{L_{\dot{\alpha}}}, c_{L_q}$)

A closed form relationship for $c_{L_{\dot{\alpha}}}$ is given by

$$c_{L_{\dot{\alpha}}} \approx 2 c_{L_{\alpha_H}} \eta_H \frac{S_H}{S} \left(\bar{x}_{AC_H} - \bar{x}_{CG}\right) \frac{d\varepsilon}{d\alpha}$$

The aerodynamic parameters ($c_{L_{\alpha_H}}, \frac{d\varepsilon}{d\alpha}$) have been calculated in Section 1.2. The geometric parameter $\bar{x}_{AC_H}$ can be extracted from the wing-horizontal tail geometry.

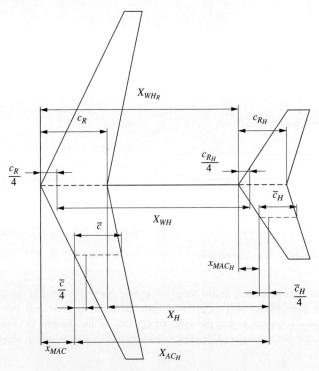

Figure CS.6 Wing-Horizontal Tail Geometric Distances

In fact, from Figure CS.6:

$$X_{WH_R} + x_{MAC_H} + \frac{\bar{c}_H}{4} = x_{MAC} + X_{AC_H}$$

solving for X_{AC_H}:

$$X_{AC_H} = X_{WH_R} + x_{MAC_H} + \frac{\bar{c}_H}{4} - x_{MAC} = 23.3 + 1.914 + \frac{3.531}{4} - 0.767 = 25.33 \text{ ft}$$

leading to $\bar{x}_{AC_H} = \dfrac{X_{AC_H}}{\bar{c}} = \dfrac{25.33}{5.856} = 4.325$.

Thus, the values for $c_{L_{\dot{\alpha}}}$ over the given Mach range are given by

Mach	0.15	0.20	0.25	0.30	0.35	0.40	0.45	0.50	0.55	0.60	0.65	0.70
$c_{L_{\dot{\alpha}}}$	1.65	1.67	1.70	1.73	1.78	1.83	1.90	1.98	2.07	2.19	2.34	2.53

They are also shown in Figure CS.7.

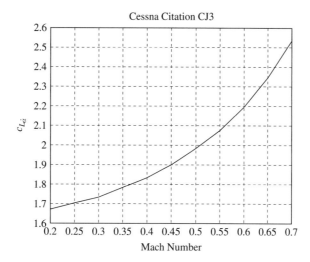

Figure CS.7 Estimates of $c_{L_{\dot{\alpha}}}$ vs. Mach

A closed-form relationship for c_{L_q} is given by:[1]

$$c_{L_q} = c_{L_{q_W}} + c_{L_{q_H}}$$

where

$$c_{L_{q_W}} = \left[\frac{AR + 2\cos\Lambda_{c/4}}{AR \cdot B + 2\cos\Lambda_{c/4}}\right] \cdot c_{L_{q_W}}\bigg|_{Mach=0}$$

with

$$B = \sqrt{1 - Mach^2(\cos\Lambda_{c/4})^2} = f(Mach) \text{ with } \Lambda_{0.25} = \Lambda_{\bar{c}/4} = 0.015 \text{ rad}$$

$$c_{L_{q_W}}\big|_{Mach=0} = \left(\frac{1}{2} + 2 \cdot |\bar{x}_{AC_W} - \bar{x}_{CG}|\right) c_{L_{\alpha_W}}\bigg|_{Mach=0}$$

$$c_{L_{q_H}} \approx 2c_{L_{\alpha_H}}\eta_H \frac{S_H}{S}(\bar{x}_{AC_H} - \bar{x}_{CG})$$

In the preceding relationship, the only unknown is the parameter $(\bar{x}_{AC_W})$, which is the location of the aerodynamic center for the wing only with respect to the leading edge of the MAC, as shown in Figure CS.8.

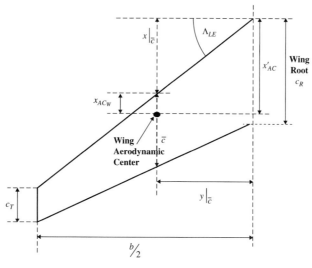

Figure CS.8 Location of the Wing Aerodynamic Center

$\bar{x}_{AC_W}$ is calculated using the closed-form relationship:

$$\bar{x}_{AC_W} = K_1 \left(\frac{x'_{AC}}{c_R} - K_2 \right)$$

where the coefficients K_1, K_2 are shown in Figure 2.28 and Figure 2.29. The geometric parameter $\left(\frac{x'_{AC}}{c_R}\right)$ is shown in Figure 2.27. The following parameters are required in Figure 2.27:

Mach	0.15	0.20	0.25	0.30	0.35	0.40	0.45	0.50	0.55	0.60	0.65	0.70
$\dfrac{\tan \Lambda_{LE}}{\sqrt{1-M^2}}$	0.071	0.071	0.072	0.073	0.075	0.076	0.078	0.081	0.084	0.087	0.092	0.098
$AR \cdot \tan \Lambda_{LE}$							0.676					

Interpolating between the curves of the plots of Figure 2.27 using the values in the above table, we have $\frac{x'_{AC}}{c_R} = 0.267$. Note that this value can be accurately assumed to be constant since it has very small variations with Mach.

From interpolation of Figure 2.28, using $\lambda = 0.305 \Rightarrow K_1 = 1.397$.

From interpolation of Figure 2.29, using

$$\Lambda_{LE} = 4°, \ \lambda = 0.305, \ AR = 9.67 \Rightarrow K_2 = 0.095$$

Therefore, the location of the wing aerodynamic center is estimated at

$$\bar{x}_{AC_W} = K_1 \left(\frac{x'_{AC}}{c_R} - K_2 \right) = 1.397 \cdot (0.267 - 0.095) = 0.240$$

At this point, with all the required values available, the calculation of c_{L_q} can be finalized using the table shown.

Mach	0.15	0.20	0.25	0.30	0.35	0.40	0.45	0.50	0.55	0.60	0.65	0.70
$c_{L_q\,H}$	8.212	8.259	8.320	8.396	8.490	8.603	8.737	8.895	9.082	9.302	9.564	9.877
B	0.989	0.980	0.968	0.954	0.937	0.917	0.893	0.866	0.835	0.800	0.760	0.714
$c_{L_{qW}}\big|_{Mach=0}$							2.782					
$c_{L_q\,W}$	2.808	2.829	2.857	2.892	2.936	2.988	3.052	3.129	3.222	3.334	3.472	3.645
c_{L_q}	11.02	11.09	11.18	11.29	11.43	11.59	11.79	12.02	12.30	12.64	13.04	13.52

The results for c_{L_q} are also shown in Figure CS.9.

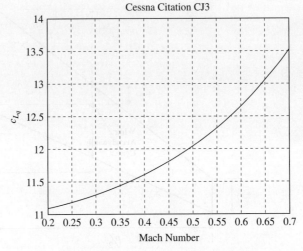

Figure CS.9 Estimates of c_{L_q} vs. Mach

1.5 Modeling of $(c_{L_{\delta_E}}, c_{L_{i_H}})$

Closed form relationship for $(c_{L_{\delta_E}}, c_{L_{i_H}})$ are given by

$$c_{L_{\delta_E}} = \eta_H \frac{S_H}{S} c_{L_{\alpha_H}} \tau_E, \quad c_{L_{i_H}} = \eta_H \frac{S_H}{S} c_{L_{\alpha_H}}$$

The elevator effectiveness factor τ_E can be found from Figure 3.7 using the size of the elevator and stabilator chords from the drawings in Appendix C. Specifically, using $\bar{c}_{Elevator}, \bar{c}_H$ the value of the elevator's effectiveness factor is found to be $\tau_E \approx 0.5$.

Therefore, the numerical values of $(c_{L_{\delta_E}}, c_{L_{i_H}})$ can be calculated as shown:

Mach	0.15	0.20	0.25	0.30	0.35	0.40	0.45	0.50	0.55	0.60	0.65	0.70
$c_{L_{\alpha_H}}$	4.657	4.683	4.718	4.761	4.814	4.878	4.954	5.044	5.150	5.275	5.423	5.601
$c_{L_{\delta_E}}$	0.504	0.507	0.510	0.515	0.521	0.528	0.536	0.546	0.557	0.571	0.587	0.606
$c_{L_{i_H}}$	1.008	1.013	1.021	1.030	1.042	1.055	1.072	1.091	1.114	1.141	1.173	1.212

The results for $c_{L_{\delta_E}}$ are also shown in Figure CS.10.

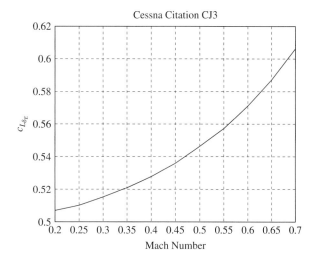

Figure CS.10 Estimates of $c_{L_{\delta_E}}$ vs. Mach

The results for $c_{L_{i_H}}$ are also shown in Figure CS.11.

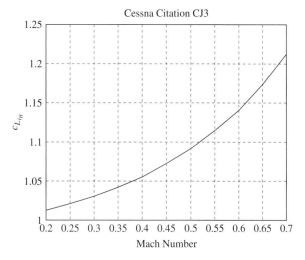

Figure CS.11 Estimates of $c_{L_{i_H}}$ vs. Mach

2 Modeling of the pitching aerodynamic coefficients

This section provides the modeling of the following pitching stability and control derivatives:

$$\left(c_{m_1}, c_{m_\alpha}, c_{m_u}, c_{m_{\dot\alpha}}, c_{m_q}, c_{m_{\delta_E}}, c_{m_{i_H}}\right)$$

2.1 Modeling of (c_{m_1})

c_{m_1} is the aircraft dimensionless pitching coefficient at steady-state conditions. By definition of steady-state conditions we have:

$$c_{m_1} = 0$$

2.2 Modeling of (c_{m_α})

A general expression for c_{m_α} is given by

$$c_{m_\alpha} = c_{L_{\alpha_W}}\left(\bar{x}_{CG} - \bar{x}_{AC_{WB}}\right) - c_{L_{\alpha_H}} \eta_H \frac{S_H}{S}\left(1 - \frac{d\varepsilon}{d\alpha}\right)\left(\bar{x}_{AC_H} - \bar{x}_{CG}\right)$$

The geometric and inertial parameters $\left(\bar{x}_{CG}, \frac{S_H}{S}\right)$ are known from Appendix C; the geometric and aerodynamic parameters $(c_{L_{\alpha_W}}, c_{L_{\alpha_H}}, \frac{d\varepsilon}{d\alpha}, \bar{x}_{AC_H})$ have been evaluated in Section 1. The only unknown value in the preceding relationship for c_{m_α} is $\bar{x}_{AC_{WB}}$, that is the aerodynamic center for the (wing + body) configuration. A general relationship for $\bar{x}_{AC_{WB}}$ is given by

$$\bar{x}_{AC_{WB}} = \bar{x}_{AC_W} + \Delta\bar{x}_{AC_B}$$

The parameter $\bar{x}_{AC_W}$ has been explained and calculated in Section 1. Next, the attention shifts on the calculation of $\Delta\bar{x}_{AC_{WB}}$. Using Munk's theory, the aerodynamic center shift due to the presence of the fuselage is calculated using the following relationship:

$$\Delta\bar{x}_{AC_B} = \frac{-\left(\frac{dM}{d\alpha}\right)}{\bar{q}\, S\, \bar{c}\, c_{L_{\alpha\,W}}} = -\frac{\left(\frac{\bar{q}}{36.5} \frac{c_{L_{\alpha\,W}}}{0.08}\right)}{\bar{q}\, S\, \bar{c}\, c_{L_{\alpha\,W}}} \sum_{i=1}^{13} w_{B_i}^2 \left(\frac{d\bar{\varepsilon}}{d\alpha}\right)_i \Delta x_i$$

$$= -\frac{1}{2.92 \cdot S\, \bar{c}} \sum_{i=1}^{13} w_{B_i}^2 \left(\frac{d\bar{\varepsilon}}{d\alpha}\right)_i \Delta x_i$$

where $w_{B_i}, \Delta x_i$ are geometric parameters for discretized aircraft sections. Particularly, w_{B_i} and Δx_i represent the width and the length of a generic i-th section of the fuselage, as shown in Figure CS.12.

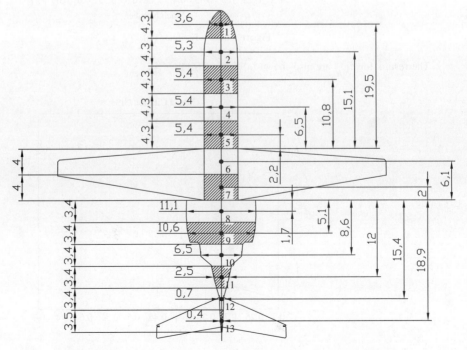

Figure CS.12 Aircraft sections for calculations of $\Delta\bar{x}_{AC_B}$ (Dimensions in ft)

The parameter x_i represents the longitudinal distance between the midpoint of the i-th section and the leading edge of the root chord (c_{Root}) for the stations of the fuselage ahead of the leading edge of the root chord. Instead, for the stations of the fuselage behind the leading edge of the root chord, x_i represents the distance between the mid-point of the i-th section and the trailing edge of the root chord. Figure CS.12 shows all the preceding parameters for one station in front and one station behind the leading edge of the root chord (the fourth and ninth stations, respectively). Finally, the parameter x_H represents the longitudinal distance between the aerodynamic center of the horizontal tail (typically located around $0.25\ \bar{c}_H$) and the trailing edge of the root chord.

The parameter $(d\bar{\varepsilon}/d\alpha)_i$ is calculated through the two curves in Figure CS.13.

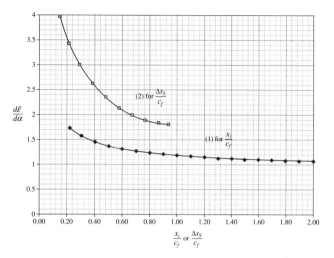

Figure CS.13 Downwash Calculations for $\Delta \bar{x}_{AC_{WB}}{}^5$

Consider first that the fuselage stations ahead of the wing will be subjected to an upwash angle effect (indicated as $\bar{\varepsilon}$) while the fuselage stations behind the wing will be exposed to a downwash effect (indicated as ε). Therefore, consider the integers $N = 6$, $M = 7$ as the number of stations in front and behind the leading edge of the root chord, exposed respectively to the upwash and the downwash effects.

Specifically, $(d\bar{\varepsilon}/d\alpha)_i$ for $i = 1, 2, 3, 4$ is obtained using curve (1) using the required values for x_i with $c_f = c_{Root} = 8.2$ ft. For $i = N - 1 = 5$, $(d\bar{\varepsilon}/d\alpha)_i$ is obtained using curve (2) and taking into account that $\Delta x_5 = 4.3$ ft.

Finally, for $i = (N = 6), \ldots, (N + M) = 13$,

$$\left(\frac{d\bar{\varepsilon}}{d\alpha}\right)_i = \left(\frac{x_i}{x_H}\right)\left(1 - \frac{d\varepsilon}{d\alpha}\bigg|_{m=0}\right)$$

The value of $d\varepsilon/d\alpha|_{m=0}$ is the downwash effect evaluated with $m = 0$, thus:

$$K_{mr} = \frac{1 - \dfrac{m}{2}}{(r)^{0.33}} = \frac{1 - 0}{(0.841)^{0.33}} = 1.059$$

$$\left(\frac{d\varepsilon}{d\alpha}\right)\bigg|_{Mach=0,\ m=0} = 4.44\left(K_{AR}K_\lambda K_{mr}\sqrt{\cos(\Lambda_{0.25})}\right)^{1.19}$$

$$= 4.44\left(0.083 \cdot 1.298 \cdot 1.059\sqrt{\cos(0.015)}\right)^{1.19} = 0.334$$

Finally, we have for $(d\varepsilon/d\alpha)|_{Mach,\ m=0}$:

$$\left(\frac{d\varepsilon}{d\alpha}\right)\bigg|_{Mach,\ m=0} = \left(\frac{d\varepsilon}{d\alpha}\right)\bigg|_{Mach=0,\ m=0} \frac{c_{L_{\alpha w}}|_{Mach}}{c_{L_{\alpha w}}|_{Mach=0}}$$

The following spreadsheet presents the values of $(d\varepsilon/d\alpha)|_{Mach,\ m=0}$.

Mach	0.15	0.20	0.25	0.30	0.35	0.40	0.45	0.50	0.55	0.60	0.65	0.70
$\left(\frac{d\varepsilon}{d\alpha}\right)\big|_{Mach,m=0}$	0.270	0.272	0.274	0.277	0.281	0.286	0.292	0.299	0.307	0.317	0.329	0.344

Note that an average value of $(d\varepsilon/d\alpha)|_{Mach, m=0} = 0.307$ is used for the calculation of $(d\bar{\varepsilon}/d\alpha)_i$. Next, the value of x_H, shown in Figure CS.14, is needed.

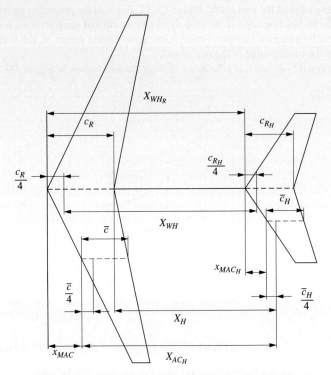

Figure CS.14 Wing-Horizontal Tail Geometric Distances

From Figure CS.14: $c_R + x_H = x_{MAC} + X_{AC_H}$.
Therefore: $x_H = x_{MAC} + X_{AC_H} - c_R = 0.767 + 25.33 - 8.2 = 17.896$.
The values for $(d\bar{\varepsilon}/d\alpha)_i$ and the final results are summarized here.

i	$x_i[ft]$	$\Delta x_i[ft]$	$\dfrac{x_i}{c_f}$	$w_i[ft]$	$\left(\dfrac{d\bar{\varepsilon}}{d\alpha}\right)_i$	$w_{bi}^2\left(\dfrac{d\bar{\varepsilon}}{d\alpha}\right)_i \Delta x_i$
1	19.5	4.3	2.38	3.6	1.07	59.629
2	15.1	4.3	1.84	5.3	1.08	129.991
3	10.8	4.3	1.32	5.4	1.13	141.595
4	6.5	4.3	0.79	5.4	1.24	156.098
5	2.2	4.3	0.27	5.4	2.25	282.229
6	6.1	4	0.74	5.4	0.24	27.560
7	2	4	0.24	5.4	0.08	9.036
8	1.7	3.4	0.21	11.1	0.07	27.585
9	5.1	3.4	0.62	10.6	0.20	75.468
10	8.6	3.4	1.05	6.5	0.33	47.853
11	12	3.4	1.46	2.5	0.46	9.877
12	15.4	3.4	1.88	0.7	0.60	0.994
13	18.9	3.5	2.30	0.4	0.73	0.410

$$\Delta x_{AC_B} = -\frac{1}{2.92 \cdot S\,\bar{c}} \sum_{i=1}^{i=13} w_{bi}^2 \left(\frac{d\bar{\varepsilon}}{d\alpha}\right)_i \Delta x_i = -0.193$$

Therefore, we have

$$\bar{x}_{AC_{WB}} = \bar{x}_{AC_W} + \Delta \bar{x}_{AC_B} = 0.24 - 0.193 = 0.047$$

Once a numerical value for $\bar{x}_{AC_{WB}}$ has been calculated, all the other parameters are known and summarized:

$$\bar{x}_{AC_H} = X_{AC_H}\bar{c} = \frac{25.33}{5.856} = 4.325, \quad \bar{x}_{CG} = 0.25, \quad \eta_H = 0.9, \quad \frac{S_H}{S} = 0.24$$

the estimates for the value of the important derivative (c_{m_α}) are shown:

$$c_{m_\alpha} = c_{L_{\alpha W}}\left(\bar{x}_{CG} - \bar{x}_{AC_{WB}}\right) - c_{L_{\alpha H}}\eta_H \frac{S_H}{S}\left(1 - \frac{d\varepsilon}{d\alpha}\right)\left(\bar{x}_{AC_H} - \bar{x}_{CG}\right) = A + B$$

Mach	0.15	0.20	0.25	0.30	0.35	0.40	0.45	0.50	0.55	0.60	0.65	0.70
A	0.874	0.881	0.889	0.899	0.912	0.927	0.946	0.968	0.995	1.027	1.066	1.114
B	−3.281	−3.294	−3.311	−3.331	−3.356	−3.385	−3.419	−3.459	−3.503	−3.554	−3.611	−3.674
c_{m_α}	−2.407	−2.414	−2.422	−2.432	−2.444	−2.458	−2.474	−2.491	−2.509	−2.527	−2.545	−2.560

The results for c_{m_α} are also shown in Figure CS.15.

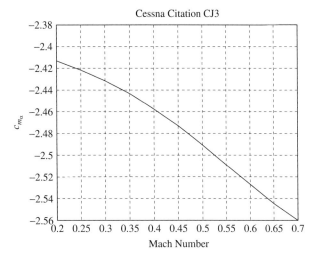

Figure CS.15 Estimates of c_{m_α} vs. Mach

2.3 Modeling of (c_{m_u})

c_{m_u} is negligible at the low subsonic conditions associated with incompressible aerodynamics; however, it takes on significant positive values as the airspeed increases. Further, its value becomes extremely critical in the transonic region. In general, it can be modeled using $c_{m_u} = -c_{L_1}\frac{\partial \bar{x}_{AC_W}}{\partial \text{Mach}}$.

Given the operational airspeeds of the Cessna CJ3 aircraft, as an approximation assume that for the Mach range of interest we have $c_{m_u} \approx 0$.

2.4 Modeling of $(c_{m_{\dot{\alpha}}}, c_{m_q})$

A closed form relationship for $c_{m_{\dot{\alpha}}}$ is given by

$$c_{m_{\dot{\alpha}}} = c_{m_{\dot{\alpha}W}} + c_{m_{\dot{\alpha}H}} \approx c_{m_{\dot{\alpha}H}} = -2c_{L_{\alpha H}}\eta_H \frac{S_H}{S}\left(\bar{x}_{AC_H} - \bar{x}_{CG}\right)^2 \frac{d\varepsilon}{d\alpha}$$

In the preceding relationship, all the parameters have been previously calculated. Therefore:

Mach	0.15	0.20	0.25	0.30	0.35	0.40	0.45	0.50	0.55	0.60	0.65	0.70
$c_{L_{\alpha_H}} \dfrac{d\varepsilon}{d\alpha}$	0.201	0.202	0.204	0.207	0.209	0.213	0.217	0.222	0.228	0.236	0.245	0.256
$c_{m_{\dot\alpha}}$	−6.72	−6.81	−6.92	−7.07	−7.25	−7.47	−7.73	−8.06	−8.46	−8.94	−9.54	−10.30

The results for $c_{m_{\dot\alpha}}$ are also shown in Figure CS.16.

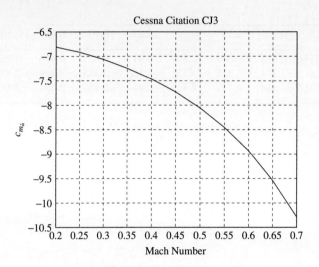

Figure CS.16 Estimates of $c_{m_{\dot\alpha}}$ vs. Mach

A closed-form relationship for c_{m_q} is given by

$$c_{m_q} \approx c_{m_{q_W}} + c_{m_{q_H}}$$

where

$$c_{m_{q_W}} = \left[\frac{\dfrac{AR^3 \tan^2 \Lambda_{c/4}}{AR \cdot B + 6 \cos \Lambda_{c/4}} + \dfrac{3}{B}}{\dfrac{AR^3 \tan^2 \Lambda_{c/4}}{AR + 6 \cos \Lambda_{c/4}} + 3} \right] \cdot c_{m_{q_W}} \bigg|_{Mach=0}$$

$$B = \sqrt{1 - Mach^2 (\cos \Lambda_{c/4})^2} = f(Mach) \text{ with } \Lambda_{0.25} = \Lambda_{\bar c/4} = 0.015 \text{ rad}$$

$$c_{m_{q_W}} \bigg|_{Mach=0} = -K_q c_{L_{\alpha_W}} \bigg|_{Mach=0} \cos \Lambda_{c/4} \cdot C \text{ with } K_q \text{ from Figure C.6}$$

$$C = \left\{ \frac{AR\left(0.5 |\bar x_{AC_W} - \bar x_{CG}| + 2|\bar x_{AC_W} - \bar x_{CG}|^2\right)}{AR + 2 \cos \Lambda_{c/4}} + \frac{1}{24} \left(\frac{AR^3 \tan \Lambda_{c/4}}{AR + 6 \cos \Lambda_{c/4}} \right) + \frac{1}{8} \right\}$$

$$c_{m_{q_H}} \approx -2 c_{L_{\alpha_H}} \eta_H \frac{S_H}{S} \left(\bar x_{AC_H} - \bar x_{CG} \right)^2$$

The preceding relationship for C contains all known and/or previously calculated values. Therefore, the direct numerical evaluation of the expressions leads to

$$C = \left\{ \frac{AR\left(0.5 |\bar x_{AC_W} - \bar x_{CG}| + 2|\bar x_{AC_W} - \bar x_{CG}|^2\right)}{AR + 2 \cos \Lambda_{c/4}} + \frac{1}{24} \left(\frac{AR^3 \tan \Lambda_{c/4}}{AR + 6 \cos \Lambda_{c/4}} \right) + \frac{1}{8} \right\} = 0.13$$

With $K_q = 0.891$ from Figure C.6, $c_{L_{\alpha_W}} \big|_{Mach=0} = 5.355$, $\cos (\Lambda_{\bar c/4}) = 0.999$, and $C = 0.13$

we have $c_{m_{qW}}|_{Mach=0} = -K_q c_{L_{\alpha W}}|_{Mach=0} \cos \Lambda_{c/4} \cdot C = -0.62$.

Since $B = \sqrt{1 - Mach^2 (\cos \Lambda_{c/4})^2} = f(Mach)$, we have

$$D = \left[\frac{\dfrac{AR^3 \tan^2 \Lambda_{c/4}}{AR \cdot B + 6 \cos \Lambda_{c/4}} + \dfrac{3}{B}}{\dfrac{AR^3 \tan^2 \Lambda_{c/4}}{AR + 6 \cos \Lambda_{c/4}} + 3} \right] = f(Mach)$$

On the other side, the contribution to c_{m_q} from the horizontal tail will only feature previously calculated parameters:

$$c_{m_{q_H}} \approx -2 c_{L_{\alpha_H}} \eta_H \frac{S_H}{S} \left(\bar{x}_{AC_H} - \bar{x}_{CG} \right)^2$$

Mach	0.15	0.20	0.25	0.30	0.35	0.40	0.45	0.50	0.55	0.60	0.65	0.70
$c_{L_{\alpha_H}}$	4.657	4.683	4.718	4.761	4.814	4.878	4.954	5.044	5.150	5.275	5.423	5.601
$c_{m_{q_H}}$	−33.47	−33.66	−33.91	−34.22	−34.60	−35.06	−35.61	−36.25	−37.01	−37.91	−38.98	−40.25
C							0.130					
K_q							0.891					
$c_{m_{qW}}|_{Mach=0}$							−0.619					
D	1.01	1.02	1.03	1.05	1.07	1.09	1.12	1.15	1.20	1.25	1.32	1.40
$c_{m_{qW}}$	−0.63	−0.63	−0.64	−0.65	−0.66	−0.68	−0.69	−0.71	−0.74	−0.77	−0.81	−0.87
c_{m_q}	−34.09	−34.29	−34.54	−34.87	−35.26	−35.73	−36.30	−36.97	−37.75	−38.68	−39.79	−41.12

Note that, despite a great effort in its modeling, it can be seen that its contribution of $c_{m_{qW}}$ to c_{m_q} is very marginal. The results for c_{m_q} are also shown in Figure CS.17.

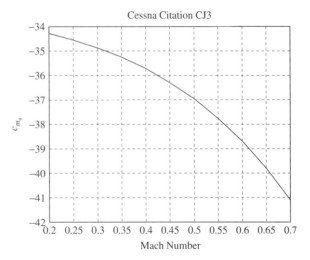

Figure CS.17 Estimates of c_{m_q} vs. Mach

2.5 Modeling of $(c_{m_{\delta_E}}, c_{m_{i_H}})$

Closed-form relationships for $c_{m_{\delta_E}}, c_{m_{i_H}}$ are given by

$$c_{m_{\delta_E}} = -c_{L_{\alpha_H}} \eta_H \frac{S_H}{S} \left(\bar{x}_{AC_H} - \bar{x}_{CG} \right) \tau_E, \quad c_{m_{i_H}} = -c_{L_{\alpha_H}} \eta_H \frac{S_H}{S} \left(\bar{x}_{AC_H} - \bar{x}_{CG} \right)$$

In the preceding relationship, all the parameters are known since they have been previously calculated. Recall that $\tau_E \approx 0.5$. Therefore, we have:

Mach	0.15	0.20	0.25	0.30	0.35	0.40	0.45	0.50	0.55	0.60	0.65	0.70
	4.657	4.683	4.718	4.761	4.814	4.878	4.954	5.044	5.150	5.275	5.423	5.601
$c_{m_{\delta_E}}$	−2.053	−2.065	−2.080	−2.099	−2.123	−2.151	−2.184	−2.224	−2.270	−2.326	−2.391	−2.469
$c_{m_{i_H}}$	−4.106	−4.129	−4.160	−4.198	−4.245	−4.301	−4.368	−4.448	−4.541	−4.651	−4.782	−4.938

The results for $c_{m_{\delta_E}}$, $c_{m_{i_H}}$ are also shown in Figures CS.18 and CS.19, respectively.

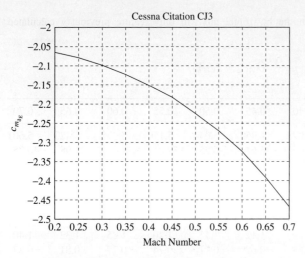

Figure CS.18 Estimates of $c_{m_{\delta_E}}$ vs. Mach

Figure CS.19 Estimates of $c_{m_{i_H}}$ vs. Mach

3 Modeling of the DRAG aerodynamic coefficients

This section provides the conceptual modeling of the following drag stability and control derivatives:

$$(c_{D_1}, c_{D_\alpha}, c_{D_u}, c_{D_{\dot\alpha}}, c_{D_q}, c_{D_{\delta_E}}, c_{D_{i_H}})$$

3.1 Modeling of (c_{D_1})

c_{D_1} is the aircraft dimensionless drag coefficient at steady-state conditions, relative to the condition:

$$c_{D_1} \bar{q} S = T \quad \rightarrow \quad c_{D_1} = \frac{T}{\bar{q} S}$$

It is considered a design parameter and is related to the thrust level at the given flight condition.

3.2 Modeling of (c_{D_α})

c_{D_α} is one of the aerodynamic coefficients for which a DATCOM analysis provides very approximated results. This is due to the difficulty in the modeling of the wing+body integration, in addition to the complexity in the modeling of the aerodynamic drag associated with the propulsion system and the drag associated with objects under the wing. In the special case when the aircraft drag polar curve can be approximated to be parabolic, the following approximate relationship can be derived:

$$c_{D_\alpha} \alpha = \frac{c_{L_1}^2}{\pi A R e}$$

leading to

$$c_{D_\alpha} \approx \frac{\partial \left(\frac{c_{L_1}^2}{\pi A R e} \right)}{\partial \alpha} = \left(\frac{2 c_{L_1}}{\pi A R e} \right) c_{L_\alpha}$$

Appendix B provides the following range of values at Mach = 0.7 for the similar size Learjet 24 aircraft:

$$c_{D_\alpha} \in [0.22 - 0.3]$$

3.3 Modeling of (c_{D_u})

This derivative is negligible at the low subsonic conditions associated with incompressible aerodynamics; however, it takes on significant positive values as the airspeed approaches transonic conditions. In general, it can be modeled using

$$c_{D_u} = Mach \cdot \frac{\partial c_D}{\partial Mach}$$

An empirical modeling of this derivative using the build-up DATCOM-based approach is very unpractical. A more suitable approach is to derive the slope $\partial c_D / \partial Mach$ once experimental drag results are available from wind tunnel analysis. Appendix B provides the following value at Mach = 0.7 for the similar size Learjet 24 aircraft:

$$c_{D_u} = 0.104$$

3.4 Modeling of ($c_{D_{\dot{\alpha}}}, c_{D_q}$)

At subsonic conditions, it is customary to assume that $c_{D_{\dot{\alpha}}} \approx 0$, $c_{D_q} \approx 0$.

3.5 Modeling of ($c_{D_{\delta_E}}, c_{D_{i_H}}$)

For most aircraft, because the horizontal tail does not provide a substantial wetted surface, it is customary to approximate $c_{D_{\delta_E}} = c_{D_{i_H}} \approx 0$. Appendix B provides the following values at Mach = 0.7 for the similar size Learjet 24 aircraft:

$$c_{D_{\delta_E}} = c_{D_{i_H}} = 0$$

SHORT PROBLEMS

Short Problem 3.1

Identify which aircraft geometric parameters affect the values of the longitudinal stability derivatives $c_{L_\alpha}, c_{m_\alpha}$. Next, explain the effect of an increase of each of the geometric parameters on each of the above stability derivatives.

Short Problem 3.2

The modeling of the aerodynamic longitudinal forces and moments has been introduced for a conventional subsonic aircraft with a wing and a horizontal tail. Provide closed-form relationships for $c_{L_\alpha}, c_{m_\alpha}$ if the aircraft features a pair of canards with a fixed surface, along with a portion of the surface that can be deflected by the pilot.

Short Problem 3.3

Consider an aircraft with a horizontal featuring both stabilators and elevators. Identify which aircraft geometric parameters affect the values of the longitudinal stability derivatives $c_{L_{\delta E}}, c_{m_{\delta E}}, c_{L_{iH}}, c_{m_{iH}}$. Next, explain the effect of an increase of each of the geometric parameters on each of the preceding stability derivatives.

Short Problem 3.4

The modeling of the aerodynamic longitudinal forces and moments has been introduced for a conventional subsonic aircraft with a wing and a horizontal tail. Assume that the aircraft features a pair of canards with a fixed surface, along with a portion of the surface that can be deflected by the pilot by an angle δ_C (positive for down deflections). Provide closed-form relationships for $c_{L_{\delta C}}, c_{m_{\delta C}}$.

Short Problem 3.5

Identify which aircraft geometric parameters affect the values of the longitudinal stability derivatives $c_{L_{\dot{\alpha}}}, c_{m_{\dot{\alpha}}}, c_{L_q}, c_{m_q}$. Next, explain the effect of an increase of each of the geometric parameters on each of the preceding stability derivatives.

Short Problem 3.6

Consider the General Dynamics F-111 aircraft shown in Appendix C. This aircraft was one of the first aircraft designed with variable wing sweep angle. Explain which longitudinal stability derivatives are affected by an increase in the wing sweep angle and provide a brief description of the expected trends.

Short Problem 3.7

The Cessna T37 and the SIAI Marchetti S211, shown in Appendix C, are both basic military jet trainers. Appendix B provides sets of values of the longitudinal stability derivatives for both aircraft at different flight conditions. Consider the cruise flight condition for the Cessna T37 and the high cruise flight condition for the SIAI Marchetti S211. Based on the actual values of the stability derivatives $c_{L_\alpha}, c_{m_\alpha}$ for both aircraft, provide qualitative comments on the origin of the differences in the values due to differences in the geometric characteristics of the two aircraft.

Short Problem 3.8

Repeat Short Problem 3.7 for the following longitudinal control derivatives:

$$c_{L_{\delta_E}}, c_{m_{\delta_E}}$$

Note that the Cessna T-37 aircraft only features elevators and does not feature stabilators.

Short Problem 3.9

Repeat Short Problem 3.7 for the following longitudinal stability derivatives:

$$c_{L_{\dot\alpha}}, c_{m_{\dot\alpha}}, c_{L_q}, c_{m_q}$$

PROBLEMS

Problem 3.1

Consider the data relative to the Aeritalia G91 aircraft in Appendix C. Use the provided drawings to extract all the relevant geometric characteristics of the aircraft. Assume that this aircraft features both stabilators and elevators for the control of the longitudinal dynamics. Also assume $\tau_E = 0.38$, $\eta_H = 0.88$. Using the modeling outlined in this chapter, find a numerical value for the following aerodynamic parameters:

- $\dfrac{d\varepsilon}{d\alpha}, \bar{x}_{AC_W}, \Delta\bar{x}_{AC_{WB}}$;
- $c_{L_\alpha}, c_{L_{\delta_E}}, c_{L_{i_H}}$;
- $\bar{x}_{AC}, c_{m_\alpha}, c_{m_{\delta_E}}, c_{m_{i_H}}$.

Problem 3.2

Consider again the data relative to the Aeritalia G91 aircraft in Appendix C. Use the provided drawings to extract all the relevant geometric characteristics of the aircraft. Assume $\eta_H = 0.88$. Using the modeling outlined in this chapter, find a numerical value for the following aerodynamic parameters:

- $c_{L_{\dot\alpha}}, c_{L_q}$;
- $c_{m_{\dot\alpha}}, c_{m_q}$.

Problem 3.3

Consider the data relative to the Beech 99 aircraft in Appendixes B and C. Use the provided drawings to extract all the relevant geometric characteristics of the aircraft. This aircraft features stabilators for trimming and elevators for maneuver purposes. Assume $\tau_E = 0.45$, $\eta_H = 0.85$. Using the modeling outlined in this chapter, find a numerical value for the following aerodynamic parameters:

- $\dfrac{d\varepsilon}{d\alpha}, \bar{x}_{AC_W}, \Delta\bar{x}_{AC_B}$;
- $c_{L_\alpha}, c_{L_{\delta_E}}, c_{L_{i_H}}$;
- $\bar{x}_{AC}, c_{m_\alpha}, c_{m_{\delta_E}}, c_{m_{i_H}}$.

Next, compare the obtained values with the true values listed in Appendix B (Aircraft 3) and evaluate the error associated with the use of the empirical modeling approach.

Problem 3.4

Consider again the data relative to the Beech 99 aircraft in Appendixes B and C. Use the provided drawings to extract all the relevant geometric characteristics of the aircraft. Assume $\eta_H = 0.85$. Using the modeling outlined in this chapter, find a numerical value for the following aerodynamic parameters:

- $c_{L_{\dot\alpha}}, c_{L_q}$;
- $c_{m_{\dot\alpha}}, c_{m_q}$.

Next, compare the obtained values with the true values listed in Appendix B (Aircraft 3) and evaluate the error associated with the use of the empirical modeling approach.

Problem 3.5

Consider the data relative to the Cessna T37 aircraft in Appendixes B and C. Use the provided drawings to extract all the relevant geometric characteristics of the aircraft. This aircraft features elevators for both trimming and maneuver purposes. Assume $\tau_E = 0.43$, $\eta_H = 0.9$. Using the modeling outlined in this chapter, find a numerical value for the following aerodynamic parameters:

- $\dfrac{d\varepsilon}{d\alpha}, \bar{x}_{AC_W}, \Delta\bar{x}_{AC_{WB}}$;
- $c_{L_\alpha}, c_{L_{\delta_E}}, c_{L_{i_H}}$;
- $\bar{x}_{AC}, c_{m_\alpha}, c_{m_{\delta_E}}, c_{m_{i_H}}$.

Next, compare the obtained values with the true values listed in Appendix B (Aircraft 4) and evaluate the error associated with the use of the empirical modeling approach.

Problem 3.6

Consider again the data relative to the Cessna T37 aircraft in Appendixes B and C. Use the provided drawings to extract all the relevant geometric characteristics of the aircraft. This aircraft features elevators for both trimming and maneuver purposes. Assume $\eta_H = 0.9$. Using the modeling outlined in this chapter, find a numerical value for the following aerodynamic parameters:

- $c_{L_{\dot\alpha}}, c_{L_q}$;
- $c_{m_{\dot\alpha}}, c_{m_q}$.

Next, compare the obtained values with the true values listed in Appendix B (Aircraft 4) and evaluate the error associated with the use of the empirical modeling approach.

Problem 3.7

Consider the data relative to the McDonnell Douglas DC-9 Series 10 aircraft in Appendix C. See Figure P.3.7.1. Use the provided drawings to extract all the relevant geometric characteristics of the aircraft. This aircraft features stabilators for trimming and elevators for maneuver purposes. Assume $\tau_E = 0.5$. Consider as baseline the following numerical estimates of the longitudinal stability and control derivatives for the DC 9 Series 10 aircraft (see Student Sample Problem 1):

$$c_{L_\alpha}, c_{L_{\delta_E}}, c_{L_{i_H}}, c_{m_\alpha}, c_{m_{\delta_E}}, c_{m_{i_H}}$$

Next, consider the later versions of the DC-9 aircraft which were introduced later in its operational life.

130 Chapter 3 Modeling of Longitudinal Aerodynamic Forces and Moments

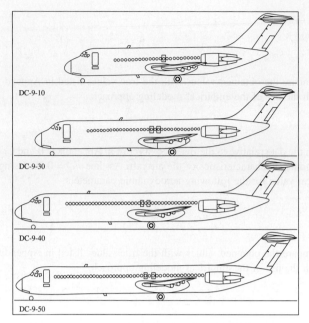

Figure P3.7.1   Lateral Views McDonnell Douglas DC 9 Series 10/30/40/50
(Source: http://www.aviastar.org/index2.html)

Ref.[9] provides detailed information about the differences in the geometric parameters. The geometric parameters for the DC 9 Series 10/30/40/50 aircraft are summarized in Tables P.3.7.1–P.3.7.3 and Figure P3.7.2.

Wing Geometric Parameters (as shown in SSP1 in Chapter 2)

Table P3.7.1   Wing Geometric Parameters

	DC9-10	DC9-30	DC9-40	DC9-50
b [ft]	89.4	93.3	93.3	93.32
c_T [ft]	3.9	3.9	3.9	3.9
c_R [ft]	15.6	16.1	16.1	16.1
Λ_{LE} [rad]		0.489		
Λ_{LE} [deg]		28		
λ	0.250	0.242	0.242	0.242
S [ft^2]	928.00	996.00	996.00	996.00
AR	8.612	8.740	8.740	8.744
$\bar{c}$ [ft]	10.92	11.240	11.240	11.240
x_{MAC} [ft]	9.51	9.880	9.880	9.882

Horizontal Tail Geometric Parameters (as shown in SSP1 in Chapter 2)

Table P3.7.2   Horizontal Tail Geometric Parameters

	DC9-10	DC9-30	DC9-40	DC9-50
b_H [ft]	36.900	36.800	36.800	36.850
c_{T_H} [ft]	4.000	4.000	4.000	4.000
c_{R_H} [ft]	10.900	10.900	10.900	10.900
Λ_{LE_H} [rad]		0.611		
Λ_{LE_H} [deg]		35		
λ_H	0.367	0.367	0.367	0.367
S_H [ft^2]	274.91	274.16	274.16	274.53
AR_H	4.95	4.94	4.94	4.95
$\bar{c}_H$ [ft]	7.98	7.98	7.98	7.98
x_{MAC_H} [ft]	5.46	5.45	5.45	5.45

Wing-Tail Geometric Parameters (as shown in SSP1 in Chapter 2)

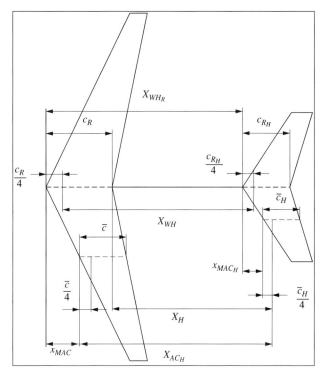

Figure P3.7.2 Wing-Horizontal Tail Geometric Distances

Table P3.7.3 Wing Geometric Parameters

	DC9-10	DC9-30	DC9-40	DC9-50
X_{WH_R} [ft]	49.500	54.900	58.300	61.500
Z_{WH} [ft]	20	20	20	20
X_{WH} [ft]	48.33	53.60	57.00	60.20
r	1.081	1.149	1.222	1.290
m	0.447	0.429	0.429	0.429
X_{AC_H} [ft]	47.451	52.463	55.863	59.068
$\bar{x}_{AC_H}$	4.345	4.667	4.970	5.255
x_H [ft]	41.358	46.243	49.643	52.851

Using the previous geometric information derived from Ref.,[9] provide estimates for the following parameters for the DC-9 30, DC-9 40, and DC-9 50:

$$c_{L_\alpha}, c_{L_{\delta_E}}, c_{L_{i_H}}, c_{m_\alpha}, c_{m_{\delta_E}}, c_{m_{i_H}}$$

for the McDonnell Douglas DC 9 Series 30, 40, and 50. Also estimate the location of the aircraft aerodynamic center $\bar{x}_{AC}$ for all the different versions of the aircraft.

Problem 3.8

Consider again the data relative to the McDonnell Douglas DC-9 Series 10 aircraft in Appendix C. Use the provided drawings to extract all the relevant geometric characteristics of the aircraft. This aircraft features stabilators for trimming and elevators for maneuver purposes. Assume $\tau_E = 0.5$. Consider as baseline the following numerical estimates of the longitudinal stability and control derivatives for the DC 9 Series 10 aircraft (see Student Sample Problem 1):

$$c_{L_\alpha}, c_{L_{\delta_E}}, c_{L_{i_H}}, c_{m_\alpha}, c_{m_{\delta_E}}, c_{m_{i_H}}$$

132 Chapter 3 Modeling of Longitudinal Aerodynamic Forces and Moments

Next, consider the versions of the DC-9 aircraft which were introduced later in its operational life. See Figure P.3.8.1.

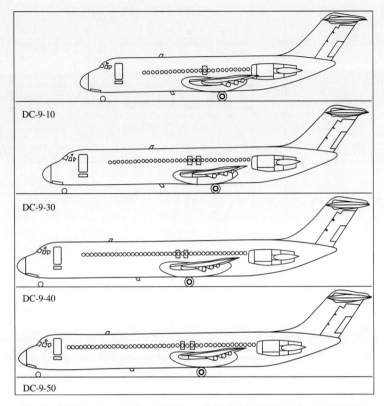

Figure P3.8.1 Lateral Views McDonnell Douglas DC 9 Series 10/30/40/50
(Source: http://www.aviastar.org/index2.html)

Ref.[9] provides detailed information about the differences in the geometric parameters. The geometric parameters for the DC 9 Series 10/30/40/50 aircraft are summarized in Tables P.3.8.1–P.3.8.3 and Figure P.3.8.2.

Wing Geometric Parameters (as shown in SSP1 in Chapter 2)

Table P3.8.1 Wing Geometric Parameters

	DC9-10	DC9-30	DC9-40	DC9-50
b [ft]	89.4	93.3	93.3	93.32
c_T [ft]	3.9	3.9	3.9	3.9
c_R [ft]	15.6	16.1	16.1	16.1
Λ_{LE} [rad]			0.489	
Λ_{LE} [deg]			28	
λ	0.250	0.242	0.242	0.242
S [ft^2]	928.00	996.00	996.00	996.00
AR	8.612	8.740	8.740	8.744
$\bar{c}$ [ft]	10.92	11.240	11.240	11.240
x_{MAC} [ft]	9.51	9.880	9.880	9.882

Horizontal Tail Geometric Parameters (as shown in SSP1 in Chapter 2)

Table P3.8.2 Horizontal Tail Geometric Parameters

	DC9-10	DC9-30	DC9-40	DC9-50
b_H [ft]	36.900	36.800	36.800	36.850
c_{T_H} [ft]	4.000	4.000	4.000	4.000
c_{R_H} [ft]	10.900	10.900	10.900	10.900
Λ_{LE_H} [rad]		0.611		
Λ_{LE_H} [deg]		35		
λ_H	0.367	0.367	0.367	0.367
S_H [ft^2]	274.91	274.16	274.16	274.53
AR_H	4.95	4.94	4.94	4.95
$\bar{c}_H$ [ft]	7.98	7.98	7.98	7.98
x_{MAC_H} [ft]	5.46	5.45	5.45	5.45

Wing-Tail Geometric Parameters (as shown in SSP1 in Chapter 2)

Table P3.8.3 Wing Geometric Parameters

	DC9-10	DC9-30	DC9-40	DC9-50
X_{WH_R} [ft]	49.500	54.900	58.300	61.500
Z_{WH} [ft]	20	20	20	20
X_{WH} [ft]	48.33	53.60	57.00	60.20
r	1.081	1.149	1.222	1.290
m	0.447	0.429	0.429	0.429
X_{AC_H} [ft]	47.451	52.463	55.863	59.068
$\bar{x}_{AC_H}$	4.345	4.667	4.970	5.255
x_H [ft]	41.358	46.243	49.643	52.851

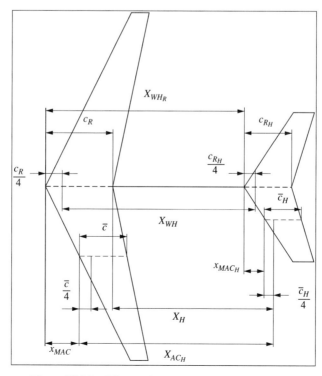

Figure P3.8.2 Wing-Horizontal Tail Geometric Distances

Using the preceding geometric information derived from Reference,[9] provide estimates for the following parameters for the DC-9 30, DC-9 40, and DC-9 50:

$$c_{L_{\dot\alpha}}, c_{L_q}, c_{m_{\dot\alpha}}, c_{m_q}$$

Problem 3.9

Consider the data relative to the Boeing B747 200 aircraft in Appendix B at high cruise conditions (Mach = 0.9) and Appendix C. Use the provided drawings to extract all the relevant geometric characteristics of the aircraft. Assume that this aircraft features stabilators for trimming and elevators for maneuver purposes. Assume $\tau_E = 0.47$, $\eta_H = 0.85$. Using the modeling outlined in this chapter, find a numerical value for the following aerodynamic parameters:

- $\dfrac{d\varepsilon}{d\alpha}, \bar{x}_{AC_W}, \Delta\bar{x}_{AC_B}$;
- $c_{L_\alpha}, c_{L_{\delta_E}}, c_{L_{i_H}}$;
- $\bar{x}_{AC}, c_{m_\alpha}, c_{m_{\delta_E}}, c_{m_{i_H}}$.

Next, compare the obtained values with the true values listed in Appendix B (Aircraft 7) and evaluate the error associated with the use of the empirical modeling approach.

Problem 3.10

Consider again the data relative to the Boeing B747 200 aircraft in Appendix B at high cruise conditions (Mach = 0.9) and Appendix C. Use the provided drawings to extract all the relevant geometric characteristics of the aircraft. Assume $\eta_H = 0.85$. Using the modeling outlined in this chapter, find a numerical value for the following aerodynamic parameters:

- $c_{L_{\dot\alpha}}, c_{L_q}$;
- $c_{m_{\dot\alpha}}, c_{m_q}$.

Next, compare the obtained values with the true values listed in Appendix B (Aircraft 7) and evaluate the error associated with the use of the empirical modeling approach.

Chapter 4

Modeling of Lateral Directional Aerodynamic Forces and Moments

TABLE OF CONTENTS

4.1 Introduction
4.2 Modeling of $F_{A_{Y_1}}$
    4.2.1 Conceptual Modeling of c_{Y_β}
    4.2.2 Mathematical Modeling of c_{Y_β}
    4.2.3 Modeling of $c_{Y_{\delta_A}}$
    4.2.4 Modeling of $c_{Y_{\delta_R}}$
4.3 Modeling of L_{A_1}
    4.3.1 Conceptual Modeling of $c_{l\beta}$
    4.3.2 Mathematical Modeling of $c_{l\beta}$
    4.3.3 Modeling of $c_{l_{\delta_A}}$
    4.3.4 Modeling of $c_{l_{\delta_R}}$
4.4 Modeling of N_{A_1}
    4.4.1 Conceptual Modeling of $c_{n\beta}$
    4.4.2 Mathematical Modeling of $c_{n\beta}$
    4.4.3 Modeling of $c_{n_{\delta_A}}$
    4.4.4 Modeling of $c_{n_{\delta_R}}$
4.5 Summary of the Lateral Directional Steady-State Force and Moments
4.6 Modeling of the Small Perturbation Lateral Directional Aerodynamic Force and Moments
    4.6.1 Modeling of $c_{Y_\beta}, c_{l_\beta}, c_{n_\beta}$
    4.6.2 Modeling of c_{Y_p}
    4.6.3 Modeling of c_{l_p}
    4.6.4 Modeling of c_{n_p}
    4.6.5 Modeling of c_{Y_r}
    4.6.6 Modeling of c_{l_r}
    4.6.7 Modeling of c_{n_r}
4.7 Summary of Longitudinal and Lateral Directional Aerodynamic Stability and Control Derivatives
4.8 Final Overview and Ranking of the Importance of the Aerodynamic Coefficients
4.9 Summary of the Modeling of the Longitudinal and Lateral-Directional Aerodynamic Forces and Moments
    References
    Student Sample Problems
    Case Study
    Short Problems
    Problems

4.1 INTRODUCTION

The goal of this chapter is to provide tools for modeling the aerodynamic lateral force and moments at steady-state and perturbed conditions. Toward this goal, it is assumed that the aircraft features conventional control surfaces, which include ailerons (δ_A) for lateral control and rudder (δ_R) for directional control. The steady-state rolling and yawing moments and the lateral force are shown in Figure 4.1.

Chapter 4 Modeling of Lateral Directional Aerodynamic Forces and Moments

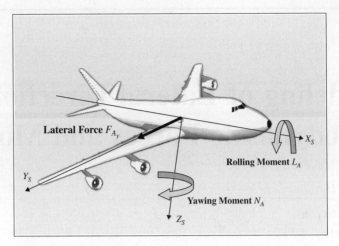

Figure 4.1 Steady-State Lateral Directional Force and Moments

A modern aircraft typically features a number of additional control surfaces, which can be deployed either symmetrically or asymmetrically. A list of available control surfaces is shown in Figure 4.2. The only aerodynamic control surfaces not shown in Figure 4.2 are the canards; canards are essentially small-size mini-wing surfaces located in front of the wing and can be deployed symmetrically or asymmetrically. Only a few commercial aircraft feature this control surface; canards are used on military experimental aircraft and a few production fighter aircraft.

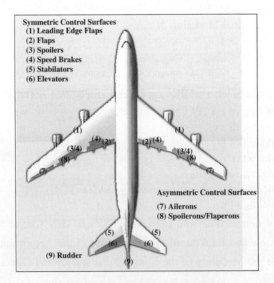

Figure 4.2 Aircraft Control Surfaces

The same role played by the angle of attack α for the modeling of the longitudinal forces and moment is played by the lateral angle of attack β, known as the sideslip angle. This angle measures essentially the deviation of the airspeed V_∞ with respect to the longitudinal stability axes; β is considered positive when the lateral component of the airspeed is in the opposite direction of the lateral stability axis, as shown in Figure 4.3.

Following the introduction of β, the sidewash angle σ is the lateral counterpart of the longitudinal downwash angle ε. As the downwash angle, the sidewash angle σ is essentially a penalty in the effective sideslip angle reaching the aircraft horizontal while vertical tail; therefore, we have

$$\beta_{(H,V)} = (\beta - \sigma)$$

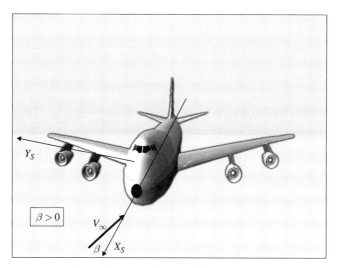

Figure 4.3 Aerodynamic Sideslip Angle β

The modeling of the sidewash angle σ is identical to the modeling for the downwash angle ε. Using a first order Taylor expansion for $\sigma = \sigma(\beta)$ we have

$$\sigma(\beta) = \sigma\big|_{\beta=0°} + \left(\frac{d\sigma}{d\beta}\right)\beta$$

The sidewash effect is negligible at zero sideslip angles $\left(\sigma\big|_{\beta=0°} = \sigma_0 = 0\right)$ leading to

$$\beta_{(H,V)} = (\beta - \sigma) = \beta - \left(\frac{d\sigma}{d\beta}\right)\beta = \left(1 - \frac{d\sigma}{d\beta}\right)\beta$$

In general, the sidewash effect is substantially smaller than its longitudinal counterpart described in Chapter 2, Section 2.4.

This chapter starts with the modeling of the steady-state lateral force followed by the steady-state rolling and yawing moments. Next, the aerodynamic modeling associated with the small perturbation lateral force, rolling, and yawing moments will be described. Additional sections will provide a ranking of all the aircraft aerodynamic coefficients, along with a summary for the modeling of all the aerodynamic forces and moments acting on the aircraft.

4.2 MODELING OF $F_{A_{Y_1}}$

The modeling of the steady-state lateral force starts from the basic relationship:

$$F_{A_{Y_1}} = F_{A_Y} = c_{Y_1}\,\bar{q}S$$

The functionality of the lateral force coefficient c_{Y_1} is given by

$$c_{Y_1} = f(\beta, \delta_A, \delta_R)$$

The introduction of a first order Taylor expansion for c_{Y_1} leads to

$$c_{Y_1} = c_{Y_0} + c_{Y_\beta}\beta + c_{Y_{\delta_A}}\delta_A + c_{Y_{\delta_R}}\delta_R$$

Due to the aircraft symmetry with respect to the XZ plane we have

$$c_{Y_0} = c_Y\big|_{\beta=\delta_A=\delta_R=0°} = 0$$

The next sections will outline the modeling of the aerodynamic coefficients $c_{Y_\beta}, c_{Y_{\delta_A}}, c_{Y_{\delta_R}}$. Because of the aerodynamic complexity, the conceptual and the mathematical modeling for c_{Y_β} are discussed separately.

4.2.1 Conceptual Modeling of c_{Y_β}

Considering the individual contributions from the different components of the aircraft, a basic relationship for c_{Y_β} is given by

$$c_{Y\beta} = c_{Y\beta_{WB}} + c_{Y\beta_H} + c_{Y\beta_V}$$

The lateral velocity field acting on the aircraft—from which the coefficient $c_{Y\beta_{WB}}$ originates—is shown in Figure 4.4.

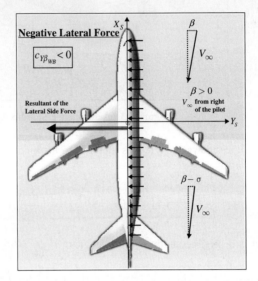

Figure 4.4 Wing Body Lateral Force Associated with β

The figure shows that the lateral force acts in the opposite direction of the lateral axis. The magnitude of this force depends on the geometric dihedral angle of the wing (shown in Figure 4.5) and the shape of the fuselage cross sections.

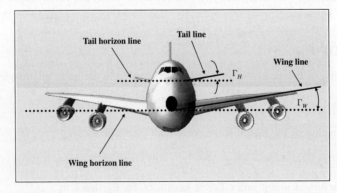

Figure 4.5 Geometric Dihedral Angles for Wing and Horizontal Tail

From the sign convention in Figure 4.4, it can be seen that

$$c_{Y\beta_{WB}} < 0$$

It is intuitive that the contribution to the lateral force associated with the sideslip angle acting on the horizontal tail is, in general, negligible. Therefore

$$c_{Y\beta_H} \approx 0$$

A special case is represented by the McDonnell Douglas F-4 aircraft. In fact, due to the substantial anhedral (or negative dihedral) angle for its horizontal tail, a non-negligible lateral force will be generated by the lateral flow on the horizontal tail, as shown in Figure 4.6.

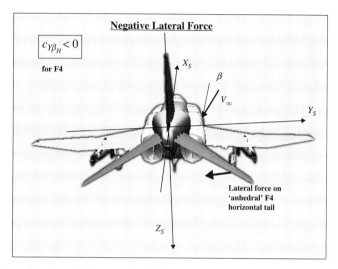

Figure 4.6 Lateral Force on the Horizontal Tail of the F-4 Associated with β

Therefore, for the F-4 we will have

$$c_{Y\beta_H} < 0$$

A substantial contribution to $c_{Y\beta}$ is instead provided by the vertical tail. The lateral velocity acting on the vertical tail, leading to the coefficient $c_{Y\beta_V}$, is shown in Figure 4.7. From the sign convention shown in Figure 4.7, it can be seen that

$$c_{Y\beta_V} < 0$$

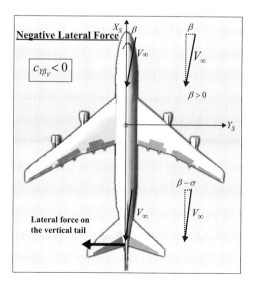

Figure 4.7 Lateral Force on the Vertical tail Associated with β

4.2.2 Mathematical Modeling of c_{Y_β}

Starting from the previous relationship:

$$c_{Y_\beta} = c_{Y_{\beta_{WB}}} + c_{Y_{\beta_H}} + c_{Y_{\beta_V}}$$

For a detailed modeling of c_{Y_β}, it is useful to separate the contributions from the wing and the fuselage. Therefore, we have

$$c_{Y_\beta} = \left(c_{Y_{\beta_W}} + c_{Y_{\beta_B}}\right) + c_{Y_{\beta_H}} + c_{Y_{\beta_V}}$$

The coefficient $c_{Y_{\beta_W}}$ is significant only when the wing has a positive dihedral angle Γ_W.
A relationship for $c_{Y_{\beta_W}}$ is given by

$$c_{Y_{\beta_W}} \approx -0.0001 \, |\Gamma_W| \cdot 57.3 \ (\text{rad}^{-1})$$

A relationship for $c_{Y_{\beta_B}}$ is instead given by

$$c_{Y_{\beta_W}} \approx -2 \cdot K_{\text{int}} \cdot \frac{S_{P \to V}}{S} \ (\text{rad}^{-1})$$

The coefficient K_{int} is an interference factor associated with the wing-fuselage interface (high-wing or low-wing). The value for K_{int} can be extracted from Figure 4.8.

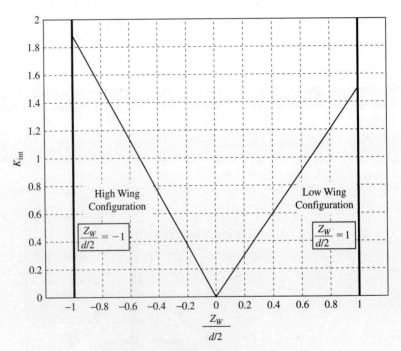

Figure 4.8 Wing-Body Interference Factor[5]

The geometric parameters Z_W and d are shown in Figures 4.9 and 4.10. Specifically, d is the maximum fuselage height at the wing-body intersection, while Z_W is the vertical distance from the fuselage centerline to 25 percent of the wing root.

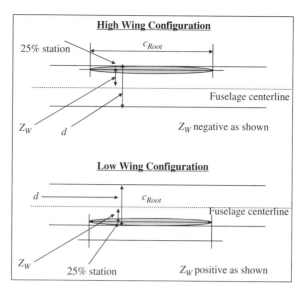

Figure 4.9 Geometric Parameters for Wing-Fuselage Integration (side view)

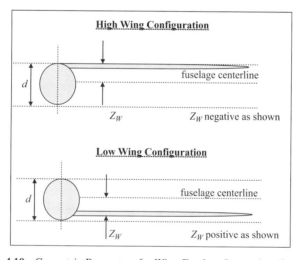

Figure 4.10 Geometric Parameters for Wing-Fuselage Integration (front view)

$S_{P \to V}$ is the cross section at the location of the fuselage—called X_0—where the flow ceases to be potential. It turns out[5] that the value of X_0 is a linear function of the specific location of the fuselage—called X_1—where the gradient $\partial S(x)/\partial S$ reaches its maximum negative value. The locations of X_0 and X_1 are shown in Figure 4.11.

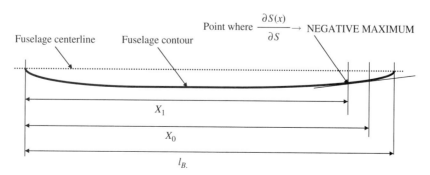

Figure 4.11 Contour of the Fuselage along the X-Axis[5]

The determination of $S_{P \to V}$ starts with the evaluation of $\partial S(x)/\partial S|_{\text{MAX.NEG}}$ and its coordinate X_1. Once X_1 has been identified, the station X_0 is found using Figure 4.12.

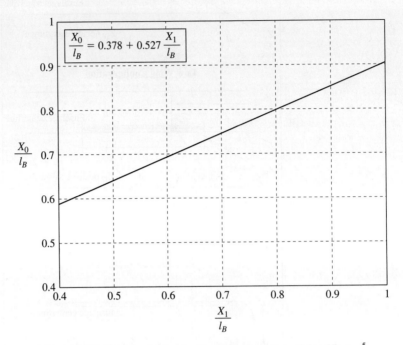

Figure 4.12 Fuselage Station for Transition from Potential to Viscous[5]

The contribution from the horizontal tail $c_{Y\beta_H}$ can be considered negligible for most aircraft. As mentioned previously, a special case is represented by the McDonnell Douglas F-4 aircraft, whose horizontal tail is characterized by a fairly large negative geometric dihedral angle Γ_H, also known as anhedral angle. In this case, a relationship for $c_{Y\beta_H}$ is given by

$$c_{Y\beta_H} \approx -0.0001 \, |\Gamma_H| \cdot 57.3 \cdot \eta_H \cdot \left(1 + \frac{d\sigma}{d\beta}\right) \frac{S_H}{S} \quad (\text{rad}^{-1})$$

An empirical relationship for $\eta_H \cdot \left(1 + \frac{d\sigma}{d\beta}\right)$ is given by[5]

$$\eta_H \cdot \left(1 + \frac{d\sigma}{d\beta}\right) = 0.724 + 3.06 \frac{S_H/S}{1 + \cos(\Lambda_{c/4})} + 0.4 \frac{Z_W}{d} + 0.009 \cdot AR$$

Finally, the contribution from the vertical tail $c_{Y\beta_V}$ is modeled through the relationship

$$c_{Y\beta_V} = -k_{Y_V} \cdot \left|c_{L\alpha \cdot V}\right| \eta_V \cdot \left(1 + \frac{d\sigma}{d\beta}\right) \frac{S_V}{S} \quad (\text{rad}^{-1})$$

where k_{Y_V} is an empirical factor shown in Figure 4.13, which is function of the geometric parameters (b_V, r_1) shown in Figure 4.14.

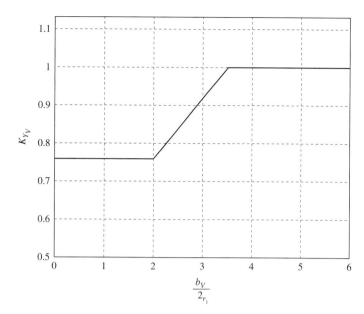

Figure 4.13 Empirical Factor for the Lateral Force at the Vertical Tail due to β[5]

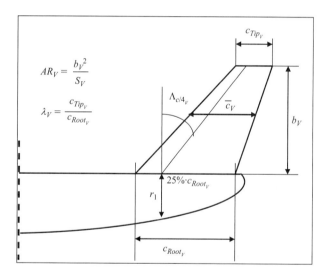

Figure 4.14 Geometric Parameters of the Vertical Tail

For the vertical tail we have

$$\eta_V \cdot \left(1 + \frac{d\sigma}{d\beta}\right) = 0.724 + 3.06 \frac{S_V/S}{1 + \cos\left(\Lambda_{c/4}\right)} + 0.4\frac{Z_W}{d} + 0.009 \cdot AR$$

$|c_{L_{\alpha V}}|$ is the lift-curve slope coefficient for the vertical tail; the numerical value for $|c_{L_{\alpha V}}|$ can be found using the Polhamus formula in Chapter 2 with the geometric parameters of the vertical tail shown in Figure 4.14. However, to account for the aerodynamic interference due to the presence of the horizontal tail, the application of the Polhamus formula requires the effective aspect ratio of the vertical tail, defined by

$$AR_{V_{eff}} = c_1 \cdot AR_V \cdot [1 + K_{HV}(c_2 - 1)]$$

where c_1 is an aspect ratio factor coefficient associated with the interference of the fuselage (shown in Figure 4.15); c_2 is an aspect ratio factor coefficient associated with the interference of the horizontal tail (shown in Figure 4.16, with the relative geometric parameters shown in Figure 4.17); finally, K_{HV} is a factor accounting for the relative size of the horizontal and vertical tails (shown in Figure 4.18).

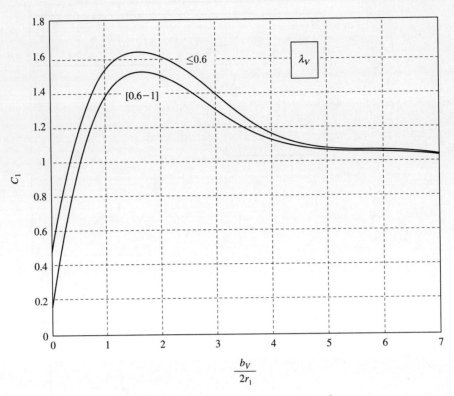

Figure 4.15 c_1 for the Evaluation of $AR_{V_{eff}}^5$

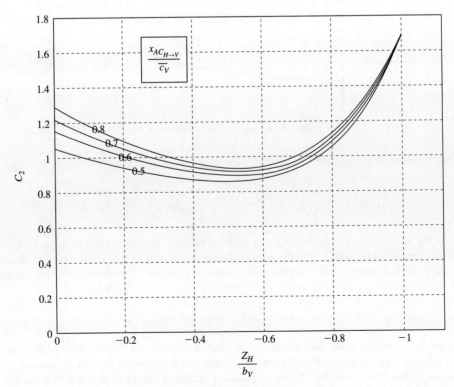

Figure 4.16 c_2 for the Evaluation of $AR_{V_{eff}}^5$

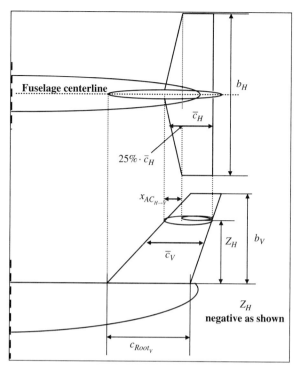

Figure 4.17 Geometric Parameters of the Interface between the Horizontal and Vertical Tail

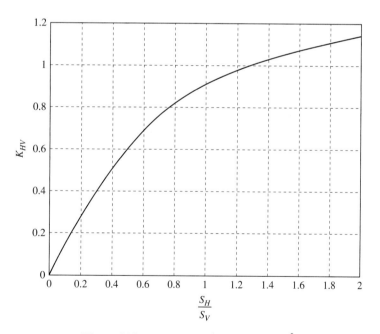

Figure 4.18 Factor for S_H/S_V Relative Size[5]

A special case is represented by the use of twin vertical tails. In this case, considering two identical vertical panels each with surface S_V, as shown in Figure 4.19, a relationship for $c_{Y\beta_V}$ is given by

$$c_{Y\beta_V} = -2 \cdot c_{VT} \cdot c_{Y\beta_{V_{eff}}} \cdot \frac{S_V}{S}$$

where $c_{Y\beta_{V_{eff}}}$ is found from Figure 4.20 using $AR_{V_{eff}}$ from Figure 4.21 and c_{VT} is a (wing + body)-horizontal tail interference coefficient found from Figure 4.22.

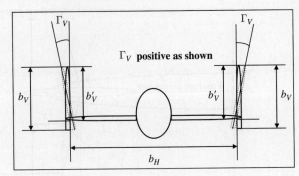

Figure 4.19 Geometric Parameters of the Twin Vertical Tails[5]

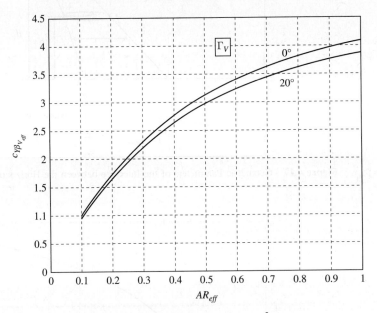

Figure 4.20 Evaluation of $c_{Y\beta_{V_{eff}}}$[5]

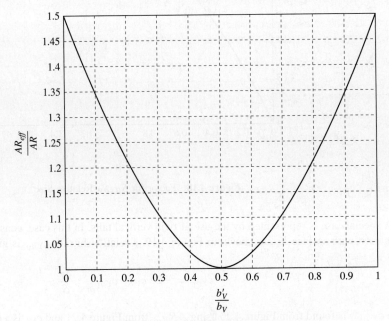

Figure 4.21 Evaluation of $AR_{V_{eff}}$[5]

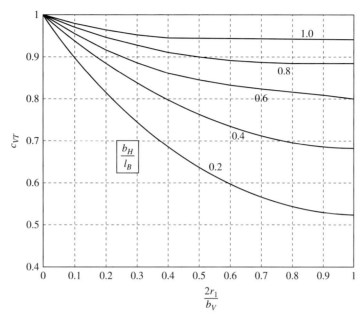

Figure 4.22 Evaluation of c_{VT}[5]

4.2.3 Modeling of $c_{Y\delta_A}$

Figure 4.23 shows the aerodynamic forces associated with a positive deflection of the conventional ailerons on a B747.

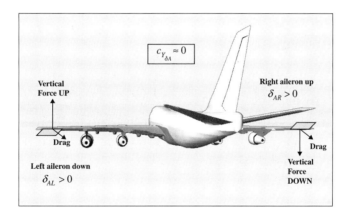

Figure 4.23 Lateral Force Associated with Deflections of Ailerons

The forces associated with the asymmetric deflections of the left and right ailerons act along the vertical and horizontal directions; therefore, their components along the lateral direction are negligible. Thus, we have

$$c_{Y\delta_A} \approx 0$$

4.2.4 Modeling of $c_{Y\delta_R}$

The contribution to the lateral force coefficient associated with the deflection of the rudder is shown in Figure 4.24. According to the sign convention, we have

$$c_{Y\delta_R} > 0$$

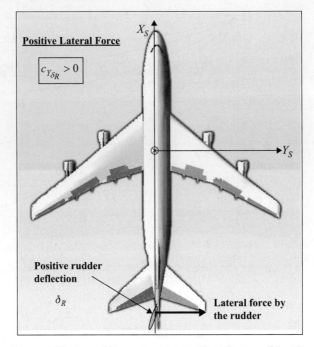

Figure 4.24 Lateral Force Associated with Deflection of Rudder

The geometric parameters of the rudder are shown in Figure 4.25.

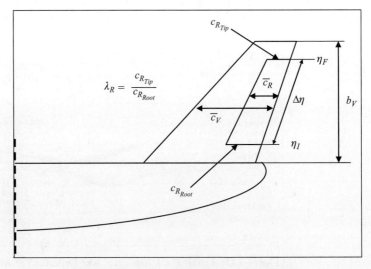

Figure 4.25 Geometric Parameters of the Rudder

A relationship for $c_{Y\delta_R}$ is given by

$$c_{Y\delta_R} = \left|c_{L\alpha_V}\right| \eta_V \frac{S_V}{S} \Delta(K_R) \tau_R$$

where $\left|c_{L\alpha_V}\right|$ is the lift-curve slope coefficient for the vertical tail calculated as for the modeling of $c_{Y\beta_V}$ (using $AR_{V_{eff}}$). τ_R is a control surface effectiveness factor for the rudder and is found using Figure 4.26 (see also Chapter 2, Section 2.4). Finally, K_R is a correction factor associated with the span of the rudder within the span of the vertical tail calculated using Figure 4.27.

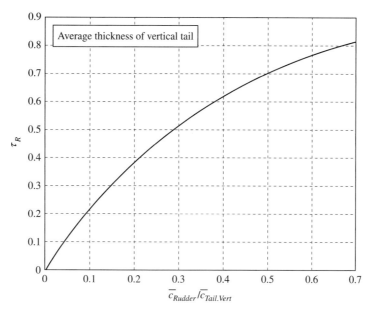

Figure 4.26 Effectiveness of the Rudder τ_R as Function of $\overline{c}_{Rudder}/\overline{c}_{Vert.Tail}$

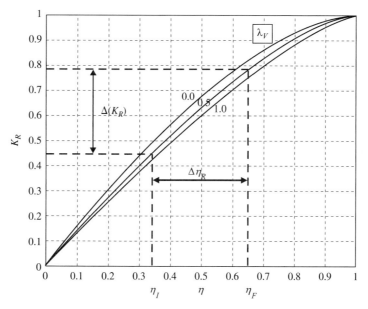

Figure 4.27 Span Factor between Rudder and Vertical Tail[5]

4.3 MODELING OF L_{A_1}

The modeling of the steady-state rolling moment starts with the relationship

$$L_{A_1} = L_A = c_{l_1} \bar{q} S b$$

where the functionality of the rolling moment coefficient is expressed by

$$c_{l_1} = f(\beta, \delta_A, \delta_R)$$

The use of a first order approximation for the Taylor expansion leads to:

$$c_{l_1} = c_{l_0} + c_{l\beta}\beta + c_{l\delta_A}\delta_A + c_{l\delta_R}\delta_R$$

The symmetry of the aircraft with respect to the XZ plane leads to the property

$$c_{l_0} = c_l\big|_{\beta=\delta_A=\delta_R=0°} = 0$$

Next, the attention shifts on the analysis of the coefficients $c_{l\beta}$, $c_{l\delta_A}$, $c_{l\delta_R}$. Because of the aerodynamic complexity of $c_{l\beta}$, the conceptual and mathematical modeling for this important aerodynamic coefficient are discussed separately.

4.3.1 Conceptual Modeling of $c_{l\beta}$

This coefficient, well known as the dihedral effect, has major implications on the lateral directional stability of the aircraft. In terms of contributions from different components of a conventional subsonic configuration, the starting relationship for $c_{l\beta}$ is given by

$$c_{l\beta} = c_{l\beta_{WB}} + c_{l\beta_H} + c_{l\beta_V}$$

The modeling starts from $c_{l\beta_{WB}}$, the wing-body component of the dihedral effect.

This modeling is fairly complex since it is function of several geometric parameters. In general, for most aircraft we can identify three main contributions of different aerodynamic nature:

$$c_{l\beta_{WB}} \approx c_{l\beta_{WB}}\big|_\mathrm{I} + c_{l\beta_{WB}}\big|_\mathrm{II} + c_{l\beta_{WB}}\big|_\mathrm{III}$$

Additional contributions, which will be outlined in the next section, are typically present but less significant.

The first (wing + body) dihedral effect contribution is due to the wing geometric dihedral angle, Γ_W, assumed to be positive with the tip of the wing pointing upward, as shown in Figure 4.28. A similar angle is also introduced for the horizontal tail Γ_H.

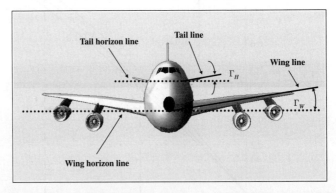

Figure 4.28 Geometric Dihedral Angles for Wing and Horizontal Tail

The sideslip angle β is positive with the velocity vector from the right of the pilot; therefore, a positive β on a wing with a positive geometric dihedral angle Γ_W leads to a $\Delta\alpha$ on the right wing modeled through

$$\Delta\alpha\bigg|_{\substack{\text{Right wing} \\ \text{w.r.t. pilot}}} = \sin(\beta)\sin(\Gamma) \approx \beta\,\Gamma$$

According to the sign convention for the rolling moment, this positive $\Delta\alpha$ generates a positive additional lift ΔL on the right wing, leading to a negative rolling moment ΔL_A, as shown in Figure 4.29.

Thus, the first contribution of the (wing + body) dihedral effect is negative:

$$c_{l\beta_{WB}}\bigg|_\mathrm{I} < 0$$

The second (wing + body) dihedral effect contribution is due to a specific aerodynamic phenomenon associated with high-wing aircraft. As shown in Figure 4.30, the presence of a positive sideslip angle leads to the production of a number of small vortices for the flow "trapped" under the right wing for a high-wing configuration.

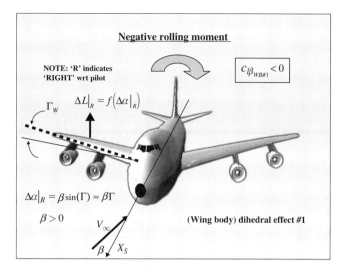

Figure 4.29 Wing-Body Dihedral Effect 1

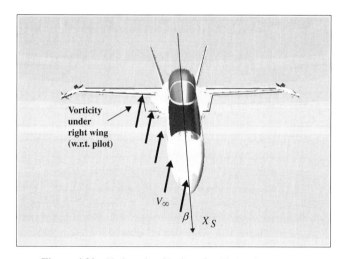

Figure 4.30 Under-wing Vortices for High-Wing Aircraft

The circulation associated with these vortices ultimately leads to the generation of a lift force acting on the right wing only, which will then create a rolling moment. This rolling moment is also negative, as shown in Figure 4.31.

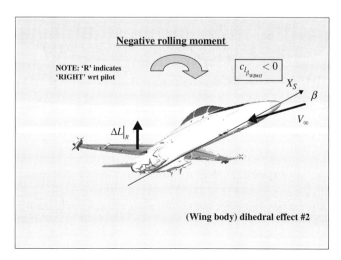

Figure 4.31 Wing-Body Dihedral Effect 2

Thus, the second contribution of the (wing + body) dihedral effect is also negative:

$$c_{l\beta_{WB}}\Big|_{II} < 0$$

It should be clear that the first (wing + body) dihedral effect is associated with the low-wing configuration whereas the second (wing + body) dihedral effect is only present with the high-wing configuration. Since low-wing aircraft typically feature positive wing dihedral angles while high-wing aircraft feature negative or negligible wing dihedral angles, the dihedral effects $c_{l\beta_{WB}}\big|_{I}$, $c_{l\beta_{WB}}\big|_{II}$ are in general mutually exclusive.

The third contribution to the (wing + body) dihedral effect is due to the geometric wing sweep angle Λ_{LE}. The origin of this effect is fairly clear. According to the cosine principle for swept wings, discussed in Chapter 2, the component of the airspeed responsible for creating lift is the normal velocity component V_{∞_n}. Without the presence of a sideslip angle V_{∞_n} is given by $V_{\infty_n} = V_{\infty} \cos(\Lambda_{LE})$. As shown in Figures 4.32 and 4.33, in the presence of a sideslip angle, the normal components of the velocity will take on different values on the different sides of the wing.

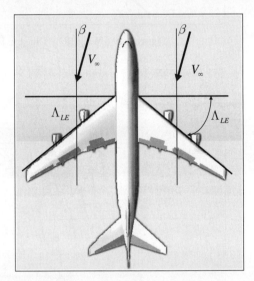

Figure 4.32 Direction of Airspeed for Swept Wings with Positive β

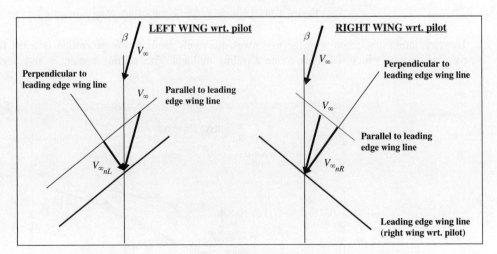

Figure 4.33 Normal Components of the Airspeed on Left and Right Wings

Particularly, the normal components of the airspeed on the left and right wings take on the following values:

$$V_{\infty_{nR}} = V_{\infty} \cos(\Lambda_{LE} - \beta)$$
$$V_{\infty_{nL}} = V_{\infty} \cos(\Lambda_{LE} + \beta)$$

From, Figure 4.33 it is clear that $V_{\infty_{nR}} > V_{\infty_{nL}}$

The difference in the values of the normal components of the velocity ultimately leads to a difference in the values of the lift force on the right and left wings, with

$$L_R > L_L$$

The difference in the values of the lift force on the two sides of the wing creates another—once again negative—rolling moment, as shown in Figure 4.34.

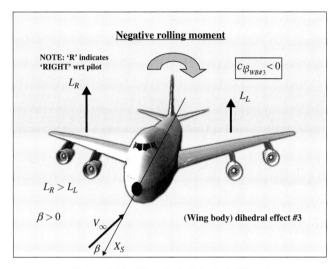

Figure 4.34 Wing-Body Dihedral Effect 3

Like the previous two contributions, the third contribution of the (wing + body) dihedral effect is also negative:

$$c_{l\beta_{WB}}|_{III} < 0$$

From the previous discussion it is clear that on different types of aircraft the total value of the (wing + body) dihedral effect will have different origins. Table 4.1 summarizes the trend for different types of aircraft, depending on the wing configurations.

Table 4.1 Wing + Body Dihedral Effect for Different Classes of Aircraft

Aircraft Type	WB Dihedral Effect 1	WB Dihedral Effect 2	WB Dihedral Effect 3
Commercial and military cargo jetliners (low-wing)	YES	NO	YES
Commercial and military cargo jetliners (high-wing)	NO	YES	YES
General aviation (low-wing with rectangular wings)	YES	NO	NO
General aviation (high-wing with rectangular wings)	NO	YES	NO
Fighter aircraft (low-wing)	YES	NO	YES
Fighter aircraft (high-wing)	NO	YES	YES

Next, the attention shifts to the tail dihedral effect $c_{l\beta_H}$ for conventional (wing + horizontal tail) aircraft configurations. In general, the tail can be considered as a smaller wing operating at a lower dynamic pressure ($\eta_H = \bar{q}_H/\bar{q} < 1$), with a smaller surface S_H, a smaller wing span b_H, and a different sideslip angle $\beta_H = (\beta - \sigma)$.

As discussed in the previous section, in the analysis of the aircraft dihedral effects for both the (wing + body) and the tail components, a special case is represented by the McDonnell Douglas F-4 aircraft. A trademark of this aircraft is the large negative geometric dihedral angle Γ_H (anhedral angle) of its horizontal tail, to the point that it is no longer appropriate to refer to it as the horizontal tail. The magnitude of this angle is evident in Figure 4.35.

Figure 4.35 F-4 Horizontal Tail Anhedral Angle Γ_H

Thus, for the F-4 aircraft, this anhedral angle Γ_H generates a substantial positive $c_{l\beta_{H_I}}$ effect, which compensates for an excessively large negative value for the $c_{l\beta_{WB}}$ coefficient.

The analysis of the dihedral effect is completed by the modeling of the contribution from the vertical tail through the coefficient $c_{l\beta_V}$, which is conceptually related to the previously described coefficient $c_{Y\beta_V}$. The velocity field along the fuselage leading to the coefficient $c_{l\beta_V}$ is shown in Figure 4.36.

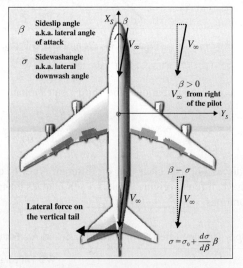

Figure 4.36 Airspeed at the Vertical Tail

As shown, the lateral flow will create a lateral force acting on the vertical tail, which, due to its moment arm, will create a rolling moment, as shown in Figures 4.37 and 4.38.

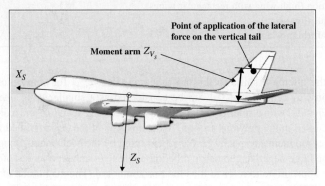

Figure 4.37 Moment Arm of the Lateral Force on the Vertical Tail

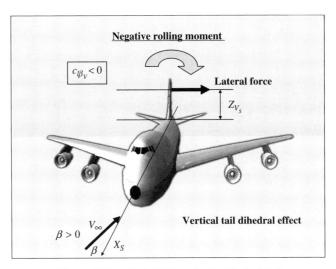

Figure 4.38 Dihedral Effect for the Vertical Tail

The previous analysis of $c_{l\beta_V}$ concludes the documentation of the overall dihedral effect $c_{l\beta}$. Summarizing, an approximate conceptual modeling of the aircraft dihedral effect is given by

$$c_{l\beta} \approx \left(c_{l\beta_{WB}}\Big|_I + c_{l\beta_{WB}}\Big|_{II} + c_{l\beta_{WB}}\Big|_{III} \right) + c_{l\beta_H} + c_{l\beta_V}$$

A detailed mathematical modeling for this important coefficient is described next.

4.3.2 Mathematical Modeling of $c_{l\beta}$

Starting from the initial breakdown for $c_{l\beta}$:

$$c_{l\beta} = c_{l\beta_{WB}} + c_{l\beta_H} + c_{l\beta_V}$$

A closed-form expression for the modeling for $c_{l\beta_{WB}}$ is given by[5]

$$c_{l\beta_{WB}} = 57.3 \cdot c_{L_1} \left[\left(\frac{c_{l\beta}}{c_{L_1}} \right)_{\Lambda_{c/2}} K_{M_\Lambda} K_f + \left(\frac{c_{l\beta}}{c_{L_1}} \right)_{AR} \right]$$

$$+ 57.3 \left\{ \Gamma_W \left[\frac{c_{l\beta}}{\Gamma_W} K_{M_\Gamma} + \frac{\Delta c_{l\beta}}{\Gamma_W} \right] + \left(\Delta c_{l\beta} \right)_{Z_W} + \varepsilon_W \tan \Lambda_{c/4} \left(\frac{\Delta c_{l\beta}}{\varepsilon_W \tan \Lambda_{c/4}} \right) \right\} \left(\frac{1}{\text{rad}} \right)$$

A detailed description of the coefficients in the preceding relationship is provided next.

$\left(\frac{c_{l\beta}}{c_{L_1}} \right)_{\Lambda_{c/2}}$ represents the contribution associated with the wing sweep angle (see Figure 4.39); K_{M_Λ} is a correction factor associated with the Mach number and the wing sweep angle (see Figure 4.40); K_f is a correction factor associated with the length of the forward portion of the fuselage (see Figure 4.41); $\left(\frac{c_{l\beta}}{c_{L_1}} \right)_{AR}$ represents the contribution associated with the wing aspect ratio (see Figure 4.42); $\left(\frac{c_{l\beta}}{\Gamma_W} \right)$ represents the contribution associated with the wing dihedral angle (see Figure 4.43); K_{M_Γ} is a correction factor associated with the Mach number and the wing dihedral angle (see Figure 4.44).

$\left(\dfrac{\Delta c_{l_\beta}}{\Gamma_W}\right)$ is a correction factor associated with the size of the fuselage modeled using

$$\left(\dfrac{\Delta c_{l_\beta}}{\Gamma_W}\right) = -0.0005 \cdot AR \cdot \left(\dfrac{d_B}{b}\right)^2 \quad \left(\dfrac{1}{\deg^2}\right)$$

where d_B is the diameter of the fuselage defined as

$$S_{f_{AVG}} = \pi R^2 = \pi \dfrac{d_B^2}{4} \quad \rightarrow \quad d_B = \sqrt{\dfrac{S_{f_{AVG}}}{\pi/4}} = \sqrt{\dfrac{S_{f_{AVG}}}{0.7854}}$$

with $S_{f_{AVG}}$ defined as the average cross-sectional area along the fuselage.

$(\Delta c_{l_\beta})_{Z_W}$ is a correction factor associated with the location of the fuselage with respect to the wing (high wing or low wing) modeled using

$$\left(\Delta c_{l_\beta}\right)_{Z_W} = \dfrac{1.2\sqrt{AR}}{57.3} \cdot \dfrac{Z_W}{b} \cdot \left(\dfrac{2d_B}{b}\right) \quad \left(\dfrac{1}{\deg}\right)$$

where Z_W is the vertical distance between the wing root (at the 25 percent station) and the fuselage centerline. This parameter is considered to be positive for a low-wing configuration, as shown in Figure 4.45.

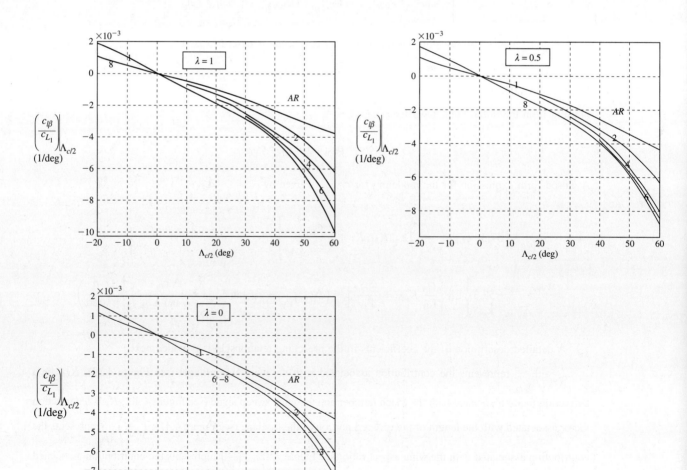

Figure 4.39 Contribution to $c_{l_{\beta_{WB}}}$ Due to Wing Sweep Angle[5]

4.3   Modeling of L_{A_1} 157

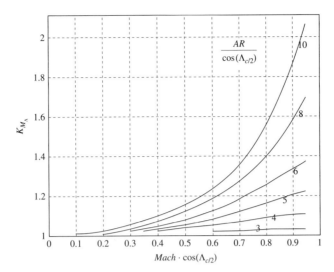

Figure 4.40   Compressibility Correction Factor for $c_{l\beta_{WB}}$ Due to Wing Sweep Angle[5]

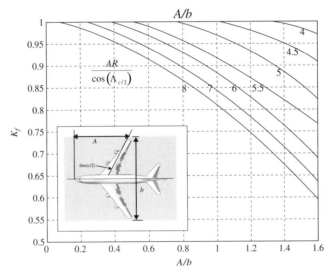

Figure 4.41   Fuselage Correction Factor for $c_{l\beta_{WB}}$ Due to Wing Sweep Angle[5]

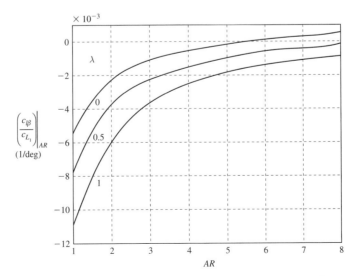

Figure 4.42   Contribution to $c_{l\beta_{WB}}$ Due to Wing Aspect Ratio[5]

158 Chapter 4 Modeling of Lateral Directional Aerodynamic Forces and Moments

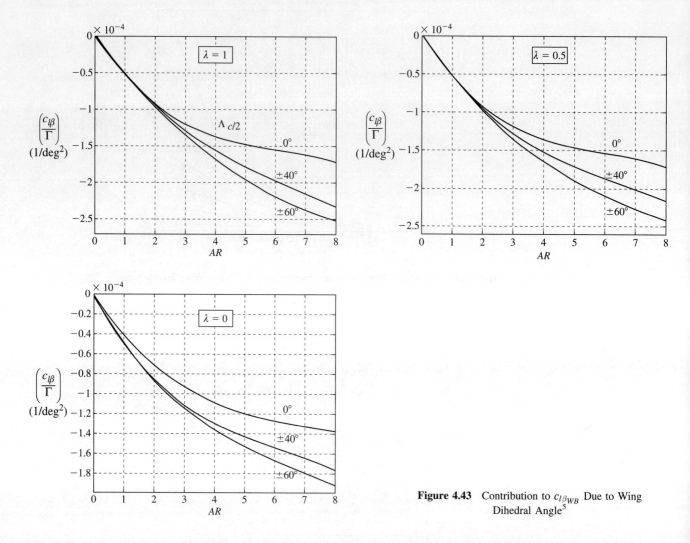

Figure 4.43 Contribution to $c_{l\beta_{WB}}$ Due to Wing Dihedral Angle[5]

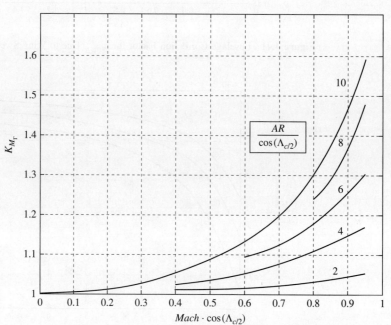

Figure 4.44 Compressibility Correction Factor to Contribution to $c_{l\beta_{WB}}$ Due to Wing Dihedral[5]

4.3 Modeling of L_{A_1} 159

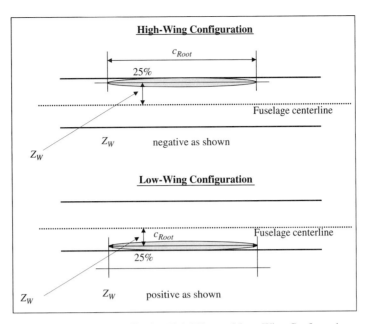

Figure 4.45 Parameter Z_W for High-Wing and Low-Wing Configurations

$\left(\dfrac{\Delta c_{l_\beta}}{\varepsilon_W \tan \Lambda_{c/4}}\right)$ is a correction factor associated with the twist angle ε_W between the zero-lift lines of the wing sections at the tip and at the root stations. This factor is shown in Figure 4.46; the twist angle ε_W is shown in Figure 4.47.

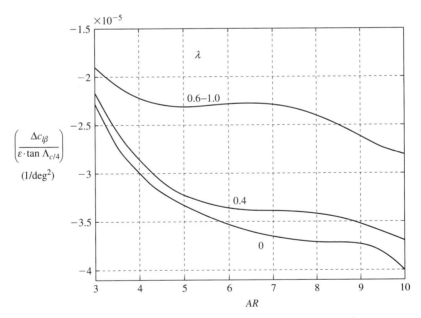

Figure 4.46 Contribution to $c_{l_{\beta_{WB}}}$ Due to Wing Twist Angle[5]

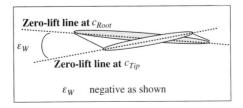

Figure 4.47 Wing Twist Angle ε_W[5]

Next, the attention shifts to the modeling of the dihedral effect associated with the horizontal tail $c_{l\beta_H}$. In general, the horizontal tail can be considered as a wing operating at a lower dynamic pressure ($\eta_H = \bar{q}_H/\bar{q} < 1$), with a smaller surface S_H, and a smaller wing span b_H. Therefore, a relationship for $c_{l\beta_H}$ is given by

$$c_{l\beta_H} = c_{l\beta_{WB}}\bigg|_H \eta_H \frac{S_H}{S} \frac{b_H}{b}$$

where $c_{l\beta_{WB}}\big|_H$ is the previously introduced $c_{l\beta_{WB}}$ evaluated with the geometric parameters of the horizontal tail.

With the exception of horizontal tails with significant dihedral or anhedral angles, for most aircraft we have the condition $c_{l\beta_H} \approx 0$.

Finally, the modeling of $c_{l\beta}$ is completed by the modeling for the dihedral effect associated with the vertical tail $c_{l\beta_V}$. It can be visualized that $c_{l\beta_V}$ is the aerodynamic moment coefficient associated with the aerodynamic force coefficient $c_{Y\beta_V}$ through a moment arm due to the location of the vertical tail with respect to the aircraft center of gravity. The geometric description of this moment arm is shown in Figure 4.48.

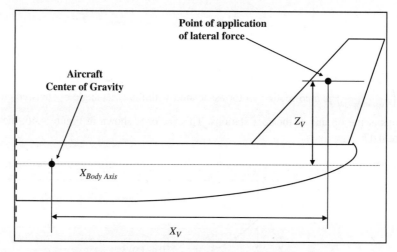

Figure 4.48 Geometric Parameters of the Vertical Tail

Considering that the modeling is performed along the aircraft stability axes, the dimensionless moment arm associated with the dihedral effect from the vertical tail is given by

$$\frac{(Z_V \cos\alpha_1 - X_V \sin\alpha_1)}{b}$$

leading to

$$c_{l\beta_V} = c_{Y\beta_V} \cdot \frac{(Z_V \cos\alpha_1 - X_V \sin\alpha_1)}{b}$$

$$= -k_{Y_V} \cdot \left|c_{L_{\alpha \cdot V}}\right|\eta_V \cdot \left(1 + \frac{d\sigma}{d\beta}\right) \frac{S_V}{S} \cdot \frac{(Z_V \cos\alpha_1 - X_V \sin\alpha_1)}{b} \quad (\text{rad}^{-1})$$

The magnitude of the contribution of the dihedral effect from the vertical tail can be quite substantial.

4.3.3 Modeling of $c_{l\delta_A}$

Next, the attention shifts to the analysis of the rolling moment coefficient associated with the deflection of the ailerons $c_{l\delta_A}$.

The ailerons are asymmetric control surfaces. A positive deflection of the ailerons implies a trailing-edge down deflection of the left aileron (w.r.t. pilot) and a trailing-edge up deflection of the right aileron. The combined result of these deflections is a positive rolling moment, as shown in Figure 4.49.

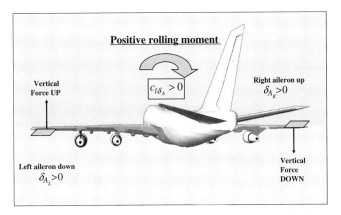

Figure 4.49 Rolling Moment Associated with Deflections of Ailerons

Despite its simple aerodynamic origin, the modeling of $c_{l\delta_A}$ is quite complex. The control authority of the ailerons in providing rolling moments is dependent mainly on the size and location of the ailerons along the aircraft wing span, as shown in Figure 4.50, and on the characteristics of the specific wing profile for that section of the wing. The following step-by-step procedure is suggested for the mathematical modeling of $c_{l\delta_A}$.

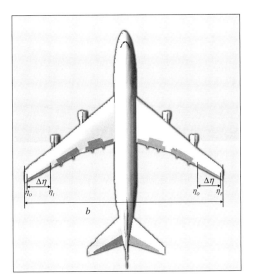

Figure 4.50 Location of the Ailerons along the Wing Span

Step 1: Determine the in-board and out-board locations of the individual left and right ailerons y_{A_I} and y_{A_O} as fractions of the semi-wing span $b/2$ leading to

$$\eta_I = \eta_{Inboard} = \frac{y_{A_I}}{b/2}, \quad \eta_O = \eta_{Outboard} = \frac{y_{A_O}}{b/2},$$

162 Chapter 4 Modeling of Lateral Directional Aerodynamic Forces and Moments

Step 2: Using η_I, η_O, find the Δ in rolling moment effectiveness (*RME*) parameter from Figures 4.51 to 4.53, assuming that the entire wing section from y_{A_I} to y_{A_O} will deflect. *RME* is a function of the parameter:

$$\Lambda_\beta = \tan^{-1}\left(\frac{\tan \Lambda_{c/4}}{\beta}\right) = \tan^{-1}\left(\frac{\tan \Lambda_{c/4}}{\sqrt{1 - Mach^2}}\right)$$

The plots in Figures 4.51 to 4.53 are given for different values of the parameter $\beta \cdot AR/k$, where

$$k = \frac{(c_{l_\alpha})_{Wing\ Section}|_{Mach} \cdot \beta}{2\pi}$$

Within a reasonable approximation the wing planform lift curve slope $c_{L_{\alpha W}}|_{Mach}$ can be used in lieu of $(c_{l_\alpha})_{Wing\ Section}|_{Mach}$. By definition, *RME* is given by $RME = \dfrac{c_{l_\delta} \cdot \beta}{k}$

The use of Figure 4.54 shows the calculation of $\Delta(RME)$:

$$\Delta(RME) = RME|_{\eta_O} - RME|_{\eta_I}$$

Step 3: With $\Delta(RME)$ from Step 2, c_{l_δ}' can be found using

$$c_{l_\delta}' = \frac{\Delta(RME) \cdot k}{\beta}$$

Again, c_{l_δ}' is relative to two full-chord sections of the wing being deflected asymmetrically.

Step 4: Considering that only a section of the chord (c_A) is deflected, c_{l_δ} is found using

$$c_{l_\delta} = \tau_A \cdot c_{l_\delta}'$$

where τ_A is found using Figure 4.55 (see Section 2.4).

Step 5: The effect of the asymmetric nature of the deflections of the ailerons is accounted for by considering $c_{l_{\delta_{R,L}}}$ as 50 percent of the previously calculated c_{l_δ} value. According to the sign convention for a positive deflection of ailerons, the left aileron is deflected down (positive) while the right aileron is deflected up (negative). Therefore, the rolling coefficient relative to the deflections of the ailerons is given by

$$c_l = \left[\left(\frac{c_{l_\delta}}{2}\right)\bigg|_{Left} + \left(\frac{c_{l_\delta}}{2}\right)\bigg|_{Right}\right]\left(\delta_{A_{Left}} - \delta_{A_{Right}}\right)$$

Since $\delta_A = \frac{1}{2}\left(\delta_{A_{Left}} - \delta_{A_{Right}}\right)$ we have

$$c_l = \left[\left(\frac{c_{l_\delta}}{2}\right)\bigg|_{Left} + \left(\frac{c_{l_\delta}}{2}\right)\bigg|_{Right}\right] \cdot \delta_A \rightarrow c_{l_{\delta_A}} = \frac{1}{2}\left[(c_{l_\delta})\big|_{Left} + (c_{l_\delta})\big|_{Right}\right]$$

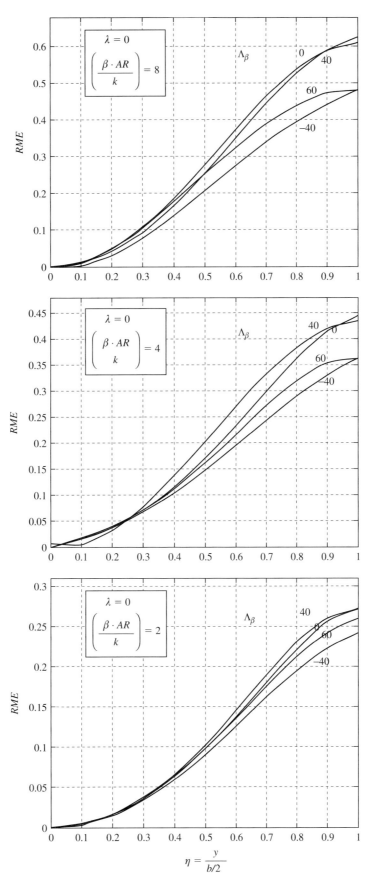

Figure 4.51 Rolling Moment Effectiveness for Different Geometries of the Wing $(\lambda = 0)^5$

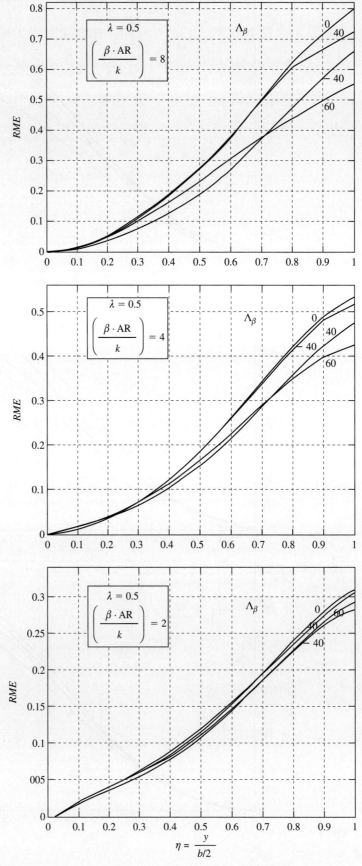

Figure 4.52 Rolling Moment Effectiveness for Different Geometries of the Wing ($\lambda = 0.5$)[5]

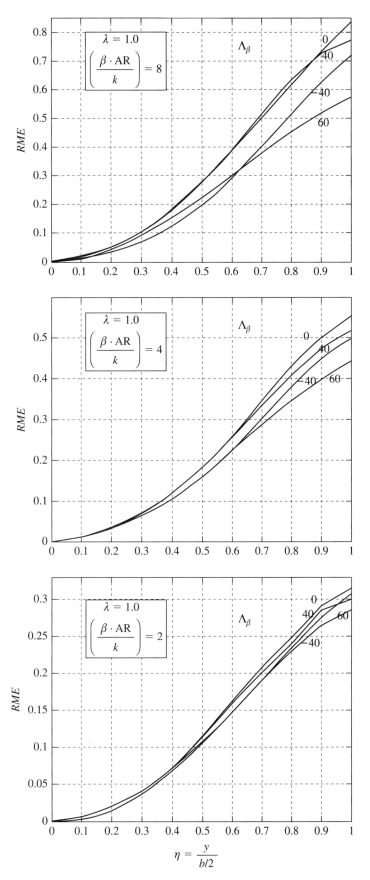

Figure 4.53 Rolling Moment Effectiveness for Different Geometries of the Wing $(\lambda = 1)^5$

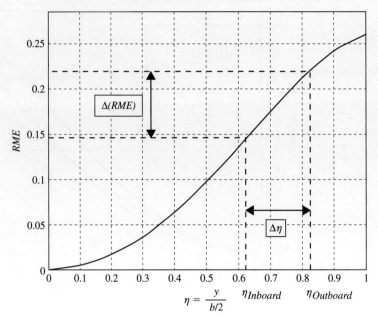

Figure 4.54 Evaluation of $\Delta(RME)$ for Given $\Delta\eta$[5]

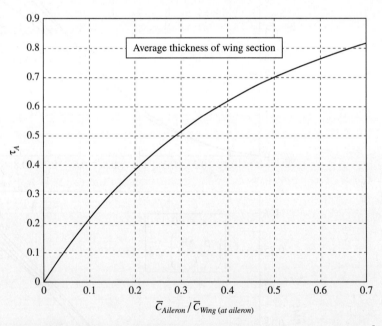

Figure 4.55 Effectiveness of the Aileron τ_A as Function of $\bar{c}_{Aileron}/\bar{c}_{Wing(Aileron)}$[5]

4.3.4 Modeling of $c_{l\delta_R}$

The analysis of the rolling moment coefficient is completed by the modeling of the rolling moment coefficient associated with the deflection of the rudder, $c_{l\delta_R}$. This moment originates from the lateral force associated with the deflection of the rudder through its moment arm with respect to the aircraft center of gravity, as shown in Figure 4.56.

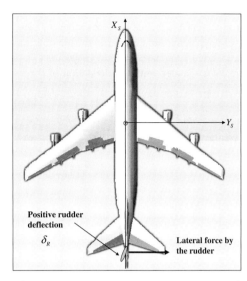

Figure 4.56 Lateral Force Associated with Rudder Deflection

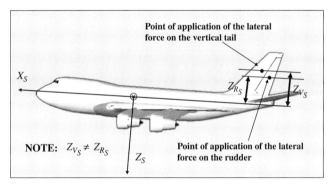

Figure 4.57 Moment Arm for Lateral Force Acting on Rudder and Vertical Tail

Although the difference between the two moment arms is fairly small, the conceptual difference between Z_{R_S} and Z_{V_S} should be pointed out. The different moment arms for the force associated with the vertical tail and the deflection of the rudder are shown in Figure 4.57. The $c_{l\delta_R}$ contribution is positive, as shown in Figure 4.58.

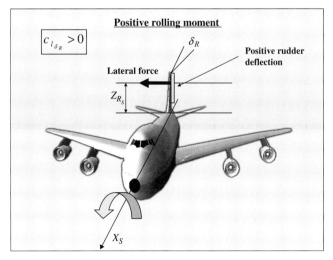

Figure 4.58 Rolling Moment Coefficient with Rudder Deflection

168 Chapter 4 Modeling of Lateral Directional Aerodynamic Forces and Moments

The mathematical modeling for $c_{l\delta_R}$ is quite intuitive. In fact, it can be visualized that the modeling of $c_{l\delta_R}$ relates to the modeling of $c_{Y\delta_R}$ in the same way that the previous modeling of $c_{l\beta_V}$ relates to the modeling of $c_{Y\beta_V}$. Considering that the modeling of the aerodynamic coefficients is performed along the stability axis, the moment arm for the rolling moment associated with the deflection of the rudder is shown in Figure 4.59 and is given by

$$\frac{Z_R \cos\alpha_1 - X_R \sin\alpha_1}{b}$$

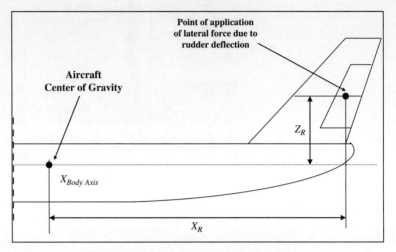

Figure 4.59 Geometric Parameters of the Rudder

Therefore, we have

$$c_{l\delta_R} = c_{Y\delta_R} \left(\frac{Z_R \cos\alpha_1 - X_R \sin\alpha_1}{b} \right)$$

$$= \left| c_{L\alpha_V} \right| \eta_V \frac{S_V}{S} K_R \tau_R \left(\frac{Z_R \cos\alpha_1 - X_R \sin\alpha_1}{b} \right)$$

where $\left|c_{L\alpha_V}\right|$, η_V, $\frac{S_V}{S}$, K_R, τ_R have been described in Section 4.2.4 for $c_{Y\delta_R}$.

4.4 MODELING OF N_{A_1}

As for the previous rolling moment, the modeling of the steady-state yawing moment starts with the relationship

$$N_{A_1} = N_A = c_{n_1} \bar{q} S b$$

where the functionality of the yawing moment coefficient is expressed by

$$c_{n_1} = f(\beta, \delta_A, \delta_R)$$

The use of a first order approximation for the Taylor expansion of c_{n_1} leads to

$$c_{n_1} = c_{n_0} + c_{n\beta}\beta + c_{n\delta_A}\delta_A + c_{n\delta_R}\delta_R$$

The symmetry of the aircraft with respect to the XZ plane leads to the property

$$c_{n_0} = c_n\big|_{\beta=\delta_A=\delta_R=0°} = 0$$

Next, the attention shifts to the analysis of the coefficients $c_{n\beta}$, $c_{n\delta_A}$, $c_{n\delta_R}$. Due to its aerodynamic complexity, the conceptual and the mathematical modeling for $c_{n\beta}$ are discussed separately.

4.4.1 Conceptual Modeling of $c_{n\beta}$

The aerodynamic coefficient $c_{n\beta}$ is also known as weathercock effect. This coefficient—along with the previous dihedral effect $c_{l\beta}$—is very important and has major implications on the lateral directional stability of the aircraft.

In terms of contributions from different components of a conventional subsonic configuration, the starting relationship for $c_{n\beta}$ is given by

$$c_{n\beta} = c_{n\beta_W} + c_{n\beta_B} + c_{n\beta_H} + c_{n\beta_V}$$

The aerodynamic origin of the components $c_{n\beta_W}, c_{n\beta_B}$ is shown in Figure 4.60.

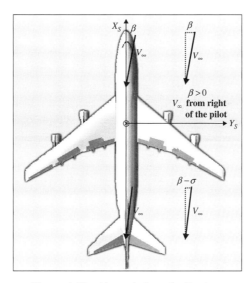

Figure 4.60 Airspeed along the Fuselage

Due to the aerodynamic flow associated with a positive sideslip angle on the right side of the pilot, the lateral forces will act on the right side of the aircraft, as shown in Figure 4.61.

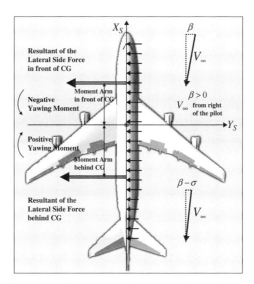

Figure 4.61 Lateral Forces and Moments Acting on the Fuselage T

The lateral force acting on the portion of the fuselage in front of the aircraft CG generates a moment in the opposite direction of the moment generated by the lateral force acting on the portion of the fuselage behind the aircraft CG. Note that the lateral force acting toward the aircraft tail is slightly smaller due to the reduction in the sideslip angle associated with the sidewash effect; on the other side, "long" aircraft—such as jetliners—have longer segments of the fuselage behind the CG than ahead of the CG. In general, these two moments with opposite directions have similar magnitudes; therefore, this coefficient can be either positive or negative and is generally small, as shown in Figure 4.62.

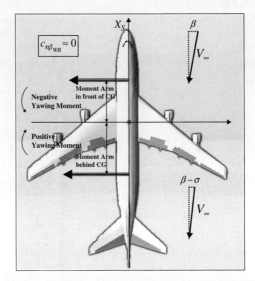

Figure 4.62 Yawing Moment Contribution Associated with β on the Wing Body

The next contribution to the important $c_{n\beta}$ is associated with the horizontal tail, which is modeled through the coefficient $c_{n\beta_H}$. The lateral force exerted on the horizontal tail by a lateral aerodynamic flow can be considered negligible due to the lack of a wetted area for the horizontal tail exposed to the lateral flow. Therefore, for the majority of aircraft with conventional configurations we have

$$c_{n\beta_H} \approx 0$$

A special case is once again represented by the McDonnell Douglas F-4 aircraft. Particularly, as shown in Figure 4.63, the horizontal tail of the F-4 is characterized by a negative dihedral angle and therefore a substantial area is exposed to the lateral flow in the presence of a sideslip angle. This in turn creates a lateral force which, ultimately, leads to a positive yawing moment, as shown in Figure 4.63.

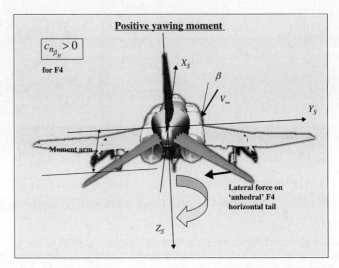

Figure 4.63 Yawing Moment Associated with Anhedral Angle on F-4 Horizontal Tail

The final contribution to $c_{n\beta}$ is provided by the coefficient $c_{n\beta_V}$ associated with the vertical tail. The origin of this contribution is conceptually similar to its counterpart for the rolling moment $c_{l\beta_V}$, with the difference being essentially the moment arm located on the horizontal axis rather than the vertical axis. Consider once again the lateral force acting on the vertical tail, as shown in Figure 4.64.

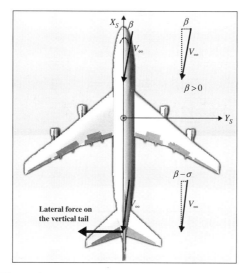

Figure 4.64 Lateral Force Acting on the Vertical Tail

The moment arm of the lateral force with respect to the aircraft center of gravity is shown in Figure 4.65, and the resulting yawing moment is shown in Figure 4.66.

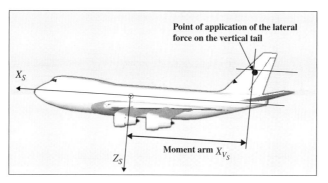

Figure 4.65 Moment Arm of the Lateral Force on the Vertical Tail

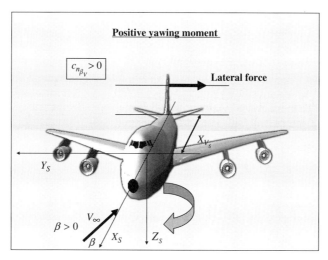

Figure 4.66 Yawing Moment Associated with the Lateral Force on the Vertical Tail

The figure shows that this contribution to the yawing moment is positive.

4.4.2 Mathematical Modeling of $c_{n\beta}$

Starting from the initial breakdown for $c_{n\beta}$

$$c_{n\beta} = c_{n\beta_W} + c_{n\beta_B} + c_{n\beta_H} + c_{n\beta_V}$$

The aerodynamic contribution to $c_{n\beta}$ from the wing is negligible for all configurations. Therefore, we have $c_{n\beta_W} \approx 0$.

The aerodynamic coefficient $c_{n\beta_B}$ can be estimated using the relationship

$$c_{n\beta_B} = -57.3 \cdot K_N \, K_{R_l} \frac{S_{B_S}}{S} \frac{l_B}{b}$$

where the geometric coefficients S_{B_S}, l_B are defined in Figure 4.67. The coefficient K_N is an empirical factor, estimated through Figure 4.68, related to the geometric coefficients of the axial cross section of the fuselage, whereas K_{R_l} is a factor, estimated through Figure 4.69, related to the Reynolds number.

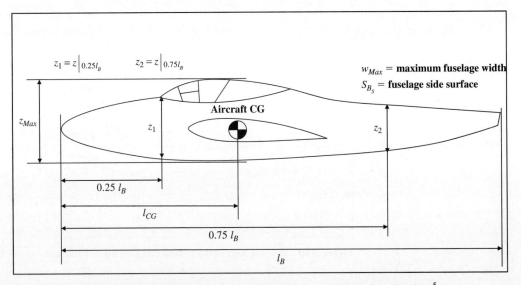

Figure 4.67 Geometric Parameters of the XZ Cross Section of the Fuselage[5]

As discussed previously, the contribution to $c_{n\beta}$ from the horizontal tail (with the rare exceptions of aircraft with very significant dihedral/anhedral angles for the tail) is negligible. Therefore, we have $c_{n\beta_H} \approx 0$.

The most significant contribution to $c_{n\beta}$ is provided by the vertical tail. As for $c_{l\beta_V}$, $c_{n\beta_V}$ is the aerodynamic moment coefficient associated with the aerodynamic force coefficient $c_{Y\beta_V}$ through a moment arm due to the location of the vertical tail with respect to the aircraft center of gravity.

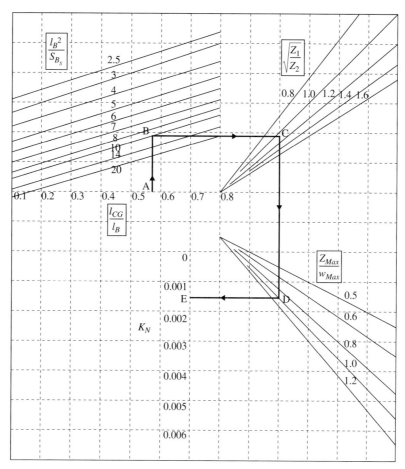

Figure 4.68 Empirical Factor K_N for Wing-Body Interface[5]

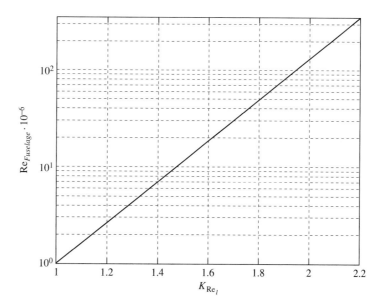

Figure 4.69 Effect of Reynolds Number on Wing-Body Interface[5]

The geometric description of this moment arm is shown in Figure 4.70.

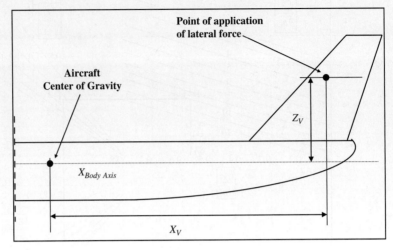

Figure 4.70 Geometric Parameters for the Vertical Tail

Considering that the modeling is performed along the aircraft stability axes, the dimensionless moment arm associated with the dihedral effect from the vertical tail is given by

$$\frac{(X_V \cos \alpha_1 + Z_V \sin \alpha_1)}{b}$$

leading to

$$c_{n\beta_V} = -c_{Y\beta_V} \cdot \frac{(X_V \cos \alpha_1 + Z_V \sin \alpha_1)}{b}$$

$$= k_{Y_V} \cdot \left|c_{L_{\alpha \cdot V}}\right| \eta_V \cdot \left(1 + \frac{d\sigma}{d\beta}\right) \frac{S_V}{S} \cdot \frac{(X_V \cos \alpha_1 + Z_V \sin \alpha_1)}{b} \quad \left(\frac{1}{\text{rad}}\right)$$

where the modeling of $c_{Y\beta_V}$ has been discussed in Section 4.2.2. It should be pointed out that the contribution from the vertical tail to the overall value of $c_{n\beta}$ is so significant that for a preliminary analysis it can be often approximated that $c_{n\beta} \approx c_{n\beta_V}$.

4.4.3 Modeling of $c_{n\delta_A}$

Next, the attention shifts to the yawing moment associated with deflections of the ailerons, modeled through the coefficient $c_{n\delta_A}$. The modeling of this coefficient requires a detailed analysis of the aerodynamic forces generated by the asymmetric deflection of the ailerons on the wings. In fact, in addition to the vertical forces with opposite directions (which generate the rolling moment), the asymmetric deflections of the left and right ailerons also generate small but not negligible drag forces leading to a small, negative yawing moment, as shown in Figure 4.71.

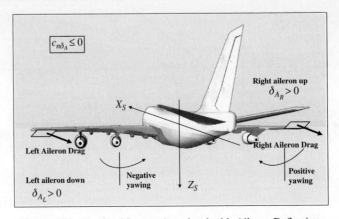

Figure 4.71 Yawing Moment Associated with Aileron Deflections

A relationship for modeling $c_{n\delta_A}$ is given by

$$c_{n\delta_A} = \Delta(K_{n_A}) c_{L_1} c_{l_{\delta_A}}$$

where the modeling of $c_{l_{\delta_A}}$ is provided in Section 4.3.3, c_{L_1} is the steady-state lift coefficient, K_{n_A} is an empirical coefficient evaluated using Figure 4.72. Note that Figure 4.72 is entered using $\eta_I = \frac{y_{A_I}}{b/2}$, $\eta_O = \frac{y_{A_O}}{b/2}$. Therefore, $\Delta(K_{n_A})$ is evaluated as the difference between the values of K_{n_A} associated with $\eta_I = \frac{y_{A_I}}{b/2}$, $\eta_O = \frac{y_{A_O}}{b/2}$, as shown in Figure 4.73.

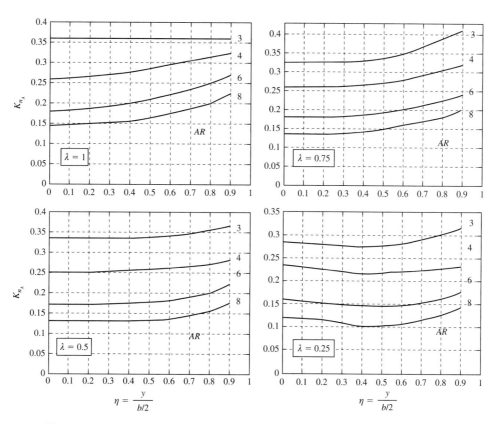

Figure 4.72 Correlation Coefficient for Yawing Moment Due to Deflection of Ailerons

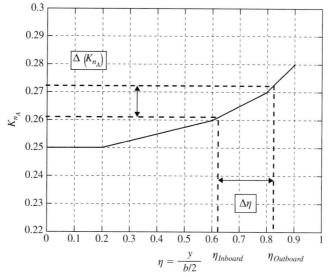

Figure 4.73 Calculation of $\Delta(K_{n_A})$ Using $\eta_I = \frac{y_{A_I}}{b/2}$, $\eta_O = \frac{y_{A_O}}{b/2}$ 5

176 Chapter 4   Modeling of Lateral Directional Aerodynamic Forces and Moments

4.4.4 Modeling of $c_{n\delta_R}$

Finally, the last yawing moment coefficient is associated with the deflection of the rudder. The positive deflection of the rudder is shown in Figure 4.74.

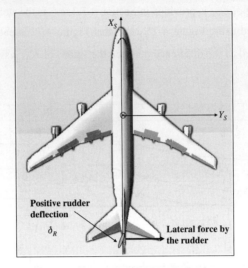

Figure 4.74 Positive Rudder Deflection

The moment arm for this lateral force is given by X_{R_S}, as shown in Figure 4.75.

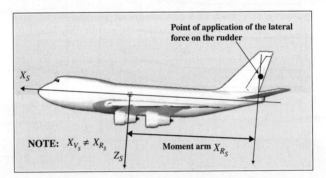

Figure 4.75 Geometric Parameters of the Rudder

As for the rolling coefficient counterpart, the difference between the moment arm X_{V_S}, associated with the vertical tail, and X_{R_S}, associated with the rudder, needs to be pointed out. The yawing moment associated with the rudder deflection is shown in Figure 4.76. Thus, the resulting yawing moment is negative, according to the sign convention.

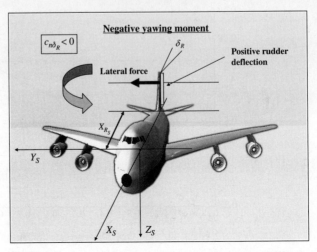

Figure 4.76 Yawing Moment Associated with Rudder Deflection

As for $c_{l\delta_R}$, the mathematical modeling for $c_{n\delta_R}$ is quite direct. Specifically, the modeling of $c_{n\delta_R}$ relates to the modeling of $c_{Y\delta_R}$ in the same way that the previous modeling of $c_{n\beta_V}$ relates to the modeling of $c_{Y\beta_V}$. Considering that the modeling of the aerodynamic coefficients is performed along the stability axis, the moment arm for the rolling moment associated with the deflection of the rudder is shown in Figure 4.77 below and it is given by

$$\frac{X_R \cos \alpha_1 + Z_R \sin \alpha_1}{b}$$

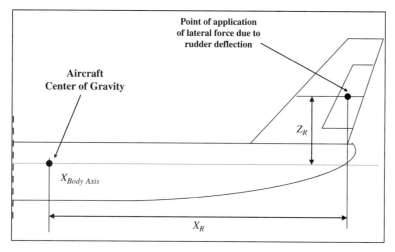

Figure 4.77 Geometric Parameters of the Rudder

Therefore, we have

$$c_{n\delta_R} = -c_{Y\delta_R}\left(\frac{X_R \cos \alpha_1 + Z_R \sin \alpha_1}{b}\right)$$

$$= -\left|c_{L\alpha_V}\right|\eta_V \frac{S_V}{S} K_R \tau_R \left(\frac{X_R \cos \alpha_1 + Z_R \sin \alpha_1}{b}\right)$$

where $\left|c_{L\alpha_V}\right|, \eta_V, \frac{S_V}{S}, K_R, \tau_R$ have been described in Section 4.2.4 for $c_{Y\delta_R}$.

The coefficient $c_{n\delta_R}$ has major implications on the directional control of the aircraft, especially for four-engine jetliners. Directional control through rudder deflection is a critical issue in the event of a single- and a dual-engine failure on the same wing, which is the worst-case scenario. In fact, the FAA requirement, which mandates the continuation of a flight with only 50 percent of the installed thrust, is essentially the main design requirement for the vertical tail and the rudder of wide-body jetliners.

4.5 SUMMARY OF THE LATERAL DIRECTIONAL STEADY-STATE FORCE AND MOMENTS

A summary of the lateral directional steady-state moments and force in a matrix format is provided:

$$\begin{Bmatrix} F_{A_{1_S}} \\ L_{A_{1_S}} \\ N_{A_{1_S}} \end{Bmatrix} = \begin{Bmatrix} F_{A_Y} \\ L_A \\ N_A \end{Bmatrix} = \begin{Bmatrix} \bar{q} S c_Y \\ \bar{q} S b c_l \\ \bar{q} S b c_n \end{Bmatrix}$$

where

$$\begin{Bmatrix} c_Y \\ c_l \\ c_n \end{Bmatrix} = \begin{bmatrix} c_{Y_\beta} & c_{Y_{\delta_A}} & c_{Y_{\delta_R}} \\ c_{l_\beta} & c_{l_{\delta_A}} & c_{l_{\delta_R}} \\ c_{n_\beta} & c_{n_{\delta_A}} & c_{n_{\delta_R}} \end{bmatrix} \begin{Bmatrix} \beta \\ \delta_A \\ \delta_R \end{Bmatrix}$$

4.6 MODELING OF THE SMALL PERTURBATION LATERAL DIRECTIONAL AERODYNAMIC FORCE AND MOMENTS

Following the modeling of the steady-state forces and moments for the lateral directional dynamics, the next objective is the modeling for the small perturbation components of the lateral force and the rolling and yawing moments. The origin of these components is associated with the perturbation terms in all the components of the linear and angular velocities, shown in Figure 4.78.

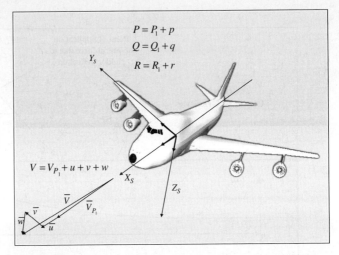

Figure 4.78 Small Perturbations in Linear and Angular Velocities

The perturbation term for the lateral velocity component can also be interpreted as perturbation in the sideslip aerodynamic angle. Therefore, using the assumption of small perturbations, we have

$$\tan \beta = \frac{\sin \beta}{\cos \beta} \approx \frac{v}{V_{P_1}} \rightarrow \beta \approx \frac{v}{V_{P_1}}, \quad \dot{\beta} \approx \frac{\dot{v}}{V_{P_1}}$$

Taking advantage of the decoupling between longitudinal and lateral directional dynamics, the functionalities of the lateral directional small perturbation moments l_A, n_A and force f_{A_Y} are described by

$$f_{A_Y} = f_{A_Y}(v, \dot{v}, p, r, \delta_A, \delta_R)$$

$$l_A = l_A(v, \dot{v}, p, r, \delta_A, \delta_R)$$

$$n_A = n_A(v, \dot{v}, p, r, \delta_A, \delta_R)$$

As for the longitudinal case, since the lateral aerodynamic coefficients are evaluated with respect to the sideslip angle β, the preceding relationships will be replaced by

$$f_{A_Y} = f_{A_Y}(\beta, \dot{\beta}, p, r, \delta_A, \delta_R)$$

$$l_A = l_A(\beta, \dot{\beta}, p, r, \delta_A, \delta_R)$$

$$n_A = n_A(\beta, \dot{\beta}, p, r, \delta_A, \delta_R)$$

Next, the modeling of the forces and moments is set up as a first order Taylor expansion with respect to the lateral directional set of independent variables $(\beta, \dot{\beta}, p, r, \delta_A, \delta_R)$.

4.6 Modeling of the Small Perturbation Lateral Directional Aerodynamic Force and Moments

$$f_{A_Y} = \frac{\partial F_{A_Y}}{\partial \beta}\bigg|_{SS} \beta + \frac{\partial F_{A_Y}}{\partial \dot{\beta}}\bigg|_{SS} \dot{\beta} + \frac{\partial F_{A_Y}}{\partial p}\bigg|_{SS} p + \frac{\partial F_{A_Y}}{\partial r}\bigg|_{SS} r + \frac{\partial F_{A_Y}}{\partial \delta_A}\bigg|_{SS} \delta_A + \frac{\partial F_{A_Y}}{\partial \delta_R}\bigg|_{SS} \delta_R$$

$$l_A = \frac{\partial L_A}{\partial \beta}\bigg|_{SS} \beta + \frac{\partial L_A}{\partial \dot{\beta}}\bigg|_{SS} \dot{\beta} + \frac{\partial L_A}{\partial p}\bigg|_{SS} p + \frac{\partial L_A}{\partial r}\bigg|_{SS} r + \frac{\partial L_A}{\partial \delta_A}\bigg|_{SS} \delta_A + \frac{\partial L_A}{\partial \delta_R}\bigg|_{SS} \delta_R$$

$$n_A = \frac{\partial N_A}{\partial \beta}\bigg|_{SS} \beta + \frac{\partial N_A}{\partial \dot{\beta}}\bigg|_{SS} \dot{\beta} + \frac{\partial N_A}{\partial p}\bigg|_{SS} p + \frac{\partial N_A}{\partial r}\bigg|_{SS} r + \frac{\partial N_A}{\partial \delta_A}\bigg|_{SS} \delta_A + \frac{\partial N_A}{\partial \delta_R}\bigg|_{SS} \delta_R$$

where the angles (β, δ_A, δ_R) are expressed in radians and are therefore dimensionless. Note that the gradients in the preceding expressions have different dimensions. An ad-hoc non-dimensionalization process is introduced with the goal of retaining the same dimensions on both sides of the equations. In this case, the goal can be achieved through using the following transformations of variables:

$$\dot{\beta}\left[\frac{\text{rad}}{\text{sec}}\right] \to \frac{\dot{\beta}b}{2V_{P_1}}, \quad p\left[\frac{\text{rad}}{\text{sec}}\right] \to \frac{pb}{2V_{P_1}}, \quad r\left[\frac{\text{rad}}{\text{sec}}\right] \to \frac{rb}{2V_{P_1}}$$

The resulting expressions are given by

$$f_{A_Y} = \frac{\partial F_{A_Y}}{\partial \beta}\bigg|_{SS} \beta + \frac{\partial F_{A_Y}}{\partial \left(\frac{\dot{\beta}b}{2V_{P_1}}\right)}\bigg|_{SS} \left(\frac{\dot{\beta}b}{2V_{P_1}}\right) + \frac{\partial F_{A_Y}}{\partial \left(\frac{pb}{2V_{P_1}}\right)}\bigg|_{SS} \left(\frac{pb}{2V_{P_1}}\right) + \frac{\partial F_{A_Y}}{\partial \left(\frac{rb}{2V_{P_1}}\right)}\bigg|_{SS} \left(\frac{rb}{2V_{P_1}}\right)_{SS}$$

$$+ \frac{\partial F_{A_Y}}{\partial \delta_A}\bigg|_{SS} \delta_A + \frac{\partial F_{A_Y}}{\partial \delta_R}\bigg|_{SS} \delta_R$$

$$l_A = \frac{\partial L_A}{\partial \beta}\bigg|_{SS} \beta + \frac{\partial L_A}{\partial \left(\frac{\dot{\beta}b}{2V_{P_1}}\right)}\bigg|_{SS} \left(\frac{\dot{\beta}b}{2V_{P_1}}\right) + \frac{\partial L_A}{\partial \left(\frac{pb}{2V_{P_1}}\right)}\bigg|_{SS} \left(\frac{pb}{2V_{P_1}}\right) + \frac{\partial L_A}{\partial \left(\frac{rb}{2V_{P_1}}\right)}\bigg|_{SS} \left(\frac{rb}{2V_{P_1}}\right)_{SS}$$

$$+ \frac{\partial L_A}{\partial \delta_A}\bigg|_{SS} \delta_A + \frac{\partial L_A}{\partial \delta_R}\bigg|_{SS} \delta_R$$

$$n_A = \frac{\partial N_A}{\partial \beta}\bigg|_{SS} \beta + \frac{\partial N_A}{\partial \left(\frac{\dot{\beta}b}{2V_{P_1}}\right)}\bigg|_{SS} \left(\frac{\dot{\beta}b}{2V_{P_1}}\right) + \frac{\partial N_A}{\partial \left(\frac{pb}{2V_{P_1}}\right)}\bigg|_{SS} \left(\frac{pb}{2V_{P_1}}\right) + \frac{\partial N_A}{\partial \left(\frac{rb}{2V_{P_1}}\right)}\bigg|_{SS} \left(\frac{rb}{2V_{P_1}}\right)_{SS}$$

$$+ \frac{\partial N_A}{\partial \delta_A}\bigg|_{SS} \delta_A + \frac{\partial N_A}{\partial \delta_R}\bigg|_{SS} \delta_R$$

Finally, the resulting expressions for the lateral directional small perturbation moments and force are given by

$$f_{A_Y} = \bar{q}S\left\{c_{Y_\beta}\beta + c_{Y_{\dot\beta}}\left(\frac{\dot\beta b}{2V_{P_1}}\right) + c_{Y_p}\left(\frac{pb}{2V_{P_1}}\right) + c_{Y_r}\left(\frac{rb}{2V_{P_1}}\right) + c_{Y_{\delta_A}}\delta_A + c_{Y_{\delta_R}}\delta_R\right\}$$

$$l_A = \bar{q}Sb\left\{c_{l_\beta}\beta + c_{l_{\dot\beta}}\left(\frac{\dot\beta b}{2V_{P_1}}\right) + c_{l_p}\left(\frac{pb}{2V_{P_1}}\right) + c_{l_r}\left(\frac{rb}{2V_{P_1}}\right) + c_{l_{\delta_A}}\delta_A + c_{l_{\delta_R}}\delta_R\right\}$$

$$n_A = \bar{q}Sb\left\{c_{n_\beta}\beta + c_{n_{\dot\beta}}\left(\frac{\dot\beta b}{2V_{P_1}}\right) + c_{n_p}\left(\frac{pb}{2V_{P_1}}\right) + c_{n_r}\left(\frac{rb}{2V_{P_1}}\right) + c_{n_{\delta_A}}\delta_A + c_{n_{\delta_R}}\delta_R\right\}$$

An analysis of the preceding relationships shows that the modeling of the small perturbation lateral directional force and moments contain nine coefficients ($c_{Y_\beta}, c_{Y_{\delta_A}}, c_{Y_{\delta_R}}, c_{l_\beta}, c_{l_{\delta_A}}, c_{l_{\delta_R}}, c_{n_\beta}, c_{n_{\delta_A}}, c_{n_{\delta_R}}$), which were introduced for the modeling of the steady-state lateral directional force and moments. The remaining nine coefficients ($c_{Y_{\dot\beta}}, c_{Y_p}, c_{Y_r}, c_{l_{\dot\beta}}, c_{l_p}, c_{l_r}, c_{n_{\dot\beta}}, c_{n_p}, c_{n_r}$) will be discussed and modeled separately.

4.6.1 Modeling of $c_{Y_{\dot\beta}}, c_{l_{\dot\beta}}, c_{n_{\dot\beta}}$

The $\dot\beta$-derivatives ($c_{Y_{\dot\beta}}, c_{l_{\dot\beta}}, c_{n_{\dot\beta}}$) are conceptually similar to the previously introduced $\dot\alpha$-derivatives ($c_{D_{\dot\alpha}}, c_{L_{\dot\alpha}}, c_{m_{\dot\alpha}}$). However, due to the lower magnitude of the sidewash effect with respect to the downwash effect, for most aircraft the importance of these derivatives is negligible. Non-negligible but still small values for these derivatives are known to exist only for aircraft with high ratios of the geometric parameter l_f/b (for example, the Boeing B-52 aircraft).

4.6.2 Modeling of c_{Y_p}

The coefficient c_{Y_p} models the contribution to the lateral force due to the roll rate. The vertical tail is the only component of the aircraft contributing to c_{Y_p}. A relationship providing an estimate for this coefficient is given by

$$c_{Y_p} \approx c_{Y_{p_V}} \approx 2\,c_{Y_{\beta_V}}\frac{(Z_V\cos\alpha_1 - X_V\sin\alpha_1)}{b}$$

where the modeling of $c_{Y_{\beta_V}}$ has been described in Section 4.2.2; the geometric parameters Z_V, X_V are shown in Figure 4.79.

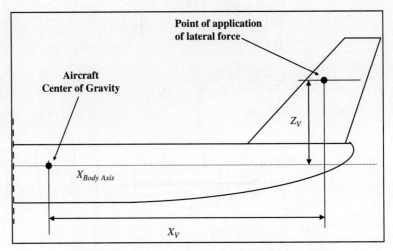

Figure 4.79 Geometric Parameters of the Vertical Tail

4.6.3 Modeling of c_{l_p}

The coefficient c_{l_p} models the contribution to the rolling moment due to the roll rate. The effect of this coefficient on the value of a key aircraft handling quality (the roll time constant) will be described in Chapter 7. The modeling for this important coefficient can be broken down into the following:

$$c_{lp} = c_{lp_{WB}} + c_{lp_H} + c_{lp_V}$$

Considering that the contribution to this coefficient from the fuselage is negligible, a relationship for $c_{lp_{WB}}$ is given by

$$c_{lp_{WB}} \approx c_{lp_W} = RDP \cdot \frac{k}{\beta}$$

where $\beta = \sqrt{1 - Mach^2}$. The parameter RDP, the rolling damping parameter, is evaluated using Figures 4.80 and 4.81.

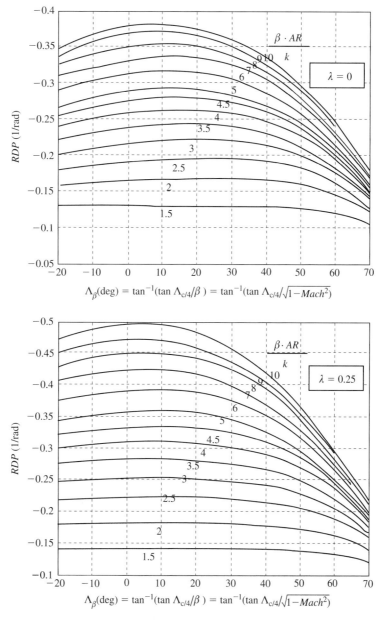

Figure 4.80 Rolling Damping Parameters for Different Wing Geometry ($\lambda = 0, \lambda = 0.25$)[5]

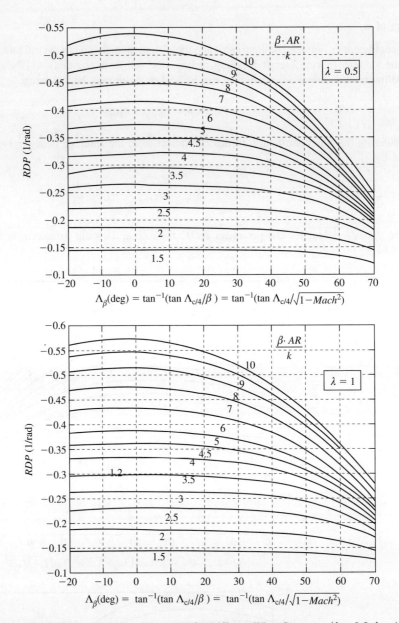

Figure 4.81 Rolling Damping Parameters for Different Wing Geometry ($\lambda = 0.5, \lambda = 1$)[5]

As for the modeling of $c_{l\delta_A}$ in Section 4.2.3, the plots in Figures 4.80 and 4.81 are given for different values of the parameter $\beta \cdot AR/k$ where

$$k = \frac{(c_{l_\alpha})_{Wing\ Section}|_{Mach} \cdot \beta}{2\pi}$$

Within a reasonable approximation the wing planform lift curve slope $c_{L_{\alpha_W}}|_{Mach}$ can be used in lieu of $(c_{l_\alpha})_{Wing\ Section}|_{Mach}$.

Next, the contribution from the horizontal tail can be modeled using

$$c_{lp_H} = \frac{1}{2}(c_{lp_W})\big|_H \frac{S_H}{S}\left(\frac{b_H}{b}\right)^2$$

where $(c_{lp_W})|_H$ is the value of the previously introduced c_{lp_W} evaluated using the parameters of the horizontal tail. For most subsonic configurations the geometric parameters of the horizontal tail and the wing are similar; therefore, it is acceptable to approximate that $(c_{lp_W})|_H \approx c_{lp_W}$. However, the value of c_{lp_H} is often negligible due to the low numerical value of the product of $S_H/S \cdot (b_H/b)^2$.

The contribution from the vertical tail can be approximated by the following relationship:

$$c_{l p_V} \approx 2\, c_{Y_{\beta_V}} \left(\frac{Z_V}{b}\right)^2$$

where the modeling of $c_{Y_{\beta_V}}$ has been described in Section 4.2.2 and the geometric parameter Z_V is shown in Figure 4.79.

4.6.4 Modeling of c_{n_p}

The coefficient c_{n_p} models the contribution to the yawing moment due to the roll rate. The modeling for this coefficient starts from the following:

$$c_{np} = c_{np_{WB}} + c_{np_H} + c_{np_V}$$

Because the fuselage and the horizontal tail do not significantly contribute to this coefficient, we have

$$c_{np} \approx c_{np_W} + c_{np_V}$$

The modeling for c_{np_W} is quite complex; a relationship for c_{np_W} is given by

$$c_{np_W} = -c_{lp_W} \tan(\alpha_1) + c_{lp} \tan(\alpha_1) + \left(\frac{c_{np}}{c_{L_1}}\right)\Bigg|_{\substack{Mach \\ C_L=0}} c_{L_1} + \left(\frac{\Delta c_{np}}{\varepsilon_W}\right)\varepsilon_W$$

where c_{lp_W}, c_{lp} are described in the previous section and c_{L_1} is the steady-state lift-coefficient. At small steady state α_1 we have

$$c_{np_W} \approx \cancel{-c_{lp_W} \tan(\alpha_1)} + \cancel{c_{lp} \tan(\alpha_1)} + \left(\frac{c_{np}}{c_{L_1}}\right)\Bigg|_{\substack{Mach \\ C_L=0}} c_{L_1} + \left(\frac{\Delta c_{np}}{\varepsilon_W}\right)\varepsilon_W$$

$$= \left(\frac{c_{np}}{c_{L_1}}\right)\Bigg|_{\substack{Mach \\ C_L=0}} c_{L_1} + \left(\frac{\Delta c_n}{\varepsilon_W}\right)\varepsilon_W$$

The coefficient $\left(\frac{c_{np}}{c_{L_1}}\right)\Big|_{\substack{Mach \\ C_L=0}}$ is given by

$$\left(\frac{c_{np}}{c_{L_1}}\right)\Bigg|_{\substack{Mach \\ C_L=0}} = C \left(\frac{c_{np}}{c_{L_1}}\right)\Bigg|_{\substack{Mach=0 \\ C_L=0}}$$

where

$$C = \frac{[AR + 4\cos(\Lambda_{c/4})]}{[AR \cdot B + 4\cos(\Lambda_{c/4})]} \cdot \left\{ \frac{AR \cdot B + \frac{1}{2}[AR \cdot B + 4\cos(\Lambda_{c/4})] \cdot \tan^2(\Lambda_{c/4})}{AR + \frac{1}{2}[AR + 4\cos(\Lambda_{c/4})] \cdot \tan^2(\Lambda_{c/4})} \right\}$$

with

$$B = \sqrt{1 - Mach^2 \cos^2(\Lambda_{c/4})}$$

$$\left(\frac{c_{np}}{c_{L_1}}\right)\Bigg|_{\substack{Mach=0 \\ C_L=0}}$$

$$= -\frac{1}{6} \cdot \frac{AR + 6(AR + \cos(\Lambda_{c/4})) \cdot \left[(\bar{x}_{CG} - \bar{x}_{AC})\frac{\tan(\Lambda_{c/4})}{AR} + \frac{\tan^2(\Lambda_{c/4})}{12}\right]}{(AR + \cos(\Lambda_{c/4}))} \quad (\text{rad}^{-1})$$

$\left(\frac{\Delta c_{np}}{\varepsilon_W}\right)$ is the contribution associated with the wing twist angle ε_W (shown again in Figure 4.82) evaluated using Figure 4.83.

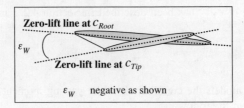

Figure 4.82 Wing Twist Angle ε_W[5]

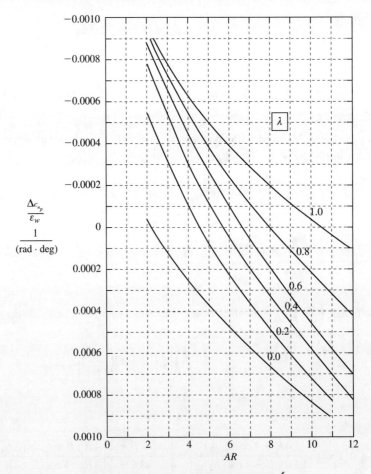

Figure 4.83 Effect of Wing Twist on c_{n_p}[5]

Next, a relationship for c_{np_V} is given by

$$c_{np_V} \approx -2 \cdot c_{Y_{\beta_V}} \cdot \frac{(X_V \cos\alpha_1 + Z_V \sin\alpha_1)}{b} \cdot \frac{(Z_V \cos\alpha_1 - X_V \sin\alpha_1 - Z_V)}{b}$$

where the modeling of $c_{Y_{\beta_V}}$ has been described in Section 4.2.2; the geometric parameters Z_V, X_V are shown in Figure 4.84.

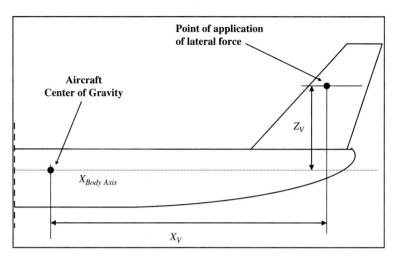

Figure 4.84 Geometric Parameters of the Vertical Tail

4.6.5 Modeling of c_{Y_r}

The coefficient c_{Y_r} models the contribution to the lateral force due to the yaw rate. As for c_{Y_p}, the vertical tail is the only component of the aircraft contributing to c_{Y_r}. A relationship for an estimate of this coefficient is given by

$$c_{Y_r} \approx c_{Y_{r_V}} \approx -2\, c_{Y_{\beta_V}} \frac{(X_V \cos\alpha_1 + Z_V \sin\alpha_1)}{b}$$

where the modeling of $c_{Y_{\beta_V}}$ has been described in Section 4.2.2; essentially the difference between c_{Y_p} and c_{Y_r} is only in the moment arm. The geometric parameters Z_V, X_V are shown in Figure 4.84.

4.6.6 Modeling of c_{l_r}

The coefficient c_{l_r} models the contribution to the rolling moment due to the yaw rate. The modeling for this coefficient starts from the following:

$$c_{l_r} = c_{l_{r_{WB}}} + c_{l_{r_H}} + c_{l_{r_V}}$$

Since the fuselage and the horizontal tail do not significantly contribute to this coefficient, we have

$$c_{l_r} \approx c_{l_{r_W}} + c_{l_{r_V}}$$

A relationship for $c_{l_{r_W}}$ is given by

$$c_{l_{r_W}} \approx \left(\frac{c_{l_r}}{c_{L_1}}\right)\bigg|_{\substack{Mach \\ C_L=0}} \cdot c_{L_1} + \left(\frac{\Delta c_{l_r}}{\Gamma}\right) \cdot \Gamma + \left(\frac{\Delta c_{l_r}}{\varepsilon_W}\right) \cdot \varepsilon_W \quad (\text{rad}^{-1})$$

The coefficient $\left(\dfrac{c_{l_r}}{c_{L_1}}\right)\bigg|_{\substack{Mach \\ C_L=0}}$ is given by

$$\left(\frac{c_{l_r}}{c_{L_1}}\right)\bigg|_{\substack{Mach \\ C_L=0}} = D \cdot \left(\frac{c_{l_r}}{c_{L_1}}\right)\bigg|_{\substack{Mach=0 \\ C_L=0}}$$

where

$$D = \frac{1 + \dfrac{AR(1-B^2)}{2B[AR \cdot B + 2\cos(\Lambda_{c/4})]} + \dfrac{[AR \cdot B + 2\cos(\Lambda_{c/4})]}{[AR \cdot B + 4\cos(\Lambda_{c/4})]} \cdot \dfrac{\tan^2(\Lambda_{c/4})}{8}}{1 + \dfrac{[AR + 2\cos(\Lambda_{c/4})]}{[AR + 4\cos(\Lambda_{c/4})]} \cdot \dfrac{\tan^2(\Lambda_{c/4})}{8}}$$

$$B = \sqrt{1 - Mach^2 \cos^2(\Lambda_{c/4})}$$

$\left(\dfrac{c_{l_r}}{c_{L_1}}\right)\bigg|_{\substack{Mach=0 \\ C_L=0}}$ is evaluated through Figure 4.85 using the wing geometric parameters.

$\left(\dfrac{\Delta c_{l_r}}{\Gamma}\right)$ is a factor due to the wing dihedral angle Γ modeled using

$$\left(\frac{\Delta c_{lr}}{\Gamma}\right) \approx \left(\frac{1}{12}\right) \cdot \frac{[\pi \cdot AR \cdot \sin(\Lambda_{c/4})]}{(AR + 4 \cdot \cos(\Lambda_{c/4}))} \quad (\text{rad}^{-2})$$

$\left(\frac{\Delta c_{lr}}{\varepsilon_W}\right)$ is a factor due to the wing twist angle ε_W, shown in Figure 4.86, modeled using Figure 4.87.

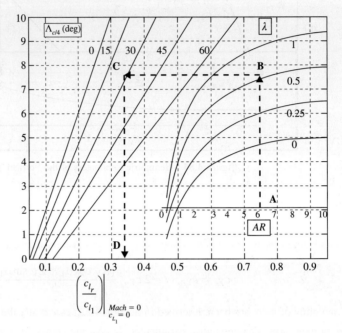

Figure 4.85 Evaluation of $(c_{lr}/c_{L_1})\big|_{\substack{Mach=0 \\ C_L=0}}$[5]

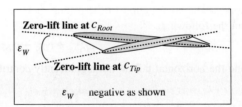

Figure 4.86 Wing Twist Angle ε_W[5]

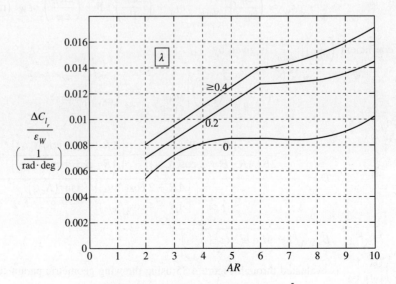

Figure 4.87 Effect of Wing Twist on c_{l_r}[5]

4.6 Modeling of the Small Perturbation Lateral Directional Aerodynamic Force and Moments

Next, a relationship for $c_{l_{r_V}}$ is given by

$$c_{l_{r_V}} \approx -2 \cdot c_{Y_{\beta_V}} \cdot \frac{(X_V \cos \alpha_1 + Z_V \sin \alpha_1)}{b} \cdot \frac{(Z_V \cos \alpha_1 - X_V \sin \alpha_1)}{b}$$

where the modeling of $c_{Y_{\beta_V}}$ has been described in Section 4.2.2; the geometric parameters Z_V, X_V are shown in Figure 4.88.

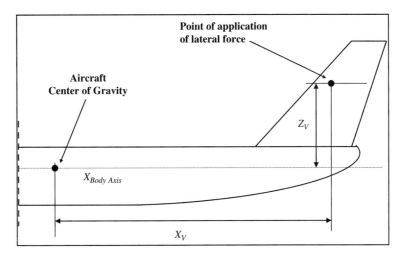

Figure 4.88 Geometric Parameters of the Vertical Tail

4.6.7 Modeling of c_{n_r}

The coefficient c_{n_r} models the contribution to the yawing moment due to the yaw rate. The importance of this coefficient on key aircraft parameters for handling qualities will be discussed in Chapter 7. The modeling for this important coefficient can be broken down into the following:

$$c_{n_r} = c_{n_{r_{WB}}} + c_{n_{r_H}} + c_{n_{r_V}}$$

The fuselage and the horizontal tail do not significantly contribute to this coefficient; therefore, we have

$$c_{n_r} \approx c_{n_{r_W}} + c_{n_{r_V}}$$

A relationship for $c_{n_{r_W}}$ is given

$$c_{n_{r_W}} \approx \left(\frac{c_{n_r}}{c_{L_1}}\right) \cdot c_{L_1}^2 + \left(\frac{c_{n_r}}{\overline{c}_{D_0}}\right) \cdot \overline{c}_{D_0} \quad (\text{rad}^{-1})$$

where $\overline{c}_{D_0}$ is the parasite drag coefficient (described in Chapter 3) which is evaluated through the drag polar curve for the total aircraft. The gradients $\left(\frac{c_{n_r}}{c_{L_1}}\right)$ and $\left(\frac{c_{n_r}}{\overline{c}_{D_0}}\right)$ are evaluated using Figure 4.89 and Figure 4.90, respectively. For most configurations, the contribution associated with $\left(\frac{c_{n_r}}{\overline{c}_{D_0}}\right)$ is negligible with respect to the contribution associated with $\left(\frac{c_{n_r}}{c_{L_1}}\right)$. Therefore, we have

$$c_{n_{r_W}} = \left(\frac{c_{n_r}}{c_{L_1}^2}\right) \cdot c_{L_1}^2 + \left(\frac{c_{n_r}}{\overline{c}_{D_0}}\right) \cdot \overline{c}_{D_0} \approx \left(\frac{c_{n_r}}{c_{L_1}^2}\right) \cdot c_{L_1}^2$$

188 Chapter 4 Modeling of Lateral Directional Aerodynamic Forces and Moments

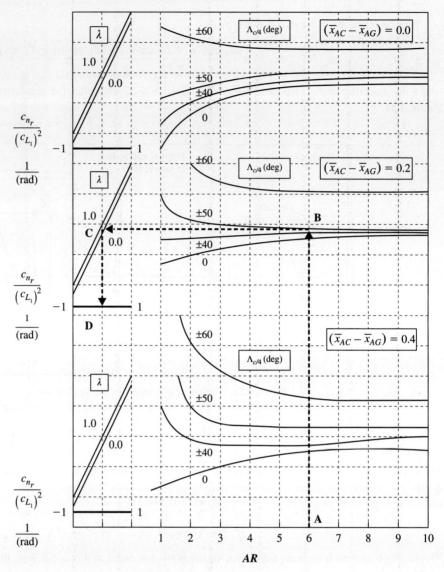

Figure 4.89 Effect of Lift on c_{n_r}[5]

Finally, a relationship for $c_{n_{r_V}}$ is given by

$$c_{n_{r_V}} \approx 2 \cdot c_{Y_{\beta_V}} \cdot \frac{(X_V \cos\alpha_1 + Z_V \sin\alpha_1)^2}{b^2}$$

where the modeling of $c_{Y_{\beta_V}}$ has been described in, Section 4.2.2; the geometric parameters Z_V, X_V are shown in Figure 4.88.

4.7 Summary of Longitudinal and Lateral Directional Aerodynamic Stability and Control Derivatives

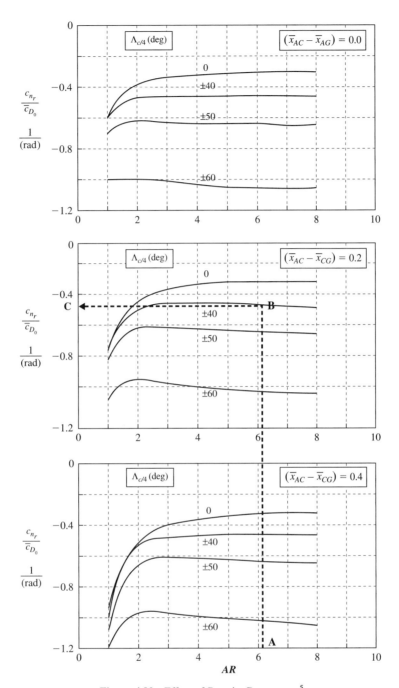

Figure 4.90 Effect of Parasite Drag on $c_{n_r}{}^5$

4.7 SUMMARY OF LONGITUDINAL AND LATERAL DIRECTIONAL AERODYNAMIC STABILITY AND CONTROL DERIVATIVES

In the following pages Figures 4.91 to 4.96 will provide a comprehensive graphical summary of all the longitudinal and lateral directional aerodynamic stability and control derivatives introduced in Chapters 3 and 4.

$$f_{A_X} = \bar{q}S\left\{-\left[c_{D_u} + 2c_{D_1}\right]\left(\frac{u}{V_{P_1}}\right) + \left[-c_{D_\alpha} + c_{L_1}\right]\alpha - c_{D_{\dot\alpha}}\left(\frac{\dot\alpha \bar{c}}{2V_{P_1}}\right) - c_{D_q}\left(\frac{q\bar{c}}{2V_{P_1}}\right) - c_{D_{\delta_E}}\delta_E - c_{D_{i_H}}i_H\right\}$$

$c_{D_1} = \dfrac{T_1}{\bar{q}_1 S} = c_{D_0} + C_{D_\alpha}\alpha_1 + C_{D_{\delta_E}}\delta_{E_1} + c_{D_{i_H}}i_{H_1}$

$c_{L_1} = \dfrac{W}{\bar{q}_1 S}$ ←—— $L_1 = c_{L_1}\bar{q}S = W$

$c_{D_0} = c_D\big|_{\alpha = \delta_E = i_H = 0°}$ ←—— [5],[6] or wind tunnel analysis

c_{D_α} ←—— [5],[6] or wind tunnel analysis

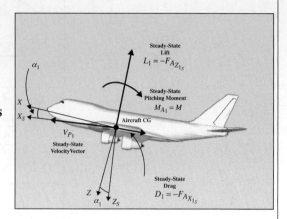

SPECIAL CASE : 'uncambered' aircraft (with parabolic 'drag' polar curve)

$\bar{c}_{D_0} = c_{D_0}$

$c_{D_\alpha} \approx \left(\dfrac{2c_{L_1}}{\pi AR\, e}\right) c_{L_\alpha}$

$c_{D_u} = Mach \cdot \dfrac{\partial c_D}{\partial Mach} \approx 0$ at low subsonic

$c_{D_{\dot\alpha}} \approx 0$

$c_{D_q} \approx 0$

$\left.\begin{array}{l}c_{D_{\delta_E}} \approx 0\\ c_{D_{i_H}} \approx 0\end{array}\right\}$ ... except for large horizontal tails with large dihedral/anhedral angles

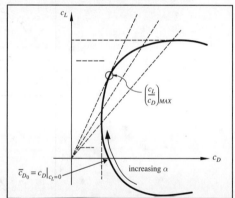

Figure 4.91 Summary of the c_D Stability and Control Derivatives

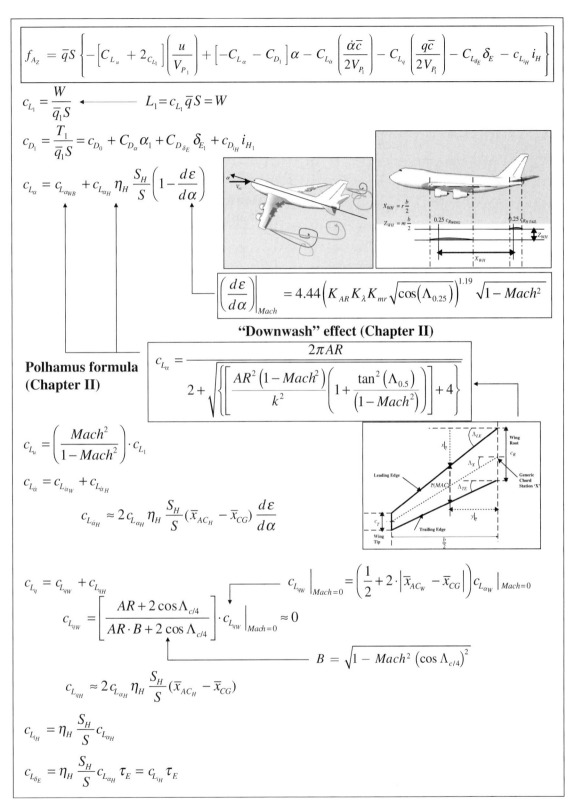

Figure 4.92 Summary of the c_L Stability and Control Derivatives

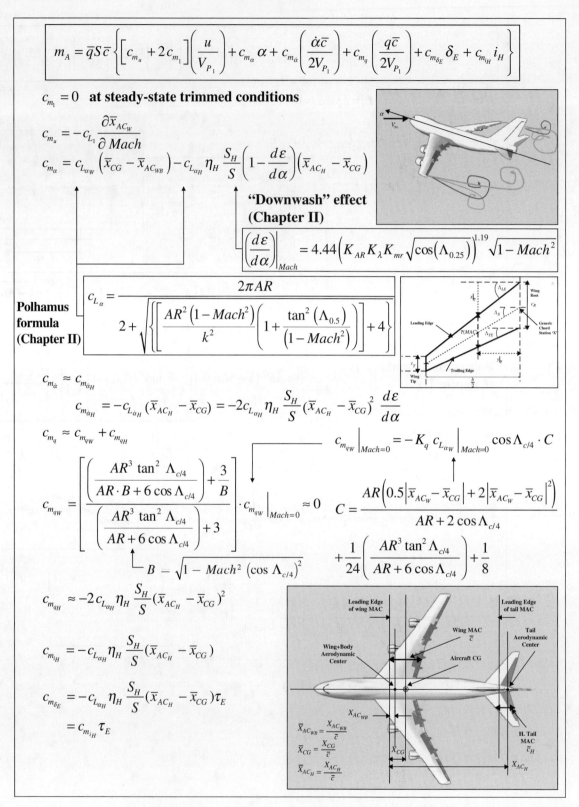

$$m_A = \bar{q}S\bar{c}\left\{\left[c_{m_u} + 2c_{m_1}\right]\left(\frac{u}{V_{P_1}}\right) + c_{m_\alpha}\alpha + c_{m_{\dot{\alpha}}}\left(\frac{\dot{\alpha}\bar{c}}{2V_{P_1}}\right) + c_{m_q}\left(\frac{q\bar{c}}{2V_{P_1}}\right) + c_{m_{\delta_E}}\delta_E + c_{m_{i_H}}i_H\right\}$$

$c_{m_1} = 0$ at steady-state trimmed conditions

$$c_{m_u} = -c_{L_1}\frac{\partial \bar{x}_{AC_W}}{\partial \text{Mach}}$$

$$c_{m_\alpha} = c_{L_{\alpha_W}}(\bar{x}_{CG} - \bar{x}_{AC_{WB}}) - c_{L_{\alpha_H}}\eta_H \frac{S_H}{S}\left(1 - \frac{d\varepsilon}{d\alpha}\right)(\bar{x}_{AC_H} - \bar{x}_{CG})$$

"Downwash" effect (Chapter II)

$$\left(\frac{d\varepsilon}{d\alpha}\right)\bigg|_{\text{Mach}} = 4.44\left(K_{AR}K_\lambda K_{mr}\sqrt{\cos(\Lambda_{0.25})}\right)^{1.19}\sqrt{1 - \text{Mach}^2}$$

Polhamus formula (Chapter II)

$$c_{L_\alpha} = \frac{2\pi AR}{2 + \sqrt{\left[\frac{AR^2(1-\text{Mach}^2)}{k^2}\left(1 + \frac{\tan^2(\Lambda_{0.5})}{(1-\text{Mach}^2)}\right)\right] + 4}}$$

$c_{m_{\dot{\alpha}}} \approx c_{m_{\dot{\alpha}H}}$

$$c_{m_{\dot{\alpha}H}} = -c_{L_{\dot{\alpha}H}}(\bar{x}_{AC_H} - \bar{x}_{CG}) = -2c_{L_{\alpha_H}}\eta_H\frac{S_H}{S}(\bar{x}_{AC_H} - \bar{x}_{CG})^2\frac{d\varepsilon}{d\alpha}$$

$c_{m_q} \approx c_{m_{qW}} + c_{m_{qH}}$

$$c_{m_{qW}} = \left[\frac{\left(\frac{AR^3\tan^2\Lambda_{c/4}}{AR\cdot B + 6\cos\Lambda_{c/4}}\right) + \frac{3}{B}}{\left(\frac{AR^3\tan^2\Lambda_{c/4}}{AR + 6\cos\Lambda_{c/4}}\right) + 3}\right]\cdot c_{m_{qW}}\bigg|_{\text{Mach}=0}$$

$B = \sqrt{1 - \text{Mach}^2(\cos\Lambda_{c/4})^2}$

$$c_{m_{qW}}\bigg|_{\text{Mach}=0} = -K_q\, c_{L_{\alpha_W}}\bigg|_{\text{Mach}=0}\cos\Lambda_{c/4}\cdot C$$

$$C = \frac{AR\left(0.5|\bar{x}_{AC_W} - \bar{x}_{CG}| + 2|\bar{x}_{AC_W} - \bar{x}_{CG}|^2\right)}{AR + 2\cos\Lambda_{c/4}} + \frac{1}{24}\left(\frac{AR^3\tan^2\Lambda_{c/4}}{AR + 6\cos\Lambda_{c/4}}\right) + \frac{1}{8}$$

$$c_{m_{qH}} \approx -2c_{L_{\alpha_H}}\eta_H\frac{S_H}{S}(\bar{x}_{AC_H} - \bar{x}_{CG})^2$$

$$c_{m_{i_H}} = -c_{L_{\alpha_H}}\eta_H\frac{S_H}{S}(\bar{x}_{AC_H} - \bar{x}_{CG})$$

$$c_{m_{\delta_E}} = -c_{L_{\alpha_H}}\eta_H\frac{S_H}{S}(\bar{x}_{AC_H} - \bar{x}_{CG})\tau_E = c_{m_{i_H}}\tau_E$$

$\bar{x}_{AC_{WB}} = \frac{X_{AC_{WB}}}{\bar{c}}$

$\bar{x}_{CG} = \frac{X_{CG}}{\bar{c}}$

$\bar{x}_{AC_H} = \frac{X_{AC_H}}{\bar{c}}$

Figure 4.93 Summary of the c_m Stability and Control Derivatives

$$f_{A_Y} = \bar{q}S\left\{c_{Y_\beta}\beta + c_{Y_{\dot\beta}}\left(\frac{\dot\beta b}{2V_{P_1}}\right) + c_{Y_r}\left(\frac{pb}{2V_{P_1}}\right) + c_{Y_r}\left(\frac{rb}{2V_{P_1}}\right) + c_{Y_{\delta_A}}\delta_A + c_{Y_{\delta_R}}\delta_R\right\}$$

$c_{Y\beta} = (c_{Y\beta_W} + c_{Y\beta_B}) + c_{Y\beta_H} + c_{Y\beta_V}$

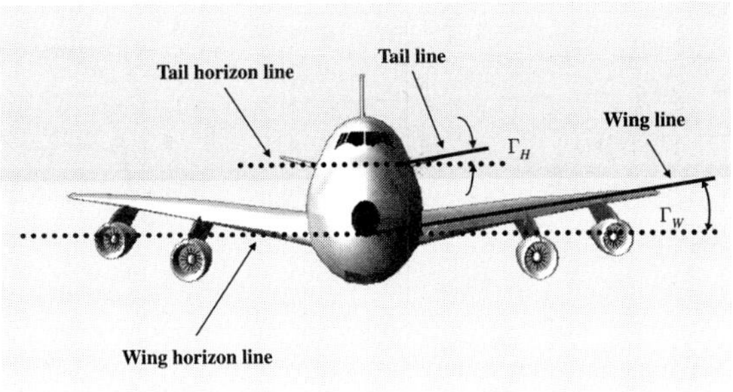

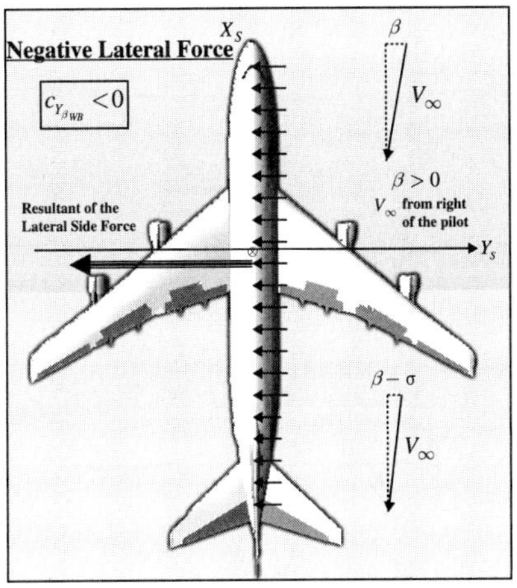

$c_{Y\beta_W} \approx -0.0001|\Gamma_W|\cdot 57.3$

$c_{Y\beta_W} \approx -2\cdot K_{int}\cdot \dfrac{S_{P\to V}}{S}$

$c_{Y\beta_H} \approx -0.0001|\Gamma_H|\cdot 57.3 \cdot \eta_H \cdot\left(1+\dfrac{d\sigma}{d\beta}\right)\dfrac{S_H}{S}$

$\quad \llcorner\; \eta_H \cdot\left(1+\dfrac{d\sigma}{d\beta}\right) = 0.724 + 3.06\dfrac{S_H/S}{1+\cos(\Lambda_{c/4})} + 0.4\dfrac{Z_W}{d} + 0.009\cdot AR$

$c_{Y\beta_V} \approx -k_{Y_V}\cdot|c_{L_{\alpha_V}}|\eta_V\cdot\left(1+\dfrac{d\sigma}{d\beta}\right)\dfrac{S_V}{S}$

$\quad \llcorner\; \eta_V \cdot\left(1+\dfrac{d\sigma}{d\beta}\right) = 0.724 + 3.06\dfrac{S_V/S}{1+\cos(\Lambda_{c/4})} + 0.4\dfrac{Z_W}{d} + 0.009\cdot AR$

— **Polhamus formula with** $AR_{V_{eff}} = c_1\cdot AR_V\cdot\left[1 + K_H(c_2 - 1)\right]$

$c_{Y_{\dot\beta}} \approx 0$

$c_{Y_p} \approx c_{Y_{p_V}} = 2\, c_{Y_{\beta_V}}\dfrac{(Z_V\cos\alpha_1 - X_V\sin\alpha_1)}{b}$

$c_{Y_r} \approx c_{Y_{r_V}} = -2\, c_{Y_{\beta_V}}\dfrac{(X_V\cos\alpha_1 + Z_V\sin\alpha_1)}{b}$

$c_{Y_{\delta_A}} \approx 0$

$c_{Y_{\delta_R}} = |c_{L_{\alpha_V}}|\eta_V\dfrac{S_V}{S}K_R\tau_R$

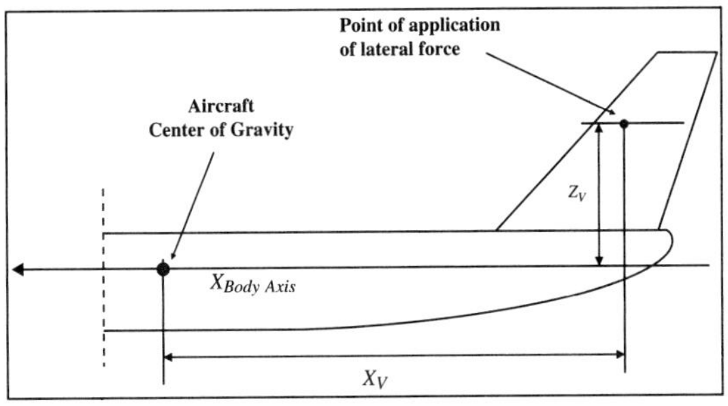

Figure 4.94 Summary of the 'c_Y' Stability and Control Derivatives

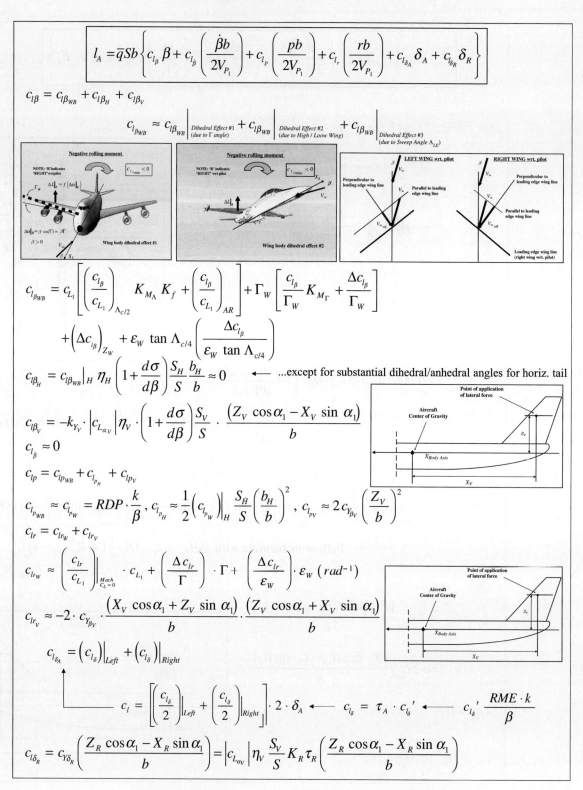

Figure 4.95 Summary of the c_l Stability and Control Derivatives

$$n_A = \bar{q}Sb\left\{c_{n_\beta}\beta + c_{n_{\dot\beta}}\left(\frac{\dot\beta b}{2V_{P_1}}\right) + c_{n_p}\left(\frac{pb}{2V_{P_1}}\right) + c_{n_r}\left(\frac{rb}{2V_{P_1}}\right) + c_{n_{\delta_A}}\delta_A + c_{n_{\delta_R}}\delta_R\right\}$$

$c_{n\beta} = c_{n\beta_W} + c_{n\beta_B} + c_{n\beta_H} + c_{n\beta_V}$

$c_{n\beta_B} = -57.3\, K_N\, K_{R_l}\, \dfrac{S_{B_S}}{S}\, \dfrac{l_B}{b}$

$c_{n\beta_H} \approx 0$

$c_{n\beta_W} \approx 0$

$c_{n\beta_V} = k_{Y_V} \cdot \left|c_{L_{\alpha V}}\right| \eta_V \cdot \left(1 + \dfrac{d\sigma}{d\beta}\right)\dfrac{S_V}{S} \cdot \dfrac{(X_V \cos\alpha_1 + Z_V \sin\alpha_1)}{b}$

$c_{n_{\dot\beta}} \approx 0$

$c_{n_p} = c_{n_{p_W}} + c_{n_{p_V}}$

$c_{n_{p_W}} \approx \left(\dfrac{c_{np}}{c_{L_1}}\right)\bigg|_{\substack{Mach\\ C_L=0}} c_{L_1} + \left(\dfrac{\Delta c_{np}}{\varepsilon_W}\right)\varepsilon_W$

$c_{n_{p_V}} \approx -2 \cdot c_{Y_{\beta_V}} \cdot \dfrac{(X_V \cos\alpha_1 + Z_V \sin\alpha_1)}{b} \cdot \dfrac{(Z_V \cos\alpha_1 - X_V \sin\alpha_1 - Z_V)}{b}$

$c_{n_r} \approx c_{n_{r_W}} + c_{n_{r_V}}$

$c_{n_{r_W}} \approx \left(\dfrac{c_{n_r}}{c_{L_1}^2}\right) \cdot c_{L_1}^2 + \left(\dfrac{c_{nr}}{\bar{c}_{D_0}}\right)\cdot \bar{c}_{D_0}$

$c_{n_{r_V}} \approx 2 \cdot c_{Y_{\beta_V}} \cdot \dfrac{(X_V \cos\alpha_1 + Z_V \sin\alpha_1)}{b^2}$

$c_{n_{\delta_A}} = K_{n_A} c_{L_1} C_{l_{\delta_A}}$

$c_{n_{\delta_R}} = -\left|c_{L_{\alpha V}}\right| \eta_V \dfrac{S_V}{S} K_R \tau_R \left(\dfrac{X_R \cos\alpha_1 + Z_R \sin\alpha_1}{b}\right)$

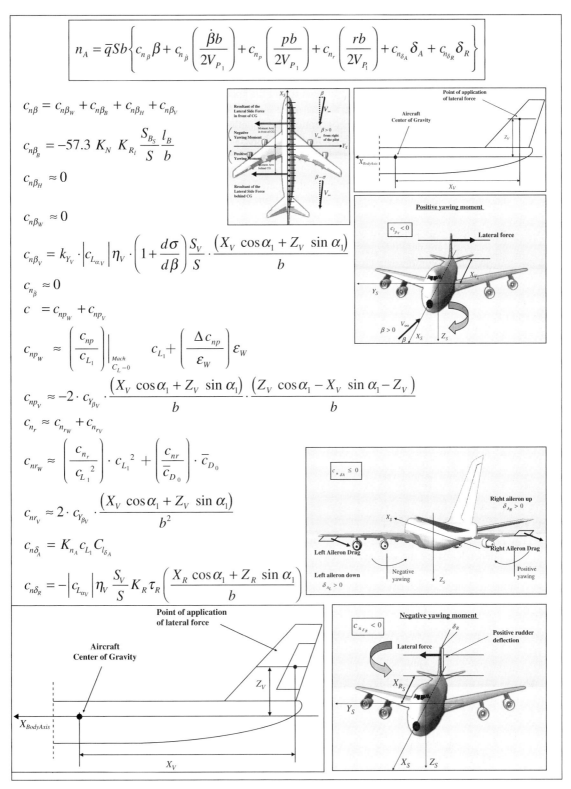

Figure 4.96 Summary of the c_n Stability and Control Derivatives

4.8 FINAL OVERVIEW AND RANKING OF THE IMPORTANCE OF THE AERODYNAMIC COEFFICIENTS

From the preceding analysis it is clear that the overall aerodynamic behavior of the aircraft is dependent on the values of a large number of coefficients. It is very critical for an aerospace engineering student to understand the relative importance of these coefficients and the specific signature of these coefficients on the overall aircraft dynamics. This important aspect will be more evident in Chapter 7.

The first step is to provide a conceptual classification of the preceding coefficients in two distinct classes, the control coefficients (also known as the control derivatives) and the stability coefficients (also known as the aerodynamic stability derivatives). Both classes are discussed separately.

Recall that only the primary longitudinal and lateral directional control surfaces have been considered in the modeling described previously. In general, depending on the specific class of aircraft, additional control surfaces might be available, including but not limited to the following:

- Leading-edge flaps, trailing-edge flaps, spoilers, or speedbrakes (δ_{LEF}, δ_{TEF}, δ_{SP}). These control surfaces on the wing are deployed symmetrically for longitudinal control at low speeds for approach and landing. The associated control derivatives are given by

$$(c_{D_{\delta_{LEF}}}, c_{L_{\delta_{LEF}}}, c_{m_{\delta_{LEF}}}, c_{D_{\delta_{TEF}}}, c_{L_{\delta_{TEF}}}, c_{m_{\delta_{TEF}}}, c_{D_{\delta_{SP}}}, c_{L_{\delta_{SP}}}, c_{m_{\delta_{SP}}}, c_{D_{\delta_{SB}}}, c_{L_{\delta_{SB}}}, c_{m_{\delta_{SB}}})$$

Since these control surfaces are located at a small longitudinal distance from the aircraft center of gravity, they provide substantial drag and lift but a limited pitching moment.

- Canards (δ_C). These control surfaces located in front of the center of gravity are deployed symmetrically for longitudinal control. The associated control derivatives are given by $(c_{D_{\delta_C}}, c_{L_{\delta_C}}, c_{m_{\delta_C}})$. As discussed in Chapter 3, these surfaces are rarely used on production aircraft. A few experimental aircraft are known to feature differential canards deployed to augment rolling capabilities. Unlike the previous control surfaces, the deflection of canards provides substantial pitching moment in the opposite direction of the pitching moment from the horizontal tail.

- Spoilerons, flaperons, and ruddervators (δ_{SPL}, δ_{FPL}, δ_{RV}). These control surfaces on the wing are deployed asymmetrically for lateral directional control. The associated control derivatives are given by

$$(c_{Y_{\delta_{SPL}}}, c_{l_{\delta_{SPL}}}, c_{n_{\delta_{SPL}}}, c_{Y_{\delta_{FPL}}}, c_{l_{\delta_{FPL}}}, c_{n_{\delta_{FPL}}}, c_{Y_{\delta_{RV}}}, c_{l_{\delta_{RV}}}, c_{n_{\delta_{RV}}})$$

For several jetliners these control surfaces might not be deflected directly by the pilot but are deployed independently by the stability augmentation systems or by the command augmentation systems.

A key aspect of the control coefficients or aerodynamic control derivatives is that their value does not affect the stability characteristics of the aircraft but only the pilot's ability to control the aircraft and to provide enough lift in the take-off and landing phases of the flight. A summary of the control derivatives is provided in Table 4.2.

Table 4.2 Summary of the Longitudinal and Lateral Directional Control Derivatives

Longitudinal Control Derivatives	Elevator: $c_{D_{\delta_E}}$, $c_{L_{\delta_E}}$, $c_{m_{\delta_E}}$
	Stabilator: $c_{D_{i_H}}$, $c_{L_{i_H}}$, $c_{m_{i_H}}$
	Leading Edge Flaps: $c_{D_{\delta_{LEF}}}$, $c_{L_{\delta_{LEF}}}$, $c_{m_{\delta_{LEF}}}$
	Trailing Edge Flaps: $c_{D_{\delta_{TEF}}}$, $c_{L_{\delta_{TEF}}}$, $c_{m_{\delta_{TEF}}}$
	Spoilers: $c_{D_{\delta_{SP}}}$, $c_{L_{\delta_{SP}}}$, $c_{m_{\delta_{SP}}}$
	Speed Brakes: $c_{D_{\delta_{SB}}}$, $c_{L_{\delta_{SB}}}$, $c_{m_{\delta_{SB}}}$
	Canards: $c_{D_{\delta_C}}$, $c_{L_{\delta_C}}$, $c_{m_{\delta_C}}$
Lateral Directional Control Derivatives	Ailerons: $c_{Y_{\delta_A}}$, $c_{l_{\delta_A}}$, $c_{n_{\delta_A}}$
	Rudder: $c_{Y_{\delta_R}}$, $c_{l_{\delta_R}}$, $c_{n_{\delta_R}}$
	Spoilerons: $c_{Y_{\delta_{SPL}}}$, $c_{l_{\delta_{SPL}}}$, $c_{n_{\delta_{SPL}}}$
	Flaperons: $c_{Y_{\delta_{FPL}}}$, $c_{l_{\delta_{FPL}}}$, $c_{n_{\delta_{FPL}}}$
	Ruddervators: $c_{Y_{\delta_{RV}}}$, $c_{l_{\delta_{RV}}}$, $c_{n_{\delta_{RV}}}$

4.8 Final Overview and Ranking of the Importance of the Aerodynamic Coefficients

Although they do not affect the aircraft stability, the correct design and sizing of the aircraft control surfaces—and therefore the appropriate numerical ranges for the values for all the above coefficients—are critical for the different phases leading to the certification of the aircraft.

Next, the attention shifts to a qualitative analysis of the relative importance of the aerodynamic stability derivatives. An approach toward providing a qualitative ranking of the specific importance of these coefficients is to introduce different groups with different levels of criticality, as shown in Table 4.3 with decreasing ranking within an arbitrary scale between 0 and 10.

Table 4.3 Relative Importance of the Longitudinal and Lateral Directional Stability Derivatives

Relative Importance	Ranking (0 to 10)	Stability Derivatives
Group #1	10	$c_{L_\alpha}, c_{m_\alpha}, c_{l_\beta}, c_{n_\beta}$
Group #2	9	$c_{m_{\dot\alpha}}, c_{m_q}, c_{l_p}, c_{n_r}$
Group #3	7–8	c_{D_0}, c_{D_α}
Group #4	6	c_{L_0}, c_{m_0}
Group #5	4	$c_{L_{\dot\alpha}}, c_{L_q}$
Group #6	3	$c_{Y_\beta}, c_{Y_p}, c_{Y_r}, c_{n_p}, c_{l_r}$
Group #7	0–1	$c_{D_{\dot\alpha}} \approx 0, c_{D_q} \approx 0,$ $c_{Y_{\dot\beta}} \approx 0, c_{l_{\dot\beta}} \approx 0, c_{n_{\dot\beta}} \approx 0$

As also outlined previously, Group #1 stability derivatives $(c_{L_\alpha}, c_{m_\alpha}, c_{l_\beta}, c_{n_\beta})$ are the most important aerodynamic coefficients. Each has major implications on the dynamic stability of the aircraft. Therefore, the appropriate design for these values is a key component for the overall design. Given the importance, it is worth to review them separately once again. Closed-form relationships for the Group #1 longitudinal stability derivatives $(c_{L_\alpha}, c_{m_\alpha})$ are given by

$$c_{L_\alpha} = c_{L_{\alpha_W}} + c_{L_{\alpha_H}} \eta_H \frac{S_H}{S} \left(1 - \frac{d\varepsilon}{d\alpha}\right)$$

$$c_{m_\alpha} = c_{L_{\alpha_W}} (\bar{x}_{CG} - \bar{x}_{AC_{WB}}) - c_{L_{\alpha_H}} \eta_H \frac{S_H}{S} \left(1 - \frac{d\varepsilon}{d\alpha}\right)(\bar{x}_{AC_H} - \bar{x}_{CG})$$

$$= c_{L_\alpha} (\bar{x}_{CG} - \bar{x}_{AC}) = c_{L_\alpha} SM$$

As previously discussed, the value of c_{L_α} mainly depends on the aerodynamic design of the wing through the wing lift-slope coefficient $c_{L_{\alpha_W}}$ with the horizontal tail providing a limited contribution (approximatley 10 to 12 percent) because of the lower dynamic pressure η_H, lower surface S_H/S, and downwash effect $\left(1 - \frac{d\varepsilon}{d\alpha}\right)$. For a given c_{L_α}, the numerical value of c_{m_α} ultimately depends on the moment arm of the horizontal tail $\bar{x}_{AC_H}$. Therefore, the location of the horizontal tail is the most critical parameter for the numerical value of c_{m_α}. The use of a simple empirical tool such as the Polhamus formula can provide approximate but realistic estimates for the values of $c_{L_\alpha}, c_{m_\alpha}$. Hoak and Roskam[5,6] review the modeling for these critical parameters, leading to realistic estimates for their values. Wind tunnel analysis and CFD codes allow more detailed measurements for these parameters.

The starting relationships for the Group #1 lateral directional stability derivatives $(c_{l_\beta}, c_{n_\beta})$ are instead given by

$$c_{l_\beta} = c_{l_{\beta_{WB}}} + c_{l_{\beta_H}} + c_{l_{\beta_V}}$$

with $c_{l_{\beta_{WB}}} \approx c_{l_{\beta_{WB}}}|_I + c_{l_{\beta_{WB}}}|_{II} + c_{l_{\beta_{WB}}}|_{III}$, $c_{l_{\beta_H}} \approx 0$

$$c_{n_\beta} = c_{n_{\beta_{WB}}} + c_{n_{\beta_H}} + c_{n_{\beta_V}}$$

with $c_{n_{\beta_{WB}}} \approx 0, c_{n_{\beta_H}} \approx 0$.

The value of the well-known dihedral effect coefficient c_{l_β} depends on the wing configuration (low-wing versus high-wing), on the wing design in terms of geometric dihedral angle (Γ_W) and sweep angle (Λ_{LE}), as well as the size of the vertical tail in terms of (Z_{V_S}). Again, Hoak and Roskam[5,6] review the modeling for these critical parameters and provide valuable tools for a quantitative analysis. Wind tunnel analysis and CFD codes are recommended for accurate estimates of c_{l_β}. Using the approximation $c_{n\beta} \approx c_{n\beta_V}$, at small α_1 an estimate of the numerical value for c_{n_β} is given by

$$c_{n\beta} \approx k_{Y_V} \cdot |c_{L_\alpha \cdot V}| \eta_V \cdot \left(1 + \frac{d\sigma}{d\beta}\right) \frac{S_V}{S} \cdot \frac{X_{V_S}}{b}$$

Clearly, the size of the vertical tail (S_V/S) and its location on the longitudinal axis with respect to the center of gravity (X_{V_S}) are the critical parameters affecting the numerical value for c_{n_β}. Later chapters will show the critical role played by c_{l_β}, c_{n_β} for the aircraft lateral directional static and dynamic stability.

The values of Group #1 stability derivatives greatly affect the overall stability of the aircraft, whereas the values of Group #2 stability derivatives $(c_{m_{\dot\alpha}}, c_{m_q}, c_{l_p}, c_{n_r})$ mainly affect the handling qualities and the dynamic characteristics of the aircraft as perceived by the pilot. These key concepts will be formally introduced in Chapter 7. The modeling for the preceding coefficients (outlined in Chapter 3 for $c_{m_{\dot\alpha}}, c_{m_q}$ and Sections 4.6.3 and 4.6.7 for c_{l_p}, c_{n_r} respectively) is too long for a quick summary. However, it is important to recall that the size of the horizontal tail (S_H/S) and its location $\bar{x}_{AC_H}$ are the most critical geometric parameters affecting $c_{m_{\dot\alpha}}, c_{m_q}$. Similarly, (S_V/S) and the geometric parameters X_{V_S}, Z_{V_S} are the most significant parameter affecting the numerical values of the coefficients (c_{l_p}, c_{n_r}). While the outlined methodologies provide fairly accurate estimates for all the coefficients, a complete wind tunnel analysis using a scale model is recommended for accurate estimates of these parameters.

The stability derivatives in Group #3 are fairly peculiar. Although technically they are classified as stability derivatives, they do not really influence the aircraft stability. In fact, their values strongly affect the overall drag at trim conditions (described in Chapter 6) and, ultimately, the necessary thrust force for steady-state flight with direct implications on the optimization of fuel costs for civilian and general aviation aircraft. Wind tunnel analysis and CFD methods are typically used for accurate estimates of these important drag coefficients while[5,6] provide simpler methods. As expected, the values for c_{D_0}, c_{D_α} depend on the cross sections of the wetted area, with the fuselage and the wing providing the most contribution. Note that c_{D_0} does not appear in the relationship for f_{A_X}.

The stability derivatives in Group #4 include the longitudinal bias terms c_{L_0}, c_{m_0}, which also affect the trimmability of the aircraft, as described in Chapter 6. Wind tunnel analysis, CFD methods, or specific modeling procedures described in[5,6] are used for estimating these coefficients. Note that these coefficients do not appear in the relationships for f_{A_Z}, m_A.

Finally, the stability derivatives in Groups #5, #6, and #7 have marginal-to-moderate roles in the aircraft aerodynamic modeling, with the coefficients in Group 7 being essentially negligible for most aircraft. Wind tunnel analysis and CFD methods are needed for obtaining reliable estimates for the preceding coefficients.

4.9 SUMMARY OF THE MODELING OF THE LONGITUDINAL AND LATERAL-DIRECTIONAL AERODYNAMIC FORCES AND MOMENTS

Chapters 3 and 4 have described basic tools for the modeling of the aerodynamic forces and moments acting on the aircraft. Following the breakdown used in Chapter 1, the modeling has been divided into the modeling of steady-state and small perturbation components, grouped into longitudinal and lateral directional forces and moments, respectively. It should be emphasized that this modeling has been developed along the stability axes (X_S, Y_S, Z_S), also known in the technical literature as wind axes, whereas the aircraft equations of motions in Chapter 1 have been introduced in terms of the body axes (X, Y, Z).

Without loss of generality, the modeling has been developed for a conventional subsonic aircraft with elevators/stabilators being used for longitudinal control and conventional ailerons/rudder being used for lateral directional control.

The associated modeling is summarized below.

4.9 Summary of the Modeling of the Longitudinal and Lateral-Directional Aerodynamic Forces and Moments 199

$$\begin{Bmatrix} F_{A_{X_{1_S}}} \\ F_{A_{Y_{1_S}}} \\ M_{A_{1_S}} \end{Bmatrix} = \begin{Bmatrix} F_{A_{X_1}} \\ F_{A_{Z_1}} \\ M_{A_1} \end{Bmatrix} = \begin{Bmatrix} -(\bar{q}Sc_{D_1}) \\ -(\bar{q}Sc_{L_1}) \\ \bar{q}S\bar{c}c_{m_1} \end{Bmatrix} \text{ with: } \begin{Bmatrix} c_{D_1} \\ c_{L_1} \\ c_{m_1} \end{Bmatrix} = \begin{bmatrix} c_{D_0} & c_{D_\alpha} & \cancel{c_{D_{\delta_E}}} & \cancel{c_{D_{i_H}}} \\ c_{L_0} & c_{L_\alpha} & c_{L_{\delta_E}} & c_{L_{i_H}} \\ c_{m_0} & c_{m_\alpha} & c_{m_{\delta_E}} & c_{m_{i_H}} \end{bmatrix} \begin{Bmatrix} 1 \\ \alpha \\ \delta_E \\ i_H \end{Bmatrix}$$

$$\begin{Bmatrix} F_A \\ L_{A_{1_S}} \\ N_{A_{1_S}} \end{Bmatrix} = \begin{Bmatrix} F_{A_{Y_1}} \\ L_{A_1} \\ N_{A_1} \end{Bmatrix} = \begin{Bmatrix} \bar{q}Sc_{Y_1} \\ \bar{q}Sbc_{l_1} \\ \bar{q}Sbc_{n_1} \end{Bmatrix} \text{ with: } \begin{Bmatrix} c_{Y_1} \\ c_{l_1} \\ c_{n_1} \end{Bmatrix} = \begin{bmatrix} c_{Y_\beta} & c_{Y_{\delta_A}} & c_{Y_{\delta_R}} \\ c_{l_\beta} & c_{l_{\delta_A}} & c_{l_{\delta_R}} \\ c_{n_\beta} & c_{n_{\delta_A}} & c_{n_{\delta_R}} \end{bmatrix} \begin{Bmatrix} \beta \\ \delta_A \\ \delta_R \end{Bmatrix}$$

Steady-State Aerodynamic Forces and Moments

$$\begin{cases} f_{A_X} = \bar{q}S \left\{ -[c_{D_u} + 2c_{D_1}]\left(\dfrac{u}{V_{P_1}}\right) + [-c_{D_\alpha} + c_{L_1}]\alpha - c_{D_{\dot{\alpha}}}\left(\dfrac{\dot{\alpha}\bar{c}}{2V_{P_1}}\right) - c_{D_q}\left(\dfrac{q\bar{c}}{2V_{P_1}}\right) - c_{D_{\delta_E}}\delta_E - c_{D_{i_H}}i_H \right\} \\[2mm]
f_{A_Z} = \bar{q}S \left\{ -[c_{L_u} + 2c_{L_1}]\left(\dfrac{u}{V_{P_1}}\right) - [c_{L_\alpha} + c_{D_1}]\alpha - c_{L_{\dot{\alpha}}}\left(\dfrac{\dot{\alpha}\bar{c}}{2V_{P_1}}\right) - c_{L_q}\left(\dfrac{q\bar{c}}{2V_{P_1}}\right) - c_{L_{\delta_E}}\delta_E - c_{L_{i_H}}i_H \right\} \\[2mm]
m_A = \bar{q}S\bar{c} \left\{ [c_{m_u} + 2c_{m_1}]\left(\dfrac{u}{V_{P_1}}\right) + c_{m_\alpha}\alpha + c_{m_{\dot{\alpha}}}\left(\dfrac{\dot{\alpha}\bar{c}}{2V_{P_1}}\right) + c_{m_q}\left(\dfrac{q\bar{c}}{2V_{P_1}}\right) + c_{m_{\delta_E}}\delta_E + c_{m_{i_H}}i_H \right\} \\[2mm]
f_{A_Y} = \bar{q}S \left\{ c_{Y_\beta}\beta + c_{Y_{\dot\beta}}\left(\dfrac{\dot{\beta}b}{2V_{P_1}}\right) + c_{Y_p}\left(\dfrac{pb}{2V_{P_1}}\right) + c_{Y_r}\left(\dfrac{rb}{2V_{P_1}}\right) + c_{Y_{\delta_A}}\delta_A + c_{Y_{\delta_R}}\delta_R \right\} \\[2mm]
l_A = \bar{q}Sb \left\{ c_{l_\beta}\beta + c_{l_{\dot\beta}}\left(\dfrac{\dot{\beta}b}{2V_{P_1}}\right) + c_{l_p}\left(\dfrac{pb}{2V_{P_1}}\right) + c_{l_r}\left(\dfrac{rb}{2V_{P_1}}\right) + c_{l_{\delta_A}}\delta_A + c_{l_{\delta_R}}\delta_R \right\} \\[2mm]
n_A = \bar{q}Sb \left\{ c_{n_\beta}\beta + c_{n_{\dot\beta}}\left(\dfrac{\dot{\beta}b}{2V_{P_1}}\right) + c_{n_p}\left(\dfrac{pb}{2V_{P_1}}\right) + c_{n_r}\left(\dfrac{rb}{2V_{P_1}}\right) + c_{n_{\delta_A}}\delta_A + c_{n_{\delta_R}}\delta_R \right\} \end{cases}$$

Small Perturbation Aerodynamic Forces and Moments

It should be emphasized that the steady—state aerodynamic forces and moments will not actually be used for the solution of the equations of motion developed in Chapter 1. However, they were needed for the derivation of the small perturbation terms, as shown in Chapter 2, Section 3.9 (for the longitudinal modeling) and Section 4.6 (for the lateral directional modeling). The small perturbation aerodynamic forces and moments $(f_{A_X}, f_{A_Z}, m_A, f_{A_Y}, l_A, n_A)$ are then provided as inputs to the small perturbation equations of motion shown:

$$\begin{cases} m[\dot{u} + Q_1 w + qW_1 - R_1 v - rV_1] = -mg\theta\cos\Theta_1 + (f_{A_X} + f_{T_X}) \\[2mm]
m[\dot{w} + P_1 v + pV_1 - Q_1 u - U_1 q] = -mg\theta\cos\Phi_1\sin\Theta_1 - mg\phi\sin\Phi_1\cos\Theta_1 + (f_{A_Z} + f_{T_Z}) \\[2mm]
\dot{q}I_{YY} + (P_1 r + pR_1)(I_{XX} - I_{ZZ}) + (2P_1 p - 2R_1 r)I_{XZ} = (m_A + m_T) \end{cases}$$

$$\begin{cases} m[\dot{v} + U_1 r + u R_1 - P_1 w - p W_1] = -mg\theta \sin\Phi_1 \sin\Theta_1 + mg\phi\cos\Phi_1\cos\Theta_1 + \left(f_{A_Y} + f_{T_Y}\right) \\ \dot{p}I_{XX} - \dot{r}I_{XZ} - (P_1 q + Q_1 p)I_{XZ} + (R_1 q + Q_1 r)(I_{ZZ} - I_{YY}) = (l_A + l_T) \\ \dot{r}I_{ZZ} - \dot{p}I_{XZ} + (P_1 q + p Q_1)(I_{YY} - I_{XX}) + (Q_1 r + R_1 q)I_{XZ} = (n_A + n_T) \end{cases}$$

The solution of the preceding equations will be discussed in Chapter 7.

REFERENCES

1. Anderson, J. D. *Fundamentals of Aerodynamics*. Mc-Graw Hill, Inc., 1984.
2. Philips, W.*Mechanics of Flight*. John Wiley and Sons, 2004.
3. Abbott, I. H., Von Doenhoff A. E.*Theory of Wing Sections*. Dover Publications Inc., 1959.
4. Roskam, J. *Airplane Flight Dynamics and Automatic Flight Controls – Part I*. Design, Analysis, and Research Corporation, Lawrence, KS, 1995 .
5. Hoak, D. E., et al. "The USAF Stability and Control DATCOM." Air Force Wright Aeronautical Laboratories. TR-83-2048, October 1960 (revised in 1978).
6. Roskam J. *Airplane Design. Part I–Part VIII*. Roskam Aviation and Engineering Corporation, Lawrence, KS, 1990.
7. Napolitano, M. R. *WVU MAE 242 Flight Testing – Class Notes*. West Virginia University, Morgantown, WV, 2001.
8. Multhopp, H. *Aerodynamics of the Fuselage*. NASA Technical Memorandum No.1036, December 1942.
9. "DC-9 Airplane Characteristics for Airport Planning," June 1984, Douglas Aircraft Company, downloaded from *http://www.boeing.com/commercial/airports/dc9.htm* (.PDF format).
10. *http://textron.vo.llnwd.net/o25/CES/cessna_aircraft_docs/citation/cj3/cj3_s&d.pdf*

STUDENT SAMPLE PROBLEMS

Student Sample Problem 4.1

Consider the data relative to the McDonnell Douglas DC-9 aircraft in Appendix C. Use the provided drawings to extract all the relevant geometric characteristics of the aircraft. Using the modeling approach outlined in this chapter, the dihedral effect c_{l_β} can be modeled as

$$c_{l\beta} = c_{l\beta_{WB}} + c_{l\beta_H} + c_{l\beta_V}$$

Estimate the value of c_{l_β} for this aircraft. Also, evaluate the percentage of each of the different contributions with respect to the total c_{l_β} value.

Solution of Student Sample Problem 4.1

All the relevant geometric parameters used in this problem were extracted from the images in Appendix C. Also, note that several geometric parameters related to the wing and the horizontal tail were calculated previously for the DC-9 problems in Chapters 2 and 3. From the drawings in Appendix C, the following geometric parameters were identified

$$b_{2V} = 2 \cdot 13.6 = 27.2 \text{ ft}, \ c_{T_V} = 10.9 \text{ ft},$$
$$c_{R_V} = 13.7 \text{ ft}, \ \Lambda_{LE_V} = 45° = 0.785 \text{ rad}$$

Note that b_{2V} is actually the same of the span of the vertical tail; this value is used so that the geometric parameters of the vertical tail can be found in a similar manner as the parameters for the wing. Next, using the assumption of straight wings, the vertical tail geometric parameters were derived using the values from the drawings in Appendix C.

Vertical Tail Geometric Parameters

The following vertical tail parameters are required to determine the stability derivatives.

$$\lambda_V = \frac{c_{T_V}}{c_{R_V}} = \frac{10.9}{13.7} = 0.796$$

$$S_{2V} = \frac{b_{2V}}{2} c_{R_V} (1 + \lambda_V) = \frac{27.2}{2} \cdot 13.7 \cdot (1 + 0.796) = 334.56 \text{ ft}^2$$

Again, it should be emphasized that S_{2V} is twice the value of the area of the vertical tail; however, the use of S_{2V} allows finding the following geometric parameters:

$$AR_V = \frac{b_{2V}^2}{S_{2V}} = \frac{27.2^2}{334.56} = 2.211;$$

$$\bar{c}_V = \frac{2}{3} c_{R_V} \frac{(1 + \lambda_V + \lambda_V^2)}{(1 + \lambda_V)} = \frac{2}{3} \cdot 13.7 \cdot \frac{(1 + 0.796 + 0.796^2)}{(1 + 0.796)} = 12.35 \text{ ft}$$

$$x_{MAC_V} = \frac{b_{2V}}{6} \frac{(1 + 2\lambda_V)}{(1 + \lambda_V)} \tan(\Lambda_{LE_V}) = \frac{27.2}{6} \cdot \frac{(1 + 2 \cdot 0.796)}{(1 + 0.796)} \tan(0.785) = 6.54 \text{ ft}$$

$$y_{MAC_V} = \frac{b_{2V}}{6} \frac{(1 + 2\lambda_V)}{(1 + \lambda_V)} = \frac{27.2}{6} \cdot \frac{(1 + 2 \cdot 0.796)}{(1 + 0.796)} = 6.54 \text{ ft}$$

$$\tan \Lambda_{0.5_V} = \tan \Lambda_{LE_V} - \frac{4 \cdot 0.5 \cdot (1 - \lambda_V)}{AR_V (1 + \lambda_V)} = \tan(0.82) - \frac{4 \cdot 0.5 \cdot (1 - 0.796)}{2.211 \cdot (1 + 0.796)} \Rightarrow \Lambda_{0.5_V} = 0.731$$

Wing-Tail Geometric Parameters

Figure SSP4.4.1 was derived to illustrate the wing-tail geometric distances. The distances X_{V_S}, Z_{V_S}, X_{R_S} and Z_{R_S} are needed to determine the stability derivatives related with the vertical tail and the rudder.

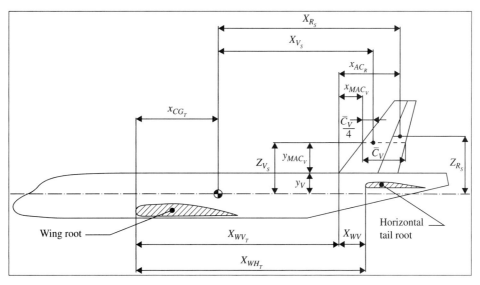

Figure SSP4.1.1 Wing-Vertical Tail Geometric Distances

From Figure SSP4.1.1:

$$x_{CG_r} + X_{V_S} = X_{WV_r} + x_{MAC_V} + \frac{\bar{c}_V}{4} \Rightarrow X_{V_S} = X_{WV_r} + \frac{\bar{c}_V}{4} + x_{MAC_V} - x_{CG_r}$$

where $X_{WH_r} = X_{WV_r} + X_{HV} \Rightarrow X_{WV_r} = X_{WH_r} - X_{HV} = 49.5 - 11.9 = 37.6$ ft.
From Figure SSP4.1.2:

$$x_{CG_r} = x_{MAC} + x_{CG} = x_{MAC} + \bar{x}_{CG} \cdot \bar{c} = 9.436 + 0.3 \cdot 11.586 = 12.912 \text{ ft}$$

Therefore:

$$X_{V_S} = X_{WV_r} + \frac{\bar{c}_V}{4} + x_{MAC_V} - x_{CG_r} = 35.7 + \frac{13.037}{4} + 7.363 - 12.912 = 33.411 \text{ ft}$$

From Figure SSP4.1.1:

$$Z_{V_S} = y_V + y_{MAC_V} = 5.6 + 6.542 = 12.142 \text{ ft}$$

$$x_{CG_r} + X_{R_S} = X_{WV_r} + x_{AC_R} \Rightarrow X_{R_S} = X_{WV_r} + x_{AC_R} - x_{CG_r} = 37.6 + 15 - 12.783 = 39.817 \text{ ft}$$

Finally, from the drawings in Appendix C: $Z_{R_S} = 9$ ft

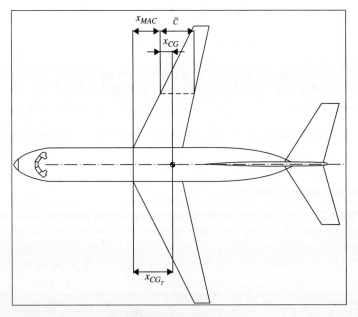

Figure SSP4.1.2 CG Location

Vertical Tail Lift Slope Coefficient

This coefficient is calculated using the Polhamus Formula with $AR_{V_{eff}}$, defined as

$$AR_{V_{eff}} = c_1 \cdot AR_V \cdot [1 + K_{HV}(c_2 - 1)]$$

The parameter c_1 is found from Figure 4.15 using $\lambda_V = 0.796$ along with the parameter:

$$\frac{b_V}{2r_1} = \frac{13.6}{2 \cdot 7.7} = 0.883$$

Using $\lambda_V = 0.796$ and $b_V/2r_1 = 0.883$ Figure 4.15 provides $c_1 = 1.284$.
  Next, the parameter c_2 is found from Figure 4.16 using

$$\frac{Z_H}{b_V} = \frac{-16}{13.6} = -1.176, \quad \frac{x_{AC_{H \to V}}}{\bar{c}_V} = \frac{7.458}{12.35} = 0.604$$

Note that Z_H is considered negative when above the CG. Using the preceding values, Figure 4.26 provides $c_2 = 2.451$. Finally, using S_H/S_V in Figure 4.18, we have

$$\frac{S_H}{S_V} = \frac{274.91}{0.5 \cdot 334.56} = 1.643 \Rightarrow K_{HV} = 1.081$$

leading to

$$AR_{V_{eff}} = c_1 \cdot AR_V \cdot [1 + K_{HV}(c_2 - 1)] = 1.284 \cdot 2.211 \cdot [1 + 1.081(2.451 - 1)] = 7.293$$

$$\tan \Lambda_{0.5_V} = \tan \Lambda_{LE_V} - \frac{4 \cdot 0.5 \cdot (1 - \lambda_V)}{AR_V(1 + \lambda_V)} = \tan(0.785) - \frac{4 \cdot 0.5 \cdot (1 - 0.796)}{2.211 \cdot (1 + 0.796)} \Rightarrow \Lambda_{0.5_V} = 0.731$$

Using $AR_{V_{eff}} = 7.293$, the application of the Polhamus formula leads to

$$k = 1 + \frac{[(8.2 - 2.3\Lambda_{LE_V}) - AR_{V_{eff}}(0.22 - 0.153\Lambda_{LE_V})]}{100}$$

$$= 1 + \frac{[(8.2 - 2.3 \cdot 0.785) - 7.293 \cdot (0.22 - 0.153 \cdot 0.785)]}{100} = 1.057$$

$$c_{L_{\alpha V,eff}} = \frac{2\pi AR_{V_{eff}}}{2 + \sqrt{\left\{\left[\frac{AR_{V_{eff}}^2(1 - Mach^2)}{k^2}\right]\left(1 + \frac{\tan^2(\Lambda_{0.5})}{(1 - Mach^2)}\right)\right] + 4\right\}}}$$

$$= \frac{2\pi \cdot 7.293}{2 + \sqrt{\left\{\left[\frac{7.293^2(1 - 0.7^2)}{1.057^2}\right]\left(1 + \frac{\tan^2(0.731)}{(1 - 0.7^2)}\right)\right] + 4\right\}}} = 4.509 \left(\frac{1}{\text{rad}}\right)$$

Since the sweep angle and the aspect ratio are outside the Polhamus range, it is reasonable to expect that the estimate exceeds somewhat the true value. Therefore, considering a 20 percent overestimate, the value of $c_{L_{\alpha V,eff}} = 3.607 \left(\frac{1}{\text{rad}}\right)$ is considered instead of the previous value.

(Wing + Body) Contribution

Using the expression for $c_{l_{\beta WB}}$:

$$c_{l_{\beta WB}} = 57.3 \cdot c_{L_1} \left[\left(\frac{c_{l_\beta}}{c_{L_1}}\right)_{\Lambda_{c/2}} K_{M_\Lambda} K_f + \left(\frac{c_{l_\beta}}{c_{L_1}}\right)_{AR}\right]$$

$$+ 57.3 \left\{\Gamma_W \left[\frac{c_{l_\beta}}{\Gamma_W} K_{M_\Gamma} + \frac{\Delta c_{l_\beta}}{\Gamma_W}\right] + \left(\Delta c_{l_\beta}\right)_{z_W} + \varepsilon_W \tan \Lambda_{c/4} \left(\frac{\Delta c_{l_\beta}}{\varepsilon_W \tan \Lambda_{c/4}}\right)\right\}$$

Next, the following individual contributions to the dihedral effect in the relationship are evaluated separately:

- Wing contribution due to the geometric dihedral angle;
- Wing contribution due to the wing-fuselage positions;
- Wing contribution due to the sweep angle;
- Wing contribution due to the aspect ratio;
- Wing contribution due to the twist angle;
- Body (fuselage) contribution.

(1) Wing contribution to the dihedral effect due to the geometric dihedral angle

Using $\lambda = 0.25$ and $\Lambda_{0.5} = 21.4°$ Figure 4.43 provides: $\frac{c_{l_\beta}}{\Gamma_W} = -0.000183 \, (1/\text{deg}^2)$

Using $Mach \cdot \cos(\Lambda_{0.5}) = 0.7 \cdot \cos(0.374) = 0.652$, $\frac{AR}{\cos(\Lambda_{0.5})} = \frac{8.612}{\cos(0.374)} = 9.252$

Figure 4.44 provides $K_{M_\Gamma} = 1.153$.

Therefore

$$c_{l\beta_{WB}}\bigg|_{\substack{Geometric \\ dihedral \\ angle}} = 57.3\, \Gamma_W \left[\frac{c_{l_\beta}}{\Gamma_W} K_{M_\Gamma}\right] = 57.3 \cdot 2.2 \cdot [-0.000183 \cdot 1.153] = -0.0266 \left(\frac{1}{\text{rad}}\right)$$

(2) Wing contribution to the dihedral effect due to wing-fuselage position

The fuselage's diameter d_B is required for the evaluation of this term. d_B is found using

$$d_B = \sqrt{\frac{S_{f_{AVG}}}{0.7854}}$$

where $S_{f_{AVG}}$ is the average cross-sectional area of the fuselage calculated using

$$S_{f_{AVG}} = \pi \left(\frac{\text{Fuselage average height}}{2}\right)\left(\frac{\text{Fuselage average width}}{2}\right)$$

$$= \pi \left(\frac{10.074}{2}\right)\left(\frac{9.14}{2}\right) = 72.32 \text{ ft}^2$$

where the average height and width of the fuselage were obtained from the drawings in Appendix C. Therefore, we have

$$d = \sqrt{\frac{S_{f_{AVG}}}{0.7854}} = \sqrt{\frac{72.32}{0.7854}} = 9.596 \text{ ft}$$

Leading to the following value for $c_{l\beta_{WB}}\big|_{\substack{High/low \\ wing}}$:

$$c_{l\beta_{WB}}\bigg|_{\substack{High/low \\ wing}} = 57.3 (\Delta c_{l_\beta})_{Z_W} = 57.3 \frac{1.2\sqrt{AR}}{57.3} \cdot \frac{Z_W}{b} \cdot \left(\frac{2d}{b}\right) = 1.2\sqrt{8.612} \cdot \frac{3.1}{89.4} \cdot \left(\frac{2 \cdot 9.596}{89.4}\right)$$

$$= 0.026 \left(\frac{1}{\text{rad}}\right)$$

(3) Wing contribution to the dihedral effect due to wing sweep angle

The value of c_{L_1} is required to calculated $c_{l\beta_{WB}}\big|_{\substack{Wing \\ sweep \\ angle}}$. This value is calculated using the given flight conditions and data from Standard Atmosphere tables:

$$c_{L_1} = \frac{W}{\frac{1}{2}\rho V^2 S} = 0.448$$

Using $\lambda = 0.25$, $AR = 8.612$ and $\Lambda_{c/2} = 21.4°$ Figure 4.39 provides

$$\left(\frac{c_{l_\beta}}{c_{L_1}}\right)_{\Lambda_{c/2}} = -0.00192 \left(\frac{1}{\text{deg}}\right)$$

Using $Mach \cdot \cos(\Lambda_{0.5}) = 0.652$, $\frac{AR}{\cos(\Lambda_{0.5})} = 9.252$ Figure 4.40 provides $K_{M_\Lambda} = 1.26$. Using $\frac{AR}{\cos(\Lambda_{0.5})} = 9.252$, $\frac{A}{b} = \frac{60.7}{89.4} = 0.68$ Figure 4.41 provides $K_f = 0.888$.
Therefore

$$c_{l\beta_{WB}}\bigg|_{\substack{Wing \\ sweep \\ angle}} = 57.3 \cdot c_{L_1}\left[\left(\frac{c_{l_\beta}}{c_{L_1}}\right)_{\Lambda_{c/2}} K_{M_\Lambda} K_f\right] = 57.3 \cdot 0.448[-0.00192 \cdot 1.26 \cdot 0.888] = -0.055 \left(\frac{1}{\text{rad}}\right)$$

(4) Wing contribution to the dihedral effect due to wing aspect ratio

Using $\lambda = 0.25$ and $AR = 8.612$ Figure 4.42 provides

$$\left(\frac{c_{l_\beta}}{C_{L_1}}\right)_{AR} = 0.00029 \left(\frac{1}{\deg}\right) 57.3 \cdot c_{L_1} \left[\left(\frac{c_{l_\beta}}{C_{L_1}}\right)_{AR}\right]$$

$$= 57.3 \cdot 0.448 \cdot 0.00029 = 0.00744 \left(\frac{1}{\text{rad}}\right)$$

(5) Wing contribution to the dihedral effect due to wing twist angle

Because of lack of specific information, the twist angle is arbitrarily assumed to be $\varepsilon_W = 2° = 0.0345$ rad.
Using $\lambda = 0.25$ and $AR = 8.612$ Figure 4.46 provides

$$\left(\frac{\Delta c_{l_\beta}}{\varepsilon_W \tan \Lambda_{c/4}}\right) = -3.57 \times 10^{-5} \left(\frac{1}{\deg^2}\right) 57.3 \cdot \varepsilon_W \tan \Lambda_{c/4} \left(\frac{\Delta c_{l_\beta}}{\varepsilon_W \tan \Lambda_{c/4}}\right)$$

$$= 57.3 \cdot 2 \cdot \tan 0.433 \cdot (-3.57 \times 10^{-5})$$

$$= -0.00189 \left(\frac{1}{\text{rad}}\right)$$

(6) Fuselage contribution to the dihedral effect

This contribution is given by

$$\left(\frac{\Delta c_{l_\beta}}{\Gamma_W}\right) = -0.0005 \cdot AR \cdot \left(\frac{d}{b}\right)^2 = -0.0005 \cdot 8.612 \cdot \left(\frac{9.596}{89.4}\right)^2 = -0.00005 \left(\frac{1}{\deg^2}\right)$$

leading to

$$57.3 \cdot \Gamma_W \left[\frac{\Delta c_{l_\beta}}{\Gamma_W}\right] = 57.3 \cdot 2.2 \cdot (-0.00005) = -0.006 \left(\frac{1}{\text{rad}}\right)$$

Adding together the previous contributions we have

$$c_{l_{\beta_{WB}}} = (-0.0266 + 0.026 - 0.055 + 0.00744 - 0.00189 - 0.006) = -0.0562 \left(\frac{1}{\text{rad}}\right)$$

Horizontal Tail Contribution

Since the horizontal tail of this aircraft does not have significant dihedral or anhedral angles, we have the condition: $c_{l_{\beta_H}} \approx 0$

Vertical Tail Contribution

The value of this contribution is given by

$$c_{l_{\beta_V}} = -k_{Y_V} \cdot |c_{L_\alpha \cdot V}| \eta_V \cdot \left(1 + \frac{d\sigma}{d\beta}\right) \frac{S_V}{S} \cdot \frac{(Z_V \cos \alpha_1 - X_V \sin \alpha_1)}{b} \quad (\text{rad}^{-1})$$

Using $\frac{b_V}{2r_1} = \frac{13.6}{2 \cdot 7.7} = 0.883$ Figure 4.13 $k_{Y_V} = 0.761$

Next, the term $\eta_V \cdot \left(1 + \frac{d\sigma}{d\beta}\right)$ is modeled using

$$\eta_V \cdot \left(1 + \frac{d\sigma}{d\beta}\right) = 0.724 + 3.06 \frac{S_V/S}{1 + \cos(\Lambda_{c/4})} + 0.4 \frac{Z_W}{d} + 0.009 \cdot AR$$

$$= 0.724 + 3.06 \frac{0.5 \cdot 334.56/928}{1 + \cos(0.433)} + 0.4 \frac{3.1}{9.596} + 0.009 \cdot 8.612 = 1.22$$

Thus,

$$c_{l\beta_V} = -k_{Y_V} \cdot |c_{L_\alpha \cdot V}| \eta_V \cdot \left(1 + \frac{d\sigma}{d\beta}\right) \frac{S_V}{S} \cdot \frac{(Z_V \cos\alpha_1 - X_V \sin\alpha_1)}{b}$$

$$= -0.761 \cdot |3.607| \, 1.22 \, \frac{0.5 \cdot 334.56}{928} \cdot \frac{(12.1 \cos 0.035 - 34.4 \sin 0.035)}{89.4}$$

$$= -0.074 \quad (\text{rad}^{-1})$$

Therefore, the final total estimated value of c_{l_β} for the DC-9 aircraft is

$$c_{l_\beta} \approx c_{l\beta_{WB}} + c_{l\beta_V} = -0.0562 - 0.074 = -0.130 \left(\frac{1}{\text{rad}}\right)$$

Table SSP4.1.1 shows the percentage of each of the coefficients with respect to the total c_{l_β} value.

Table SSP4.1.1 Percentage of Each of the Contributions to the Total c_{l_β} Value

Term	Value [1/rad]	% of c_{l_β}
Wing contribution due to the geometric dihedral angle	−0.0266	20.5%
Wing contribution due to the wing-fuselage position	0.026	−20.2%
Wing contribution due to the sweep angle	−0.055	42.4%
Wing contribution due to the aspect ratio	0.00744	−5.7%
Wing contribution due to the twist angle	−0.00189	1.5%
Body (fuselage) contribution	−0.006	4.8%
Horizontal tail contribution	0	0.0%
Vertical tail contribution	−0.074	56.8%
Total (c_{l_β})	−0.144	100%

The table suggest a few interesting observations. First, the anhedral effect from the low wing contribution virtually cancels out the dihedral effect due to the wing geometric dihedral angle. Second, the main contributions to the dihedral effect for this aircraft are due to the wing sweep angle and, largely, the vertical tail, with the fuselage providing very small contribution.

Student Sample Problem 4.2

Consider the data relative to the SIAI Marchetti S211 aircraft in Appendix B and Appendix C. Use the provided drawings to extract all the relevant geometric characteristics of the aircraft. Using the modeling approach outlined in this chapter, the dihedral effect c_{l_β} can be modeled as

$$c_{l\beta} = c_{l\beta_{WB}} + c_{l\beta_H} + c_{l\beta_V}$$

Estimate the value of c_{l_β} for this aircraft. Also, evaluate the percentage of each of the different contributions with respect to the total c_{l_β} value. Next, compare the obtained c_{l_β} value with the true value listed in Appendix B (Aircraft 8) and evaluate the modeling error.

Solution of Student Sample Problem 4.2

All the relevant geometric parameters used in this problem were extracted from the images in Appendix C. Also, note that several geometric parameters related to the wing and the horizontal tail were calculated previously for the S211 problems in Chapters 2 and 3. From the drawings in Appendix C, the following geometric parameters were identified:

$$b_{2V} = 2 \cdot 5.8 = 11.6 \text{ ft}, \quad c_{T_V} = 2 \text{ ft}, \, c_{R_V} = 5.7 \text{ ft}, \quad \Lambda_{LE_V} = 40° = 0.698 \text{ rad}$$

Note that b_{2V} is actually the same of the span of the vertical tail; this value is used so that the geometric parameters of the vertical tail can be found in a similar manner as the parameters for the wing. Next, using the assumption of straight wings, the vertical tail geometric parameters were derived using the values from the drawings in Appendix C.

Vertical Tail Geometric Parameters

The following vertical tail parameters are required to determine the stability derivatives.

$$\lambda_V = \frac{c_{T_V}}{c_{R_V}} = \frac{2}{5.7} = 0.351;$$

$$S_{2V} = \frac{b_{2V}}{2} c_{R_V}(1 + \lambda_V) = \frac{11.6}{2} \cdot 5.7 \cdot (1 + 0.351) = 44.66 \text{ ft}^2$$

Again, it should be emphasized that S_{2V} is twice the value of the area of the vertical tail; however, the use of S_{2V} allows finding the following geometric parameters:

$$AR_V = \frac{b_{2V}^2}{S_{2V}} = \frac{11.6^2}{44.66} = 3.013$$

$$\bar{c}_V = \frac{2}{3} c_{R_V} \frac{(1 + \lambda_V + \lambda_V^2)}{(1 + \lambda_V)} = \frac{2}{3} \cdot 5.7 \cdot \frac{(1 + 0.351 + 0.351^2)}{(1 + 0.351)} = 4.146 \text{ ft}$$

$$x_{MAC_V} = \frac{b_{2V}}{6} \frac{(1 + 2\lambda_V)}{(1 + \lambda_V)} \tan(\Lambda_{LE_V}) = \frac{11.6}{6} \cdot \frac{(1 + 2 \cdot 0.351)}{(1 + 0.351)} \tan(0.698) = 2.044 \text{ ft}$$

$$y_{MAC_V} = \frac{b_{2V}}{6} \frac{(1 + 2\lambda_V)}{(1 + \lambda_V)} = \frac{11.6}{6} \cdot \frac{(1 + 2 \cdot 0.351)}{(1 + 0.351)} = 2.435 \text{ ft}$$

$$\tan\Lambda_{0.5_V} = \tan\Lambda_{LE_V} - \frac{4 \cdot 0.5 \cdot (1 - \lambda_V)}{AR_V(1 + \lambda_V)} = \tan(0.698) - \frac{4 \cdot 0.5 \cdot (1 - 0.351)}{3.013 \cdot (1 + 0.351)} \Rightarrow \Lambda_{0.5_V} = 0.48$$

Wing-Tail Geometric Parameters

The distances $X_{V_S}, Z_{V_S}, X_{R_S}$ and Z_{R_S} are evaluated as in SSP4.2. They are needed to determine the stability derivatives related with the vertical tail and the rudder. They are given by

$$X_{V_S} = 9.349 \text{ ft}, Z_{V_S} = 5.435 \text{ ft}, X_{R_S} = 11.469 \text{ ft}, Z_{R_S} = 5.1 \text{ ft}$$

Technically speaking the parameters X_{R_S} and Z_{R_S}—relative to the rudder—are not needed in this problem. However, they were evaluated since they are conceptually similar to X_{V_S} and Z_{V_S}.

Vertical Tail Lift Slope Coefficient

This coefficient is calculated using the Polhamus Formula with $AR_{V_{eff}}$, defined as

$$AR_{V_{eff}} = c_1 \cdot AR_V \cdot [1 + K_{HV}(c_2 - 1)]$$

The parameter c_1 is found from Figure 4.15 using $\lambda_V = 0.351$ along with the parameter

$$\frac{b_V}{2r_1} = \frac{5.8}{2 \cdot 4.1} = 0.707$$

Using $\lambda_V = 0.351$ and $b_V/2r_1 = 0.707$ Figure 4.15 provides $c_1 = 1.335$.

Next, the parameter c_2 is found from Figure 4.16 using

$$\frac{Z_H}{b_V} = \frac{-2.9}{5.8} = -0.5$$

The geometric parameter $x_{AC_{H \to V}}$, shown in Figure 4.17, is calculated using

$$x_{AC_{H \to V}} = x_{MAC_H} + 0.25 \cdot \bar{c}_H = 0.969 + 0.25 \cdot 2.65 = 1.632 \text{ft} \Rightarrow \frac{x_{AC_{H \to V}}}{\bar{c}_V} = \frac{1.632}{4.146} = 0.394$$

Note that Z_H is considered negative when above the CG. Using the above $x_{AC_{H \to V}}$ value, Figure 4.16 provides $c_2 = 0.815$. Finally, using S_H/S_V in Figure 4.18, we have

$$\frac{S_H}{S_V} = \frac{33.583}{0.5 \cdot 44.66} = 1.504 \Rightarrow K_{HV} = 1.05$$

leading to

$$AR_{V_{eff}} = c_1 \cdot AR_V \cdot [1 + K_{HV}(c_2 - 1)]$$

$$= 1.335 \cdot 3.013 \cdot [1 + 1.05(0.815 - 1)] = 3.241$$

$$\tan \Lambda_{0.5_V} = \tan \Lambda_{LE_V} - \frac{4 \cdot 0.5 \cdot (1 - \lambda_V)}{AR_V(1 + \lambda_V)}$$

$$= \tan(0.698) - \frac{4 \cdot 0.5 \cdot (1 - 0.351)}{3.013 \cdot (1 + 0.351)} \Rightarrow \Lambda_{0.5_V} = 0.48$$

Using $AR_{V_{eff}} = 3.241 \leq 4$, the application of the Polhamus formula leads to

$$k_V = 1 + \frac{AR_{V_{eff}}(1.87 - 0.000233 \Lambda_{LE_V})}{100}$$

$$= 1 + \frac{3.241(1.87 - 0.000233 \cdot 0.698)}{100} = 1.061$$

$$c_{L_{\alpha V,eff}} = \frac{2\pi AR_{V_{eff}}}{2 + \sqrt{\left\{\left[\frac{AR_{V_{eff}}(1 - Mach^2)}{k^2}\right]\left(1 + \frac{\tan^2(\Lambda_{0.5})}{(1 - Mach^2)}\right)\right] + 4\right\}}}$$

$$= \frac{2\pi \cdot 3.241}{2 + \sqrt{\left\{\left[\frac{3.241^2(1 - 0.6^2)}{1.061^2}\right]\left(1 + \frac{\tan^2(0.48)}{(1 - 0.6^2)}\right)\right] + 4\right\}}} = 3.678 \left(\frac{1}{\text{rad}}\right)$$

(Wing + Body) Contribution

The modeling for $c_{l\beta_{WB}}$ is given by

$$c_{l_\beta \; WB} = 57.3 \cdot c_{L_1} \left[\left(\frac{c_{l_\beta}}{c_{L_1}}\right)_{\Lambda_{c/2}} K_{M_\Lambda} K_f + \left(\frac{c_{l_\beta}}{c_{L_1}}\right)_{AR}\right]$$

$$+ 57.3 \left\{\Gamma_W \left[\frac{c_{l_\beta}}{\Gamma_W} K_{M_\Gamma} + \frac{\Delta c_{l_\beta}}{\Gamma_W}\right] + \left(\Delta c_{l_\beta}\right)_{Z_W} + \varepsilon_W \tan \Lambda_{c/4} \left(\frac{\Delta c_{l_\beta}}{\varepsilon_W \tan \Lambda_{c/4}}\right)\right\}$$

Next, the following individual contributions to the dihedral effect in the relationship are evaluated separately:

- Wing contribution due to the geometric dihedral angle;
- Wing contribution due to the wing-fuselage positions;
- Wing contribution due to the sweep angle;
- Wing contribution due to the aspect ratio;
- Wing contribution due to the twist angle;
- Body (fuselage) contribution.

(1) Wing contribution to the dihedral effect due to the geometric dihedral angle

Using $\lambda = 0.517$ and $\Lambda_{0.5} = 12.9°$ Figure 4.43 provides

$$\frac{c_{l_\beta}}{\Gamma_W} = -0.000155 \; (1/\text{deg}^2)$$

Using $Mach \cdot \cos(\Lambda_{0.5}) = 0.6 \cdot \cos(0.225) = 0.585$, $\frac{AR}{\cos(\Lambda_{0.5})} = \frac{5.086}{\cos(0.225)} = 5.217$

Figure 4.44 provides $K_{M_\Gamma} = 1.05$. Therefore

$$c_{l\beta_{WB}}\bigg|_{\substack{Geometric \\ dihedral \\ angle}} = 57.3\, \Gamma_W \left[\frac{c_{l\beta}}{\Gamma_W} K_{M_\Gamma}\right] = 57.3 \cdot (-1.9) \cdot [-0.000155 \cdot 1.05] = 0.0177 \left(\frac{1}{rad}\right)$$

(2) Wing contribution to the dihedral effect due to wing-fuselage position

The fuselage's diameter d_B is required for the evaluation of this term. d_B is found using

$$d_B = \sqrt{\frac{S_{f_{AVG}}}{0.7854}}$$

where $S_{f_{AVG}}$ is the average cross-sectional area of the fuselage calculated using

$$S_{f_{AVG}} = \pi \left(\frac{\text{Fuselage average height}}{2}\right)\left(\frac{\text{Fuselage average width}}{2}\right)$$

$$= \pi \left(\frac{3.75}{2}\right)\left(\frac{3.15}{2}\right) = 9.278 \text{ ft}^2$$

where the average height and width of the fuselage were obtained from the drawings in Appendix C. Therefore, we have

$$d = \sqrt{\frac{S_{f_{AVG}}}{0.7854}} = \sqrt{\frac{9.278}{0.7854}} = 3.437 \text{ ft}$$

leading to the following value for $c_{l\beta_{WB}}\big|_{\substack{High/low \\ wing}}$:

$$c_{l\beta_{WB}}\bigg|_{\substack{High/low \\ wing}} = 57.3 \left(\Delta c_{l\beta}\right)_{Z_W} = 57.3 \frac{1.2\sqrt{AR}}{57.3} \cdot \frac{Z_W}{b} \cdot \left(\frac{2d}{b}\right)$$

$$= \frac{1.2\sqrt{5.086}}{57.3} \cdot \frac{(-1.1)}{26.3} \cdot \left(\frac{2 \cdot 3.437}{26.3}\right) = -0.030 \left(\frac{1}{rad}\right)$$

(3) Wing contribution to the dihedral effect due to wing sweep angle

The value of c_{L_1} is required to calculated $c_{l\beta_{WB}}\big|_{\substack{Wing \\ sweep \\ angle}}$. This value is calculated using the given flight conditions and data from Standard Atmosphere tables:

$$c_{L_1} = \frac{W}{\frac{1}{2}\rho V^2 S} = 0.149$$

Using $\lambda = 0.517$, $AR = 5.086$ and $\Lambda_{c/2} = 12.9°$ Figure 4.39 provides:

$$\left(\frac{c_{l\beta}}{c_{L_1}}\right)_{\Lambda_{c/2}} = -0.000977 \left(\frac{1}{\deg}\right)$$

Using $Mach \cdot \cos(\Lambda_{0.5}) = 0.585$, $\frac{AR}{\cos(\Lambda_{0.5})} = 5.217$ Figure 4.40 provides $K_{M_\Lambda} = 1.084$. Using $\frac{AR}{\cos(\Lambda_{0.5})} = 5.217$, $\frac{A}{b} = \frac{18.4}{26.3} = 0.70$ Figure 4.41 provides $K_f = 0.997$.
Therefore

$$c_{l\beta_{WB}}\bigg|_{\substack{Wing \\ sweep \\ angle}} = 57.3 \cdot c_{L_1}\left[\left(\frac{c_{l\beta}}{c_{L_1}}\right)_{\Lambda_{c/2}} K_{M_\Lambda} K_f\right]$$

$$= 57.3 \cdot 0.149 \, [-0.000977 \cdot 1.084 \cdot 0.997] = -0.009 \left(\frac{1}{rad}\right)$$

(4) Wing contribution to the dihedral effect due to wing aspect ratio

Using $\lambda = 0.517$ and $AR = 5.086$ Figure 4.42 provides

$$\left(\frac{c_{l_\beta}}{c_{L_1}}\right)_{AR} = -0.00094 \left(\frac{1}{\deg}\right) 57.3 \cdot c_{L_1} \left[\left(\frac{c_{l_\beta}}{c_{L_1}}\right)_{AR}\right]$$

$$= 57.3 \cdot 0.149 \cdot (-0.00094) = -0.00801 \left(\frac{1}{\text{rad}}\right)$$

(5) Wing contribution to the dihedral effect due to wing twist angle

Because of lack of specific information, the twist angle is assumed to be $\varepsilon_W = 2° \simeq 0.0345$.
Using $\lambda = 0.517$ and $AR = 5.086$ Figure 4.46 provides

$$\left(\frac{\Delta c_{l_\beta}}{\varepsilon_W \tan \Lambda_{c/4}}\right) = -3.2 \times 10^{-5} \left(\frac{1}{\deg^2}\right) 57.3 \cdot \varepsilon_W \tan \Lambda_{c/4} \left(\frac{\Delta c_{l_\beta}}{\varepsilon_W \tan \Lambda_{c/4}}\right)$$

$$= 57.3 \cdot 2 \cdot \tan 0.284 \cdot (-3.2 \times 10^{-5}) = -0.00107 \left(\frac{1}{\text{rad}}\right)$$

(6) Fuselage contribution to the dihedral effect

This contribution is given by

$$\left(\frac{\Delta c_{l_\beta}}{\Gamma_W}\right) = -0.0005 \cdot AR \cdot \left(\frac{d}{b}\right)^2 = -0.0005 \cdot 5.086 \cdot \left(\frac{3.437}{26.3}\right)^2 = -0.00004 \left(\frac{1}{\deg^2}\right)$$

leading to

$$57.3 \cdot \Gamma_W \left[\frac{\Delta c_{l_\beta}}{\Gamma_W}\right] = 57.3 \cdot (-1.9) \cdot (-0.00004) = 0.005 \left(\frac{1}{\text{rad}}\right)$$

Adding together the previous contributions we have

$$c_{l_{\beta_{WB}}} = (0.0177 - 0.030 - 0.009 - 0.00801 - 0.00107 + 0.005) = -0.0252 \left(\frac{1}{\text{rad}}\right)$$

Horizontal Tail Contribution

The horizontal tail of this aircraft does not have significant dihedral or anhedral angles, so we have the condition $c_{l_{\beta_H}} \approx 0$.

Vertical Tail Contribution

The value of this contribution is given by

$$c_{l_{\beta_V}} = -k_{Y_V} \cdot |c_{L_{\alpha \cdot V}}| \eta_V \cdot \left(1 + \frac{d\sigma}{d\beta}\right) \frac{S_V}{S} \cdot \frac{(Z_V \cos \alpha_1 - X_V \sin \alpha_1)}{b} \quad (\text{rad}^{-1})$$

Using $\frac{b_V}{2r_1} = \frac{5.8}{2 \cdot 4.1} = 0.707$ Figure 4.13 provides $k_{Y_V} = 0.761$

Next, the term $\eta_V \cdot \left(1 + \frac{d\sigma}{d\beta}\right)$ is modeled using

$$\eta_V \cdot \left(1 + \frac{d\sigma}{d\beta}\right) = 0.724 + 3.06 \frac{S_V/S}{1 + \cos(\Lambda_{c/4})} + 0.4 \frac{Z_W}{d} + 0.009 \cdot AR$$

$$= 0.724 + 3.06 \frac{0.5 \cdot 44.66/136}{1 + \cos(0.284)} + 0.4 \frac{(-1.1)}{3.437} + 0.009 \cdot 5.086 = 0.898$$

Thus,

$$c_{l\beta_V} = -k_{Y_V} \cdot |c_{L_{\alpha \cdot V}}| \eta_V \cdot \left(1 + \frac{d\sigma}{d\beta}\right) \frac{S_V}{S} \cdot \frac{(Z_V \cos \alpha_1 - X_V \sin \alpha_1)}{b}$$

$$= -0.761 \cdot |3.678| 0.898 \, \frac{0.5 \cdot 44.66}{136} \cdot \frac{(5.435 \cos(0) - 9.349 \sin(0))}{26.3}$$

$$= -0.085 \, \left(\text{rad}^{-1}\right)$$

Note that for this aircraft $\alpha_1 = 0°$ at low cruise conditions (see Appendix B, Aircraft 8).

Therefore, the final total estimated value of c_{l_β} for the S211 aircraft is

$$c_{l_\beta} \approx c_{l\beta_{WB}} + c_{l\beta_V} = -0.0252 - 0.085 = -0.111 \, \left(\frac{1}{\text{rad}}\right)$$

Table SSP4.2.1 shows the percentage of each of the coefficients with respect to the total c_{l_β} value.

Table SSP4.2.1 Percentage of Each of the Contributions to the Total c_{l_β} Value

Term	Value [1/rad]	% of c_{l_β}
Wing contribution due to the geometric dihedral angle	0.0177	−16%
Wing contribution due to the wing-fuselage position	−0.030	26.8%
Wing contribution due to the sweep angle	−0.009	8.1%
Wing contribution due to the aspect ratio	−0.00801	7.2%
Wing contribution due to the twist angle	−0.00107	1.0%
Body (fuselage) contribution	0.005	−4.3%
Horizontal tail contribution	0	0%
Vertical tail contribution	−0.085	77.2%
Total (c_{l_β})	−0.111	100%

The table suggests a few interesting observations. First, the anhedral angle of the wing naturally leads to an anhedral effect (rather than dihedral); however, a substantial contribution to the dihedral effect is given by the high wing configuration and, to a much lower extent, by the (limited) wing sweep angle. Second, the other and more substantial contribution to the dihedral effect for this aircraft comes from the vertical tail.

Comparison between True Values and Empirical Values

A comparison between the true value of c_{l_β} and its estimate using the empirical aerodynamic modeling approach, along with a calculation of the percentage of the error, is shown in Table SSP4.2.2:

Table SSP4.2.2 Comparative Analysis with Percentages of the Error

	Empirical Values	True Values	Percent Error [%]
c_{l_β}	−0.111	−0.11	1.0 %

The table result suggests that the mathematical modeling provides a very accurate estimate of this important aerodynamic coefficient.

Student Sample Problem 4.3

Consider the data relative to the McDonnell Douglas F-4 (Phantom) aircraft at subsonic conditions in Appendix B and Appendix C. It has been pointed out that the geometry of the horizontal tail of this aircraft makes it a special case. Use the provided drawings and table to extract all the relevant geometric characteristics of the aircraft. Using the modeling approach outlined in this chapter, the dihedral effect c_{l_β} can be modeled as

$$c_{l\beta} = c_{l\beta_{WB}} + c_{l\beta_H} + c_{l\beta_V}$$

Estimate the value of c_{l_β} for this aircraft. Also, evaluate the percentage of each of the different contributions with respect to the total c_{l_β} value. Next, compare the obtained c_{l_β} value with the true value listed in Appendix B (Aircraft 10) and evaluate the modeling error.

Solution of Student Sample Problem 4.3

All the relevant geometric parameters used in this problem were extracted from the images and the tables in Appendix C. Also, note that several geometric parameters related to the wing and the horizontal tail were calculated previously for the McDonnell Douglas F-4 problems in Chapters 2 and 3. From the drawings in Appendix C, the following geometric parameters were identified:

$$b_{2V} = 2 \cdot 6.2 = 12.4 \text{ ft}, \quad c_{T_V} = 3.9 \text{ ft}, \quad c_{R_V} = 16.4 \text{ ft}, \quad \Lambda_{LE_V} = 63° = 1.1 \text{ rad}$$

Note that b_{2V} is actually the same of the span of the vertical tail; this value is used so that the geometric parameters of the vertical tail can be found in a similar manner as the parameters for the wing. Next, using the assumption of straight wings, the vertical tail geometric parameters were derived using the values from the drawings in Appendix C.

Vertical Tail Geometric Parameters

The following vertical tail parameters are required to determine the stability derivatives.

$$\lambda_V = \frac{c_{T_V}}{c_{R_V}} = \frac{3.9}{16.4} = 0.238;$$

$$S_{2V} = \frac{b_{2V}}{2} c_{R_V} (1 + \lambda_V) = \frac{12.4}{2} \cdot 16.4 \cdot (1 + 0.238) = 125.86 \text{ ft}^2$$

Note that S_{2V} is twice the value of the area of the vertical tail; the use of S_{2V} allows finding the following geometric parameters:

$$AR_V = \frac{b_{2V}^2}{S_{2V}} = \frac{12.4^2}{125.86} = 1.222,$$

$$\bar{c}_V = \frac{2}{3} c_{R_V} \frac{(1 + \lambda_V + \lambda_V^2)}{(1 + \lambda_V)} = \frac{2}{3} \cdot 5.7 \cdot \frac{(1 + 0.351 + 0.351^2)}{(1 + 0.351)} = 4.146 \text{ ft}$$

$$x_{MAC_V} = \frac{b_{2V}}{6} \frac{(1 + 2\lambda_V)}{(1 + \lambda_V)} \tan(\Lambda_{LE_V}) = \frac{11.6}{6} \cdot \frac{(1 + 2 \cdot 0.238)}{(1 + 0.238)} \tan(1.1) = 11.433 \text{ ft}$$

$$y_{MAC_V} = \frac{b_{2V}}{6} \frac{(1 + 2\lambda_V)}{(1 + \lambda_V)} = \frac{12.4}{6} \cdot \frac{(1 + 2 \cdot 0.238)}{(1 + 0.238)} = 2.464 \text{ ft}$$

$$\tan \Lambda_{0.5_V} = \tan \Lambda_{LE_V} - \frac{4 \cdot 0.5 \cdot (1 - \lambda_V)}{AR_V(1 + \lambda_V)} = \tan(1.1) - \frac{4 \cdot 0.5 \cdot (1 - 0.238)}{1.222 \cdot (1 + 0.238)} \Rightarrow \Lambda_{0.5_V} = 0.762$$

Wing-Tail Geometric Parameters

The derivation of the distances $X_{V_S}, Z_{V_S}, X_{R_S}$ and Z_{R_S} is shown in SSP4.2. These parameters are needed to determine the stability derivatives related with the vertical tail and the rudder. From the drawings in Appendix C they are given by

$$X_{V_S} = 13.754 \text{ ft}, \quad Z_{V_S} = 7.664, \quad X_{R_S} = 21.16 \text{ ft}, \quad Z_{R_S} = 7.4 \text{ ft}$$

Technically speaking the parameters X_{R_S} and Z_{R_S}—relative to the rudder—are not needed in this problem. However, they were evaluated because they are conceptually similar to X_{V_S} and Z_{V_S}.

Vertical Tail Lift Slope Coefficient

This coefficient is calculated using the Polhamus formula with $AR_{V_{eff}}$, defined as

$$AR_{V_{eff}} = c_1 \cdot AR_V \cdot [1 + K_{HV}(c_2 - 1)]$$

The parameter c_1 is found from Figure 4.15 using $\lambda_V = 0.238$ along with the parameter:

$$\frac{b_V}{2r_1} = \frac{6.2}{2 \cdot 6.2} = 0.5$$

The unusually large value of r_1 is due to the large size of the engine exhaust on this aircraft. Using $\lambda_V = 0.238$ and $b_V/2r_1 = 0.5$ Figure 4.15 provides $c_1 = 1.156$.

Next, the parameter c_2 is found from Figure 4.16 using

$$\frac{Z_H}{b_V} = \frac{-3.2}{6.2} = -0.516$$

The geometric parameter $x_{AC_H \to V}$, shown in Figure 4.17, is calculated using

$$x_{AC_H \to V} = x_{MAC_H} + 0.25 \cdot \bar{c}_H = 3.121 + 0.25 \cdot 5.396 = 4.47 \text{ ft} \Rightarrow \frac{x_{AC_H \to V}}{\bar{c}_V} = \frac{4.47}{11.433} = 0.391$$

Note that Z_H is considered negative when above the CG. Using the $x_{AC_H \to V}$ value, Figure 4.16 provides $c_2 = 0.822$.

Finally, using S_H/S_V in Figure 4.18, we have

$$\frac{S_H}{S_V} = \frac{80.36}{0.5 \cdot 125.86} = 1.277 \Rightarrow K_{HV} = 0.989$$

leading to

$$AR_{V_{eff}} = c_1 \cdot AR_V \cdot [1 + K_{HV}(c_2 - 1)] = 1.156 \cdot 1.222 \cdot [1 + 0.989(0.822 - 1)] = 1.164$$

$$\tan \Lambda_{0.5_V} = \tan \Lambda_{LE_V} - \frac{4 \cdot 0.5 \cdot (1 - \lambda_V)}{AR_V(1 + \lambda_V)} = \tan(1.1) - \frac{4 \cdot 0.5 \cdot (1 - 0.238)}{1.222 \cdot (1 + 0.238)} \Rightarrow \Lambda_{0.5_V} = 0.762$$

Using $AR_{V_{eff}} = 1.164 \leq 4$, the application of the Polhamus formula leads to

$$k_V = 1 + \frac{AR_{V_{eff}}(1.87 - 0.000233\Lambda_{LE_V})}{100}$$

$$= 1 + \frac{1.164(1.87 - 0.000233 \cdot 1.1)}{100} = 1.023$$

$$c_{L_{\alpha V,eff}} = \frac{2\pi AR_{V_{eff}}}{2 + \sqrt{\left\{\left[\frac{AR_{V_{eff}}^2(1 - Mach^2)}{k^2}\right]\left(1 + \frac{\tan^2(\Lambda_{0.5})}{(1 - Mach^2)}\right) + 4\right\}}}$$

$$= \frac{2\pi \cdot 1.164}{2 + \sqrt{\left\{\left[\frac{1.164^2(1 - 0.9^2)}{1.023^2}\right]\left(1 + \frac{\tan^2(0.762)}{(1 - 0.9^2)}\right) + 4\right\}}} = 1.689 \left(\frac{1}{\text{rad}}\right)$$

Note that in this case the application of the Polhamus formula will lead to approximate results since the parameters λ, Λ_{LE}, AR and *Mach* are outside the allowed ranges.

(Wing + Body) Contribution

The modeling for $c_{l\beta_{WB}}$ is given by

$$c_{l\beta_{WB}} = 57.3 \cdot c_{L_1} \left[\left(\frac{c_{l_\beta}}{c_{L_1}}\right)_{\Lambda_{c/2}} K_{M_\Lambda} K_f + \left(\frac{c_{l_\beta}}{c_{L_1}}\right)_{AR}\right]$$

$$+ 57.3 \left\{\Gamma_W \left[\frac{c_{l_\beta}}{\Gamma_W} K_{M_\Gamma} + \frac{\Delta c_{l_\beta}}{\Gamma_W}\right] + \left(\Delta c_{l_\beta}\right)_{Z_W} + \varepsilon_W \tan \Lambda_{c/4} \left(\frac{\Delta c_{l_\beta}}{\varepsilon_W \tan \Lambda_{c/4}}\right)\right\}$$

Next, the following individual contributions to the dihedral effect in the above relationship are evaluated separately:

- Wing contribution due to the geometric dihedral angle;
- Wing contribution due to the wing-fuselage positions;
- Wing contribution due to the sweep angle;

214 Chapter 4 Modeling of Lateral Directional Aerodynamic Forces and Moments

- Wing contribution due to the aspect ratio;
- Wing contribution due to the twist angle;
- Body (fuselage) contribution.

(1) Wing contribution to the dihedral effect due to the geometric dihedral angle

A closed look at the front drawing of the F-4 wing in Appendix C reveals that that the wing has a negligible dihedral angle for approximately 60 percent of the wing span and a $\Gamma_W \approx 4°$ for the remaining 40 percent portion of the wing span toward the wing tip. Therefore, for the purpose of estimating $c_{l_{\beta_{WB}}}\big|_{\substack{Geometric \\ dihedral \\ angle}}$ we consider an average $\Gamma_{W_{Average}} \approx 2°$.

Using $\lambda = 0.261$ and $\Lambda_{0.5} = 35.6°$ Figure 4.43 provides

$$\frac{c_{l_\beta}}{\Gamma_W} = -0.0001138\,(1/\text{deg}^2)$$

Using $Mach \cdot \cos(\Lambda_{0.5}) = 0.9 \cdot \cos(0.621) = 0.732$, $\dfrac{AR}{\cos(\Lambda_{0.5})} = \dfrac{2.826}{\cos(0.621)} = 3.475$

Figure 4.44 provides $K_{M_\Gamma} = 1.068$. Therefore

$$c_{l_{\beta_{WB}}}\big|_{\substack{Geometric \\ dihedral \\ angle}} = 57.3\,\Gamma_{W_{Average}} \left[\frac{c_{l_\beta}}{\Gamma_W} K_{M_\Gamma}\right] = 57.3 \cdot (2) \cdot [-0.0001138 \cdot 1.068] = -0.014\left(\frac{1}{\text{rad}}\right)$$

(2) Wing contribution to the dihedral effect due to wing-fuselage position

The fuselage's diameter d_B is required for the evaluation of this term. d_B is found using

$$d_B = \sqrt{\frac{S_{f_{AVG}}}{0.7854}}$$

where $S_{f_{AVG}}$ is the average cross-sectional area of the fuselage calculated using

$$S_{f_{AVG}} = \pi\left(\frac{\text{Fuselage average height}}{2}\right)\left(\frac{\text{Fuselage average width}}{2}\right)$$

$$= \pi\left(\frac{5.6}{2}\right)\left(\frac{5.6}{2}\right) = 24.63\,\text{ft}^2$$

where the average height and width of the fuselage were obtained from the drawings in Appendix C. Therefore, we have

$$d = \sqrt{\frac{S_{f_{AVG}}}{0.7854}} = \sqrt{\frac{24.63}{0.7854}} = 5.6\,\text{ft}$$

leading to the following value for $c_{l_{\beta_{WB}}}\big|_{\substack{High/low \\ wing}}$:

$$c_{l_{\beta_{WB}}}\big|_{\substack{High/low \\ wing}} = 57.3\left(\Delta c_{l_\beta}\right)_{Z_W} = 57.3\frac{1.2\sqrt{AR}}{57.3}\cdot\frac{Z_W}{b}\cdot\left(\frac{2d}{b}\right) = \frac{1.2\sqrt{2.826}}{57.3}\cdot\frac{(1.9)}{38.7}\cdot\left(\frac{2\cdot 5.6}{38.7}\right) = 0.029\left(\frac{1}{\text{rad}}\right)$$

(3) Wing contribution to the dihedral effect due to wing sweep angle

The value of c_{L_1} is required to calculated $c_{l_{\beta_{WB}}}\big|_{\substack{Wing \\ sweep \\ angle}}$. This value is calculated using the given flight conditions and data from Standard Atmosphere tables:

$$c_{L_1} = \frac{W}{\frac{1}{2}\rho V^2 S} = 0.261$$

Using $\lambda = 0.261$, $AR = 2.826$ and $\Lambda_{c/2} = 35.6°$ Figure 4.39 provides

$$\left(\frac{c_{l_\beta}}{c_{L_1}}\right)_{\Lambda_{c/2}} = -0.000286 \left(\frac{1}{\deg}\right)$$

Using $Mach \cdot \cos(\Lambda_{0.5}) = 0.732$, $\frac{AR}{\cos(\Lambda_{0.5})} = 3.475$ Figure 4.40 provides $K_{M_\Lambda} = 1.048$. Using $\frac{AR}{\cos(\Lambda_{0.5})} = 3.475$, $\frac{A}{b} = \frac{41.5}{38.7} = 1.07$ Figure 4.41 provides $K_f = 1$.

Therefore

$$c_{l\beta_{WB}}\bigg|_{\substack{\text{Wing} \\ \text{sweep} \\ \text{angle}}} = 57.3 \cdot c_{L_1}\left[\left(\frac{c_{l_\beta}}{c_{L_1}}\right)_{\Lambda_{c/2}} K_{M_\Lambda} K_f\right]$$

$$= 57.3 \cdot 0.261[-0.000286 \cdot 1.048 \cdot 1] = -0.045 \left(\frac{1}{\text{rad}}\right)$$

(4) Wing contribution to the dihedral effect due to wing aspect ratio

Using $\lambda = 0.261$ and $AR = 2.826$ Figure 4.42 provides

$$\left(\frac{c_{l_\beta}}{c_{L_1}}\right)_{AR} = -0.00184 \left(\frac{1}{\deg}\right) 57.3 \cdot c_{L_1}\left[\left(\frac{c_{l_\beta}}{c_{L_1}}\right)_{AR}\right]$$

$$= 57.3 \cdot 0.261 \cdot (-0.00184) = -0.027 \left(\frac{1}{\text{rad}}\right)$$

(5) Wing contribution to the dihedral effect due to wing twist angle

Because of lack of specific information, the twist angle is assumed to be $\varepsilon_W = 2° \simeq 0.0345$.

Using $\lambda = 0.261$ and $AR = 2.826$ Figure 4.46 provides

$$\left(\frac{\Delta c_{l_\beta}}{\varepsilon_W \tan \Lambda_{c/4}}\right) = -2.08 \times 10^{-5} \left(\frac{1}{\deg^2}\right) 57.3 \cdot \varepsilon_W \tan \Lambda_{c/4} \left(\frac{\Delta c_{l_\beta}}{\varepsilon_W \tan \Lambda_{c/4}}\right)$$

$$= 57.3 \cdot 2 \cdot \tan 0.745 \cdot (-2.08 \times 10^{-5}) = -0.0022 \left(\frac{1}{\text{rad}}\right)$$

(6) Fuselage contribution to the dihedral effect

This contribution is given by

$$\left(\frac{\Delta c_{l_\beta}}{\Gamma_W}\right) = -0.0005 \cdot AR \cdot \left(\frac{d}{b}\right)^2 = -0.0005 \cdot 2.826 \cdot \left(\frac{5.6}{38.7}\right)^2 = -0.00003 \left(\frac{1}{\deg^2}\right)$$

leading to

$$57.3 \cdot \Gamma_W \left[\frac{\Delta c_{l_\beta}}{\Gamma_W}\right] = 57.3 \cdot (4) \cdot (-0.00003) = -0.007 \left(\frac{1}{\text{rad}}\right)$$

Adding together the previous contributions we have

$$c_{l\beta_{WB}} = (-0.014 + 0.029 - 0.045 - 0.027 - 0.002 - 0.007) = -0.066 \left(\frac{1}{\text{rad}}\right)$$

Horizontal Tail Contribution

The horizontal tail for this aircraft is quite large and has a considerable anhedral angle, so this contribution is modeled in a similar way as the wing contribution.

(1) Horizontal tail contribution to the dihedral effect due to the geometric dihedral angle

Using $\lambda = 0.289$ and $\Lambda_{0.5} = 31.1°$ Figure 4.43 provides

$$\frac{c_{l_\beta}}{\Gamma_H} = -0.0001256 (1/\deg^2)$$

Using $Mach \cdot \cos(\Lambda_{0.5}) = 0.9 \cdot \cos(0.543) = 0.77$, $\frac{AR}{\cos(\Lambda_{0.5})} = \frac{3.347}{\cos(0.543)} = 3.91$

Figure 4.44 provides $K_{M_\Gamma} = 1.093$. Therefore

$$c_{l\beta_{WB}}\Big|_{\substack{Geometric \\ dihedral \\ angle}} = 57.3 \, \Gamma_H \left[\frac{c_{l\beta}}{\Gamma_H} K_{M_\Gamma}\right] = 57.3 \cdot (-27) \cdot [-0.0001256 \cdot 1.093] = 0.212 \left(\frac{1}{rad}\right)$$

(2) Horizontal tail contribution to the dihedral effect due to wing-fuselage position

$$c_{l\beta_{HB}}\Big|_{\substack{High/low \\ wing}} = 57.3 \left(\Delta c_{l\beta}\right)_{Z_H} = 57.3 \frac{1.2\sqrt{AR}}{57.3} \cdot \frac{Z_H}{b_H} \cdot \left(\frac{2d}{b_H}\right)$$

$$= 1.2\sqrt{3.347} \cdot \frac{(-2.1)}{16.4} \cdot \left(\frac{2 \cdot 5.6}{16.4}\right) = -0.192 \left(\frac{1}{rad}\right)$$

(3) Horizontal tail contribution to the dihedral effect due to wing sweep angle

Using $\lambda = 0.289$, $AR = 3.347$ and $\Lambda_{c/2} = 31.1°$ Figure 4.39 provides

$$\left(\frac{c_{l\beta}}{c_{L_1}}\right)_{\Lambda_{c/2}} = -0.00208 \left(\frac{1}{deg}\right)$$

Using $Mach \cdot \cos(\Lambda_{0.5}) = 0.9 \cdot \cos(0.543) = 0.77$, $\frac{AR}{\cos(\Lambda_{0.5})} = \frac{3.347}{\cos(0.543)} = 3.91$ Figure 4.40 provides $K_{M_\Lambda} = 3.081$.
Using $\frac{AR}{\cos(\Lambda_{0.5})} = 3.91$, $\frac{A}{b} = \frac{41.5}{38.7} = 1.07$ Figure 4.41 provides $K_f \simeq 1$. Therefore

$$c_{l\beta_{WB}}\Big|_{\substack{Wing \\ sweep \\ angle}} = 57.3 \cdot c_{L_1} \left[\left(\frac{c_{l\beta}}{c_{L_1}}\right)_{\Lambda_{c/2}} K_{M_\Lambda} K_f\right] = 57.3 \cdot 0.261 \cdot (-0.00208) \cdot 3.081 \cdot 1 = -0.096 \left(\frac{1}{rad}\right)$$

(4) Horizontal tail contribution to the dihedral effect due to wing aspect ratio

Using $\lambda = 0.289$ and $AR = 3.347$ Figure 4.42 provides

$$\left(\frac{c_{l\beta}}{c_{L_1}}\right)_{AR} = -0.00149 \left(\frac{1}{deg}\right) 57.3 \cdot c_{L_1} \left[\left(\frac{c_{l\beta}}{c_{L_1}}\right)_{AR}\right]$$

$$= 57.3 \cdot 0.261 \cdot (-0.00149) = -0.022 \left(\frac{1}{rad}\right)$$

(5) Fuselage contribution to the dihedral effect

This contribution is given by

$$\left(\frac{\Delta c_{l\beta}}{\Gamma_H}\right) = -0.0005 \cdot AR_H \cdot \left(\frac{d}{b_H}\right)^2 = -0.0005 \cdot 3.347 \cdot \left(\frac{5.6}{16.4}\right)^2 = -0.0002 \left(\frac{1}{deg^2}\right)$$

leading to

$$57.3 \cdot \Gamma_H \left[\frac{\Delta c_{l\beta}}{\Gamma_W}\right] = 57.3 \cdot (-27) \cdot (-0.0002) = 0.302 \left(\frac{1}{rad}\right)$$

Adding together the previous five contributions we have

$$c_{l\beta_{WB_H}} = (0.212 - 0.192 - 0.096 - 0.022 + 0.302) = 0.204 \left(\frac{1}{rad}\right)$$

Due to the lower surface and wing span, assuming $\eta_H = 0.9$, the total dihedral effect for the horizontal tail is given by

$$c_{l\beta_H} = c_{l\beta_{WB}}\Big|_H \eta_H \frac{S_H}{S} \frac{b_H}{b} = 0.204 \cdot 0.9 \cdot \frac{80.36}{530} \cdot \frac{16.4}{38.7} = 0.0118$$

Vertical Tail Contribution

The value of this contribution is given by

$$c_{l\beta_V} = -k_{Y_V} \cdot |c_{L_{\alpha \cdot V}}| \eta_V \cdot \left(1 + \frac{d\sigma}{d\beta}\right) \frac{S_V}{S} \cdot \frac{(Z_V \cos\alpha_1 - X_V \sin\alpha_1)}{b} \quad (\text{rad}^{-1})$$

Using $\frac{b_V}{2r_1} = \frac{6.2}{2 \cdot 6.2} = 0.5$ Figure 4.13 provides $k_{Y_V} = 0.761$

Next, the term $\eta_V \cdot \left(1 + \frac{d\sigma}{d\beta}\right)$ is modeled using

$$\eta_V \cdot \left(1 + \frac{d\sigma}{d\beta}\right) = 0.724 + 3.06 \frac{S_V/S}{1 + \cos(\Lambda_{c/4})} + 0.4 \frac{Z_W}{d} + 0.009 \cdot AR$$

$$= 0.724 + 3.06 \frac{0.5 \cdot 125.86/530}{1 + \cos(0.745)} + 0.4 \frac{(1.9)}{5.6} + 0.009 \cdot 2.826 = 1.095$$

Thus,

$$c_{l\beta\ V} = -k_{Y_V} \cdot |c_{L_{\alpha \cdot V}}| \eta_V \cdot \left(1 + \frac{d\sigma}{d\beta}\right) \frac{S_V}{S} \cdot \frac{(Z_V \cos\alpha_1 - X_V \sin\alpha_1)}{b}$$

$$= -0.761 \cdot |1.689| 1.095 \frac{0.5 \cdot 125.86}{530} \cdot \frac{(7.664 \cos(0.045) - 13.754 \sin(0.045))}{38.7}$$

$$= -0.030 \ (\text{rad}^{-1})$$

Note that for this aircraft $\alpha_1 = 2.6° = 0.045$ rad at low cruise conditions (see Appendix B, Aircraft 10). Therefore, the final total estimated value of $c_{l\beta}$ for the F-4 aircraft is

$$c_{l\beta} \approx c_{l\beta_{WB}} + c_{l\beta_H} + c_{l\beta_V} = -0.066 + 0.0118 - 0.030 = -0.0842 \ \left(\frac{1}{\text{rad}}\right)$$

Table SSP4.3.1 shows the percentage of each of the coefficients with respect to the total $c_{l\beta}$ value.

Table SSP4.3.1 Percentage of Each of the Contributions to the Total $c_{l\beta}$ Value

Term	Value [1/rad]	% of $c_{l\beta}$
Wing contribution due to the geometric dihedral angle	−0.014	16.6%
Wing contribution due to the wing-fuselage position	0.029	−34.4%
Wing contribution due to the sweep angle	−0.045	53.4%
Wing contribution due to the aspect ratio	−0.027	32.1%
Wing contribution due to the twist angle	−0.002	2.6%
Body (fuselage) contribution	−0.007	8.3%
Horizontal tail contribution	0.0118	−14%
Vertical tail contribution	−0.030	35.6%
Total ($c_{l\beta}$)	−0.0842	100

The above table suggests a few interesting observations. The substantial negative dihedral angle of the horizontal tail and the low wing-body configuration provide a substantial anhedral (negative dihedral) effect, which is counteracted mainly by larger dihedral effects due to the vertical tail and the wing sweep angle.

Comparison between True Values and Empirical Values

A comparison between the true value of $c_{l\beta}$ and its estimate using the empirical aerodynamic modeling approach, along with a calculation of the percentage of the error, is shown in the Table SSP.4.3.2:

Table SSP4.3.2 Comparative Analysis with Percentages of the Error

	Empirical Values	True Values	Percent Error [%]
$c_{l\beta}$	−0.0842	−0.080	5.2 %

218 Chapter 4 Modeling of Lateral Directional Aerodynamic Forces and Moments

The result suggests that the mathematical modeling provides an accurate estimate of this important aerodynamic coefficient.

Student Sample Problem 4.4

Consider the data relative to the McDonnell Douglas DC-9 aircraft in Appendix C. Use the provided drawings to extract all the relevant geometric characteristics of the aircraft.

Assume $\eta_H = \eta_V = 0.9$. Using the modeling approach outlined in this chapter, provide estimates for the following stability derivatives:

$$c_{Y_\beta}, c_{n_\beta}$$

For both aerodynamic parameters highlight the contribution of the vertical tail versus the total value.

Solution of Student Sample Problem 4.4

All the relevant geometric parameters used in this problem were extracted from the images in Appendix C. Also, note that several geometric parameters related to the wing, horizontal tail, and vertical tail were calculated previously for the DC-9 problems in Chapters 2 and 3 and in SSP4.1.

Calculation of c_{Y_β}

The general modeling of this derivative is given by

$$c_{Y_\beta} = \left(c_{Y_{\beta_W}} + c_{Y_{\beta_B}}\right) + c_{Y_{\beta_H}} + c_{Y_{\beta_V}}$$

Since the dihedral angle of the horizontal tail angle is negligible, $c_{Y_{\beta_H}} \approx 0$. Therefore

$$c_{Y_\beta} \approx \left(c_{Y_{\beta_W}} + c_{Y_{\beta_B}}\right) + c_{Y_{\beta_V}}$$

Wing Contribution

$c_{Y_{\beta_W}}$ is given by

$$c_{Y_{\beta_W}} \approx -0.0001 \, |\Gamma_W| \cdot 57.3 = -0.0001 \, |2.2| \cdot 57.3 = -0.013 \, (\text{rad}^{-1})$$

Body Contribution

$c_{Y_{\beta_B}}$ is given by

$$c_{Y_{\beta_B}} \approx -2 \cdot K_{\text{int}} \cdot \frac{S_{P \to V}}{S} \, (\text{rad}^{-1})$$

Using $\frac{Z_W}{d/2} = \frac{3.1}{11.5/2} = 0.539$ Figure 4.8 provides $K_{\text{int}} = 0.8$

As described in Section 4.2.1 $S_{P \to V}$ is instead the cross section at the location of the fuselage where the flow transitions from potential to viscous. From the drawings in Appendix C, using the approach described in Section 4.2.2 $S_{P \to V} = 89.54 \, \text{ft}^2$. Therefore

$$c_{Y_{\beta_B}} \approx -2 \cdot K_{\text{int}} \cdot \frac{S_{P \to V}}{S} = -2 \cdot 0.8 \cdot \frac{89.54}{928} = -0.154 \, (\text{rad}^{-1})$$

Horizontal Tail Contribution

Since the dihedral angle of the horizontal tail angle is negligible, $c_{Y_{\beta_H}} \approx 0$.

Vertical Contribution

$c_{Y_{\beta_V}}$ is found using

$$c_{Y_{\beta_V}} \approx -k_{Y_V} \cdot \left|c_{L_{\alpha \cdot V}}\right| \eta_V \cdot \left(1 + \frac{d\sigma}{d\beta}\right) \frac{S_V}{S} \, (\text{rad}^{-1})$$

Again, the term $\eta_V \cdot \left(1 + \frac{d\sigma}{d\beta}\right)$ is modeled using

$$\eta_V \cdot \left(1 + \frac{d\sigma}{d\beta}\right) = 0.724 + 3.06 \frac{S_V/S}{1+\cos(\Lambda_{c/4})} + 0.4\frac{Z_W}{d} + 0.009 \cdot AR$$

$$= 0.724 + 3.06 \frac{0.5 \cdot 334.56/928}{1+\cos(0.433)} + 0.4 \frac{3.1}{9.596} + 0.009 \cdot 8.612 = 1.22$$

leading to

$$c_{Y\beta_V} \approx -k_{Y_V} \cdot |c_{L_{\alpha \cdot V}}|\eta_V \cdot \left(1 + \frac{d\sigma}{d\beta}\right)\frac{S_{2V}}{S} \ (\text{rad}^{-1})$$

$$= -0.761 \cdot |3.607|1.22\frac{334.56}{928} = -1.207 \ (\text{rad}^{-1})$$

Finally the total value of $c_{Y\beta}$ is given by

$$c_{Y\beta} \approx (c_{Y\beta_W} + c_{Y\beta_B}) + c_{Y\beta_V}$$
$$= -0.013 - 0.154 - 1.207 = -1.374 \ (\text{rad}^{-1})$$

Calculation of $c_{n\beta}$

The general modeling of this derivative is

$$c_{n\beta} = c_{n\beta_W} + c_{n\beta_B} + c_{n\beta_H} + c_{n\beta_V} \approx c_{n\beta_B} + c_{n\beta_V}$$

Wing Contribution

At small angles of attack it can be approximated that: $c_{n\beta_W} \approx 0$.

Body Contribution

$c_{n\beta_B}$ is given by

$$c_{n\beta_B} = -57.3 \cdot K_N K_{R_l} \frac{S_{B_S}}{S}\frac{l_B}{b}$$

To obtain K_N from Figure 4.68 the following parameters from Figure 4.67 related to dimensions of the DC-9 aircraft from Appendix C are required.

$$\frac{l_{CG}}{l_B} = \frac{49.2}{92.1} = 0.534; \quad \frac{l_B^2}{S_{B_S}} = \frac{92.1^2}{896.55} = 9.461$$

$$\sqrt{\frac{z_1}{z_2}} = \sqrt{\frac{11.5}{10.7}} = 1.044; \quad \frac{z_{\text{Max}}}{w_{\text{Max}}} = \frac{11.7}{11} = 1.064$$

Using these parameters Figure 4.68 provides: $K_N = 0.00152$

Next, the following parameter is required to obtain K_{Re_l}.

$$Re_{Fuselage} = \frac{V \cdot l_B}{\nu} = \frac{(0.7 \cdot 981.66) \cdot 92.1}{3.823 \times 10^{-4}} = 1.6554 \times 10^8$$

The values of the speed of sound and the kinematic viscosity were obtained from the Standard Atmosphere tables, using the given flight altitude. Finally, using $Re_{Fuselage}$, Figure 4.69 provides $K_{Re_l} = 2.05$, leading to

$$c_{n\beta_B} = -57.3 \cdot K_N K_{R_l} \frac{S_{B_S}}{S}\frac{l_B}{b} = -57.3 \cdot 0.00152 \cdot 2.05 \cdot \frac{896.55}{928} \cdot \frac{92.1}{89.4} = -0.178 \ \left(\frac{1}{\text{rad}}\right)$$

Horizontal Tail Contribution

Since the dihedral angle of the horizontal tail angle is negligible, $c_{n\beta_H} \approx 0$.

Vertical Tail Contribution

$c_{n\beta_V}$ is given by

$$c_{n\beta_V} = -c_{Y\beta_V} \cdot \frac{(X_V \cos\alpha_1 + Z_V \sin\alpha_1)}{b}$$

$$= -(-1.207)\frac{(34.4 \cos(0.035) + 12.1 \sin(0.035))}{89.4} = 0.471 \left(\frac{1}{\text{rad}}\right)$$

X_V and Z_V were calculated in SSP4.1.

Therefore, the total value of $c_{n\beta}$ is given by

$$c_{n\beta} = c_{n\beta_B} + c_{n\beta_V} \cong -0.178 + 0.471 = 0.293 \ (\text{rad}^{-1})$$

Student Sample Problem 4.5

Consider the data relative to the SIAI Marchetti S211 aircraft in Appendix B and Appendix C. Use the provided drawings to extract all the relevant geometric characteristics of the aircraft. Using the modeling approach outlined in this chapter, provide estimates for the following stability derivatives:

$$c_{Y\beta}, c_{n\beta}$$

For both aerodynamic parameters highlight the contribution of the vertical tail versus the total value. Next, compare the obtained values with the true values listed in Appendix B (Aircraft 8) and evaluate the modeling error.

Solution of Student Sample Problem 4.5

All the relevant geometric parameters used in this problem were extracted from the images in Appendix C. Also, note that several geometric parameters related to the wing, horizontal tail, and vertical tail were calculated previously for the S211 problems in Chapters 2 and 3 and in SSP4.2.

Calculation of $c_{Y\beta}$

The general modeling of this derivative is given by

$$c_{Y\beta} = \left(c_{Y\beta_W} + c_{Y\beta_B}\right) + c_{Y\beta_H} + c_{Y\beta_V}$$

Since the dihedral angle of the horizontal tail angle is negligible, $c_{Y\beta_H} \approx 0$. Therefore

$$c_{Y\beta} \approx \left(c_{Y\beta_W} + c_{Y\beta_B}\right) + c_{Y\beta_V}$$

Wing Contribution

The coefficient $c_{Y\beta_W}$ is significant only when the wing has a positive dihedral angle Γ_W. Since for this aircraft $\Gamma_W \leq 0$ we have $c_{Y\beta_W} \approx 0$.

Body Contribution

$c_{Y\beta_B}$ is given by

$$c_{Y\beta_B} \approx -2 \cdot K_{\text{int}} \cdot \frac{S_{P \to V}}{S} \ (\text{rad}^{-1})$$

Using $\dfrac{Z_W}{d/2} = \dfrac{-1.1}{5/2} = -0.44$ Figure 4.8 provides $K_{\text{int}} = 0.830$.

As described in Section 4.2.2 $S_{P \to V}$ is the cross section at the location of the fuselage where the flow transitions from potential to viscous. From the drawings in Appendix C, using the approach described in Section 4.2.2 $S_{P \to V} = 4.85$ ft^2. Therefore

$$c_{Y\beta_B} \approx -2 \cdot K_{\text{int}} \cdot \frac{S_{P \to V}}{S} = -2 \cdot 0.830 \cdot \frac{4.85}{136} = -0.059 \ (\text{rad}^{-1})$$

Vertical Tail Contribution

$c_{Y\beta_V}$ is found using

$$c_{Y\beta_V} \simeq -k_{Y_V} \cdot \left|c_{L_\alpha \cdot V}\right| \eta_V \cdot \left(1 + \frac{d\sigma}{d\beta}\right) \frac{S_V}{S} \quad (\text{rad}^{-1})$$

Again, the term $\eta_V \cdot \left(1 + \frac{d\sigma}{d\beta}\right)$ is modeled using

$$\eta_V \cdot \left(1 + \frac{d\sigma}{d\beta}\right) = 0.724 + 3.06 \frac{S_V/S}{1 + \cos(\Lambda_{c/4})} + 0.4 \frac{Z_W}{d} + 0.009 \cdot AR$$

$$= 0.724 + 3.06 \frac{0.5 \cdot 44.66/136}{1 + \cos(0.284)} + 0.4 \frac{(-1.1)}{3.437} + 0.009 \cdot 5.086 = 0.898$$

leading to

$$c_{Y\beta_V} \approx -k_{Y_V} \cdot \left|c_{L_\alpha \cdot V}\right| \eta_V \cdot \left(1 + \frac{d\sigma}{d\beta}\right) \frac{S_V}{S} \quad (\text{rad}^{-1})$$

$$= -0.761 \cdot |3.678| 0.898 \frac{44.66}{136} = -0.826 \quad (\text{rad}^{-1})$$

Finally, the total value of $c_{Y\beta}$ is given by

$$c_{Y\beta} \approx c_{Y\beta_B} + c_{Y\beta_V} = -0.059 - 0.826 = -0.885 \quad (\text{rad}^{-1})$$

Calculation of c_{n_β}

The general modeling of this derivative is

$$c_{n\beta} = c_{n\beta_W} + c_{n\beta_B} + c_{n\beta_H} + c_{n\beta_V}$$

Wing Contribution

At small angles of attack it can be approximated that $c_{n\beta_W} \simeq 0$.

Body Contribution

$c_{n\beta_B}$ is given by

$$c_{n\beta_B} = -57.3 \cdot K_N K_{R_l} \frac{S_{B_S}}{S} \frac{l_B}{b}$$

To obtain K_N from Figure 4.68 the following parameters from Figure 4.67 related to dimensions of the S211 aircraft from Appendix C are required.

$$\frac{l_{CG}}{l_B} = \frac{16.6}{30.9} = 0.537, \quad \frac{l_B^2}{S_{B_S}} = \frac{30.9^2}{116.009} = 8.231,$$

$$\sqrt{\frac{z_1}{z_2}} = \sqrt{\frac{4.5}{4.3}} = 1.023; \quad \frac{z_{\text{Max}}}{w_{\text{Max}}} = \frac{5.1}{5.9} = 0.864$$

Using the above parameters Figure 4.68 provides $K_N = 0.00138$.
Next, the following parameter is required to obtain K_{Re_l}.

$$\text{Re}_{\text{Fuselage}} = \frac{V \cdot l_B}{\nu} = \frac{(0.6 \cdot 1015.98) \cdot 30.9}{3.019 \times 10^{-4}} = 6.239 \times 10^7$$

The values of the speed of sound and the kinematic viscosity were obtained from the Standard Atmosphere tables, using the given flight altitude. Finally, using $\text{Re}_{\text{Fuselage}}$, Figure 4.69 provides $K_{Re_l} = 1.85$ leading to

$$c_{n\beta_B} = -57.3 \cdot K_N K_{R_l} \frac{S_{B_S}}{S} \frac{l_B}{b} = -57.3 \cdot 0.00138 \cdot 1.85 \cdot \frac{116.009}{136} \cdot \frac{30.9}{26.3} = -0.147 \left(\frac{1}{\text{rad}}\right)$$

Horizontal Tail Contribution

Since the dihedral angle of the horizontal tail angle is negligible, $c_{n\beta_H} \approx 0$.

Vertical Tail Contribution

$c_{n\beta_V}$ is given by

$$c_{n\beta_V} = -c_{Y\beta_V} \cdot \frac{(X_V \cos\alpha_1 + Z_V \sin\alpha_1)}{b}$$

$$= -(-0.826)\frac{(9.349 \cos(0) + 5.435 \sin(0))}{26.3} = 0.294 \left(\frac{1}{\text{rad}}\right)$$

X_V and Z_V were calculated in SSP4.2. Thus, the total value of $c_{n\beta}$ is given by

$$c_{n\beta} \approx c_{n\beta_B} + c_{n\beta_V} \cong -0.147 + 0.294 = 0.147 \ (\text{rad}^{-1})$$

Comparison between True Values and Empirical Values

A comparison between the true value of $c_{Y\beta}$ and $c_{n\beta}$ and their estimates along with a calculation of the percentage of the error is shown in Table SSP4.5.1:

Table SSP4.5.1 Comparative Analysis with Percentages of the Error

	Empirical Values	True Values	Percent Error [%]
$c_{Y\beta}$	−0.885	−1.00	11.5%
$c_{n\beta}$	0.147	0.170	13.5%

The results in the table suggest a reasonable accuracy for the calculation of the aerodynamic coefficients.

Student Sample Problem 4.6

Consider the data relative to the McDonnell Douglas DC-9 aircraft in Appendix C. Use the provided drawings to extract all the relevant geometric characteristics of the aircraft.

Assume $\eta_H = \eta_V = 0.9$. Using the modeling approach outlined in this chapter, provide estimates for the following control derivatives:

- $c_{Y_{\delta_R}}, c_{l_{\delta_R}}, c_{n_{\delta_R}}$;
- $c_{l_{\delta_A}}$.

Solution of Student Sample Problem 4.6

All the relevant geometric parameters used in this problem were extracted from the drawings in Appendix C. Additionally, all the geometric parameters related to the wing and the horizontal and vertical tail were calculated previously in the DC-9 problems solved in Chapters 2 and 3 and in SSP4.1.

Calculation of $c_{Y_{\delta_R}}$

The modeling of $c_{Y_{\delta_R}}$ is given by

$$c_{Y_{\delta_R}} = \left|c_{L_{\alpha_V}}\right| \eta_V \frac{S_V}{S} \Delta(K_R) \tau_R$$

where $c_{L_{\alpha_V}}$ was found and shown in SSP4.1. To find $\Delta(K_R)$ the parameters η_I and η_F and their associated K_R (shown in Section 4.2.4, Figure 4.27) are required. Using $\lambda_V = 0.796$, Figure 4.27 provides:

$$\eta_I = \frac{y_{R_I}}{b_V} = \frac{0}{13.6} = 0 \Rightarrow K_{R_I} = 0$$

$$\eta_F = \frac{y_{R_F}}{b_V} = \frac{10.4}{13.6} = 0.765 \Rightarrow K_{R_F} = 0.866$$

$$\Delta(K_R) = K_{R_F} - K_{R_I} = 0.866 - 0 = 0.866$$

Using $\dfrac{\bar{c}_{Rudder}}{\bar{c}_{Vert.Tail}} = \dfrac{6.2}{12.35} = 0.50$ from the drawings in Appendix C, Figure 4.26 provides $\tau_R = 0.7$.

Finally, the value of $c_{Y\delta_R}$ is given by

$$c_{Y\delta_R} = \left|c_{L_{\alpha_V}}\right| \eta_V \dfrac{S_V}{S} \Delta(K_R)\tau_R$$

$$= |3.607|0.9 \, \dfrac{0.5 \cdot 334.56}{928} \, 0.866 \cdot 0.7 = 0.355 \, \left(\dfrac{1}{rad}\right)$$

Calculation of $c_{l\delta_R}$

The modeling of $c_{l\delta_R}$ is given by

$$c_{l\delta_R} = c_{Y\delta_R}\left(\dfrac{Z_R \cos\alpha_1 - X_R \sin\alpha_1}{b}\right)$$

$$= 0.355 \, \dfrac{(9\cos(0.035) - 39.8\sin(0.035))}{89.4} = 0.030 \, \left(\dfrac{1}{rad}\right)$$

where X_R and Z_R were found and shown in SSP4.1.

Calculation of $c_{n\delta_R}$

The modeling of $c_{n\delta_R}$ is given by

$$c_{n\delta_R} = -c_{Y\delta_R}\left(\dfrac{X_R \cos\alpha_1 + Z_R \sin\alpha_1}{b}\right)$$

$$= -0.355 \, \dfrac{(39.8\cos(0.035) + 9\sin(0.035))}{89.4} = -0.159 \, \left(\dfrac{1}{rad}\right)$$

Calculation of $c_{l\delta_R}$

As shown in Section 4.3.3, the following parameters are required for the modeling of $c_{l\delta_A}$:

$$\eta_I = \eta_{Inboard} = \dfrac{y_{A_I}}{b/2} = \dfrac{27.1}{89.4/2} = 0.606;$$

$$\eta_O = \eta_{Outboard} = \dfrac{y_{A_O}}{b/2} = \dfrac{38}{89.4/2} = 0.85$$

$$\beta = \sqrt{1 - Mach^2} = \sqrt{1 - 0.7^2} = 0.714;$$

$$\Lambda_\beta = \tan^{-1}\left(\dfrac{\tan \Lambda_{c/4}}{\beta}\right) = \tan^{-1}\left(\dfrac{\tan 0.433}{0.714}\right) = 32.9°$$

$$k = \dfrac{(c_{l\alpha})_{Wing\,Section}|_{Mach} \cdot \beta}{2\pi} = \dfrac{4.85 \cdot 0.714}{2\pi} = 0.551;$$

$$\dfrac{\beta \cdot AR}{k} = \dfrac{0.714 \cdot 8.612}{0.551} = 11.158$$

From Figure 4.51, using $\Lambda_\beta = 32.9°$, $\dfrac{\beta \cdot AR}{k} = 11.158$ and $\lambda = 0.25$, we have

$$\eta_I = 0.606 \Rightarrow RME_I = 0.46$$

$$\eta_O = 0.85 \Rightarrow RME_O = 0.74$$

$$\Delta(RME) = RME|_{\eta_O} - RME|_{\eta_I} = 0.74 - 0.46 = 0.28$$

leading to

$$c_{l_\delta} = \frac{\Delta(RME) \cdot k}{\beta} = \frac{0.28 \cdot 0.551}{0.714} = 0.216$$

Using $\dfrac{\bar{c}_{Aileron}}{\bar{c}_{Wing(at\_aileron)}} = \dfrac{2.2}{7.4} = 0.297$ Figure 4.55 provides $\tau_A = 0.513$ leading to

$$c_{l_\delta} = \tau_A \cdot c_{l_\delta} = 0.513 \cdot 0.216 = 0.111$$

Finally, the value of $c_{l_{\delta_A}}$ is obtained as

$$c_{l_{\delta_A}} = \frac{1}{2}\left[(c_{l_\delta})|_{Left} + (c_{l_\delta})|_{Right}\right] = 0.111 \ (\text{rad}^{-1})$$

Student Sample Problem 4.7

Consider the data relative to the SIAI Marchetti S211 aircraft in Appendix B and Appendix C. Use the provided drawings to extract all the relevant geometric characteristics of the aircraft.

Using the modeling approach outlined in this chapter, provide estimates for the following control derivatives:

- $c_{Y_{\delta_R}}, c_{l_{\delta_R}}, c_{n_{\delta_R}}$;
- $c_{l_{\delta_A}}$.

Next, compare the obtained values with the true values listed in Appendix B (Aircraft 8) and evaluate the errors associated with the use of the empirical modeling approach.

Solution of Student Sample Problem 4.7

All the relevant geometric parameters used in this problem were extracted from the drawings in Appendix C. Additionally, all the geometric parameters related to the wing and the horizontal and vertical tail were calculated previously in the S211 problems solved in Chapters 2 and 3 and in SSP4.2.

Calculation of $c_{Y_{\delta_R}}$

The modeling of $c_{Y_{\delta_R}}$ is given by

$$c_{Y_{\delta_R}} = \left|c_{L_{\alpha_V}}\right|\eta_V \frac{S_V}{S}\Delta(K_R)\tau_R$$

where $c_{L_{\alpha_V}}$ was found and shown in SSP4.2. To find $\Delta(K_R)$ the parameters η_I and η_F and their associated K_R (shown in Section 4.2.4, Figure 4.27) are required. Using $\lambda_V = 0.351$, Figure 4.27 provides

$$\eta_I = \frac{y_{R_I}}{b_V} = \frac{0}{14.6} = 0 \Rightarrow K_{R_I} = 0$$

$$\eta_F = \frac{y_{R_F}}{b_V} = \frac{4.8}{5.8} = 0.828 \Rightarrow K_{R_F} = 0.936$$

$$\Delta(K_R) = K_{R_F} - K_{R_I} = 0.936 - 0 = 0.936$$

Using $\dfrac{\bar{c}_{Rudder}}{\bar{c}_{Vert.Tail}} = \dfrac{1.7}{4.146} = 0.41$ from the drawings in Appendix C, Figure 4.26 provides

$$\tau_R = 0.624$$

Finally, the value of $c_{Y_{\delta_R}}$ is given by

$$c_{Y_{\delta_R}} = \left|c_{L_{\alpha_V}}\right|\eta_V \frac{S_V}{S}\Delta(K_R)\tau_R$$

$$= |3.678|0.88 \cdot 0.164 \cdot 0.936 \cdot 0.624 = 0.310 \ \left(\frac{1}{\text{rad}}\right)$$

Calculation of $c_{l_{\delta_R}}$

The modeling of $c_{l_{\delta_R}}$ is given by

$$c_{l\delta_R} = c_{Y\delta_R}\left(\frac{Z_R \cos\alpha_1 - X_R \sin\alpha_1}{b}\right)$$

$$= 0.310 \frac{(5.1\cos(0) - 11.469\sin(0))}{26.3} = 0.060 \left(\frac{1}{\text{rad}}\right)$$

where X_R and Z_R were found and shown in SSP4.2.

Calculation of $c_{n\delta_R}$

The modeling of $c_{n\delta_R}$ is given by

$$c_{n\delta_R} = -c_{Y\delta_R}\left(\frac{X_R \cos\alpha_1 + Z_R \sin\alpha_1}{b}\right)$$

$$= -0.310 \frac{(11.469\cos(0) + 5.1\sin(0))}{26.3} = -0.135 \left(\frac{1}{\text{rad}}\right)$$

Calculation of $c_{l\delta_A}$

As shown in Section 4.3.3, the following parameters are required for the modeling of $c_{l\delta_A}$:

$$\eta_I = \eta_{Inboard} = \frac{y_{A_I}}{b/2} = \frac{7.6}{26.3/2} = 0.578;$$

$$\eta_O = \eta_{Outboard} = \frac{y_{A_O}}{b/2} = \frac{12.6}{26.3/2} = 0.958$$

$$\beta = \sqrt{1 - Mach^2} = \sqrt{1 - 0.6^2} = 0.8;$$

$$\Lambda_\beta = \tan^{-1}\left(\frac{\tan \Lambda_{c/4}}{\beta}\right) = \tan^{-1}\left(\frac{\tan 0.284}{0.8}\right) = 20°$$

$$k = \frac{(c_{l_\alpha})_{Wing\ Section}|_{Mach} \cdot \beta}{2\pi} = \frac{4.957 \cdot 0.8}{2\pi} = 0.631;$$

$$\frac{\beta \cdot AR}{k} = \frac{0.8 \cdot 5.086}{0.631} = 6.45$$

From Figure 4.51, using $\Lambda_\beta = 20°$, $\frac{\beta \cdot AR}{k} = 6.45$ and $\lambda = 0.517$, we have

$$\eta_I = 0.578 \Rightarrow RME_I = 0.313$$

$$\eta_O = 0.958 \Rightarrow RME_O = 0.650$$

$$\Delta(RME) = RME|_{\eta_O} - RME|_{\eta_I} = 0.650 - 0.313 = 0.337$$

leading to

$$c_{l_\delta}' = \frac{\Delta(RME) \cdot k}{\beta} = \frac{0.337 \cdot 0.631}{0.8} = 0.266$$

Using $\frac{\bar{c}_{Aileron}}{\bar{c}_{Wing(at\_aileron)}} = \frac{0.9}{4} = 0.225$ Figure 4.55 provides $\tau_A = 0.422$ leading to

$$c_{l_\delta} = \tau_A \cdot c_{l_\delta}' = 0.422 \cdot 0.266 = 0.112$$

Finally, the value of $c_{l\delta_A}$ is obtained as

$$c_{l\delta_A} = \frac{1}{2}\left[(c_{l_\delta})|_{Left} + (c_{l_\delta})|_{Right}\right] = 0.112\ (\text{rad}^{-1})$$

Comparison between True Values and Empirical Values

A comparison between the true value of $c_{Y_{\delta_R}}, c_{l_{\delta_R}}, c_{n_{\delta_R}}, c_{l_{\delta_A}}$ and $c_{n_{\delta_A}}$ and their estimates using the empirical aerodynamic modeling approach, along with a calculation of the percentage of the error, is shown in Table SSP4.7.1:

Table SSP4.7.1 Comparative Analysis with Percentages of the Error

	Empirical Values	True Values	Percent Error [%]
$c_{Y_{\delta_R}}$	0.31	0.28	10%
$c_{l_{\delta_R}}$	0.06	0.050	20%
$c_{n_{\delta_R}}$	−0.135	−0.120	13%
$c_{l_{\delta_A}}$	0.112	0.10	12%

The table results suggest a reasonable accuracy in the estimates of the aerodynamic coefficients.

Student Sample Problem 4.8

Consider the data relative to the McDonnell Douglas DC-9 aircraft in Appendix C. Use the provided drawings to extract all the relevant geometric characteristics of the aircraft.

Using the modeling approach outlined in this chapter, provide estimates for the following stability derivatives:

- $c_{Y_p}, c_{l_p}, c_{n_p}$;
- $c_{Y_r}, c_{l_r}, c_{n_r}$.

Solution of Student Sample Problem 4.8

All the relevant geometric parameters used in this problem were extracted from the drawings in Appendix C. Additionally, all the geometric parameters related to the wing and the horizontal and vertical tail were calculated previously in the DC-9 problems solved in Chapters 2 and 3 and in SSP4.1.

Calculation of c_{Y_p}

As shown in Section 4.6.2, the modeling for c_{Y_p} is given by

$$c_{Y_p} \approx c_{Y_{p_V}} \approx 2\, c_{Y_{\beta_V}} \frac{(Z_V \cos\alpha_1 - X_V \sin\alpha_1)}{b}$$

$$= 2(-1.207) \frac{(12.1 \cos(0.035) - 34.4 \sin(0.035))}{89.4} = -0.295 \left(\frac{1}{\text{rad}}\right)$$

where $c_{Y_{\beta_V}}$ was found in SSP4.4 while X_V and Z_V were shown in SSP4.1.

Calculation of c_{l_p}

As shown in Section 4.6.3, the modeling for c_{l_p} is given by

$$c_{l_p} = c_{l_{p_{WB}}} + c_{l_{p_H}} + c_{l_{p_V}}$$

$c_{l_{p_{WB}}}$ is given by $c_{l_{p_{WB}}} \approx c_{l_{p_W}} = RDP \cdot \frac{k}{\beta}$ where $\beta = 0.714$, $k = 0.551$ (from SSP 4.6).

The parameters $\frac{\beta \cdot AR}{k} = 11.158$ and $\Lambda_\beta = 32.9°$ (from SSP 4.6) are needed to evaluate RDP. Using the preceding parameters, along with $\lambda = 0.25$, Figures 4.80 and 4.81 provide $RDP = -0.465$, leading to

$$c_{l_{p_{WB}}} \approx c_{l_{p_W}} = RDP \cdot \frac{k}{\beta} = -0.465 \cdot \frac{0.551}{0.714} = -0.359$$

The contribution from the horizontal is given by

$$c_{l_{p_H}} \approx \frac{1}{2} \left(c_{l_{p_W}}\right)\big|_H \frac{S_H}{S} \left(\frac{b_H}{b}\right)^2$$

where $(c_{l_{pW}})|_H = RDP_H \cdot \frac{k_H}{\beta_H}$. It should be clear that this contribution is expected to be negligible due to the small values of $\frac{S_H}{S}$, $\left(\frac{b_H}{b}\right)^2$. The parameters k_H, β_H are given by

$$\beta_H = \sqrt{1 - Mach^2} = \sqrt{1 - 0.7^2} = 0.714$$

$$k_H = \frac{\left(c_{l_{\alpha H}}\right)_{Wing\ Section}\big|_{Mach} \cdot \beta_H}{2\pi} = \frac{3.793 \cdot 0.714}{2\pi} = 0.431$$

Finally, the following parameters are required to obtain RDP_H:

$$\frac{\beta \cdot AR}{k} = \frac{0.714 \cdot 4.953}{0.431} = 8.205$$

$$\Lambda_\beta = \tan^{-1}\left(\frac{\tan \Lambda_{c/4H}}{\beta_H}\right) = \tan^{-1}\left(\frac{\tan 0.545}{0.714}\right) = 40.4°$$

Using $\Lambda_{\beta_H} = 40.4°$, $\frac{\beta_H \cdot AR_H}{k_H} = 8.205$, $\lambda_H = 0.367$ Figures 4.80 and 4.81 provide $RDP_H = -0.409$.

$$c_{l_{pH}} \simeq RDP_H \cdot \frac{k_H}{\beta_H} = -0.409 \cdot \frac{0.431}{0.714} = -0.006$$

The contribution from the vertical is given by

$$c_{l_{pV}} \approx 2\, c_{Y_{\beta_V}} \left(\frac{Z_V}{b}\right)^2 = 2 \cdot (-1.384)\left(\frac{12.366}{88}\right)^2 = -0.054$$

where $c_{Y_{\beta_V}}$ was obtained from SSP4.4.

Finally the total value of c_{lp} is given by

$$c_{lp} = c_{l_{pWB}} + c_{l_{pH}} + c_{l_{pV}} = -0.359 - 0.006 - 0.045 = -0.410$$

Calculation of c_{n_p}

As shown in Section 4.6.4, the modeling for c_{n_p} is given by

$$c_{np} \approx c_{np_W} + c_{n_{pV}}$$

The contribution from the wing is given by

$$c_{np_W} \cong \left(\frac{c_{np}}{c_{L_1}}\right)\bigg|_{\substack{Mach \\ C_L=0}} c_{L_1} + \left(\frac{\Delta c_{np}}{\varepsilon_W}\right)\varepsilon_W$$

As shown in Section 4.6.4, the coefficient $\left(\frac{c_{np}}{c_{L_1}}\right)\big|_{\substack{Mach \\ C_L=0}}$ is given by

$$\left(\frac{c_{np}}{c_{L_1}}\right)\bigg|_{\substack{Mach \\ C_L=0}} = C\left(\frac{c_{np}}{c_{L_1}}\right)\bigg|_{\substack{Mach=0 \\ C_L=0}}$$

where C is given by

$$C = \frac{[AR + 4\cos(\Lambda_{c/4})]}{[AR \cdot B + 4\cos(\Lambda_{c/4})]} \cdot \left\{ \frac{AR \cdot B + \frac{1}{2}[AR \cdot B + 4\cos(\Lambda_{c/4})] \cdot \tan^2(\Lambda_{c/4})}{AR + \frac{1}{2}[AR + 4\cos(\Lambda_{c/4})] \cdot \tan^2(\Lambda_{c/4})} \right\}$$

with B given by

$$B = \sqrt{1 - Mach^2 \cos^2(\Lambda_{c/4})} = \sqrt{1 - 0.7^2 \cos^2(0.433)} = 0.772$$

leading to

$$C = \frac{[8.612 + 4\cos(0.433)]}{[8.612 \cdot 0.772 + 4\cos(0.433)]}$$

$$\cdot \left\{ \frac{8.612 \cdot 0.772 + \frac{1}{2}[8.612 \cdot 0.772 + 4\cos(0.433)] \cdot \tan^2(0.433)}{8.612 + \frac{1}{2}[8.612 + 4\cos(0.433)] \cdot \tan^2(0.433)} \right\} = 0.930$$

$\left(\frac{c_{np}}{c_{L_1}}\right)\Big|_{\substack{Mach=0 \\ C_L=0}}$ is instead given by

$$\left(\frac{c_{np}}{c_{L_1}}\right)\Big|_{\substack{Mach=0 \\ C_L=0}} = -\frac{1}{6} \cdot \frac{AR + 6(AR + \cos(\Lambda_{c/4})) \cdot \left[(\overline{x}_{CG} - \overline{x}_{AC})\frac{\tan(\Lambda_{c/4})}{AR} + \frac{\tan^2(\Lambda_{c/4})}{12}\right]}{(AR + \cos(\Lambda_{c/4}))}$$

$$= -\frac{1}{6} \cdot \frac{8.612 + 6(8.612 + \cos(0.433)) \cdot \left[(0.3 - 0.602)\frac{\tan(0.433)}{8.612} + \frac{\tan^2(0.433)}{12}\right]}{(8.612 + \cos(0.433))}$$

$$= -0.152 \quad (\text{rad}^{-1})$$

where $\overline{x}_{AC}$ was obtained in SSP3.1. Using the previously calculated values

$$\left(\frac{c_{np}}{c_{L_1}}\right)\Big|_{\substack{Mach=0 \\ C_L=0}} = C\left(\frac{c_{np}}{c_{L_1}}\right)\Big|_{\substack{Mach=0 \\ C_L=0}} = 0.930 \cdot (-0.152) = -0.142$$

Next, using $\lambda = 0.25$ and $AR = 8.612$, Figure 4.83 provides

$$\left(\frac{\Delta c_{np}}{\varepsilon_W}\right) = 0.000520 \quad \left(\frac{1}{\text{rad} \cdot \text{deg}}\right)$$

Assuming a wing twist angle of 2°:

$$c_{np_W} \cong \left(\frac{c_{np}}{c_{L_1}}\right)\Big|_{\substack{Mach=0 \\ C_L=0}} c_{L_1} + \left(\frac{\Delta c_{np}}{\varepsilon_W}\right)\varepsilon_W = -0.142 \cdot 0.448 + 0.000520 \cdot 2 = -0.062$$

The contribution from the vertical tail is given by

$$c_{np_V} \approx -2 \cdot c_{Y_{\beta_V}} \cdot \frac{(X_V \cos\alpha_1 + Z_V \sin\alpha_1)}{b} \cdot \frac{(Z_V \cos\alpha_1 - X_V \sin\alpha_1 - Z_V)}{b}$$

$$= -2 \cdot (-1.207)\frac{(34.4\cos(0.035) + 12.1\sin(0.035))}{89.4} \cdot \frac{(12.1\cos(0.035) - 34.4\sin(0.035) - 12.1)}{89.4}$$

$$= 0.013 \quad \left(\frac{1}{\text{rad}}\right)$$

where, again, the value of $c_{Y_{\beta_V}}$ was found in SSP4.4. Finally, the total value of c_{np} is

$$c_{np} \approx c_{n p_W} + c_{n p_V} = -0.062 - 0.013 = -0.075$$

Calculation of c_{Y_r}

As shown in Section 4.6.5, the modeling for c_{Y_r} is given by

$$c_{Y_r} \approx c_{Y_{r_V}} \approx -2c_{Y_{\beta_V}} \frac{(X_V \cos\alpha_1 + Z_V \sin\alpha_1)}{b}$$

$$= -2 \cdot (-1.207) \frac{(34.4 \cos(0.035) + 12.1 \sin(0.035))}{89.4} = 0.941 \left(\frac{1}{\text{rad}}\right)$$

Calculation of c_{l_r}

As shown in Section 4.6.6, the modeling for c_{l_r} is given by

$$c_{l_r} \approx c_{l_{r_W}} + c_{l_{r_V}}$$

A relationship for $c_{l_{r_W}}$ is given by

$$c_{l_{r_W}} \approx \left(\frac{c_{l_r}}{c_{L_1}}\right)\bigg|_{\substack{\text{Mach} \\ C_L=0}} \cdot c_{L_1} + \left(\frac{\Delta c_{l_r}}{\Gamma}\right) \cdot \Gamma + \left(\frac{\Delta c_{l_r}}{\varepsilon_W}\right) \cdot \varepsilon_W \quad (\text{rad}^{-1})$$

The coefficient $\left(\frac{c_{l_r}}{c_{L_1}}\right)\bigg|_{\substack{\text{Mach} \\ C_L=0}}$ is given by

$$\left(\frac{c_{l_r}}{c_{L_1}}\right)\bigg|_{\substack{\text{Mach} \\ C_L=0}} = D \cdot \left(\frac{c_{l_r}}{c_{L_1}}\right)\bigg|_{\substack{\text{Mach}=0 \\ C_L=0}}$$

To obtain D the following parameter is required:

$$B = \sqrt{1 - \text{Mach}^2 \cos^2(\Lambda_{c/4})} = \sqrt{1 - 0.7^2 \cos^2(0.433)} = 0.772$$

where D is given by

$$D = \frac{1 + \dfrac{AR(1-B^2)}{2B[AR \cdot B + 2\cos(\Lambda_{c/4})]} + \dfrac{[AR \cdot B + 2\cos(\Lambda_{c/4})]}{[AR \cdot B + 4\cos(\Lambda_{c/4})]} \cdot \dfrac{\tan^2(\Lambda_{c/4})}{8}}{1 + \dfrac{[AR + 2\cos(\Lambda_{c/4})]}{[AR + 4\cos(\Lambda_{c/4})]} \cdot \dfrac{\tan^2(\Lambda_{c/4})}{8}}$$

To parameter B is given by

$$B = \sqrt{1 - \text{Mach}^2 \cos^2(\Lambda_{c/4})} = \sqrt{1 - 0.7^2 \cos^2(0.429)} = 0.771$$

leading to

$$D = \frac{1 + \dfrac{8.372(1 - 0.771^2)}{2B[8.372 \cdot 0.771 + 2\cos(0.429)]} + \dfrac{[8.372 \cdot 0.771 + 2\cos(0.429)]}{[8.372 \cdot 0.771 + 4\cos(0.429)]} \cdot \dfrac{\tan^2(0.429)}{8}}{1 + \dfrac{[8.372 + 2\cos(0.429)]}{[8.372 + 4\cos(0.429)]} \cdot \dfrac{\tan^2(0.429)}{8}} = 1.259$$

Next, using Figure 4.85 with the parameters $\lambda = 0.25$, $AR = 8.612$, and $\Lambda_{c/4} = 24.8°$ we have $\left(\frac{c_{l_r}}{c_{L_1}}\right)\bigg|_{\substack{\text{Mach}=0 \\ C_L=0}} = 0.29$

leading to

$$\left.\left(\frac{c_{l_r}}{c_{L_1}}\right)\right|_{\substack{Mach \\ C_L=0}} = D \cdot \left.\left(\frac{c_{l_r}}{c_{L_1}}\right)\right|_{\substack{Mach=0 \\ C_L=0}} = 1.259 \cdot 0.29 = 0.365$$

Next, the term $\left(\frac{\Delta c_{l_r}}{\Gamma}\right)$ is given by

$$\left(\frac{\Delta c_{l_r}}{\Gamma}\right) \approx \left(\frac{1}{12}\right) \cdot \frac{[\pi \cdot AR \cdot \sin(\Lambda_{c/4})]}{(AR + 4 \cdot \cos(\Lambda_{c/4}))} = \left(\frac{1}{12}\right) \cdot \frac{[\pi \cdot 8.612 \cdot \sin(0.433)]}{(8.612 + 4 \cdot \cos(0.433))} = 0.077 \; (\text{rad}^{-2})$$

Also, from Figure 4.87 with $\lambda = 0.25$, $AR = 8.612$

$$\left(\frac{\Delta c_{l_r}}{\varepsilon_W}\right) = 0.0138 \left(\frac{1}{\text{rad} \cdot \text{deg}}\right)$$

Adding the previous terms, the value of $c_{l_{r_W}}$ is given by

$$c_{l_{r_W}} \approx \left.\left(\frac{c_{l_r}}{c_{L_1}}\right)\right|_{\substack{Mach \\ C_L=0}} \cdot c_{L_1} + \left(\frac{\Delta c_{l_r}}{\Gamma}\right) \cdot \Gamma + \left(\frac{\Delta c_{l_r}}{\varepsilon_W}\right) \cdot \varepsilon_W = 0.365 \cdot 0.448 + 0.077 \cdot 0.038 + 0.0138 \cdot 2 = 0.194 \; (\text{rad}^{-1})$$

A relationship for $c_{l_{r_V}}$ is given by

$$c_{l_{r_V}} \approx -2 \cdot c_{Y_{\beta_V}} \cdot \frac{(X_V \cos \alpha_1 + Z_V \sin \alpha_1)}{b} \cdot \frac{(Z_V \cos \alpha_1 - X_V \sin \alpha_1)}{b}$$

$$= -2 \cdot (-1.207) \frac{(34.4 \cos(0.035) + 12.1 \sin(0.035))}{89.4} \cdot \frac{(12.1 \cos(0.035) - 34.4 \sin(0.035))}{89.4}$$

$$= 0.115 \left(\frac{1}{\text{rad}}\right)$$

Thus, the total value of c_{l_r} is given by

$$c_{l_r} \approx c_{l_{r_W}} + c_{l_{r_V}} = 0.194 + 0.115 = 0.309 \left(\frac{1}{\text{rad}}\right)$$

Calculation of c_{n_r}

As shown in Section 4.6.7, the modeling for c_{n_r} is given by

$$c_{n_r} \approx c_{n_{r_W}} + c_{n_{r_V}}$$

A relationship for $c_{n_{r_W}}$ is given $c_{n_{r_W}} \cong \left(\frac{c_{n_r}}{c_{L_1}}\right) \cdot c_{L_1}^2$

Using Figure 4.89 with $\lambda = 0.25$, $AR = 8.612$, $\Lambda_{c/4} = 24.8°$, and $\bar{x}_{AC} - \bar{x}_{CG} = 0.602 - 0.3 = 0.30$ we will have

$$\left(\frac{c_{n_r}}{c_{L_1}^2}\right) = 0.018 \left(\frac{1}{\text{rad}}\right)$$

Leading to

$$c_{n_{r_W}} \cong \left(\frac{c_{n_r}}{c_{L_1}}\right) \cdot c_{L_1} = -0.018 \cdot 0.448^2 = -0.004 \left(\frac{1}{\text{rad}}\right)$$

A relationship for $c_{n_{r_W}}$ is given by

$$c_{n_{r_W}} \approx 2 \cdot c_{Y_{\beta_V}} \cdot \frac{(X_V \cos \alpha_1 + Z_V \sin \alpha_1)^2}{b^2}$$

$$= 2 \cdot (-1.207) \frac{(34.4 \cos 0.035 + 12.1 \sin 0.035)^2}{89.4^2} = -0.367 \left(\frac{1}{\text{rad}}\right)$$

Thus, the total value of c_{n_r} is given by

$$c_{n_r} \approx c_{n_{r_W}} + c_{n_{r_V}} = -0.004 - 0.367 = -0.363 \ \left(\frac{1}{\text{rad}}\right)$$

Student Sample Problem 4.9

Consider the data relative to the SIAI Marchetti S211 aircraft in Appendixes B and C. Use the provided drawings to extract all the relevant geometric characteristics of the aircraft.

Using the modeling approach outlined in this chapter, provide estimates for the following stability derivatives:

- $c_{Y_p}, c_{l_p}, c_{n_p}$;
- $c_{Y_r}, c_{l_r}, c_{n_r}$.

Next, compare the obtained values with the true values listed in Appendix B (Aircraft 8) and evaluate the errors associated with the use of the empirical modeling approach.

Solution of Student Sample Problem 4.9

All the relevant geometric parameters used in this problem were extracted from the drawings in Appendix C. Additionally, all the geometric parameters related to the wing and the horizontal and vertical tail were calculated previously in the S211 problems solved in Chapters 2 and 3 and in SSP4.2.

Calculation of c_{Y_p}

As shown in Section 4.6.2, the modeling for c_{Y_p} is given by

$$c_{Y_p} \approx c_{Y_{p_V}} \approx 2\, c_{Y_{\beta_V}} \frac{(Z_V \cos\alpha_1 - X_V \sin\alpha_1)}{b}$$

$$= 2(-0.826)\frac{(5.435\cos(0) - 9.349\sin(0))}{26.3} = -0.342 \ \left(\frac{1}{\text{rad}}\right)$$

where $c_{Y_{\beta_V}}$ was found in SSP4.6 while X_V and Z_V were shown in SSP4.2.

Calculation of c_{l_p}

As shown in Section 4.6.3, the modeling for c_{l_p} is given by

$$c_{l_p} = c_{l_{p_{WB}}} + c_{l_{p_H}} + c_{l_{p_V}}$$

$c_{l_{p_{WB}}}$ is given by $c_{l_{p_{WB}}} \approx c_{l_{p_W}} = RDP \cdot \frac{k}{\beta}$ where $\beta = 0.8$, $k = 0.631$.

The parameters $\frac{\beta \cdot AR}{k} = 6.446$ and $\Lambda_\beta = 20°$ are needed to evaluate RDP. Using the preceding parameters, along with $\lambda = 0.517$, Figures 4.80 and 4.81 provide $RDP = -0.423$, leading to

$$c_{l_{p_{WB}}} \approx c_{l_{p_W}} = RDP \cdot \frac{k}{\beta} = -0.423 \cdot \frac{0.631}{0.8} = -0.334$$

The contribution from the horizontal is given by

$$c_{l_{p_H}} \approx \frac{1}{2}\left(c_{l_{p_W}}\right)\bigg|_H \frac{S_H}{S}\left(\frac{b_H}{b}\right)^2$$

where $\left(c_{l_{p_W}}\right)\bigg|_H = RDP_H \cdot \frac{k_H}{\beta_H}$. This contribution is expected to be negligible due to the small values of $S_H/S, \left(\frac{b_H}{b}\right)^2$.

The parameters k_H, β_H are given by

$$\beta_H = \sqrt{1 - Mach^2} = \sqrt{1 - 0.6^2} = 0.8$$

$$k_H = \frac{(c_{l_{\alpha_H}})_{\text{Wing Section}}\big|_{Mach} \cdot \beta_H}{2\pi} = \frac{5.070 \cdot 0.8}{2\pi} = 0.646$$

Finally, the following parameters are required to obtain RDP_H:

$$\frac{\beta \cdot AR}{k} = \frac{0.8 \cdot 5.267}{0.646} = 6.528$$

$$\Lambda_\beta = \tan^{-1}\left(\frac{\tan \Lambda_{c/4_H}}{\beta_H}\right) = \tan^{-1}\left(\frac{\tan 0.256}{0.8}\right) = 18.1°$$

Using $\Lambda_{\beta_H} = 18.1°$, $\frac{\beta_H \cdot AR_H}{k_H} = 6.528$, $\lambda_H = 0.443$ Figures 4.80 and 4.81 provide $RDP_H = -0.422$.

$$c_{lp_H} \approx RDP_H \cdot \frac{k_H}{\beta_H} = -0.422 \cdot \frac{0.646}{0.8} = -0.011$$

The contribution from the vertical is given by

$$c_{lp_V} \approx 2\, c_{Y_{\beta_V}}\left(\frac{Z_V}{b}\right)^2 = 2 \cdot (-0.826)\left(\frac{5.435}{26.3}\right)^2 = -0.07$$

where $c_{Y_{\beta_V}}$ was obtained from SSP4.5.

Finally, the total value of c_{lp} is given by

$$c_{lp} = c_{lp_{WB}} + c_{lp_H} + c_{lp_V} = -0.334 - 0.011 - 0.07 = -0.415$$

Calculation of c_{n_p}

As shown in Section 4.6.4, the modeling for c_{n_p} is given by

$$c_{np} \approx c_{np_W} + c_{np_V}$$

The contribution from the wing is given by

$$c_{np_W} \cong \left.\left(\frac{c_{np}}{c_{L_1}}\right)\right|_{\substack{\text{Mach} \\ C_L=0}} c_{L_1} + \left(\frac{\Delta c_{np}}{\varepsilon_W}\right)\varepsilon_W$$

As shown in Section 4.6.4, the coefficient $\left.\left(\frac{c_{np}}{c_{L_1}}\right)\right|_{\substack{\text{Mach} \\ C_L=0}}$ is given by

$$\left.\left(\frac{c_{np}}{c_{L_1}}\right)\right|_{\substack{\text{Mach} \\ C_L=0}} = C\left.\left(\frac{c_{np}}{c_{L_1}}\right)\right|_{\substack{\text{Mach} \\ C_L=0}}$$

where C is given by

$$C = \frac{[AR + 4\cos(\Lambda_{c/4})]}{[AR \cdot B + 4\cos(\Lambda_{c/4})]} \cdot \left\{\frac{AR \cdot B + \frac{1}{2}[AR \cdot B + 4\cos(\Lambda_{c/4})] \cdot \tan^2(\Lambda_{c/4})}{AR + \frac{1}{2}[AR + 4\cos(\Lambda_{c/4})] \cdot \tan^2(\Lambda_{c/4})}\right\}$$

with B given by

$$B = \sqrt{1 - \text{Mach}^2 \cos^2(\Lambda_{c/4})} = \sqrt{1 - 0.6^2 \cos^2(0.284)} = 0.817$$

leading to

$$C = \frac{[5.086 + 4\cos(0.284)]}{[5.086 \cdot 0.817 + 4\cos(0.284)]} \cdot \left\{\frac{5.086 \cdot 0.817 + \frac{1}{2}[5.086 \cdot 0.817 + 4\cos(0.284)] \cdot \tan^2(0.284)}{5.086 + \frac{1}{2}[5.086 + 4\cos(0.284)] \cdot \tan^2(0.284)}\right\} = 0.918$$

$\left(\dfrac{c_{np}}{c_{L_1}}\right)\bigg|_{\substack{\text{Mach}\\C_L=0}}$ is instead given by

$$\left(\dfrac{c_{np}}{c_{L_1}}\right)\bigg|_{\substack{\text{Mach}\\C_L=0}} = -\dfrac{1}{6} \cdot \dfrac{AR + 6(AR + \cos(\Lambda_{c/4})) \cdot \left[(\overline{x}_{CG} - \overline{x}_{AC})\dfrac{\tan(\Lambda_{c/4})}{AR} + \dfrac{\tan^2(\Lambda_{c/4})}{12}\right]}{(AR + \cos(\Lambda_{c/4}))}$$

$$= -\dfrac{1}{6} \cdot \dfrac{5.086 + 6(5.086 + \cos(0.284)) \cdot \left[(0.25 - 0.278)\dfrac{\tan(0.284)}{5.086} + \dfrac{\tan^2(0.284)}{12}\right]}{(5.086 + \cos(0.284))} = -0.146 \ (\text{rad}^{-1})$$

where $\overline{x}_{AC}$ was obtained in SSP3.2. Using the previously calculated values

$$\left(\dfrac{c_{np}}{c_{L_1}}\right)\bigg|_{\substack{\text{Mach}\\C_L=0}} = C\left(\dfrac{c_{np}}{c_{L_1}}\right)\bigg|_{\substack{\text{Mach}=0\\C_L=0}} = 0.916 \cdot (-0.146) = -0.134$$

Next, using $\lambda = 0.517$ and $AR = 5.086$, Figure 4.83 provides

$$\left(\dfrac{\Delta c_{np}}{\varepsilon_W}\right) = -0.000189 \ \left(\dfrac{1}{\text{rad} \cdot \text{deg}}\right)$$

Assuming a wing twist angle of 2°

$$c_{np_W} \approx \left(\dfrac{c_{np}}{c_{L_1}}\right)\bigg|_{\substack{\text{Mach}\\C_L=0}} c_{L_1} + \left(\dfrac{\Delta c_{np}}{\varepsilon_W}\right)\varepsilon_W = -0.134 \cdot 0.149 - 0.000189 \cdot 2 = -0.02$$

The contribution from the vertical tail is given by

$$c_{np_V} \approx -2 \cdot c_{Y_{\beta_V}} \cdot \dfrac{(X_V \cos\alpha_1 + Z_V \sin\alpha_1)}{b} \cdot \dfrac{(Z_V \cos\alpha_1 - X_V \sin\alpha_1 - Z_V)}{b}$$

$$= -2 \cdot (-0.826)\dfrac{(9.349\cos(0) + 5.435\sin(0))}{26.3} \cdot \dfrac{(5.435\cos(0) - 9.349\sin(0) - 5.435)}{26.3} = 0 \ \left(\dfrac{1}{\text{rad}}\right)$$

Where the value of $c_{Y_{\beta_V}}$ was found in SSP4.5. Finally, the total value of c_{np} is

$$c_{np} \approx c_{np_W} + c_{np_V} = -0.02 + 0 = -0.02$$

Calculation of c_{Y_r}

As shown in Section 4.6.5, the modeling for c_{Y_r} is given by

$$c_{Y_r} \approx c_{Y_{r_V}} \approx -2\,c_{Y_{\beta_V}}\dfrac{(X_V\cos\alpha_1 + Z_V\sin\alpha_1)}{b}$$

$$= -2 \cdot (-0.826)\dfrac{(9.349\cos(0) + 5.435\sin(0))}{26.3} = 0.586 \ \left(\dfrac{1}{\text{rad}}\right)$$

Calculation of c_{l_r}

As shown in Section 4.6.6, the modeling for c_{l_r} is given by

$$c_{l_r} \approx c_{l_{r_W}} + c_{l_{r_V}}$$

A relationship for $c_{l_{r_W}}$ is given by

$$c_{l_{r_W}} \approx \left(\frac{c_{l_r}}{c_{L_1}}\right)\Bigg|_{\substack{Mach \\ C_L=0}} \cdot c_{L_1} + \left(\frac{\Delta c_{l_r}}{\Gamma}\right) \cdot \Gamma + \left(\frac{\Delta c_{l_r}}{\varepsilon_W}\right) \cdot \varepsilon_W \ (\text{rad}^{-1})$$

The coefficient $\left(\frac{c_{l_r}}{c_{L_1}}\right)\Big|_{\substack{Mach \\ C_L=0}}$ is given by

$$\left(\frac{c_{l_r}}{c_{L_1}}\right)\Bigg|_{\substack{Mach \\ C_L=0}} = D \cdot \left(\frac{c_{l_r}}{c_{L_1}}\right)\Bigg|_{\substack{Mach \\ C_L=0}}$$

To obtain D the following parameter is required:

$$B = \sqrt{1 - Mach^2 \cos^2(\Lambda_{c/4})} = \sqrt{1 - 0.6^2 \cos^2(0.284)} = 0.817$$

where D is given by

$$D = \frac{1 + \dfrac{AR(1-B^2)}{2B[AR \cdot B + 2\cos(\Lambda_{c/4})]} + \dfrac{[AR \cdot B + 2\cos(\Lambda_{c/4})]}{[AR \cdot B + 4\cos(\Lambda_{c/4})]} \cdot \dfrac{\tan^2(\Lambda_{c/4})}{8}}{1 + \dfrac{[AR + 2\cos(\Lambda_{c/4})]}{[AR + 4\cos(\Lambda_{c/4})]} \cdot \dfrac{\tan^2(\Lambda_{c/4})}{8}}$$

leading to

$$D = \frac{1 + \dfrac{5.086(1 - 0.817^2)}{2 \cdot 0.817[5.086 \cdot 0.817 + 2\cos(0.284)]} + \dfrac{[5.086 \cdot 0.817 + 2\cos(0.284)]}{[5.086 \cdot 0.817 + 4\cos(0.284)]} \cdot \dfrac{\tan^2(0.284)}{8}}{1 + \dfrac{[5.086 + 2\cos(0.284)]}{[5.086 + 4\cos(0.284)]} \cdot \dfrac{\tan^2(0.284)}{8}} = 1.168$$

Next, using Figure 4.85 with the parameters $\lambda = 0.517$, $AR = 5.086$, and $\Lambda_{c/4} = 16.2°$ we have $\left(\frac{c_{l_r}}{c_{L_1}}\right)\Big|_{\substack{Mach=0 \\ C_L=0}} = 0.28$

leading to

$$\left(\frac{c_{l_r}}{c_{L_1}}\right)\Bigg|_{\substack{Mach \\ C_L=0}} = D \cdot \left(\frac{c_{l_r}}{c_{L_1}}\right)\Bigg|_{\substack{Mach=0 \\ C_L=0}} = 1.168 \cdot 0.28 = 0.327$$

Next, the term $\left(\frac{\Delta c_{l_r}}{\Gamma}\right)$ is given by

$$\left(\frac{\Delta c_{l_r}}{\Gamma}\right) \approx \left(\frac{1}{12}\right) \cdot \frac{[\pi \cdot AR \cdot \sin(\Lambda_{c/4})]}{(AR + 4 \cdot \cos(\Lambda_{c/4}))} = \left(\frac{1}{12}\right) \cdot \frac{[\pi \cdot 5.086 \cdot \sin(0.284)]}{(5.086 + 4 \cdot \cos(0.284))} = 0.042 \ (\text{rad}^{-2})$$

Also, from Figure 4.87 with $\lambda = 0.517$, $AR = 5.086$

$$\left(\frac{\Delta c_{l_r}}{\varepsilon_W}\right) = 0.01259 \ \left(\frac{1}{\text{rad} \cdot \text{deg}}\right)$$

Adding the previous terms, the value of $c_{l_{r_W}}$ is given by

$$c_{l_{r_W}} \approx \left(\frac{c_{l_r}}{c_{L_1}}\right)\Bigg|_{\substack{Mach \\ C_L=0}} \cdot c_{L_1} + \left(\frac{\Delta c_{l_r}}{\Gamma}\right) \cdot \Gamma + \left(\frac{\Delta c_{l_r}}{\varepsilon_W}\right) \cdot \varepsilon_W$$

$$= 0.327 \cdot 0.149 + 0.042 \cdot (-0.033) + 0.01259 \cdot 2 = 0.072 \ (\text{rad}^{-1})$$

A relationship for $c_{l_{r_V}}$ is given by

$$c_{l_{r_V}} \approx -2 \cdot c_{Y_{\beta_V}} \cdot \frac{(X_V \cos \alpha_1 + Z_V \sin \alpha_1)}{b} \cdot \frac{(Z_V \cos \alpha_1 - X_V \sin \alpha_1)}{b}$$

$$= -2 \cdot (-0.826) \frac{\big(9.349 \cos(0) + 5.435 \sin(0)\big)}{26.3} \cdot \frac{\big(5.435 \cos(0) - 9.349 \sin(0)\big)}{26.3} = 0.122 \left(\frac{1}{\text{rad}}\right)$$

Thus, the total value of c_{l_r} is given by

$$c_{l_r} \approx c_{l_{r_W}} + c_{l_{r_V}} = 0.072 + 0.122 = 0.194 \left(\frac{1}{\text{rad}}\right)$$

Calculation of c_{n_r}

As shown in Section 4.6.7, the modeling for c_{n_r} is given by

$$c_{n_r} \approx c_{n_{r_W}} + c_{n_{r_V}}$$

A relationship for $c_{n_{r_W}}$ is given $c_{n_{r_W}} \cong \left(\frac{c_{n_r}}{c_{L_1}}\right) \cdot c_{L_1}^2$

Using Figure 4.89 with $\lambda = 0.517$, $AR = 5.086$, $\Lambda_{c/4} = 16.2°$ and $\bar{x}_{AC} - \bar{x}_{CG} = 0.278 - 0.25 = 0.028$ we will have

$$\left(\frac{c_{n_r}}{c_{L_1}}\right) = -0.0181 \left(\frac{1}{\text{rad}}\right)$$

leading to

$$c_{n_{r_W}} \approx \left(\frac{c_{n_r}}{c_{L_1}^2}\right) \cdot c_{L_1} = -0.0181 \cdot 0.149^2 = -0.004 \left(\frac{1}{\text{rad}}\right)$$

A relationship for $c_{n_{r_V}}$ is given by

$$c_{n_{r_V}} \approx 2 \cdot c_{Y_{\beta_V}} \cdot \frac{(X_V \cos \alpha_1 + Z_V \sin \alpha_1)^2}{b^2}$$

$$= 2 \cdot (-0.826) \frac{(9.349 \cos 0 + 5.435 \sin 0)^2}{26.3^2} = -0.208 \left(\frac{1}{\text{rad}}\right)$$

Thus, the total value of c_{n_r} is given by

$$c_{n_r} \approx c_{n_{r_W}} + c_{n_{r_V}} = -0.004 - 0.208 = -0.212 \left(\frac{1}{\text{rad}}\right)$$

Comparison between True Values and Empirical Values

A comparison between the true values of $c_{Y_p}, c_{l_p}, c_{n_p}, c_{Y_r}, c_{l_r}$ and c_{n_r} and their estimates along with a calculation of the percentage of the error are shown in Table SSP4.9.1.

Table SSP4.9.1 Comparative Analysis with Percentages of the Error

	Empirical Values	True Values	Percent Error [%]
c_{Y_p}	−0.342	−0.140	144%
c_{l_p}	−0.415	−0.390	6%
c_{n_p}	−0.02	0.09	−123%
c_{Y_r}	0.586	0.61	4%
c_{l_r}	0.194	0.28	30%
c_{n_r}	−0.212	−0.26	18%

The table results suggest the following two conclusions. The modeling does not provide a desirable level of accuracy for the coefficients c_{Y_p} and c_{n_p}. However, these coefficients are considered of secondary importance, as explained in Section 4.8. On the other side, the modeling is fairly accurate for the coefficients c_{l_p} and c_{n_r}, which are very important for shaping the aircraft dynamic response, as it will be shown in Chapter 7.

CASE STUDY

Modeling of the Lateral Directional Aerodynamics for the Cessna CJ3

The objective of this Case Study is to show a detailed DATCOM-based modeling for the lateral directional aerodynamic coefficients of the Cessna CJ3 aircraft over a [0.2–0.7] Mach range[10].

1 Modeling of the Side Force Aerodynamic Coefficients

This section provides the modeling of the following side force stability and control derivatives:

$$\left(c_{y_\beta}, c_{y_p}, c_{y_r}, c_{y_{\delta_A}}, c_{y_{\delta_R}}\right)$$

1.1 Modeling of (c_{y_β})

From the drawings in Appendix C, the following geometric parameters were identified:

$$b_{2V} = 2 \cdot 7.6 = 15.2 \text{ ft}, \ c_{T_V} = 4.9 \text{ ft}, \ c_{R_V} = 8.5 \text{ ft}, \ \Lambda_{LE_V} = 52° = 0.908 \text{ rad}$$

Next, using the assumption of straight wings, the vertical tail geometric parameters were derived using the values from the drawings in Appendix C.

Vertical Tail Geometric Parameters

The following vertical tail parameters are required:

$$\lambda_V = \frac{c_{T_V}}{c_{R_V}} = \frac{4.9}{8.5} = 0.576$$

$$S_{2V} = \frac{b_{2V}}{2} c_{R_V} (1 + \lambda_V) = \frac{15.2}{2} \cdot 8.5 \cdot (1 + 0.576) = 101.84 \text{ ft}^2$$

It should be emphasized that S_{2V} is twice the value of the area of the vertical tail; the use of S_{2V} allows finding the following geometric parameters:

$$AR_V = \frac{b_{2V}^2}{S_{2V}} = \frac{15.2^2}{101.84} = 2.269,$$

$$\bar{c}_V = \frac{2}{3} c_{R_V} \frac{(1 + \lambda_V + \lambda_V^2)}{(1 + \lambda_V)} = \frac{2}{3} \cdot 8.5 \cdot \frac{(1 + 0.576 + 0.576^2)}{(1 + 0.576)} = 6.86 \text{ ft}$$

$$x_{MAC_V} = \frac{b_{2V}}{6} \frac{(1 + 2\lambda_V)}{(1 + \lambda_V)} \tan(\Lambda_{LE_V}) = \frac{15.2}{6} \cdot \frac{(1 + 2 \cdot 0.576)}{(1 + 0.576)} \tan(0.908) = 4.43 \text{ ft}$$

$$y_{MAC_V} = \frac{b_{2V}}{6} \frac{(1 + 2\lambda_V)}{(1 + \lambda_V)} = \frac{15.2}{6} \cdot \frac{(1 + 2 \cdot 0.576)}{(1 + 0.576)} = 3.46 \text{ ft}$$

$$\tan \Lambda_{0.5_V} = \tan \Lambda_{LE_V} - \frac{4 \cdot 0.5 \cdot (1 - \lambda_V)}{AR_V(1 + \lambda_V)} = \tan(0.908) - \frac{4 \cdot 0.5 \cdot (1 - 0.576)}{2.269 \cdot (1 + 0.576)} \Rightarrow \Lambda_{0.5_V} = 0.806$$

Wing-Tail Geometric Parameters

Figures CS.1 and CS.2 illustrate the wing-tail geometric distances.

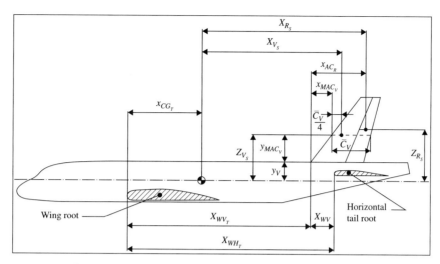

Figure CS.1 Wing-Vertical Tail Geometric Distances (Side view)

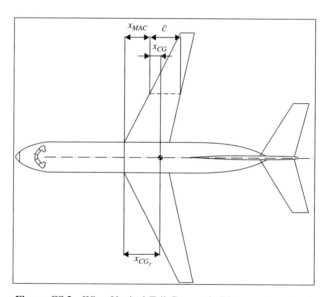

Figure CS.2 Wing-Vertical Tail Geometric Distances (top view)

The distances X_{V_S}, Z_{V_S}, X_{R_S} and Z_{R_S} are needed to determine the directional stability and control derivatives. From Figure CS.1:

$$x_{CG_T} + X_{V_S} = X_{WV_r} + x_{MAC_V} + \frac{\bar{c}_V}{4} \Rightarrow X_{V_S} = X_{WV_r} + \frac{\bar{c}_V}{4} + x_{MAC_V} - x_{CG_T}$$

with $X_{WH_r} = X_{WV_r} + X_{HV} \Rightarrow X_{WV_r} = X_{WH_r} - X_{HV} = 23.3 - 8.4 = 14.9$ ft
From Figure CS.2:

$$x_{CG_T} = x_{MAC} + x_{CG} = x_{MAC} + \bar{x}_{CG} \cdot \bar{c} = 0.767 + 0.25 \cdot 5.856 = 2.2 \text{ ft}$$

Therefore $X_{V_S} = X_{WV_r} + \frac{\bar{c}_V}{4} + x_{MAC_V} - x_{CG_T} = 14.9 + \frac{6.86}{4} + 4.43 - 2.2 = 18.8$ ft

From Figure CS.1 again:

$$Z_{V_S} = y_V + y_{MAC_V} = 1.8 + 3.46 = 5.3 \text{ ft}$$

$$x_{CG_T} + X_{R_S} = X_{WV_r} + x_{AC_R} \Rightarrow X_{R_S} = X_{WV_r} + x_{AC_R} - x_{CG_T} = 14.9 + 9.3 - 2.2 = 22 \text{ ft}$$

From the drawings in Appendix C: $Z_{R_S} = 4.9$ ft
Therefore, the values for the distances X_{V_S}, Z_{V_S}, X_{R_S} and Z_{R_S} are given by

$$X_{V_S} = 18.8 \text{ ft}, \ Z_{V_S} = 5.3 \text{ ft}, \ X_{R_S} = 22 \text{ ft}, \ Z_{R_S} = 4.9 \text{ ft}$$

The second subscript 's' stands for stability axes. From now on the second subscript will be no longer used and the above parameters will be referred as

$$X_V = 18.8 \text{ ft}, \ Z_V = 5.3 \text{ ft}, \ X_R = 22 \text{ ft}, \ Z_R = 4.9 \text{ ft}$$

Vertical Tail Lift Slope Coefficient

This coefficient is calculated using the Polhamus formula with $AR_{V_{eff}}$, defined as

$$AR_{V_{eff}} = c_1 \cdot AR_V \cdot [1 + K_{HV}(c_2 - 1)]$$

The parameter c_1 is found from Figure 4.15 using $\lambda_V = 0.576$ along with the parameter

$$\frac{b_V}{2r_1} = \frac{7.6}{2 \cdot 3.6} = 1.056$$

Using $\lambda_V = 0.576$ and $b_V/2r_1 = 1.056$ Figure 4.15 provides $c_1 = 1.522$.

Next, the parameter c_2 is found from Figure 4.16 using

$$\frac{Z_H}{b_V} = \frac{-8.8}{7.6} = -1.158$$

The geometric parameter $x_{AC_H \to V}$, shown in Figure 4.17, is calculated using

$$x_{AC_H \to V} = x_{MAC_H} + 0.25 \cdot \bar{c}_H = 2.796 \text{ ft} \Rightarrow \frac{x_{AC_H \to V}}{\bar{c}_V} = \frac{2.796}{6.86} = 0.408$$

Note that Z_H is considered negative when above the CG. Using the above $x_{AC_H \to V}$ value, Figure 4.16 provides $c_2 = 2.375$. Finally, using S_H/S_V in Figure 4.18, we have

$$\frac{S_H}{S_V} = \frac{70.7}{0.5 \cdot 101.84} = 1.388 \Rightarrow K_{HV} = 1.018$$

leading to

$$AR_{V_{eff}} = c_1 \cdot AR_V \cdot [1 + K_{HV}(c_2 - 1)] = 1.522 \cdot 2.269 \cdot [1 + 1.018(2.375 - 1)] = 8.286$$

Also, knowing that

$$\tan \Lambda_{0.5_V} = \tan \Lambda_{LE_V} - \frac{4 \cdot 0.5 \cdot (1 - \lambda_V)}{AR_V(1 + \lambda_V)} = \tan(0.908) - \frac{4 \cdot 0.5 \cdot (1 - 0.576)}{2.269 \cdot (1 + 0.576)} \Rightarrow \Lambda_{0.5_V} = 0.806$$

The application of the Polhamus formula leads to

$$c_{L_{\alpha V, eff}} = \frac{2\pi AR_{V_{eff}}}{2 + \sqrt{\left\{\left[\frac{AR_{V_{eff}}^2(1 - Mach^2)}{k_V^2}\right]\left(1 + \frac{\tan^2(\Lambda_{0.5})}{(1 - Mach^2)}\right)\right] + 4\right\}}}$$

where

$$\tan \Lambda_{0.5_V} = \tan \Lambda_{LE_V} - \frac{4 \cdot 0.5 \cdot (1 - \lambda_V)}{AR_V(1 + \lambda_V)} = \tan(0.908) - \frac{4 \cdot 0.5 \cdot (1 - 0.576)}{2.269 \cdot (1 + 0.576)} \Rightarrow \Lambda_{0.5_V} = 0.806$$

with $k_V = 1 + \dfrac{\left[(8.2 - 2.3\Lambda_{LE_V}) - AR_{V_{Eff}}(0.22 - 0.153\Lambda_{LE_V})\right]}{100} = 1.054$

(since $AR_{V_{eff}} = 8.286 > 4$).

Once again, the results of the Polhamus formula have been multiplied by an empirical correction factor of 0.8 since the aspect ratio is above the allowed range. The results are shown in this table:

Mach	0.15	0.20	0.25	0.30	0.35	0.40	0.45	0.50	0.55	0.60	0.65	0.70
$c_{L_{\alpha V, eff}}$ (80%)	3.092	3.103	3.117	3.135	3.156	3.181	3.210	3.244	3.282	3.326	3.376	3.433

Using all the parameters calculated previously, the general modeling for c_{Y_β} is given by

$$c_{Y_\beta} = \left(c_{Y_{\beta_W}} + c_{Y_{\beta_B}}\right) + c_{Y_{\beta_H}} + c_{Y_{\beta_V}}$$

Wing Contribution

$c_{Y_{\beta_W}}$ is given by

$$c_{Y_{\beta_W}} \approx -0.0001 \, |\Gamma_W| \cdot 57.3 = -0.0001 \, |5| \cdot 57.3 = -0.029 \, \left(\text{rad}^{-1}\right)$$

Body Contribution

$c_{Y_{\beta_B}}$ is given by

$$c_{Y_{\beta_B}} \approx -2 \cdot K_{\text{int}} \cdot \frac{S_{P \to V}}{S} \, \left(\text{rad}^{-1}\right)$$

Using $\frac{Z_W}{d/2} = \frac{2.9}{6.2/2} = 0.935$ (where Z_W, d are shown in Figures 4.45 and 4.10) Figure 4.8 provides $K_{\text{int}} = 1.4$.

$S_{P \to V}$ is the cross section at the location of the fuselage where the flow transitions from potential to viscous. From the drawings in Appendix C $S_{P \to V} = 12.86 \text{ ft}^2$. Therefore

$$c_{Y_{\beta_B}} \approx -2 \cdot K_{\text{int}} \cdot \frac{S_{P \to V}}{S} = -2 \cdot 1.4 \cdot \frac{12.86}{294.1} = -0.122 \, \left(\text{rad}^{-1}\right)$$

Horizontal Tail Contribution

$c_{Y_{\beta_H}}$ is found using

$$c_{Y_{\beta_H}} \approx -0.0001 \, |\Gamma_H| \cdot 57.3 \cdot \eta_H \cdot \frac{S_H}{S} = 0$$

where the horizontal tail dihedral angle is shown in Appendix C ($\Gamma_H = 0$).

Vertical Tail Contribution

$c_{Y_{\beta_V}}$ is found using

$$c_{Y_{\beta_V}} \approx -k_{Y_V} \cdot \left|c_{L_{\alpha \cdot V}}\right| \cdot \eta_V \cdot \left(1 + \frac{d\sigma}{d\beta}\right) \cdot \frac{S_{2V}}{S} \, \left(\text{rad}^{-1}\right)$$

where $\left|c_{L_{\alpha \cdot V}}\right|$ has been found previously. Using $\frac{b_V}{2r_1} = 1.056$ Figure 4.13 provides $k_{Y_V} = 0.761$

The term $\eta_V \cdot \left(1 + \frac{d\sigma}{d\beta}\right)$ is modeled using

$$\eta_V \cdot \left(1 + \frac{d\sigma}{d\beta}\right) = 0.724 + 3.06 \, \frac{S_V/S}{1 + \cos(\Lambda_{c/4})} + 0.4 \, \frac{Z_W}{d} + 0.009 \cdot AR = 1.317$$

Therefore, we have

Mach	0.15	0.20	0.25	0.30	0.35	0.40	0.45	0.50	0.55	0.60	0.65	0.70
$c_{Y_{\beta_V}}$	−1.073	−1.077	−1.082	−1.088	−1.095	−1.104	−1.114	−1.126	−1.139	−1.154	−1.171	−1.191

Finally, the total value of c_{Y_β} is given by

Mach	0.15	0.20	0.25	0.30	0.35	0.40	0.45	0.50	0.55	0.60	0.65	0.70
$c_{Y_{\beta_W}}$	−0.029	−0.029	−0.029	−0.029	−0.029	−0.029	−0.029	−0.029	−0.029	−0.029	−0.029	−0.029
$c_{Y_{\beta_B}}$	−0.122	−0.122	−0.122	−0.122	−0.122	−0.122	−0.122	−0.122	−0.122	−0.122	−0.122	−0.122
$c_{Y_{\beta_H}}$	0.000	0.000	0.000	0.000	0.000	0.000	0.000	0.000	0.000	0.000	0.000	0.000
$c_{Y_{\beta_V}}$	−1.073	−1.077	−1.082	−1.088	−1.095	−1.104	−1.114	−1.126	−1.139	−1.154	−1.171	−1.191
c_{Y_β}	−1.224	−1.228	−1.233	−1.239	−1.246	−1.255	−1.265	−1.277	−1.290	−1.305	−1.323	−1.342

$$c_{Y_\beta} = c_{Y_{\beta_W}} + c_{Y_{\beta_B}} + c_{Y_{\beta_H}} + c_{Y_{\beta_V}}$$

The results for $c_{Y\beta}$ are also shown in Figure CS.3.

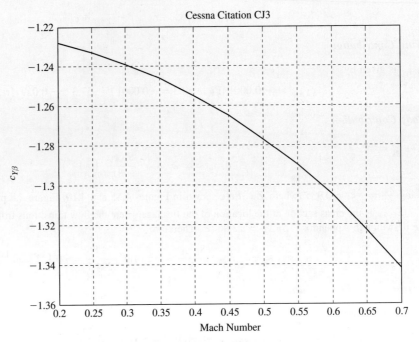

Figure CS.3 Estimates of $c_{Y\beta}$ vs. Mach

1.2 Modeling of (c_{y_p})

The modeling for c_{Y_p} is given by

$$c_{Y_p} \approx c_{Y_{p_V}} \approx 2\, c_{Y_{\beta_V}} \frac{(Z_V \cos\alpha_1 - X_V \sin\alpha_1)}{b}$$

where $c_{Y_{\beta_V}}$ was found in Section 1.1 while X_V and Z_V are shown in Figure CS.4. The values for X_V and Z_V were shown previously.

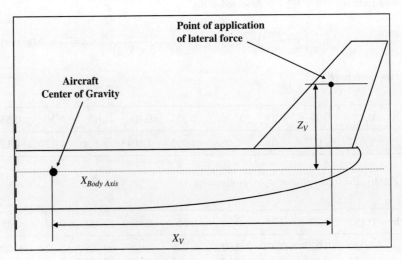

Figure CS.4 Geometric Parameters of the Vertical Tail

Assuming $\alpha_1 = 2° = 0.035$ rad and considering the previously introduced values of $X_V = 18.8$, $Z_V = 5.3$, $b = 53.33$, we have

$$c_{Y_p} \approx 2\, c_{Y_{\beta_V}} \frac{(5.3 \cos(0.035) - 18.8 \sin(0.035))}{53.3}$$

Mach	0.15	0.20	0.25	0.30	0.35	0.40	0.45	0.50	0.55	0.60	0.65	0.70
c_{Y_p}	−0.185	−0.186	−0.187	−0.188	−0.189	−0.190	−0.192	−0.194	−0.196	−0.199	−0.202	−0.205

The results for c_{Y_p} are also shown in Figure CS.5.

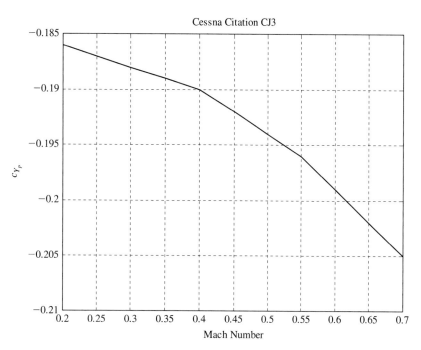

Figure CS.5 Estimates of c_{Y_p} vs. Mach

1.3 Modeling of (c_{y_r})

The modeling for c_{Y_r} is given by

$$c_{Y_r} \approx c_{Y_{r_V}} \approx -2\, c_{Y_{\beta_V}} \frac{(X_V \cos \alpha_1 + Z_V \sin \alpha_1)}{b}$$

$$= -2 c_{Y_{\beta_V}} \frac{(18.8 \cos(0.035) + 5.3 \sin(0.035))}{53.33}$$

where $c_{Y_{\beta_V}}$ was found in Section 1.1 while X_V and Z_V are shown in Figure CS.4. The values for X_V and Z_V were shown previously. The relationship leads to

Mach	0.15	0.20	0.25	0.30	0.35	0.40	0.45	0.50	0.55	0.60	0.65	0.70
c_{Y_r}	0.764	0.767	0.770	0.774	0.780	0.786	0.793	0.801	0.811	0.822	0.834	0.848

The results for c_{Y_r} are also shown below in Figure CS.6.

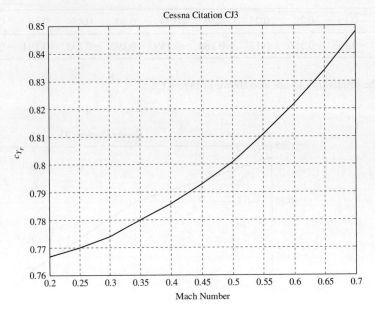

Figure CS.6 Estimates of c_{Y_r} vs. Mach

1.4 Modeling of ($c_{y_{\delta_A}}$)

The forces associated with the asymmetric deflections of the left and right ailerons act along the vertical and horizontal directions; therefore, their components along the lateral direction are negligible. Thus, it can be safely approximated that

$$c_{Y\delta_A} \approx 0$$

1.5 Modeling of ($c_{y_{\delta_R}}$)

The modeling of $c_{Y\delta_R}$ is given by

$$c_{Y\delta_R} = \left|c_{L_{\alpha_V}}\right| \eta_V \frac{S_V}{S} \Delta(K_R) \tau_R$$

where $c_{L_{\alpha_V}}$ was found previously. To find $\Delta(K_R)$ the parameters η_I and η_F and their associated K_R (shown in Figure 4.27) are required. The geometric meaning of the parameters η_I and η_F is shown in Figure CS.7.

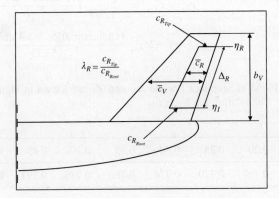

Figure CS.7 Geometric Parameters of the Rudder

Using $\lambda_V = 0.576$ and the geometric data provided in Appendix C, Figure 4.27 provides

$$\eta_I = \frac{y_{R_I}}{b_V} = \frac{0}{7.6} = 0. \Rightarrow K_{R_I} = 0$$

$$\eta_F = \frac{y_{R_F}}{b_V} = \frac{6.3}{7.6} = 0.829 \Rightarrow K_{R_F} = 0.926$$

$$\Delta(K_R) = K_{R_F} - K_{R_I} = 0.926 - 0 = 0.926$$

Using $\dfrac{\overline{c}_{Rudder}}{\overline{c}_{Vert.Tail}} = \dfrac{2.8}{6.86} = 0.41$ from the drawings in Appendix C, Figure 4.26 provides $\tau_R = 0.624$.

Using the preceding parameters, the values of $c_{Y\delta_R}$ over the Mach range are given by the following:

Mach	0.15	0.20	0.25	0.30	0.35	0.40	0.45	0.50	0.55	0.60	0.65	0.70
$c_{Y\delta_R}$	0.278	0.279	0.281	0.282	0.284	0.286	0.289	0.292	0.296	0.299	0.304	0.309

The results for $c_{Y\delta_R}$ are also shown below in Figure CS.8.

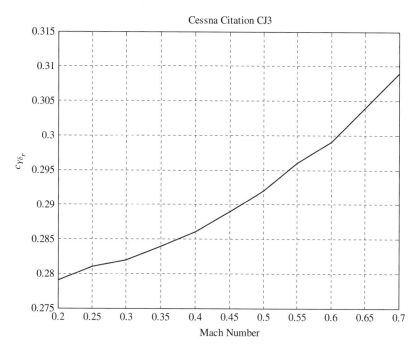

Figure CS.8 Estimates of $c_{Y\delta_R}$ vs. Mach

2 Modeling of the Rolling Aerodynamic Coefficients

This section provides the modeling of the following rolling stability and control derivatives:

$$\left(c_{l_\beta},\ c_{l_p},\ c_{l_r},\ c_{l_{\delta_A}},\ c_{l_{\delta_R}}\right)$$

2.1 Modeling of (c_{l_β})

The starting modeling relationship is given by

$$c_{l\beta} = c_{l\beta_{WB}} + c_{l\beta_H} + c_{l\beta_V}$$

Next, each contribution will be discussed separately.

2.1.1 Wing and Body Contribution

The modeling for $c_{l\beta_{WB}}$ is given by

$$c_{l_{\beta_{WB}}} = 57.3 \left\{ \Gamma_W \left[\dfrac{c_{l_\beta}}{\Gamma_W} K_{M_\Gamma} + \dfrac{\Delta c_{l_\beta}}{\Gamma_W} \right] + \left(\Delta c_{l_\beta}\right)_{Z_W} + \varepsilon_W \tan \Lambda_{c/4} \left(\dfrac{\Delta c_{l_\beta}}{\varepsilon_W \tan \Lambda_{c/4}} \right) \right\}$$
$$+ 57.3 \cdot c_{L_1} \left[\left(\dfrac{c_{l_\beta}}{c_{L_1}}\right)_{\Lambda_{c/2}} K_{M_\Lambda} K_f + \left(\dfrac{c_{l_\beta}}{c_{L_1}}\right)_{AR} \right]$$

244 Chapter 4 Modeling of Lateral Directional Aerodynamic Forces and Moments

Next, the following individual contributions to the dihedral effect in the preceding relationship are evaluated separately for the Cessna CJ3 aircraft:

- Wing contribution due to the geometric dihedral effect;
- Body contribution to the geometric dihedral effect;
- Body contribution due to high/low wing configuration;
- Wing contribution due to the twist angle;
- Wing contribution due to the sweep angle;
- Wing contribution due to the aspect ratio.

(1) Wing contribution due to the geometric dihedral effect

Figures 4.43 and 4.44 are used for the calculation of this contribution.

Using λ and $\Lambda_{0.5}$ Figure 4.43 provides $\frac{c_{l_\beta}}{\Gamma_W} = -0.0001764 \, (1/\text{deg}^2)$.

Using $Mach \cdot \cos(\Lambda_{0.5})$, $\frac{AR}{\cos(\Lambda_{0.5})} = 9.678$, Figure 4.44 provides $K_{M_\Gamma} = f(Mach)$. The following spreadsheet has been prepared for $c_{l_{\beta_{WB\#1}}}$:

Mach	0.15	0.20	0.25	0.30	0.35	0.40	0.45	0.50	0.55	0.60	0.65	0.70
Γ_W [deg]						5						
$Mach \cdot \cos(\Lambda_{0.5})$	0.150	0.200	0.250	0.300	0.350	0.400	0.450	0.500	0.550	0.600	0.649	0.699
$\frac{AR}{\cos(\Lambda_{0.5})}$						9.678						
$\frac{c_{l_\beta}}{\Gamma_W}$						-1.764×10^{-4}						
K_{M_Γ}	1.007	1.011	1.018	1.027	1.038	1.052	1.069	1.088	1.110	1.135	1.162	1.192
$c_{l_{\beta_{WB\#1}}}$	−0.051	−0.051	−0.051	−0.052	−0.052	−0.053	−0.054	−0.055	−0.056	−0.057	−0.059	−0.060

(2) Fuselage contribution to the geometric dihedral effect

This contribution is given by

$$\left(\frac{\Delta c_{l_\beta}}{\Gamma_W}\right) = -0.0005 \cdot AR \cdot \left(\frac{d}{b}\right)^2 = -0.0005 \cdot 9.67 \cdot \left(\frac{4.817}{53.33}\right)^2 = -0.00004 \left(\frac{1}{\text{deg}^2}\right)$$

leading to

$$c_{l_{\beta_{WB\#2}}} = 57.3 \cdot \Gamma_W \left[\frac{\Delta c_{l_\beta}}{\Gamma_W}\right] = 57.3 \cdot (5) \cdot (-0.00004) = -0.011 \left(\frac{1}{\text{rad}}\right)$$

Note that this contribution is constant throughout the range of Mach.

Mach	0.15	0.20	0.25	0.30	0.35	0.40	0.45	0.50	0.55	0.60	0.65	0.70
$c_{l_{\beta_{WB\#2}}}$	−0.011	−0.011	−0.011	−0.011	−0.011	−0.011	−0.011	−0.011	−0.011	−0.011	−0.011	−0.011

(3) Body contribution due to high/low wing configuration

The fuselage's diameter d_B is required for the evaluation of this term. d_B is found using

$$d_B = \sqrt{\frac{S_{f_{AVG}}}{0.7854}}$$

where $S_{f_{AVG}}$ is the average cross-sectional area of the fuselage calculated using

$$S_{f_{AVG}} = \pi \left(\frac{\text{Fuselage average height}}{2}\right)\left(\frac{\text{Fuselage average width}}{2}\right)$$

$$= \pi \left(\frac{4.55}{2}\right)\left(\frac{5.1}{2}\right) = 18.23 \text{ ft}^2$$

where the average height and width of the fuselage were obtained from the drawings in Appendix C. Therefore, we have

$$d = \sqrt{\frac{S_{f_{AVG}}}{0.7854}} = \sqrt{\frac{18.23}{0.7854}} = 4.817 \text{ ft}$$

leading to the following value for $c_{l_{\beta_{WB\#3}}} = c_{l_{\beta_{WB}}}\big|_{\substack{High/low \\ wing}}$:

$$c_{l_{\beta_{WB\#3}}} = c_{l_{\beta_{WB}}}\big|_{\substack{High/low \\ wing}} = 57.3 \left(\Delta c_{l_\beta}\right)_{Z_W} = 57.3 \frac{1.2\sqrt{AR}}{57.3} \cdot \frac{Z_W}{b} \cdot \left(\frac{2d}{b}\right)$$

$$= 1.2\sqrt{9.67} \cdot \frac{(2.9)}{53.33} \cdot \left(\frac{2 \cdot 4.817}{53.33}\right) = 0.037 \left(\frac{1}{\text{rad}}\right)$$

Note that this contribution is also constant over the range of Mach.

Mach	0.15	0.20	0.25	0.30	0.35	0.40	0.45	0.50	0.55	0.60	0.65	0.70
$c_{l_{\beta_{WB\#3}}}$	0.037	0.037	0.037	0.037	0.037	0.037	0.037	0.037	0.037	0.037	0.037	0.037

(4) Wing contribution due to wing twist angle

Due to a lack of specific information, the wing twist angle is assumed to be $\varepsilon_W = 2° = 0.0345$ rad. The wing twist angle is shown in Figure CS.9.

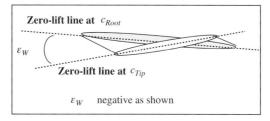

Figure CS.9 Wing Twist Angle ε_W

Using $\lambda = 0.305$ and $AR = 9.67$ Figure 4.46 provides

$$\left(\frac{\Delta c_{l_\beta}}{\varepsilon_W \tan \Lambda_{c/4}}\right) = -3.686 \times 10^{-5} \left(\frac{1}{\deg^2}\right)$$

$$c_{l_{\beta_{WB\#4}}} = 57.3 \cdot \varepsilon_W \tan \Lambda_{c/4} \left(\frac{\Delta c_{l_\beta}}{\varepsilon_W \tan \Lambda_{c/4}}\right) = 57.3 \cdot 2 \cdot \tan(0.015) \cdot \left(-3.686 \times 10^{-5}\right) = -0.00006 \left(\frac{1}{\text{rad}}\right)$$

Note that this contribution is also constant over the range of Mach.

Mach	0.15	0.20	0.25	0.30	0.35	0.40	0.45	0.50	0.55	0.60	0.65	0.70
$c_{l_{\beta_{WB\#4}}}$	-6.0×10^{-5}	-6.0×10^{-5}	-6.0×10^{-5}	-6.0×10^{-5}	-6.0×10^{-5}	-6.0×10^{-5}	-6.0×10^{-5}	-6.0×10^{-5}	-6.0×10^{-5}	-6.0×10^{-5}	-6.0×10^{-5}	-6.0×10^{-5}

(5) Wing contribution due to wing sweep angle

Using $\lambda = 0.305$, $AR = 9.67$ and $\Lambda_{c/2} = -2.3°$ the interpolation from Figure 4.39 provides

$$\left(\frac{c_{l_\beta}}{C_{L_1}}\right)_{\Lambda_{c/2}} = 0.0002 \left(\frac{1}{\deg}\right)$$

246 Chapter 4 Modeling of Lateral Directional Aerodynamic Forces and Moments

The value of c_{L_1} is required to calculate this contribution. Starting from the basic definition $c_{L_1} = \frac{W}{\bar{q}S} = \frac{W}{\frac{1}{2}\rho V^2 S}$ assumptions were made for the calculation of the density. A fairly linear trend was assumed for the altitude versus airspeed profile for the purpose of using realistic c_{L_1} values from Mach = 0.15 to Mach = 0.7. Next, using different values for $Mach \cdot \cos(\Lambda_{0.5})$ along with $\frac{AR}{\cos(\Lambda_{0.5})} = 9.678$ Figure 4.40 provides $K_{M_\Lambda} = f(Mach)$. Using again $\frac{AR}{\cos(\Lambda_{0.5})} = 9.678$ along with $\frac{A}{b} = 0.46$ Figure 4.41 provides $K_f = 0.9$.

Therefore

$$c_{l\beta_{WB\#5}} = c_{l\beta_{WB}}\bigg|_{\substack{Wing \\ sweep \\ angle}} = 57.3 \cdot c_{L_1} \left[\left(\frac{c_{l\beta}}{c_{L_1}}\right)_{\Lambda_{c/2}} K_{M_\Lambda} K_f \right]$$

The calculations are shown in the following spreadsheet.

Mach	0.15	0.20	0.25	0.30	0.35	0.40	0.45	0.50	0.55	0.60	0.65	0.70
c_{L_1}	1.715	1.072	0.726	0.536	0.420	0.344	0.293	0.257	0.232	0.214	0.203	0.197
K_f						0.9						
$\left(\frac{c_{l\beta}}{c_{L_1}}\right)_{\Lambda_{c/2}}$						2×10^{-4}						
K_{M_Λ}	1.010	1.026	1.041	1.056	1.074	1.095	1.120	1.150	1.185	1.228	1.278	1.338
$c_{l\beta_{WB\#5}}$	0.018	0.011	0.008	0.006	0.005	0.004	0.003	0.003	0.003	0.003	0.003	0.003

(6) Wing contribution due to wing aspect ratio

Using $\lambda = 0.305$ and $AR = 9.67$ the extrapolation from Figure 4.42 provides

$$\left(\frac{c_{l\beta}}{c_{L_1}}\right)_{AR} = 0.0004 \left(\frac{1}{\deg}\right)$$

Therefore

$$c_{l\beta_{WB\#6}} = 57.3 \cdot c_{L_1} \left[\left(\frac{c_{l\beta}}{c_{L_1}}\right)_{AR}\right]$$

The calculations are shown in the following spreadsheet.

Mach	0.15	0.20	0.25	0.30	0.35	0.40	0.45	0.50	0.55	0.60	0.65	0.70
c_{L_1}	1.715	1.072	0.726	0.536	0.420	0.344	0.293	0.257	0.232	0.214	0.203	0.197
$\left(\frac{c_{l\beta}}{c_{L_1}}\right)_{AR}$						4×10^{-4}						
$c_{l\beta\ WB\#6}$	0.039	0.025	0.017	0.012	0.010	0.008	0.007	0.006	0.005	0.005	0.005	0.005

The sum of all the different contributions to $c_{l\beta_{WB}}$ is shown in the following spreadsheet.

Mach	0.15	0.20	0.25	0.30	0.35	0.40	0.45	0.50	0.55	0.60	0.65	0.70
$c_{l\beta_{WB\#1}}$	−0.051	−0.051	−0.051	−0.052	−0.052	−0.053	−0.054	−0.055	−0.056	−0.057	−0.059	−0.060
$c_{l\beta_{WB\#2}}$	−0.011	−0.011	−0.011	−0.011	−0.011	−0.011	−0.011	−0.011	−0.011	−0.011	−0.011	−0.011
$c_{l\beta_{WB\#3}}$	0.037	0.037	0.037	0.037	0.037	0.037	0.037	0.037	0.037	0.037	0.037	0.037
$c_{l\beta_{WB\#4}}$	-6.0×10^{-5}	-6.0×10^{-5}	-6.0×10^{-5}	-6.0×10^{-5}	-6.0×10^{-5}	-6.0×10^{-5}	-6.0×10^{-5}	-6.0×10^{-5}	-6.0×10^{-5}	-6.0×10^{-5}	-6.0×10^{-5}	-6.0×10^{-5}
$c_{l\beta_{WB\#5}}$	0.018	0.011	0.008	0.006	0.005	0.004	0.003	0.003	0.003	0.003	0.003	0.003
$c_{l\beta_{WB\#6}}$	0.039	0.025	0.017	0.012	0.010	0.008	0.007	0.006	0.005	0.005	0.005	0.005
$c_{l\beta_{WB}}$	0.032	0.010	−0.002	−0.008	−0.013	−0.016	−0.019	−0.021	−0.023	−0.024	−0.026	−0.028

Case Study   **247**

2.1.2 Horizontal Tail Contribution

From the drawings in Appendix C, it is clear that the CJ3 aircraft has a negligible dihedral angle for the horizontal tail. Therefore, the only contribution to the dihedral effect from the horizontal tail comes from the sweep angle.

Using $\lambda_H = 0.426$, $AR_H = 6.137$ and $\Lambda_{c/2_H} = 16.3°$ the interpolation of Figure 4.39 provides

$$\left(\frac{c_{l_\beta}}{c_{L_1}}\right)_{\Lambda_{c/2}} = -0.0015 \left(\frac{1}{\text{deg}}\right)$$

Using $\text{Mach} \cdot \cos(\Lambda_{0.5_H})$, $\frac{AR_H}{\cos(\Lambda_{0.5_H})} = 6.39$ Figure 4.40 provides an approximated constant value $K_{M_\Lambda} = 1.1$.

Using $\frac{AR_H}{\cos(\Lambda_{0.5_H})} = 6.39$, $\frac{A}{b_H} \approx 5$ the extrapolation of Figure 4.41 provides an approximated value of $K_f = 0.96$

Therefore

$$c_{l\beta_{WB}}\Big|_{\substack{\text{Wing}\\\text{sweep}\\\text{angle}}} = 57.3 \cdot c_{L_1} \left[\left(\frac{c_{l_\beta}}{c_{L_1}}\right)_{\Lambda_{c/2}} K_{M_\Lambda} K_f\right]$$

The preceding relationship provides the contribution of the horizontal tail assumed to be as a wing. Because of the smaller size and lower wing span, the true dihedral contribution from the horizontal is found using

$$c_{l\beta_H} = c_{l\beta_{WB}}\Big|_H \eta_H \frac{S_H}{S} \frac{b_H}{b} = c_{l\beta_{WB}}\Big|_H \cdot 0.9 \cdot \frac{70.7}{294.4} \cdot \frac{20.83}{53.3}$$

All the preceding calculations are summarized in the following spreadsheet.

Mach	0.15	0.20	0.25	0.30	0.35	0.40	0.45	0.50	0.55	0.60	0.65	0.70
c_{L_1}	1.715	1.072	0.726	0.536	0.420	0.344	0.293	0.257	0.232	0.214	0.203	0.197
K_f						0.96						
$\left(\frac{c_{l_\beta}}{c_{L_1}}\right)_{\Lambda_{c/2}}$						2×10^{-4}						
K_{M_Λ}						1.1						
$c_{l\beta_{WB}}\Big|_{\substack{\text{Wing}\\\text{sweep}\\\text{angle}}}$	−0.156	−0.097	−0.066	−0.049	−0.038	−0.031	−0.027	−0.023	−0.021	−0.019	−0.018	−0.018
$c_{l\beta_H}$	−0.013	−0.008	−0.006	−0.004	−0.003	−0.003	−0.002	−0.002	−0.002	−0.002	−0.002	−0.002

2.1.3 Vertical Tail Contribution

A relationship for this contribution is given by

$$c_{l\beta_V} = -k_{Y_V} \cdot \left|c_{L_{\alpha_V}}\right| \eta_V \cdot \left(1 + \frac{d\sigma}{d\beta}\right) \frac{S_V}{S} \cdot \frac{(Z_V \cos\alpha_1 - X_V \sin\alpha_1)}{b} \quad (\text{rad}^{-1})$$

Using $\frac{b_V}{2r_1} = \frac{7.6}{2 \cdot 3.6} = 1.055$ with the data from Appendix C, Figure 4.13 provides $k_{Y_V} = 0.761$. Next, the term $\eta_V \cdot \left(1 + \frac{d\sigma}{d\beta}\right)$ is modeled using

$$\eta_V \cdot \left(1 + \frac{d\sigma}{d\beta}\right) = 0.724 + 3.06 \frac{S_V/S}{1 + \cos(\Lambda_{c/4})} + 0.4 \frac{Z_W}{d} + 0.009 \cdot AR = 1.317$$

Thus

$$c_{l\beta_V} = -k_{Y_V} \cdot \left|c_{L_{\alpha_V}}\right| \cdot \eta_V \cdot \left(1 + \frac{d\sigma}{d\beta}\right) \cdot \frac{S_V}{S} \cdot \frac{(Z_V \cos\alpha_1 - X_V \sin\alpha_1)}{b}$$

$$= -0.761 \cdot \left|c_{L_{\alpha_V}}\right| \cdot 1.317 \cdot \frac{0.5 \cdot 101.84}{294.1} \cdot \frac{(5.3\cos(0.035) - 18.8\sin(0.035))}{53.3}$$

Note that $\alpha_1 = 2° = 0.035$ rad is assumed. The preceding calculation is summarized in the following spreadsheet.

Mach	0.15	0.20	0.25	0.30	0.35	0.40	0.45	0.50	0.55	0.60	0.65	0.70
$c_{l\beta_V}$	−0.046	−0.046	−0.047	−0.047	−0.047	−0.048	−0.048	−0.049	−0.049	−0.050	−0.051	−0.051

2.1.4 Total Dihedral Effect

The starting point for the modeling of $c_{l\beta}$ was given by

$$c_{l\beta} = c_{l\beta_{WB}} + c_{l\beta_H} + c_{l\beta_V}$$

The following spreadsheet shows the breakdown.

Mach	0.15	0.20	0.25	0.30	0.35	0.40	0.45	0.50	0.55	0.60	0.65	0.70
$c_{l\beta_{WB}}$	0.032	0.010	−0.002	−0.008	−0.013	−0.016	−0.019	−0.021	−0.023	−0.024	−0.026	−0.028
$c_{l\beta_H}$	−0.013	−0.008	−0.006	−0.004	−0.003	−0.003	−0.002	−0.002	−0.002	−0.002	−0.002	−0.002
$c_{l\beta_V}$	−0.046	−0.046	−0.047	−0.047	−0.047	−0.048	−0.048	−0.049	−0.049	−0.050	−0.051	−0.051
$c_{l\beta}$	−0.028	−0.045	−0.054	−0.059	−0.063	−0.066	−0.069	−0.071	−0.074	−0.076	−0.078	−0.081

The results for $c_{l\beta}$ are also shown in Figure CS.10.

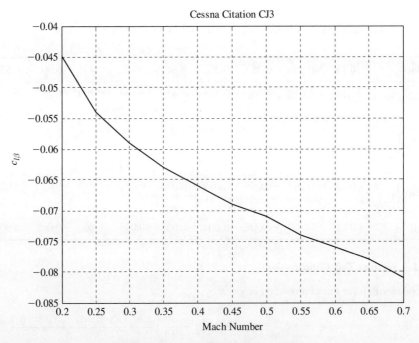

Figure CS.10 Estimates of $c_{l\beta}$ vs. Mach

From a design point of view, an interesting analysis is the breakdown of the contributions to the important dihedral effect from all the different sources previously explained. The breakdown is shown in the following spreadsheet.

Mach	0.15	0.20	0.25	0.30	0.35	0.40	0.45	0.50	0.55	0.60	0.65	0.70
% $c_{l\beta_{WB\#1}}(\Gamma_W)$	182.7	114.7	95.4	87.2	82.8	80.2	78.4	77.2	76.3	75.6	75.1	74.7
% $c_{l\beta_{WB\#2}}$ (fuselage)	40.6	25.4	21.0	19.0	17.8	17.0	16.4	15.9	15.4	14.9	14.5	14.0
% $c_{l\beta_{WB\#3}}$ (low wing)	−131.6	−82.3	−68.0	−61.6	−57.9	−55.3	−53.2	−51.4	−49.8	−48.3	−46.9	−45.5
% $c_{l\beta_{WB\#4}}(\varepsilon_W)$	0.2	0.1	0.1	0.1	0.1	0.1	0.1	0.1	0.1	0.1	0.1	0.1
% $c_{l\beta_{WB\#5}}(\Lambda_W)$	−64.1	−25.4	−14.5	−9.8	−7.3	−5.9	−4.9	−4.3	−3.9	−3.6	−3.4	−3.4
% $c_{l\beta_{WB\#6}}(AR_W)$	−141.1	−55.1	−30.9	−20.6	−15.2	−11.9	−9.7	−8.3	−7.2	−6.5	−5.9	−5.6
% $c_{l\beta_H}$	47.2	18.4	10.3	6.9	5.1	4.0	3.3	2.8	2.4	2.2	2.0	1.9
% $c_{l\beta_V}$	166.1	104.2	86.5	78.9	74.6	71.8	69.7	68.1	66.8	65.6	64.6	63.7
Total (100%)	100.0	100.0	100.0	100.0	100.0	100.0	100.0	100.0	100.0	100.0	100.0	100.0

The main conclusions from an analysis of the above spreadsheet are the following:

- The low wing configuration for this aircraft produces a very significant negative dihedral (anhedral) effect across the range of Mach;
- The contribution to the dihedral effect due to the wing sweep angle is very limited due to the very low value of this geometric parameter;
- The main contributions to the dihedral effect across the range of Mach are from the wing geometric dihedral and from the vertical tail.

2.2 Modeling of (c_{l_p})

The modeling for c_{l_p} is given by

$$c_{l_p} = c_{l_{p_{WB}}} + c_{l_{p_H}} + c_{l_{p_V}}$$

$c_{l_{p_{WB}}}$ is given by $c_{l_{p_{WB}}} \approx c_{l_{p_W}} = RDP \cdot \frac{k}{\beta}$

where $\beta = \sqrt{1 - Mach^2} = 0.760$, $k = \frac{(c_{l_\alpha})_{Wing\ Section}|_{Mach} \cdot \beta}{2\pi}$ within a reasonable approximation the wing planform lift curve slope $c_{L_{\alpha_W}}|_{Mach}$ can be used in lieu of $(c_{l_\alpha})_{Wing\ Section}|_{Mach}$. Therefore, we have $k = \frac{c_{L_{\alpha_W}}|_{Mach} \cdot \beta}{2\pi}$.

The parameters $\frac{\beta \cdot AR}{k}$ and $\Lambda_\beta = \tan^{-1}\left(\frac{\tan \Lambda_{c/4}}{\beta}\right)$ are needed to evaluate RDP. Using the above parameters, along with $\lambda = 0.305$, Figures 4.80 and 4.81 provide $RDP = f(Mach)$.

The contribution from the horizontal is given by

$$c_{l_{p_H}} \approx \frac{1}{2}\left(c_{l_{p_W}}\right)\bigg|_H \frac{S_H}{S}\left(\frac{b_H}{b}\right)^2$$

where $\left(c_{l_{p_W}}\right)\big|_H = RDP_H \cdot \frac{k_H}{\beta_H}$. This contribution is expected to be small due to the low values of $\frac{S_H}{S}, \left(\frac{b_H}{b}\right)^2$. As for the wing, the parameters k_H, β_H are given by

$$\beta_H = \sqrt{1 - Mach^2}, k_H = c_{L_{\alpha_H}}\bigg|_{Mach} \cdot \frac{\beta_H}{2\pi}$$

The following parameters are required to obtain RDP_H from Figures 4.80 and 4.81:

$$\frac{\beta_H \cdot AR_H}{k_H}, \Lambda_{\beta_H} = \tan^{-1}\left(\frac{\tan \Lambda_{c/4_H}}{\beta_H}\right), \lambda_H = 0.426$$

Finally, Figures 4.80 and 4.81 provide $RDP_H = f(Mach)$ leading to

$$\left(c_{l_{p_W}}\right)_H = RDP_H \cdot \frac{k_H}{\beta_H} = f(Mach)$$

Next, the above contribution is normalized with respect to the wing parameters using:

$$c_{l_{p_H}} = \frac{1}{2}\left(c_{l_{p_W}}\right)\bigg|_H \frac{S_H}{S}\left(\frac{b_H}{b}\right)^2 = \frac{1}{2}\left(c_{l_{p_W}}\right)\bigg|_H \frac{70.7}{294.4}\left(\frac{20.83}{53.3}\right)^2 = f(Mach)$$

The contribution from the vertical is given by

$$c_{l_{p_V}} \approx 2\, c_{Y_{\beta_V}}\left(\frac{Z_V}{b}\right)^2 = 2 \cdot c_{Y_{\beta_V}}\left(\frac{5.3}{53.3}\right)^2 = f(Mach)$$

where $c_{Y_{\beta_V}}$ was obtained in Section 1.2.

All the preceding modeling has been summarized in the following spreadsheet

Mach	0.15	0.20	0.25	0.30	0.35	0.40	0.45	0.50	0.55	0.60	0.65	0.70
β	0.989	0.980	0.968	0.954	0.937	0.917	0.893	0.866	0.835	0.800	0.760	0.714
k	0.680	0.679	0.677	0.675	0.672	0.669	0.665	0.660	0.654	0.646	0.637	0.626
$\frac{\beta \cdot AR}{k}$	14.06	13.96	13.83	13.67	13.48	13.25	12.99	12.70	12.36	11.97	11.53	11.03
RDP	−0.608	−0.605	−0.602	−0.598	−0.593	−0.588	−0.581	−0.574	−0.565	−0.555	−0.544	−0.532
Λ_β [deg]	0.86	0.87	0.88	0.89	0.91	0.93	0.95	0.98	1.02	1.06	1.12	1.19
$c_{l_{p_W}}$	−0.418	−0.419	−0.421	−0.423	−0.426	−0.429	−0.432	−0.437	−0.442	−0.449	−0.457	−0.466
$(c_{l_{p_W}})\vert_H$	−0.385	−0.386	−0.388	−0.390	−0.392	−0.394	−0.397	−0.400	−0.404	−0.407	−0.411	−0.414
β_H	0.989	0.980	0.968	0.954	0.937	0.917	0.893	0.866	0.835	0.800	0.760	0.714
k_H	0.733	0.730	0.727	0.723	0.718	0.712	0.704	0.695	0.685	0.672	0.656	0.637
$\frac{\beta_H \cdot AR_H}{k_H}$	8.280	8.234	8.173	8.099	8.009	7.904	7.783	7.645	7.487	7.310	7.110	6.885
Λ_{β_H} [deg]	19.9	20.1	20.3	20.6	21.0	21.4	21.9	22.5	23.2	24.2	25.3	26.7
RDP_H	−0.520	−0.518	−0.516	−0.514	−0.511	−0.508	−0.504	−0.499	−0.493	−0.485	−0.476	−0.464
$c_{l_{p_H}}$	−0.007	−0.007	−0.007	−0.007	−0.007	−0.007	−0.007	−0.007	−0.007	−0.007	−0.008	−0.008
$c_{l_{p_V}}$	−0.021	−0.021	−0.021	−0.021	−0.021	−0.021	−0.022	−0.022	−0.022	−0.022	−0.023	−0.023
c_{l_p}	−0.446	−0.447	−0.449	−0.451	−0.454	−0.457	−0.461	−0.466	−0.472	−0.479	−0.487	−0.497

The results for c_{l_p} are also shown in Figure CS.11.

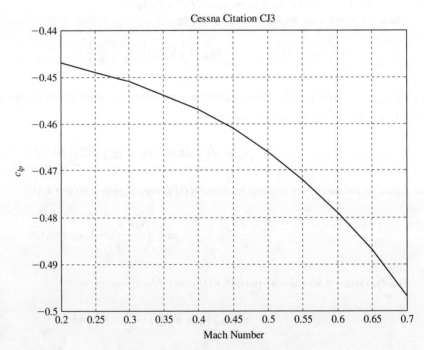

Figure CS.11 Estimates of c_{l_p} vs. Mach

2.3 Modeling of (c_{l_r})

The modeling for c_{l_r} is given by

$$c_{l_r} \approx c_{l_{r_W}} + c_{l_{r_V}}$$

A relationship for $c_{l_{r_W}}$ is given by

$$c_{l_{r_W}} \approx \left(\frac{c_{l_r}}{c_{L_1}}\right)\bigg|_{\substack{Mach=0 \\ C_L=0}} \cdot c_{L_1} + \left(\frac{\Delta c_{l_r}}{\Gamma}\right) \cdot \Gamma + \left(\frac{\Delta c_{l_r}}{\varepsilon_W}\right) \cdot \varepsilon_W \quad (\text{rad}^{-1})$$

The coefficient $\left(\frac{c_{l_r}}{c_{L_1}}\right)\bigg|_{\substack{Mach \\ C_L=0}}$ is given by

$$\left(\frac{c_{lr}}{C_{L_1}}\right)\bigg|_{\substack{Mach \\ C_L=0}} = D \cdot \left(\frac{c_{lr}}{C_{L_1}}\right)\bigg|_{\substack{Mach=0 \\ C_L=0}}$$

To obtain D the following parameter is required

$$B = \sqrt{1 - Mach^2 \cos^2(\Lambda_{c/4})}$$

where D is given by

$$D = \frac{1 + \dfrac{AR(1-B^2)}{2B[AR \cdot B + 2\cos(\Lambda_{c/4})]} + \dfrac{[AR \cdot B + 2\cos(\Lambda_{c/4})]}{[AR \cdot B + 4\cos(\Lambda_{c/4})]} \cdot \dfrac{\tan^2(\Lambda_{c/4})}{8}}{1 + \dfrac{[AR + 2\cos(\Lambda_{c/4})]}{[AR + 4\cos(\Lambda_{c/4})]} \cdot \dfrac{\tan^2(\Lambda_{c/4})}{8}} = f(Mach)$$

with $AR = 9.67$ and $\Lambda_{c/4} = 0.86° = 0.015$ rad.

Next, using Figure 4.85 with the parameters $\lambda = 0.305$, $AR = 9.67$, and $\Lambda_{c/4} = 0.86°$ we have $\left(\dfrac{c_{lr}}{C_{L_1}}\right)\bigg|_{\substack{Mach=0 \\ C_L=0}} = 0.223$ leading to

$$\left(\frac{c_{lr}}{C_{L_1}}\right)\bigg|_{\substack{Mach \\ C_L=0}} = D \cdot \left(\frac{c_{lr}}{C_{L_1}}\right)\bigg|_{\substack{Mach=0 \\ C_L=0}} = f(Mach)$$

Next, the term $\left(\dfrac{\Delta c_{lr}}{\Gamma}\right)$ is given by

$$\left(\frac{\Delta c_{lr}}{\Gamma}\right) \approx \left(\frac{1}{12}\right) \cdot \frac{[\pi \cdot AR \cdot \sin(\Lambda_{c/4})]}{(AR + 4 \cdot \cos(\Lambda_{c/4}))} = \left(\frac{1}{12}\right) \cdot \frac{[\pi \cdot 9.67 \cdot \sin(0.015)]}{(9.67 + 4 \cdot \cos(0.015))} = 0.003 \; (\text{rad}^{-2})$$

Also, from Figure 4.87 with $\lambda = 0.305$, $AR = 9.67$:

$$\left(\frac{\Delta c_{lr}}{\varepsilon_W}\right) = 0.01542 \left(\frac{1}{\text{rad} \cdot \text{deg}}\right)$$

Note that a positive 2° twist angle ε_W is here assumed.

Adding the previous terms, the value of c_{lr_W} is given by:

$$c_{lr_W} = \left(\frac{c_{lr}}{C_{L_1}}\right)\bigg|_{\substack{Mach \\ C_L=0}} \cdot C_{L_1} + \left(\frac{\Delta c_{lr}}{\Gamma}\right) \cdot \Gamma + \left(\frac{\Delta c_{lr}}{\varepsilon_W}\right) \cdot \varepsilon_W$$

$$= \left(\frac{c_{lr}}{C_{L_1}}\right)\bigg|_{\substack{Mach \\ C_L=0}} \cdot C_{L_1} + \left(0.003 \cdot \frac{5}{57.3}\right) + (0.01542 \cdot 2)$$

A relationship for c_{lr_V} is given by

$$c_{lr_V} \approx -2 \cdot c_{Y_{\beta_V}} \cdot \frac{(X_V \cos\alpha_1 + Z_V \sin\alpha_1)}{b} \cdot \frac{(Z_V \cos\alpha_1 - X_V \sin\alpha_1)}{b}$$

$$= -2 \cdot c_{Y_{\beta_V}} \frac{(18.8 \cos(0.035) + 5.3 \sin(0.035))}{53.3} \cdot \frac{(5.3 \cos(0.035) - 18.8 \sin(0.035))}{53.3}$$

with $c_{Y_{\beta_V}} = f(Mach)$ from the previous section. Thus, the total value of c_{lr} is given by

$$c_{lr} \approx c_{lr_W} + c_{lr_V}$$

The preceding calculations are summarized in the following spreadsheet.

Mach	0.15	0.20	0.25	0.30	0.35	0.40	0.45	0.50	0.55	0.60	0.65	0.70
c_{L_1}	1.715	1.072	0.726	0.536	0.420	0.344	0.293	0.257	0.232	0.214	0.203	0.197
$\left(\dfrac{c_{l_r}}{c_{L_1}}\right)\bigg\|_{\substack{Mach \\ C_L=0}}$	0.225	0.227	0.229	0.232	0.236	0.240	0.246	0.253	0.262	0.273	0.287	0.306
$\left(\dfrac{\Delta c_{l_r}}{\Gamma}\right)$						0.003						
$\left(\dfrac{\Delta c_{l_r}}{\varepsilon_W}\right)$						0.01542						
$c_{l_{r_W}}$	0.417	0.274	0.197	0.155	0.130	0.114	0.103	0.096	0.092	0.090	0.089	0.091
$c_{Y_{\beta_V}}$	−1.073	−1.077	−1.082	−1.088	−1.095	−1.104	−1.114	−1.126	−1.139	−1.154	−1.171	−1.191
$c_{l_{r_V}}$	0.066	0.066	0.066	0.067	0.067	0.068	0.068	0.069	0.070	0.071	0.072	0.073
c_{l_r}	0.483	0.340	0.264	0.222	0.197	0.182	0.172	0.165	0.162	0.160	0.161	0.164

The results for c_{l_r} are also shown in Figure CS.12.

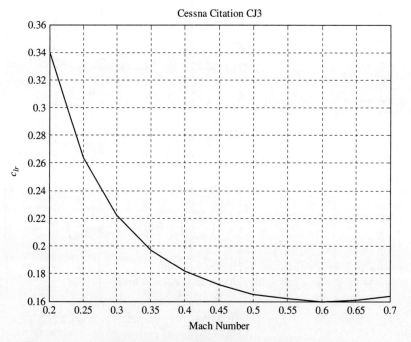

Figure CS.12 Estimates of c_{l_r} vs. Mach

2.4 Modeling of ($c_{l_{\delta_A}}$)

From the data in Appendix C, the following parameters are required for the modeling of $c_{l_{\delta_A}}$:

$$\eta_I = \eta_{Inboard} = \frac{y_{A_I}}{b/2} = \frac{18.8}{26.65} = 0.705,$$

$$\eta_O = \eta_{Outboard} = \frac{y_{A_O}}{b/2} = \frac{26.2}{26.65} = 0.983$$

$$\beta = \sqrt{1 - Mach^2} = f(Mach);$$

$$\Lambda_\beta = \tan^{-1}\left(\frac{\tan \Lambda_{c/4}}{\beta}\right) = f(Mach)$$

$$k = \frac{(c_{L_\alpha})_W|_{Mach} \cdot \beta}{2\pi} = f(Mach), \quad \frac{\beta \cdot AR}{k} = f(Mach)$$

From Figures 4.51 to 4.53, using Λ_β, $\frac{\beta \cdot AR}{k}$ and $\lambda = 0.305$, we have

$$\eta_I = 0.705 \Rightarrow RME_I = f(Mach)$$
$$\eta_O = 0.983 \Rightarrow RME_O = f(Mach)$$

Using Figure 4.54 we have

$$\Delta(RME) = RME|_{\eta_O} - RME|_{\eta_I} = f(Mach)$$

leading to

$$c_{l_\delta}' = \frac{\Delta(RME) \cdot k}{\beta} = f(Mach)$$

Using $\frac{\bar{c}_{Aileron}}{\bar{c}_{Wing(at\_aileron)}} = \frac{1.1}{3.4} = 0.324$ from the data in Appendix C, Figure 4.55 provides $\tau_A = 0.544$ leading to $c_{l_\delta}' = \tau_A \cdot c_{l_\delta} = f(Mach)$.

Finally, the value of $c_{l_{\delta_A}}$ is obtained as

$$c_{l_{\delta_A}} = \frac{1}{2}\left[(c_{l_\delta})|_{Left} + (c_{l_\delta})|_{Right}\right] = f(Mach)$$

The preceding modeling is summarized in the following spreadsheet.

Mach	0.15	0.20	0.25	0.30	0.35	0.40	0.45	0.50	0.55	0.60	0.65	0.70
y_{A_I} [ft]						18.8						
y_{A_O} [ft]						26.2						
η_I						0.705						
η_O						0.983						
Λ_β [deg]	0.86	0.87	0.88	0.89	0.91	0.93	0.95	0.98	1.02	1.06	1.12	1.19
$\frac{k}{\beta}$	0.688	0.693	0.699	0.707	0.717	0.730	0.744	0.762	0.783	0.808	0.839	0.877
$\frac{\beta \cdot AR}{k}$	14.06	13.96	13.83	13.67	13.48	13.25	12.99	12.70	12.36	11.97	11.53	11.03
RME_I	0.73	0.73	0.72	0.71	0.71	0.70	0.69	0.67	0.66	0.65	0.63	0.61
RME_O	1.07	1.07	1.06	1.05	1.04	1.03	1.01	0.99	0.97	0.95	0.93	0.90
$\Delta(RME)$	0.342	0.340	0.338	0.335	0.332	0.328	0.323	0.318	0.312	0.305	0.298	0.289
c_{l_δ}'	0.235	0.236	0.236	0.237	0.238	0.239	0.241	0.242	0.244	0.247	0.250	0.253
$\bar{c}_{Aileron}$ [ft]						1.1						
$\bar{c}_{Wing@aileron}$ [ft]						3.4						
$\bar{c}_{Ail.}/\bar{c}_{Wing@Ail.}$						0.324						
τ_A						0.544						
c_{l_δ}	0.128	0.128	0.129	0.129	0.130	0.130	0.131	0.132	0.133	0.134	0.136	0.138

The results for $c_{l_{\delta_A}}$ are also shown in Figure CS.13.

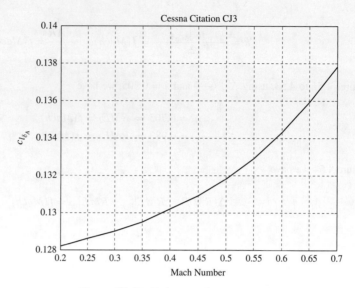

Figure CS.13 Estimates of $c_{l_{\delta_A}}$ vs. Mach

2.5 Modeling of ($c_{l_{\delta_R}}$)

The modeling of $c_{l_{\delta_R}}$ is given by

$$c_{l_{\delta_R}} = c_{Y_{\delta_R}} \left(\frac{Z_R \cos \alpha_1 - X_R \sin \alpha_1}{b} \right)$$

$$= c_{Y_{\delta_R}} \frac{(4.9 \cos(0.035) - 22 \sin(0.035))}{53.3} = f(Mach)$$

where $c_{Y_{\delta_R}}$, X_R, and Z_R were found and shown in Section 1.

The preceding modeling is summarized in the following spreadsheet.

Mach	0.15	0.20	0.25	0.30	0.35	0.40	0.45	0.50	0.55	0.60	0.65	0.70
$c_{Y_{\delta_R}}$	0.688	0.693	0.699	0.707	0.717	0.730	0.744	0.762	0.783	0.808	0.839	0.877
$c_{l_{\delta_R}}$	0.02156	0.02164	0.02174	0.02186	0.02201	0.02218	0.02239	0.02262	0.02289	0.02319	0.02354	0.02394

The results for $c_{l\delta_R}$ are also shown in Figure CS.14.

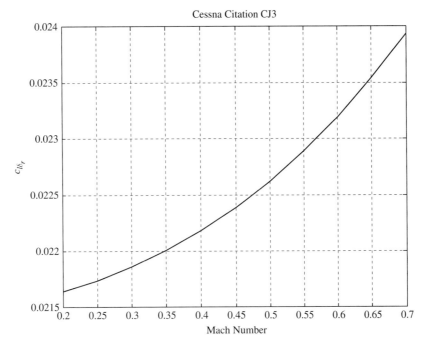

Figure CS.14 Estimates of $c_{l\delta_R}$ vs. Mach

3 Modeling of the Yawing Aerodynamic Coefficients

This section provides the modeling of the following yawing stability and control derivatives:

$$\left(c_{n_\beta}, c_{n_p}, c_{n_r}, c_{n_{\delta_A}}, c_{n_{\delta_R}}\right)$$

3.1 Modeling of (c_{n_β})

The general modeling of this derivative is

$$c_{n\beta} = c_{n\beta_W} + c_{n\beta_B} + c_{n\beta_H} + c_{n\beta_V}$$

It is easy to visualize that the both the wing and the horizontal tail will not provide a significant contribution to c_{n_β}. Therefore, we have

$$c_{n\beta} \approx c_{n\beta_B} + c_{n\beta_V}$$

Body Contribution

$c_{n\beta_B}$ is given by

$$c_{n\beta_B} = -57.3 \cdot K_N \, K_{R_l} \frac{S_{B_S}}{S} \frac{l_B}{b}$$

To obtain K_N from Figure 4.55 the following parameters from Figure 4.67 related to dimensions of the Cessna CJ3 aircraft from Appendix C are required.

$$\frac{l_{CG}}{l_B} = \frac{23.8}{45.1} = 0.528, \quad \frac{l_B^2}{S_{B_S}} = \frac{45.1^2}{205.33} = 9.906,$$

$$\sqrt{\frac{z_1}{z_2}} = \sqrt{\frac{5.3}{5.0}} = 1.029, \quad \frac{z_{\text{Max}}}{w_{\text{Max}}} = \frac{6.4}{5.4} = 1.185$$

Using the preceding parameters Figure 4.68 provides $K_N = 0.00158$.

256 Chapter 4 Modeling of Lateral Directional Aerodynamic Forces and Moments

Next, the following parameter is required to obtain K_{Re_l}.

$$Re_{Fuselage} = \frac{V \cdot l_B}{\nu} = \frac{f(Mach) \cdot 45.1}{3.823 \times 10^{-4}} = f(Mach)$$

The values of the speed of sound and the kinematic viscosity were obtained from the Standard Atmosphere tables. Finally, using $Re_{Fuselage} = f(Mach)$, Figure 4.69 provides $K_{Re_l} = f(Mach)$, leading to

$$c_{n\beta_B} = -57.3 \cdot K_N \cdot K_{Re_l} \frac{S_{B_S}}{S} \frac{l_B}{b} = -57.3 \cdot K_N \cdot K_{Re_l} \cdot \frac{205.33}{294.4} \cdot \frac{45.1}{53.3} = f(Mach)$$

Vertical Tail Contribution

$c_{n\beta_V}$ is given by

$$c_{n\beta_V} = -c_{Y\beta_V} \cdot \frac{(X_V \cos\alpha_1 + Z_V \sin\alpha_1)}{b}$$

$$= -c_{Y\beta_V} \frac{(18.8 \cos(0.035) + 5.3 \sin(0.035))}{53.3} = f(Mach)$$

$c_{y\beta_V}$, X_V and Z_V were calculated in Section 1.

Thus, the total value of $c_{n\beta}$ is given by

$$c_{n\beta} \approx c_{n\beta_B} + c_{n\beta_V}$$

The preceding modeling is summarized in the following spreadsheet.

Mach	0.15	0.20	0.25	0.30	0.35	0.40	0.45	0.50	0.55	0.60	0.65	0.70
K_N						0.00158						
Kinematic viscosity [ft²/sec]						3.823×10^{-4}						
Speed of sound [ft/sec]						981.66						
Airspeed [ft/sec]	147.2	196.3	245.4	294.5	343.6	392.7	441.7	490.8	539.9	589.0	638.1	687.2
$Re_{Fuselage}$	1.737×10^7	2.316×10^7	2.895×10^7	3.474×10^7	4.053×10^7	4.632×10^7	5.211×10^7	5.790×10^7	6.369×10^7	6.948×10^7	7.527×10^7	8.106×10^7
K_{Re_l}	1.59	1.65	1.69	1.73	1.76	1.79	1.81	1.83	1.85	1.87	1.89	1.90
$c_{n\beta_B}$	−0.085	−0.088	−0.090	−0.092	−0.094	−0.096	−0.097	−0.098	−0.099	−0.100	−0.101	−0.102
$c_{y\beta_V}$	−1.073	−1.077	−1.082	−1.088	−1.095	−1.104	−1.114	−1.126	−1.139	−1.154	−1.171	−1.191
$c_{n\beta_V}$	0.382	0.383	0.385	0.387	0.390	0.393	0.397	0.401	0.405	0.411	0.417	0.424
$c_{n\beta}$	0.297	0.295	0.295	0.295	0.296	0.297	0.300	0.303	0.306	0.311	0.316	0.322

An interesting observation is that the vertical tail has to produce enough positive and stable $c_{n\beta}$ to overcome a negative and unstable contribution from the fuselage.

The results for $c_{n\beta}$ are also shown in Figure CS.15.

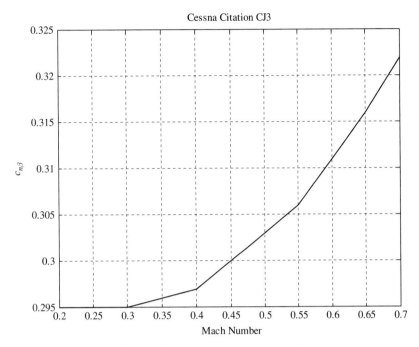

Figure CS.15 Estimates of $c_{n\beta}$ vs. Mach

3.2 Modeling of (c_{n_p})

The modeling for c_{n_p} is given by

$$c_{n_p} \approx c_{n_{p_W}} + c_{n_{p_V}}$$

The contribution from the wing is given by

$$c_{n_{p_W}} \approx \left(\frac{c_{n_p}}{c_{L_1}}\right)\bigg|_{\substack{Mach \\ C_L=0}} c_{L_1} + \left(\frac{\Delta c_{n_p}}{\varepsilon_W}\right)\varepsilon_W$$

The coefficient $\left(\frac{c_{n_p}}{c_{L_1}}\right)\bigg|_{\substack{Mach \\ C_L=0}}$ is given by

$$\left(\frac{c_{n_p}}{c_{L_1}}\right)\bigg|_{\substack{Mach \\ C_L=0}} = C\left(\frac{c_{n_p}}{c_{L_1}}\right)\bigg|_{\substack{Mach=0 \\ C_L=0}}$$

where C is given by

$$C = \frac{[AR + 4\cos(\Lambda_{c/4})]}{[AR \cdot B + 4\cos(\Lambda_{c/4})]} \cdot \left\{ \frac{AR \cdot B + \frac{1}{2}[AR \cdot B + 4\cos(\Lambda_{c/4})] \cdot \tan^2(\Lambda_{c/4})}{AR + \frac{1}{2}[AR + 4\cos(\Lambda_{c/4})] \cdot \tan^2(\Lambda_{c/4})} \right\}$$

with B given by

$$B = \sqrt{1 - Mach^2 \cos^2(\Lambda_{c/4})} = f(Mach), \quad AR = 9.67, \quad \Lambda_{c/4} = 0.015 \text{ rad}$$

leading to

$$C = \frac{[9.67 + 4\cos(0.015)]}{[9.67 \cdot B + 4\cos(0.015)]} \cdot \left\{ \frac{9.67 \cdot B + \frac{1}{2}[9.67 \cdot B + 4\cos(0.015)] \cdot \tan^2(0.015)}{9.67 + \frac{1}{2}[9.67 + 4\cos(0.015)] \cdot \tan^2(0.015)} \right\}$$

$$= f(Mach)$$

$\left(\dfrac{c_{np}}{c_{L_1}}\right)\bigg|_{\substack{Mach=0 \\ C_L=0}}$ is instead given by

$$\left(\frac{c_{np}}{c_{L_1}}\right)\bigg|_{\substack{Mach=0 \\ C_L=0}} = -\frac{1}{6} \cdot \frac{AR + 6(AR + \cos(\Lambda_{c/4})) \cdot \left[(\bar{x}_{CG} - \bar{x}_{AC})\dfrac{\tan(\Lambda_{c/4})}{AR} + \dfrac{\tan^2(\Lambda_{c/4})}{12}\right]}{(AR + \cos(\Lambda_{c/4}))}$$

$$= -\frac{1}{6} \cdot \frac{9.67 + 6(9.67 + \cos(0.015)) \cdot \left[(0.25 - 0.457)\dfrac{\tan(0.015)}{9.67} + \dfrac{\tan^2(0.015)}{12}\right]}{(9.67 + \cos(0.015))} = -0.150 \; (\text{rad}^{-1})$$

where $\bar{x}_{AC}$ was obtained in Chapter 3 by setting $\bar{x}_{AC} = \bar{x}_{CG}|_{c_{m_\alpha}=0}$. Using the previously calculated values:

$$\left(\frac{c_{np}}{c_{L_1}}\right)\bigg|_{\substack{Mach \\ C_L=0}} = C\left(\frac{c_{np}}{c_{L_1}}\right)\bigg|_{\substack{Mach=0 \\ C_L=0}} = -0.150 \cdot C = f(Mach)$$

Next, using $\lambda = 0.305$ and $AR = 9.67$, Figure 4.83 provides

$$\left(\frac{\Delta c_{np}}{\varepsilon_W}\right) = -0.00061 \left(\frac{1}{\text{rad} \cdot \text{deg}}\right)$$

Note that a positive 2° twist angle ε_W is here assumed. Therefore

$$c_{np_W} \cong \left(\frac{c_{np}}{c_{L_1}}\right)\bigg|_{\substack{Mach \\ C_L=0}} c_{L_1} + \left(\frac{\Delta c_{np}}{\varepsilon_W}\right)\varepsilon_W = -0.150 \cdot C \cdot c_{L_1} - 0.00061 \cdot 2 = f(Mach)$$

The contribution from the vertical tail is given by

$$c_{np_V} = -2 \cdot c_{Y_{\beta_V}} \cdot \frac{(X_V \cos\alpha_1 + Z_V \sin\alpha_1)}{b} \cdot \frac{(Z_V \cos\alpha_1 - X_V \sin\alpha_1 - Z_V)}{b}$$

$$= -2 \cdot c_{Y_{\beta_V}} \frac{(18.8\cos(0.035) + 5.3\sin(0.035))}{53.3} \cdot \frac{(5.3\cos(0.035) - 18.8\sin(0.035) - 33.7)}{53.3} = f(Mach)$$

where the value of $c_{Y_{\beta_V}}$ was found in Section 1.
Finally, the total value of c_{np} is

$$c_{np} \approx c_{np_W} + c_{np_V}$$

The preceding modeling is summarized in the following spreadsheet.

Mach	0.15	0.20	0.25	0.30	0.35	0.40	0.45	0.50	0.55	0.60	0.65	0.70
c_{L_1}	1.715	1.072	0.726	0.536	0.420	0.344	0.293	0.257	0.232	0.214	0.203	0.197
$\left(\frac{c_{np}}{c_{L_1}}\right)\bigg\|_{\substack{Mach=0 \\ C_L=0}}$						-0.150						
$\left(\frac{\Delta c_{np}}{\varepsilon_W}\right)$						-0.00061						
$\left(\frac{c_{np}}{c_{L_1}}\right)\bigg\|_{\substack{Mach \\ C_L=0}}$	-0.150	-0.149	-0.149	-0.148	-0.147	-0.146	-0.145	-0.144	-0.142	-0.140	-0.138	-0.135
c_{np_W}	-0.256	-0.159	-0.107	-0.078	-0.061	-0.049	-0.041	-0.036	-0.032	-0.029	-0.027	-0.025
$c_{y\beta_V}$	-1.073	-1.077	-1.082	-1.088	-1.095	-1.104	-1.114	-1.126	-1.139	-1.154	-1.171	-1.191
c_{np_V}	-0.009	-0.009	-0.010	-0.010	-0.010	-0.010	-0.010	-0.010	-0.010	-0.010	-0.010	-0.010
c_{np}	-0.265	-0.168	-0.116	-0.088	-0.070	-0.059	-0.051	-0.046	-0.042	-0.039	-0.037	-0.036

The results for c_{np} are also shown in Figure CS.16.

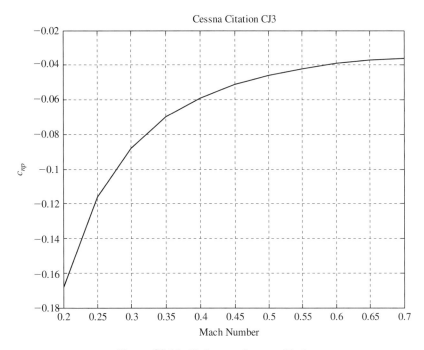

Figure CS.16 Estimates of c_{np} vs. Mach

3.3 Modeling of (c_{n_r})

The modeling for c_{n_r} is given by

$$c_{n_r} \approx c_{n_{r_W}} + c_{n_{r_V}}$$

A relationship for $c_{n_{r_W}}$ is given by: $c_{n_{r_W}} \cong \left(\frac{c_{n_r}}{c_{L_1}^2}\right) \cdot c_{L_1}^2$.

Using Figure 4.89 with $\lambda = 0.305$, $AR = 9.67$, $\Lambda_{c/4} = 0.86°$ and $\bar{x}_{AC} - \bar{x}_{CG} = 0.457 - 0.25 = 0.207$ we will have $\left(\frac{c_{n_r}}{c_{L_1}^2}\right) = f(Mach)$.

A relationship for $c_{n r_V}$ is given by

$$c_{n r_V} \approx 2 \cdot c_{Y_{\beta_V}} \cdot \frac{(X_V \cos\alpha_1 + Z_V \sin\alpha_1)^2}{b^2}$$

$$= 2 \cdot c_{Y_{\beta_V}} \frac{(18.8 \cos(0.035) + 5.3 \sin(0.035))^2}{53.3^2} = f(Mach)$$

Thus, the total value of c_{n_r} is given by

$$c_{n_r} \approx c_{n_{r_W}} + c_{n_{r_V}}$$

The preceding modeling is summarized in the following spreadsheet.

Mach	0.15	0.20	0.25	0.30	0.35	0.40	0.45	0.50	0.55	0.60	0.65	0.70
c_{L_1}	1.715	1.072	0.726	0.536	0.420	0.344	0.293	0.257	0.232	0.214	0.203	0.197
$\left(\frac{c_{n_r}}{c_{L_1}}\right)$	−0.144	−0.144	−0.143	−0.142	−0.142	−0.141	−0.140	−0.138	−0.137	−0.135	−0.133	−0.130
$c_{n_{r_W}}$	−0.423	−0.165	−0.075	−0.041	−0.025	−0.017	−0.012	−0.009	−0.007	−0.006	−0.005	−0.005
$c_{y_{\beta_V}}$	−1.073	−1.077	−1.082	−1.088	−1.095	−1.104	−1.114	−1.126	−1.139	−1.154	−1.171	−1.191
$c_{n_{r_V}}$	−0.272	−0.273	−0.274	−0.276	−0.278	−0.280	−0.282	−0.285	−0.289	−0.293	−0.297	−0.302
c_{n_r}	−0.695	−0.438	−0.350	−0.317	−0.303	−0.296	−0.294	−0.294	−0.296	−0.299	−0.302	−0.307

The results for c_{n_r} are also shown in Figure CS.17.

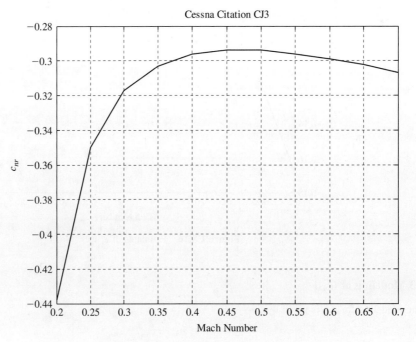

Figure CS.17 Estimates of c_{n_r} vs. Mach

3.4 Modeling of $(c_{n_{\delta_A}})$

The modeling of this coefficient requires the analysis of the aerodynamic forces on the horizontal axis exerted by the asymmetric deflection of the ailerons on the wings. In fact, in addition to the vertical forces with opposite directions, which will eventually generate the rolling moment, the asymmetric deflections of the left and right ailerons also generate some small but not negligible drag forces leading to a small negative yawing moment.

Because of lack of specific information about the drag characteristics of the left and right ailerons for the Cessna CJ3 aircraft, it is simply assumed that the two associated yawing moments will have similar size and opposite directions and will therefore cancel each other. Therefore, we can safely assume

$$c_{n_{\delta_A}} \approx 0$$

3.5 Modeling of ($c_{n_{\delta_R}}$)

The modeling of $c_{n\delta_R}$ is given by

$$c_{n\delta_R} = -c_{Y\delta_R}\left(\frac{X_R \cos\alpha_1 + Z_R \sin\alpha_1}{b}\right)$$

$$= -c_{Y\delta_R} \cdot \frac{(22\cos(0.035) + 4.9\sin(0.035))}{53.3} = f(Mach)$$

where $c_{Y\delta_R}, X_R, Z_R$ were shown in Section 1.

The preceding modeling is summarized in the following spreadsheet.

Mach	0.15	0.20	0.25	0.30	0.35	0.40	0.45	0.50	0.55	0.60	0.65	0.70
$c_{Y\delta_R}$	0.688	0.693	0.699	0.707	0.717	0.730	0.744	0.762	0.783	0.808	0.839	0.877
$c_{n\delta_R}$	−0.115	−0.116	−0.116	−0.117	−0.118	−0.119	−0.120	−0.121	−0.123	−0.124	−0.126	−0.128

The results for $c_n\delta_R$ are also shown in Figure CS.18.

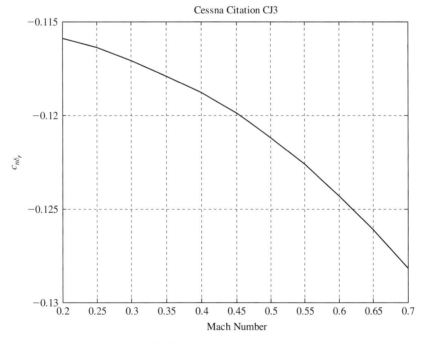

Figure CS.18 Estimates of $c_{n\delta_R}$ vs. Mach

SHORT PROBLEMS

Short Problem 4.1

Identify which aircraft geometric parameters affect the values of the lateral-directional stability derivatives:

$$c_{Y_\beta}, c_{l_\beta}, c_{n_\beta}$$

Next, explain the effect of an increase of each of the geometric parameters on each of the preceding stability derivatives. Organize your results in a table.

Short Problem 4.2

Repeat Short Problem 4.1 for the following lateral directional control derivatives:

$$c_{l_{\delta_A}}, c_{Y_{\delta_R}}, c_{l_{\delta_R}}, c_{n_{\delta_R}}$$

Short Problem 4.3

Repeat Short Problem 4.1 for the following lateral directional stability derivatives:

$$c_{Y_p}, c_{Y_r}, c_{l_p}, c_{l_r}, c_{n_p}, c_{n_r}$$

Short Problem 4.4

Consider the General Dynamics F-111 aircraft shown in Appendix C. This aircraft was one of the first aircraft designed with variable wing sweep angle. Describe which of the lateral directional stability derivatives change as the aircraft configuration changes from lower to higher sweep angles.

Short Problem 4.5

Consider an aircraft with an identical planform for the wing and the horizontal tail. Assuming negligible downwash on the horizontal tail, $\frac{S_H}{S} = 0.2$, $\eta_H = 0.95$, how much negative dihedral (anhedral) angle for the horizontal tail Γ_H will be necessary just to compensate for $\Gamma_W = 3°$?

Short Problem 4.6

Consider the Boeing B747-200 and the McDonnell Douglas C-17 aircraft in Appendix C. These aircraft have similar dimensions. Based on the different geometry of the aircraft, provide qualitative comments for a comparison of the different contributions to the overall dihedral effect c_{l_β}.

Short Problem 4.7

The Cessna T37 and the SIAI Marchetti S211, shown in Appendix C, are both basic military jet trainers. Appendix B provides sets of values of the lateral directional stability derivatives for both aircraft at different flight conditions. Consider the cruise flight condition for the Cessna T37 and the high cruise flight condition for the SIAI Marchetti S211. Based on the actual values of the stability derivatives $c_{y_\beta}, c_{l_\beta}, c_{n_\beta}$ for both aircraft, provide qualitative comments on the origin of the differences in the values due to differences in the geometric characteristics of the two aircraft.

Short Problem 4.8

Repeat Short Problem 4.7 for the following lateral directional control derivatives:

$$c_{l_{\delta_A}}, c_{Y_{\delta_R}}, c_{l_{\delta_R}}, c_{n_{\delta_R}}$$

Short Problem 4.9

Repeat Short Problem 4.7 for the following lateral directional stability derivatives:

$$c_{Y_p}, c_{Y_r}, c_{l_p}, c_{n_r}$$

PROBLEMS

Problem 4.1

Consider the data relative to the McDonnell Douglas DC 8 aircraft in Appendix C. Use the provided drawings to extract all the relevant geometric characteristics of the aircraft. Using the modeling approach outlined in this chapter, the dihedral effect c_{l_β} can be modeled as

$$c_{l\beta} = c_{l\beta_{WB}} + c_{l\beta_H} + c_{l\beta_V}$$

Estimate the value of c_{l_β} for this aircraft. Also, evaluate the percentage of each of the different contributions with respect to the total c_{l_β} value.

Problem 4.2

Consider the data relative to the McDonnell Douglas DC 8 aircraft in Appendix C. Use the provided drawings to extract all the relevant geometric characteristics of the aircraft.

Using the modeling approach outlined in this chapter, provide estimates for the following stability derivatives:

$$c_{Y_\beta}, c_{n_\beta}$$

For both aerodynamic parameters highlight the contribution of the vertical tail versus the total value.

Problem 4.3

Consider the data relative to the McDonnell Douglas DC 8 aircraft in Appendix C. Use the provided drawings to extract all the relevant geometric characteristics of the aircraft.

Using the modeling approach outlined in this chapter, provide estimates for the following control derivatives:

- $c_{Y_{\delta_R}}, c_{l_{\delta_R}}, c_{n_{\delta_R}}$;
- $c_{l_{\delta_A}}$.

Problem 4.4

Consider the data relative to the McDonnell Douglas DC 8 aircraft in Appendix C. Use the provided drawings to extract all the relevant geometric characteristics of the aircraft.

Using the modeling approach outlined in this chapter, provide estimates for the following stability derivatives:

$$c_{l_p}, c_{n_r}$$

Problem 4.5

Consider the data relative to the Aeritalia G91 aircraft in Appendix C. Use the provided drawings to extract all the relevant geometric characteristics of the aircraft. Using the modeling approach outlined in this chapter, the dihedral effect c_{l_β} can be modeled as

$$c_{l\beta} = c_{l\beta_{WB}} + c_{l\beta_H} + c_{l\beta_V}$$

Estimate the value of c_{l_β} for this aircraft. Also, evaluate the percentage of each of the different contributions with respect to the total c_{l_β} value.

Problem 4.6

Consider the data relative to the Aeritalia G91 aircraft in Appendix C. Use the provided drawings to extract all the relevant geometric characteristics of the aircraft.

Using the modeling approach outlined in this chapter, provide estimates for the following stability derivatives:

$$c_{Y_\beta}, c_{n_\beta}$$

For both aerodynamic parameters highlight the contribution of the vertical tail versus the total value.

Problem 4.7

Consider the data relative to the Aeritalia G91 aircraft in Appendix C. Use the provided drawings to extract all the relevant geometric characteristics of the aircraft.

Using the modeling approach outlined in this chapter, provide estimates for the following control derivatives:

- $c_{Y_{\delta_R}}$, $c_{l_{\delta_R}}$, $c_{n_{\delta_R}}$;
- $c_{l_{\delta_A}}$.

Problem 4.8

Consider the data relative to the Aeritalia G91 aircraft in Appendix C. Use the provided drawings to extract all the relevant geometric characteristics of the aircraft.

Using the modeling approach outlined in this chapter, provide estimates for the following stability derivatives:

- c_{Y_p}, c_{l_p}, c_{n_p};
- c_{Y_r}, c_{l_r}, c_{n_r}.

Problem 4.9

Consider the data relative to the Beech 99 aircraft in Appendixes B and C. Use the provided drawings to extract all the relevant geometric characteristics of the aircraft. Using the modeling approach outlined in this chapter, the dihedral effect c_{l_β} can be modeled as

$$c_{l\beta} = c_{l\beta_{WB}} + c_{l\beta_H} + c_{l\beta_V}$$

Estimate the value of c_{l_β} for this aircraft. Also, evaluate the percentage of each of the different contributions with respect to the total c_{l_β} value. Next, compare the obtained c_{l_β} value with the true value listed in Appendix B (Aircraft 3) and evaluate the errors associated with the use of the empirical modeling approach.

Problem 4.10

Consider the data relative to the Beech 99 aircraft in Appendixes B and C. Use the provided drawings to extract all the relevant geometric characteristics of the aircraft.

Using the modeling approach outlined in this chapter, provide estimates for the following stability derivatives:

$$c_{Y_\beta}, c_{n_\beta}$$

For both aerodynamic parameters highlight the contribution of the vertical tail versus the total value. Next, compare the obtained values with the true values listed in Appendix B (Aircraft 3) and evaluate the errors associated with the use of the empirical modeling approach.

Problem 4.11

Consider the data relative to the Beech 99 aircraft in Appendixes B and C. Use the provided drawings to extract all the relevant geometric characteristics of the aircraft.

Using the modeling approach outlined in this chapter, provide estimates for the following stability derivatives:

- c_{Y_p}, c_{l_p}, c_{n_p};
- c_{Y_r}, c_{l_r}, c_{n_r}.

Next, compare the obtained values with the true values listed in Appendix B (Aircraft 3) and evaluate the errors associated with the use of the empirical modeling approach.

Problem 4.12

Consider the data relative to the Cessna T37 aircraft in Appendixes B and C. Use the provided drawings to extract all the relevant geometric characteristics of the aircraft.

Using the modeling approach outlined in this chapter, the dihedral effect c_{l_β} can be modeled as

$$c_{l\beta} = c_{l\beta_{WB}} + c_{l\beta_H} + c_{l\beta_V}$$

Estimate the value of c_{l_β} for this aircraft. Also, evaluate the percentage of each of the different contributions with respect to the total c_{l_β} value. Next, compare the obtained c_{l_β} value with the true value listed in Appendix B (Aircraft 4) and evaluate the errors associated with the use of the empirical modeling approach.

Problem 4.13

Consider the data relative to the Cessna T37 aircraft in Appendixes B and C. Use the provided drawings to extract all the relevant geometric characteristics of the aircraft.

Using the modeling approach outlined in this chapter, provide estimates for the following stability derivatives:

$$c_{Y_\beta}, c_{n_\beta}$$

For both aerodynamic parameters highlight the contribution of the vertical tail versus the total value. Next, compare the obtained values with the true values listed in Appendix B (Aircraft 4) and evaluate the errors associated with the use of the empirical modeling approach.

Problem 4.14

Consider the data relative to the Cessna T37 aircraft in Appendixes B and C. Use the provided drawings to extract all the relevant geometric characteristics of the aircraft.

Using the modeling approach outlined in this chapter, provide estimates for the following control derivatives:

- $c_{Y_{\delta_R}}, c_{l_{\delta_R}}, c_{n_{\delta_R}}$;
- $c_{l_{\delta_A}}$.

Next, compare the obtained values with the true values listed in Appendix B (Aircraft 4) and evaluate the errors associated with the use of the empirical modeling approach.

Problem 4.15

Consider the data relative to the Cessna T37 aircraft in Appendixes B and C. Use the provided drawings to extract all the relevant geometric characteristics of the aircraft.

Using the modeling approach outlined in this chapter, provide estimates for the following stability derivatives:

- $c_{Y_p}, c_{l_p}, c_{n_p}$;
- $c_{Y_r}, c_{l_r}, c_{n_r}$.

Next, compare the obtained values with the true values listed in Appendix B (Aircraft 4) and evaluate the errors associated with the use of the empirical modeling approach.

Problem 4.16

Consider the data relative to the Boeing B747 200 aircraft in Appendixes B and C. Use the provided drawings to extract all the relevant geometric characteristics of the aircraft. Using the modeling approach outlined in this chapter, the so-called dihedral effect c_{l_β} can be modeled as

$$c_{l_\beta} = c_{l_{\beta_{WB}}} + c_{l_{\beta_H}} + c_{l_{\beta_V}}$$

Estimate the value of c_{l_β} for this aircraft. Also, evaluate the percentage of each of the different contributions with respect to the total c_{l_β} value. Next, compare the obtained c_{l_β} value with the true value listed in Appendix B (Aircraft 7) and evaluate the errors associated with the use of the empirical modeling approach.

Problem 4.17

Consider the data relative to the Boeing B747 200 aircraft in Appendixes B and C. Use the provided drawings to extract all the relevant geometric characteristics of the aircraft.

Using the modeling approach outlined in this chapter, provide estimates for the following stability derivatives:

$$c_{Y_\beta}, c_{n_\beta}$$

266 Chapter 4 Modeling of Lateral Directional Aerodynamic Forces and Moments

For both aerodynamic parameters highlight the contribution of the vertical tail versus the total value. Next, compare the obtained values with the true values listed in Appendix B (Aircraft 7) and evaluate the errors associated with the use of the empirical modeling approach.

Problem 4.18

Consider the data relative to the Boeing B747 200 aircraft in Appendixes B and C. Use the provided drawings to extract all the relevant geometric characteristics of the aircraft. Using the modeling approach outlined in this chapter, provide estimates for the following control derivatives:

- $c_{Y_{\delta_R}}, c_{l_{\delta_R}}, c_{n_{\delta_R}}$;
- $c_{l_{\delta_A}}$.

Next, compare the obtained values with the true values listed in Appendix B (Aircraft 7) and evaluate the errors associated with the use of the empirical modeling approach.

Problem 4.19

Consider the data relative to the Boeing B747 200 aircraft in Appendixes B and C. Use the provided drawings to extract all the relevant geometric characteristics of the aircraft. Using the modeling approach outlined in this chapter, provide estimates for the following stability derivatives:

- $c_{Y_p}, c_{l_p}, c_{n_p}$;
- $c_{Y_r}, c_{l_r}, c_{n_r}$.

Next, compare the obtained values with the true values listed in Appendix B (Aircraft 7) and evaluate the errors associated with the use of the empirical modeling approach.

Problem 4.20

Consider the data relative to the McDonnell Douglas DC-9 Series 10 aircraft in Appendix C. Use the provided drawings to extract all the relevant geometric characteristics of the aircraft. Student Sample Problem 4.1 has shown an estimate of the dihedral effect c_{l_β} for the DC-9 Series 10. Next, consider the later versions of the Mc Donnell Douglas DC-9 aircraft which were introduced later in its operational life.

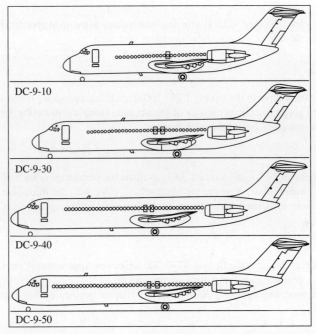

Figure P4.20.1 Lateral Views McDonnell Douglas DC 9 Series 10/30/40/50

(Source: http://www.aviastar.org/index2.html)

Reference[9] provides all the key geometric parameters for all the different versions of the DC-9 aircraft. The relevant geometric parameters for the lateral directional modeling were derived from[9] and are summarized in the following tables.

Table P4.20.1 Vertical Tail Dimensions

	DC-9-10	DC-9-30	DC-9-40	DC-9-50
b_{2V} [ft]	27.2	27.2	27.2	27.2
c_{T_V} [ft]	10.9	10.9	10.9	10.9
c_{R_V} [ft]	13.7	13.7	13.7	13.7
Λ_{LE_V} [rad]	0.785	0.785	0.785	0.785
Λ_{LE_V} [deg]	45	45	45	45

Table P4.20.2 Vertical Tail Geometric Parameters

	DC-9-10	DC-9-30	DC-9-40	DC-9-50
λ_V	0.796	0.796	0.796	0.796
S_{2V} [ft^2]	334.56	334.56	334.56	334.56
AR_V	2.211	2.211	2.211	2.211
$\bar{c}_V$ [ft]	12.353	12.353	12.353	12.353
y_{MAC_V} [ft]	6.542	6.542	6.542	6.542
x_{MAC_V} [ft]	6.542	6.542	6.542	6.542
$\Lambda_{0.5_V}$ [rad]	0.731	0.731	0.731	0.731

Table P4.20.3 Wing-Tail Geometric Parameters

	DC-9-10	DC-9-30	DC-9-40	DC-9-50
Z_{V_S} [ft]	12.142	12.142	12.142	12.142
Z_{R_S} [ft]	9.000	9.000	9.000	9.000
X_{V_S} [ft]	34.447	39.378	42.778	45.976
X_{R_S} [ft]	39.817	44.748	48.148	51.345

Using the data in the preceding tables, provide an estimate of c_{l_β} for the DC-9 Series 30, 40, and 50 aircraft.

Problem 4.21

Repeat Problem 4.20 for the estimates of the following stability derivative:

$$c_{n_\beta}$$

Problem 4.22

Repeat Problem 4.20 for the estimates of the following control derivatives:

$$c_{l_{\delta_R}}, c_{n_{\delta_R}}$$

Problem 4.23

Repeat Problem 4.20 for the estimates of the following stability derivatives:

$$c_{l_p}, c_{n_r}$$

Problem 4.24

Consider the data relative to the Rockwell B-1 aircraft in Appendix C. Use the provided drawings to extract all the relevant geometric characteristics of the aircraft. Using the modeling approach outlined in this chapter, provide estimates of the dihedral effect c_{l_β} in both configurations (minimum and maximum Λ_{LE}).

Chapter 5

Review of Basic Aircraft Performance and Modeling of Thrust Forces and Moments

TABLE OF CONTENTS

5.1 Introduction
5.2 Review of Different Aircraft Propulsion Systems
    5.2.1 Piston Engine (Propeller) Aircraft Engines
    5.2.2 Turboprop Aircraft Engines
    5.2.3 Turbojet Aircraft Engines
    5.2.4 Turbofan Aircraft Engines
    5.2.5 Ramjet Aircraft Engines
5.3 Review of Basic Aircraft Performance
5.4 Power at Level Flight
    5.4.1 Maximum Aerodynamic Efficiency
    5.4.2 Minimum Aerodynamic Drag
    5.4.3 Minimum Power Required
5.5 Determination of Power Required
5.6 Determination of Power Available
5.7 Modeling of the Thrust Forces and Moments
    5.7.1 Modeling of the Steady-State Thrust Forces and Moments
    5.7.2 Modeling of the Small Perturbation Thrust Forces and Moments
5.8 Summary
References
Student Sample Problems
Problems

5.1 INTRODUCTION

The performances of an aircraft depend on its weight, its aerodynamic characteristics, and the installed thrust in the power plant. The selection of the type of propulsion system is dictated mainly by the operational flight envelope of the given aircraft. This chapter will first provide a brief conceptual review of the different types of propulsion systems; next, a brief review of the key aircraft equations in the vertical plane is presented, followed by an analysis of the key aircraft propulsive performance. Finally, the modeling of the forces and moments associated with the thrust is described, with the goal of completing the modeling for all the forces and moments acting on the aircraft prior to the solution of the equations of motion to be discussed in Chapter 7.

5.2 REVIEW OF DIFFERENT AIRCRAFT PROPULSION SYSTEMS

The matching between an aircraft and its engine is a key design issue. There are major design implications associated with the selection of the power plant in terms of the overall aircraft aerodynamic behavior, structural strength, as well as inertial characteristics. In addition to providing enough power and thrust according to the design specifications, a propulsion system is required to be

- Reliable (at all different conditions in the operational envelope);
- Fuel efficient (for minimizing operational costs as well as maximizing range and minimizing take-off weight);
- Low maintenance (for minimizing operational costs and ground time).

Phillips[1] provides an interesting and enjoyable review of the history of the development of different propulsion systems, along with advances in aerodynamics leading to exponentially increasing performance of the aircraft systems in the 20th century. A classification of the different propulsion systems along with a brief description of their key characteristics follows.[2]

5.2.1 Piston Engine (Propeller) Aircraft Engines

The combination of a piston engine (also known as reciprocating engine) with a multiblade propeller was the first type of propulsion system; to date, it is still the most common type of aircraft propulsion. Phillips[1] provides an excellent review of the propeller blade theory. A simple conceptual diagram of the system is shown in Figure 5.1.

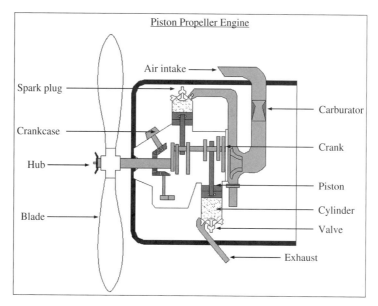

Figure 5.1 Simplified Diagram of a Piston Propeller Aircraft Engine

The reciprocating engine is the most efficient propulsion system in terms of fuel consumption, which is typically expressed using lbs. of fuel per hour per brake horsepower.[1,2] Brake horsepower is here referred to as the value of the engine horsepower without the power loss due to the propeller. This propulsion system is extensively used for aircraft with maximum operational speed up to Mach ∼ 0.3; at higher speeds the compressibility effects on the blades lead to excessive loss of the efficiency of the propeller. The historical development of this propulsion system started with "in-line" engines, in which multiple cylinders are lined up along one row aligned with the longitudinal axis of the aircraft with the goal of minimizing the cross section for drag reduction purposes. Within in-line engines the most common configuration is the inverted in-line configuration in which the engine crankshaft is located above the cylinders; this configuration allows larger propellers because it increases the ground clearance.

Although appropriate for small-size engines, the design of in-line piston engines has two inherent drawbacks. One drawback is that the crankcase and the crankshaft are fairly long and heavy, leading to high fuel consumption and low power/weight ratios. Another drawback is the inherent difficulty in cooling the rear cylinders of the engine, which are not exposed to the airflow, leading to the need of liquid cooling and therefore additional weight and complexity. Nevertheless, despite the drawbacks, this propulsion system has allowed the birth of aviation.

As the size, the speed, and the weight of aircraft increased, the in-line design was abandoned and other design configurations were adopted. One configuration is the V engines, in which the cylinders are located in two in-line groups with a tilt angle of either 30 or 60 degrees. Due to the need of cooling the rear cylinders, those engines were liquid-cooled. With respect to the in-line design, the V design allowed better power/weight ratios while maintaining the advantage of minimizing the cross-sectional area for drag purposes.

Another design approach is the rotary design, in which the crankshaft remains stationary (attached to the aircraft structure) while the entire block of cylinders rotates around the crankshaft with the propeller bolted in front of the crankcase. This engine design had a number of advantages, including a fairly high power/weight ratios (due to the fact that there are no reciprocating parts, and the rotating masses of the cylinders acted as a flywheel, eliminating the need for a flywheel) and no need for cooling systems (because the rotating cylinder block created its own cooling airflow even at rest or low airspeeds). The main drawbacks of this design were the difficulty by the pilots in providing appropriate fuel/air mixtures for different throttling at the different flight conditions and its limitation on breathing capacity, which ultimately capped its torque performance. Additionally, these engines had fairly high fuel consumption and created substantial gyroscopic effects (due to the rotating mass in one direction only) with undesirable consequences on the aircraft dynamics in the high angular rates conditions associated with duel combat situations.

Perhaps the most common design configuration for aircraft piston propeller systems is the radial configuration. Rotary and radial aircraft engines look very similar when they are not running, because both have cylinders arranged radially around a central crankshaft. However, unlike the rotary engines, radial engines use a conventional rotating crankshaft in a fixed engine block. In a radial engine, the pistons are connected to the crankshaft with a master-and-articulating rod assembly. Typically, only one piston has a master rod with a direct attachment to the crankshaft, while the remaining pistons from the other cylinders have the connecting rods pinned to rings around the edge of the master rod. The schematic of a radial engine is shown in Figure 5.2.

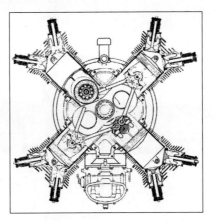

Figure 5.2 Caminez 447 Engine
(Source: http://www.aqpl43.dsl.pipex.com)

The main advantages of radial engines were their simplicity (there was no need for liquid cooling), their reliability, and their lower weight as opposed to the heavier in-line engines. The main drawback was their large cross-section, producing substantial aerodynamic drag (later somewhat alleviated by aerodynamically shaped engine cowlings) and low visibility for the pilots, especially critical for landing on carriers.

Radial piston engines powered several aircraft (including a number of World War II aircraft) and represented the most common design approach until the introduction of turbine engines.

5.2.2 Turboprop Aircraft Engines

A turboprop engine is essentially a propulsion system featuring a gas turbine driving a propeller. Specifically, the output of the turbine is used to drive the propeller with the engine exhaust gas containing limited energy, compared to a jet engine. As shown in Figure 5.3, a turboprop engine consists of an intake, a compressor, a combustor, a turbine, and a nozzle.

As the airflow enters the engine, it is somewhat compressed by the ram effect; however, most of the compression is performed in the compressor. Fuel is mixed and burned in the combustor while the hot gases turn the turbine to drive the compressor and the propeller. The propeller is connected with the turbine through a reduction gear converting the high RPM and low torque outputs of the turbine to low RPM and high torque.

The main advantages of turboprop engines are their reliability and their high power/weight ratios due to the large airflow through the gas turbine engine. Also, the power of turboprop engines increases somewhat with airspeed (up to Mach ~ 0.6) due to the ram effect mentioned previously. The main drawback of turboprop engines is their high price, which makes them suitable when specific performances are required (such as for short take-off and landing [STOL] aircraft) or when their high reliability and low maintenance justify high acquisition costs (such as for commercial commuter aircraft).

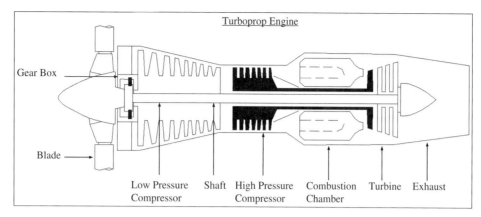

Figure 5.3 Schematic of a Turboprop Engine

5.2.3 Turbojet Aircraft Engines

As the aircraft airspeed increased during the 1930s and 1940s, a clear need originated for a propulsion system with higher efficiency for airspeeds, which were approaching the Mach ~ 1 condition. The turbojet is the first type of aircraft jet engines. The turbojet is a gas turbine in which the thrust force is produced by the velocity of the exhaust gas. Specifically, the airflow enters the rotating compressor through the intake and is compressed by a compressor before it enters into the combustion chamber. In the chamber, the fuel is mixed with the airflow and ignited. The high-temperature gases leaving the combustion chamber expand in the turbine, where a small amount of power is extracted to drive the compressor. The high-pressure exhaust gases exiting the turbine are accelerated through the nozzle, producing a high-velocity reaction jet, which ultimately creates a forward thrust force. The schematic of a turbojet engine is shown in Figure 5.4.

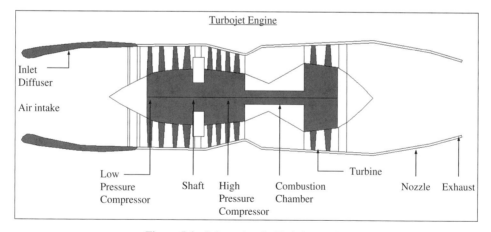

Figure 5.4 Schematic of a Turbojet Engine

Early generations of turbojet engines featured centrifugal compressors, later replaced by axial compressors to reduce the overall diameters of the engine housing and the weight of the system. The most significant design limitation in the design of a turbojet is the maximum temperature allowed in the turbine, which ultimately depends on the characteristics of the material used for the manufacturing of the turbine blades. Unlike piston engines, fuel consumption for turbojet engines is measured using lbs. of fuel per hour per lbs. of thrust; this parameter decreases with increasing compressor pressure ratio.

For supersonic aircraft, turbojets are augmented with afterburner systems. Afterburners are essentially extended ducts at the end of a turbojet. In the afterburner the hot exhaust gases exiting the turbine are mixed with additional fuel and ignited again before exiting through the nozzle. Since afterburners are used only at transonic and supersonic speeds, variable area nozzles are required for the different operating conditions of the turbojet. While reducing the overall efficiency and increasing substantially the fuel consumption, afterburners provide substantial additional thrust (up to 100 percent more thrust) and allow airspeeds up to Mach $\sim [2.5-3.0]$.

272 Chapter 5 Review of Basic Aircraft Performance and Modeling of Thrust Forces and Moments

5.2.4 Turbofan Aircraft Engines

Turbofan engines are the propulsion systems for modern commercial aviation. A turbofan engine is essentially a hybrid between a turbojet and a ducted turboprop. In a turbofan a percentage of the airflow (referred to in the literature as the secondary airflow) only goes through the fan and exits the engine, providing a certain level of thrust. The rest of the airflow (the primary airflow) goes through the entire system as in a turbojet. The ratio of the secondary flow rate over the primary flow rate is known as the by-pass ratio (BPR); using this parameter, turbofan engines are therefore classified into high BPR and low BPR. High BPR turbofan engines (up to 10:1) are more fuel efficient; however, they require larger cross-sectional areas due to the need for large fans. The size of this cross-sectional areas essentially limits the use of high BPR turbofans to high subsonic airspeeds (essentially the airspeed of commercial jet liners); on the other side, low BRP turbofans, although less fuel efficient, are suitable for aircraft with supersonic speed. The schematic of high BPR and low BPR are shown in Figures 5.5 and 5.6, respectively.

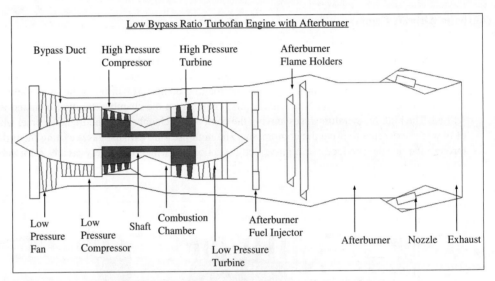

Figure 5.5 Schematic of a Low BPR Turbofan Engine

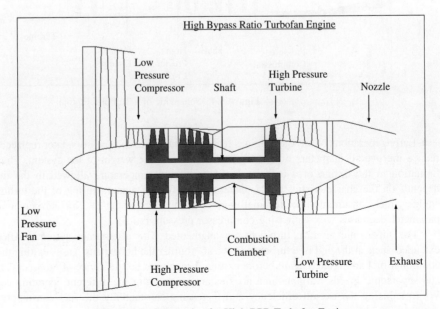

Figure 5.6 Schematic of a High BPR Turbofan Engine

5.2.5 Ramjet Aircraft Engines

At supersonic speed the airflow entering the engine intake is fast enough to produce enough compression, eliminating the need for a compressor. The airflow is then mixed to the fuel sprayed through a circular ring and ignited to produce hot exhaust gases, which expand through the nozzle. The main advantage of a ramjet is its simplicity, since it requires neither a turbine nor a compressor. The obvious disadvantage is that the system requires a very high speed before providing thrust. Therefore, this type of propulsion system is mainly used on missiles, where rocket engines provide the necessary thrust to reach airspeeds at which a ramjet can start to operate (Mach ~ 3). The schematic of a ramjet engine is shown in Figure 5.7.

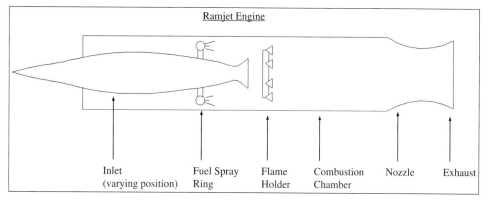

Figure 5.7 Schematic of a Ramjet Engine

5.3 REVIEW OF BASIC AIRCRAFT PERFORMANCE

The previous section provided a brief review of different aircraft propulsion systems. Clearly, the interface of the propulsion system with the rest of the aircraft has major aerodynamic, structural, and inertial implications. Nevertheless, within this context, the presence of a propulsion system simply implies the existence of a thrust vector (or a set of thrust vectors for multiengine aircraft), which ultimately dictates all the aircraft dynamic performance.

For this purpose, consider the motion of an aircraft in the vertical plane, as shown in Figure 5.8.

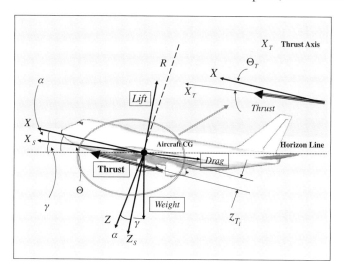

Figure 5.8 Forces Acting on Aircraft on Vertical Plane

The forces acting on the aircraft are

- Drag, aligned with the airspeed but in the opposite direction;
- Lift, perpendicular to the drag;
- Weight, pointing toward the center of the earth (parallel to the Z' axis);
- Thrust, with an angle Θ_T with respect to the flight path.

The angle γ is referred to as the flight path angle; it is also referred to as the climb or angle depending on its sign. The pitch angle θ is given by

$$\theta = \alpha + \gamma$$

Finally, R is the radius of curvature of the flight path in the vertical plane.

The equations invoking equilibrium of the forces along the directions of the drag and the lift (which are essentially the simplified form of two of the Conservation of Linear Momentum Equations [CLMEs], discussed in Chapter 1) are given by

$$T\cos\Theta_T - D - W\sin\gamma = \frac{W}{g}\frac{dV}{dt}$$

$$T\sin\Theta_T + L - W\cos\gamma = \frac{W}{g}\frac{V^2}{R}$$

The preceding general equations can take on specific formats under specific conditions. For example, assuming that the installed thrust is aligned with the direction of the airspeed (that is, $\Theta_T \approx 0$, $\cos\Theta_T \approx 1$, $\sin\Theta_T \approx 0$), we have

$$T - D - W\sin\gamma = \frac{W}{g}\frac{dV}{dt}$$

$$L - W\cos\gamma = \frac{W}{g}\frac{V^2}{R}$$

Additionally, if the aircraft is flying along a rectilinear flight path, we have $R \to \infty$, $\frac{dV}{dt} \approx 0$, leading to the following equations:

$$T - D - W\sin\gamma = 0$$

$$L - W\cos\gamma = 0$$

Additionally, consider the Flight Path Equations (FPEs) introduced in Chapter 1; considering the direction of the flight path and its perpendicular direction, the equations of the distances along those directions with respect to the ground (that is, the first and third of the FPEs) are given by

$$\dot{x} = V\cos\gamma$$

$$\dot{h} = V\sin\gamma$$

The next sections focus on basic aircraft performance, followed by a detailed analysis of two specific propulsion-related performances: power required and power available. A detailed review of all different aircraft performance is provided in.[3]

5.4 POWER AT LEVEL FLIGHT

Consider an aircraft on a level flight path ($\gamma \approx 0$); therefore, the above CLMEs are simply given by

$$T - D = 0 \to T = D$$

$$L - W = 0 \to L = W$$

Next, consider the generic relationships for drag and lift introduced in Chapter 3:

$$D = \frac{1}{2}\rho V^2 S c_D, \quad L = \frac{1}{2}\rho V^2 S c_L$$

Using the Prandtl relationship for c_D we have

$$c_D = c_{D_O} + \frac{c_L^2}{\pi ARe}$$

where $c_{D_O} = c_D|_{c_L=0}$ is known as the parasite drag and e is known as the wing efficiency factor or Oswald efficiency factor (as discussed in Chapter 2, Section 2.4).

5.4.1 Maximum Aerodynamic Efficiency

Given the definition of Aerodynamic Efficiency

$$E = \frac{L}{D} = \frac{\frac{1}{2}\rho V^2 S c_L}{\frac{1}{2}\rho V^2 S c_D} = \frac{c_L}{c_D}$$

This parameter can be maximized using the following inverse relationship:

$$E_{Max} = \left.\frac{L}{D}\right|_{Max} = \left.\frac{D}{L}\right|_{Min} = \left.\frac{c_D}{c_L}\right|_{Min}$$

Of specific interest is the value of the lift coefficient c_L associated with the condition of maximum efficiency; this value can be found using the relationship

$$\frac{d}{dc_L}\left(\frac{c_D}{c_L}\right) = \frac{d}{dc_L}\left(\frac{c_{D_O} + \frac{c_L^2}{\pi ARe}}{c_L}\right) = 0$$

Therefore, we have

$$\frac{d}{dc_L}\left(\frac{c_{D_O} + \frac{c_L^2}{\pi ARe}}{c_L}\right) = \frac{\left(\frac{2c_L^2}{\pi ARe} - \left(c_{D_O} + \frac{c_L^2}{\pi ARe}\right)\right)}{c_L^2} = \frac{\left(\frac{c_L^2}{\pi ARe} - c_{D_O}\right)}{c_L^2} = 0$$

$$\left(\frac{c_L^2}{\pi ARe} - c_{D_O}\right) = 0 \longrightarrow c_L^2 = c_{D_O}(\pi ARe)$$

leading to

$$c_L|_{E_{Max}} = \sqrt{c_{D_O}(\pi ARe)}$$

The associated maximum value of the aerodynamic efficiency is given by

$$E_{Max} = \left.\frac{c_D}{c_L}\right|_{Min} = \left.\frac{c_{D_O} + \frac{c_{D_O}(\pi ARe)}{(\pi ARe)}}{\sqrt{c_{D_O}(\pi ARe)}}\right|_{Min} = \frac{2c_{D_O}}{\sqrt{c_{D_O}(\pi ARe)}} = 2\sqrt{\frac{c_{D_O}}{(\pi ARe)}}$$

5.4.2 Minimum Aerodynamic Drag

Consider again the simplified CLMEs at level flight:

$$T = D$$
$$L = W$$

Solving for c_L we have

$$\frac{1}{2}\rho V^2 S c_L = W \rightarrow c_L = \frac{2W}{\rho V^2 S}, \quad c_L^2 = \frac{4W^2}{\rho^2 V^4 S^2}$$

Therefore, we have

$$D = \frac{1}{2}\rho V^2 S \left(c_{D_O} + \frac{c_L^2}{\pi A R e} \right) = \frac{1}{2}\rho V^2 S \left(c_{D_O} + \frac{1}{(\pi A R e)} \cdot \frac{4W^2}{\rho^2 V^4 S^2} \right)$$

$$= \frac{1}{2}\rho V^2 S c_{D_O} + \frac{1}{(\pi A R e)} \cdot \frac{2W^2}{\rho V^2 S} = D_{Parasite} + D_{Induced}$$

Thus, the total aircraft drag is the sum of the parasite drag and the induced drag. The induced drag can be considered as the penalty associated with the generation of lift. The parasite drag is proportional to the square of the airspeed whereas the induced drag is inversely proportional to the square of the airspeed. Of specific interest is the value of the speed at which the total aircraft drag is minimized, also referred to as $V|_{D_{MINIMUM}}$. This value can be found by setting the derivative $dD/dV = 0$ and solving for the associated value of V. Therefore, we have

$$\frac{dD}{dV} = \frac{d}{dV}\left(\frac{1}{2}\rho V^2 S c_{D_O} + \frac{1}{(\pi A R e)} \cdot \frac{2W^2}{\rho V^2 S} \right) = \rho V S c_{D_O} - \frac{1}{(\pi A R e)} \cdot \frac{4W^2}{\rho V^3 S} = 0$$

leading to:

$$\rho V S c_{D_O} = \frac{1}{(\pi A R e)} \cdot \frac{4W^2}{\rho V^3 S} \rightarrow \rho S c_{D_O} = \frac{1}{(\pi A R e)} \cdot \frac{4W^2}{\rho V^4 S}$$

and finally to

$$V|_{D_{Minimum}}^4 = \frac{1}{(\pi A R e)(\rho S c_{D_O})} \frac{4W^2}{\rho S} \rightarrow V|_{D_{Minimum}}^2 = \frac{2W}{\rho S} \cdot \sqrt{\frac{1}{(\pi A R e) c_{D_O}}}$$

$$\rightarrow V|_{D_{Minimum}} = \sqrt{\frac{2W}{\rho S}} \cdot \sqrt[4]{\frac{1}{(\pi A R e) c_{D_O}}}$$

The trend of the total aircraft drag versus airspeed is shown in Figure 5.9.

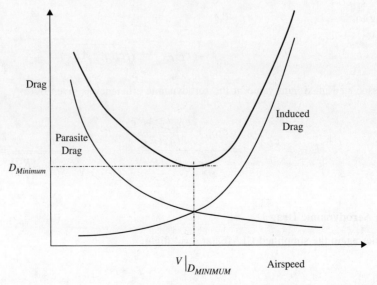

Figure 5.9 Trend of the Aircraft Drag (Parasite + Induced) vs. Airspeed

From Figure 5.9 it can be shown (see Student Sample Problem 5.1) that $V|_{D_{MINIMUM}}$ is associated with the condition:

$$D_{Parasite}\Big|_{V|_{D_{MINIMUM}}} = D_{Induced}\Big|_{V|_{D_{MINIMUM}}}$$

Therefore, we have

$$D_{Minimum} = 2 \cdot D_{Parasite}\Big|_{V|_{D_{MINIMUM}}} = 2 \cdot D_{Induced}\Big|_{V|_{D_{MINIMUM}}}$$

It can also be shown (see Student Sample Problem 5.1) that

$$D_{Minimum} = 2W\sqrt{\frac{c_{D_0}}{(\pi ARe)}}$$

It is interesting to point out that at low subsonic speed, the aerodynamic coefficient c_{D_0} can be considered to be constant; however, c_{D_0} substantially increases at high subsonic and even more at supersonic conditions due to the generation of the drag associated with the shock waves.

5.4.3 Minimum Power Required

The power required for sustaining level flight is given by

$$P_{required} = T \cdot V = D \cdot V = \frac{1}{2}\rho V^3 S c_{D_O} + \frac{1}{(\pi ARe)} \cdot \frac{2W^2}{\rho VS}$$

$$= P_{required}\Big|_{\substack{Parasite \\ Drag}} + P_{required}\Big|_{\substack{Induced \\ Drag}}$$

As performed previously for the drag, it is important to find the airspeed at which the required power for level flight is minimized. This value can be found through setting

$$\frac{dP_{required}}{dV} = \frac{d}{dV}\left(\frac{1}{2}\rho V^3 S c_{D_O} + \frac{1}{(\pi ARe)} \cdot \frac{2W^2}{\rho VS}\right) = 0$$

and solving for the associated value $V|_{P_{REQUIRED_{Minimum}}}$.

It can be shown (see Student Sample Problem 5.2) that

$$V|_{P_{REQUIRED_{Minimum}}} = \frac{1}{\sqrt[4]{3}}\sqrt{\frac{2W}{\rho S}} \cdot \sqrt[4]{\frac{1}{(\pi ARe)c_{D_O}}} = \frac{1}{\sqrt[4]{3}}V|_{D_{Minimum}} \approx 0.758 V|_{D_{Minimum}}$$

The general trend for the power required is shown in Figure 5.10. It is interesting to point out the difference between $P_{required_{Minimum}}$ and $P_{required_{DRAGMinimum}}$, which are associated with the values of $V|_{P_{REQUIRED_{Minimum}}}$ and $V|_{D_{Minimum}}$, respectively.

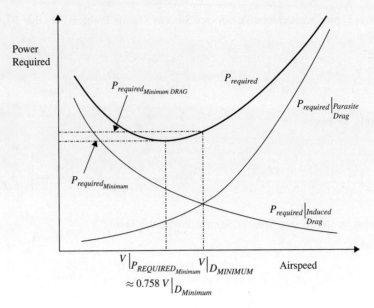

Figure 5.10 Typical Trend of Power Required vs. Airspeed

The value of the lift coefficient associated with the minimum power required condition is given by

$$c_L\big|_{P_{REQUIRED_{Minimum}}} = \frac{2W}{\rho V\big|^2_{P_{REQUIRED_{Minimum}}} S}$$

Using the relationship for $V\big|^2_{P_{REQUIRED_{Minimum}}}$ (see Student Sample Problem 5.2)

$$V\big|^2_{P_{REQUIRED_{Minimum}}} = \frac{2W}{\rho S}\sqrt{\frac{1}{3(\pi ARe)c_{D_O}}}$$

we have

$$c_L\big|_{P_{REQUIRED_{Minimum}}} = \frac{2W}{\rho \dfrac{2W}{\rho S}\sqrt{\dfrac{1}{3(\pi ARe)c_{D_O}}} S} = \frac{\cancel{2W}}{\cancel{\rho}\dfrac{\cancel{2W}}{\cancel{\rho S}}\sqrt{\dfrac{1}{3(\pi ARe)c_{D_O}}} S}$$

$$= \sqrt{\frac{1}{3(\pi ARe)c_{D_O}}} = \sqrt{3(\pi ARe)c_{D_O}}$$

Recall that the lift coefficient in the condition of maximum aerodynamic efficiency is given by

$$c_L\big|_{E_{Max}} = \sqrt{c_{D_O}(\pi ARe)}$$

Therefore, we have

$$c_L\big|_{P_{REQUIRED_{Minimum}}} = \sqrt{3(\pi ARe)c_{D_O}} = \sqrt{3}\, c_L\big|_{E_{Max}}$$

5.5 DETERMINATION OF POWER REQUIRED

Assuming the availability of engine performance charts from the engine manufacturer, a specific approach allows the determination of the power required for a propeller aircraft using flight data. Additionally, two important aerodynamic parameters—the parasite drag coefficient c_{D_O} and the Oswald efficiency factor e—can also be evaluated through this approach.

Starting again from

$$P_{Required} = T \cdot V = D \cdot V = \frac{1}{2}\rho V^3 S c_{D_O} + \frac{1}{(\pi AR e)} \cdot \frac{2W^2}{\rho V S}$$

The first step consists of expressing the relationship for power in terms of horsepower (HP) using the conversion factor:

$$P_{HP_{Required}} = \frac{D \cdot V}{550}$$

Therefore, in terms of HP we have

$$P_{HP\ Required} = \frac{1}{1100}\rho V^3 S c_{D_O} + \frac{4}{1100} \frac{1}{(\pi AR e)} \cdot \frac{W^2}{\rho V S}$$

Assuming the availability of $P_{HP_{Required}}$ and the measurements of the (true) airspeed $V = V_{True}$, the preceding relationship does not allow the determination of the aerodynamic coefficients (c_{D_O}, e) since it is a function of aircraft weight and altitude (through the density). Further, it would not be possible to extract multiple aerodynamic coefficients from a single relationship. Thus, a critical task is the evaluation of the coefficients c_{D_O}, e assuming that the engine charts are available from the engine manufacturer.

A first conceptual challenge is to properly address the dependency from both altitude and weight. An approach to deal with this problem is given by the Equivalent Weight (EW) approach, described in.[4] Within this approach the first parameter is the Equivalent Weight Velocity (EWV), defined as

$$V_{EW} = V\sqrt{\frac{W_0}{W} \cdot \frac{\rho}{\rho_0}}$$

where W_0 can be selected to be the maximum take-off weight for the specific aircraft while ρ_0 is the sea level density. The procedure outlined allows the determination of the coefficients c_{D_O}, e using the EW approach.

Using again:

$$P_{HP_{Required}} = \frac{D \cdot V}{550} = \frac{\rho V^3 S c_D}{1100}$$

Using the expression for V from the relationship $L = W$:

$$\frac{1}{2}\rho V^2 S c_L = W \rightarrow V = \sqrt{\frac{2W}{\rho S c_L}}, \quad V^3 = \left(\frac{2W}{\rho S c_L}\right)\sqrt{\frac{2W}{\rho S c_L}}$$

leading to

$$P_{HP_{Required}} = \frac{\rho S c_D}{1100} V^3 = \frac{\rho S c_D}{1100} \left(\frac{2W}{\rho S c_L}\right)\sqrt{\frac{2W}{\rho S c_L}}$$

$$= \frac{\rho S c_D}{1100} \left(\frac{2W}{\rho S c_L}\right)\sqrt{\frac{2W}{\rho S c_L}} = \frac{1}{550}\sqrt{\frac{2W^3 c_D^3}{\rho S c_L^3}}$$

Next, the EW power required $P_{HP_{Required_{EW}}}$ is defined as

$$P_{HP_{Required_{EW}}} = \frac{1}{550}\sqrt{\frac{2W_0^3 c_D^2}{\rho_0 S c_L^3}}$$

leading to

$$P_{HP_{Required_{EW}}} = P_{HP_{Required}}\sqrt{\frac{W_0^3 \rho}{W^3 \rho_0}}$$

where $P_{HP_{Required}}$ is given by

$$P_{HP_{Required}} = \frac{1}{1100}\rho V^3 S c_{D_O} + \frac{4}{1100}\frac{1}{(\pi AR e)} \cdot \frac{W^2}{\rho V S}$$

Next, the brake horsepower is here introduced as

$$B_{HP} = \frac{P_{HP}}{\eta_P}$$

where η_P is known as the propeller efficiency.[1] This parameter is a function of the geometric parameters as well as the operating conditions of the propeller. Harmann and Bierman[5] provide an approach for the experimental evaluation of η_P. Using the preceding definitions for EW, it can be shown (see Student Sample Problem 5.3) that

$$V_{EW} \cdot B_{HP_{Required_{EW}}} = \frac{1}{1100}\frac{S c_{D_O} \rho_0}{\eta_P} \cdot V_{EW}^4 + \frac{4}{1100}\frac{1}{(\pi AR e)S\eta_P} \cdot \frac{W_0^2}{\rho_0}$$

This relationship is very important for graphical interpretation purposes. In fact, $V_{EW} \cdot B_{HP_{Required_{EW}}}$ can be rewritten as

$$V_{EW} \cdot B_{HP_{Required_{EW}}} = c_1 \cdot V_{EW}^4 + c_2$$

where $c_1 = \frac{1}{1100}\frac{S c_{D_O} \rho_0}{\eta_P}$, $c_2 = \frac{4}{1100}\frac{1}{(\pi AR e)S\eta_P} \cdot \frac{W_0^2}{\rho_0}$.

The relationship can be set to

$$y = c_1 \cdot x + c_2$$

where

$$y = V_{EW} \cdot B_{HP_{Required_{EW}}}, \quad x = V_{EW}^4$$

Therefore, imposing a linear interpolation for the flight data, the following aerodynamic parameters can be evaluated:

$$c_{D_O} = \frac{1100 \, c_1 \eta_P}{S\rho_0}$$

$$e = \frac{4}{1100}\frac{1}{(\pi AR c_2)S\eta_P} \cdot \frac{W_0^2}{\rho_0}$$

Thus, the introduction of the EW approach allows the determination of the coefficients c_{D_O}, e through a simple linear relationship between $V_{EW} \cdot B_{HP_{Required_{EW}}}$ and V_{EW}.

A sample of the determination of c_{D_O}, e from flight data is shown in Figure 5.12 using the WVU Cessna 206 aircraft. A set of drawings for this aircraft are shown in Figure 5.11.

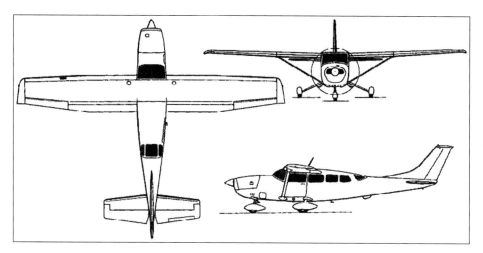

Figure 5.11 3D View of Cessna U206
(Source: http://www.aviastar.org/index2.html)

The value $\eta_P = 0.725$ was calculated for this aircraft using the method outlined in.[5] Note that for the Cessna 206 the values for the wing surface and the wing aspect ratio[4] are $S = 175.5 \text{ ft}^2$, $AR = 7.62$, respectively.

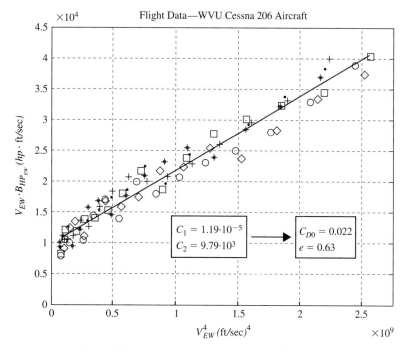

Figure 5.12 Determination of c_{D_O}, e from Flight Data

Once c_{D_O}, e are experimentally evaluated using the EW approach, for a known value of η_P, the brake horsepower required for level flight can be evaluated using

$$B_{HP_{Required}} = \frac{1}{1100} \frac{\rho V^3 S c_{D_O}}{\eta_P} + \frac{4}{1100} \frac{1}{(\pi AR e)} \cdot \frac{W^2}{\rho V S} \frac{1}{\eta_P}$$

From the preceding relationship $B_{HP_{Required}}$ is primarily a function of weight and altitude (through the air density ρ). Figures 5.13 and 5.14 show the trend for $B_{HP_{Required}}$ for the WVU Cessna 206 aircraft for different weight and altitudes.

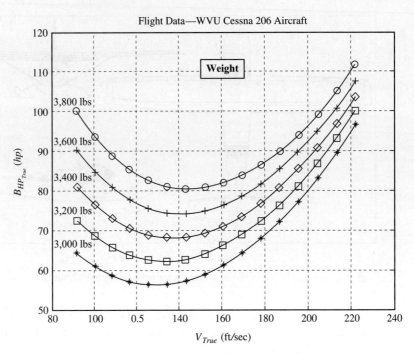

Figure 5.13 $B_{HP_{Required}}$ for Different Weight Configurations

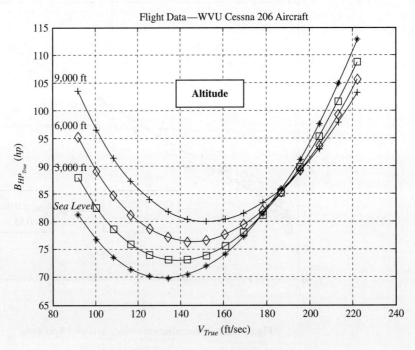

Figure 5.14 $B_{HP_{Required}}$ for Different Altitudes

As expected, the minimum value of $B_{HP_{Required}}$ is a function of the specific flight condition (in terms of weight and altitude).

5.6 DETERMINATION OF POWER AVAILABLE

Following the determination of the power required for level flight, another critical propulsive parameter for a given propulsion system is the power available, the total amount of power and thrust being provided by the propulsion system at the given flight condition. The detailed evaluation of this parameter will then allow the determination of additional aircraft performance parameters, such as excess power, best rate of climb, and maximum climb angle for an aircraft.

A general approach for the determination of the power available (PA) is based on the total energy (TE) method.

The TE method focuses on the analysis of the total energy of the aircraft system. Starting from general terms, TE can be considered to be the sum of the potential energy (PE) and the kinetic energy (KE).

$$TE = PE + KE$$

where the potential energy can be in general terms expressed as

$$PE = mgh = Wh$$

while the kinetic energy can be expressed as

$$KE = \frac{1}{2}mV^2 = \frac{1}{2}\frac{W}{g}V^2$$

leading to the general relationship

$$TE = Wh + \frac{1}{2}\frac{W}{g}V^2$$

Next, the change in total energy over time for a given climbing maneuver is expressed as

$$\frac{d(TE)}{dt} = h\frac{dW}{dt} + W\frac{dh}{dt} + \frac{1}{2}\frac{V^2}{g}\frac{dW}{dt} + \frac{W}{g}V\frac{dV}{dt}$$

A critical assumption is made such that the change in the aircraft weight over a small amount of time can be considered negligible. This leads to the simplification

$$\frac{d(TE)}{dt} = h\cancel{\frac{dW}{dt}} + W\frac{dh}{dt} + \cancel{\frac{1}{2}\frac{V^2}{g}\frac{dW}{dt}} + \frac{W}{g}V\frac{dV}{dt}$$

$$= W\frac{dh}{dt} + \frac{W}{g}V\frac{dV}{dt}$$

Next, simplified relationships are needed for the aircraft acceleration (dV/dt) and for the aircraft rate of climb (dh/dt). A sketch of the acting forces (assuming no moment arms associated with the thrust) is shown in Figure 5.15, where γ is the aircraft climb angle.

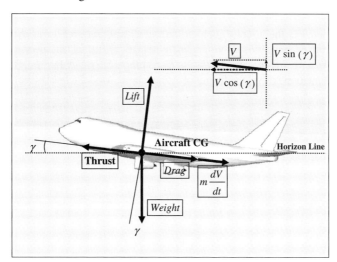

Figure 5.15 Simplified Sketch of the Longitudinal Aerodynamic and Thrust Forces

From Figure 5.15, an expression for (dV/dt) can be found from the balance of forces acting on the longitudinal axis using

$$\frac{dV}{dt} = \frac{g}{W}(T - D - W\sin(\gamma))$$

Next, the aircraft rate of climb can be generically expressed as

$$\frac{dh}{dt} = V\sin(\gamma)$$

Using the preceding relationships, we have

$$\frac{d(TE)}{dt} = W\frac{dh}{dt} + \frac{W}{g}V\frac{dV}{dt} = WV\sin(\gamma) + \frac{W}{g}V\frac{g}{W}(T - D - W\sin(\gamma))$$

$$= WV\sin(\gamma) + V(T - D - W\sin(\gamma)) = VT - VD$$

Given the basic definition of power required (P_R)

$$P_{Required} = VD$$

at maximum throttle setting we have the power available (P_A) defined as

$$P_{Available} = VT$$

Therefore, the following introduction of excess power (P_E) can be introduced:

$$P_{Excess} = P_{Available} - P_{Required} = \frac{d(TE)}{dt} = W\frac{dh}{dt} + \frac{W}{g}V\frac{dV}{dt}$$

The experimental evaluation of the P_E using the total energy method requires one term associated with the aircraft acceleration (dV/dt) and one term associated with the aircraft rate of climb (dh/dt). Therefore, conceptually there are two potential simplifications. The first option is to perform flight experiments at maximum throttle setting at constant altitude ($dh/dt \approx 0$). However, this is virtually impossible because during the acceleration at maximum throttle setting, the aircraft will exhibit a natural tendency to pitch up and climb, due to the increased production of lift. Therefore, a more realistic flight condition to evaluate P_A and P_E is through a set of steady-state climbs at constant airspeed ($dV/dt \approx 0$); under these conditions the previous relationship reduces to

$$P_{Excess} = W\frac{dh}{dt} + \frac{W}{g}V\frac{dV}{dt} = W\frac{dh}{dt}$$

Next, note that h is the aircraft geometric altitude, also referred to as the true altitude. This parameter is not measured through conventional aircraft instrumentation; in fact, the pressure altitude H_P is the only observable altitude parameter being measured. However, an approach for evaluating ($dh/dt \approx 0$) is provided by the use of the Hydrostatic Equation within the Standard Atmospheric Model (SAM) (see Appendix A). In fact, first we can set

$$\frac{dh}{dt} = \frac{dH_P}{dt} \cdot \frac{dh}{dH_P}$$

Next, at standard and actual atmospheric conditions the hydrostatic relationship takes on the following forms:

$$dp = -\rho_S g dH_P, \quad dp = -\rho_T g dh$$

leading to

$$-\rho_S g dH_P = -\rho_{True} g \, dh \rightarrow \frac{dh}{dH_P} = \frac{\rho_S}{\rho_{True}}$$

The air density cannot be commonly measurable by the on-board instrumentation. However, the true air temperature T_{True} is typically measured. Therefore, from the Equation of State in the SAM (see Appendix A), we have

$$\rho_S = \frac{P}{RT_S}, \rho_{True} = \frac{P}{RT_{True}} \rightarrow \frac{\rho_S}{\rho_{True}} = \frac{T_{True}}{T_S}$$

Finally, introducing the rate of climb as $RC = \Delta(Altitude)/\Delta(Time)$, we have

$$(RC)_{True} = \frac{dh}{dt} = \frac{dH_P}{dt} \cdot \frac{dh}{dH_P} = (RC)_{Obs} \frac{T_{True}}{T_S}$$

$$\rightarrow P_E = W(RC)_{True} = W(RC)_{Obs} \frac{T_{True}}{T_S} = P_A - P_R$$

The preceding relationship is very generic in nature and does not account for different take-off weights as well as weight variations due to fuel consumption. An approach for a comprehensive solution is to evaluate the rate of climb parameter $(RC)_{True}$ over all the weight configurations of a given aircraft, as well as to account for the changes in weight due to the fuel consumption.[4] Starting from a generic i-th flight condition, the resulting parameter is known as $(RC)_{True_{ref_i}}$. It can be shown (see Student Sample Problem 5.4) that an expression for $(RC)_{True_{ref_i}}$ is given by

$$(RC)_{True_{ref_i}} = (RC)_{True_i} \frac{W_i}{W_{ref}} - \frac{2\,W_{ref}}{\rho\,V_i S(\pi\,AR\,e)} \left(1 - \frac{W_i^2}{W_{ref}^2}\right)$$

where W_{ref} is selected to be the average of all possible operational take-off weights for the given aircraft:

$$W_{ref} = \frac{W_{TO_1} + W_{TO_2} + \ldots + W_{TO_n}}{n}$$

Note that in the preceding relationship V_i is the true airspeed at the generic i-th flight condition, while ρ is considered as the sea level air density (since the effects of the altitude on the air density are factored in the true airspeed). The evaluation of $(RC)_{True_{ref_i}}$ allows to account for all the different operational take-off weights as well as for the fuel consumption during the flight. A sample of the calculation $(RC)_{True_{ref_i}}$ using flight data from the WVU Cessna 206 aircraft (generated through a set of climbs performed at different airspeed at maximum throttle setting) is shown in Figure 5.16. The actual flight data are shown along with the best polynomial fit; in this specific case a third order polynomial provided the best statistical fit; therefore

$$(RC)_{True_{ref_{BEST\,FIT}}} = c_3\,V_{True}^3 + c_2\,V_{True}^2 + c_1\,V_{True} + c_0$$

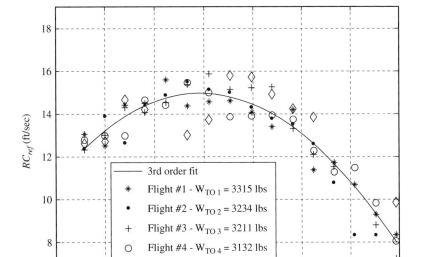

Figure 5.16 Determination of $(RC)_{True_{ref_i}}$ from Flight Data

It can be shown (see Student Sample Problem 5.5) that the $(RC)_{True_{ref\,BEST\,FIT}}$ polynomial allows the calculation of closed-form expressions for two important climbing parameters, the climb angle at maximum rate of climb $(\gamma_{RC_{MAX}})$ and the maximum climb angle (γ_{MAX}). See Figure 5.17.

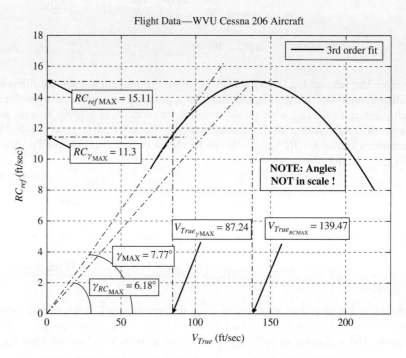

Figure 5.17 Determination of γ_{MAX}, $\gamma_{RC_{MAX}}$ Angles from Flight Data

Once $(RC)_{True_{ref}}$ is available, the available horsepower in terms of brake horsepower $B_{HP} = P/(550 \cdot \eta_P)$ can be evaluated using

$$B_{HP_{Available}} = B_{HP_{Required}} + B_{HP_{Excess}}$$

where

$$B_{HP_{Required}} = \frac{1}{1100} \frac{\rho V^3 S c_{D_O}}{\eta_P} + \frac{4}{1100} \frac{1}{(\pi\,AR\,e)} \cdot \frac{W_{ref}^2}{\rho V S} \frac{1}{\eta_P}$$

$$B_{HP_{Excess}} = \frac{1}{550} \frac{W_{ref}(RC)_{True_{ref}}}{\eta_P}$$

using the values for c_{D_O}, e, η_P as evaluated in the previous section:

$$c_{D_O} = 0.022,\ e = 0.62,\ \eta_P = 0.725$$

The experimental determination of the $B_{HP_{Available}}$, $B_{HP_{Required}}$, $B_{HP_{Excess}}$ curves for the WVU Cessna U206 is shown in Figures 5.18 and 5.19.

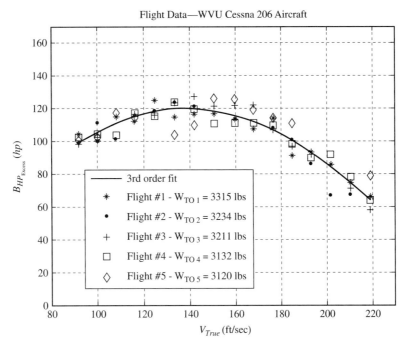

Figure 5.18 Determination of $B_{HP_{Excess}}$ from Flight Data

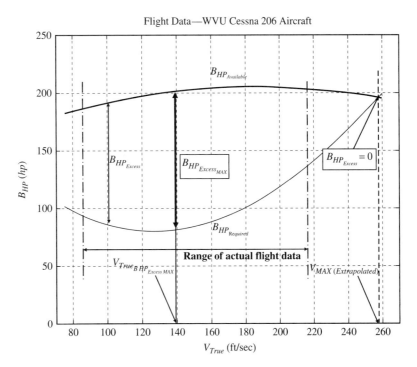

Figure 5.19 Determination of $B_{HP_{Available}}$ from Flight Data

5.7 MODELING OF THE THRUST FORCES AND MOMENTS

Although conceptually different, the modeling of the thrust forces and moments will be introduced in a similar fashion as the modeling of the aerodynamic forces and moments. Using the small perturbation theory from Chapter 1, the starting set of relationships for the thrust forces and moments is given by

Thrust Longitudinal Forces and Moment

$$\text{Along X-axis} \quad F_{T_X} = F_{T_{X_1}} + f_{T_X}$$

$$\text{Along Z-axis} \quad F_{T_Z} = F_{T_{Z_1}} + f_{T_Z}$$

$$\text{Around Y-axis} \quad M_T = M_{T_1} + m_T$$

Thrust Lateral Directional Force and Moments

$$\text{Around X-axis} \quad L_T = L_{T_1} + l_T$$

$$\text{Around Z-axis} \quad N_T = N_{T_1} + n_T$$

$$\text{Along Y-axis} \quad F_{T_Y} = F_{T_{Y_1}} + f_{T_Y}$$

The steady-state modeling will be described first, followed by the small perturbation modeling.

5.7.1 Modeling of the Steady-State Thrust Forces and Moments

The modeling of the thrust effects assumes the knowledge of the installed thrust T as provided by the engine manufacturer. The modeling of the installed thrust requires knowledge of the specific propulsion system (either propeller-based or jet-based). In general terms, the installed thrust is a function of the following variables:

- Altitude and Mach number;
- Thrust and fuel-mixture settings;
- Inlet conditions (for jet) or propeller settings (for propellers);
- Aerodynamic flow angles (α, β);
- Temperature and humidity.

The modeling of the thrust effects clearly depends on the number and the location of the engines on the aircraft. In the case of multiple engines, for modeling purposes it will be assumed that at nominal flight conditions the pilot operates all the engines at the same throttle setting.

The operational use of multiple engines at different throttle settings is known as differential thrust and is a procedure used by pilots only in emergency conditions to provide control authority in the event of failures to primary control surfaces. As will be evident from the modeling discussed below, differential thrust can be effective only when the thrust force associated with the engines has a sizable moment arm with respect to the center of gravity for creating a compensating moment. The modeling assumes a thrust vector whose magnitude can be varied (depending on the throttle setting) but with a fixed orientation. The capability of changing the orientation of the thrust for the propulsion system is instead known as thrust vectoring. This capability is currently a feature of only a few military aircraft (such as the YF-23 and the Harrier), and it has been explored for flight control purposes on a selected number of research aircraft used as a test bed for high angles of attack aerodynamic research.

In the general case of a multiengine aircraft, each installed thrust can be considered as a generic i-th vector T_i (with $i = 1, \ldots, N$, where N is the number of engines). This vector is aligned on a thrust axis, which has a different orientation with respect to both the body axes X, Y, Z and the stability axes X_S, Y_S, Z_S. Therefore, to ensure consistency in the modeling, the vector T_i needs to be expressed with respect to the same stability axes X_S, Y_S, Z_S along which the modeling of the aerodynamic forces and moments has been provided. The orientation of the T_i vector with respect to both X, Y, Z and X_S, Y_S, Z_S axes is shown in Figure 5.20.

Although it can be assumed to be small, it is possible for the engine to be installed with a non-negligible angle Θ_{T_i} with respect to the body axis X. The angle Θ_T, shown in Figure 5.20, is known as the engine toe-up angle. Similarly, the thrust axis could have a lateral orientation, as shown in Figure 5.21.

The angle Ψ_T is known as the engine toe-in angle. Clearly, engine toe-in angles on the right wing have opposite orientation with respect to the engine toe-in angles on the left wing. Small but not negligible Θ_T and Ψ_T angles have been shown to provide a small but non-negligible reduction in the drag associated with the airflow around the nacelles for wing-mounted engines. In the specific case of the McDonnell Douglas DC 9 aircraft, the use of the engine toe-up Θ_T angle actually allowed a significant improvement in the take-off performance of the aircraft.

5.7 Modeling of the Thrust Forces and Moments 289

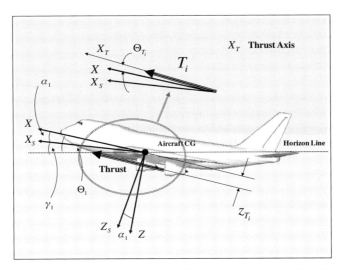

Figure 5.20 Direction of the Generic T_i with Respect to Body and Stability Axes

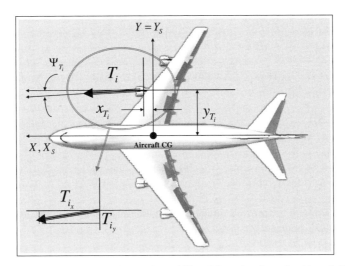

Figure 5.21 Directional Orientation of T_i and Location from the Aircraft CG

Based on the preceding modeling [6], the components of T_i along the X, Y, Z body axes are given by

$$T_{i_x} = T_i \cos(\Theta_{T_i}) \cos(\Psi_{T_i})$$
$$T_{i_y} = T_i \cos(\Theta_{T_i}) \sin(\Psi_{T_i})$$
$$T_{i_z} = T_i \sin(\Theta_{T_i})$$

Considering the N engines, the components of thrust along the X, Y, Z body axes are given by

$$F_{T_X} = \sum_{i=1}^{N} T_{i_X} = \sum_{i=1}^{N} \left(T_i \cos(\Theta_{T_i}) \cos(\Psi_{T_i}) \right)$$

$$F_{T_Y} = \sum_{i=1}^{N} T_{i_Y} = \sum_{i=1}^{N} \left(T_i \cos(\Theta_{T_i}) \sin(\Psi_{T_i}) \right)$$

$$F_{T_Z} = \sum_{i=1}^{N} T_{i_Z} = \sum_{i=1}^{N} \left(T_i \sin(\Theta_{T_i}) \right)$$

Next, the components of the forces need to be expressed at steady-state conditions with respect to the X_S, Y_S, Z_S stability axes for ensuring consistency with the previous modeling of the steady-state aerodynamic forces along the stability axes. Therefore, we have

$$F_{T_{X_{1_S}}} = F_{T_{X_1}} = \left(\sum_{i=1}^{N} T_i \cos(\Theta_{T_i}) \cos(\Psi_{T_i})\right) \cos(\alpha_1) - \left(\sum_{i=1}^{N} T_i \sin(\Theta_{T_i})\right) \sin(\alpha_1)$$

$$F_{T_{Y_{1_S}}} = F_{T_{Y_1}} = \left(\sum_{i=1}^{N} T_i \cos(\Theta_{T_i}) \sin(\Psi_{T_i})\right)$$

$$F_{T_{Z_{1_S}}} = F_{T_{Z_1}} = -\left(\sum_{i=1}^{N} T_i \sin(\Theta_{T_i})\right) \cos(\alpha_1) - \left(\sum_{i=1}^{N} T_i \cos(\Theta_{T_i}) \cos(\Psi_{T_i})\right) \sin(\alpha_1)$$

The associated thrust moments at steady-state conditions around the X_S, Y_S, Z_S stability axes are instead given by

$$L_{T_{1_S}} = L_{T_1} = \left(\sum_{i=1}^{N} T_i (z_{T_i} \cos(\Theta_{T_i}) \sin(\Psi_{T_i}) - y_{T_i} \sin(\Psi_{T_i}))\right) \cos(\alpha_1)$$
$$+ \left(\sum_{i=1}^{N} T_i (x_{T_i} \cos(\Theta_{T_i}) \sin(\Psi_{T_i}) - y_{T_i} \cos(\Theta_{T_i}) \cos(\Psi_{T_i}))\right) \sin(\alpha_1)$$

$$M_{T_{1_S}} = M_{T_1} = \left(\sum_{i=1}^{N} T_i (z_{T_i} \cos(\Theta_{T_i}) \cos(\Psi_{T_i}))\right) + \left(\sum_{i=1}^{N} T_i (x_{T_i} \sin(\Psi_{T_i}))\right)$$

$$N_{T_{1_S}} = N_{T_1} = \left(\sum_{i=1}^{N} T_i (x_{T_i} \cos(\Theta_{T_i}) \sin(\Psi_{T_i}) - y_{T_i} \cos(\Theta_{T_i}) \cos(\Psi_{T_i}))\right) \cos(\alpha_1)$$
$$+ \left(\sum_{i=1}^{N} T_i (z_{T_i} \cos(\Theta_{T_i}) \sin(\Psi_{T_i}) - y_{T_i} \sin(\Psi_{T_i}))\right) \sin(\alpha_1)$$

To be consistent with the previous breakdown in longitudinal and lateral directional components, the steady-state thrust forces and moments are given by

Steady-State Longitudinal Thrust Forces and Moment (N engines)

Along X_S: $\quad F_{T_{X_1}} = \left(\sum_{i=1}^{N} T_i \cos(\Theta_{T_i}) \cos(\Psi_{T_i})\right) \cos(\alpha_1) - \left(\sum_{i=1}^{N} T_i \sin(\Theta_{T_i})\right) \sin(\alpha_1)$

Around Y_S: $\quad M_{T_1} = \left(\sum_{i=1}^{N} T_i (z_{T_i} \cos(\Theta_{T_i}) \cos(\Psi_{T_i}))\right) + \left(\sum_{i=1}^{N} T_i (x_{T_i} \sin(\Psi_{T_i}))\right)$

Along Z_S: $\quad F_{T_{Z_1}} = -\left(\sum_{i=1}^{N} T_i \sin(\Theta_{T_i})\right) \cos(\alpha_1) - \left(\sum_{i=1}^{N} T_i \cos(\Theta_{T_i}) \cos(\Psi_{T_i})\right) \sin(\alpha_1)$

Steady-State Lateral Directional Thrust Force and Moments (N engines)

Around X_S: $\quad L_{T_1} = \left(\sum_{i=1}^{N} T_i (z_{T_i} \cos(\Theta_{T_i}) \sin(\Psi_{T_i}) - y_{T_i} \sin(\Psi_{T_i}))\right) \cos(\alpha_1)$

$\qquad\qquad\qquad + \left(\sum_{i=1}^{N} T_i (x_{T_i} \cos(\Theta_{T_i}) \sin(\Psi_{T_i}) - y_{T_i} \cos(\Theta_{T_i}) \cos(\Psi_{T_i}))\right) \sin(\alpha_1)$

Along Y_S: $\quad F_{T_{Y_1}} = \left(\sum_{i=1}^{N} T_i \cos(\Theta_{T_i}) \sin(\Psi_{T_i})\right)$

$$\text{Around } Z_S: \quad N_{T_1} = \left(\sum_{i=1}^{N} T_i\left(x_{T_i}\cos(\Theta_{T_i})\sin(\Psi_{T_i}) - y_{T_i}\cos(\Theta_{T_i})\cos(\Psi_{T_i})\right)\right)\cos(\alpha_1)$$
$$+ \left(\sum_{i=1}^{N} T_i\left(z_{T_i}\cos(\Theta_{T_i})\sin(\Psi_{T_i}) - y_{T_i}\sin(\Psi_{T_i})\right)\right)\sin(\alpha_1)$$

A few assumptions can substantially simplify the preceding relationships when all the engines are operating at nominal conditions with the same throttle setting. For example, while both engine toe-up and engine toe-in angles (Θ_T and Ψ_T) are small, typically we have $\Theta_T > \Psi_T$ with $\cos(\Theta_T) < \cos(\Psi_T) \approx 1$. Additionally, since the steady-state angle is α_1 we can assume

$$\sin(\Theta_T) \approx \Theta_T, \ \sin(\Psi_T) \approx \Psi_T, \ \sin(\alpha_1) \approx \alpha_1, \quad \alpha_1\Psi_T \approx 0, \ \alpha_1\Theta_T \approx 0$$

Further, to ensure symmetry of forces with respect to the XZ plane and equilibrium conditions around the aircraft center of gravity, all the engine toe-in angles (and the relative $\sin(\Psi_{T_i})$) will have opposite signs on the two semiwings leading to a cancellation of the involved terms. Therefore, we have

Simplified Steady-State Longitudinal Thrust Forces and Moment (*N* engines)

$$\text{Along } X_S: \quad F_{T_{X_1}} = \left(\sum_{i=1}^{N} T_i\cos(\Theta_{T_i})\right)\cos(\alpha_1)$$

$$\text{Around } Y_S: \quad M_{T_1} = \left(\sum_{i=1}^{N} T_i(z_{T_i}\cos(\Theta_{T_i}))\right)$$

$$\text{Along } Z_S: \quad F_{T_{Z_1}} = -\left(\sum_{i=1}^{N} T_i\Theta_{T_i}\right)\cos(\alpha_1) - \left(\sum_{i=1}^{N} T_i\cos(\Theta_{T_i})\right)\sin(\alpha_1)$$

Simplified Steady-State Lateral Directional Thrust Force and Moments (*N* engines)

$$\text{Around } X_S: \quad L_{T_1} = 0$$
$$\text{Along } Y_S: \quad F_{T_{Y_1}} = 0$$
$$\text{Around } Z_S: \quad N_{T_1} = 0$$

The previous modeling does not account for the aerodynamic interaction between the vortices associated with the propellers or the jet exhaust with the aircraft surfaces. These effects are very specific and require detailed localized modeling.

The preceding relationships, in the regular and nonsimplified form, become very important for the modeling of a specific flight condition known as OEI (one-engine lnoperative). Unless the OEI condition involves the failure of the central (tail) engine for a three-engine aircraft, any OEI leads to substantial yawing and rolling moments, in addition to a loss of thrust in the longitudinal axis and a pitching moment. Furthermore, a drag ΔD_i is typically associated with the inoperative engine for propeller-based propulsion. The additional drag associated with the OEI should be properly modeled for a detailed simulation of the aircraft dynamics at postengine failure conditions, in addition to the net loss of thrust (and relative change in longitudinal and lateral-directional moments) associated with the OEI condition.

5.7.2 Modeling of the Small Perturbation Thrust Forces and Moments

The small perturbation thrust forces and moments—divided into longitudinal and lateral-directional—are given by

$$f_{T_X}, f_{T_Z}, m_T$$
$$l_T, f_{T_Y}, n_T$$

The modeling of the small perturbation thrust forces and moments is conceptually similar to the modeling of the small perturbation aerodynamic forces and moments. However, it turns out that only a limited number of small perturbations in the motion variables have significant effects on the thrust characteristics. As outlined in Chapter 3 and 4 for the aerodynamic forces and moments, the nondimensional small perturbation terms are used. Specifically, we have

$$f_{T_X} = f\left(\frac{u}{V_{P_1}}, \alpha\right), \quad f_{T_Z} = f\left(\frac{u}{V_{P_1}}, \alpha\right), \quad m_T = f\left(\frac{u}{V_{P_1}}, \alpha\right)$$

$$f_{T_Y} = f(\beta), \quad l_T = f(\beta), \quad n_T = f(\beta)$$

Next, by calculating the gradient evaluated at steady-state conditions, we have

$$f_{T_X} = \left.\frac{\partial F_{T_X}}{\partial\left(\frac{u}{V_{P_1}}\right)}\right|_{SS} \left(\frac{u}{V_{P_1}}\right) + \left.\frac{\partial F_{T_X}}{\partial \alpha}\right|_{SS} \alpha$$

$$f_{T_Z} = \left.\frac{\partial F_{T_Z}}{\partial\left(\frac{u}{V_{P_1}}\right)}\right|_{SS} \left(\frac{u}{V_{P_1}}\right) + \left.\frac{\partial F_{T_Z}}{\partial \alpha}\right|_{SS} \alpha$$

$$m_T = \left.\frac{\partial M_T}{\partial\left(\frac{u}{V_{P_1}}\right)}\right|_{SS} \left(\frac{u}{V_{P_1}}\right) + \left.\frac{\partial M_T}{\partial \alpha}\right|_{SS} \alpha$$

$$f_{T_Y} = \left.\frac{\partial F_{T_Y}}{\partial \beta}\right|_{SS} \beta, \quad l_T = \left.\frac{\partial L_T}{\partial \beta}\right|_{SS} \beta, \quad n_T = \left.\frac{\partial N_T}{\partial \beta}\right|_{SS} \beta$$

where the above forces and moments can be generically expressed using

$$F_{T_X} = c_{T_X} \bar{q} S, \quad F_{T_Z} = c_{T_Z} \bar{q} S, \quad M_T = c_{m_T} \bar{q} S \bar{c}$$

$$F_{T_Y} = c_{T_Y} \bar{q} S, \quad L_T = c_{l_T} \bar{q} S b, \quad N_T = c_{n_T} \bar{q} S b$$

The derivation of each of the gradient terms will be briefly discussed next.
Starting from the terms in f_{T_X} we have

$$\left.\frac{\partial F_{T_X}}{\partial\left(\frac{u}{V_{P_1}}\right)}\right|_{SS} = (\bar{q}_1 S) \left.\frac{\partial c_{T_X}}{\partial\left(\frac{u}{V_{P_1}}\right)}\right|_{SS} + (S c_{T_X}) \left.\frac{\partial \bar{q}}{\partial\left(\frac{u}{V_{P_1}}\right)}\right|_{SS}$$

At small perturbation conditions, with the presence of small perturbation linear velocity components (u,v,w), the dynamic pressure $\bar{q}$ is given by

$$\bar{q} = \frac{1}{2}\rho V^2 = \frac{1}{2}\rho\left[(V_{P_1} + u)^2 + v^2 + w^2\right]$$

Therefore: $\left.\dfrac{\partial \bar{q}}{\partial\left(\frac{u}{V_{P_1}}\right)}\right|_{SS} = V_{P_1} \left.\dfrac{\partial \bar{q}}{\partial u}\right|_{SS}$

Using the previous relationship for $\bar{q}$, the gradient with respect to u leads to

$$V_{P_1}\left.\frac{\partial \bar{q}}{\partial u}\right|_{SS} = V_{P_1} \frac{\partial\left(\frac{1}{2}\rho\left[(V_{P_1}+u)^2+v^2+w^2\right]\right)}{\partial u}\bigg|_{SS} = V_{P_1}\frac{\partial\left(\frac{1}{2}\rho\left[(V_{P_1}^2+2uV_{P_1}+u^2)\right]\right)}{\partial u}\bigg|_{SS}$$

$$= V_{P_1}\left[\frac{1}{2}\rho(2V_{P_1}+2u)\right]\bigg|_{SS} = V_{P_1}\left[\frac{1}{2}\rho\,2(V_{P_1}+u)\right]\bigg|_{SS} = \rho V_{P_1}^2 = 2\bar{q}_1$$

Back to the previous gradient calculation we will have

$$\left.\frac{\partial F_{T_X}}{\partial\left(\frac{u}{V_{P_1}}\right)}\right|_{SS} = \bar{q}_1 S\, c_{T_{X_U}} + 2\bar{q}_1 S\, c_{T_{X_1}}$$

Similarly, for the term relative to the small perturbation α, we have

$$\left.\frac{\partial F_{T_X}}{\partial \alpha}\right|_{SS}\alpha = \bar{q}_1 S c_{T_{X_\alpha}}$$

Therefore, a relationship for f_{T_X} is given by

$$\boxed{f_{T_X} = \left.\frac{\partial F_{T_X}}{\partial\left(\frac{u}{V_{P_1}}\right)}\right|_{SS}\left(\frac{u}{V_{P_1}}\right) + \left.\frac{\partial F_{T_X}}{\partial\alpha}\right|_{SS}\alpha = \left[\bar{q}_1 S\left(c_{T_{X_U}}+2c_{T_{X_1}}\right)\right]\left(\frac{u}{V_{P_1}}\right) + \left(\bar{q}_1 S c_{T_{X_\alpha}}\right)\alpha}$$

Thus, the attention shifts to the analysis of the thrust coefficients $c_{T_{X_1}}, c_{T_{X_U}}, c_{T_{X_\alpha}}$.

$c_{T_{X_1}}$ is the coefficient opposing the drag coefficient c_{D_1} at steady-state conditions to ensure the classic longitudinal equilibrium condition $T = D$.

$c_{T_{X_U}}$ is the coefficient modeling the variation of the thrust along X_S associated with small variations in the linear speed in the forward direction. The quantification of this effect is very dependent on the specific type of propulsion system. Roskam[6,7] provides an overview of the modeling for these coefficients for the following types of propulsion system:

- Rocket propulsion;
- Jet propulsion;
- Fixed-pitch propellers;
- Variable-pitch propellers.

$c_{T_{X_\alpha}}$ is the coefficient modeling the variation of the thrust along X_S associated with small variation in the longitudinal angle of attack. This coefficient is typically negligible for most aircraft.

Using the same modeling procedure introduced for f_{T_X}, a relationship for f_{T_Z} is given by

$$\boxed{f_{T_Z} = \left.\frac{\partial F_{T_Z}}{\partial\left(\frac{u}{V_{P_1}}\right)}\right|_{SS}\left(\frac{u}{V_{P_1}}\right) + \left.\frac{\partial F_{T_Z}}{\partial\alpha}\right|_{SS}\alpha = \left[\bar{q}_1 S\left(c_{T_{Z_U}}+2c_{T_{Z_1}}\right)\right]\left(\frac{u}{V_{P_1}}\right) + \left(\bar{q}_1 S c_{T_{Z_\alpha}}\right)\alpha}$$

Next, the attention shifts to the analysis of the thrust coefficients $c_{T_{Z_1}}, c_{T_{Z_U}}, c_{T_{Z_\alpha}}$. The physical meaning of these coefficients is similar to the previous $c_{T_{X_1}}, c_{T_{X_U}}, c_{T_{X_\alpha}}$ with the only difference being the relationship along Z_S in lieu of X_S. Unless a thrust vectoring system is implemented, it turns out that for most aircraft these coefficients are all negligible.

Similarly, a relationship for m_T is given by

$$m_T = \left.\frac{\partial M_T}{\partial \left(\frac{u}{V_{P_1}}\right)}\right|_{SS} \left(\frac{u}{V_{P_1}}\right) + \left.\frac{\partial M_T}{\partial \alpha}\right|_{SS} \alpha = \left[\bar{q}_1 S \bar{c}\left(c_{m_{T_U}} + 2c_{m_{T1}}\right)\right]\left(\frac{u}{V_{P_1}}\right) + \left(\bar{q}_1 S \bar{c}\, c_{m_{T_\alpha}}\right)\alpha$$

Next, the attention shifts to the analysis of the coefficients $c_{m_{T_1}}$, $c_{m_{T_U}}$, $c_{m_{T_\alpha}}$. The value of the coefficient $c_{m_{T_1}}$ depends on the general aircraft equilibrium conditions $M_A + M_T = 0$. At trim conditions we have $c_{m_{T_1}} + c_{m_1} = 0$. If the thrust axis is aligned with X_S we will have the condition $c_{m_{T_1}} = 0$. Otherwise, the aerodynamic pitching moment at steady-state conditions will have to be enough to compensate for the $c_{m_{T_1}} \neq 0$ coefficient.

$c_{m_{T_U}}$ depends instead on the value of the coefficient $c_{T_{X_U}}$. In fact, a relationship is given by

$$c_{m_{T_U}} = \pm c_{T_{X_U}} \frac{z_T}{\bar{c}}$$

where z_T is the overall moment arm of the thrust with respect to the aircraft center of gravity along the Z_S axis.

$c_{m_{T_\alpha}}$ models instead the variation in pitching moment associated with a small perturbation in the longitudinal angle of attack. It turns out that this effect is not negligible. Roskam[6,7] provides estimates for this coefficient for both propeller-based and jet-based propulsion systems.

Switching to the modeling of the small perturbation thrust lateral directional forces and moments we have

$$f_{T_Y} = \left.\frac{\partial F_{T_Y}}{\partial \beta}\right|_{SS} \beta = \bar{q}_1 S c_{T_{Y_\beta}}$$

$$l_T = \left.\frac{\partial L_T}{\partial \beta}\right|_{SS} \beta = \bar{q}_1 S b c_{l_{T_\beta}}$$

$$n_T = \left.\frac{\partial N_T}{\partial \beta}\right|_{SS} \beta = \bar{q}_1 S b c_{n_{T_\beta}}$$

It turns out that the coefficients $c_{T_{Y_\beta}}$, $c_{l_{T_\beta}}$, $c_{n_{T_\beta}}$ are all negligible for most aircraft propulsive configurations.

5.8 SUMMARY

This chapter provided a brief review of aircraft propulsion systems followed by a review of basic aircraft performance parameters, including the analysis of the power required and the power available leading to the modeling of the thrust forces and moments at steady-state and small perturbation conditions.

With the modeling of the thrust forces and moments, the overall modeling task has been completed. A final summary of the aerodynamic and thrust forces and moments is provided here.

Steady State Aerodynamic Forces and Moments

$$\begin{Bmatrix} F_A \\ F_A \\ M_{A_{1S}} \end{Bmatrix} = \begin{Bmatrix} F_{A_{X_1}} \\ F_{A_{Z_1}} \\ M_{A_1} \end{Bmatrix} = \begin{Bmatrix} -(\bar{q}Sc_{D_1}) \\ -(\bar{q}Sc_{L_1}) \\ \bar{q}S\bar{c}c_{m_1} \end{Bmatrix} \text{ with } \begin{Bmatrix} c_{D_1} \\ c_{L_1} \\ c_{m_1} \end{Bmatrix} = \begin{bmatrix} c_{D_0} & c_{D_\alpha} & \cancel{c_{D_{\delta_E}}} & \cancel{c_{D_{i_H}}} \\ c_{L_0} & c_{L_\alpha} & c_{L_{\delta_E}} & c_{L_{i_H}} \\ c_{m_0} & c_{m_\alpha} & c_{m_{\delta_E}} & c_{m_{i_H}} \end{bmatrix} \begin{Bmatrix} 1 \\ \alpha \\ \delta_E \\ i_H \end{Bmatrix}$$

$$\begin{Bmatrix} F_A \\ L_{A_{1S}} \\ N_{A_{1S}} \end{Bmatrix} = \begin{Bmatrix} F_{A_{Y_1}} \\ L_{A_1} \\ N_{A_1} \end{Bmatrix} = \begin{Bmatrix} \bar{q}Sc_{Y_1} \\ \bar{q}Sbc_{l_1} \\ \bar{q}Sbc_{n_1} \end{Bmatrix} \text{ with } \begin{Bmatrix} c_{Y_1} \\ c_{l_1} \\ c_{n_1} \end{Bmatrix} = \begin{bmatrix} c_{Y_\beta} & c_{Y_{\delta_A}} & c_{Y_{\delta_R}} \\ c_{l_\beta} & c_{l_{\delta_A}} & c_{l_{\delta_R}} \\ c_{n_\beta} & c_{n_{\delta_A}} & c_{n_{\delta_R}} \end{bmatrix} \begin{Bmatrix} \beta \\ \delta_A \\ \delta_R \end{Bmatrix}$$

Small Perturbation Aerodynamic Forces and Moments

$$\begin{cases} f_{A_X} = \bar{q}S\left\{-[c_{D_u} + 2c_{D_1}]\left(\dfrac{u}{V_{P_1}}\right) + [-c_{D_\alpha} + c_{L_1}]\alpha - c_{D_{\dot{\alpha}}}\left(\dfrac{\dot{\alpha}\bar{c}}{2V_{P_1}}\right) - c_{D_q}\left(\dfrac{q\bar{c}}{2V_{P_1}}\right) - c_{D_{\delta_E}}\delta_E - c_{D_{i_H}}i_H\right\} \\[2mm] f_{A_z} = \bar{q}S\left\{-[c_{L_u} + 2c_{L_1}]\left(\dfrac{u}{V_{P_1}}\right) + [-c_{L_\alpha} - c_{D_1}]\alpha - c_{L_{\dot{\alpha}}}\left(\dfrac{\dot{\alpha}\bar{c}}{2V_{P_1}}\right) - c_{L_q}\left(\dfrac{q\bar{c}}{2V_{P_1}}\right) - c_{L_{\delta_E}}\delta_E - c_{L_{i_H}}i_H\right\} \\[2mm] m_A = \bar{q}S\bar{c}\left\{[c_{m_u} + 2c_{m_1}]\left(\dfrac{u}{V_{P_1}}\right) + c_{m_\alpha}\alpha + c_{m_{\dot{\alpha}}}\left(\dfrac{\dot{\alpha}\bar{c}}{2V_{P_1}}\right) + c_{m_q}\left(\dfrac{q\bar{c}}{2V_{P_1}}\right) + c_{m_{\delta_E}}\delta_E + c_{m_{i_H}}i_H\right\} \end{cases}$$

$$\begin{cases} f_{A_Y} = \bar{q}S\left\{c_{Y_\beta}\beta + c_{Y_{\dot{\beta}}}\left(\dfrac{\dot{\beta}b}{2V_{P_1}}\right) + c_{Y_p}\left(\dfrac{pb}{2V_{P_1}}\right) + c_{Y_r}\left(\dfrac{rb}{2V_{P_1}}\right) + c_{Y_{\delta_A}}\delta_A + c_{Y_{\delta_R}}\delta_R\right\} \\[2mm] l_A = \bar{q}Sb\left\{c_{l_\beta}\beta + c_{l_{\dot{\beta}}}\left(\dfrac{\dot{\beta}b}{2V_{P_1}}\right) + c_{l_p}\left(\dfrac{pb}{2V_{P_1}}\right) + c_{l_r}\left(\dfrac{rb}{2V_{P_1}}\right) + c_{l_{\delta_A}}\delta_A + c_{l_{\delta_R}}\delta_R\right\} \\[2mm] n_A = \bar{q}Sb\left\{c_{n_\beta}\beta + c_{n_{\dot{\beta}}}\left(\dfrac{\dot{\beta}b}{2V_{P_1}}\right) + c_{n_p}\left(\dfrac{pb}{2V_{P_1}}\right) + c_{n_r}\left(\dfrac{rb}{2V_{P_1}}\right) + c_{n_{\delta_A}}\delta_A + c_{n_{\delta_R}}\delta_R\right\} \end{cases}$$

Steady-State Thrust Forces and Moments (at nominal engine operating conditions)

$$\begin{cases} F_{T_{X_1}} = \left(\sum_{i=1}^{N} T_i \cos(\Theta_{T_i}) \cos(\Psi_{T_i})\right) \cos(\alpha_1) - \left(\sum_{i=1}^{N} T_i \sin(\Theta_{T_i})\right) \sin(\alpha_1) \\[2mm] F_{T_{Z_1}} = -\left(\sum_{i=1}^{N} T_i \sin(\Theta_{T_i})\right) \cos(\alpha_1) - \left(\sum_{i=1}^{N} T_i \cos(\Theta_{T_i}) \cos(\Psi_{T_i})\right) \sin(\alpha_1) \\[2mm] M_{T_1} = \left(\sum_{i=1}^{N} T_i(z_{T_i} \cos(\Theta_{T_i}) \cos(\Psi_{T_i}))\right) + \left(\sum_{i=1}^{N} T_i(x_{T_i} \sin(\Psi_{T_i}))\right) \end{cases}$$

$$\begin{cases} F_{T_{Y_1}} = \left(\sum_{i=1}^{N} T_i \cos(\Theta_{T_i}) \sin(\Psi_{T_i})\right) \\[2mm] L_{T_1} = \left(\sum_{i=1}^{N} T_i(z_{T_i} \cos(\Theta_{T_i}) \sin(\Psi_{T_i}) - y_{T_i} \sin(\Psi_{T_i}))\right) \cos(\alpha_1) \\[2mm] \qquad + \left(\sum_{i=1}^{N} T_i(x_{T_i} \cos(\Theta_{T_i}) \sin(\Psi_{T_i}) - y_{T_i} \cos(\Theta_{T_i}) \cos(\Psi_{T_i}))\right) \sin(\alpha_1) \\[2mm] N_{T_1} = \left(\sum_{i=1}^{N} T_i(x_{T_i} \cos(\Theta_{T_i}) \sin(\Psi_{T_i}) - y_{T_i} \cos(\Theta_{T_i}) \cos(\Psi_{T_i}))\right) \cos(\alpha_1) \\[2mm] \qquad + \left(\sum_{i=1}^{N} T_i(z_{T_i} \cos(\Theta_{T_i}) \sin(\Psi_{T_i}) - y_{T_i} \sin(\Psi_{T_i}))\right) \sin(\alpha_1) \end{cases}$$

Small Perturbation Thrust Forces and Moments (at nominal engine operating conditions)

$$\begin{cases} f_{T_X} = \bar{q}_1 S\left[\left(c_{T_{X_U}} + 2c_{T_{X_1}}\right)\left(\dfrac{u}{V_{P_1}}\right) + \cancel{(c_{T_{X_\alpha}})}\alpha\right] \\[6pt] f_{T_Z} = \bar{q}_1 S\left[\cancel{\left(c_{T_{Z_U}} + 2c_{T_{Z_1}}\right)}\left(\dfrac{u}{V_{P_1}}\right) + \cancel{(c_{T_{Z_\alpha}})}\alpha\right] \\[6pt] m_T = \bar{q}_1 S\bar{c}\left[\left(c_{m_{T_U}} + 2c_{m_{T1}}\right)\left(\dfrac{u}{V_{P_1}}\right) + \cancel{(c_{m_{T_\alpha}})}\alpha\right] \end{cases}$$

$$\begin{cases} f_{T_Y} = \bar{q}_1 S \cancel{c_{T_{Y_\beta}}} \\ l_T = \bar{q}_1 S b \cancel{c_{l_{T_\beta}}} \\ n_T = \bar{q}_1 S b \cancel{c_{n_{T_\beta}}} \end{cases}$$

REFERENCES

1. Phillips, Warren F. *Mechanics of Flight*. John Wiley & Sons, 2004.
2. Yechout, T. R., Morris, S. L., Bossert, D. E., and Hallgren, W. F. *Introduction to Aircraft Flight Mechanics*. AIAA (Education Series), 2003.
3. Pamadi, B. N. *Performance, Stability, Dynamics, and Control of Airplanes*. AIAA (Education Series), 2004.
4. Napolitano, M. R. *WVU MAE 242 Flight Testing—Class Notes*. Department of Mechanical and Aerospace Engineering, West Virginia University, Morgantown, WV, 2001.
5. Hartmann, E. P., Biermann, D. NACA Report No. 640, "The Aerodynamic Characteristics of Full-Scale Propellers Having 2, 3, and 4 Blades of Clark Y and R.A.F. 6 Airfoil Sections."1938.
6. Roskam, J. *Airplane Flight Dynamics and Automatic Flight Controls—Part I*. Design, Analysis, and Research Corporation, Lawrence, KS, 1995.
7. Roskam, J. Airplane Design. *Part I–Part VIII*. Roskam Aviation and Engineering Corporation, Lawrence, KS, 1990.

STUDENT SAMPLE PROBLEMS

Student Sample Problem 5.1

The total aircraft drag (sum of parasite and induced drag) is shown in Figure SSP5.1.1.

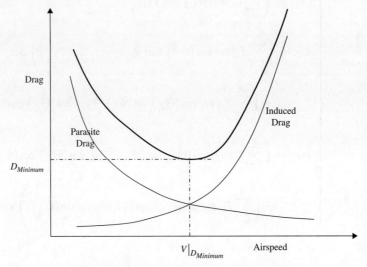

Figure SSP5.1.1 Generic Drag vs. Velocity Plot

Knowing that $V|_{D_{Minimum}} = \sqrt{\dfrac{2W}{\rho S} \cdot \sqrt{\dfrac{1}{\pi AR e c_{D_O}}}} \rightarrow V|_{D_{Minimum}} = \sqrt{\dfrac{2W}{\rho S}} \cdot \sqrt[4]{\dfrac{1}{\pi AR e c_{D_O}}}$, demonstrate that

(1) $D_{Parasite}|_{V|_{D_{MINIMUM}}} = D_{Induced}|_{V|_{D_{MINIMUM}}}$

(2) $D_{Minimum} = 2 \cdot D_{Parasite}|_{V|_{D_{MINIMUM}}} = 2 \cdot D_{Induced}|_{V|_{D_{MINIMUM}}} = 2W\sqrt{\dfrac{c_{D_0}}{(\pi AR e)}}$

Solution of Student Sample Problem 5.1

Starting from the general relationship for the drag of the entire aircraft

$$D = \dfrac{1}{2}\rho V^2 S \left(c_{D_O} + \dfrac{c_L^2}{\pi AR e} \right) = \dfrac{1}{2}\rho V^2 S \left(c_{D_O} + \dfrac{1}{\pi AR e} \cdot \dfrac{4W^2}{\rho^2 V^4 S^2} \right)$$

$$= \dfrac{1}{2}\rho V^2 S c_{D_O} + \dfrac{1}{\pi AR e} \cdot \dfrac{2W^2}{\rho V^2 S} = D_{Parasite} + D_{Induced}$$

Next, $D_{Minimum}$ is found using

$$D_{Minimum} = \dfrac{1}{2}\rho \left(V|_{D_{MINIMUM}} \right)^2 S c_{D_O} + \dfrac{1}{\pi AR e} \cdot \dfrac{2W^2}{\rho \left(V|_{D_{MINIMUM}} \right)^2 S}$$

$$= \dfrac{1}{2}\rho \dfrac{2W}{\rho S} \cdot \sqrt{\dfrac{1}{\pi AR e c_{D_O}}} S c_{D_O} + \dfrac{1}{\pi AR e} \cdot \dfrac{2W^2}{\rho \dfrac{2W}{\rho S} \cdot \sqrt{\dfrac{1}{\pi AR e c_{D_O}}} S}$$

$$= \dfrac{1}{\cancel{2}} \cancel{\rho} \dfrac{\cancel{2}W}{\cancel{\rho} \cancel{S}} \cdot \sqrt{\dfrac{1}{\pi AR e c_{D_O}}} \cancel{S} c_{D_O} + \dfrac{1}{\pi AR e} \cdot \dfrac{2W^{\cancel{2}}}{\cancel{\rho} \dfrac{2W}{\cancel{\rho} \cancel{S}} \cdot \sqrt{\dfrac{1}{\pi AR e c_{D_O}}} \cancel{S}}$$

leading to

$$D_{Minimum} = W c_{D_O} \sqrt{\dfrac{1}{(\pi AR e) c_{D_O}}} + \dfrac{1}{(\pi AR e)} \cdot \dfrac{W}{\sqrt{\dfrac{1}{(\pi AR e) c_{D_O}}}}$$

$$= W\sqrt{\dfrac{c_{D_O}^2}{(\pi AR e) c_{D_O}}} + \dfrac{W}{\sqrt{\dfrac{(\pi AR e)^2}{(\pi AR e) c_{D_O}}}} = W\sqrt{\dfrac{c_{D_O}}{(\pi AR e)}} + W\sqrt{\dfrac{c_{D_O}}{(\pi AR e)}}$$

Therefore, it is clear that $D_{Parasite}|_{V|_{D_{MINIMUM}}} = D_{Induced}|_{V|_{D_{MINIMUM}}}$

Furthermore, $D_{Minimum} = 2 \cdot W\sqrt{\dfrac{c_{D_O}}{(\pi AR e)}}$

Student Sample Problem 5.2

Starting from the relationship

$$P_{required} = D \cdot V = \dfrac{1}{2}\rho V^3 S c_{D_O} + \dfrac{1}{(\pi AR e)} \cdot \dfrac{2W^2}{\rho V S} = P_{required}\big|_{\substack{Parasite \\ Drag}} + P_{required}\big|_{\substack{Induced \\ Drag}}$$

Show that

$$V\big|_{P_{REQUIRED_{Minimum}}} = \frac{1}{\sqrt[4]{3}} V\big|_{D_{Minimum}} = \frac{1}{\sqrt[4]{3}} \sqrt{\frac{2W}{\rho S}} \cdot \sqrt[4]{\frac{1}{(\pi ARe)c_{D_O}}} \approx 0.76 \sqrt{\frac{2W}{\rho S}} \cdot \sqrt[4]{\frac{1}{(\pi ARe)c_{D_O}}}$$

Solution of Student Sample Problem 5.2

From

$$P_{required} = D \cdot V = \frac{1}{2}\rho V^3 S c_{D_O} + \frac{1}{(\pi ARe)} \cdot \frac{2W^2}{\rho VS} = P_{required}\big|_{\substack{Parasite\\Drag}} + P_{required}\big|_{\substack{Induced\\Drag}}$$

$$\frac{dP_{required}}{dV} = \frac{d}{dV}\left(\frac{1}{2}\rho V^3 S c_{D_O} + \frac{1}{(\pi ARe)} \cdot \frac{2W^2}{\rho VS}\right)$$

$$= \frac{3}{2}\rho V^2 S c_{D_O} - \frac{1}{(\pi ARe)} \cdot \frac{2W^2}{\rho V^2 S} = 0$$

$$\rightarrow \frac{3}{2}\rho V^2 S c_{D_O} = \frac{1}{(\pi ARe)} \cdot \frac{2W^2}{\rho V^2 S}$$

Multiplying both sides by V^2 we have

$$\frac{3}{2}\rho V^4 S c_{D_O} = \frac{1}{(\pi ARe)} \cdot \frac{2W^2}{\rho S}$$

Leading to

$$V^4\big|_{\substack{Power\\required_{Minimum}}} = \frac{1}{(\pi ARe)} \cdot \frac{2W^2}{\rho S} \cdot \frac{2}{3\rho S c_{D_O}} = \frac{4W^2}{3\rho^2 S^2(\pi ARe)c_{D_O}}$$

From which

$$V^2\big|_{\substack{Power\\required_{Minimum}}} = \sqrt{\frac{4W^2}{3\rho^2 S^2(\pi ARe)c_{D_O}}} = \frac{2W}{\rho S}\sqrt{\frac{1}{3(\pi ARe)c_{D_O}}}$$

Leading finally to

$$V\big|_{\substack{Power\\required_{Minimum}}} = \sqrt{\frac{2W}{\rho S}} \sqrt[4]{\frac{1}{3(\pi ARe)c_{D_O}}}$$

From Student Sample Problem 5.1 we recall

$$V\big|_{D_{Minimum}} = \sqrt{\frac{2W}{\rho S}} \cdot \sqrt[4]{\frac{1}{(\pi ARe)c_{D_O}}}$$

Therefore, it is clear that

$$V\big|_{\substack{Power\\required_{Minimum}}} = \sqrt{\frac{2W}{\rho S}} \sqrt[4]{\frac{1}{3(\pi ARe)c_{D_O}}} = \sqrt[4]{\frac{1}{3}} V\big|_{D_{Minimum}} \approx 0.758 V\big|_{D_{Minimum}}$$

Thus, the airspeed at which the power required is minimized is approximately 76 percent of the airspeed at which the drag of the aircraft is minimized.

Student Sample Problem 5.3

Starting from the definition of

$$P_{HP_{Required}} = \frac{1}{1100}\rho V^3 S c_{D_O} + \frac{4}{1100}\frac{1}{(\pi ARe)}\cdot\frac{W^2}{\rho VS}$$

Using the EW parameters

$$V_{EW} = V\sqrt{\frac{W_0}{W}\cdot\frac{\rho}{\rho_0}},\ P_{HP_{Required_{EW}}} = P_{HP_{Required}}\sqrt{\frac{W_0^3\rho}{W^3\rho_0}}$$

and the definition of brake horsepower

$$B_{HP} = \frac{P_{HP}}{\eta_P}$$

show that

$$V_{EW}\cdot B_{HP_{Required_{EW}}} = \frac{1}{1100}\frac{Sc_{D_O}\rho_0}{\eta_P}\cdot V_{EW}^4 + \frac{4}{1100}\frac{1}{(\pi ARe)S\eta_P}\cdot\frac{W_0^2}{\rho_0}$$

Solution of Student Sample Problem 5.3

Using the preceding relationship for $P_{HP_{Required}}$, using the definitions for V_{EW}, $P_{HP_{Required_{EW}}}$ we have

$$P_{HP_{Required_{EW}}} = \frac{1}{1100}\rho V_{EW}^3 S c_{D_O}\cdot\left(\sqrt{\frac{W_0^3\rho}{W^3\rho_0}}\right) + \frac{4}{1100}\frac{1}{(\pi ARe)}\cdot\frac{W^2}{\rho V_{EW}S}\cdot\left(\sqrt{\frac{W_0^3\rho}{W^3\rho_0}}\right)$$

Using the previous definition for V_{EW}

$$V_{EW} = V\sqrt{\frac{W_0}{W}\cdot\frac{\rho}{\rho_0}}$$

we have for the two right-hand side terms of $P_{HP_{Required_{EW}}}$

1st: $\dfrac{1}{1100}\rho V^3 S c_{D_O}\cdot\left(\sqrt{\dfrac{W_0^3\rho}{W^3\rho_0}}\right) = \dfrac{1}{1100}\rho V^3 S c_{D_O}\cdot\left(\sqrt{\dfrac{W_0^3\rho}{W^3\rho_0}}\right)\cdot\left(\dfrac{\rho_0}{\rho_0}\right) = \dfrac{1}{1100}Sc_{D_O}\rho_0 V_{EW}^3$

2nd: $\dfrac{4}{1100}\dfrac{1}{(\pi ARe)}\cdot\dfrac{W^2}{\rho VS}\cdot\left(\sqrt{\dfrac{W_0^3\rho}{W^3\rho_0}}\right)$

$= \dfrac{4}{1100}\dfrac{1}{(\pi ARe)}\cdot\dfrac{W^2}{\rho VS}\cdot\left(\sqrt{\dfrac{W_0^3\rho}{W^3\rho_0}}\right)\dfrac{\left(\sqrt{\dfrac{W_0\rho}{W\rho_0}}\right)}{\left(\sqrt{\dfrac{W_0\rho}{W\rho_0}}\right)} = \dfrac{4}{1100}\dfrac{1}{(\pi ARe)S}\cdot\dfrac{W_0^2}{\rho_0 V_{EW}}$

leading to

$$P_{HP_{Required_{EW}}} = \frac{1}{1100}Sc_{D_O}\rho_0 V_{EW}^3 + \frac{4}{1100}\frac{1}{(\pi ARe)S}\cdot\frac{W_0^2}{\rho_0 V_{EW}}$$

Next, the definition of brake horsepower is introduced as

$$B_{HP} = \frac{P_{HP}}{\eta_P}$$

where η_P is known as the propeller efficiency. Using the previous relationship for brake horsepower applied at the EW condition, following a pre-multiplication by V_{EW}, we finally have

$$V_{EW} \cdot B_{HP_{Required_{EW}}} = \frac{1}{1100} \frac{S c_{D_O} \rho_0}{\eta_P} \cdot V_{EW}^4 + \frac{4}{1100} \frac{1}{(\pi ARe) S \eta_P} \cdot \frac{W_0^2}{\rho_0}$$

Student Sample Problem 5.4

Starting from a generic *i-th* flight condition at maximum throttle setting, we have

$$(RC)_{True_i} = (RC)_{Obs_i} \frac{T_{True_i}}{T_{S_i}} = \frac{P_{E_i}}{W_i} = P_{A_i} - P_{R_i} = \frac{T_i V_i}{W_i} - \frac{D_i V_i}{W_i}$$

Show that $(RC)_{True_{ref_i}}$ is given by

$$(RC)_{True_{ref_i}} = (RC)_{True_i} \frac{W_i}{W_{ref}} - \frac{2 W_{ref}}{\rho V_i S (\pi ARe)} \left(1 - \frac{W_i^2}{W_{ref}^2}\right)$$

where

$$W_{ref} = \frac{W_{TO_1} + W_{TO_2} + \cdots + W_{TO_n}}{n}$$

Solution of Student Sample Problem 5.4

At a generic *i-th* flight condition at maximum throttle setting we have

$$(RC)_{True_i} = (RC)_{Obs_i} \frac{T_{True_i}}{T_{S_i}} = \frac{P_{E_i}}{W_i} = P_{A_i} - P_{R_i} = \frac{T_i V_i}{W_i} - \frac{D_i V_i}{W_i}$$

Of particular importance is the sensitivity of this parameter with respect to the aircraft weight. The derivative of $(RC)_{True}|_i$ with respect to the weight W leads to

$$\left.\frac{d(RC)_{True}}{dW}\right|_i = \left(-\frac{T_i V_i}{W_i^2} + \frac{D_i V_i}{W_i^2} - \frac{V_i}{W_i}\frac{dD}{dW}\bigg|_i\right)$$

The evaluation of the parameter dD/dW starts from the basic definition:

$$D = \frac{1}{2}\rho V^2 S\left(c_{D_O} + \frac{c_L^2}{\pi ARe}\right)$$

During a steady climb we have

$$L = W \cdot \cos(\gamma) = \frac{1}{2}\rho V^2 S c_L \cdot \cos(\gamma)$$

Assuming small enough climb angles (γ) we have

$$L = W = \frac{1}{2}\rho V^2 S c_L \rightarrow c_L^2 = \frac{4W^2}{\rho^2 V^4 S^2}$$

leading to

$$D = \frac{1}{2}\rho V^2 S c_{D_O} + \frac{1}{2}\rho V^2 S \left(\frac{\frac{4W^2}{\rho^2 V^4 S^2}}{\pi ARe}\right)$$

$$= \frac{1}{2}\rho V^2 S c_{D_O} + \frac{2W^2}{\rho V^2 S (\pi ARe)}$$

From the preceding relationship dD/dW can be found as

$$\frac{dD}{dW} = \frac{4W}{\rho V^2 S(\pi ARe)}$$

Next, insert the previous expressions for D and dD/dW into the relationship for $\left(d(RC)_{True}/dW\right)$ leading to

$$\left.\frac{d(RC)_{True}}{dW}\right|_i = \left(-\frac{T_i V_i}{W_i^2} + \frac{\left(\frac{1}{2}\rho V_i^2 S c_{D_O} + \frac{2W_i^2}{\rho V_i^2 S(\pi ARe)}\right)V_i}{W_i^2} - \frac{V_i}{W_i}\left(\frac{4W_i}{\rho V_i^2 S(\pi ARe)}\right)\right)\Bigg|_i$$

$$= \left(-\frac{T_i V_i}{W_i^2} + \rho S c_{D_O} \frac{V_i^3}{2W_i^2} + \frac{2\cancel{W_i^2} \cancel{V_i}}{\cancel{W_i^2} \rho V_i^{\cancel{2}} S(\pi ARe)} - \frac{4\cancel{W_i^2} \cancel{V_i}}{\cancel{W_i^2} \rho V_i^{\cancel{2}} S(\pi ARe)}\right)$$

$$= \left(-\frac{T_i V_i}{W_i^2} + \rho S c_{D_O} \frac{V_i^3}{2W_i^2} - \frac{2}{\rho V_i S(\pi ARe)}\right)$$

Next, an aircraft reference weight needs to be introduced. A suitable selection is an average of all possible take-off weights within the operational weight envelope, as shown:

$$W_{ref} = \frac{W_{TO_1} + W_{TO_2} + \ldots + W_{TO_n}}{n}$$

Next, integrating with respect to weight, we have

$$(RC)_{True_{ref_i}} - (RC)_{True_i} = \int_{W_i}^{W_{ref}} \left.\frac{d(RC)_{True}}{dW}\right|_i dW$$

$$= \int_{W_i}^{W_{ref}} \left(-\frac{T_i V_i}{W_i^2} + \rho S c_{D_O} \frac{V_i^3}{2W_i^2} - \frac{2}{\rho V_i S(\pi ARe)}\right) dW$$

The solution of the above integral leads to the following relationship:

$$(RC)_{True_{ref_i}} - (RC)_{True_i} = T_i V_i \cdot \left.\frac{1}{W}\right|_{W_i}^{W_{ref}} - \frac{1}{2}\rho S c_{D_O} V_i^3 \cdot \left.\frac{1}{W}\right|_{W_i}^{W_{ref}} - \frac{2}{\rho V_i S(\pi ARe)} \cdot W\Big|_{W_i}^{W_{ref}}$$

which ultimately leads to

$$(RC)_{True_{ref_i}} - (RC)_{True_i} = T_i V_i \cdot \left(\frac{1}{W_{ref}} - \frac{1}{W_i}\right) - \frac{1}{2}\rho S c_{D_O} V_i^3 \cdot \left(\frac{1}{W_{ref}} - \frac{1}{W_i}\right) - \frac{2}{\rho V_i S(\pi ARe)} \cdot (W_{ref} - W_i)$$

Next, recall that: $P_{Available} = P_{Required} + P_{Excess}$.
Therefore, at the generic *i-th* flight condition at maximum throttle setting, we have

$$T_i V_i = D_i V_i + W_i (RC)_{True_i} = \frac{1}{2}\rho V_i^3 S c_{D_O} + \frac{2W_i^2}{\rho V_i S(\pi ARe)} + W_i (RC)_{True_i}$$

Therefore, inserting the previous expression into the $(RC)_{True}\big|_{ref}$ relationship we have

$$(RC)_{True_{ref_i}} - (RC)_{True_i} = \left[\frac{1}{2}\rho V_i^3 S c_{D_O} + \frac{2W_i^2}{\rho V_i S(\pi ARe)} + W_i(RC)_{True_i}\right] \cdot \left(\frac{1}{W_{ref}} - \frac{1}{W_i}\right)$$

$$- \frac{1}{2}\rho S c_{D_O} V_i^3 \cdot \left(\frac{1}{W_{ref}} - \frac{1}{W_i}\right) - \frac{2}{\rho V_i S(\pi ARe)} \cdot (W_{ref} - W_i)$$

By grouping common terms and rearranging, we have

$$(RC)_{True_{ref_i}} = (RC)_{True_i} + W_i(RC)_{True_i}\frac{1}{W_{ref}} - \cancel{W_i(RC)_{True_i}\frac{1}{W_i}}$$

$$+ \cancel{\frac{1}{2}\rho V_i^3 S c_{D_O}\left(\frac{1}{W_{ref}} - \frac{1}{W_i}\right)} - \cancel{\frac{1}{2}\rho V_i^3 S c_{D_O}\left(\frac{1}{W_{ref}} - \frac{1}{W_i}\right)}$$

$$+ \frac{2W_i^2}{\rho V_i S(\pi ARe)}\left(\frac{1}{W_{ref}} - \frac{1}{W_i}\right) - \frac{2}{\rho V_i S(\pi ARe)} \cdot (W_{ref} - W_i)$$

$$= (RC)_{True_i}\frac{W_i}{W_{ref}} - \frac{2W_{ref}}{\rho V_i S(\pi ARe)}\left(1 - \frac{W_i^2}{W_{ref}^2}\right)$$

ultimately leading to

$$(RC)_{True_{ref_i}} = (RC)_{True_i}\frac{W_i}{W_{ref}} - \frac{2W_{ref}}{\rho V_i S(\pi ARe)}\left(1 - \frac{W_i^2}{W_{ref}^2}\right)$$

Student Sample Problem 5.5

Given the polynomial

$$(RC)_{True_{ref_{BESTFIT}}} = c_3 V_{True}^3 + c_2 V_{True}^2 + c_1 V_{True} + c_0$$

Find closed-form analytical relationships for the climb angle at maximum rate of climb $(\gamma_{RC_{MAX}})$ and the maximum climb angle (γ_{MAX}).

Solution of Student Sample Problem 5.5

Starting from

$$(RC)_{True_{ref_{BESTFIT}}} = c_3 V_{True}^3 + c_2 V_{True}^2 + c_1 V_{True} + c_0$$

$(\gamma_{RC_{MAX}})$ can be found by first setting $d(RC)_{True_{ref_{BESTFIT}}}/dV_{True} = 0$ and solving for the associated value of V_{True}. Therefore, we have:

$$\frac{d(RC)_{True_{ref_{BESTFIT}}}}{dV_{True}} = 3c_3 V_{True}^2 + 2c_2 V_{True} + c_1 = 0$$

The associated value of $V_{True}|_{RC_{MAX}}$ can be found solving the preceding quadratic relationship:

$$V_{True}\bigg|_{RC_{MAX}} = \frac{-2c_2 \pm \sqrt{(2c_2)^2 - 4(3c_3)(c_1)}}{2(3c_3)}$$

From the analysis of the WVU Cessna 206 flight data the following coefficients were evaluated:

$$c_3 = 7.67 \cdot 10^{-7}, \ c_2 = -1.46 \cdot 10^{-3}, \ c_1 = 0.3625, \ c_0 = -9.128$$

leading to

$$V_{True}\Big|_{RC_{MAX}} = \frac{-2c_2 \pm \sqrt{(2c_2)^2 - 4(3c_3)(c_1)}}{2(3c_3)}$$

$$= \frac{2.92 \cdot 10^{-3} \pm \sqrt{(2.92 \cdot 10^{-3})^2 - 4(2.301 \cdot 10^{-6})(0.3625)}}{4.6 \cdot 10^{-6}}$$

$$= 139.47 \ \text{ft/sec}$$

The associated value of $(RC)_{True_{ref_{BESTFIT}}MAX}$

$$(RC)_{True_{ref_{BESTFIT}}MAX} = c_3 V_{True}\Big|_{RC_{MAX}}^3 + c_2 V_{True}\Big|_{RC_{MAX}}^2 + c_1 V_{True}\Big|_{RC_{MAX}} + c_0$$

$$= 15.11 \ \text{ft/sec}$$

leading to

$$\tan(\gamma_{RC\,MAX}) = \frac{(RC)_{True_{ref_{BESTFIT}}MAX}}{V_{True}\Big|_{RC_{MAX}}} = 0.1083 \ \rightarrow \ \gamma_{RC\,MAX} \approx 6.18°$$

Next, γ_{MAX} can be evaluated using the following approach. At γ_{MAX} conditions the $(RC)_{True_{ref_{BESTFIT}}}$ intersects a line crossing the origin described by the equation $(RC)_{True_{ref}} = a\,V_{True}$. Furthermore, at the intersection of the preceding line and curve, the condition $d(RC)_{True_{ref_{BESTFIT}}}/dV_{True} = a$ applies. Therefore, we have

$$\begin{cases} c_3 V_{True}^3 + c_2 V_{True}^2 + c_1 V_{True} + c_0 = a V_{True} \\ 3c_3 V_{True}^2 + 2c_2 V_{True} + c_1 = a \end{cases}$$

By plugging the second equation into the first equation we have

$$c_3 V_{True}^3 + c_2 V_{True}^2 + c_1 V_{True} + c_0 = (3c_3 V_{True}^2 + 2c_2 V_{True} + c_1) V_{True}$$

leading to the polynomial

$$(c_3 - 3c_3) V_{True}^3 + (c_2 - 2c_2) V_{True}^2 + \cancel{(c_1 - c_1)} V_{True} + c_0 = 0$$

The solution of this polynomial leads to three different solutions; the solution in the area of interest is given by

$$V_{True}\Big|_{\gamma_{MAX}} = 82.74 \ \text{ft/sec}$$

Next, using the second equation, the slope a can be found:

$$3c_3 V_{True}\Big|_{\gamma_{MAX}}^2 + 2c_2 V_{True}\Big|_{\gamma_{MAX}} + c_1 = a = 0.1366$$

Leading to

$$(RC)_{True_{ref_{BESTFIT}\gamma MAX}} = a\,V_{True}\Big|_{\gamma_{MAX}}$$

Finally, we have

$$\tan(\gamma_{MAX}) = \frac{(RC)_{True_{ref_{BESTFIT_{\gamma MAX}}}}}{V_{True}\big|_{\gamma_{MAX}}} = 0.1365 \rightarrow \gamma_{MAX} \approx 7.77°$$

Both maximum rate of climb $(\gamma_{RC_{MAX}})$ and the maximum climb angle (γ_{MAX}) are shown in Figure SSP5.5.1.

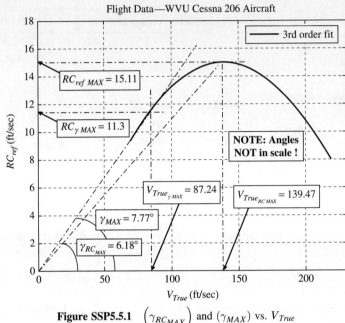

Figure SSP5.5.1 $\left(\gamma_{RC_{MAX}}\right)$ and (γ_{MAX}) vs. V_{True}

PROBLEMS

Problem 5.1

Consider a general aviation aircraft with the following geometric parameters:

$$S = 180 \text{ ft}^2, AR = 7.5$$

The aircraft has a reference weight (maximum take-off weight) $W_0 = 3{,}800$ *lbs* and a propeller efficiency $\eta_P = 0.7$.
Using the EW method, the statistical analysis of the flight data provides the following $V_{EW} \cdot B_{HP_{Required_{EW}}}$ vs. V_{EW}^4 relationship:

$$V_{EW} \cdot B_{HP_{Required_{EW}}} = 1.3 \cdot 10^{-5} \cdot V_{EW}^4 + 11{,}946$$

Find the numerical values for the coefficients c_{D_O}, e.

Problem 5.2

Consider a general aviation aircraft. The statistical analysis of the flight data provides the following $(RC)_{True_{ref_{BESTFIT}}}$ polynomial:

$$(RC)_{True_{ref_{BEST\ FIT}}} = -1.11 \cdot 10^{-3} V_{True}^2 + 0.3089\, V_{True} - 6.5712$$

Find the numerical values for the climb angle at maximum rate of climb $(\gamma_{RC_{MAX}})$ and the maximum climb angle (γ_{MAX}).

Chapter 6

Aircraft Stability and Design for Trim Conditions

TABLE OF CONTENTS

6.1 Introduction
6.2 Concept of Aircraft Stability
6.3 Criteria for Aircraft Static Stability
    6.3.1 Static Stability Criteria #1 (SSC #1)
    6.3.2 Static Stability Criteria #2 (SSC #2)
    6.3.3 Static Stability Criteria #3 (SSC #3)
    6.3.4 Static Stability Criteria #4 (SSC #4)
    6.3.5 Static Stability Criteria #5 (SSC #5)
    6.3.6 Static Stability Criteria #5, #6, and #7 (SSC #5, SSC #6, SSC #7)
    6.3.7 Static Stability Criteria #9 (SSC #9)
    6.3.8 Static Stability Criteria #10 (SSC #10)
6.4 Longitudinal Analysis of Steady-State Straight Flight
6.5 Lift Chart and Trim Diagram
    6.5.1 Lift Chart
    6.5.2 Trim Diagram
    6.5.3 Trim Diagrams for Different Classes of Aircraft
    6.5.4 Trim Diagrams for Thrust Axis Above/Below Center of Gravity
6.6 Lateral Directional Analysis of Steady-State Straight Flight
6.7 Summary
    References
    Student Sample Problems
    Problems

6.1 INTRODUCTION

This chapter reviews basic concepts related to the stability of the aircraft system. Emphasis will be placed on the concept of static stability. The conditions for dynamic stability will be discussed in Chapter 7. The equations of the aircraft at steady-state rectilinear conditions will be discussed, leading to the analysis of the longitudinal trim conditions and to the introduction of the trim diagram, an important tool for aircraft design purposes. Next, the analysis of the lateral set of the steady-state rectilinear equations yields to the lateral directional trim conditions. The chapter concludes with a detailed review of the important engines-out flight condition.

6.2 CONCEPT OF AIRCRAFT STABILITY

This section to provides a review of the definition of the important concepts of aircraft static and dynamic stability. In general terms, a dynamic system is said to be stable if the response of the system is bounded following a bounded input. The aircraft response is defined in this context as the time histories of the motion variables starting from a steady-state condition following an external disturbance or a pilot's maneuver. The perturbation terms of the motion variables considered for this analysis are

- linear velocity: u, v, w;
- angular velocity: p, q, r;
- aerodynamic angles: α, β.

The starting conditions for the analysis of the stability of the aircraft system are the following:[1]

1. Given a flight envelope, the aircraft has enough control authority—from a propulsive and aerodynamic point of view—to be able to go from any arbitrary point to any arbitrary point in the flight envelope. Furthermore, the aircraft must be able to maintain a steady-state rectilinear flight condition at any point in its operational flight envelope.[2,3,4,5,6]
2. The forces to be applied by the pilot on the control commands connected to the aerodynamic control surfaces need to be within specific ranges for each of the control commands, as defined by the appropriate rules of the certifying agency for the specific aircraft class.[2,3,4,5,6]

For an aircraft system with the preceding characteristics, the following definition of *Aircraft static stability* is provided.

An aircraft is *statically stable* if it features the capability of developing aerodynamic and thrust forces and moments which counteract a perturbation in a motion variable starting from a steady-state flight condition.

The aircraft static stability implies the capability for the aircraft to oppose any perturbation from a preexisting condition. Clearly, the analysis of the aircraft static stability does not depend on the Earth-fixed reference frame $X'\,Y'\,Z'$.

The next section will introduce a number of criteria to be satisfied for ensuring the aircraft static stability.

Aircraft dynamic stability is defined as the following:

An aircraft is dynamically stable if the perturbations of the motion variables—associated with either an external disturbance or a pilot's maneuver—decrease with time to zero or to a new set of steady-state values after either the disturbance or the pilot's maneuver has stopped.

The specific conditions for guaranteeing the aircraft dynamic stability will be discussed in Chapter 7. For the current analysis, the conditions for dynamic stability are more restrictive than the conditions for static stability. Particularly, the aircraft static stability is *a necessary but not sufficient condition* for dynamic stability. The aircraft dynamic stability requires the aircraft static stability in addition to other conditions, to be discussed in Chapter 7. This concept is graphically shown in Figure 6.1.

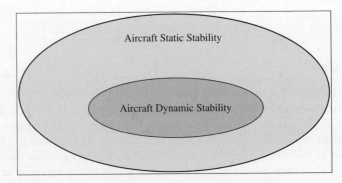

Figure 6.1 Relationship between Aircraft Static and Dynamic Stability

The next section outlines a list of specific criteria to be satisfied for ensuring aircraft static stability.

6.3 CRITERIA FOR AIRCRAFT STATIC STABILITY

An aircraft is classified as statically stable if its dynamic response satisfies the following general static stability criteria (SSC):[7,8,9]

SSC #1: $\dfrac{\partial(F_{A_X}+F_{T_X})}{\partial u} < 0$

SSC #2: $\dfrac{\partial(F_{A_Y}+F_{T_Y})}{\partial v} < 0$

SSC #3: $\dfrac{\partial(F_{A_Z}+F_{T_Z})}{\partial w} < 0$

SSC #4: $\dfrac{\partial(M_A+M_T)}{\partial \alpha} < 0$

SSC #5: $\dfrac{\partial(N_A+N_T)}{\partial \beta} > 0$

SSC #6: $\dfrac{\partial(L_A+L_T)}{\partial p} < 0$

SSC #7: $\dfrac{\partial(M_A+M_T)}{\partial q} < 0$

SSC #8: $\dfrac{\partial(N_A+N_T)}{\partial r} > 0$

SSC #9: $\dfrac{\partial(L_A+L_T)}{\partial \beta} > 0$

SSC #10: $\dfrac{\partial(M_A+M_T)}{\partial u} > 0$

The aerodynamic implications for the preceding criteria are discussed next.

6.3.1 Static Stability Criteria #1 (SSC #1)

From the aerodynamic modeling in Chapter 3, the relationship for F_{A_X}, F_{T_X} along the stability axes $X_S Y_S Z_S$ is given by

$$F_{A_X} + F_{T_X} = (-c_D + c_{T_X}) \bar{q} S$$

Taking the derivative with respect to the perturbation in the forward speed u the application of the SSC #1 leads to the following relationship along the stability axes $X_S Y_S Z_S$:

$$\left(c_{T_{X_u}} - c_{D_u}\right) + \dfrac{2}{V_{P_1}} \left(c_{T_{X_1}} - c_{D_1}\right) < 0$$

At steady state condition trim conditions we have $c_{T_{X_1}} = c_{D_1}$. Therefore

$$\boxed{\text{SSC \#1} \quad \dfrac{\partial(F_{A_X}+F_{T_X})}{\partial u} < 0 \Rightarrow \left(c_{T_{X_u}} - c_{D_u}\right) < 0}$$

From an engineering point of view, this criteria describes the desirable balance between drag and thrust following a small perturbation in the aircraft forward linear velocity. This particular aspect becomes critical during the terminal phases of the flight—final descent, approach, and landing—due to the fact that a number of control surfaces are typically deployed for generating the necessary lift and also the necessary drag for decreasing the airspeed prior to landing. In those specific conditions, assuming that the aircraft is experiencing a perturbation causing an increase in drag, the creation of additional thrust by the propulsion system is required to recover from the perturbation and to maintain static stability. The capability of generating additional thrust within a short amount of time is clearly dependent on the characteristic of the propulsion system.

6.3.2 Static Stability Criteria #2 (SSC #2)

From the aerodynamic modeling in Chapter 4, a relationship for F_{A_Y}, F_{T_Y} along the stability axes $X_S Y_S Z_S$ is given by

$$F_{A_Y} + F_{T_Y} = (c_Y + c_{T_Y})\,\bar{q}\,S$$

Using the relationship

$$\beta \approx \frac{v}{V_{P_1}} \rightarrow v \approx \beta V_{P_1}$$

and using the following expression for the derivative of a function

$$\frac{\partial(F_{A_Y} + F_{T_Y})}{\partial v} = \frac{\partial(F_{A_Y} + F_{T_Y})}{\partial \beta} \frac{\partial \beta}{\partial v}$$

the application of the SSC #2 simply leads to the relationship:

$$\left(c_{Y_\beta} + c_{T_{Y_\beta}} \right) < 0$$

Unless the propulsion system features a thrust vectoring mechanism, we have the condition $c_{T_{Y_\beta}} = 0$. Therefore

$$\boxed{\text{SSC \#2} \quad \frac{\partial(F_{A_Y} + F_{T_Y})}{\partial v} < 0 \Rightarrow c_{Y_\beta} < 0}$$

According to the modeling in, Chapter 4, this condition is trivially satisfied,[1,7,8,9] as shown in Figure 6.2.

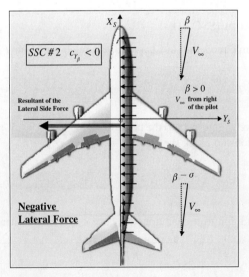

Figure 6.2 Aerodynamic Implication of SSC #2

6.3.3 Static Stability Criteria #3 (SSC #3)

From the aerodynamic modeling in Chapter 3, an expression for F_{A_Z}, F_{T_Z} along the stability axes $X_S Y_S Z_S$ is given by

$$F_{A_Z} + F_{T_Z} = (-c_L + c_{T_Z})\,\bar{q}\,S$$

Using the relationship: $\alpha \approx w/V_{P_1} \rightarrow w \approx \alpha V_{P_1}$ and using the following expression for the derivative of a function

$$\frac{\partial(F_{A_Z} + F_{T_Z})}{\partial w} = \frac{\partial(F_{A_Z} + F_{T_Z})}{\partial \alpha} \frac{\partial \alpha}{\partial w}$$

the application of the SSC #3 simply leads to the relationship:

$$\left(c_{T_{Z_\alpha}} - c_{L_\alpha}\right) < 0$$

For most propulsion systems the engine performance is mostly independent of the value of the angles of attack α, at least for the linear regions. Therefore, the approximation $c_{T_{Z_\alpha}} \approx 0$ can be introduced. Thus:

$$\boxed{\text{SSC \#3} \quad \frac{\partial(F_{A_Z} + F_{T_Z})}{\partial w} < 0 \Rightarrow c_{L_\alpha} > 0}$$

As shown in Figure 6.3, there will always be a positive lift associated with a positive α. Therefore, like the previous static stability criteria, this condition is trivially satisfied.[1,7,8,9]

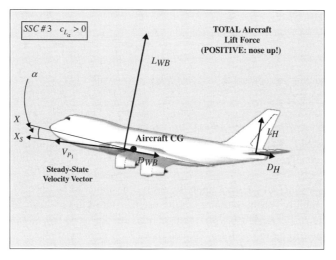

Figure 6.3 Aerodynamic Implication of SSC #3

6.3.4 Static Stability Criteria #4 (SSC #4)

From the aerodynamic modeling in Chapter 3, an expression for M_A, M_T along the stability axes $X_S Y_S Z_S$ is given by

$$M_A + M_T = (c_m + c_{m_T}) \bar{q} S \bar{c}$$

Thus, the application of the SSC #4 trivially leads to the condition $(c_{m_\alpha} + c_{m_{T_\alpha}}) < 0$.

Because $c_{m_{T_\alpha}}$ is the moment coefficient associated with the previous coefficient $c_{T_{Z_\alpha}}$, the assumption $c_{T_{Z_\alpha}} \approx 0$ will also lead to the assumption $c_{m_{T_\alpha}} \approx 0$. Thus,

$$\boxed{\text{SSC \#4} \quad \frac{\partial(M_A + M_T)}{\partial \alpha} < 0 \Rightarrow c_{m_\alpha} < 0}$$

The aerodynamic implications of SSC #4 are visualized in Figure 6.4. It is clear that because of the presence of the horizontal tail, an aircraft will naturally develop a negative pitching moment (nose down) with a positive angle of attack.[1,7,8,9]

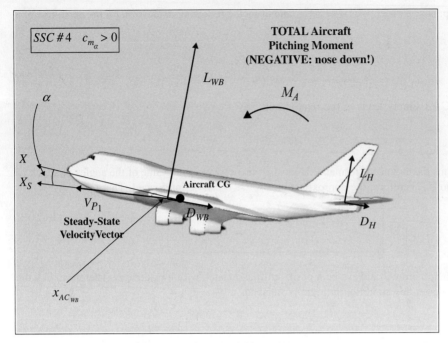

Figure 6.4 Aerodynamic Implication of SSC #4

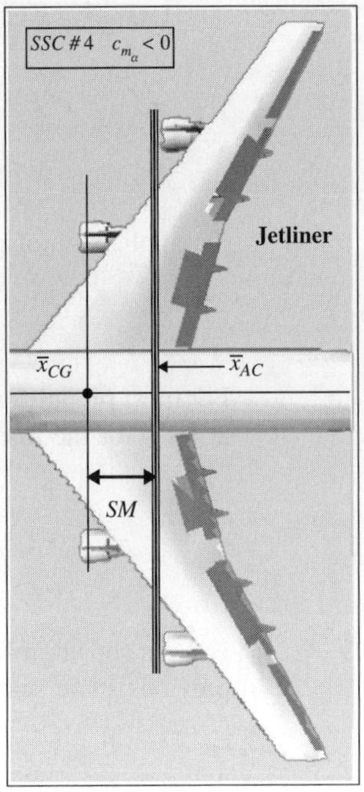

Figure 6.5 Aircraft Static Margin (SM) for a Statically Stable Aircraft

From the modeling introduced in Chapter 3 we also recall

$$c_{m_\alpha} = c_{L_\alpha} \, SM = c_{L_\alpha}(\bar{x}_{CG} - \bar{x}_{AC})$$

Therefore, the SSC #4 implies that the aircraft center of gravity is located in front of the aircraft aerodynamic center. This condition is satisfied for aircraft designed to be open-loop longitudinally stable, as shown in Figure 6.5.

Conversely, SSC #4 does not apply to aircraft designed to be open-loop unstable. For this class of aircraft, longitudinal stability is achieved through the use of control laws whose deflections are directly activated by the flight computer. This design has the goal of maximizing specific aircraft characteristics, such as agility and maneuverability. Refer to[10,11] for a detailed discussion of these topics.

6.3.5 Static Stability Criteria #5 (SSC #5)

From the aerodynamic modeling in Chapter 4, an expression for N_A, N_T along the stability axes $X_S Y_S Z_S$ is given by

$$N_A + N_T = (c_n + c_{n_T}) \, \bar{q} \, S \, b$$

Thus, the application of the SSC #5 trivially leads to the condition $(c_{n_\beta} + c_{n_{T_\beta}}) > 0$. It can be easily visualized that at nominal conditions $c_{n_{T_\beta}} \approx 0$. Therefore,

$$\boxed{\text{SSC \#5} \quad \frac{\partial (N_A + N_T)}{\partial \beta} > 0 \Rightarrow c_{n_\beta} > 0}$$

Once again, the aerodynamic implications of the static stability criteria are clear. Figure 6.6 shows that an aircraft will naturally develop a positive yawing moment following a perturbation in the sideslip angle. The modeling in Chapter 4 has shown that this effect, the weathercock effect, is largely due to the presence of the vertical tail.[1,7,8,9]

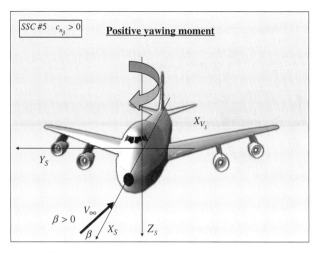

Figure 6.6 Aerodynamic Implication of SSC #5

6.3.6 Static Stability Criteria #5, #6, and #7 (SSC #5, SSC #6, SSC #7)

These SSCs are related to the damping derivatives along the roll, pitch, and yaw axes, respectively. The importance of these derivatives will be clearly outlined during the analysis of the aircraft dynamic stability in Chapter 7. These SSCs are trivially satisfied for all aircraft configurations. Therefore, we will have

$$\text{SSC \#6} \quad \frac{\partial(L_A + L_T)}{\partial p} < 0 \Rightarrow c_{l_p} < 0$$

$$\text{SSC \#7} \quad \frac{\partial(M_A + M_T)}{\partial q} < 0 \Rightarrow c_{m_q} < 0$$

$$\text{SSC \#8} \quad \frac{\partial(N_A + N_T)}{\partial r} > 0 \Rightarrow c_{n_r} < 0$$

Specifically, c_{l_p}, the derivative associated with the SSC #6, will be responsible for the important rolling dynamic mode; c_{m_q}, the derivative associated with the SSC #7, will be responsible for the damping coefficient of the important short period mode; finally c_{n_r}, the derivative associated with the SSC #8, will be responsible for the damping coefficient of the important dutch roll mode.

6.3.7 Static Stability Criteria #9 (SSC #9)

From the aerodynamic modeling in Chapter 4, an expression for L_A, L_T along the stability axes $X_S Y_S Z_S$ is given by

$$L_A + L_T = (c_l + c_{l_T})\,\bar{q}\,S\,b$$

Similar to the SSC #5 relative to the yawing moment, the application of the SSC #9 leads to the condition $(c_{l_\beta} + c_{l_{T_\beta}}) > 0$. At nominal conditions $c_{l_{T_\beta}} \approx 0$. Therefore,

$$\text{SSC \#9} \quad \frac{\partial(L_A + L_T)}{\partial \beta} > 0 \Rightarrow c_{l_\beta} < 0$$

The modeling in Chapter 4 demonstrated how this stability criterion is trivially satisfied for virtually all aircraft configuration.[1,7,8,9] Particularly, this cumulative effect, Dihedral Effect, is given by the following contributions:

$$c_{l\beta} = c_{l\beta_{WB}} + c_{l\beta_H} + c_{l\beta_V} \approx \left(c_{l\beta_{WB}}\big|_{\text{I}} + c_{l\beta_{WB}}\big|_{\text{II}} + c_{l\beta_{WB}}\big|_{\text{III}}\right) + c_{l\beta_H} + c_{l\beta_V}$$

Recall that, with the exception of $c_{l\beta_H}$ being negligible for most aircraft, each of the preceding contributions is <0, with the mutual exclusion of the $c_{l\beta_{WB}}\big|_{\text{I}}$, $c_{l\beta_{WB}}\big|_{\text{II}}$ effects, as shown in Figure 6.7.

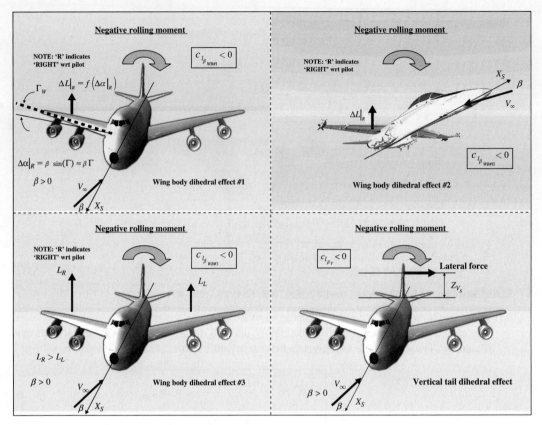

Figure 6.7 Aerodynamic Implication of SSC #9

6.3.8 Static Stability Criteria #10 (SSC #10)

From the aerodynamic modeling in Chapter 3, an expression for M_A, M_T along the stability axes $X_S Y_S Z_S$ is given by

$$M_A + M_T = (c_m + c_{m_T})\, \bar{q}\, S\, \bar{c}$$

Taking the derivative with respect to the perturbation in the forward speed u, the application of SSC #10 leads to the following relationship along the stability axes $X_S Y_S Z_S$:

$$(c_{m_u} + c_{m_{T_u}}) + \frac{2}{V_{P_1}}(c_{m_1} + c_{m_{T_1}}) > 0$$

By definition of trim conditions we have $(c_{m_1} + c_{m_{T_1}}) = 0$. Additionally, for most propulsion systems we have $c_{m_{T_u}}$. Therefore,

$$\boxed{\text{SSC \#10} \quad \frac{\partial(M_A + M_T)}{\partial u} > 0 \Rightarrow c_{m_u} > 0}$$

It turns out that the SSC #10 is not trivially satisfied, especially for jetliners at transonic flight conditions. Specifically, c_{m_u}, known as the tuck derivative, tends to have small positive values at low subsonic speeds; however, as the speed approaches the transonic region, this coefficient tends to decrease to very low values and eventually, for some specific aircraft, can turn negative, leading to a loss of static stability. Ultimately, this phenomenon is due to a sudden shift of the aircraft aerodynamic center from approximately 0.25 of the MAC to approximately 0.5 of the MAC in supersonic. Therefore, the value of c_{m_u} is dependent on the specific wing sections used for the wing design. Earlier classes of jetliners were quite sensitive to the tuck problem, while newer jetliners feature a more sophisticated aerodynamic design preventing the occurrence of the tuck problem. (See Student Sample Problems 6.1 and 6.2.)

6.4 LONGITUDINAL ANALYSIS OF STEADY-STATE STRAIGHT FLIGHT

This section analyzes the steady-state straight flight conditions from a longitudinal point of view. Once these equations are properly set it is then possible to solve for the aircraft trim conditions.

In the most general form the solution for the aircraft trim conditions is a nonlinear problem, since all the aerodynamic coefficients are or can be nonlinear functions of the angle of attack, Mach, or other variables. In this case, the problem of finding the design trim conditions can be solved using a nonlinear multivariable root finder. Clearly, a solution for trim conditions outside the linear region is not an acceptable outcome from a design point of view and will necessarily lead to a modification of the design. For example, it would not be acceptable for the design trim conditions for a general aviation aircraft to be with an angle of attack above or around stall conditions or with a deflection of the control surface exceeding or close to the limit deflection.

Once the design for an aircraft has reached a mature stage and the designed trim conditions are expected to be well within the linear aerodynamic range, it is appropriate to assume linearity. Therefore, the use of the stability and control derivatives discussed so far is recommended.

Prior to this analysis, it is important to specify the selected configuration for the horizontal tail. Without loss of generality, consider the most general case of a horizontal tail, shown in Figure 6.8, featuring a stabilator (for trimming the aircraft) and an elevator (for maneuvering the aircraft starting from a trim condition).

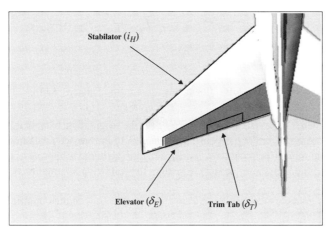

Figure 6.8 (Stabilator + Elevator) Horizontal Tail Configuration

Considering the longitudinal steady-state equations developed in Chapter 1 along with the aerodynamic modeling in Chapters 3 and 4, with respect to the stability axes shown in Figure 6.9, the longitudinal equations describing steady-state conditions[1,7,8,9] are given by

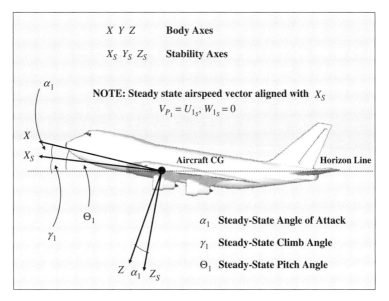

Figure 6.9 Body and Stability Axes

$$\begin{cases} mg\sin(\gamma_1) = -\bar{q}_1 S \left(c_{D_O} + c_{D_\alpha}\alpha_1 + c_{D_{\delta_E}}\delta_{E_1} + c_{D_{i_H}}i_{H_1}\right) + T_1\cos(\Theta_T + \alpha_1) \\ mg\cos(\gamma_1) = \bar{q}_1 S \left(c_{L_O} + c_{L_\alpha}\alpha_1 + c_{L_{\delta_E}}\delta_{E_1} + c_{L_{i_H}}i_{H_1}\right) + T_1\sin(\Theta_T + \alpha_1) \\ 0 = \bar{q}_1 S \bar{c}\left(c_{m_O} + c_{m_\alpha}\alpha_1 + c_{m_{\delta_E}}\delta_{E_1} + c_{m_{i_H}}i_{H_1}\right) - T_1 z_T \end{cases}$$

where the propulsive geometric parameters Θ_T, z_T are shown in Figure 6.10.

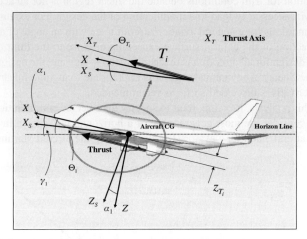

Figure 6.10 Longitudinal Geometric Parameters of the Propulsion System

It is assumed that the stabilators i_{H_1} are used for the longitudinal trim; thus, it can be assumed that the deflections of the elevators are set to zero or to a constant value δ_{E_1}. Therefore, the preceding system can be considered as a system of three equations in the unknowns γ_1, α_1, i_{H_1}, T; thus, an iterative approach needs to be introduced for the solution of the system. Within a first level of approximation, the following assumptions can be made:

- γ_1 is small enough such that $\cos(\gamma_1) \approx 1$, $\sin(\gamma_1) \approx 0$, with the condition $\gamma_1 = 0°$ identifying the condition of rectilinear flight;
- $(\Theta_T + \alpha_1)$ is small enough such that $T_1\sin(\Theta_T + \alpha_1) \approx 0$;
- At any point in the aircraft flight envelope there is enough installed thrust to overcome the aerodynamic drag;
- z_T is at least set to be zero such that $M_{T_1} = T_1 z_T = 0$. However, this assumption can be later revisited, and it will be shown how the solutions can be modified in the more general case of $M_{T_1} = T_1 z_T \leq 0$ or ≥ 0, depending on whether the thrust axis is located above or below the aircraft center of gravity.

Under these assumptions, the previous system of three equations will reduce itself to the following system of two equations:

$$\begin{cases} mg = \bar{q}_1 S \left(c_{L_O} + c_{L_\alpha}\alpha_1 + c_{L_{\delta_E}}\delta_{E_1} + c_{L_{i_H}}i_{H_1}\right) \\ 0 = \bar{q}_1 S \bar{c}\left(c_{m_O} + c_{m_\alpha}\alpha_1 + c_{m_{\delta_E}}\delta_{E_1} + c_{m_{i_H}}i_{H_1}\right) \end{cases}$$

in the unknowns α_1, i_{H_1}. By imposing the relationship $mg = \bar{q}_1 S c_{L_1} \Rightarrow c_{L_1} = \dfrac{mg}{\bar{q}_1 S}$, the previous simplified longitudinal equations can be rearranged in the following matrix format:

$$\begin{bmatrix} c_{L_\alpha} & c_{L_{i_H}} \\ c_{m_\alpha} & c_{m_{i_H}} \end{bmatrix} \begin{Bmatrix} \alpha_1 \\ i_{H_1} \end{Bmatrix} = \begin{Bmatrix} c_{L_1} - c_{L_O} - c_{L_{\delta_E}}\delta_{E_1} \\ -c_{m_O} - c_{m_{\delta_E}}\delta_{E_1} \end{Bmatrix}$$

where the quantities $\left(c_{L_{\delta_E}}\delta_{E_1}, c_{m_{\delta_E}}\delta_{E_1}\right)$ are either known or zero, as previously discussed. The application of Cramer's rule leads to the following solutions:

$$\alpha_1 = \dfrac{\begin{vmatrix} (c_{L_1} - c_{L_O} - c_{L_{\delta_E}}\delta_{E_1}) & c_{L_{i_H}} \\ (-c_{m_O} - c_{m_{\delta_E}}\delta_{E_1}) & c_{m_{i_H}} \end{vmatrix}}{\begin{vmatrix} c_{L_\alpha} & c_{L_{i_H}} \\ c_{m_\alpha} & c_{m_{i_H}} \end{vmatrix}} = \dfrac{c_{m_{i_H}}\left(c_{L_1} - c_{L_O} - c_{L_{\delta_E}}\delta_{E_1}\right) - c_{L_{i_H}}\left(-c_{m_O} - c_{m_{\delta_E}}\delta_{E_1}\right)}{\left(c_{L_\alpha}c_{m_{i_H}} - c_{m_\alpha}c_{L_{i_H}}\right)}$$

$$i_{H_1} = \frac{\begin{vmatrix} c_{L_\alpha} & (c_{L_1} - c_{L_O} - c_{L_{\delta_E}}\delta_{E_1}) \\ c_{m_\alpha} & (-c_{m_O} - c_{m_{\delta_E}}\delta_{E_1}) \end{vmatrix}}{\begin{vmatrix} c_{L_\alpha} & c_{L_{i_H}} \\ c_{m_\alpha} & c_{m_{i_H}} \end{vmatrix}} = \frac{c_{L_\alpha}(-c_{m_O} - c_{m_{\delta_E}}\delta_{E_1}) - c_{m_\alpha}(c_{L_1} - c_{L_O} - c_{L_{\delta_E}}\delta_{E_1})}{(c_{L_\alpha}c_{m_{i_H}} - c_{m_\alpha}c_{L_{i_H}})}$$

The values obtained for α_1, i_{H_1} from the preceding equations are essentially a description of the longitudinal trim conditions of the aircraft. In the configuration in which the aircraft does not feature stabilators and elevators are used for both trimming and maneuvering purposes, the alternative system of equations will be used:

$$\begin{bmatrix} c_{L_\alpha} & c_{L_{\delta_E}} \\ c_{m_\alpha} & c_{m_{\delta_E}} \end{bmatrix} \begin{Bmatrix} \alpha_1 \\ \delta_{E_1} \end{Bmatrix} = \begin{Bmatrix} c_{L_1} - c_{L_O} \\ -c_{m_O} \end{Bmatrix}$$

leading to the following solutions in terms of α_1, δ_{E_1}:

$$\alpha_1 = \frac{\begin{vmatrix} (c_{L_1} - c_{L_O}) & c_{L_{\delta_E}} \\ (-c_{m_O}) & c_{m_{\delta_E}} \end{vmatrix}}{\begin{vmatrix} c_{L_\alpha} & c_{L_{\delta_E}} \\ c_{m_\alpha} & c_{m_{\delta_E}} \end{vmatrix}} = \frac{c_{m_{\delta_E}}(c_{L_1} - c_{L_O}) - c_{L_{\delta_E}}(-c_{m_O})}{(c_{L_\alpha}c_{m_{\delta_E}} - c_{m_\alpha}c_{L_{\delta_E}})}$$

$$\delta_{E_1} = \frac{\begin{vmatrix} c_{L_\alpha} & (c_{L_1} - c_{L_O}) \\ c_{m_\alpha} & (-c_{m_O}) \end{vmatrix}}{\begin{vmatrix} c_{L_\alpha} & c_{L_{\delta_E}} \\ c_{m_\alpha} & c_{m_{\delta_E}} \end{vmatrix}} = \frac{c_{L_\alpha}(-c_{m_O}) - c_{m_\alpha}(c_{L_1} - c_{L_O})}{(c_{L_\alpha}c_{m_{\delta_E}} - c_{m_\alpha}c_{L_{\delta_E}})}$$

The resulting values for (α_1, i_{H_1}) or (α_1, δ_{E_1}) have major implications on the aircraft design. Particularly, the value of α_1 is critically important for the minimization of the entire drag force, which is directly related to the necessary thrust (or power) for steady-state flight, which is consequently related to fuel consumption, fuel weight, and aircraft autonomy. An expected range for α_1 is between $[-1°, 3°]$ at cruise conditions; clearly, higher α_1 values are needed during approach and descent. It should be clear that a difference in a fraction of a degree for α_1 can lead to a difference in fuel consumption up to thousands of pounds of fuel for a widebody jetliner during a typical oversea flight. The value for (i_{H_1}, δ_{E_1}) has instead implications from a pilot point of view in terms of the control authority of the horizontal tail and its suitability for trimming the aircraft at every point in the flight envelope. Low values in the range of $[-3°, 3°]$ are desirable for the purpose of avoiding potential saturation of the control surface, especially in the configuration in which the elevator is used for both trimming and maneuvering purposes[1]. A few numerical examples confirm the accuracy of this simplified approach.

Consider, for example, two different types of aircraft from Appendix B, the Cessna 182 (general aviation) and the Learjet 24 (business jet).

For the Cessna 182, which features elevators for trimming and maneuvering purposes, at cruise steady-state conditions ($Alt. = 5{,}000$ ft, $V_{P_1} = 220$ ft/sec) we have

$$\alpha_1 = \frac{c_{m_{\delta_E}}(c_{L_1} - c_{L_O}) - c_{L_{\delta_E}}(-c_{m_O})}{(c_{L_\alpha}c_{m_{\delta_E}} - c_{m_\alpha}c_{L_{\delta_E}})} = \frac{-1.122 \cdot (0.307 - 0.307) - 0.43 \cdot (-0.04)}{((4.41 \cdot -1.122) - (-0.613 \cdot 0.43))}$$
$$= -0.0037 \text{ rad} \approx -0.2°$$

compared to the true provided value $\alpha_{1_{TRUE}} \approx 0°$. Therefore, the preceding approximation can be considered accurate. The associated value for δ_{E_1} is given by

$$\delta_{E_1} = \frac{c_{L_\alpha}(-c_{m_O}) - c_{m_\alpha}(c_{L_1} - c_{L_O})}{(c_{L_\alpha}c_{m_{\delta_E}} - c_{m_\alpha}c_{L_{\delta_E}})} = \frac{4.41 \cdot (-0.04) + 0.613(0.307 - 0.307)}{((4.41 \cdot -1.122) - (-0.613 \cdot 0.43))} = 0.038 \text{ rad} \approx 2.1°$$

This result implies that at cruise conditions this aircraft will produce enough lift that the pilot needs to provide a down deflection (positive) for the elevators to generate a nose down pitching moment to maintain the aircraft at steady-state rectilinear conditions.

For the Learjet 24 which features stabilators for trimming purposes, at cruise steady-state conditions with maximum weight (Alt. = 40,000 ft, $V_{P_1} = 677$ ft/sec) we have

$$\alpha_1 = \frac{c_{m_{i_H}}(c_{L_1} - c_{L_O}) - c_{L_{i_H}}(-c_{m_O})}{(c_{L_\alpha} c_{m_{i_H}} - c_{m_\alpha} c_{L_{i_H}})} = \frac{-2.5 \cdot (0.41 - 0.13) - 0.94 \cdot (-0.05)}{((5.84 \cdot -2.5) - (-0.64 \cdot 0.94))} = 0.0467 \text{ rad} \approx 2.67°$$

compared to the true provided value $\alpha_{1_{TRUE}} = 2.7°$ showing that the preceding approximation can be considered accurate. Assuming zero elevator deflections, the associated value for i_{H_1} is given by

$$i_{H_1} = \frac{c_{L_\alpha}(-c_{m_O} - \cancel{c_{m_{\delta_E}} \delta_{E_1}}) - c_{m_\alpha}(c_{L_1} - c_{L_O} - \cancel{c_{L_{\delta_E}} \delta_{E_1}})}{(c_{L_\alpha} c_{m_{i_H}} - c_{m_\alpha} c_{L_{i_H}})} = 0.0081 \text{ rad} \approx 0.46°$$

Consider again the Learjet 24 at cruise steady-state conditions (Alt. = 40,000 ft, $V_{P_1} = 677$ ft/sec); however, this time consider the low weight configuration. This implies a lower value for $c_{L_1} = m\, g/\bar{q}_1\, S$. In this case, using the data in Appendix B we have

$$\alpha_1 = \frac{c_{m_{i_H}}(c_{L_1} - c_{L_O}) - c_{L_{i_H}}(-c_{m_O})}{(c_{L_\alpha} c_{m_{i_H}} - c_{m_\alpha} c_{L_{i_H}})} = \frac{-2.5 \cdot (0.28 - 0.13) - 0.94 \cdot (-0.05)}{((5.84 \cdot -2.5) - (-0.64 \cdot 0.94))} = 0.0234 \text{ rad} \approx 1.4°$$

Therefore, an important result is that the aircraft require lower values of the angle of attack when flying at lower mass configurations, since a lower amount of lift is required to offset the lower weight.

Assuming again zero elevator deflections, the associated value for i_{H_1} is given by

$$i_{H_1} = \frac{c_{L_\alpha}(-c_{m_O} - \cancel{c_{m_{\delta_E}} \delta_{E_1}}) - c_{m_\alpha}(c_{L_1} - c_{L_O} - \cancel{c_{L_{\delta_E}} \delta_{E_1}})}{(c_{L_\alpha} c_{m_{i_H}} - c_{m_\alpha} c_{L_{i_H}})} = \frac{5.84 \cdot (-0.05) + 0.64 \cdot (0.28 - 0.13)}{((5.84 \cdot -2.5) - (-0.64 \cdot 0.94))}$$

$$= 0.014 \text{ rad} \approx 0.8°$$

Thus, the effect of the lower mass on the i_{H_1} trim value is fairly small; in fact, only a slightly higher positive (down) stabilator deflection is needed to generate a nose down pitching moment.

The preceding analysis leads to the following gradients for a sensitivity analysis of the trim conditions:

$$\left(\frac{\partial \alpha_1}{\partial c_{L_1}}, \frac{\partial i_{H_1}}{\partial c_{L_1}}\right) \text{ for aircraft featuring stabilators for trimming purposes}$$

$$\left(\frac{\partial \alpha_1}{\partial c_{L_1}}, \frac{\partial \delta_{E_1}}{\partial c_{L_1}}\right) \text{ for aircraft featuring elevators for trimming purposes}$$

Since $c_{L_1} = mg/\bar{q}_1\, S = W/\bar{q}_1\, S$ where $\bar{q}_1 = 0.5\, \rho V_{P_1}^2$, we have $c_{L_1} = \dfrac{2W}{\rho V_{P_1}^2\, S}$. Therefore, the value of c_{L_1} is function of the following:

- Aircraft weight;
- Airspeed;
- Altitude (through the density).

Expressions for the preceding gradients are given by

$$\frac{\partial \alpha_1}{\partial c_{L_1}} = \frac{c_{m_{i_H}}}{(c_{L_\alpha} c_{m_{i_H}} - c_{m_\alpha} c_{L_{i_H}})} \quad \text{or} \quad \frac{\partial \alpha_1}{\partial c_{L_1}} = \frac{c_{m_{\delta_E}}}{(c_{L_\alpha} c_{m_{\delta_E}} - c_{m_\alpha} c_{L_{\delta_E}})}$$

$$\frac{\partial i_{H_1}}{\partial c_{L_1}} = \frac{-c_{m_\alpha}}{(c_{L_\alpha} c_{m_{i_H}} - c_{m_\alpha} c_{L_{i_H}})} \quad \text{or} \quad \frac{\partial \delta_{E_1}}{\partial c_{L_1}} = \frac{-c_{m_\alpha}}{(c_{L_\alpha} c_{m_{\delta_E}} - c_{m_\alpha} c_{L_{\delta_E}})}$$

A first important step is the analysis of the signs for the preceding gradients. For this purpose, it is important to recall that for all aircraft we have that $|c_{L_\alpha} c_{m_{i_H}}| \gg |c_{m_\alpha} c_{L_{i_H}}|$ or $|c_{L_\alpha} c_{m_{\delta_E}}| \gg |c_{m_\alpha} c_{L_{\delta_E}}|$. Therefore, the sign of the denominators is dictated by the signs of the products $(c_{L_\alpha} c_{m_{i_H}})$ or $(c_{L_\alpha} c_{m_{\delta_E}})$, which are negative due to the negative sign for $c_{m_{i_H}}$ and $c_{m_{\delta_E}}$. Because c_{m_α} is also negative for statically stable aircraft, we have

$$\frac{\partial \alpha_1}{\partial c_{L_1}} = \frac{c_{m_{i_H}}}{\left(c_{L_\alpha} c_{m_{i_H}} - c_{m_\alpha} c_{L_{i_H}}\right)} > 0 \quad \text{or} \quad \frac{\partial \alpha_1}{\partial c_{L_1}} = \frac{c_{m_{\delta_E}}}{\left(c_{L_\alpha} c_{m_{\delta_E}} - c_{m_\alpha} c_{L_{\delta_E}}\right)} > 0$$

$$\frac{\partial i_{H_1}}{\partial c_{L_1}} = \frac{-c_{m_\alpha}}{\left(c_{L_\alpha} c_{m_{i_H}} - c_{m_\alpha} c_{L_{i_H}}\right)} < 0 \quad \text{or} \quad \frac{\partial \delta_{E_1}}{\partial c_{L_1}} = \frac{-c_{m_\alpha}}{\left(c_{L_\alpha} c_{m_{\delta_E}} - c_{m_\alpha} c_{L_{\delta_E}}\right)} < 0$$

It is critical to point out that for longitudinally static unstable aircraft (an aircraft with the center of gravity located beyond the aircraft aerodynamic center) we have

$$c_{m_\alpha} > 0 \Rightarrow \frac{\partial i_{H_1}}{\partial c_{L_1}} = \frac{-c_{m_\alpha}}{\left(c_{L_\alpha} c_{m_{i_H}} - c_{m_\alpha} c_{L_{i_H}}\right)} > 0$$

$$c_{m_\alpha} > 0 \Rightarrow \frac{\partial \delta_{E_1}}{\partial c_{L_1}} = \frac{-c_{m_\alpha}}{\left(c_{L_\alpha} c_{m_{\delta_E}} - c_{m_\alpha} c_{L_{\delta_E}}\right)} > 0$$

The following analysis is provided for the different parameters affecting the value of c_{L_1}.

As the aircraft weight decreases (for example, due to fuel consumption for civilian aircraft or dropping of payloads for military aircraft), the value of c_{L_1} will also decrease. Given the positive value of the gradient $\partial \alpha_1 / \partial c_{L_1}$, a lower value of the angle of attack α_1 will therefore be required to trim the aircraft. In terms of the deflections for the control surfaces, given the negative values of the gradients $\left(\partial i_{H_1}/\partial c_{L_1}, \partial \delta_{E_1}/\partial c_{L_1}\right)$, higher positive (down) deflections will be required to generate an upward force at the horizontal tail, leading to a negative nose down pitching moment, associated with the lower value of α_1. The functionality is summarized as in the following:

$$\Delta W < 0 \Rightarrow \Delta c_{L_1} = \frac{2\Delta W}{\rho V_{P_1}^2 S} < 0 \Rightarrow \Delta \alpha_1 = \frac{\partial \alpha_1}{\partial c_{L_1}} \Delta c_{L_1} < 0 \Rightarrow \text{Nose} \downarrow$$

$$\Delta W < 0 \Rightarrow \Delta c_{L_1} = \frac{2\Delta W}{\rho V_{P_1}^2 S} < 0 \Rightarrow \begin{cases} \Delta i_{H_1} = \dfrac{\partial i_{H_1}}{\partial c_{L_1}} \cdot \Delta c_{L_1} > 0 \Rightarrow \text{Stab.} \downarrow \Rightarrow \text{Force} \uparrow \Rightarrow \text{Nose} \downarrow \\ \Delta \delta_{E_1} = \dfrac{\partial \delta_{E_1}}{\partial c_{L_1}} \cdot \Delta c_{L_1} > 0 \Rightarrow \text{Elev.} \downarrow \Rightarrow \text{Force} \uparrow \Rightarrow \text{Nose} \downarrow \end{cases}$$

The trend is shown in Figure 6.11.

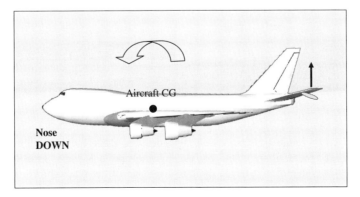

Figure 6.11 Boeing 747 with a Nose Down Pitching Moment

As the airspeed decreases, the value of c_{L_1} increases in a nonlinear fashion given the relationship $c_{L_1} = 2W/\rho V_{P_1}^2 S$. In this case, given the positive value of the gradient $\partial \alpha_1/\partial c_{L_1}$, a higher value of the angle of attack α_1 is therefore required to trim the aircraft. In fact, values of α_1 up to 8 to10 degrees might be required to trim the aircraft during the last phases of an approach. In terms of the deflections for the control surfaces, given the negative values of the gradients $\left(\frac{\partial i_{H_1}}{\partial c_{L_1}}, \frac{\partial \delta_{E_1}}{\partial c_{L_1}}\right)$, higher negative (up) deflections are required to generate a downward force at the horizontal tail leading to a positive nose up pitching moment, associated with the higher value for α_1. The functionality is summarized as in the following:

$$\Delta V_{P_1} < 0 \Rightarrow \Delta c_{L_1} = \frac{2W}{\rho (\Delta V_{P_1})^2 S} > 0 \Rightarrow \Delta \alpha_1 = \frac{\partial \alpha_1}{\partial c_{L_1}} \cdot \Delta c_{L_1} > 0 \Rightarrow Nose \uparrow$$

$$\Delta V_{P_1} < 0 \Rightarrow \Delta c_{L_1} = \frac{2W}{\rho (\Delta V_{P_1})^2 S} > 0 \Rightarrow \begin{cases} \Delta i_{H_1} = \frac{\partial i_{H_1}}{\partial c_{L_1}} \cdot \Delta c_{L_1} < 0 \Rightarrow Stab.\uparrow \Rightarrow Force \downarrow \Rightarrow Nose \uparrow \\ \Delta \delta_{E_1} = \frac{\partial \delta_{E_1}}{\partial c_{L_1}} \cdot \Delta c_{L_1} < 0 \Rightarrow Elev.\uparrow \Rightarrow Force \downarrow \Rightarrow Nose \uparrow \end{cases}$$

The trend is shown in Figure 6.12.

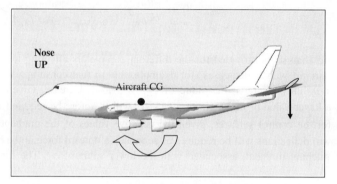

Figure 6.12 Boeing 747 with a Nose up Pitching Moment

Given the positive value of the gradient $\partial \alpha_1 / \partial c_{L_1}$, a lower value of the angle of attack α_1 is therefore required to trim the aircraft. In terms of the deflections for the control surfaces, given the negative values of the gradients $\left(\frac{\partial i_{H_1}}{\partial c_{L_1}}, \frac{\partial \delta_{E_1}}{\partial c_{L_1}}\right)$, higher positive (down) deflections are required to generate an upward force at the horizontal tail, leading to a negative nose down pitching moment, associated with the lower value of α_1. This functionality is summarized as in the following:

$$\Delta W < 0 \Rightarrow \Delta c_{L_1} = \frac{2 \Delta W}{\rho V_{P_1}^2} < 0 \Rightarrow \Delta \alpha_1 = \frac{\partial \alpha_1}{\partial c_{L_1}} \cdot \Delta c_{L_1} < 0 \Rightarrow Nose \downarrow$$

$$\Delta W < 0 \Rightarrow \Delta c_{L_1} = \frac{2 \Delta W}{\rho V_{P_1}^2} < 0 \Rightarrow \begin{cases} \Delta i_{H_1} = \frac{\partial i_{H_1}}{\partial c_{L_1}} \cdot \Delta c_{L_1} > 0 \Rightarrow Stab.\downarrow \Rightarrow Force \uparrow \Rightarrow Nose \downarrow \\ \Delta \delta_{E_1} = \frac{\partial \delta_{E_1}}{\partial c_{L_1}} \cdot \Delta c_{L_1} > 0 \Rightarrow Elev.\downarrow \Rightarrow Force \uparrow \Rightarrow Nose \downarrow \end{cases}$$

Finally, as the altitude decreases during a descent, the value of the air density ρ will increase (due to the $\rho = \rho(Altitude)$ relationship in the Standard Atmospheric Model). This, in turn, leads to a mild nonlinear decrease for the value of c_{L_1}. Given the positive value of the gradient $\partial \alpha_1 / \partial c_{L_1}$, a lower value of the angle of attack α_1 is therefore required to trim the aircraft at a lower altitude. In terms of the deflections for the control surfaces, given the negative values of the gradients $\left(\frac{\partial i_{H_1}}{\partial c_{L_1}}, \frac{\partial \delta_{E_1}}{\partial c_{L_1}}\right)$, slightly higher positive (down) deflections will be required at a lower altitude to generate an upward force at the horizontal tail, leading to a negative nose down pitching moment, associated with the slightly lower value of α_1. This functionality is summarized as in the following:

$$\Delta(Alt.) < 0 \Rightarrow \Delta\rho > 0 \Rightarrow \Delta c_{L_1} = \frac{2W}{(\Delta\rho) V_{P_1}^2} < 0 \Rightarrow \Delta\alpha_1 = \frac{\partial \alpha_1}{\partial c_{L_1}} \cdot \Delta c_{L_1} < 0 \Rightarrow Nose\downarrow$$

$$\Delta(Alt.) < 0 \Rightarrow \Delta\rho > 0 \Rightarrow \Delta c_{L_1} = \frac{2W}{(\Delta\rho) V_{P_1}^2} < 0$$

$$\Rightarrow \begin{cases} \Delta i_{H_1} = \dfrac{\partial i_{H_1}}{\partial c_{L_1}} \cdot \Delta c_{L_1} > 0 \Rightarrow Stab.\downarrow \Rightarrow Force\uparrow \Rightarrow Nose\downarrow \\[2mm] \Delta \delta_{E_1} = \dfrac{\partial \delta_{E_1}}{\partial c_{L_1}} \cdot \Delta c_{L_1} > 0 \Rightarrow Elev.\downarrow \Rightarrow Force\uparrow \Rightarrow Nose\downarrow \end{cases}$$

It should be emphasized that the effect of the altitude on the trim conditions is substantially less significant than the effects of the aircraft weight and airspeed.

In addition to varying conditions in terms of weight, airspeed, and altitude, it is critical from a design point of view to evaluate the relationship between the trim conditions (α_1, i_{H_1}) or (α_1, δ_{E_1}) and the aircraft center of gravity.

As described in the aerodynamic modeling in Chapter 3, the aerodynamic stability derivatives $(c_{L_\alpha}, c_{L_{i_H}}, c_{L_{\delta_E}})$ are not functions of the location of the aircraft center of gravity; however, $(c_{m_\alpha}, c_{m_{i_H}}, c_{m_{\delta_E}})$, especially c_{m_α}, are strongly affected by the aircraft CG location, as recalled here:

$$c_{m_\alpha} = c_{L_\alpha}(\bar{x}_{CG} - \bar{x}_{AC}) = c_{L_\alpha} SM$$

$$c_{m_{i_H}} = -c_{L_{\alpha_H}} \eta_H \frac{S_H}{S}(\bar{x}_{AC_H} - \bar{x}_{CG}) = -c_{L_{i_H}}(\bar{x}_{AC_H} - \bar{x}_{CG})$$

$$c_{m_{\delta_E}} = -c_{L_{\alpha_H}} \eta_H \frac{S_H}{S}(\bar{x}_{AC_H} - \bar{x}_{CG})\tau_E = -c_{L_{\delta_E}}(\bar{x}_{AC_H} - \bar{x}_{CG})$$

Considering the configuration where the stabilators are used for trimming purposes, an approach for modeling these functionalities is given by the introduction of the following relationships:

$$\frac{\partial \alpha_1}{\partial c_{L_1}} = \frac{\left[-c_{L_{i_H}}(\bar{x}_{AC_H} - \bar{x}_{CG})\right]}{\left[c_{L_\alpha}\left(-c_{L_{i_H}}(\bar{x}_{AC_H} - \bar{x}_{CG})\right) - (c_{L_\alpha}(\bar{x}_{CG} - \bar{x}_{AC}))c_{L_{i_H}}\right]}$$

$$\frac{\partial i_{H_1}}{\partial c_{L_1}} = \frac{-(c_{L_\alpha}(\bar{x}_{CG} - \bar{x}_{AC}))}{\left[c_{L_\alpha}\left(-c_{L_{i_H}}(\bar{x}_{AC_H} - \bar{x}_{CG})\right) - (c_{L_\alpha}(\bar{x}_{CG} - \bar{x}_{AC}))c_{L_{i_H}}\right]}$$

An example of the application of the preceding relationships is provided next. Consider again the Learjet 24 aircraft at cruise steady-state conditions ($Alt. = 40{,}000$ ft, $V_{P_1} = 677$ ft/sec). This aircraft features the stabilators for trimming purposes. From Appendix B, the values for the key aerodynamic stability derivatives are given by

$$c_{m_\alpha} = -0.64, \quad c_{L_\alpha} = 5.84, \quad c_{m_{i_H}} = -2.5, \quad c_{L_{i_H}} = 0.94$$

From this data the static margin can be evaluated using

$$c_{m_\alpha} = c_{L_\alpha}(\bar{x}_{CG} - \bar{x}_{AC}) = c_{L_\alpha} SM \Rightarrow SM = \frac{c_{m_\alpha}}{c_{L_\alpha}} = \frac{-0.64}{5.84} \approx -0.11$$

The reference center of gravity for this aircraft is provided as $\bar{x}_{CG} = 0.32$.

Therefore, the location of the aircraft aerodynamic center can be evaluated using

$$\bar{x}_{AC} = \bar{x}_{CG} - SM = 0.32 - (-0.11) = 0.43$$

Next, knowing that $c_{m_{i_H}} = -c_{L_{i_H}}(\bar{x}_{AC_H} - \bar{x}_{CG})$, the geometric parameter $\bar{x}_{AC_H}$ is evaluated using

$$(\bar{x}_{AC_H} - \bar{x}_{CG}) = -\frac{c_{m_{i_H}}}{c_{L_{i_H}}} = -\frac{-2.5}{0.94} = 2.66 \Rightarrow \bar{x}_{AC_H} = 2.66 + \bar{x}_{CG} = 2.66 + 0.32 = 2.98$$

320 Chapter 6 Aircraft Stability and Design for Trim Conditions

leading to the following linear slopes:

$$\frac{\partial \alpha_1}{\partial c_{L_1}} = \frac{\left[-c_{L_{i_H}}(\bar{x}_{AC_H} - \bar{x}_{CG})\right]}{\left[c_{L_\alpha}\left(-c_{L_{i_H}}(\bar{x}_{AC_H} - \bar{x}_{CG})\right) - (c_{L_\alpha}(\bar{x}_{CG} - \bar{x}_{AC}))c_{L_{i_H}}\right]} = \frac{-0.94(2.98 - \bar{x}_{CG})}{[-5.49(2.98 - \bar{x}_{CG}) - 5.49(\bar{x}_{CG} - 0.43)]}$$

$$\frac{\partial i_{H_1}}{\partial c_{L_1}} = \frac{-[c_{L_\alpha}(\bar{x}_{CG} - \bar{x}_{AC})]}{\left[c_{L_\alpha}\left(-c_{L_{i_H}}(\bar{x}_{AC_H} - \bar{x}_{CG})\right) - (c_{L_\alpha}(\bar{x}_{CG} - \bar{x}_{AC}))c_{L_{i_H}}\right]} = \frac{-5.84(\bar{x}_{CG} - 0.43)}{[-5.49(2.98 - \bar{x}_{CG}) - 5.49(\bar{x}_{CG} - 0.43)]}$$

These relationships are shown in Figures 6.13 and 6.14.

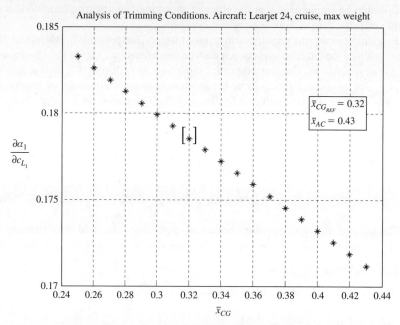

Figure 6.13 Analysis of Trim Conditions for the Learjet 24

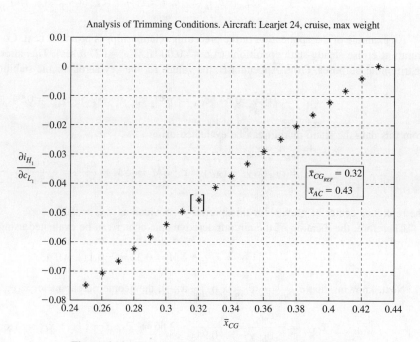

Figure 6.14 Analysis of Trim Conditions for the Learjet 24

The analysis of Figure 6.14 provides an important consideration. In fact, the gradient $\frac{\partial i_{H_1}}{\partial c_{L_1}}$ goes to zero when the aircraft center of gravity is located at the aircraft aerodynamic center. From a pilot point of view this implies that at that specific location for the aircraft center of gravity, no stabilator deflection can change the trim value for c_{L_1}, which makes the stabilators totally ineffective. For this reason the aircraft aerodynamic center is also known as the neutral point. Note that this analysis also applies to an aircraft trimmed through the elevators, in lieu of the stabilators.

Another important trim parameter is the trim stabilator versus airspeed gradient or the trim elevator versus airspeed gradient, as denoted by

$$\frac{\partial i_{H_1}}{\partial V_{P_1}}, \frac{\partial \delta_{E_1}}{\partial V_{P_1}}$$

These gradients have more practical implications than the previous gradients versus c_{L_1}. In fact, pilots have a much better feel of the variation of the trim conditions with variations of the airspeed, rather than variations in the coefficient c_{L_1}. Without any loss of generality, consider an aircraft where stabilators are used for trimming purposes. An expression for the trim stabilator versus airspeed gradient can be derived using the following composite derivative relationship:

$$\frac{\partial i_{H_1}}{\partial V_{P_1}} = \frac{\partial i_{H_1}}{\partial c_{L_1}} \cdot \frac{\partial c_{L_1}}{\partial V_{P_1}} = \left[\frac{-c_{m_\alpha}}{\left(c_{L_\alpha} c_{m_{i_H}} - c_{m_\alpha} c_{L_{i_H}}\right)}\right] \cdot \frac{\partial c_{L_1}}{\partial V_{P_1}}$$

Starting from $c_{L_1} = 2W/\rho V_{P_1}^2 S$, an expression for $\left(\frac{\partial c_{L_1}}{\partial V_{P_1}}\right)$ can be derived as

$$\left(\frac{\partial c_{L_1}}{\partial V_{P_1}}\right) = \left(\frac{\partial \left(2W/\rho V_{P_1}^2 S\right)}{\partial V_{P_1}}\right) = \frac{2W}{\rho S} \frac{\partial \left(1/V_{P_1}^2\right)}{\partial V_{P_1}} = \frac{2W}{\rho S} \frac{-2}{V_{P_1}^3} = -\frac{4W}{\rho S V_{P_1}^3}$$

leading to

$$\frac{\partial i_{H_1}}{\partial V_{P_1}} = \frac{\partial i_{H_1}}{\partial c_{L_1}} \cdot \frac{\partial c_{L_1}}{\partial V_{P_1}} = \left[\frac{-c_{m_\alpha}}{\left(c_{L_\alpha} c_{m_{i_H}} - c_{m_\alpha} c_{L_{i_H}}\right)}\right] \left(-\frac{4W}{\rho S V_{P_1}^3}\right) = \left(\frac{4W}{\rho S V_{P_1}^3}\right) \frac{c_{m_\alpha}}{\left(c_{L_\alpha} c_{m_{i_H}} - c_{m_\alpha} c_{L_{i_H}}\right)}$$

The sign of this gradient is positive since both numerator and denominator are negative. Therefore,

$$\frac{\partial i_{H_1}}{\partial V_{P_1}} = \frac{4W}{\rho S V_{P_1}^3} \frac{c_{m_\alpha}}{\left(c_{L_\alpha} c_{m_{i_H}} - c_{m_\alpha} c_{L_{i_H}}\right)} > 0$$

The implications for this sign are clear. Assuming an increase in airspeed, the aircraft will generate additional lift causing the aircraft to pitch up; therefore, the pilot will require a positive (down) trim deflection of the stabilators, generating an upward force on the horizontal tail and, thus, causing a nose down pitching moment. Conversely, a decrease in airspeed will require a negative (up) trim deflection of the stabilators, generating a downward force and, thus, a nose up pitching moment. A similar analysis can be conducted with the elevators as the trim control surface.

Like $\frac{\partial i_{H_1}}{\partial c_{L_1}}$, the value of $\frac{\partial i_{H_1}}{\partial V_{P_1}}$ depends on the location of the aircraft center of gravity. Specifically, $\frac{\partial i_{H_1}}{\partial V_{P_1}} \to 0$ as $\bar{x}_{CG} \to 0 \Rightarrow c_{m_\alpha} \to 0$. Thus $\frac{\partial i_{H_1}}{\partial V_{P_1}}$ is negative only for a statically stable aircraft.

The functionality of this gradient with the center of gravity location is given by

$$\frac{\partial i_{H_1}}{\partial V_{P_1}} = \frac{-4W}{\rho S V_{P_1}^3} \frac{\partial i_{H_1}}{\partial c_{L_1}} = \frac{4W}{\rho S V_{P_1}^3} \frac{[c_{L_\alpha}(\bar{x}_{CG} - \bar{x}_{AC})]}{\left[c_{L_\alpha}\left(-c_{L_{i_H}}(\bar{x}_{AC_H} - \bar{x}_{CG})\right) - \left(c_{L_\alpha}(\bar{x}_{CG} - \bar{x}_{AC})\right) c_{L_{i_H}}\right]}$$

Consider again Aircraft 6 (Learjet 24) from Appendix B at cruise steady-state conditions (Alt. = 40,000 ft, $V_{P_1} = 677$ ft/sec). The values for the key aerodynamic stability derivatives are given by $c_{m_\alpha} = -0.64$, $c_{L_\alpha} = 5.84$, $c_{m_{i_H}} = -2.5$, $c_{L_{i_H}} = 0.94$. Using the previously calculated $\bar{x}_{AC} = 0.43$, $\bar{x}_{AC_H} = 2.98$ and the provided values $S = 230$ ft^2, $W = 13,000$ lbs with the associated value for $\rho(Alt)$, $\dfrac{\partial i_{H_1}}{\partial V_{P_1}}$ can be found using the numerical relationship:

$$\frac{\partial i_{H_1}}{\partial V_{P_1}} = \frac{4 \cdot 13,000}{5.88 \cdot 10^{-4} \cdot 230 \cdot (677)^3} \frac{5.84(\bar{x}_{CG} - 0.43)}{[-5.49(2.98 - \bar{x}_{CG}) - 5.49(\bar{x}_{CG} - 0.43)]}$$

The trend is shown in Figure 6.15.

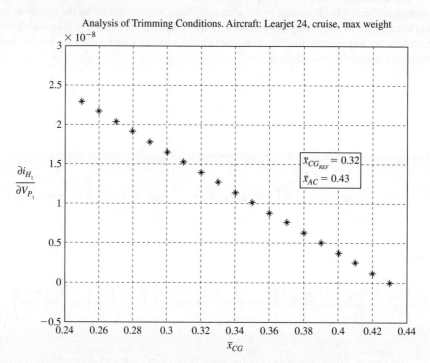

Figure 6.15 Analysis of Trim Conditions for the Learjet 24

(See Student Sample Problems 6.3.)

6.5 LIFT CHART AND TRIM DIAGRAM

Without loss of generality, consider again an aircraft featuring stabilators for trimming purposes and elevators for maneuvering. The graphic tools introduced in this section can be similarly developed if elevators are used for trimming purposes.

6.5.1 Lift Chart

Following a preliminary aircraft design and a preliminary sizing of the longitudinal control surfaces, the designer can develop two important graphic tools. The first tool is known as the c_L vs. α chart. A typical c_L vs. α chart is shown in Figure 6.16.

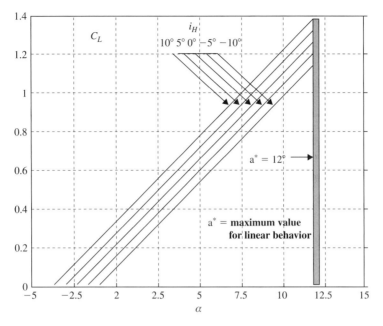

Figure 6.16 Typical c_L vs. α Chart

In the chart the value of α is capped at the end of the linear behavior for the lift force (in this case $\alpha = 12°$). The chart shows different lines for different deflections of the stabilators. Consider now only the line associated with $i_H = 0°$, as shown in Figure 6.17.

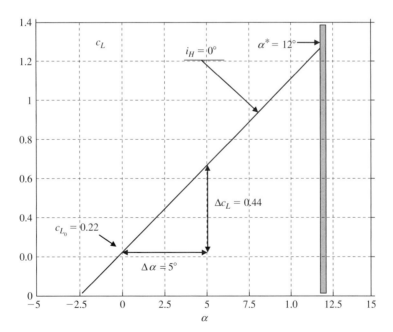

Figure 6.17 Typical c_L vs. α chart ($i_H = 0°$)

The following data can be extracted from the chart: $c_{L_0} = c_L|_{\alpha=0°}$, c_{L_α}. For this generic aircraft $c_{L_0} = 0.22$, as read from the intersection of the c_L line with the vertical axis associated with the condition $\alpha = 0°$. The value of c_{L_α} can be instead calculated as the generic ratio of two $\Delta's$; considering, without loss of generality, $\Delta\alpha = 5°$ and the associated $\Delta c_L = 0.44$, we have

$$c_{L_\alpha} = \frac{\Delta c_L}{\Delta \alpha} = \frac{0.44}{5°} \approx 0.088 \,\frac{1}{\text{deg}} \approx 5.04 \,\frac{1}{\text{rad}}$$

Consider now the lines associated with two specific values of the deflection of the stabilators, that is $i_H = 0°$, $i_H = 10°$, as shown in Figure 6.18.

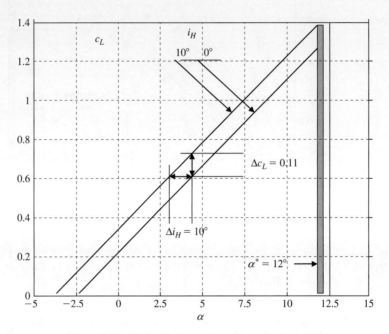

Figure 6.18 Typical c_L vs. α chart ($i_H = 0°$, $i_H = 10°$)

The value of $c_{L_{i_H}}$ can be evaluated from the chart as the generic ratio of two $\Delta's$; considering, without loss of generality, $\Delta i_H = 10°$ and the associated $\Delta c_L = 0.11$, we have

$$c_{L_{i_H}} = \frac{\Delta c_L}{\Delta i_H} = \frac{0.11}{10°} \approx 0.011 \,\frac{1}{\text{deg}} \approx 0.63 \,\frac{1}{\text{rad}}$$

6.5.2 Trim Diagram

The second—and much more important—graphic tool is the c_L vs. c_m chart, better known as the aircraft trim diagram (TD).[1] A typical shape for a trim diagram is shown in Figure 6.19.

A general definition of the trim diagram is the following.

> The Trim Diagram is the 2D representation in the c_L vs. c_m plane of all the possible set of values α_1, i_{H_1} at which it is possible to trim an aircraft at every possible condition in its flight envelope and at every possible location of its center of gravity.

A first characteristic of the TD is that on the x-axis the values of c_m are indicated in reverse order, positive values on the LHS and negative values on the RHS. Also, the TD must be derived with respect to a reference location of the aircraft center of gravity. This location is arbitrary but it can be, in general, selected as either the empty weight (EW) configuration or the center of gravity location associated with the average loading configuration for the aircraft. In the TD in Figure 6.19 $\bar{x}_{CG_{REF}} = 0.25$ is shown.

The TD is shaped as an irregular triangle; therefore, it features three segments. The first segment, denoted as **AC**, is associated with the highest possible value of $\bar{x}_{CG}$ at which the aircraft can be trimmed by the pilot.

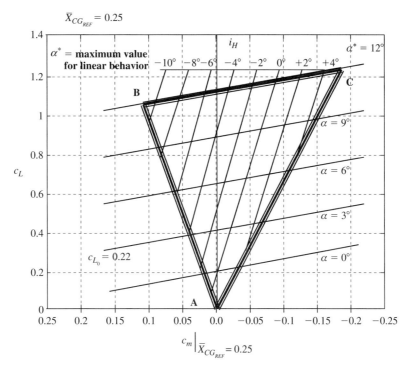

Figure 6.19 Typical c_L vs. c_m Chart (Trim Diagram)

By definition, this value is the aircraft aerodynamic center ($\bar{x}_{AC}$). The second segment, denoted as **AB**, is associated with the most forward position of the center of gravity at which the aircraft can be trimmed. The third segment, denoted as **BC**, is associated with the highest possible value of the angle of attack α at which the aircraft can be trimmed (in this case $\alpha_{MAX} = 12°$).

The following can be deduced from the preceding TD, shown again in Figure 6.20.

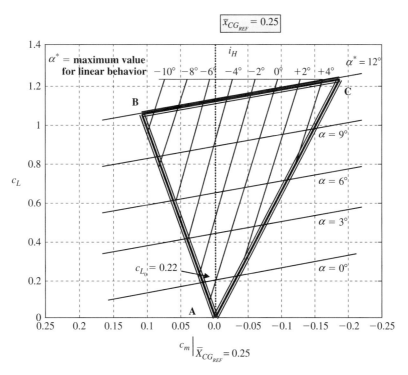

Figure 6.20 Typical c_L vs. c_m Chart (Trim Diagram) $(\bar{x}_{CG} = \bar{x}_{CG_{REF}})$

The thick, dotted line on the y-axis denotes all the possible α_1, i_{H_1} trim conditions associated with $\bar{x}_{CG} = \bar{x}_{CG_{REF}}$. As expected, negative (up) deflections for the stabilators are required as the angle of attack α increases to provide downward forces generating nose up pitching moments associated with increases in the values of c_{L_1} (associated, for example, with decreases in airspeed).

As stated previously, the aircraft aerodynamic center can also be considered as the maximum $\bar{x}_{CG}$ value at which the aircraft can be trimmed. This value can be derived from the TD, as shown in Figure 6.21.

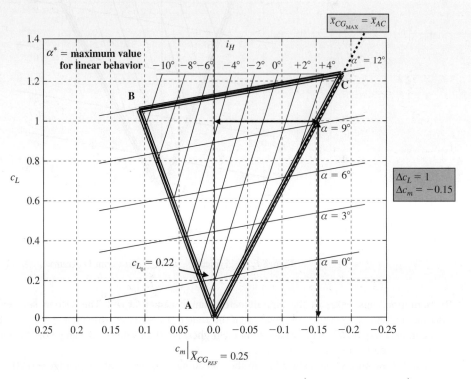

Figure 6.21 Typical c_L vs. c_m Chart (Trim Diagram) $(\bar{x}_{CG} = \bar{x}_{CG_{MAX}} = \bar{x}_{AC})$

Starting from the basic relationship $c_{m_\alpha} = c_{L_\alpha}(\bar{x}_{CG} - \bar{x}_{AC})$, removing the functionality of α we have $c_m = c_L(\bar{x}_{CG} - \bar{x}_{AC})$. At this point, starting from the y-axis line associated with $\bar{x}_{CG} = \bar{x}_{CG_{REF}}$, the aircraft aerodynamic center can be derived from the maximum deviation from $\bar{x}_{CG_{REF}}$. We have

$$\Delta c_m = \Delta c_L(\bar{x}_{CG_{REF}} - \bar{x}_{AC}) \Rightarrow (\bar{x}_{CG_{REF}} - \bar{x}_{AC}) = \frac{\Delta c_m}{\Delta c_L}$$

Without loss of generality, for simplicity purposes select $\Delta c_L = 1$ associated with $\Delta c_m = -0.15$ (as shown), leading to

$$(\bar{x}_{CG_{REF}} - \bar{x}_{AC}) = \frac{\Delta c_m}{\Delta c_L} = -0.15 \Rightarrow \bar{x}_{AC} = \bar{x}_{CG_{REF}} - \frac{\Delta c_m}{\Delta c_L} = \bar{x}_{CG_{REF}} - (-0.15) = 0.4$$

An identical approach can be used for the evaluation of $\bar{x}_{CG_{MIN}}$, as shown in Figure 6.22.

Without loss of generality, for simplicity purposes select $\Delta c_L = 1$ associated with $\Delta c_m = 0.1$, as shown; therefore, for the evaluation of $\bar{x}_{CG_{MIN}}$ we have

$$(\bar{x}_{CG_{REF}} - \bar{x}_{CG_{MIN}}) = \frac{\Delta c_m}{\Delta c_L} = 0.1 \Rightarrow \bar{x}_{CG_{MIN}} = \bar{x}_{CG_{REF}} - \frac{\Delta c_m}{\Delta c_L} = \bar{x}_{CG_{REF}} - 0.1 = 0.15$$

In general the value of $\bar{x}_{CG_{MIN}}$ is dictated by specific requirements for the aircraft trimmability during descent and approach. It is clear that a forward $\bar{x}_{CG}$ location can present maneuvering challenges for the pilot due to the aircraft tendency to nose down during a flight phase where the lower values of the airspeed require larger values of the angles of attack and, therefore, a nose-up pitching moment.

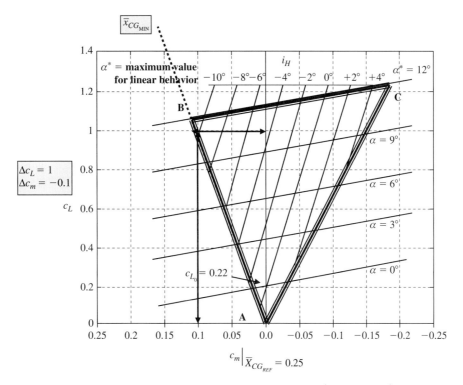

Figure 6.22 Typical c_L vs. c_m Chart (Trim Diagram) $\left(\bar{x}_{CG} = \bar{x}_{CG_{\text{MIN}}}\right)$

Finally, the upper limit of the TD, denoted by the segment **BC**, is associated with the maximum value of the angle of attack at which the aircraft can be trimmed for any $\bar{x}_{CG}$ location, as shown in Figure 6.23. Clearly, a desirable design ensures trimmability of the aircraft at low values of the angle of attack for drag (and fuel consumption) minimization purposes, with the exception of trimming the aircraft in the approach and descent phases.

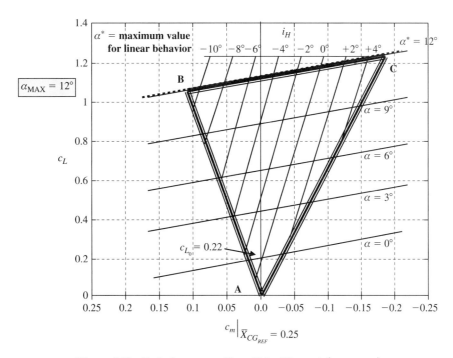

Figure 6.23 Typical c_L vs. c_m Chart (Trim Diagram) $(\alpha = \alpha_{\text{MAX}})$

In the plot the maximum value of the angle of attack is given by $\alpha_{MAX} = 12°$. An interesting property of the TD is that it is intersected by parallel lines associated with constant values of α.

A typical use of the TD is providing the aircraft designer a set of (α_1, i_{H_1}) for a given $\bar{x}_{CG}$ location. For example, consider an aircraft with a $\bar{x}_{CG} = 0.3$ location; assume that the aerodynamic and thrust analysis has revealed that the optimal value for the angle of attack is $\alpha_1 = 2°$. The TD can then be used as in the following. First, a line is drawn associated with the $\bar{x}_{CG} = 0.3$. Because $\bar{x}_{CG_{REF}} = 0.25$, the line can be found by connecting the origin with the point associated with $\Delta c_L = 1$, $\Delta c_m = -0.05$. Next, the line associated with $\alpha_1 = 2°$ is found through linear interpolation between the lines associated with $\alpha_1 = 3°$ and $\alpha_1 = 0°$. The intersection between the previous two lines identifies a point; finally, a line crossing that point is found from a linear interpolation between two parallel lines associated with different deflections of stabilators.

The process is shown in Figure 6.24.

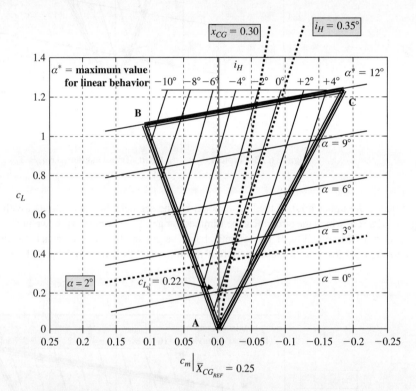

Figure 6.24 Typical c_L vs. c_m Chart (Trim Diagram) ($\bar{x}_{CG} = 0.3$, $\alpha_1 = 2°$)

In this specific case with ($\bar{x}_{CG} = 0.3$, $\alpha_1 = 2°$) the aircraft can be trimmed with a stabilator $i_{H_i} \approx 0.35°$. The preceding example represents the most important use of the TD since the value of α_1 is far from critical than the value of i_{H_i}, due to the direct relationships between α_1 and the total aerodynamic drag and, therefore, the fuel consumption.

Although less frequently, the TD can also be used to identify the value of α_1 associated with specific $(\bar{x}_{CG}, i_{H_1})$ values. For example, consider $(\bar{x}_{CG} = 0.3, i_{H_1} \approx -1°)$. From the intersection between the two lines, it can be seen that the aircraft can be trimmed at the angle of attack $\alpha_1 = 7.5°$, as shown in Figure 6.25.

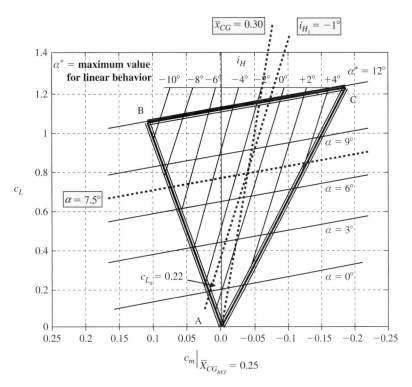

Figure 6.25 Typical c_L vs. c_m Chart (Trim Diagram) $(\bar{x}_{CG} = 0.3, i_{H_1} = -1°)$

6.5.3 Trim Diagrams for Different Classes of Aircraft

The TD shown in the previous sections is a generic diagram for a generic aircraft. The shape of this important tool depends on the specific aircraft belonging to a specific class. The following trends typically apply. General aviation aircraft tend to have fairly flexible loading conditions and, therefore, a wide range of possible $\bar{x}_{CG}$ locations; also, these aircraft are usually trimmable at large values of the angle of attack. Military cargo aircraft and commercial jetliners also tend to have a wide range of center of gravity locations (although not as wide as for general aviation); however, their range of trim conditions in terms of the angle of attack is much more limited. Finally, jet fighter aircraft have a smaller range for the center of gravity locations but can be trimmed up to higher values of the angle of attack.

The preceding trends are shown in Figure 6.26. The trends outlined previously are to be considered as general trends; specific exceptions do exist for each class of aircraft.

6.5.4 Trim Diagrams for Thrust Axis Above/Below Center of Gravity

Consider again the generic TD for a generic aircraft introduced in Section 6.5.2, shown again in Figure 6.27.

The TD is associated with the condition of the thrust aligned with the longitudinal stability axis X_S, implying $z_T = 0$ in Figure 6.10. Although not frequently, there are aircraft configurations where the installed thrust force generates substantial pitching moments at steady-state flight conditions.[1] Specifically, this can lead to positive Δc_m if the thrust line is above X_S or negative Δc_m if the thrust line is below X_S. Both situations are outlined in Figure 6.28. The reported Δc_m values are unusually high (implying large $\pm z_T$ values) for clarity purposes. Note that the trim values for the stabilators in Figure 6.28 should be extended and interpolated to include these specific conditions. (See Student Sample Problem 6.4.)

330 Chapter 6 Aircraft Stability and Design for Trim Conditions

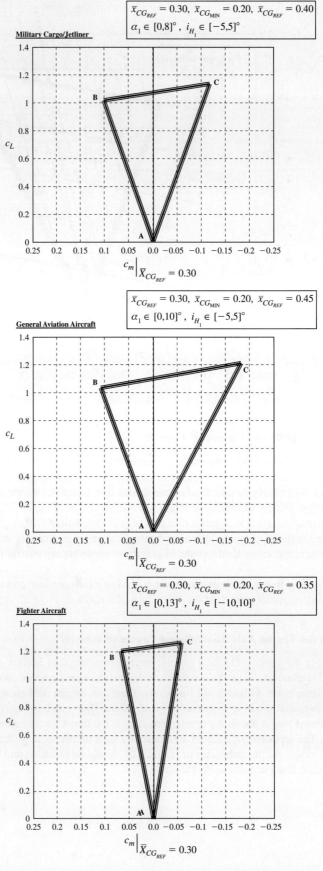

Figure 6.26 Typical c_L vs. c_m Charts (Trim Diagrams) for Different Classes of Aircraft

6.5 Lift Chart and Trim Diagram

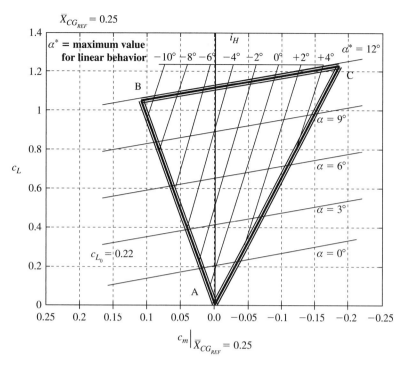

Figure 6.27 Typical c_L vs. c_m Chart (Trim Diagram)

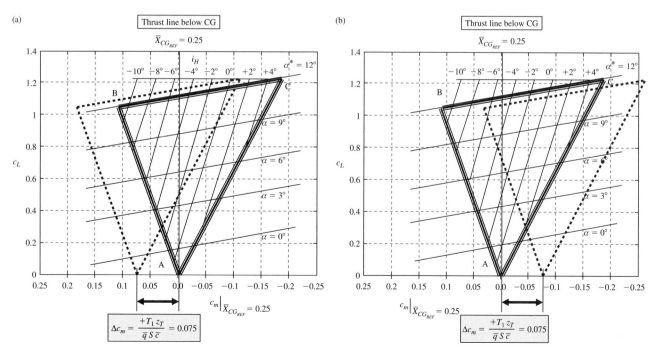

Figure 6.28 Trim Diagrams for Thrust Line Above/Below CG

6.6 LATERAL DIRECTIONAL ANALYSIS OF STEADY-STATE STRAIGHT FLIGHT

This section analyzes the steady-state straight flight conditions from a lateral directional point of view. Once these equations are properly set, then it is possible to solve for the aircraft trim conditions.

The same considerations introduced for the longitudinal trim conditions in Section 6.4 also apply to the lateral directional trim problem. In the most general form, the solution for the aircraft trim conditions is a nonlinear problem as all the aerodynamic coefficients are or can be nonlinear functions of the sideslip angle, Mach, or other variables. Thus, the problem of finding the design trim conditions can be solved using a nonlinear multivariable root finder. Clearly, a solution for trim conditions outside the linear region is not an acceptable outcome from a design view and will necessarily lead to a modification of the design. However, the design for lateral directional trim conditions is greatly simplified by the fact that the lateral directional aerodynamic biases $c_{Y_0}, c_{l_0}, c_{n_0}$ are all zero. Therefore, the designed trim conditions are expected to be well within the linear aerodynamic range. In this case, it is appropriate to assume linearity, and therefore the use of the stability and control derivatives discussed so far is recommended. The only flight condition which can substantially affect the lateral directional trim conditions is the engine out condition, which will be discussed separately.

Consider again an aircraft with the body and stability axes shown in Figure 6.29.

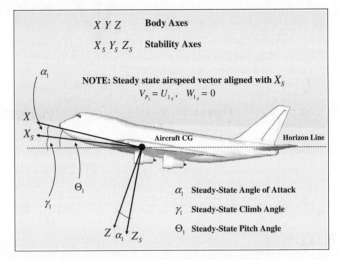

Figure 6.29 Body and Stability Axes

The aircraft features ailerons and rudder as conventional primary control surfaces for the lateral and the directional dynamics, respectively. Using the modeling for the aerodynamic and thrust steady-state forces and moments from Chapter 4, the steady-state lateral-directional equations of motion along the stability axes are given by:[1]

$$\begin{cases} -mg\cos(\gamma_1)\sin(\Phi_1) = \overline{q}_1 S \left(c_{Y_\beta}\beta_1 + c_{Y_{\delta_A}}\delta_{A_1} + c_{Y_{\delta_R}}\delta_{R_1}\right) + F_{T_{Y_1}} \\ 0 = \overline{q}_1 S b \left(c_{l_\beta}\beta_1 + c_{l_{\delta_A}}\delta_{A_1} + c_{l_{\delta_R}}\delta_{R_1}\right) + L_{T_1} \\ 0 = \overline{q}_1 S b \left(c_{n_\beta}\beta_1 + c_{n_{\delta_A}}\delta_{A_1} + c_{n_{\delta_R}}\delta_{R_1}\right) + N_{T_1} \end{cases}$$

These equations are in the most general form. Unless thrust vectoring or differential thrust is implemented and unless engine failure conditions exist, the following applies: $F_{T_{Y_1}} = L_{T_1} = N_{T_1} = 0$.

The analysis of the equations also reveals that the system features three equations and four unknowns $(\Phi_1, \beta_1, \delta_{A_1}, \delta_{R_1})$. For the purpose of solving this system using tools from linear algebra, the value for Φ_1 is set to zero or to a constant and known value. This approach is also dictated by the need for modeling specific conditions associated with the requirements for coping with engine out conditions. Therefore, with the assumption of known value for Φ_1, the preceding equations can be reduced into the following matrix format:

$$\begin{bmatrix} c_{Y_\beta} & c_{Y_{\delta_A}} & c_{Y_{\delta_R}} \\ c_{l_\beta} & c_{l_{\delta_A}} & c_{l_{\delta_R}} \\ c_{n_\beta} & c_{n_{\delta_A}} & c_{n_{\delta_R}} \end{bmatrix} \begin{Bmatrix} \beta_1 \\ \delta_{A_1} \\ \delta_{R_1} \end{Bmatrix} = \begin{Bmatrix} -\left[mg\cos(\gamma_1)\sin(\Phi_1) + F_{T_{Y_1}}\right]/\overline{q}_1 S \\ -L_{T_1}/\overline{q}_1 S b \\ -N_{T_1}/\overline{q}_1 S b \end{Bmatrix}$$

6.6 Lateral Directional Analysis of Steady-State Straight Flight

The application of Cramer's rule leads to the following general solutions as ratios of determinants:

$$\beta_1 = \frac{\begin{vmatrix} -[mg\cos(\gamma_1)\sin(\Phi_1) + F_{T_{Y_1}}]/\bar{q}_1 S & c_{Y_{\delta_A}} & c_{Y_{\delta_R}} \\ -L_{T_1}/\bar{q}_1 S b & c_{l_{\delta_A}} & c_{l_{\delta_R}} \\ -N_{T_1}/\bar{q}_1 S b & c_{n_{\delta_A}} & c_{n_{\delta_R}} \end{vmatrix}}{\begin{vmatrix} c_{Y_\beta} & c_{Y_{\delta_A}} & c_{Y_{\delta_R}} \\ c_{l_\beta} & c_{l_{\delta_A}} & c_{l_{\delta_R}} \\ c_{n_\beta} & c_{n_{\delta_A}} & c_{n_{\delta_R}} \end{vmatrix}}$$

$$\delta_{A_1} = \frac{\begin{vmatrix} c_{Y_\beta} & -[mg\cos(\gamma_1)\sin(\Phi_1) + F_{T_{Y_1}}]/\bar{q}_1 S & c_{Y_{\delta_R}} \\ c_{l_\beta} & -L_{T_1}/\bar{q}_1 S b & c_{l_{\delta_R}} \\ c_{n_\beta} & -N_{T_1}/\bar{q}_1 S b & c_{n_{\delta_R}} \end{vmatrix}}{\begin{vmatrix} c_{Y_\beta} & c_{Y_{\delta_A}} & c_{Y_{\delta_R}} \\ c_{l_\beta} & c_{l_{\delta_A}} & c_{l_{\delta_R}} \\ c_{n_\beta} & c_{n_{\delta_A}} & c_{n_{\delta_R}} \end{vmatrix}}$$

$$\delta_{R_1} = \frac{\begin{vmatrix} c_{Y_\beta} & c_{Y_{\delta_A}} & -[mg\cos(\gamma_1)\sin(\Phi_1) + F_{T_{Y_1}}]/\bar{q}_1 S \\ c_{l_\beta} & c_{l_{\delta_A}} & -L_{T_1}/\bar{q}_1 S b \\ c_{n_\beta} & c_{n_{\delta_A}} & -N_{T_1}/\bar{q}_1 S b \end{vmatrix}}{\begin{vmatrix} c_{Y_\beta} & c_{Y_{\delta_A}} & c_{Y_{\delta_R}} \\ c_{l_\beta} & c_{l_{\delta_A}} & c_{l_{\delta_R}} \\ c_{n_\beta} & c_{n_{\delta_A}} & c_{n_{\delta_R}} \end{vmatrix}}$$

where

$$\begin{vmatrix} c_{Y_\beta} & c_{Y_{\delta_A}} & c_{Y_{\delta_R}} \\ c_{l_\beta} & c_{l_{\delta_A}} & c_{l_{\delta_R}} \\ c_{n_\beta} & c_{n_{\delta_A}} & c_{n_{\delta_R}} \end{vmatrix} = c_{Y_\beta}\left(c_{l_{\delta_A}} c_{n_{\delta_R}} - c_{n_{\delta_A}} c_{l_{\delta_R}}\right) + c_{Y_{\delta_A}}\left(c_{l_{\delta_R}} c_{n_\beta} - c_{l_\beta} c_{n_{\delta_R}}\right) + c_{Y_{\delta_R}}\left(c_{l_\beta} c_{n_{\delta_A}} - c_{n_\beta} c_{l_{\delta_A}}\right)$$

leading to the following general solutions:

$$\beta_1 = \frac{A\left(c_{l_{\delta_A}} c_{n_{\delta_R}} - c_{n_{\delta_A}} c_{l_{\delta_R}}\right) + B\left(c_{n_{\delta_A}} c_{Y_{\delta_R}} - c_{Y_{\delta_A}} c_{n_{\delta_R}}\right) + C\left(c_{Y_{\delta_A}} c_{l_{\delta_R}} - c_{l_{\delta_A}} c_{Y_{\delta_R}}\right)}{c_{Y_\beta}\left(c_{l_{\delta_A}} c_{n_{\delta_R}} - c_{n_{\delta_A}} c_{l_{\delta_R}}\right) + c_{Y_{\delta_A}}\left(c_{l_{\delta_R}} c_{n_\beta} - c_{l_\beta} c_{n_{\delta_R}}\right) + c_{Y_{\delta_R}}\left(c_{l_\beta} c_{n_{\delta_A}} - c_{n_\beta} c_{l_{\delta_A}}\right)}$$

$$\delta_{A_1} = \frac{c_{Y_\beta}\left(B c_{n_{\delta_R}} - C c_{l_{\delta_R}}\right) + A\left(c_{n_\beta} c_{l_{\delta_R}} - c_{l_\beta} c_{n_{\delta_R}}\right) + c_{Y_{\delta_R}}\left(c_{l_\beta} C - c_{n_\beta} B\right)}{c_{Y_\beta}\left(c_{l_{\delta_A}} c_{n_{\delta_R}} - c_{n_{\delta_A}} c_{l_{\delta_R}}\right) + c_{Y_{\delta_A}}\left(c_{l_{\delta_R}} c_{n_\beta} - c_{l_\beta} c_{n_{\delta_R}}\right) + c_{Y_{\delta_R}}\left(c_{l_\beta} c_{n_{\delta_A}} - c_{n_\beta} c_{l_{\delta_A}}\right)}$$

$$\delta_{R_1} = \frac{c_{Y_\beta}\left(c_{l_{\delta_A}} C - c_{n_{\delta_A}} B\right) + c_{Y_{\delta_A}}\left(c_{l_\beta} C - c_{n_\beta} B\right) + A\left(c_{l_\beta} c_{n_{\delta_A}} - c_{n_\beta} c_{l_{\delta_A}}\right)}{c_{Y_\beta}\left(c_{l_{\delta_A}} c_{n_{\delta_R}} - c_{n_{\delta_A}} c_{l_{\delta_R}}\right) + c_{Y_{\delta_A}}\left(c_{l_{\delta_R}} c_{n_\beta} - c_{l_\beta} c_{n_{\delta_R}}\right) + c_{Y_{\delta_R}}\left(c_{l_\beta} c_{n_{\delta_A}} - c_{n_\beta} c_{l_{\delta_A}}\right)}$$

where

$$A = -[m g \cos(\gamma_1)\sin(\Phi_1) + F_{T_{Y_1}}]/\bar{q}_1 S$$
$$B = (-L_{T_1}/\bar{q}_1 S b)$$
$$C = (-N_{T_1}/\bar{q}_1 S b)$$

It is easy to visualize that in specific case of wing level ($\Phi_1 = 0°$), rectilinear ($\gamma_1 = 0°$) flight, with symmetric thrust (no engine out conditions and differential thrust), and in the absence of thrust vectoring ($F_{T_{Y_1}} = L_{T_1} = N_{T_1} = 0$), the preceding solutions take on the trivial form: $\beta_1 = \delta_{A_1} = \delta_{R_1} = 0$.

Clearly, the solutions of the preceding equations are of specific interest to the designers to analyze the aircraft dynamics following a propulsion system malfunction.[1] However, a problem with the general solutions outlined here is that it would be difficult to understand specific trends of individual components, given the general coupling of all the aerodynamic terms along the lateral axis and around the rolling and yawing axes. A first simplifying but realistic assumption is that, following an engine-out condition, we will have

$$F_{T_{Y_1}} = 0, \quad L_{T_1} \neq 0, \quad N_{T_1} \neq 0$$

Note that in the event of a multiengine propeller aircraft, the parameter N_{T_1} models not only the yawing moment associated with the nonsymmetric thrust but also the additional yawing moment ΔN_T associated with the ΔD generated by the nonrotating propellers of the malfunctioning engine.

Consider for example a Boeing B747 with an engine-out condition, as shown in Figure 6.30.

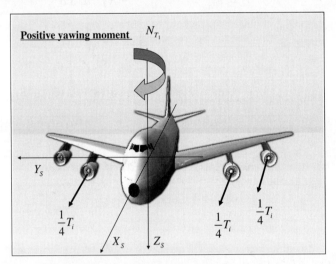

Figure 6.30 Engine Out Condition for a B747

The first step toward modeling the lateral directional dynamics following an engine-out condition is given by the analysis of the yawing moment equation:

$$0 = \overline{q}_1 S b \left(c_{n_\beta} \beta_1 + c_{n_{\delta_A}} \delta_{A_1} + c_{n_{\delta_R}} \delta_{R_1}\right) + N_{T_1}$$

Within the engine-out condition, a particularly important phase is the transient of the failure. In certain instances the pilot might be aware of the occurrence of the failure if he/she can deduce it from the readings from the instruments in the cockpit. However, in other cases the pilot might be totally unaware of the incoming engine failure, such as the case of birds being ingested in the intakes of the engine.[12] Therefore, there might be a transient of few seconds when the pilot is in the process of detecting and identifying the failure, from a combination of motion cues and instrument readings, before attempting to take the necessary corrective actions.

Assuming that in those seconds the positions of the lateral directional control surfaces remain at the trim conditions ($\delta_{A_1} \approx 0°$, $\delta_{R_1} \approx 0°$), the aircraft will begin to yaw and roll along with a sudden change in the sideslip angle from $\beta_1 \approx 0°$ to

$$\beta_1|_{ENGINE\ OUT} = \beta_1|_{EO} = -\frac{N_{T_1}}{c_{n_\beta} \overline{q}_1 S b}$$

Next, following the transient of the failure, the pilot will start to take compensating action on both the rolling and yawing axes. Starting from the rolling moment equation assuming the presence of a thrust-induced rolling moment L_{T_1}:

$$0 = \overline{q}_1 S b \left(c_{l_\beta} \beta_1 + c_{l_{\delta_A}} \delta_{A_1} + c_{l_{\delta_R}} \delta_{R_1}\right) + L_{T_1}$$

Starting from $\delta_{R_1} = 0°$, $\beta_1 = \beta_1|_{EO}$, we have

$$\delta_{A_1}\big|_{EO} = \frac{\left(-c_{l_\beta}\beta_1\big|_{EO} - \frac{L_{T_1}}{\overline{q}_1 S b}\right)}{c_{l_{\delta_A}}}$$

Next, inserting the previously calculated value of $\beta_1|_{EO}$, we have

$$\delta_{A_1}\big|_{EO} = \frac{\left(\frac{c_{l_\beta} N_{T_1}}{c_{n_\beta} \overline{q}_1 S b} - \frac{L_{T_1}}{\overline{q}_1 S b}\right)}{c_{l_{\delta_A}}}$$

Clearly, it is desirable to minimize the compensating deflections for the aileron to avoid saturation of this primary rolling control surface and to avoid stalls of one wing. A $\pm 20°$ maximum deflection is typically set for this purpose.

Next, the attention shifts on the yawing moment equation with the values of $\delta_{A_1}|_{EO}$, $\beta_1|_{EO}$:

$$0 = \overline{q}_1 S b \left(c_{n_\beta} \beta_1\big|_{EO} + c_{n_{\delta_A}} \delta_{A_1}\big|_{EO} + c_{n_{\delta_R}} \delta_{R_1}\right) + N_{T_1}$$

The preceding relationship allows the calculation of the compensating rudder deflection $\delta_{R_1}|_{EO}$ as:

$$\delta_{R_1}\big|_{EO} = \frac{-\left(c_{n_\beta} \beta_1\big|_{EO} + c_{n_{\delta_A}} \delta_{A_1}\big|_{EO} + \frac{N_{T_1}}{\overline{q}_1 S b}\right)}{c_{n_{\delta_R}}}$$

It is worthwhile to mention that the term in $\delta_{R_1}|_{EO}$ associated with the aileron compensating deflection $(-(c_{n_{\delta_A}}/c_{n_{\delta_R}})\delta_{A_1}|_{EO})$ is typically very small due to the small or negligible size of the control derivatives $c_{n_{\delta_A}}$. However, while $c_{n_{\delta_A}}$ is small or negligible, $c_{l_{\delta_R}}$ can be significant, especially for jetliners, due to the relatively large size of the rudder. Therefore, for completeness, an additional aileron correction might be required—starting from the previous deflection labeled $\delta_{A_1}|_{EO_{INITIAL}}$—for canceling the additional rolling moment associated with the rudder compensating deflection. This correction might be significant for substantial values of $\delta_{R_1}|_{EO}$. From the rolling moment equation, this additional correction can be calculated as

$$\Delta\delta_{A_1}\big|_{EO} = \frac{\left(-c_{l_{\delta_R}} \delta_{R_1}\big|_{EO}\right)}{c_{l_{\delta_A}}}$$

leading to

$$\delta_{A_1}\big|_{EO_{FINAL}} = \delta_{A_1}\big|_{EO_{INITIAL}} + \Delta\delta_{A_1}\big|_{EO} = \frac{\left(-c_{l_\beta}\beta_1\big|_{EO} - \frac{L_{T_1}}{\overline{q}_1 S b}\right)}{c_{l_{\delta_A}}} + \frac{\left(-c_{l_{\delta_R}} \delta_{R_1}\big|_{EO}\right)}{c_{l_{\delta_A}}}$$

To provide a numerical example of the lateral dynamics described, consider the aerodynamic and propulsive data of the Boeing B747 (Aircraft 10 in Appendix B) at high cruise flight conditions:

$$\text{Alt.} = 40{,}000 \text{ ft}, \quad \overline{q}_1 = 222.8 \text{ lbs/ft}^2, \quad V_{P_1} = 817 \text{ ft/sec}, \quad M = 0.9$$
$$S = 5{,}500 \text{ ft}^2, \quad b = 196 \text{ ft}$$
$$c_{T_{X_1}} = 0.045 \Rightarrow T_1 = c_{T_{X_1}} \overline{q}_1 S = 55{,}143 \text{ lbs}$$

Initial trim conditions: $\beta_1 = \delta_{A_1} = \delta_{R_1} = 0°$

$$c_{n_\beta} = 0.210, \quad c_{n_{\delta_A}} = -0.0028, \quad c_{n_{\delta_R}} = -0.095$$
$$c_{l_\beta} = -0.095, \quad c_{l_{\delta_A}} = 0.014, \quad c_{l_{\delta_R}} = 0.005$$

An approximate modeling of the thrust moments can be developed from Figure 6.31, assuming the failure of the outer engine on the right wing.

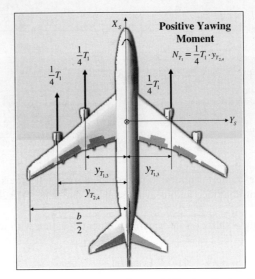

Figure 6.31 Modeling of the Thrust Moment Following Engine Out on a B747

Assuming negligible toe-in angle for the propulsion system, for the B747 aircraft we can approximate $L_{T_1} \approx 0$ while an expression for N_{T_1} can be given by

$$N_{T_1} = \frac{1}{4} T_1 \, y_{T_{2,4}} = \frac{1}{4} T_1 \cdot \left(0.75 \frac{b}{2}\right) = 13786 \cdot 73.5 \approx 1{,}013{,}252.6 \text{ lbs ft}$$

Immediately following the occurrence of the engine-out condition, the sideslip angle $\beta_1|_{EO}$ will develop starting from $\beta_1 = 0°$:

$$\beta_1\Big|_{EO} = -\frac{N_{T_1}}{c_{n_\beta} \bar{q}_1 \, S \, b} = -0.0201 \text{ rad} = -1.15°$$

Next, due to the substantial dihedral effect c_{l_β}, even with $L_{T_1} \approx 0$, the aircraft will develop a substantial rolling moment, which will be eventually compensated by the pilot with an aileron deflection $\delta_{A_1}|_{EO}$ given by

$$\delta_{A_1}\Big|_{EO} = \frac{\left(-c_{l_\beta} \beta_1\Big|_{EO} - \dfrac{L_{T_1}}{\bar{q}_1 \, S \, b}\right)}{c_{l_{\delta_A}}} = \frac{(0.095(-0.0201) - 0)}{0.014} = -0.1363 \text{ rad} = -7.78°$$

Thus, substantial aileron compensation is required, which is expected due to the significant dihedral effect for this aircraft.

Therefore, the compensating deflection for the rudder to eliminate the thrust induced yawing moment is calculated using

$$\delta_{R_1}\Big|_{EO} = \frac{-\left(c_{n_\beta} \beta_1\Big|_{EO} + c_{n_{\delta_A}} \delta_{A_1}\Big|_{EO} + \dfrac{N_{T_1}}{\bar{q}_1 \, S \, b}\right)}{c_{n_{\delta_R}}}$$

$$= \frac{-\left((0.21 \cdot -0.0201) + (-0.0028 \cdot -0.1363) + \dfrac{1{,}013{,}252.6}{(222.8 \cdot 5500 \cdot 196)}\right)}{-0.095} = 0.004 \text{ rad} \approx 0.23°$$

Due to the low value of $\delta_{R_1}|_{EO}$, the secondary compensation for the ailerons will be negligible. In this specific case we have

$$\Delta\delta_{A_1}\Big|_{EO} = \frac{\left(-c_{l_{\delta_R}}\delta_{R_1}\Big|_{EO}\right)}{c_{l_{\delta_A}}} = \frac{-0.005 \cdot 0.004}{0.014} = -0.0014 \text{ rad} \approx -0.082°$$

leading to

$$\delta_{A_1}\Big|_{EO_{FINAL}} = \delta_{A_1}\Big|_{EO_{INITIAL}} + \Delta\delta_{A_1}\Big|_{EO} = \delta_{A_1}\Big|_{EO} = -7.78° - 0.082° \approx -7.89°$$

The low value of the rudder deflection $\delta_{R_1}|_{EO}$ might indicate that the occurrence of an engine-out condition is a minor problem from a directional point of view. However, the following needs to be considered. The B747 aircraft is a four-engine jetliner; therefore, its vertical tail and its rudder are designed to handle up to two engine failures on the same wing. Therefore, the value for the control derivative $c_{n_{\delta_R}}$ is fairly large for this specific aircraft. Also, the engine-out condition is considered at a very high airspeed and dynamic pressure where the rudder is very effective. The analysis of the compensating rudder deflection throughout the range of operational airspeed is very important from a design point of view. The trend for $\delta_{R_1}|_{EO}$ versus airspeed for the B747 aircraft is shown in Figure 6.32.

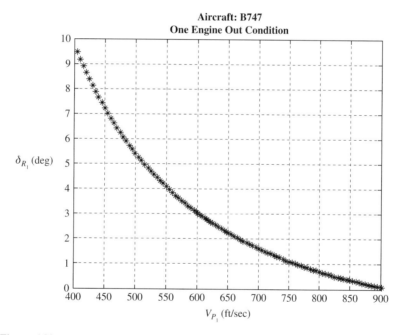

Figure 6.32 Rudder Compensation vs. Airspeed for Engine-Out Condition for the B747

It is clear that substantial compensating deflections are required for the rudder as the airspeed decreases due to the lower effectiveness of the rudder in generating yawing moments.

The availability of the data for the B747 aircraft also allows the analysis of a specific and rare condition, which is the sequential failure of two engines. According to FAA regulations, a civilian aircraft is required to be able to continue its flight following the loss of up to 50 percent of the installed thrust.[2,3] For a four-engine aircraft, the worst-case scenario from a directional point of view is represented by the condition of two engines out on the same wing, as shown in Figure 6.33.

338 Chapter 6 Aircraft Stability and Design for Trim Conditions

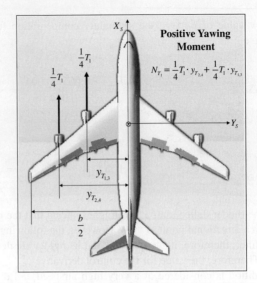

Figure 6.33 Modeling of the Thrust Moment Following Two Engines Out on a B747

Assuming $y_{T_{1,3}} = \left(0.4 \dfrac{b}{2}\right)$, the modeling for N_{T_1} in this special case can be given by

$$N_{T_1} = \frac{1}{4} T_1 \, y_{T_{2,4}} + \frac{1}{4} T_1 \, y_{T_{1,2}} = \frac{1}{4} T_1 \cdot \left(0.75 \frac{b}{2} + 0.4 \frac{b}{2}\right)$$

$$= 13786 \cdot 112.7 \approx 1{,}553{,}682.2 \text{ lbs ft}$$

The previous numerical analysis using the new N_{T_1} estimate for the case of two engines out will provide the following results in terms of trim conditions:

$$\beta_1 \Big|_{EO} = -\frac{N_{T_1}}{c_{n_\beta} \bar{q}_1 \, S \, b} \approx -1.76°$$

$$\delta_{A_1} \Big|_{EO} = \frac{\left(-c_{l_\beta} \beta_1 \Big|_{EO} - \dfrac{L_{T_1}}{\bar{q}_1 \, S \, b}\right)}{c_{l_{\delta_A}}} \approx -11.98°$$

$$\delta_{R_1} \Big|_{EO} = \frac{-\left(c_{n_\beta} \beta_1 \Big|_{EO} + c_{n_{\delta_A}} \delta_{A_1} \Big|_{EO} + \dfrac{N_{T_1}}{\bar{q}_1 \, S \, b}\right)}{c_{n_{\delta_R}}} \approx 0.35°$$

$$\Delta \delta_{A_1} \Big|_{EO} = \frac{\left(-c_{l_{\delta_R}} \delta_{R_1} \Big|_{EO}\right)}{c_{l_{\delta_A}}} \approx -0.13°$$

$$\delta_{A_1} \Big|_{EO_{FINAL}} = \delta_{A_1} \Big|_{EO_{INITIAL}} + \Delta \delta_{A_1} \Big|_{EO} = \delta_{A_1} \Big|_{EO} \approx -12.10°$$

Finally, the trend of the compensating rudder deflections with the airspeed is shown in Figure 6.34.

6.6 Lateral Directional Analysis of Steady-State Straight Flight

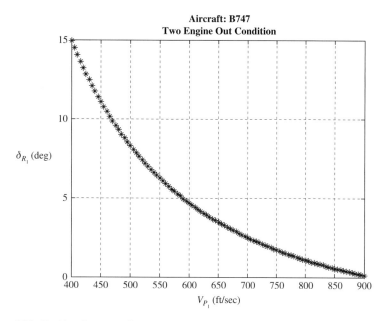

Figure 6.34 Rudder Compensation vs. Airspeed for Two Engines Out Condition for the B747

Obviously, the amount of rudder compensation $\delta_{R_1}|_{EO}$ required at lower airspeed is substantial for a two engines out condition; in fact, the capability of regaining trim conditions following propulsion failures is a key factor for the design of vertical tails for commercial aircraft. Additionally, especially at lower speed, it is very desirable to minimize the sideslip angle $\beta_1|_{EO}$ for the specific goal of minimizing the drag and therefore leaving much needed excess power from the residual thrust to the pilot in case a climb is necessary for continuing the flight.

Typically a $\pm 25°$ value is set for $\delta_{R_1}|_{EO_{MAX}}$, which leaves a small margin to avoid control saturation for this primary control surface. Assuming zero sideslip angle, the use of $\delta_{R_1}|_{EO_{MAX}}$ directly identifies a minimum control airspeed at which the aircraft can be controlled in the event of an engines-out condition. From the yawing moment equation with zero sideslip angle and negligible value for $c_{n_{\delta_A}}$

$$0 = \frac{1}{2}\rho V_{P_1}^2 S b \left(c_{n_\beta}\cancel{\beta} + \cancel{c_{n_{\delta_A}}}\delta_{A_1} + c_{n_{\delta_R}}\delta_{R_1}\right) + N_{T_1}$$

this airspeed can be derived using

$$V_{MC} = \sqrt{\frac{2 N_{T_1}}{\left(\rho S\, b c_{n_{\delta_R}}\right)\delta_{R_1}|_{EO_{MAX}}}}$$

It is intuitive that FAA regulations for civilian aircraft and MIL regulations for military aircraft[2,3,4,5,6] require V_{MC} to be larger than the stall speed for the aircraft at that specific configuration.

Finally, an important aid for the pilot following the engine-out can be given by imposing a moderate bank angle Φ_1 into the operational engines, as shown in Figure 6.35. This leads to a reduction of the compensating rudder deflection $\delta_{R_1}|_{EO}$. Typically this value for the bank angle is set to a maximum value of $\pm 5°$.

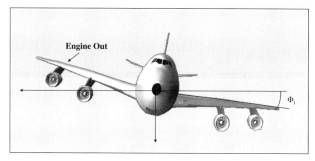

Figure 6.35 Bank Angle into Operational Engines Following Engine Out for a B747

As a final set of conclusions, the following can be considered a suitable sequence of events for a pilot trying to restore lateral directional equilibrium conditions following an engine out on a multiengine aircraft (starting from trim conditions) and a short transient allowing a failure detection and identification:

Step 1 — Aileron deflection $\delta_{A_1}|_{EO}$ to compensate for the rolling moment associated with the engine failure

Step 2 — Rudder deflection $\delta_{R_1}|_{EO}$ to regain equilibrium around the yawing axis

Step 3 — Additional aileron deflection $\Delta\delta_{A_1}|_{EO}$ to compensate for the rolling moment associated with the previous rudder deflection

Step 4 — Reduction of the sideslip angle to minimize the generation of additional drag following engine failure and thus allowing excess power potentially needed for a climb

Step 5 — Commanding a small bank angle into the operational engines if needed for decreasing the magnitude of $\delta_{R_1}|_{EO}$

The analysis of the lateral directional dynamics following engine failure starting from a maneuvered flight condition is substantially more complex. Refer to [1] for this specific analysis.

(See Student Sample Problem 6.5.)

6.7 SUMMARY

This chapter outlined some key concepts relative to the overall aircraft stability, with emphasis on introducing specific criteria to be satisfied the aircraft static stability. The equations relative to the longitudinal dynamics starting from steady-state rectilinear flight conditions have been analyzed, leading to the important discussion of the trim conditions followed by the introduction of the lift chart and the important trim diagram. The lateral directional equations from a steady-state rectilinear flight have been analyzed with an important and detailed discussion of the engine-out condition for one and two engines on a multiengine aircraft.

REFERENCES

1. Roskam, J. *Airplane Flight Dynamics and Automatic Flight Controls—Part I*. Design, Analysis, and Research Corporation, Lawrence, KS, 1995.
2. Airworthiness Requirements FAR 23.
3. Airworthiness Requirements FAR 25.
4. Airworthiness Requirements JAR-VLA.
5. Airworthiness Requirement MIL-F-8785C.
6. Airworthiness Requirement MIL-STD-1797A.
7. Russell, J. B. *Performance & Stability of Aircraft*. Arnold and John Wiley & Sons, 1996.
8. Phillips, W. F. *Mechanics of Flight*. John Wiley & Sons, 2004.
9. Pamadi, B. N. *Performance, Stability, Dynamics, and Control of Airplanes* (2nd ed.). AIAA (Education Series), 2004.
10. Stevens, B. L., Lewis, F. *Aircraft Control and Simulation* (2nd ed.). John Wiley & Sons, 2003.
11. Roskam, J. *Airplane Flight Dynamics and Automatic Flight Controls—Part II*. Design, Analysis, and Research Corporation, Lawrence, KS, 1995.
12. Web Search, "*USAirways—Flight 1549.*" February 2009.

STUDENT SAMPLE PROBLEMS

Student Sample Problem 6.1

Consider the aerodynamic data for the following aircraft from Appendix B:

- Cessna 182 (General Aviation Aircraft). Flight condition: cruise;
- Learjet 24 (Business jet). Flight condition: cruise, max. weight;
- Boeing 747 (Commercial jetliner). Flight condition: high cruise.

Verify that the static stability criteria are satisfied for those aircraft at the given flight conditions.

Solution of Student Sample Problem 6.1

Table SSP6.1.1 Static Stability Criteria for the Cessna 182, Learjet, and Boeing B747 Aircraft

Stability Criteria	Cessna 182	Learjet 24	Boeing 747
SSC #1: $\left(c_{T_{X_u}} - c_{D_u}\right) < 0$	$c_{T_{X_u}} - 0.096, c_{D_u} = 0$ YES	$c_{T_{X_u}} - 0.07, c_{D_u} = 0.104$ YES	$c_{T_{X_u}} - 0.95, c_{D_u} = 0.22$ YES
SSC #2: $c_{Y_\beta} < 0$	$c_{Y_\beta} = -0.393$ YES	$c_{Y_\beta} = -0.73$ YES	$c_{Y_\beta} = -0.9$ YES
SSC #3: $c_{L_\alpha} > 0$	$c_{L_\alpha} = 4.41$ YES	$c_{L_\alpha} = 5.84$ YES	$c_{L_\alpha} = 5.5$ YES
SSC #4: $c_{m_\alpha} < 0$	$c_{m_\alpha} = -0.613$ YES	$c_{m_\alpha} = -0.64$ YES	$c_{m_\alpha} = -1.6$ YES
SSC #5: $c_{n_\beta} > 0$	$c_{n_\beta} = 0.0587$ YES	$c_{n_\beta} = 0.127$ YES	$c_{n_\beta} = 0.16$ YES
SSC #6: $c_{l_p} < 0$	$c_{l_p} = -0.484$ YES	$c_{l_p} = -0.45$ YES	$c_{l_p} = -0.34$ YES
SSC #7: $c_{m_q} < 0$	$c_{m_q} = -12.4$ YES	$c_{m_q} = -15.5$ YES	$c_{m_q} = -25.5$ YES
SSC #8: $c_{n_r} < 0$	$c_{n_r} = -0.0937$ YES	$c_{n_r} = -0.0937$ YES	$c_{n_r} = -0.28$ YES
SSC #9: $c_{l_\beta} < 0$	$c_{l_\beta} = -0.0923$ YES	$c_{l_\beta} = -0.11$ YES	$c_{l_\beta} = -0.16$ YES
SSC #10: $c_{m_u} > 0$	$c_{m_u} = 0$ MARGINAL	$c_{m_u} = 0.05$ YES	$c_{m_u} = -0.09$ UNSTABLE

Table SSP6.1.1 analysis shows in terms of c_{m_u} a marginal stability for the Cessna 182 and instability for the Boeing 747, which is expected for this class of aircraft. The condition can be controlled by the pilot for the Cessna 182 and by a Stability Augmentation System (SAS) for the Boeing B747. Referr to [10,11] for a description of the design of these control systems.

Student Sample Problem 6.2

Consider the aerodynamic data for the following aircraft from Appendix B:

- SIAI Marchetti S211 (Fighter Trainer). Flight condition: high subsonic cruise;
- Lockheed F-104 (Fighter/Interceptor). Flight condition: supersonic cruise;
- McDonnell F-4 (Fighter/Attack). Flight condition: supersonic cruise.

Verify that the static stability criteria are satisfied for those aircraft at the given flight conditions.

Solution of Student Sample Problem 6.2

The Table SSP6.2.1 shows a marginal stability in terms of c_{m_u} for the SIAI Marchetti S211 and the Lockheed F-104. This condition can be easily controlled by the pilot and/or by a Stability Augmentation System (SAS).[10,11]

Table SSP6.2.1 Static Stability Criteria for the SIAI Marchetti S211, Lockhead F-104, and McDonnell F-4 Aircraft

Stability Criteria	SIAI Marchetti S211	Lockheed F-104	McDonnell F-4
SSC #1: $\left(c_{T_{X_u}} - c_{D_u}\right) < 0$	$c_{T_{X_u}} - 0.055, c_{D_u} = 0.05$ YES	$c_{T_{X_u}} - 0.13, c_{D_u} = -0.06$ YES	$c_{T_{X_u}} - 0.1, c_{D_u} = -0.054$ YES
SSC #2: $c_{Y_\beta} < 0$	$c_{Y_\beta} = -1.0$ YES	$c_{Y_\beta} = -1.045$ YES	$c_{Y_\beta} = -0.7$ YES
SSC #3: $c_{L_\alpha} > 0$	$c_{L_\alpha} = 5.5$ YES	$c_{L_\alpha} = 2.0$ YES	$c_{L_\alpha} = 2.8$ YES
SSC #4: $c_{m_\alpha} < 0$	$c_{m_\alpha} = -0.24$ YES	$c_{m_\alpha} = -1.31$ YES	$c_{m_\alpha} = -0.78$ YES
SSC #5: $c_{n_\beta} > 0$	$c_{n_\beta} = 0.170$ YES	$c_{n_\beta} = 0.242$ YES	$c_{n_\beta} = 0.09$ YES
SSC #6: $c_{l_p} < 0$	$c_{l_p} = -0.39$ YES	$c_{l_p} = -0.272$ YES	$c_{l_p} = -0.2$ YES
SSC #7: $c_{m_q} < 0$	$c_{m_q} = -17.7$ YES	$c_{m_q} = -4.83$ YES	$c_{m_q} = -2.0$ YES
SSC #8: $c_{n_r} < 0$	$c_{n_r} = -0.26$ YES	$c_{n_r} = -0.65$ YES	$c_{n_r} = -0.26$ YES
SSC #9: $c_{l_\beta} < 0$	$c_{l_\beta} = -0.11$ YES	$c_{l_\beta} = -0.093$ YES	$c_{l_\beta} = -0.025$ YES
SSC #10: $c_{m_u} > 0$	$c_{m_u} = 0$ MARGINAL	$c_{m_u} = 0$ MARGINAL	$c_{m_u} = 0.054$ YES

Student Sample Problem 6.3

Consider the aerodynamic data for the SIAI Marchetti S211 aircraft at high subsonic cruise (Appendix B, Aircraft 8). Note that this aircraft features a horizontal tail with elevators for maneuvering and stabilators for trimming purposes. Assume zero deflections for the elevators at trim conditions. Evaluate the following trim characteristics:

$$\alpha_1, i_{H_1}, \frac{\partial \alpha_1}{\partial c_{L_1}}, \frac{\partial i_{H_1}}{\partial c_{L_1}}, \frac{\partial i_{H_1}}{\partial V_{P_1}}$$

Plot of $\frac{\partial \alpha_1}{\partial c_{L_1}}$ vs. $\bar{x}_{CG}$, plot of $\frac{\partial i_{H_1}}{\partial c_{L_1}}$ vs. $\bar{x}_{CG}$, plot of $\frac{\partial i_{H_1}}{\partial V_{P_1}}$ vs. $\bar{x}_{CG}$

Solution of Student Sample Problem 6.3

Starting from the aircraft geometry, weight, and flight conditions:

$$S = 136 \text{ ft}^2, \ \bar{c} = 5.4 \text{ ft}, \ W = 4000 \text{ lbs}, \ \bar{x}_{CG} = 0.25$$
$$\bar{q}_1 = 125.7 \text{ lbs/ft}^2, \ Alt. = 35{,}000 \text{ ft}, \ V_{P_1} = 584 \text{ ft/sec},$$
$$\rho = 7.37 \cdot 10^{-4} \ slug/\text{ft}^3$$

Consider the following longitudinal aerodynamic stability and control derivatives:

$$c_{m_0} = -0.08, \ c_{m_\alpha} = -0.24, \ c_{m_{i_H}} = -2.3$$
$$c_{L_0} = 0.149, \ c_{L_\alpha} = 5.5, \ c_{L_{i_H}} = 0.99$$

First, evaluate the value of c_{L_1}:

$$c_{L_1} = \frac{W}{\bar{q}_1 \cdot S} = \frac{4000}{125.7 \cdot 136} = 0.234$$

Next, evaluate α_1 using

$$\alpha_1 = \frac{c_{m_{i_H}}(c_{L_1} - c_{L_0}) - c_{L_{i_H}}(-c_{m_0})}{\left(c_{L_\alpha} c_{m_{i_H}} - c_{m_\alpha} c_{L_{i_H}}\right)} = \frac{-2.3 \cdot (0.234 - 0.149) - 0.99 \cdot (-0.08)}{((5.5 \cdot -2.3) - (-0.24 \cdot 0.99))} = 0.0937 \text{ rad} \approx 0.54°$$

and evaluate i_{H_1} using

$$i_{H_1} = \frac{c_{L_\alpha}(-c_{m_0}) - c_{m_\alpha}(c_{L_1} - c_{L_0})}{\left(c_{L_\alpha} c_{m_{i_H}} - c_{m_\alpha} c_{L_{i_H}}\right)} = \frac{5.5 \cdot (+0.08) + 0.24 \cdot (0.234 - 0.149)}{((5.5 \cdot -2.3) - (-0.24 \cdot 0.99))} = -0.037 \text{ rad} \approx -2.12°$$

The gradient $\frac{\partial \alpha_1}{\partial c_{L_1}}$ is evaluated using

$$\frac{\partial \alpha_1}{\partial c_{L_1}} = \frac{c_{m_{i_H}}}{\left(c_{L_\alpha} c_{m_{i_H}} - c_{m_\alpha} c_{L_{i_H}}\right)} = \frac{-2.3}{((5.5 \cdot -2.3) - (-0.24 \cdot 0.99))} = \frac{-2.3}{-12.41} = 0.185$$

which is positive, as expected. Similarly, the gradient $\frac{\partial i_{H_1}}{\partial c_{L_1}}$ is evaluated using

$$\frac{\partial i_{H_1}}{\partial c_{L_1}} = \frac{-c_{m_\alpha}}{\left(c_{L_\alpha} c_{m_{i_H}} - c_{m_\alpha} c_{L_{i_H}}\right)} = \frac{0.24}{((5.5 \cdot -2.3) - (-0.24 \cdot 0.99))} = \frac{0.24}{-12.41} = -0.0193$$

which is negative, as expected.

Also, the gradient $\frac{\partial i_{H_1}}{\partial V_{P_1}}$ is evaluated using

$$\frac{\partial i_{H_1}}{\partial V_{P_1}} = \frac{4W}{\rho S V_{P_1}^3} \frac{c_{m_\alpha}}{\left(c_{L_\alpha} c_{m_{i_H}} - c_{m_\alpha} c_{L_{i_H}}\right)} = \left(\frac{4 \cdot 4000}{7.37 \cdot 10^{-4} \cdot 136 \cdot 584^3}\right) \frac{-0.24}{((5.5 \cdot -2.3) - (-0.24 \cdot 0.99))} = 1.55 \cdot 10^{-6}$$

which is positive, as expected.

From the preceding aerodynamic data the static margin for this aircraft at this configuration can be calculated using

$$c_{m_\alpha} = c_{L_\alpha}(\bar{x}_{CG} - \bar{x}_{AC}) = c_{L_\alpha} SM \Rightarrow SM = \frac{c_{m_\alpha}}{c_{L_\alpha}} = \frac{-0.24}{5.5} \approx -0.044$$

Starting from the reference center of gravity $\bar{x}_{CG} = 0.25$, we have

$$\bar{x}_{AC} = \bar{x}_{CG} - SM = 0.25 - (-0.044) = 0.294$$

With a similar approach, the moment arm associated with the horizontal tail can be calculated using:

$$c_{m_{i_H}} = -c_{L_{i_H}}(\bar{x}_{AC_H} - \bar{x}_{CG}) \Rightarrow (\bar{x}_{AC_H} - \bar{x}_{CG}) = -\frac{c_{m_{i_H}}}{c_{L_{i_H}}} = -\frac{-2.3}{0.99} = 2.32$$

leading therefore to $\bar{x}_{AC_H} = 2.32 + \bar{x}_{CG} = 2.32 + 0.25 = 2.57$.

These calculations allow setting the evaluation of $\frac{\partial \alpha_1}{\partial c_{L_1}} = f(\bar{x}_{CG})$ as

$$\frac{\partial \alpha_1}{\partial c_{L_1}} = \frac{\left[-c_{L_{i_H}}(\bar{x}_{AC_H} - \bar{x}_{CG})\right]}{\left[c_{L_\alpha}\left(-c_{L_{i_H}}(\bar{x}_{AC_H} - \bar{x}_{CG})\right) - (c_{L_\alpha}(\bar{x}_{CG} - \bar{x}_{AC}))c_{L_{i_H}}\right]} = \frac{-0.99(2.57 - \bar{x}_{CG})}{[-5.445(2.57 - \bar{x}_{CG}) - 5.445(\bar{x}_{CG} - 0.294)]}$$

A plot of $\frac{\partial \alpha_1}{\partial c_{L_1}} = f(\bar{x}_{CG})$ for a restricted range of $\bar{x}_{CG}$ location is shown in Figure SSP6.3.1:

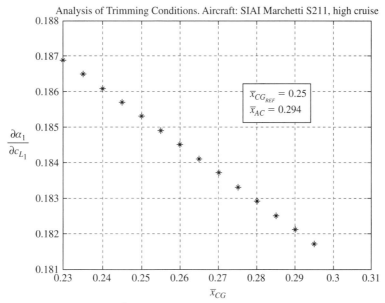

Figure SSP6.3.1 $\frac{\partial \alpha_1}{\partial c_{L_1}}$ vs. $(\bar{x}_{CG})$ for the SIAI Marchetti S211 Aircraft

Similarly, for $\frac{\partial i_{H_1}}{\partial c_{L_1}} = \frac{-c_{m_\alpha}}{\left(c_{L_\alpha}c_{m_{i_H}} - c_{m_\alpha}c_{L_{i_H}}\right)}$, a relationship $\frac{\partial i_{H_1}}{\partial c_{L_1}} = f(\bar{x}_{CG})$ can be given by

$$\frac{\partial i_{H_1}}{\partial c_{L_1}} = \frac{-[c_{L_\alpha}(\bar{x}_{CG} - \bar{x}_{AC})]}{\left[c_{L_\alpha}\left(-c_{L_{i_H}}(\bar{x}_{AC_H} - \bar{x}_{CG})\right) - (c_{L_\alpha}(\bar{x}_{CG} - \bar{x}_{AC}))c_{L_{i_H}}\right]} = \frac{-5.5(\bar{x}_{CG} - 0.294)}{[-5.445(2.57 - \bar{x}_{CG}) - 5.445(\bar{x}_{CG} - 0.294)]}$$

344 Chapter 6 Aircraft Stability and Design for Trim Conditions

A plot of $\dfrac{\partial i_{H_1}}{\partial c_{L_1}} = f(\overline{x}_{CG})$ for a restricted range of $\overline{x}_{CG}$ location is shown in Figure SSP6.3.2:

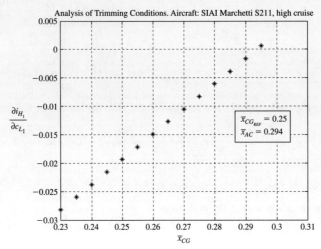

Figure SSP6.3.2 $\dfrac{\partial i_{H_1}}{\partial c_{L_1}}$ vs. $(\overline{x}_{CG})$ for the SIAI Marchetti S211 Aircraft

Finally, for $\dfrac{\partial i_{H_1}}{\partial V_{P_1}} = \dfrac{4\,W}{\rho\,S\,V_{P_1}^3}\dfrac{c_{m_\alpha}}{\left(c_{L_\alpha}c_{m_{i_H}} - c_{m_\alpha}c_{L_{i_H}}\right)}$ vs. $\overline{x}_{CG}$, a relationship $\dfrac{\partial i_{H_1}}{\partial V_{P_1}} = f(\overline{x}_{CG})$ can be given by

$$\dfrac{\partial i_{H_1}}{\partial V_{P_1}} = \dfrac{-4\,W}{\rho\,S\,V_{P_1}^3}\dfrac{\partial i_{H_1}}{\partial c_{L_1}} = \dfrac{4\,W}{\rho\,S\,V_{P_1}^3}\dfrac{[c_{L_\alpha}(\overline{x}_{CG} - \overline{x}_{AC})]}{\left[c_{L_\alpha}\left(-c_{L_{i_H}}(\overline{x}_{AC_H} - \overline{x}_{CG})\right) - (c_{L_\alpha}(\overline{x}_{CG} - \overline{x}_{AC}))c_{L_{i_H}}\right]}$$

$$= \left(\dfrac{4\cdot 4000}{7.37\cdot 10^{-4}\cdot 136\cdot 584^3}\right)\dfrac{[5.5(\overline{x}_{CG} - 0.294)]}{[-5.445(2.57 - \overline{x}_{CG}) - 5.445(\overline{x}_{CG} - 0.294)]}$$

A plot of $\dfrac{\partial i_{H_1}}{\partial V_{P_1}} = f(\overline{x}_{CG})$ for a restricted range of $\overline{x}_{CG}$ location is shown in Figure SPP6.3.3:

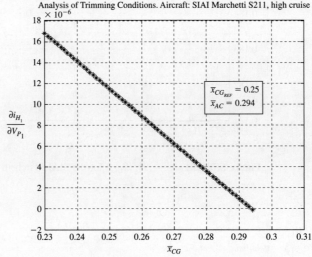

Figure SSP6.3.3 $\dfrac{\partial i_{H_1}}{\partial V_{P_1}}$ vs. $(\overline{x}_{CG})$ for the SIAI Marchetti S211 Aircraft

Student Sample Problem 6.4

Consider a generic aircraft with the TD shown in Figure SSP6.4.1.

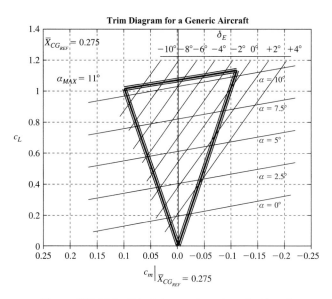

Figure SSP6.4.1 Trim Diagram for a Generic Aircraft

For the given aircraft $\bar{x}_{CG_{REF}} = 0.275$ is selected. Note that the aircraft features elevators (δ_E) for trimming purposes. Find the following:

- The elevator deflection required for trimming the aircraft at $\alpha_1 = 6°$ with $\bar{x}_{CG} = 0.35$;
- The angle of attack required for trimming the aircraft with $\delta_{E_1} = -2°$ and $\bar{x}_{CG} = 0.2$;
- The value of c_{L_0} and assessment on whether the aircraft can be trimmed with $c_{L_1} = c_{L_0}$.

Solution of Student Sample Problem 6.4

Consider the aircraft configuration with $\bar{x}_{CG} = 0.35$. Since $\bar{x}_{CG_{REF}} = 0.275$, considering an arbitrary $\Delta c_L = 1$, there will be a $\Delta c_m = -0.075$ associated with $\Delta(SM) = \bar{x}_{CG_{REF}} - 0.35 = -0.075$. Connecting the origin with this point will generate the new trim line associated with $\bar{x}_{CG} = 0.35$. Next, a line associated with the condition $\alpha_1 = 6°$ is drawn using a linear interpolation between the parallel lines associated with $\alpha_1 = 5°$ and $\alpha_1 = 7.5°$. Finally, the required δ_{E_1} is found by interpolating a line parallel to the constant elevator deflections $\delta_{E_1} = 0°$ and $\delta_{E_1} = 2°$. It turns out that the elevator deflection required for trimming the aircraft at $\alpha_1 = 6°$ with $\bar{x}_{CG} = 0.35$ is given by $\delta_{E_1} = +1°$, as shown in Figure SSP6.4.2.

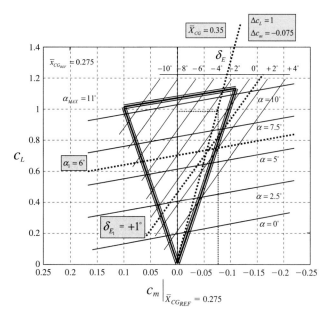

Figure SSP6.4.2 Trim Diagram for a Generic Aircraft

Next, consider the aircraft configuration with $\bar{x}_{CG} = 0.2$. Since $\bar{x}_{CG_{REF}} = 0.275$, considering an arbitrary $\Delta c_L = 1$, there will be a $\Delta c_m = +0.075$ associated with $\Delta(SM) = \bar{x}_{CG_{REF}} - 0.2 = +0.075$. Connecting the origin with this point will generate the new trim line associated with $\bar{x}_{CG} = 0.2$. Next, the line associated with the condition $\delta_{E_1} = -2°$ is selected. Finally, the required α_1 is found by interpolating a line parallel to the constant angles of attack $\alpha_1 = 2.5°$ and $\alpha_1 = 5°$. It turns out that the angle of attack required for trimming the aircraft at $\delta_{E_1} = -2°$ with $\bar{x}_{CG} = 0.2$ is given by $\alpha_1 \approx 4°$, as shown in Figure SSP6.4.3.

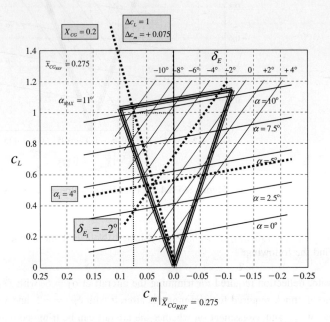

Figure SSP6.4.3 Trim Diagram for a Generic Aircraft

By definition we have $c_{L_0} = c_L|_{\alpha = \delta_E = 0°}$.

Therefore, the value of c_{L_0} can be found at the intersection of the trim lines with $\alpha_1 = 0°$ and $\delta_{E_1} = 0°$. As shown in Figure SSP6.4.4, $c_{L_0} \approx 0.16$. Because this point is located outside the trim triangle, it can be concluded that the aircraft can not be trimmed at the conditions $\alpha_1 = 0°$, $\delta_{E_1} = 0°$.

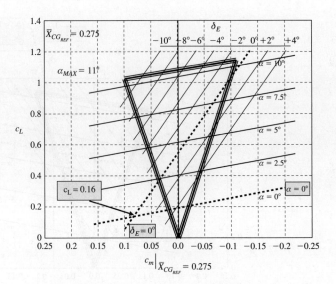

Figure SSP6.4.4 Trim Diagram for a Generic Aircraft

Student Sample Problem 6.5

Consider an engine-out condition for the Beech 99 twin-turbo prop aircraft, shown in Figure SSP6.5.1, representative of a regional commercial aircraft. Describe the solution for new steady-state conditions assuming a worst-case scenario with the engine failure occurring during the approach (approx. 3,000 ft) and $V_{P_1} = 170$ ft/sec.

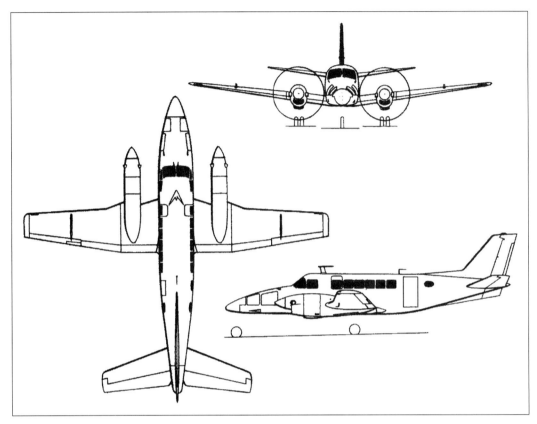

Figure SSP6.5.1 3D View of a Beech 99 Aircraft
(Source: http://www.aviastar.org/index2.html)

Starting from the aerodynamic and propulsive data of the Beech 99 (Appendix B, Aircraft 3):

$$Alt. = 3{,}000 \text{ ft}, \quad \bar{q}_1 = 34.2 \text{ lbs/ft}^2, \quad V_{P_1} = 170 \text{ ft/sec},$$
$$S = 280 \text{ ft}^2, \quad b = 46 \text{ ft}$$
$$c_{T_{X_1}} = 0.162 \Rightarrow T_1 = c_{T_{X_1}} \bar{q}_1 S = 1{,}551.3 \text{ lbs}$$
$$c_{n_\beta} = 0.12, \quad c_{n_{\delta_A}} = -0.0012, \quad c_{n_{\delta_R}} = -0.0763$$
$$c_{l_\beta} = -0.13, \quad c_{l_{\delta_A}} = 0.156, \quad c_{l_{\delta_R}} = 0.0087$$

Consider the conventional lateral-directional initial trim conditions

$$\beta_1 = \delta_{A_1} = \delta_{R_1} = 0°$$

Assuming the failure of the right engine, an approximate modeling of the thrust moments can be deduced from Figure SSP6.5.2, assuming a negligible toe-in angle and $L_{T_1} \approx 0$.

348 Chapter 6 Aircraft Stability and Design for Trim Conditions

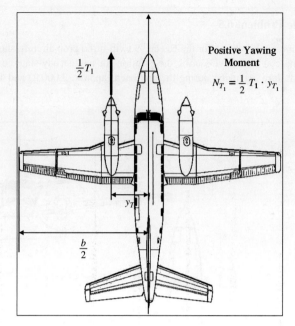

Figure SSP6.5.2 Geometric Parameters for Engine-Out Condition on a Beech 99 Aircraft

From a graphic analysis it can be assumed that

$$y_{T_1} \approx \frac{1}{3}\frac{b}{2} = \frac{b}{6} \approx 7.67 \text{ ft}$$

Therefore, it can be approximated

$$N_{T_1} = \frac{1}{2} T_1 \; y_{T_1} \approx 5{,}949.3 \text{ lbs ft}$$

Using the preceding modeling, the following sequence of compensating actions can be taken by the pilot following the detection and identification of the engine failure problem:

Step 1: Aileron compensating deflection $\delta_{A_1}|_{EO}$ to overcome the rolling moment associated with the engine failure

$$\delta_{A_1}\Big|_{EO} = \frac{\left(-c_{l_\beta}\beta_1\Big|_{EO} - \cancel{\frac{L_{T_1}}{\overline{q}_1 S b}}\right)}{c_{l_{\delta_A}}} \quad \text{where } \beta_1\Big|_{EO} = -\frac{N_{T_1}}{c_{n_\beta}\overline{q}_1 S b}$$

Numerically, for the sideslip angle we have

$$\beta_1\Big|_{EO} = -\frac{5{,}949.3}{(0.12 \cdot 34.2 \cdot 280 \cdot 46)} = -0.1125 \text{ rad} = -6.45°$$

leading to

$$\delta_{A_1}\Big|_{EO} = \frac{\left(-c_{l_\beta}\beta_1\Big|_{EO} - \frac{L_{T_1}}{\overline{q}_1 S b}\right)}{c_{l_{\delta_A}}} = \frac{(0.13 \cdot -0.1125)}{0.156} = -0.094 \text{ rad} = -5.37°$$

Step 2: Rudder compensating deflection $\delta_{R_1}|_{EO}$ to regain equilibrium around the yawing axis

$$\delta_{R_1}\Big|_{EO} = \frac{-\left(c_{n_\beta}\beta_1\Big|_{EO} + c_{n_{\delta_A}}\delta_{A_1}\Big|_{EO} + \frac{N_{T_1}}{\overline{q}_1 S b}\right)}{c_{n_{\delta_R}}}$$

$$= \frac{-\left((0.12 \cdot -0.1125) + (-0.0012 \cdot -0.094) + \frac{5{,}949.3}{(34.2 \cdot 280 \cdot 46)}\right)}{-0.0763} = 0.00148 \text{ rad} \approx 0.08°$$

The preceding value implies that due to the low throttle setting during the approach, the engine-out condition is fairly simple to control from a directional point of view. Due to the low value of $\delta_{R_1}|_{EO}$, a secondary compensation for the ailerons will not be necessary. In fact, numerically we have

$$\Delta \delta_{A_1}\Big|_{EO} = \frac{\left(-c_{l_{\delta_R}} \delta_{R_1}\Big|_{EO}\right)}{c_{l_{\delta_A}}} = \frac{-0.0087 \cdot 0.00148}{0.156} = -8.25 \cdot 10^{-5} \text{ rad} = -4.72 \cdot 10^{-3\circ} \approx 0°$$

It is important to verify that the airspeed during approach ($V_{P_1} = 170$ ft/sec) is above the minimum control airspeed provided by

$$V_{MC} = \sqrt{\frac{2N_{T_1}}{\left(\rho S \, b \, |c_{n_{\delta_R}}|\right) \delta_{R_1}\Big|_{EO_{MAX}}}}$$

Assuming a $\pm 25°$ for $\delta_{R_1}|_{EO_{MAX}}$ and calculating the air density using

$$\rho = 2\bar{q}_1/V_{P_1} \approx 2.37 \cdot 10^{-3} \text{ slug/ft}^3$$

we have

$$V_{MC} = \sqrt{\frac{2 \cdot 5,849.3}{(2.37 \cdot 10^{-3} \cdot 280 \cdot 46 \cdot 0.0763S)(25/57.3)}} \approx 103 \text{ ft/sec}$$

which is substantially lower than $V_{P_1} = 170$ ft/sec.

From the preceding analysis it appears that a pilot should be able to handle the engine-out condition without significant saturation problems on the lateral directional control surfaces. Since the aircraft is at approach and the required compensating deflection for the rudder is quite small, it would not be recommended in this case to command a small bank angle into the operational engine. However, it is recommended to the pilot to reduce $\beta_1|_{EO}$ with an additional compensating rudder deflection for a desirable alignment of the aircraft during the descent phase.

PROBLEMS

Problem 6.1

Consider the aerodynamic data for the Boeing B747 aircraft at high subsonic cruise conditions (Appendix B, Aircraft 7). This aircraft features a horizontal tail with elevators for maneuvering and stabilators for trimming purposes. Assume zero deflections for the elevators at trim conditions. Evaluate the following trim characteristics:

$$\alpha_1, i_{H_1}, \frac{\partial \alpha_1}{\partial c_{L_1}}, \frac{\partial i_{H_1}}{\partial c_{L_1}}, \frac{\partial i_{H_1}}{\partial V_{P_1}}$$

$$\text{Plot of } \frac{\partial \alpha_1}{\partial c_{L_1}}, \text{ vs. } \bar{x}_{CG}, \text{ plot of } \frac{\partial i_{H_1}}{\partial c_{L_1}} \text{ vs. } \bar{x}_{CG}, \text{ plot of } \frac{\partial i_{H_1}}{\partial V_{P_1}} \text{ vs. } \bar{x}_{CG}$$

Problem 6.2

Consider the aerodynamic data for the McDonnell Douglas F-4 aircraft at high subsonic cruise conditions (Appendix B, Aircraft 10). This aircraft features an all-moving horizontal tail (stabilators) for both maneuvering and trimming. Evaluate the following trim characteristics:

$$\alpha_1, i_{H_1}, \frac{\partial \alpha_1}{\partial c_{L_1}}, \frac{\partial i_{H_1}}{\partial c_{L_1}}, \frac{\partial i_{H_1}}{\partial V_{P_1}}$$

$$\text{Plot of } \frac{\partial \alpha_1}{\partial c_{L_1}} \text{ vs. } \bar{x}_{CG}, \text{ plot of } \frac{\partial i_{H_1}}{\partial c_{L_1}} \text{ vs. } \bar{x}_{CG}, \text{ plot of } \frac{\partial i_{H_1}}{\partial V_{P_1}} \text{ vs. } \bar{x}_{CG}$$

Problem 6.3

Consider a generic general aviation aircraft with the trim diagram shown in Figure P6.3.1.

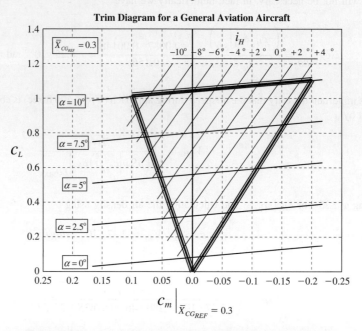

Figure P6.3.1 Trim Diagram for a Generic General Aviation Aircraft

The aircraft features stabilators (i_H) for trimming purposes. Find the following:

- The aircraft aerodynamic center $\bar{x}_{AC}$;
- The stabilator deflection required for trimming the aircraft at $\alpha_1 = 5°$ with $\bar{x}_{CG} = 0.42$;
- The angle of attack required for trimming the aircraft with $i_H = -4°$ and $\bar{x}_{CG} = 0.25$;
- The value of c_{L_0} and assessment on whether the aircraft can be trimmed with $c_{L_1} = c_{L_0}$.

Problem 6.4

Consider a generic commercial aviation jetliner with the trim diagram shown in Figure P.6.4.1.

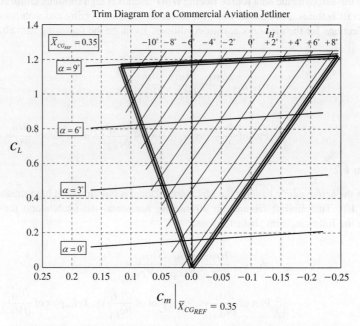

Figure P6.4.1 Trim Diagram for a Generic Commercial Aviation Jetliner

The aircraft features stabilators (i_H) for trimming purposes. Find the following:

- The aircraft aerodynamic center $\bar{x}_{AC}$;
- The stabilator deflection required for trimming the aircraft at $\alpha_1 = 3°$ with $\bar{x}_{CG} = 0.44$;
- The angle of attack required for trimming the aircraft with $i_H = -5°$ and $\bar{x}_{CG} = 0.3$;
- The value of c_{L_0} and assessment on whether the aircraft can be trimmed with $c_{L_1} = c_{L_0}$.

Problem 6.5

Consider an engine-out condition for the Cessna 310 twin-engine aircraft (Appendix B, Aircraft 2). Describe the solution for new steady-state conditions with an engine failure (right engine) occurring at approach conditions (see Appendix B) assuming

$$y_{T_1} \approx \frac{1}{3}\frac{b}{2} = \frac{b}{6}$$

Problem 6.6

Consider the data for the Cessna 620 4-engine aircraft (Appendix B, Aircraft 5). Describe the solutions for new steady-state conditions with the following:

1. Scenario 1: An engine failure for engine 1 on the right side, occurring at approach conditions (see Appendix B) assuming

$$y_{T_1} \approx 0.27 \frac{b}{2}$$

2. Scenario 2: Simultaneous failures for a engine 1 and engine 2 on the right side, occurring at approach conditions (see Appendix B) assuming

$$y_{T_1} \approx 0.27 \frac{b}{2}, \quad y_{T_2} \approx 0.6 \frac{b}{2}$$

Chapter 7

Solution of the Aircraft Equations of Motion Based on Laplace Transformations and Transfer Functions

TABLE OF CONTENTS

- 7.1 Introduction
- 7.2 Application of Laplace Transformations to the Longitudinal Small Perturbation Equations
- 7.3 Routh–Hurwitz Analysis of the Longitudinal Stability
- 7.4 Longitudinal Dynamic Modes: Short Period and Phugoid
- 7.5 Solution of the Longitudinal Equations
- 7.6 Short Period Approximation
- 7.7 Phugoid Approximation
- 7.8 Summary of the Longitudinal Equations
- 7.9 Application of Laplace Transformations to the Lateral Directional Small Perturbation Equations
- 7.10 Routh–Hurwitz Analysis of the Lateral Directional Stability
- 7.11 Lateral Directional Dynamic Modes: Rolling, Spiral, and Dutch Roll
- 7.12 Solution of the Lateral Directional Equations
- 7.13 Rolling Approximation
- 7.14 Summary of Lateral Directional Equations
- 7.15 Sensitivity Analysis for the Aircraft Dynamics
  - 7.15.1 Short Period Sensitivity Analysis
  - 7.15.2 Phugoid Sensitivity Analysis
  - 7.15.3 Sensitivity Analysis for the Lateral Directional Parameters
- 7.16 Summary
- References
- Student Sample Problems
- Problems

7.1 INTRODUCTION

This chapter provides a detailed analysis of the aircraft dynamics starting from the small perturbation equations of motion developed in Chapter 1 leading to the solution of these equations in the linearized form.

The initial point is given by the small perturbation equations—that is, Conservation of Linear Momentum (CLMEs), Conservation of Angular Momentum (CAMEs), and Inverted Kinematic Equations (IKEs)—at steady-state, wing-level rectilinear flight with respect to the aircraft body axes (X, Y, Z):

$$m[\dot{u} + qW_1] = -mg\theta \cos\Theta_1 + (f_{A_X} + f_{T_X})$$

$$m[\dot{v} + U_1 r - pW_1] = mg\phi \cos\Theta_1 + (f_{A_Y} + f_{T_Y})$$

$$m[\dot{w} - U_1 q] = -mg\theta \sin\Theta_1 + (f_{A_Z} + f_{T_Z})$$

$$\dot{p} I_{XX} - \dot{r} I_{XZ} = (l_A + l_T)$$
$$\dot{q} I_{YY} = (m_A + m_T)$$
$$\dot{r} I_{ZZ} - \dot{p} I_{XZ} = (n_A + n_T)$$
$$p = \dot{\phi} - \dot{\psi} \sin \Theta_1$$
$$q = \dot{\theta}$$
$$r = \dot{\psi} \cos \Theta_1$$

A first important observation is that the introduction of the small perturbation assumptions and the application to rectilinear wing-level flight conditions has allowed the removal of the nonlinear terms in the original CLMEs and CAMEs. Furthermore, the preceding conditions also allow decoupling the equations into two sets, the longitudinal and the lateral-directional equations, as shown:

Longitudinal Equations

$$m[\dot{u} + qW_1] = -mg\,\theta \cos \Theta_1 + \left(f_{A_X} + f_{T_X}\right)$$
$$m[\dot{w} - U_1 q] = -mg\,\theta \sin \Theta_1 + \left(f_{A_Z} + f_{T_Z}\right)$$
$$\dot{q} I_{YY} = (m_A + m_T)$$
$$q = \dot{\theta}$$

Lateral Directional Equations

$$m[\dot{v} + U_1 r - pW_1] = mg\phi \cos \Theta_1 + \left(f_{A_Y} + f_{T_Y}\right)$$
$$\dot{p} I_{XX} - \dot{r} I_{XZ} = (l_A + l_T)$$
$$\dot{r} I_{ZZ} - \dot{p} I_{XZ} = (n_A + n_T)$$
$$p = \dot{\phi} - \dot{\psi} \sin \Theta_1$$
$$r = \dot{\psi} \cos \Theta_1$$

Starting from the longitudinal equations, following the introduction of the aerodynamic modeling from Chapter 3 (with the aerodynamic coefficients expressed along the (X_S, Y_S, Z_S) stability axes) the resulting differential equations have been linearized; furthermore, they feature constant coefficients. Therefore, they can be solved using the well-known Laplace transform approach. Next, the longitudinal transfer functions are derived, leading to the analysis of longitudinal dynamic stability. The short period and phugoid dynamic modes are then introduced, with emphasis on the importance of the short period approximation. The preceding analysis is essentially repeated in the second part of the chapter for the lateral directional dynamics with the aerodynamic modeling from Chapter 4 leading to the analysis of the rolling, spiral, and dutch roll modes. Additionally, a detailed analysis of the parameters affecting the aircraft handling qualities, known as sensitivity analysis, is presented.

7.2 APPLICATION OF LAPLACE TRANSFORMATIONS TO THE LONGITUDINAL SMALL PERTURBATION EQUATIONS

Consider again the previous small perturbation longitudinal equations with respect to the aircraft body axes (X, Y, Z):

$$m[\dot{u} + qW_1] = -mg\theta \cos \Theta_1 + \left(f_{A_X} + f_{T_X}\right)$$
$$m[\dot{w} - U_1 q] = -mg\theta \sin \Theta_1 + \left(f_{A_Z} + f_{T_Z}\right)$$
$$\dot{q} I_{YY} = (m_A + m_T)$$
$$q = \dot{\theta}$$

Because of the mathematical implications associated with the introduction of the small perturbations assumptions, these differential equations are linear with constant coefficients. As reported in Chapter 1, the solution of the linearized equations of aircraft dynamics follow the theoretical work by Lanchester and Bryan in the early 1900s.[1,2] The modeling from Chapter 3 provided the following expressions for the longitudinal aerodynamic and thrust forces and moment:

$$(f_{A_X} + f_{T_X}) = \bar{q}_1 S \left[-(c_{D_u} + 2c_{D_1})\frac{u}{V_{P_1}} + (c_{T_{X_u}} + 2c_{T_{X_1}})\frac{u}{V_{P_1}} - (c_{D_\alpha} - c_{L_1})\alpha - c_{D_{\delta_E}}\delta_E \right]$$

$$(f_{A_Z} + f_{T_Z}) = \bar{q}_1 S \left[-(c_{L_u} + 2c_{L_1})\frac{u}{V_{P_1}} - (c_{L_\alpha} + c_{D_1})\alpha - c_{L_{\dot\alpha}}\frac{\dot\alpha \bar c}{2V_{P_1}} - c_{L_q}\frac{q\bar c}{2V_{P_1}} - c_{L_{\delta_E}}\delta_E \right]$$

$$(m_A + m_T) = \bar{q}_1 S \bar c \left[(c_{m_u} + 2c_{m_1})\frac{u}{V_{P_1}} + (c_{m_{T_u}} + 2c_{m_{T_1}})\frac{u}{V_{P_1}} + (c_{m_\alpha} + c_{m_{T_\alpha}})\alpha + c_{m_{\dot\alpha}}\frac{\dot\alpha \bar c}{2V_{P_1}} + c_{m_q}\frac{q\bar c}{2V_{P_1}} + c_{m_{\delta_E}}\delta_E \right]$$

Once again, recall that the aerodynamic forces and moments are expressed in the aerodynamic-based stability axes (X_S, Y_S, Z_S). A specific property of this frame is that the only nonzero component of the linear velocity is along the X_S axis with $U_{1_S} = V_{P_1}, W_{1_S} = 0$. Therefore, the previous CLMEs originally derived in the body axes (X, Y, Z) can be redefined in the stability axes (X_S, Y_S, Z_S) as shown below:

$$m\dot u = -mg\cos\Theta_1\theta + \bar{q}_1 S \left[-(c_{D_u} + 2c_{D_1})\frac{u}{V_{P_1}} + (c_{T_{X_u}} + 2c_{T_{X_1}})\frac{u}{V_{P_1}} - (c_{D_\alpha} - c_{L_1})\alpha - c_{D_{\delta_E}}\delta_E \right]$$

$$m(\dot w - V_{P_1} q) = -mg\sin\Theta_1\theta + \bar{q}_1 S \left[-(c_{L_u} + 2c_{L_1})\frac{u}{V_{P_1}} - (c_{L_\alpha} + c_{D_1})\alpha - c_{L_{\dot\alpha}}\frac{\dot\alpha \bar c}{2V_{P_1}} - c_{L_q}\frac{q\bar c}{2V_{P_1}} - c_{L_{\delta_E}}\delta_E \right]$$

$$I_{YY}\dot q = \bar{q}_1 S \bar c \left[(c_{m_u} + 2c_{m_1})\frac{u}{V_{P_1}} + (c_{m_{T_u}} + 2c_{m_{T_1}})\frac{u}{V_{P_1}} + (c_{m_\alpha} + c_{m_{T_\alpha}})\alpha + c_{m_{\dot\alpha}}\frac{\dot\alpha \bar c}{2V_{P_1}} + c_{m_q}\frac{q\bar c}{2V_{P_1}} + c_{m_{\delta_E}}\delta_E \right]$$

along with the additional IKE around the pitch axis $q = \dot\theta$.

Before starting an approach leading to the solution of these equations, an important transformation of variables is required to address an inconsistency in the setup of the equations. In fact, the second equation along the Z_S axis is formulated in the small perturbation vertical velocity w, while for aerodynamic modeling purposes the aircraft stability derivatives are expressed in terms of the longitudinal angle of attack α. Additionally, the IKE can be embedded into the pitching moment equation. Therefore, the following transformation of variables is introduced:

$$w \to \alpha, \quad q \to \theta$$

with

$$q = \dot\theta, \quad \dot q = \ddot\theta$$
$$w \approx V_{P_1}\alpha, \quad \dot w \approx V_{P_1}\dot\alpha$$

leading to a new set of equations:

$$\dot u = -g\cos\Theta_1\theta + \frac{\bar{q}_1 S}{m}\left[-(c_{D_u} + 2c_{D_1})\frac{u}{V_{P_1}} + (c_{T_{X_u}} + 2c_{T_{X_1}})\frac{u}{V_{P_1}} - (c_{D_\alpha} - c_{L_1})\alpha - c_{D_{\delta_E}}\delta_E \right]$$

7.2 Application of Laplace Transformations to the Longitudinal Small Perturbation Equations

$$\left(V_{P_1}\dot{\alpha} - V_{P_1}q\right) = -g\sin\Theta_1\theta + \frac{\bar{q}_1 S}{m}\left[-\left(c_{L_u} + 2c_{L_1}\right)\frac{u}{V_{P_1}} - \left(c_{L_\alpha} + c_{D_1}\right)\alpha - c_{L_{\dot{\alpha}}}\frac{\dot{\alpha}\bar{c}}{2V_{P_1}} - c_{L_q}\frac{q\bar{c}}{2V_{P_1}} - c_{L_{\delta_E}}\delta_E\right]$$

$$I_{YY}\ddot{\theta} = \bar{q}_1 S\bar{c}\left[\left(c_{m_u} + 2c_{m_1}\right)\frac{u}{V_{P_1}} + \left(c_{m_{T_u}} + 2c_{m_{T_1}}\right)\frac{u}{V_{P_1}} + \left(c_{m_\alpha} + c_{m_{T_\alpha}}\right)\alpha + c_{m_{\dot{\alpha}}}\frac{\dot{\alpha}\bar{c}}{2V_{P_1}} + c_{m_q}\frac{q\bar{c}}{2V_{P_1}} + c_{m_{\delta_E}}\delta_E\right]$$

Next, an inspection of the right-hand side (RHS) of the preceding equations shows a combination of terms that are functions of the aerodynamic modeling as well as the flight conditions $\bar{q}_1$, geometric characteristics ($\bar{c}$, S), and aircraft inertial characteristics (m, I_{YY}). The different dimensions for all the RHS coefficients lead to the introduction of a set of more descriptive aerodynamic and control derivatives, known as the dimensional stability and control derivatives, which combine the flight conditions, the geometric and inertial characteristics with the aerodynamic modeling. The longitudinal dimensional stability and control derivatives are shown in Table 7.1.

Table 7.1 Dimensional Longitudinal Stability Derivatives

Longitudinal Dimensional Stability and Control Derivatives	
$X_u = \dfrac{-\bar{q}_1 S(c_{D_U} + 2c_{D_1})}{mU_1} \cdot (\sec^{-1})$	$X_{T_u} = \dfrac{\bar{q}_1 S(c_{T_{xu}} + 2c_{T_{x1}})}{mU_1} \cdot (\sec^{-1})$
$X_\alpha = \dfrac{-\bar{q}_1 S(c_{D_\alpha} - c_{L_1})}{m} \cdot (\text{ft sec}^{-2})$	$X_{\delta_E} = \dfrac{-\bar{q}_1 S c_{D_{\delta E}}}{m} \cdot (\text{ft sec}^{-2})$
$Z_u = -\dfrac{\bar{q}_1 S(c_{L_U} + 2c_{L_1})}{mU_1} \cdot (\sec^{-1})$	$Z_\alpha = -\dfrac{\bar{q}_1 S(c_{L_\alpha} + c_{D_1})}{m} \cdot (\text{ft sec}^{-2})$
$Z_{\dot{\alpha}} = -\dfrac{\bar{q}_1 S \bar{c}\, c_{L_{\dot{\alpha}}}}{2mU_1} \cdot (\text{ft sec}^{-1})$	$Z_q = -\dfrac{\bar{q}_1 S \bar{c}\, c_{L_q}}{2mU_1} \cdot (\text{ft sec}^{-1})$
$Z_{\delta_E} = \dfrac{-\bar{q}_1 S c_{L_{\delta E}}}{m} \cdot (\text{ft sec}^{-2})$	
$M_u = \dfrac{\bar{q}_1 S \bar{c}(c_{m_U} + 2c_{m_1})}{U_1 I_{yy}} \cdot (\text{ft}^{-1}\sec^{-1})$	$M_{Tu} = \dfrac{\bar{q}_1 S \bar{c}(c_{m_{TU}} + 2c_{m_{T1}})}{U_1 I_{yy}} \cdot (\text{ft}^{-1}\sec^{-1})$
$M_\alpha = \dfrac{\bar{q}_1 S \bar{c}\, c_{m_\alpha}}{I_{yy}} \cdot (\sec^{-2})$	$M_{T\alpha} = \dfrac{\bar{q}_1 S \bar{c}\, c_{m_{T\alpha}}}{I_{yy}} \cdot (\sec^{-2})$
$M_{\dot{\alpha}} = \dfrac{\bar{q}_1 S \bar{c}\, c_{m_{\dot{\alpha}}}}{I_{yy}} \cdot \dfrac{\bar{c}}{2U_1} (\sec^{-1})$	$M_q = \dfrac{\bar{q}_1 S \bar{c}\, c_{m_q}}{I_{yy}} \cdot \dfrac{\bar{c}}{2U_1} (\sec^{-1})$
$M_{\delta_E} = \dfrac{\bar{q}_1 S \bar{c}\, c_{m_{\delta E}}}{I_{yy}} \cdot (\sec^{-2})$	

Using the expressions in Table 7.1 the preceding equations take on the form

$$\dot{u} = -g\cos\Theta_1\theta + (X_u + X_{T_u})u + X_\alpha\alpha + X_{\delta_E}\delta_E$$

$$V_{P_1}\dot{\alpha} = -g\sin\Theta_1\theta + Z_u u + Z_\alpha\alpha + Z_{\dot{\alpha}}\dot{\alpha} + (Z_q + V_{P_1})\dot{\theta} + Z_{\delta_E}\delta_E$$

$$\ddot{\theta} = (M_u + M_{T_u})u + (M_\alpha + M_{T_\alpha})\alpha + M_{\dot{\alpha}}\dot{\alpha} + M_q\dot{\theta} + M_{\delta_E}\delta_E$$

Next, the following Laplace transformations are applied to the RHS and LHS of the equations, assuming zero initial conditions (since the small perturbations are considered with respect to a steady-state rectilinear flight):

$$L(\delta_E) = \delta_E(s)$$

$$L(u) = u(s), \quad L(\dot{u}) = s\,u(s)$$

$$L(\alpha) = \alpha(s), \quad L(\dot{\alpha}) = s\,\alpha(s)$$

$$L(\theta) = \theta(s), \quad L(\dot{\theta}) = s\,\theta(s), \quad L(\ddot{\theta}) = s^2\,\theta(s)$$

In this system the known input is the longitudinal control surface, δ_E. Without loss of generality, the preceding modeling can be applied to any other longitudinal control surface (such as stabilators, flaps, symmetric spoilers, and so on). Next, the coefficients of the equations are grouped in terms of $(u(s), \alpha(s), \theta(s))$, leading to

$$(s - (X_u + X_{T_u}))u(s) - X_\alpha \alpha(s) + g\cos\Theta_1 \theta(s) = X_{\delta_E}\delta_E(s)$$

$$-Z_u u(s) + (s(V_{P_1} - Z_{\dot\alpha}) - Z_\alpha)\alpha(s) + (-s(Z_q + V_{P_1}) + g\sin\Theta_1)\theta(s) = Z_{\delta_E}\delta_E(s)$$

$$-(M_u + M_{T_u})u(s) - (M_{\dot\alpha}s + (M_\alpha + M_{T_\alpha}))\alpha(s) + s(s - M_q)\theta(s) = M_{\delta_E}\delta_E(s)$$

In general, given a known input $u(t)$ and its Laplace transform $U(s) = L(u(t))$, the output $y(t)$ of a generic system in the time domain can be derived as the inverse Laplace transform $y(t) = L^{-1}(Y(s)) = L^{-1}\left(\frac{Y(s)}{U(s)} \cdot U(s)\right)$ where $\frac{Y(s)}{U(s)}$ is known as the transfer function. The concept of transfer function [3,4,5] is shown in Figure 7.1.

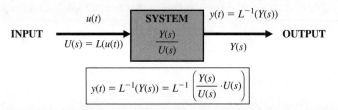

Figure 7.1 Concept of Transfer Function

For the longitudinal dynamics the input is represented by the deflection of the elevator $\delta_E(t)$, whereas the individual outputs are represented by the small perturbation variables $(u(t), \alpha(t), \theta(t))$. Therefore, using the transfer functions $\left\{\frac{u(s)}{\delta_E(s)}, \frac{\alpha(s)}{\delta_E(s)}, \frac{\theta(s)}{\delta_E(s)}\right\}$, the preceding equations can be rearranged in the following matrix format:

$$\begin{bmatrix} (s-(X_u+X_{T_u})) & -X_\alpha & +g\cos\Theta_1 \\ -Z_u & (s(V_{P_1}-Z_{\dot\alpha})-Z_\alpha) & (-s(Z_q+V_{P_1})+g\sin\Theta_1) \\ -(M_u+M_{T_u}) & -(M_{\dot\alpha}s+(M_\alpha+M_{T_\alpha})) & s(s-M_q) \end{bmatrix} \begin{Bmatrix} \frac{u(s)}{\delta_E(s)} \\ \frac{\alpha(s)}{\delta_E(s)} \\ \frac{\theta(s)}{\delta_E(s)} \end{Bmatrix} = \begin{Bmatrix} X_{\delta_E} \\ Z_{\delta_E} \\ M_{\delta_E} \end{Bmatrix}$$

The application of Cramer's rule leads to the following solutions for the preceding transfer functions:

$$\frac{u(s)}{\delta_E(s)} = \frac{\begin{vmatrix} X_{\delta_E} & -X_\alpha & +g\cos\Theta_1 \\ Z_{\delta_E} & (s(V_{P_1}-Z_{\dot\alpha})-Z_\alpha) & (-s(Z_q+V_{P_1})+g\sin\Theta_1) \\ M_{\delta_E} & -(M_{\dot\alpha}s+(M_\alpha+M_{T_\alpha})) & s(s-M_q) \end{vmatrix}}{\begin{vmatrix} (s-(X_u+X_{T_u})) & -X_\alpha & +g\cos\Theta_1 \\ -Z_u & (s(V_{P_1}-Z_{\dot\alpha})-Z_\alpha) & (-s(Z_q+V_{P_1})+g\sin\Theta_1) \\ -(M_u+M_{T_u}) & -(M_{\dot\alpha}s+(M_\alpha+M_{T_\alpha})) & s(s-M_q) \end{vmatrix}} = \frac{Num_u(s)}{D_1(s)}$$

$$\frac{\alpha(s)}{\delta_E(s)} = \frac{\begin{vmatrix} (s-(X_u+X_{Tu})) & X_{\delta_E} & +g\cos\Theta_1 \\ -Z_u & Z_{\delta_E} & (-s(Z_q+V_{P_1})+g\sin\Theta_1) \\ -(M_u+M_{Tu}) & M_{\delta_E} & s(s-M_q) \end{vmatrix}}{\begin{vmatrix} (s-(X_u+X_{Tu})) & -X_\alpha & +g\cos\Theta_1 \\ -Z_u & (s(V_{P_1}-Z_{\dot\alpha})-Z_\alpha) & (-s(Z_q+V_{P_1})+g\sin\Theta_1) \\ -(M_u+M_{Tu}) & -(M_{\dot\alpha}s+(M_\alpha+M_{T_\alpha})) & s(s-M_q) \end{vmatrix}} = \frac{Num_\alpha(s)}{\overline{D}_1(s)}$$

$$\frac{\theta(s)}{\delta_E(s)} = \frac{\begin{vmatrix} (s-(X_u+X_{Tu})) & -X_\alpha & X_{\delta_E} \\ -Z_u & (s(V_{P_1}-Z_{\dot\alpha})-Z_\alpha) & Z_{\delta_E} \\ -(M_u+M_{Tu}) & -(M_{\dot\alpha}s+(M_\alpha+M_{T_\alpha})) & M_{\delta_E} \end{vmatrix}}{\begin{vmatrix} (s-(X_u+X_{Tu})) & -X_\alpha & +g\cos\Theta_1 \\ -Z_u & (s(V_{P_1}-Z_{\dot\alpha})-Z_\alpha) & (-s(Z_q+V_{P_1})+g\sin\Theta_1) \\ -(M_u+M_{Tu}) & -(M_{\dot\alpha}s+(M_\alpha+M_{T_\alpha})) & s(s-M_q) \end{vmatrix}} = \frac{Num_\theta(s)}{\overline{D}_1(s)}$$

A detailed understanding of the time response of the aircraft system requires an analysis of the coefficients of the polynomials at the numerator and, more importantly, the denominator of the preceding transfer functions. These coefficients are derived from grouping terms with the same order from the calculation of the preceding determinants. The expressions for the coefficients of the numerator polynomials ($Num_u(s)$, $Num_\alpha(s)$, $Num_\theta(s)$) are given by

$Num_u(s) = A_u s^3 + B_u s^2 + C_u s + D_u$

$A_u = X_{\delta_E}(V_{P_1} - Z_{\dot\alpha})$

$B_u = -X_{\delta_E}[(V_{P_1} - Z_{\dot\alpha})M_q + Z_\alpha + M_{\dot\alpha}(V_{P_1} + Z_q)] + Z_{\delta_E} X_\alpha$

$C_u = X_{\delta_E}[M_q Z_\alpha + M_{\dot\alpha} g \sin\Theta_1 - (M_\alpha + M_{T_\alpha})(V_{P_1} + Z_q)]$

$D_u = g\sin\Theta_1 X_{\delta_E}(M_\alpha + M_{T_\alpha}) - g\cos\Theta_1 Z_{\delta_E}(M_\alpha + M_{T_\alpha}) + M_{\delta_E}(g\cos\Theta_1 Z_\alpha - g\sin\Theta_1 X_\alpha)$

$Num_\alpha(s) = A_\alpha s^3 + B_\alpha s^2 + C_\alpha s + D_\alpha$

$A_\alpha = Z_{\delta_E}$

$B_\alpha = X_{\delta_E} Z_u - Z_{\delta_E}((X_u + X_{Tu}) + M_q) + M_{\delta_E}(Z_q + V_{P_1})$

$C_\alpha = X_{\delta_E}[(Z_q + V_{P_1})(M_u + M_{Tu}) - M_q Z_u] + Z_{\delta_E} M_q(X_u + X_{Tu})$
$\qquad - M_{\delta_E}[g\sin\Theta_1 + (Z_q + V_{P_1})(X_u + X_{Tu})]$

$D_\alpha = -g\sin\Theta_1 X_{\delta_E}(M_u + M_{Tu}) + g\cos\Theta_1 Z_{\delta_E}(M_u + M_{Tu}) + M_{\delta_E}[g\sin\Theta_1(X_u + X_{Tu}) - g\cos\Theta_1 Z_u]$

$Num_\theta(s) = A_\theta s^2 + B_\theta s + C_\theta$

$A_\theta = Z_{\delta_E} M_{\dot\alpha} + M_{\delta_E}(V_{P_1} - Z_{\dot\alpha})$

$B_\theta = X_{\delta_E}[Z_u M_{\dot\alpha} + (V_{P_1} - Z_{\dot\alpha})(M_u + M_{Tu})] + Z_{\delta_E}[(M_\alpha + M_{T_\alpha}) - M_{\dot\alpha}(X_u + X_{Tu})]$
$\qquad - M_{\delta_E}[(V_{P_1} - Z_{\dot\alpha})(X_u + X_{Tu}) - Z_\alpha]$

$C_\theta = X_{\delta_E}[(M_\alpha + M_{T_\alpha})Z_u - (M_u + M_{Tu})Z_\alpha] - Z_{\delta_E}[(M_\alpha + M_{T_\alpha})(X_u + X_{Tu}) + X_\alpha(M_u + M_{Tu})]$
$\qquad + M_{\delta_E}[(X_u + X_{Tu})Z_\alpha - X_\alpha Z_u]$

All the transfer functions share a common denominator, which will be referred to as the longitudinal characteristic equation (CE) and indicated by $\overline{D}_1(s)$. By grouping the terms with the same exponent, an expression for the CE is given by

$$\overline{D}_1(s) = A_1 s^4 + B_1 s^3 + C_1 s^2 + D_1 s + E_1$$

$$A_1 = (V_{P_1} - Z_{\dot{\alpha}})$$

$$B_1 = -(V_{P_1} - Z_{\dot{\alpha}})(X_u + X_{T_u} + M_q) - Z_\alpha - M_{\dot{\alpha}}(Z_q + V_{P_1})$$

$$C_1 = (X_u + X_{T_u})[M_q(V_{P_1} - Z_{\dot{\alpha}}) + Z_\alpha + M_{\dot{\alpha}}(V_{P_1} + Z_q)]$$

$$+ M_q Z_\alpha - Z_u X_\alpha + M_{\dot{\alpha}} g \sin\Theta_1 - (M_\alpha + M_{T_\alpha})(V_{P_1} + Z_q)$$

$$D_1 = g \sin\Theta_1 [(M_\alpha + M_{T_\alpha}) - M_{\dot{\alpha}}(X_u + X_{T_u})] + g \cos\Theta_1 [M_{\dot{\alpha}} Z_u + (M_u + M_{T_u})(V_{P_1} - Z_{\dot{\alpha}})]$$

$$- X_\alpha (M_u + M_{T_u})(Z_q + V_{P_1}) + Z_u X_\alpha M_q + (X_u + X_{T_u})[(M_\alpha + M_{T_\alpha})(Z_q + V_{P_1}) - M_q Z_\alpha]$$

$$E_1 = g \cos\Theta_1 [Z_u(M_\alpha + M_{T_\alpha}) - Z_\alpha(M_u + M_{T_u})] + g \sin\Theta_1 [(M_u + M_{T_u})X_\alpha - (X_u + X_{T_u})(M_\alpha + M_{T_\alpha})]$$

The stability of the longitudinal dynamics is directly related to the values of the preceding coefficients. A simple tool for assessing the dynamic stability is given by the application of the Routh–Hurwitz stability criteria.[6]

7.3 ROUTH–HURWITZ ANALYSIS OF THE LONGITUDINAL STABILITY

The assessment of the dynamic stability of a system can today be easily performed in MATLAB® by verifying that the roots of the associated characteristic equation (CE) are negative—if real—or have negative real parts—if complex conjugate. However, the solution of third or higher order algebraic equations was not a trivial matter until few years ago. Without actually solving for the roots of the CE, a simple approach for assessing the dynamic stability of a system is provided by the Routh–Hurwitz stability criteria using the Routh–Hurwitz array associated with the CE.[6] An example is provided next. Given the longitudinal CE developed in the previous section

$$\overline{D}_1 = A_1 s^4 + B_1 s^3 + C_1 s^2 + D_1 s + E_1$$

The Routh–Hurwitz array associated with the above CE is given by

s^4	A_1	C_1	E_1
s^3	B_1	D_1	–
s^2	k_{11}	k_{12}	–
s	k_{21}	–	–
s^0	E_1	–	–

where

$$k_{11} = \frac{B_1 C_1 - A_1 D_1}{B_1}, \quad k_{12} = \frac{B_1 E_1}{B_1} = E_1$$

$$k_{21} = \frac{D_1 k_{11} - B_1 k_{12}}{k_{11}} = \frac{D_1 \left(\frac{B_1 C_1 - A_1 D_1}{B_1}\right) - B_1 E_1}{\frac{B_1 C_1 - A_1 D_1}{B_1}} = \frac{D_1(B_1 C_1 - A_1 D_1) - B_1^2 E_1}{B_1 C_1 - A_1 D_1}$$

Next, the attention focuses on the first column of the Routh–Hurwitz array:

7.3 Routh–Hurwitz Analysis of the Longitudinal Stability

$$\begin{array}{c|ccc} s^4 & A_1 & C_1 & E_1 \\ s^3 & B_1 & D_1 & - \\ s^2 & k_{11} & k_{12} & - \\ s & k_{21} & - & - \\ s^0 & E_1 & - & - \end{array}$$

According to the Routh–Hurwitz stability criterion,[6] a dynamic system with a given CE is dynamically stable if *all* the coefficients in the first column of the Routh–Hurwitz array have the same sign. Furthermore, the number of unstable roots of the CE (that is positive, if real, or with positive real part, if complex conjugate) is equal to the number of sign changes in the first column of the Routh–Hurwitz array. Therefore, the following conditions must be satisfied for longitudinal dynamic stability:

$$A_1 > 0, \quad B_1 > 0, \quad (B_1 C_1 - A_1 D_1) > 0, \quad \Delta_1 > 0, \quad E_1 > 0$$

where $\Delta_1 = D_1(B_1 C_1 - A_1 D_1) - B_1^2$ is known as the longitudinal Routh discriminant.

From the preceding analysis the assessment of the longitudinal stability requires the verification of each of the five preceding conditions. While some of these conditions are trivially satisfied (for example, $A_1 > 0$, $B_1 > 0$), the other conditions might be substantially more restrictive. Typically, longitudinal dynamic instability might potentially arise from the following two sources:

$$\Delta_1 = D_1(B_1 C_1 - A_1 D_1) - B_1^2 E_1 < 0$$
$$E_1 < 0$$

The condition $\Delta = D_1(B_1 C_1 - A_1 D_1) - B_1^2 E_1 < 0$ along with the condition $E_1 > 0$ is associated with two unstable roots. It will be shown later that this situation is related to the pair of unstable complex conjugate roots associated with an unstable phugoid mode.[3–15] The condition $\Delta = D_1(B_1 C_1 - A_1 D_1) - B_1^2 E_1 > 0$ along with the condition $E_1 < 0$ is associated instead with one unstable root. Recall that

$$E_1 = g\cos\Theta_1[Z_u(M_\alpha + M_{T_\alpha}) - Z_\alpha(M_u + M_{T_u})] + g\sin\Theta_1[(M_u + M_{T_u})X_\alpha - (X_u + X_{T_u})(M_\alpha + M_{T_\alpha})]$$

For small values of Θ_1 we can approximate $E_1 \approx [Z_u(M_\alpha + M_{T_\alpha}) - Z_\alpha(M_u + M_{T_u})]$.

Neglecting the contribution from the propulsive dimensional derivatives and recalling the relationships for the aerodynamic dimensional derivatives

$$Z_u = -\frac{\bar{q}_1 S(c_{L_U} + 2c_{L_1})}{mU_1} < 0, \quad Z_\alpha = -\frac{\bar{q}_1 S(c_{L_\alpha} + c_{D_1})}{m} < 0$$

$$M_u = \frac{\bar{q}_1 S\bar{c}(c_{m_U} + 2c_{m_1})}{U_1 I_{yy}} \leq \approx \geq 0, \quad M_\alpha = \frac{\bar{q}_1 S\bar{c} c_{m_\alpha}}{I_{yy}} < 0$$

with $c_{D_1}, c_{L_1}, c_{m_1}$ being the values of c_D, c_L, c_m at trim conditions (with $c_{m_1} = 0$).

It turns out that the condition $M_u = \dfrac{\bar{q}_1 S\bar{c}(c_{m_U} + 2c_{m_1})}{U_1 I_{yy}} \leq \approx \geq 0$ is the most critical condition for satisfying the Hurwitz stability criteria.[3,6] In fact, for longitudinal dynamic stability purposes it is desirable to have

$$M_u > 0 \rightarrow c_{m_u} > 0$$

The violation of this condition is associated with the aerodynamic transonic problem known as tuck,[8,13,14] which was briefly described in Chapter 6, Section 6.3. The tuck problem is essentially associated with a drastic shift in the aircraft aerodynamic center following a small variation in speed around the *Mach* ≈ 1 condition. A detailed study to assess the longitudinal dynamic stability at all possible flight conditions within the aircraft flight envelope is an important phase of the overall design and is known as sensitivity analysis. From the relationships

$$A_1 > 0, \quad B_1 > 0, \quad (B_1 C_1 - A_1 D_1) > 0, \quad \Delta > 0, \quad E_1 > 0$$

it is clear that the conditions for stability are functions of a large number of variables, including flight conditions, geometric parameters, and inertial characteristics, in addition to the aerodynamic behavior. A number of studies are typically performed to evaluate the effect of specific parameters (related to the aerodynamic behavior, flight conditions, or aircraft geometry) on the overall longitudinal stability. This issue will be specifically addressed in a later section.

7.4 LONGITUDINAL DYNAMIC MODES: SHORT PERIOD AND PHUGOID

For dynamically stable aircraft the longitudinal CE has two pairs of complex conjugate roots.[3,13,14] Each pair is associated with a specific dynamic mode related to a second order system. Refer to Appendix A.5 for a review of the characteristics of second order dynamic systems. The longitudinal dynamic modes are called short period and phugoid.[3-15] Therefore, the longitudinal CE takes on the form:

$$\overline{D}_1 = A_1 s^4 + B_1 s^3 + C_1 s^2 + D_1 s + E_1 = (s^2 + 2\varsigma_{SP}\omega_{n_{SP}}s + \omega_{n_{SP}}^2)(s^2 + 2\varsigma_{PH}\omega_{n_{PH}}s + \omega_{n_{PH}}^2)$$

These two dynamic modes provide two very distinct signatures on the aircraft longitudinal dynamics. The typical locations of the CE roots associated with these modes are shown in Figure 7.2.

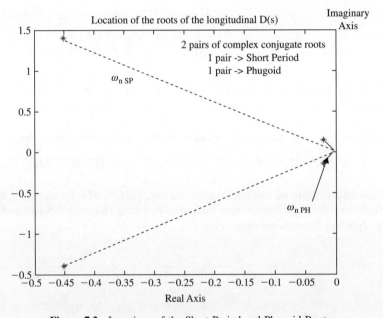

Figure 7.2 Locations of the Short Period and Phugoid Roots

The short period is characterized by fairly high values of damping coefficient ς_{SP} along with high values of natural frequency $\omega_{n_{SP}}$. Conversely, the phugoid is characterized by low values of damping coefficient ς_{Ph} along with generally low values of natural frequency $\omega_{n_{Ph}}$. The following relationships always apply to the longitudinal dynamics:

$$\omega_{n_{SP}} \gg \omega_{n_{PH}}, \quad \varsigma_{SP} \gg \varsigma_{PH}$$

The preceding longitudinal parameters are also key design objectives for satisfying requirements for the aircraft longitudinal handling qualities. The values for all the preceding parameters depend on the specific aircraft type (military versus civilian), aircraft class, and level of handling qualities, as discussed in Chapter 10. Table 7.2 shows a sample of the values for $(\varsigma_{SP}, \omega_{n_{SP}}, \varsigma_{PH}, \omega_{n_{PH}})$ for different aircraft from different classes.[3]

Table 7.2 Sample $(\varsigma_{SP}, \omega_{n_{SP}}, \varsigma_{PH}, \omega_{n_{PH}})$ for Different Aircraft

Aircraft Class	Longitudinal Parameters			
	ς_{SP}	$\omega_{n_{SP}}$	ς_{PH}	$\omega_{n_{PH}}$
General Aviation/Light Commercial Aviation				
Cessna 182 (Alt. = 5,000 ft. Mach = 0.2)	0.844	5.27	0.129	0.171
Beech 99 (Alt. = 20,000ft. Mach = 0.43)	0.485	5.0	0.062	0.095
Cessna 620 (Alt. = SL, Mach = 0.17)	0.720	2.71	0.087	0.205
Military Trainer				
SIAI Marchetti 211 (Alt. = SL, Mach = 0.11)	0.742	1.645	0.019	0.293
Cessna T37 A (Alt. = 30,000 ft. Mach = 0.46)	0.493	4.652	0.053	0.093
Business jet				
Learjet 24 (Alt. = SL, Mach = 0.15)	0.564	1.562	0.067	0.263
Fighter				
Lockheed F-104 (Alt. = SL, Mach = 0.26)	0.221	2.847	—	—
McDonnell F-4 (Alt. = 35,000 ft. Mach = 0.9)	0.307	1.468	0.138	0.148
Commercial Jetliner				
Boeing B747-200 (Alt. = 40,000 ft, Mach = 0.9)	0.353	1.321	—	—

----: 'Degenerated' phugoid with 2 real roots

The distinct dynamic signatures of the short period and the phugoid dynamic modes will be highlighted in the aircraft time responses, discussed in the next sections.

7.5 SOLUTION OF THE LONGITUDINAL EQUATIONS

Using the available previously evaluated transfer functions

$$\frac{u(s)}{\delta_E(s)} = \frac{Num_u(s)}{\overline{D}_1(s)} = \frac{A_u s^3 + B_u s^2 + C_u s + D_u}{A_1 s^4 + B_1 s^3 + C_1 s^2 + D_1 s + E_1}$$

$$\frac{\alpha(s)}{\delta_E(s)} = \frac{Num_\alpha(s)}{\overline{D}_1(s)} = \frac{A_\alpha s^3 + B_\alpha s^2 + C_\alpha s + D_\alpha}{A_1 s^4 + B_1 s^3 + C_1 s^2 + D_1 s + E_1}$$

$$\frac{\theta(s)}{\delta_E(s)} = \frac{Num_\theta(s)}{\overline{D}_1(s)} = \frac{A_\theta s^2 + B_\theta s + C_\theta}{A_1 s^4 + B_1 s^3 + C_1 s^2 + D_1 s + E_1}$$

for a known $\delta_E(s)$ associated with a given $\delta_E(t)$ we have

$$u(s) = \frac{u(s)}{\delta_E(s)} \delta_E(s) \Rightarrow u(t) = L^{-1}[u(s)]$$

$$\alpha(s) = \frac{\alpha(s)}{\delta_E(s)} \delta_E(s) \Rightarrow \alpha(t) = L^{-1}[\alpha(s)]$$

$$\theta(s) = \frac{\theta(s)}{\delta_E(s)} \delta_E(s) \Rightarrow \theta(t) = L^{-1}[\theta(s)]$$

The hand solution leading to closed-form expressions for $(u(t), \alpha(t), \theta(t))$ is not a trivial matter. A suitable approach is given by the application of the so-called partial fraction expansion (PFE) method.[6] Refer to Appendix A.4 for a review of this method. The widespread availability of MATLAB® allows simple solutions to these previously long and tedious calculations. A detailed discussion of the important MATLAB® commands associated with the solution of aircraft dynamics is provided in the Student Sample Problems section.

A MATLAB®-based solution for the longitudinal dynamics provides the time histories shown in Figure 7.3 using aerodynamic, geometry, inertial, and flight conditions data for a high-performance fighter aircraft (Lockheed F-104) at approach conditions, from Appendix B.[3]

362 Chapter 7 Solution of the Aircraft Equations of Motion Based on Laplace Transformations and Transfer Functions

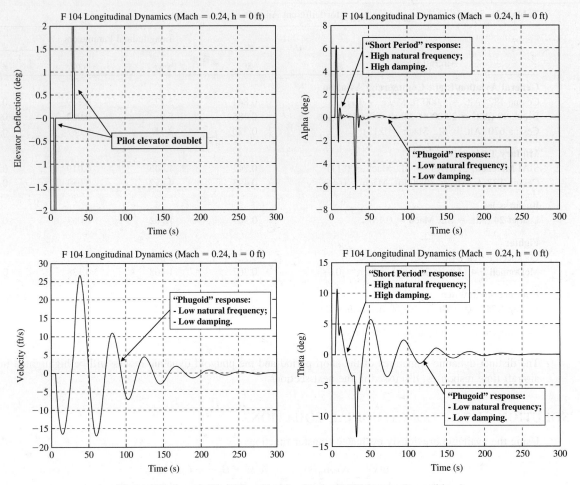

Figure 7.3 Longitudinal Time-histories for the F-104 (approach conditions)

The above plots show two distinct phases in the aircraft longitudinal responses.[3,5,8,13,14] The first phase, associated with the short period dynamics, is only evident in the immediate transient. In fact, due to the high value of the damping coefficient, this phase only lasts one or two seconds immediately following a pilot maneuver or an external disturbance (such as a gust or an atmospheric turbulence). The second phase, associated with the phugoid dynamics, becomes evident only later when the effects of the short period dynamics have vanished. The roots of the longitudinal characteristic equation associated with the short period and phugoid modes for the given aircraft at the given flight conditions are shown in Figure 7.4.

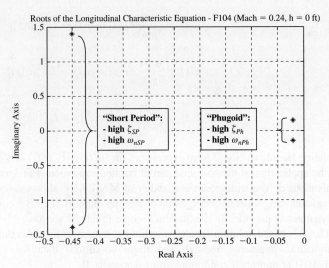

Figure 7.4 Roots of the Longitudinal Characteristic Equation for the F-104 (approach conditions)

The preceding modeling procedure can be applied without loss of generality to the more general case of multiple inputs through deflections of multiple control surfaces. Figure 7.5 shows a variety of control surfaces located on the wing of a typical jetliner.

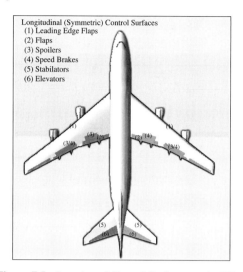

Figure 7.5 Location of Control Surfaces on the Wing

The extension of the Laplace transformation approach to the case of multiple deflections is made possible by the linearity of the constant coefficient differential equations.[6] Therefore, using the superimposition property of linear systems, the aircraft responses associated with multiple control inputs will be given by

$$u(s) = \frac{u(s)}{\delta_E(s)}\delta_E(s) + \frac{u(s)}{i_H(s)}i_H(s) + \frac{u(s)}{\delta_{THROTTLE}(s)}\delta_{THROTTLE}(s) + \frac{u(s)}{\delta_{LE\_FLAPS}(s)}\delta_{LE\_FLAPS}(s)$$

$$+ \frac{u(s)}{\delta_{TE\_FLAPS}(s)}\delta_{TE\_FLAPS}(s) + \frac{u(s)}{\delta_{SPOILERS}(s)}\delta_{SPOILERS}(s)\ldots \Rightarrow u(t) = L^{-1}[u(s)]$$

$$\alpha(s) = \frac{\alpha(s)}{\delta_E(s)}\delta_E(s) + \frac{\alpha(s)}{i_H(s)}i_H(s) + \frac{\alpha(s)}{\delta_{THROTTLE}(s)}\delta_{THROTTLE}(s) + \frac{\alpha(s)}{\delta_{LE\_FLAPS}(s)}\delta_{LE\_FLAPS}(s)$$

$$+ \frac{\alpha(s)}{\delta_{TE\_FLAPS}(s)}\delta_{TE\_FLAPS}(s) + \frac{\alpha(s)}{\delta_{SPOILERS}(s)}\delta_{SPOILERS}(s)\ldots \Rightarrow \alpha(t) = L^{-1}[\alpha(s)]$$

$$\theta(s) = \frac{\theta(s)}{\delta_E(s)}\delta_E(s) + \frac{\theta(s)}{i_H(s)}i_H(s) + \frac{\theta(s)}{\delta_{THROTTLE}(s)}\delta_{THROTTLE}(s) + \frac{\theta(s)}{\delta_{LE\_FLAPS}(s)}\delta_{LE\_FLAPS}(s)$$

$$+ \frac{\theta(s)}{\delta_{TE\_FLAPS}(s)}\delta_{TE\_FLAPS}(s). + \frac{\theta(s)}{\delta_{SPOILERS}(s)}\delta_{SPOILERS}(s)\ldots \Rightarrow \theta(t) = L^{-1}[\theta(s)]$$

7.6 SHORT PERIOD APPROXIMATION

As discussed in Section 7.2, the application of Laplace transformations to the linearized longitudinal equations leads to the solution of the longitudinal transfer functions using

$$\begin{bmatrix} (s-(X_u + X_{T_u})) & -X_\alpha & +g\cos\Theta_1 \\ -Z_u & (s(V_{P_1} - Z_{\dot\alpha}) - Z_\alpha) & (-s(Z_q + V_{P_1}) + g\sin\Theta_1) \\ -(M_u + M_{T_u}) & -(M_{\dot\alpha}s + (M_\alpha + M_{T_\alpha})) & s(s - M_q) \end{bmatrix} \begin{Bmatrix} \dfrac{u(s)}{\delta_E(s)} \\ \dfrac{\alpha(s)}{\delta_E(s)} \\ \dfrac{\theta(s)}{\delta_E(s)} \end{Bmatrix} = \begin{Bmatrix} X_{\delta_E} \\ Z_{\delta_E} \\ M_{\delta_E} \end{Bmatrix}$$

From an analysis of the longitudinal dynamics it can be seen that typically for longitudinally stable aircraft the short period dynamics is so fast that the condition $u(t) \approx 0$ can be approximated to be valid throughout the limited duration of the short period dynamics. This approximation allows a major simplification of the preceding relationship, leading to

$$\begin{bmatrix} \cancel{(s(X_u + X_{T_u}))} & \cancel{X_\alpha} & \cancel{+g\cos\Theta_1} \\ -Z_u & (s(V_{P_1} - Z_{\dot\alpha}) - Z_\alpha) & (-s(Z_q + V_{P_1}) + g\sin\Theta_1) \\ -(M_u + M_{T_u}) & -(M_{\dot\alpha}s + (M_\alpha + M_{T_\alpha})) & s(s - M_q) \end{bmatrix} \begin{Bmatrix} \cancel{\dfrac{u(s)}{\delta_E(s)}} \\ \dfrac{\alpha(s)}{\delta_E(s)} \\ \dfrac{\theta(s)}{\delta_E(s)} \end{Bmatrix} = \begin{Bmatrix} \cancel{X_{\delta_E}} \\ Z_{\delta_E} \\ M_{\delta_E} \end{Bmatrix}$$

Therefore, the previous (3 x 3) system can be reduced to the following (2 x 2) system:

$$\begin{bmatrix} s(V_{P_1} - Z_{\dot\alpha}) - Z_\alpha & -s(Z_q + V_{P_1}) + g\sin\theta_1 \\ -(M_{\dot\alpha}s + M_\alpha + M_{T_\alpha}) & s(s - M_q) \end{bmatrix} \begin{Bmatrix} \dfrac{\alpha(s)}{\delta_E(s)} \\ \dfrac{\theta(s)}{\delta_E(s)} \end{Bmatrix} = \begin{Bmatrix} Z_{\delta_E} \\ M_{\delta_E} \end{Bmatrix}$$

Additionally, the following assumptions[3,4,7,10–15] are possible:

$$Z_{\dot\alpha} \ll V_{P_1} \rightarrow Z_{\dot\alpha} \approx 0, \quad Z_q \ll V_{P_1} \rightarrow Z_q \approx 0, \quad \sin\theta_1 \approx \theta_1 \approx 0, \quad M_{T_\alpha} \approx 0$$

The above assumptions lead to the additional simplification:

$$\begin{bmatrix} sV_{P_1} - Z_\alpha & -V_{P_1}s \\ -(M_{\dot\alpha}s + M_\alpha) & s(s - M_q) \end{bmatrix} \begin{Bmatrix} \dfrac{\alpha(s)}{\delta_E(s)} \\ \dfrac{\theta(s)}{\delta_E(s)} \end{Bmatrix} = \begin{Bmatrix} Z_{\delta_E} \\ M_{\delta_E} \end{Bmatrix}$$

The application of Cramer's rule provides the following expressions for the transfer functions $\left(\dfrac{\alpha(s)}{\delta_E(s)}, \dfrac{\theta(s)}{\delta_E(s)}\right)$:

$$\dfrac{\alpha(s)}{\delta_E(s)} = \dfrac{\begin{vmatrix} Z_{\delta_E} & -V_{P_1}s \\ M_{\delta_E} & s(s-M_q) \end{vmatrix}}{\begin{vmatrix} sV_{P_1} - Z_\alpha & -V_{P_1}s \\ -(M_{\dot\alpha}s + M_\alpha) & s(s-M_q) \end{vmatrix}} = \dfrac{sZ_{\delta_E}(s-M_q) + sV_{P_1}M_{\delta_E}}{[s(sV_{P_1} - Z_\alpha)(s-M_q) - V_{P_1}s(M_{\dot\alpha}s + M_\alpha)]}$$

$$= \dfrac{\cancel{s}}{\cancel{s}}\dfrac{sZ_{\delta_E} + (V_{P_1}M_{\delta_E} - Z_{\delta_E}M_q)}{V_{P_1}\left[s^2 - \left(M_{\dot\alpha} + M_q + \dfrac{Z_\alpha}{V_{P_1}}\right)s + \left(\dfrac{Z_\alpha M_q}{V_{P_1}} - M_\alpha\right)\right]}$$

$$\dfrac{\alpha(s)}{\delta_E(s)} = \dfrac{\begin{vmatrix} sV_{P_1} - Z_\alpha & Z_{\delta_E} \\ -(M_{\dot\alpha}s + M_\alpha) & M_{\delta_E} \end{vmatrix}}{\begin{vmatrix} sV_{P_1} - Z_\alpha & -V_{P_1}s \\ -(M_{\dot\alpha}s + M_\alpha) & s(s-M_q) \end{vmatrix}} = \dfrac{M_{\delta_E}(sV_{P_1} - Z_\alpha) + Z_{\delta_E}(M_{\dot\alpha}s + M_\alpha)}{[s(sV_{P_1} - Z_\alpha)(s-M_q) - V_{P_1}s(M_{\dot\alpha}s + M_\alpha)]}$$

$$= \dfrac{(M_{\delta_E}V_{P_1} + Z_{\delta_E}M_{\dot\alpha})s + (Z_{\delta_E}M_\alpha - M_{\delta_E}Z_\alpha)}{V_{P_1}s\left[s^2 - \left(M_{\dot\alpha} + M_q + \dfrac{Z_\alpha}{V_{P_1}}\right)s + \left(\dfrac{Z_\alpha M_q}{V_{P_1}} - M_\alpha\right)\right]}$$

Next, the attention shifts to the quadratic expression in the reduced order short period characteristic equation.[3,5] By definition we have

$$s^2 - \left(M_{\dot{\alpha}} + M_q + \frac{Z_\alpha}{V_{P_1}}\right)s + \left(\frac{Z_\alpha M_q}{V_{P_1}} - M_\alpha\right) = s^2 + 2\ \zeta_{SP}\omega_{nSP}s + \omega_{nSP}^2$$

Interested students are referred to Appendix A.5 for a review of 2nd order dynamic systems. By equating the coefficients we have

$$\omega_{nSP} = \sqrt{\left(\frac{Z_\alpha M_q}{V_{P_1}} - M_\alpha\right)}$$

$$\zeta_{SP} = -\frac{\left(M_{\dot{\alpha}} + M_q + \frac{Z_\alpha}{V_{P_1}}\right)}{2\omega_{nSP}} = -\frac{\left(M_{\dot{\alpha}} + M_q + \frac{Z_\alpha}{V_{P_1}}\right)}{2\sqrt{\left(\frac{Z_\alpha M_q}{V_{P_1}} - M_\alpha\right)}}$$

The previous relationship for ζ_{SP} explains the longitudinal damping derivatives terminology typically used for the dimensionless stability derivatives $c_{m_{\dot{\alpha}}}, c_{mq}$ (see Chapter 3, Section 3.9.3) in the dimensional correspondent stability derivatives:

$$M_{\dot{\alpha}} = \frac{c_{m_{\dot{\alpha}}} \overline{q}_1 S \overline{c}^2}{2 I_{YY} V_{P_1}}, \quad M_q = \frac{c_{mq} \overline{q}_1 S \overline{c}^2}{2 I_{YY} V_{P_1}}$$

At high-speed flight conditions the following additional approximation can be made for most aircraft:

$$\left|\frac{Z_\alpha M_q}{V_{P_1}}\right| \ll |M_\alpha| \rightarrow \omega_{nSP} = \sqrt{\left(\frac{Z_\alpha M_q}{V_{P_1}} - M_\alpha\right)} \approx \sqrt{-M_\alpha} = \sqrt{\frac{c_{m_\alpha} \overline{q}_1 S \overline{c}}{I_{YY}}}$$

However, the use of this approximation is in general not recommended.

It should be emphasized again that the widespread availability of numerical codes capable of solving routinely third and higher order algebraic equations might make the use of the short period approximation unnecessary and obsolete. However, the short period approximation provides very valuable insights in the longitudinal stability with the very simple calculations shown previously. Particularly, the following conclusions can be drawn from the preceding relationships:

1. The natural frequency of the short period dynamics is a strong function of the dimensional stability derivative M_α and, therefore, of the corresponding dimensionless aerodynamic coefficient c_{m_α}. Since $c_{m_\alpha} = c_{L_\alpha}\ \overline{SM} = c_{L_\alpha}(\overline{x}_{CG} - \overline{x}_{AC})$, higher $\overline{SM}$ values lead to higher ω_{nSP} values and, therefore, faster short period dynamics. For naturally unstable aircraft (with the center gravity located after $\overline{x}_{AC}$, leading to positive $\overline{SM}$) with $\left|\frac{Z_\alpha M_q}{V_{P_1}}\right| > |M_\alpha|$, it is inappropriate to introduce $\omega_{nSP} = \sqrt{\left(\frac{Z_\alpha M_q}{V_{P_1}} - M_\alpha\right)}$ because the argument of the square root is negative.
2. Higher values of airspeed lead to higher ω_{nSP} values and therefore faster short period dynamics. The same trend applies with lower magnitude to decreasing altitude and, therefore, higher air density.
3. The moment of inertia I_{YY} plays a major role in the value for ω_{nSP}. In fact, low values for I_{YY} (associated with smaller and lighter aircraft) lead to high values of ω_{nSP}. For example, as shown in Table 7.2 ω_{nSP} for the Boeing B747-200 aircraft is $\omega_{nSP} = 1.321$ rad/sec, whereas for the Cessna 182 $\omega_{nSP} = 5.27$ rad/sec. As expected, smaller planes have faster dynamics and larger aircraft have slower dynamics, regardless of the difference in airspeed. Therefore, within aircraft of similar size and weight, the mass distribution plays a

central role. With similar payloads, the only substantial difference in the values of I_{YY} can be provided by the location of the propulsion system. A typical example is provided by a comparison between earlier versions of the Boeing B737 and McDonnell Douglas DC-9 aircraft. These jetliners had similar ranges, performance, and payload. However, the engines for the DC-9 are located on the tail, and the B737 has the engines under the wings. This naturally leads to the following relationships in terms of the moments of inertia:

$$I_{YY_{DC9}} > I_{YY_{B737}}, \quad I_{XX_{DC9}} < I_{XX_{B737}}, \quad I_{ZZ_{DC9}} \approx I_{ZZ_{B737}}$$

Therefore the DC9, whose engines are in the tail, has a lower ω_{nSP} than the B737, which features engines under the wings.

A numerical example is shown, with the goal of providing an estimate of the accuracy of the short period approximation.

Consider the Cessna 182 in Appendix B.[3] At Alt = 5,000 ft. and Mach = 0.21 the numerical expression for the full longitudinal characteristic equation (CE) is given by

$$\overline{D}_1(s) = A_1 s^4 + B_1 s^3 + C_1 s^2 + D_1 s + E_1$$
$$= 222.05 \, s^4 + 1985.95 \, s^3 + 6262.29 \, s^2 + 329.88 \, s + 180.58$$

Using MATLAB® *roots* command, the CE's roots are given by

$$-4.4498 + 2.8248i, \, -4.4498 - 2.8248i \qquad -0.0220 + 0.1697i, \, -0.0220 - 0.1697i$$

The location of these roots is associated with the following short period and phugoid characteristics (to be considered the true baseline values):

$$\zeta_{SP} = 0.844, \, \omega_{nSP} = 5.27 \, \frac{\text{rad}}{\text{sec}}, \quad \zeta_{Ph} = 0.129, \, \omega_{nPh} = 0.171 \, \frac{\text{rad}}{\text{sec}}$$

Using the short period approximation, the reduced order CE is given by

$$\overline{D}_{1_{SP}} = V_{P_1} s \left[s^2 - \left(M_{\dot{\alpha}} + M_q + \frac{Z_\alpha}{V_{P_1}} \right) s + \left(\frac{Z_\alpha M_q}{V_{P_1}} - M_\alpha \right) \right]$$
$$= 220.1 \, s \left[s^2 - \left(-2.5428 - 4.337 - \frac{464.71}{220.1} \right) s + \left(\frac{-464.71 \, -4.337}{220.1} + 19.26 \right) \right]$$

Thus, equating the preceding CE to the denominator of a second order system, we have

$$s^2 + 8.99s + 28.41 = s^2 + 2\zeta_{SP}\omega_{nSP} \, s + \omega_{nSP}^2$$

leading to the following short period approximation results:

$$\zeta_{SP_{APPROX.}} = 0.843, \, \omega_{nSP_{APPROX.}} = 5.33 \, \frac{\text{rad}}{\text{sec}}$$

The differences with respect to the above true values are around 1 percent for ω_{nSP} and totally negligible for ζ_{SP}, confirming, therefore, the validity of the short period approximation.

7.7 PHUGOID APPROXIMATION

The previous section has shown the accuracy of the short period approximation in solving for the dynamic characteristics of this dynamic mode. A conceptually similar approach can be introduced for the analysis of the phugoid mode.

Consider again the full (3 x 3) longitudinal state matrix whose determinant provides the full fourth order longitudinal characteristic equation $\overline{D}_1(s)$.

$$\begin{bmatrix} (s-(X_u+X_{T_u})) & -X_\alpha & +g\cos\Theta_1 \\ -Z_u & (s(V_{P_1}-Z_{\dot\alpha})-Z_\alpha) & (-s(Z_q+V_{P_1})+g\sin\Theta_1) \\ -(M_u+M_{T_u}) & -(M_{\dot\alpha}s+(M_\alpha+M_{T_\alpha})) & s(s-M_q) \end{bmatrix} \begin{Bmatrix} \dfrac{u(s)}{\delta_E(s)} \\ \dfrac{\alpha(s)}{\delta_E(s)} \\ \dfrac{\theta(s)}{\delta_E(s)} \end{Bmatrix} = \begin{Bmatrix} X_{\delta_E} \\ Z_{\delta_E} \\ M_{\delta_E} \end{Bmatrix}$$

It can be approximated that after the effects of the short period have vanished, the aircraft dynamic response shows no changes in the angle of attack. Mathematically this implies that the pitching moment equation can be neglected. Furthermore, all the terms associated with the perturbations in the angle of attack can be discarded. In other words, it can be approximated that during a typical phugoid oscillation we have $\alpha(t) \approx 0$. Therefore, the preceding relationship can be simplified to the solution of the following (2 x 2) system.

$$\begin{bmatrix} (s(X_u+X_{T_u})) & \cancel{-X_\alpha} & +g\cos\Theta \\ -Z_u & \cancel{(s(V_{P_1}-Z_{\dot\alpha})-Z_\alpha)} & (-s(Z_q+V_{P_1})+g\sin\Theta_1) \\ \cancel{-(M_u+M_{T_u})} & \cancel{-(M_{\dot\alpha}s+(M_\alpha+M_{T_\alpha}))} & \cancel{s(s-M_q)} \end{bmatrix} \begin{Bmatrix} \dfrac{u(s)}{\delta_E(s)} \\ \dfrac{\alpha(s)}{\delta_E(s)} \\ \cancel{\dfrac{\theta(s)}{\delta_E(s)}} \end{Bmatrix} = \begin{Bmatrix} X_{\delta_E} \\ Z_{\delta_E} \\ \cancel{M_{\delta_E}} \end{Bmatrix}$$

Additionally, since the aircraft is neither climbing or descending, we have the relationship $\alpha(t) \approx \theta(t)$. Also, it is possible to make the following assumptions:

$$Z_q \ll V_{P_1}, \Theta_1 \approx 0 \rightarrow \sin\Theta_1 \approx 0, \cos\Theta_1 \approx 1$$

Thus, the reduced order phugoid characteristic equation can be found using

$$\overline{D}_{1_{PHUGOID}}(s) = \begin{vmatrix} (s-X_U-X_{T_U}) & g \\ -Z_U & -V_{P_1}s \end{vmatrix} = -V_{P_1}s(s-X_U-X_{T_U}) + gZ_U$$

This relationship can be reduced to

$$\overline{D}_{1_{PHUGOID}}(s) = V_{P_1}s^2 - V_{P_1}(X_U+X_{T_U})s - gZ_U = V_{P_1}\left[s^2 - (X_U+X_{T_U})s - \dfrac{gZ_U}{V_{P_1}}\right]$$

By imposing the second order system characteristics

$$\overline{D}_{1_{PHUGOID}}(s) = V_{P_1}\left[s^2 - (X_U+X_{T_U})s - \dfrac{gZ_U}{V_{P_1}}\right] = V_{P_1}[s^2 + 2\zeta_{Ph}\omega_{n_{Ph}}s + \omega_{n_{Ph}}^2]$$

we have

$$\omega_{n_{Ph}} = \sqrt{-\dfrac{gZ_U}{V_{P_1}}}, \quad \zeta_{Ph} = \dfrac{-(X_U+X_{T_U})}{2\omega_{n_{Ph}}}$$

At this point recall from Table 7.1: $Z_u = -\dfrac{\overline{q}_1 S(c_{L_u}+2c_{L_1})}{mV_{P_1}}$.

Therefore, inserting the previous relationship, we have

$$\omega_{n_{Ph}} = \sqrt{\dfrac{g\dfrac{\overline{q}_1 S(c_{L_u}+2c_{L_1})}{mV_{P_1}}}{V_{P_1}}} = \sqrt{\dfrac{g\,\overline{q}_1 S(c_{L_u}+2c_{L_1})}{mV_{P_1}^2}}$$

Recall that in the low subsonic regions we have $c_{L_u} \approx 0$, $c_{L_u} \ll 2c_{L_1}$ with $c_{L_1} = \dfrac{mg}{\overline{q}_1 S}$.

Therefore, we have $\omega_{n_{Ph}} = \sqrt{\dfrac{g\,\overline{q}_1 S\left(\dfrac{2mg}{\overline{q}_1 S}\right)}{mV_{P_1}^2}} = \sqrt{\dfrac{g\,\overline{q}_1 S\left(\dfrac{2mg}{\overline{q}_1 S}\right)}{mV_{P_1}^2}} = \dfrac{g}{V_{P_1}}\sqrt{2}$

Thus, $\omega_{n_{Ph}} = \dfrac{g}{V_{P_1}}\sqrt{2}$.

The preceding relationship implies that at low subsonic speed the phugoid natural frequency is only a function of the airspeed, which is a major approximation. Next, inserting the natural frequency relationship in the expression for the damping coefficient we have

$$\zeta_{Ph} = \dfrac{-(X_U + X_{T_U})}{2\dfrac{g}{V_{P_1}}\sqrt{2}}$$

From Table 7.1 we recall that

$$X_u = -\dfrac{\overline{q}_1 S(c_{D_u}+2c_{D_1})}{mV_{P_1}},\ X_{T_u} = \dfrac{\overline{q}_1 S(c_{T_{X_u}}+2c_{T_{X_1}})}{mV_{P_1}}$$

Also, recall that at steady-state conditions $c_{T_{X_1}} = -c_{D_1}$.

Therefore, we have $\zeta_{Ph} = \dfrac{-(X_U + X_{T_U})}{2\dfrac{g}{V_{P_1}}\sqrt{2}} = \dfrac{\overline{q}_1 S(c_{D_u} - c_{T_{X_u}})}{2\dfrac{g}{V_{P_1}}\sqrt{2}(mV_{P_1})}$.

Recalling that $c_{L_1} = \dfrac{mg}{\overline{q}_1 S}$, we have

$$\zeta_{Ph} = \dfrac{\overline{q}_1 S(c_{D_u} - c_{T_{X_u}})}{2\dfrac{g}{V_{P_1}}\sqrt{2}(mV_{P_1})} = \dfrac{\overline{q}_1 S(c_{D_u} - c_{T_{X_u}})}{2\sqrt{2}(mg)} = \dfrac{(c_{D_u} - c_{T_{X_u}})}{2\sqrt{2}\ c_{L_1}}$$

leading to

$$\zeta_{Ph} = \dfrac{(c_{D_u} - c_{T_{X_u}})}{2\sqrt{2}\ c_{L_1}}$$

Therefore, the value of the phugoid damping is a function of the aircraft propulsive characteristics in addition to its aerodynamic characteristics.

Clearly, for aircraft at low subsonic conditions (low values for c_{D_u}, fairly high values of c_{L_1}) the value of the phugoid damping coefficient is lower than for aircraft at higher subsonic conditions (higher values for c_{D_u} along with lower values of c_{L_1}).

The numerical accuracy of the phugoid approximation is verified using the data of the SIAI Marchetti S211 (Appendix B, Aircraft 8) at approach conditions.

The full longitudinal characteristic equation is given by

$$\overline{D}_1(s) = A_1 s^4 + B_1 s^3 + C_1 s^2 + D_1 s + E_1$$
$$= 125.48\,s^4 + 307.66\,s^3 + 353.80\,s^2 + 30.07\,s + 29.13$$

The associated MATLAB® roots are given by

$$-1.2203 + 1.1033i,\quad -1.2203 - 1.1033i,\quad -0.0056 + 0.2928i,\ -0.0056 - 0.2928i$$

providing the following phugoid characteristics:

$$\omega_{n_{Ph}} = 0.2928 \, \frac{\text{rad}}{\text{sec}}$$

$$\zeta_{Ph} = 0.0191$$

Using instead the phugoid approximation, we have

$$\omega_{n_{Ph_{APPROX.}}} = \frac{g}{V_{P_1}}\sqrt{2} = \frac{32.17}{124}\sqrt{2} = 0.367 \, \frac{\text{rad}}{\text{sec}}$$

$$\zeta_{Ph_{APPROX.}} = \frac{(c_{D_u} - c_{T_{X_u}})}{2\sqrt{2} \, c_{L_1}} = 0 \text{ since } c_{D_u} = 0$$

It is clear that the phugoid approximation does not lead to acceptable results, especially for the estimate of the damping for aircraft at low airspeed flight conditions.

Consider now the Learjet aircraft at cruise conditions, low weight (Appendix B, Aircraft 6). The full longitudinal characteristic equation is given by

$$\overline{D}_1(s) = A_1 s^4 + B_1 s^3 + C_1 s^2 + D_1 s + E_1$$

$$= 678.26 \, s^4 + 1615.29 \, s^3 + 5892.29 \, s^2 + 184.70 \, s + 59.49$$

The associated MATLAB® roots are given by

$$-1.1763 + 2.6880i \quad -1.1763 - 2.6880i, \quad -0.0144 + 0.0999i \quad -0.0144 - 0.0999i$$

leading to the following phugoid characteristics:

$$\omega_{n_{Ph}} = 0.1 \, \frac{\text{rad}}{\text{sec}}$$

$$\zeta_{Ph} = 0.1429$$

Using instead the phugoid approximation, we have

$$\omega_{n_{Ph_{APPROX.}}} = \frac{g}{V_{P_1}}\sqrt{2} = \frac{32.17}{677}\sqrt{2} = 0.067 \, \frac{\text{rad}}{\text{sec}}$$

$$\zeta_{Ph_{APPROX.}} = \frac{(c_{D_u} - c_{T_{X_u}})}{2\sqrt{2} \, c_{L_1}} = \frac{(0.104 + 0.07)}{2\sqrt{2} \, 0.28} = \frac{0.174}{0.792} \approx 0.22$$

Once again, although it has somewhat improved from lower to higher airspeed, the phugoid approximation does not provide accurate results with error in range of 30 to 40 percent for the estimates of both natural frequency and damping.

7.8 SUMMARY OF THE LONGITUDINAL EQUATIONS

The derivation of the longitudinal equations is summarized in Figure 7.6.

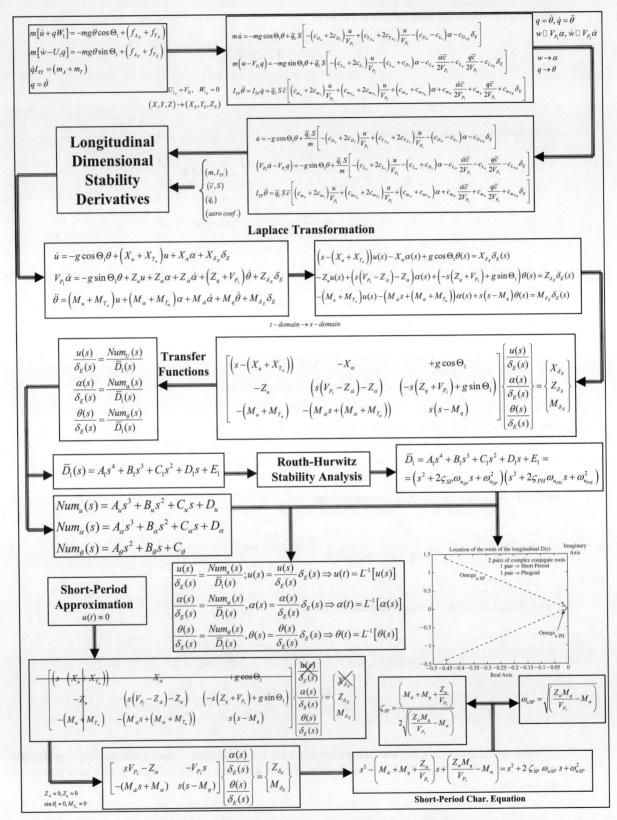

Figure 7.6 Overview of the Derivation of the Longitudinal Equations

7.9 APPLICATION OF LAPLACE TRANSFORMATIONS TO THE LATERAL DIRECTIONAL SMALL PERTURBATION EQUATIONS

Consider the small perturbation lateral directional equations with respect to the aircraft body-axes (X, Y, Z):

$$m[\dot{v} + U_1 r - p W_1] = mg\, \phi \cos\Theta_1 + (f_{A_Y} + f_{T_Y})$$

$$\dot{p}\, I_{XX} - \dot{r}\, I_{XZ} = (l_A + l_T)$$

$$\dot{r}\, I_{ZZ} - \dot{p}\, I_{XZ} = (n_A + n_T)$$

$$p = \dot{\phi} - \dot{\psi} \sin\Theta_1$$

$$r = \dot{\psi} \cos\Theta_1$$

Due to the small perturbation assumptions, the above differential equations are linear with constant coefficients. The modeling in Chapter 4 provided the following expressions for the small perturbation rolling and yawing moments and the small perturbation lateral force:

$$(f_{A_Y} + f_{T_Y}) = \bar{q}_1 S \left[\left(c_{Y_\beta} + c_{Y_{T\beta}}\right)\beta + c_{Y_p}\frac{pb}{2V_{P_1}} + c_{Y_r}\frac{rb}{2V_{P_1}} + c_{Y_{\delta_A}}\delta_A + c_{Y_{\delta_R}}\delta_R \right]$$

$$(l_A + l_T) = \bar{q}_1 S b \left[\left(c_{l_\beta} + c_{l_{T\beta}}\right)\beta + c_{l_p}\frac{pb}{2V_{P_1}} + c_{l_r}\frac{rb}{2V_{P_1}} + c_{l_{\delta_A}}\delta_A + c_{l_{\delta_R}}\delta_R \right]$$

$$(n_A + n_T) = q_1 S b \left[\left(c_{n_\beta} + c_{n_{T\beta}}\right)\beta + c_{n_p}\frac{pb}{2V_{P_1}} + c_{n_r}\frac{rb}{2V_{P_1}} + c_{n_{\delta_A}}\delta_A + c_{n_{\delta_R}}\delta_R \right]$$

Recall that the aerodynamic forces and moments are expressed in the aerodynamic-based stability axes (X_S, Y_S, Z_S). A specific property is that in this reference frame $U_{1_S} = V_{P_1}$ along the X_S axis is the only nonzero component of the linear velocity, with the lateral and vertical components being zero $(V_{1_S} = W_{1_S} = 0)$. Additionally, the thrust-induced rolling and yawing moments and lateral force are negligible, assuming that the aircraft does not have thrust vectoring or engine-out conditions (for multiengine aircraft). Finally, assuming that the starting condition is at a longitudinal trimmed condition such that $\cos\Theta_1 \approx 1$, $\sin\Theta_1 \approx \Theta_1 \approx 0$, the previous equations can be redefined in the stability axes (X_S, Y_S, Z_S) as in the following:

$$m(\dot{v} + V_{P_1} r) = mg\phi + \bar{q}_1 S \left[c_{Y_\beta}\beta + c_{Y_p}\frac{pb}{2V_{P_1}} + c_{Y_r}\frac{rb}{2V_{P_1}} + c_{Y_{\delta_A}}\delta_A + c_{Y_{\delta_R}}\delta_R \right]$$

$$I_{XX}\dot{p} - I_{XZ}\dot{r} = \bar{q}_1 S b \left[c_{l_\beta}\beta + c_{l_p}\frac{pb}{2V_{P_1}} + c_{l_r}\frac{rb}{2V_{P_1}} + c_{l_{\delta_A}}\delta_A + c_{l_{\delta_R}}\delta_R \right]$$

$$I_{ZZ}\dot{r} - I_{XZ}\dot{p} = \bar{q}_1 S b \left[c_{n_\beta}\beta + c_{n_p}\frac{pb}{2V_{P_1}} + c_{n_r}\frac{rb}{2V_{P_1}} + c_{n_{\delta_A}}\delta_A + c_{n_{\delta_R}}\delta_R \right]$$

along with the additional IKEs around the roll and yaw axes $p \approx \dot{\phi}$, $\dot{p} \approx \ddot{\phi}$, $r \approx \dot{\psi}$, $\dot{r} \approx \ddot{\psi}$.

Prior to the solution of the preceding equations, another transformation is required to address the issue that the aerodynamic rolling and yawing moments along with the lateral force are modeled with respect to the aerodynamic-based stability axes (X_S, Y_S, Z_S), while the moments of inertia and the product of inertia (I_{XX}, I_{YY}, I_{XZ}) are provided along the body-axes (X, Y, Z). As shown in Figure 7.7 these two reference frames of the axes (X, Z) and (X_S, Z_S) are rotated by the angle α_1.

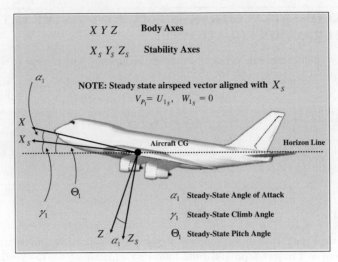

Figure 7.7 Boby-axes and Stability-axes

Therefore, the following transformation is introduced to express (I_{XX}, I_{YY}, I_{XZ}) with respect to the stability axes (X_S, Y_S, Z_S):

$$\begin{Bmatrix} I_{XX} \\ I_{ZZ} \\ I_{XZ} \end{Bmatrix}_{\text{Stability axes}} = \begin{bmatrix} \cos^2\alpha_1 & \sin^2\alpha_1 & -\sin(2\alpha_1) \\ \sin^2\alpha_1 & \cos^2\alpha_1 & \sin(2\alpha_1) \\ 0.5\sin(2\alpha_1) & -0.5\sin(2\alpha_1) & \cos(2\alpha_1) \end{bmatrix} \begin{Bmatrix} I_{XX} \\ I_{ZZ} \\ I_{XZ} \end{Bmatrix}_{\text{Body axes}}$$

From this point it is therefore assumed that (I_{XX}, I_{YY}, I_{XZ}) are expressed with respect to the frame (X_S, Y_S, Z_S). Next, a transformation of variables is introduced, allowing the previous equation in $\dot{v}$ to be expressed in terms of $\dot{\beta}$. Therefore, using

$$v \approx V_{P_1}\beta \rightarrow \beta \approx \frac{v}{V_{P_1}}, \quad \dot{v} \approx V_{P_1}\dot{\beta} \rightarrow \dot{\beta} = \frac{\dot{v}}{V_{P_1}}$$

and substituting

$$p \approx \dot{\phi}, \; \dot{p} \approx \ddot{\phi}, \; r \approx \dot{\psi}, \; \dot{r} \approx \ddot{\psi}$$

we have

$$(V_{P_1}\dot{\beta} + V_{P_1}\dot{\psi}) = g\phi + \frac{\bar{q}_1 S}{m}\left[c_{Y_\beta}\beta + c_{Y_p}\frac{b}{2V_{P_1}}\dot{\phi} + c_{Y_r}\frac{b}{2V_{P_1}}\dot{\psi} + c_{Y_{\delta_A}}\delta_A + c_{Y_{\delta_R}}\delta_R\right]$$

$$\ddot{\phi} - \frac{I_{XZ}}{I_{XX}}\ddot{\psi} = \frac{\bar{q}_1 Sb}{I_{XX}}\left[c_{l_\beta}\beta + c_{l_p}\frac{b}{2V_{P_1}}\dot{\phi} + c_{l_r}\frac{b}{2V_{P_1}}\dot{\psi} + c_{l_{\delta_A}}\delta_A + c_{l_{\delta_R}}\delta_R\right]$$

$$\ddot{\psi} - \frac{I_{XZ}}{I_{ZZ}}\ddot{\phi} = \frac{\bar{q}_1 Sb}{I_{ZZ}}\left[c_{n_\beta}\beta + c_{n_p}\frac{b}{2V_{P_1}}\dot{\phi} + c_{n_r}\frac{b}{2V_{P_1}}\dot{\psi} + c_{n_{\delta_A}}\delta_A + c_{n_{\delta_R}}\delta_R\right]$$

As for the longitudinal equations in Section 7.2 the RHS of the preceding equations features a combination of aerodynamic coefficients along with aircraft geometric (b,S) and inertial parameters (I_{XX}, I_{YY}, I_{XZ}), as well as flight conditions $(\bar{q}_1)$. To provide a unified frame for all these parameters a set of lateral-directional dimensional stability and control derivatives is introduced in Table 7.3.[3]

7.9 Application of Laplace Transformations to the Lateral Directional Small Perturbation Equations

Table 7.3 Dimensional Lateral Directional Stability Derivatives

Lateral-Directional Dimensional Stability and Control Derivatives	
$Y_\beta = \dfrac{\bar{q}_1 S c_{Y_\beta}}{m} \cdot (\text{ft sec}^{-2})$	$Y_p = \dfrac{\bar{q}_1 S c_{Y_p}}{m} \dfrac{b}{2U_1} \cdot (\text{ft sec}^{-2})$
$Y_r = \dfrac{\bar{q}_1 S c_{Y_r}}{m} \dfrac{b}{2U_1} \cdot (\text{ft sec}^{-1})$	$Y_{\delta_A} = \dfrac{\bar{q}_1 S c_{Y_{\delta A}}}{m} \cdot (\text{ft sec}^{-2})$
$Y_{\delta_R} = \dfrac{\bar{q}_1 S c_{Y_{\delta R}}}{m} \cdot (\text{ft sec}^{-2})$	
$L_\beta = \dfrac{\bar{q}_1 S c_{l_\beta} b}{I_{xx}} \cdot (\text{sec}^{-2})$	$L_p = \dfrac{\bar{q}_1 S b c_{lp}}{I_{xx}} \dfrac{b}{2U_1} \cdot (\text{sec}^{-1})$
$L_r = \dfrac{\bar{q}_1 S b c_{lr}}{I_{xx}} \dfrac{b}{2U_1} \cdot (\text{sec}^{-1})$	$L_{\delta_A} = \dfrac{\bar{q}_1 S c_{l_{\delta A}} b}{I_{xx}} \cdot (\text{sec}^{-2})$
$L_{\delta_R} = \dfrac{\bar{q}_1 S c_{l_{\delta R}} b}{I_{xx}} \cdot (\text{sec}^{-2})$	
$N_\beta = \dfrac{\bar{q}_1 S c_{n_\beta} b}{I_{zz}} \cdot (\text{sec}^{-2})$	$N_{T\beta} = \dfrac{\bar{q}_1 S c_{n_{T\beta}} b}{I_{zz}} \cdot (\text{sec}^{-2})$
$N_p = \dfrac{\bar{q}_1 S b c_{np}}{I_{zz}} \dfrac{b}{2U_1} \cdot (\text{sec}^{-1})$	$N_r = \dfrac{\bar{q}_1 S b c_{nr}}{I_{zz}} \dfrac{b}{2U_1} \cdot (\text{sec}^{-1})$
$N_{\delta_A} = \dfrac{\bar{q}_1 S c_{n_{\delta A}} b}{I_{zz}} \cdot (\text{sec}^{-2})$	$N_{\delta_R} = \dfrac{\bar{q}_1 S c_{n_{\delta R}} b}{I_{zz}} \cdot (\text{sec}^{-2})$

Using the coefficients in Table 4.3 the preceding small perturbation lateral directional equations take on the form

$$\left(V_{P_1}\dot{\beta} + V_{P_1}\dot{\psi}\right) = g\phi + Y_\beta \beta + Y_{\dot\phi}\dot\phi + Y_{\dot\psi}\dot\psi + Y_{\delta_A}\delta_A + Y_{\delta_R}\delta_R$$

$$\ddot{\phi} - \frac{I_{XZ}}{I_{XX}}\ddot{\psi} = L_\beta \beta + L_{\dot\phi}\dot\phi + L_{\dot\psi}\dot\psi + L_{\delta_A}\delta_A + L_{\delta_R}\delta_R$$

$$\ddot{\psi} - \frac{I_{XZ}}{I_{ZZ}}\ddot{\phi} = N_\beta \beta + N_{\dot\phi}\dot\phi + N_{\dot\psi}\dot\psi + N_{\delta_A}\delta_A + N_{\delta_R}\delta_R$$

where $Y_{\dot\phi} = Y_p$, $Y_{\dot\psi} = Y_r$, $L_{\dot\phi} = L_p$, $L_{\dot\psi} = L_r$, $N_{\dot\phi} = N_p$, $N_{\dot\psi} = N_r$.

Next, the following Laplace transformations are applied to the RHS and LHS of the equations, assuming zero initial conditions (because the small perturbations are considered from a steady-state rectilinear flight):

$$L(\delta_A) = \delta_A(s), \ L(\delta_R) = \delta_R(s)$$

$$L(\beta) = \beta(s), \ L(\dot\beta) = s\,\beta(s)$$

$$L(\phi) = \phi(s), \ L(\dot\phi) = L(p) = s\phi(s), L(\ddot\phi) = L(\dot p) = s^2\phi(s)$$

$$L(\psi) = \psi(s), \ L(\dot\psi) = L(r) = s\,\psi(s), \ L(\ddot\psi) = L(\dot r) = s^2\psi(s)$$

Introducing the following ratios $I_1 = \dfrac{I_{XZ}}{I_{XX}}$, $I_2 = \dfrac{I_{XZ}}{I_{ZZ}}$, the application of the Laplace transformations leads to

$$(sV_{P_1} - Y_\beta)\beta(s) - (sY_p + g\cos\Theta_1)\phi(s) + s(V_{P_1} - Y_r)\psi(s) = Y_\delta \delta(s)$$

$$-L_\beta \beta(s) + s(s - L_p)\phi(s) - s(sI_1 + L_r)\psi(s) = L_\delta \delta(s)$$

$$-N_\beta \beta(s) - s(sI_2 + N_p)\phi(s) + s(s - N_r)\psi(s) = N_\delta \delta(s)$$

In the preceding equations $\delta(s)$ denotes a generic lateral directional control surfaces (among the ailerons or the rudder). Using the same approach introduced for the longitudinal dynamics, transfer functions are now introduced for the lateral directional dynamics, with the only difference that for the lateral directional dynamics there are at least two control surfaces (ailerons and rudder) leading to the following six transfer functions:

$$\left\{\frac{\beta(s)}{\delta_A(s)}, \frac{\phi(s)}{\delta_A(s)}, \frac{\psi(s)}{\delta_A(s)}\right\}, \left\{\frac{\beta(s)}{\delta_R(s)}, \frac{\phi(s)}{\delta_R(s)}, \frac{\psi(s)}{\delta_R(s)}\right\}$$

The determination of the lateral directional transfer functions is introduced here for a generic control surface $\delta(s)$. Rearranging the previous equations in a matrix format with the transfer functions in the unknown vector we have

$$\begin{bmatrix} (sV_{P_1} - Y_\beta) & -(sY_p + g) & s(V_{P_1} - Y_r) \\ -L_\beta & s(s - L_p) & -s(sI_1 + L_r) \\ -N_\beta & -s(sI_2 + N_p) & s(s - N_r) \end{bmatrix} \begin{Bmatrix} \dfrac{\beta(s)}{\delta(s)} \\ \dfrac{\phi(s)}{\delta(s)} \\ \dfrac{\psi(s)}{\delta(s)} \end{Bmatrix} = \begin{Bmatrix} Y_\delta \\ L_\delta \\ N_\delta \end{Bmatrix}$$

where $\delta(s)$ can be considered either $\delta_A(s)$ or $\delta_R(s)$ and where the dimensional control derivatives $\{Y_\delta \ L_\delta \ N_\delta\}$ can be associated either to the ailerons or to the rudder. Next, the application of Cramer's rule leads to the following expressions for the transfer functions $\left\{\dfrac{\beta(s)}{\delta(s)}, \dfrac{\phi(s)}{\delta(s)}, \dfrac{\psi(s)}{\delta(s)}\right\}$:

$$\frac{\beta(s)}{\delta(s)} = \frac{\begin{vmatrix} Y_\delta & -(sY_p + g\cos\Theta_1) & s(V_{P_1} - Y_r) \\ L_\delta & s(s - L_p) & -s(sI_1 + L_r) \\ N_\delta & -s(sI_2 + N_p) & s(s - N_r) \end{vmatrix}}{\begin{vmatrix} (sV_{P_1} - Y_\beta) & -(sY_p + g\cos\Theta_1) & s(V_{P_1} - Y_r) \\ -L_\beta & s(s - L_p) & -s(sI_1 + L_r) \\ -N_\beta & -s(sI_2 + N_p) & s(s - N_r) \end{vmatrix}} = \frac{Num_\beta(s)}{D_2(s)}$$

$$\frac{\phi(s)}{\delta(s)} = \frac{\begin{vmatrix} (sV_{P_1} - Y_\beta) & Y_\delta & s(V_{P_1} - Y_r) \\ -L_\beta & L_\delta & -s(sI_1 + L_r) \\ -(N_\beta + N_{T_\beta}) & N_\delta & s(s - N_r) \end{vmatrix}}{\begin{vmatrix} (sV_{P_1} - Y_\beta) & -(sY_p + g\cos\Theta_1) & s(V_{P_1} - Y_r) \\ -L_\beta & s(s - L_p) & -s(sI_1 + L_r) \\ -N_\beta & -s(sI_2 + N_p) & s(s - N_r) \end{vmatrix}} = \frac{Num_\phi(s)}{D_2(s)}$$

$$\frac{\psi(s)}{\delta(s)} = \frac{\begin{vmatrix} (sV_{P_1} - Y_\beta) & -(sY_p + g\cos\Theta_1) & Y_\delta \\ -L_\beta & s(s - L_p) & L_\delta \\ -(N_\beta + N_{T_\beta}) & -s(sI_2 + N_p) & N_\delta \end{vmatrix}}{\begin{vmatrix} (sV_{P_1} - Y_\beta) & -(sY_p + g\cos\Theta_1) & s(V_{P_1} - Y_r) \\ -L_\beta & s(s - L_p) & -s(sI_1 + L_r) \\ -N_\beta & -s(sI_2 + N_p) & s(s - N_r) \end{vmatrix}} = \frac{Num_\psi(s)}{D_2(s)}$$

As for the longitudinal analysis, the lateral directional aircraft response is a function of the numerical values of the coefficients of the preceding polynomials at the numerator and, especially, the denominator. Starting from the numerator polynomials $(Num_\beta(s), Num_\phi(s), Num_\psi(s))$ we have

$$Num_\beta(s) = s(A_\beta s^3 + B_\beta s^2 + C_\beta s + D_\beta)$$

$$A_\beta = Y_\delta(1 - I_1 I_2)$$

$$B_\beta = -Y_\delta(L_p + N_r + I_1 N_p + I_2 L_r) + Y_p(L_\delta + I_1 N_\delta) + Y_r(L_\delta I_2 + N_\delta) - V_{P_1}(L_\delta I_2 + N_\delta)$$

$$C_\beta = Y_\delta(L_p N_r - N_p L_r) + Y_p(L_r N_\delta - N_r L_\delta) + g(L_\delta + I_1 N_\delta)$$
$$+ Y_r(L_\delta N_p - N_\delta L_p) - V_{P_1}(L_\delta N_p - N_\delta L_p)$$

$$D_\beta = g(L_r N_\delta - N_r L_\delta)$$

$$Num_\phi(s) = s(A_\phi s^2 + B_\phi s + C_\phi)$$

$$A_\beta = V_{P_1}(L_\delta + I_1 N_\delta)$$

$$B_\phi = V_{P_1}(L_r N_\delta - N_r L_\delta) - Y_\beta(L_\delta + I_1 N_\delta) + Y_\delta(L_\beta + I_1 N_\beta)$$

$$C_\phi = -Y_\beta(L_r N_\delta - N_r L_\delta) + Y_\delta(L_r N_\beta - L_\beta N_r)$$
$$+ V_{P_1}(L_\delta N_\beta - L_\beta N_\delta) - Y_r(L_\delta N_\beta - L_\beta N_\delta)$$

$$Num_\psi(s) = A_\psi s^3 + B_\psi s^2 + C_\psi s + D_\psi$$

$$A_\psi = V_{P_1}(N_\delta + I_2 L_\delta)$$

$$B_\psi = V_{P_1}(L_\delta N_p - N_\delta L_p) - Y_\beta(N_\delta + I_2 L_\delta) + Y_\delta(L_\beta I_2 + N_\beta)$$

$$C_\psi = -Y_\beta(L_\delta N_p - N_\delta L_p) + Y_p(L_\delta N_\beta - L_\beta N_\delta) + Y_\delta(L_\beta N_p - L_p N_\beta)$$

$$D_\psi = g(L_\delta N_\beta - L_\beta N_\delta)$$

All of the lateral directional transfer functions share the following CE $\overline{D}_2(s)$:

$$\overline{D}_2(s) = s(A_2 s^4 + B_2 s^3 + C_2 s^2 + D_2 s + E_2)$$

$$A_2 = V_{P_1}(1 - I_1 I_2)$$

$$B_2 = -Y_\beta(1 - I_1 I_2) - V_{P_1}(L_p + N_r + I_1 N_p + I_2 L_r)$$

$$C_2 = V_{P_1}(L_p N_r - N_p L_r) + Y_\beta(L_p + N_r + I_1 N_p + I_2 L_r) - Y_p(L_\beta + I_1 N_\beta)$$
$$+ V_{P_1}(L_\beta I_2 + N_\beta) - Y_r(L_\beta I_2 + N_\beta)$$

$$D_2 = -Y_\beta(L_p N_r - N_p L_r) + Y_p(L_\beta N_r - L_r N_\beta) - g(L_\beta + I_1 N_\beta)$$
$$+ V_{P_1}(L_\beta N_p - L_p N_\beta) - Y_r(L_\beta N_p - L_p N_\beta)$$

$$E_2 = g(L_\beta N_r - L_r N_\beta)$$

The stability of the lateral directional dynamics is a function of the values of the preceding coefficients. As for the longitudinal dynamics, the Routh–Hurwitz stability criteria[6] can be used for assessing the lateral directional stability.

7.10 ROUTH–HURWITZ ANALYSIS OF THE LATERAL DIRECTIONAL STABILITY

For dynamic stability purposes the roots of a CE are required to be negative, if real, or have negative real parts, if complex conjugates. The longitudinal CE is a fourth order polynomial, and the lateral directional CE is a fifth order polynomial $\overline{D}_2 = s(A_2 s^4 + B_2 s^3 + C_2 s^2 + D_2 s + E_2)$. However, excluding the trivial root at the origin ($s = 0$), the stability analysis will be similar to the longitudinal case. The application of the Routh–Hurwitz criteria to the lateral directional CE leads to the following Routh–Hurwitz array:

s^4	A_2	C_2	E_2
s^3	B_2	D_2	—
s^2	k_{11}	k_{12}	—
s	k_{21}	—	—
s^0	E_2	—	—

where

$$k_{11} = \frac{B_2 C_2 - A_2 D_2}{B_2}, \quad k_{12} = \frac{B_2 E_2}{B_2} = E_2$$

$$k_{21} = \frac{D_2 k_{11} - B_2 k_{12}}{k_{11}} = \frac{D_2 \left(\frac{B_2 C_2 - A_2 D_2}{B_2} \right) - B_2 E_2}{\frac{B_2 C_2 - A_2 D_2}{B_2}} = \frac{D_2 (B_2 C_2 - A_2 D_2) - B_2^2 E_2}{B_2 C_2 - A_2 D_2}$$

Next, the analysis focuses on the first column of the preceding array:

s^4	A_2	C_2	E_2
s^3	B_2	D_2	—
s^2	k_{11}	k_{12}	—
s	k_{21}	—	—
s^0	E_2	—	—

Recall that a system is dynamically stable if *all* the coefficients in the first column of the Routh–Hurwitz array have the same sign. Additionally, the number of unstable roots is equal to the number of sign changes in the first column of the Routh–Hurwitz array. Thus, the following conditions need to be satisfied for ensuring lateral directional dynamic stability:

$$A_2 > 0, \quad B_2 > 0, \quad (B_2 C_2 - A_2 D_2) > 0, \quad \Delta_2 > 0, \quad E_2 > 0$$

where $\Delta_2 = D_2(B_2 C_2 - A_2 D_2) - B_2^2$ is known as the lateral directional Routh Discriminant. As in the longitudinal case, dynamic instability for the lateral directional dynamics can be tracked to the following two conditions:

$$\Delta_2 = D_2(B_2 C_2 - A_2 D_2) - B_2^2 E_2 < 0$$
$$E_2 < 0$$

The condition $\Delta_2 = D_2(B_2 C_2 - A_2 D_2) - B_2^2 E_2 < 0$ along with the condition $E_2 > 0$ leads to two unstable roots (that is one pair of complex conjugate roots with positive real part). The condition $\Delta_2 = D_2(B_2 C_2 - A_2 D_2) - B_2^2 E_2 > 0$ along with the condition $E_2 < 0$ leads instead to one unstable root. The mathematical implications of $\Delta_2 < 0$ are difficult to visualize; however, it is easy to identify the source of instability associated with the condition $E_2 < 0$.

In fact, since $E_2 = g(L_\beta N_r - L_r N_\beta)$, dynamic stability is associated with the condition

$$L_\beta N_r - L_r N_\beta > 0$$

From Table 4.3 recall that

$$L_\beta = \frac{\bar{q}_1 S c_{l_\beta} b}{I_{xx}}, \quad L_r = \frac{\bar{q}_1 S b c_{l_r}}{I_{xx}} \frac{b}{2V_{P_1}}, \quad N_\beta = \frac{\bar{q}_1 S c_{n_\beta} b}{I_{zz}}, \quad N_r = \frac{\bar{q}_1 S b c_{n_r}}{I_{zz}} \frac{b}{2V_{P_1}}$$

Therefore

$$L_\beta N_r - L_r N_\beta > 0 \rightarrow c_{l_\beta} \, c_{nr} - c_{l_r} \, c_{n_\beta} > 0$$

At this point an analysis of the signs for the preceding stability derivatives is required. Recall that for static stability purposes

$$c_{l_\beta} < 0, \quad c_{n_\beta} > 0$$

Additionally, for the r derivatives the following static stability criteria apply:

$$c_{nr} < 0, \quad c_{lr} > 0$$

In terms of signs, for dynamic stability purposes we have

$$(< 0)(< 0) - (> 0)(> 0) \; > \; 0$$
$$(> 0) - (> 0) \; > \; 0$$

Therefore, for lateral directional stability purposes the following condition needs to be verified:

$$|L_\beta N_r| - |L_r N_\beta| > 0$$

This relationship highlights once again the critical importance of the aircraft dihedral effect previously discussed and modeled in Chapter 4, Section 4.3.1.

As for the longitudinal dynamics, the stability of the lateral directional dynamics is affected by a large number of variables, including flight conditions, geometric parameters, and inertial characteristics, in addition to the aerodynamic modeling. A specific study called sensitivity analysis is performed in the design phase to evaluate the effect of specific parameters (related to the aerodynamic behavior, flight conditions, or aircraft geometry) on the overall lateral directional stability. This issue will be addressed in Section 7.15.

7.11 LATERAL DIRECTIONAL DYNAMIC MODES: ROLLING, SPIRAL, AND DUTCH ROLL

For dynamically stable aircraft, the lateral directional CE has one pair of complex conjugate roots along with two real roots. The pair of complex conjugate roots is associated with a second order system, whereas each of the real roots is associated wit a first order system. Refer to Appendix A.5 for a review of the characteristics of first and second order dynamic systems. Thus, the lateral directional CE can be broken down into the following:

$$\overline{D}_2 = s(A_2 s^4 + B_2 s^3 + C_2 s^2 + D_2 s + E_2) = s(s^2 + 2\zeta_{DR}\omega_{n_{DR}} s + \omega_{n_{DR}}^2)(s + \lambda_R)(s + \lambda_S)$$

Each of these dynamic modes provides distinct signatures on the aircraft lateral directional dynamic responses. The typical locations of the CE roots associated with these modes are shown in Figure 7.8.

The dutch roll is characterized by moderate values of the damping coefficient ζ_{DR} along with moderate values of the natural frequency $\omega_{n_{DR}}$. Particularly, in most aircraft we have $(\omega_{n_{PH}} < \omega_{n_{DR}} < \omega_{n_{SP}})$ along with $(\zeta_{PH} < \zeta_{DR} < \zeta_{SP})$.

The spiral is instead a very slow first order mode, with the root at $-\lambda_S$ typically located around the origin of the s-plane, leading to a very large value of the associated time constant $T_S = -\dfrac{1}{\lambda_S}$. For many aircraft this first order mode can actually have a slightly positive root, leading to a slow-developing dynamic instability. However, this mode is easily controlled manually by the pilot or by the flight control laws of the autopilot or stability augmentation systems.

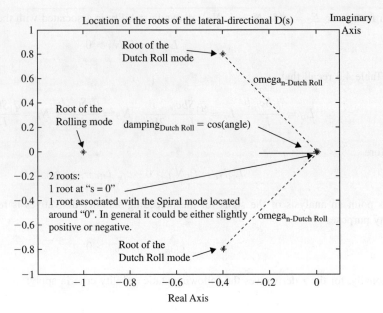

Figure 7.8 Locations of the Dutch Roll, Spiral, and Rolling Roots

Finally, the rolling is a reasonably fast first order mode, with the root at $-\lambda_R$ associated with the time constant $T_R = -\frac{1}{\lambda_R}$. This specific parameter has major importance on the maneuverability and the handling qualities of fighter aircraft.

As for the longitudinal dynamics, the preceding lateral directional parameters are major design objectives for satisfying the requirements for the aircraft handling qualities. The values for each of these parameters depend on the specific aircraft type (military versus civilian), aircraft class, and level of handling qualities, as described in Chapter 10. Table 7.4 shows a sample of the values for $(\varsigma_{DR}, \omega_{n_{DR}}, T_S, T_R)$ for different aircraft from different classes.[3]

Table 7.4 Sample $(\varsigma_{DR}, \omega_{n_{DR}}, T_S, T_R)$ for Different Aircraft

Aircraft Class	Lateral / Direct Parameters			
	ς_{DR}	$\omega_{n_{DR}}$	$T_{R(sec)}$	$T_{S(sec)}$
General Aviation/Light Commercial Aviation				
Cessna 182 (Alt. = 5,000 ft. Mach = 0.2)	0.207	3.25	0.077	55.921
Cessna 310 (Alt. = SL, Mach = 0.16)	0.105	1.94	0.584	−44.48*
Beech 99 (Alt. = 20,000ft. Mach = 0.43)	0.036	1.874	0.306	40.17
Cessna 620 (Alt. = SL, Mach = 0.17)	0.13	1.587	0.84	−47.49*
Military Trainer				
SIAI Marchetti 211 (Alt. = SL, Mach = 0.11)	0.212	1.798	0.276	−8.09*
Cessna T37 A (Alt. = 30,000 ft. Mach = 0.46)	0.047	2.409	0.79	271.31
Business jet				
Learjet 24 (Alt. = SL, Mach = 0.15)	−0.045	1.041	1.363	−34.14*
Fighter				
Lockheed F-104 (Alt. = SL, Mach = 0.26)	0.128	2.881	0.967	−966.96*
McDonnell F-4 (Alt. = 35,000 ft. Mach = 0.9)	0.048	2.396	0.748	77.02
Commercial Jetliner				
Boeing B747-200 (Alt. = 40,000 ft, Mach = 0.9)	0.064	0.911	1.689	78.26

^: 'Degenerated' unstable dutch roll mode

*: unstable spiral mode

7.12 SOLUTION OF THE LATERAL DIRECTIONAL EQUATIONS

The conceptual approach to the solution of the lateral directional equations is similar to the longitudinal case. However, a conventional aircraft features at least two lateral directional control surfaces (ailerons and rudder). Starting from the previously discussed generic transfer functions

$$\frac{\beta(s)}{\delta(s)} = \frac{Num_\beta(s)}{\overline{D}_2(s)}, \quad \frac{\phi(s)}{\delta(s)} = \frac{Num_\phi(s)}{\overline{D}_2(s)}, \quad \frac{\psi(s)}{\delta(s)} = \frac{Num_\psi(s)}{\overline{D}_2(s)}$$

For a known $\delta(s) = \delta_A(s)$ associated with a given $\delta_A(t)$ we have

$$\frac{\beta(s)}{\delta_A(s)} = \frac{Num_{\beta A}}{\overline{D}_2} = \frac{s(A_{\beta A}s^3 + B_{\beta A}s^2 + C_{\beta A}s + D_{\beta A})}{s(A_2 s^4 + B_2 s^3 + C_2 s^2 + D_2 s + E_2)}$$

$$\frac{\phi(s)}{\delta_A(s)} = \frac{Num_{\phi A}}{\overline{D}_2} = \frac{s(A_{\phi A}s^2 + B_{\phi A}s + C_{\phi A})}{s(A_2 s^4 + B_2 s^3 + C_2 s^2 + D_2 s + E_2)}$$

$$\frac{\psi(s)}{\delta_A(s)} = \frac{Num_{\psi A}}{\overline{D}_2} = \frac{A_{\psi A}s^3 + B_{\psi A}s^2 + C_{\psi A}s + D_{\psi A}}{s(A_2 s^4 + B_2 s^3 + C_2 s^2 + D_2 s + E_2)}$$

using $\{Y_{\delta_A} \ L_{\delta_A} \ N_{\delta_A}\}$ in the expressions for $(Num_{\beta A}, Num_{\phi A}, Num_{\psi A})$.

Similarly, for a known $\delta(s) = \delta_R(s)$ associated with a given $\delta_R(t)$ we have

$$\frac{\beta(s)}{\delta_R(s)} = \frac{Num_{\beta R}}{\overline{D}_2} = \frac{s(A_{\beta R}s^3 + B_{\beta R}s^2 + C_{\beta R}s + D_{\beta R})}{s(A_2 s^4 + B_2 s^3 + C_2 s^2 + D_2 s + E_2)}$$

$$\frac{\phi(s)}{\delta_R(s)} = \frac{Num_{\phi R}}{\overline{D}_2} = \frac{s(A_{\phi R}s^2 + B_{\phi R}s + C_{\phi R})}{s(A_2 s^4 + B_2 s^3 + C_2 s^2 + D_2 s + E_2)}$$

$$\frac{\psi(s)}{\delta_R(s)} = \frac{Num_{\psi R}}{\overline{D}_2} = \frac{A_{\psi R}s^3 + B_{\psi R}s^2 + C_{\psi R}s + D_{\psi R}}{s(A_2 s^4 + B_2 s^3 + C_2 s^2 + D_2 s + E_2)}$$

using $\{Y_{\delta_R} \ L_{\delta_R} \ N_{\delta_R}\}$ in the expressions for $(Num_{\beta R}, Num_{\phi R}, Num_{\psi R})$.

Using the super-imposition property for linear systems, the solution of the lateral directional equations with the aircraft subjected to both ailerons and rudder input is given by

$$\beta(s) = \frac{\beta(s)}{\delta_A(s)}\delta_A(s) + \frac{\beta(s)}{\delta_R(s)}\delta_R(s) \Rightarrow \beta(t) = L^{-1}[\beta(s)]$$

$$\phi(s) = \frac{\phi(s)}{\delta_A(s)}\delta_A(s) + \frac{\phi(s)}{\delta_R(s)}\delta_R(s) \Rightarrow \phi(t) = L^{-1}[\phi(s)]$$

$$\psi(s) = \frac{\psi(s)}{\delta_A(s)}\delta_A(s) + \frac{\psi(s)}{\delta_R(s)}\delta_R(s) \Rightarrow \psi(t) = L^{-1}[\psi(s)]$$

The hand solution using the partial fraction expansion (PFE) method leading to expressions for $(\beta(t), \phi(t), \psi(t))$ is not a trivial matter. The availability of MATLAB® has allowed quick solutions to these formerly long and tedious calculations.

A MATLAB®-based solution for the lateral directional dynamics provides the following time histories shown in Figures 7.9 and 7.10 associated with aileron and rudder maneuvers using aerodynamic, geometry, inertial, and flight conditions data for a high-performance fighter aircraft (Lockheed F-104) at approach conditions from Appendix B.[3]

The preceding time histories show fairly well behaved dutch roll dynamics with a desirable damping (ς_{DR}) along with reasonably fast rolling dynamics, associated with a fairly low value for T_R, which is expected for this class of aircraft.

380 Chapter 7 Solution of the Aircraft Equations of Motion Based on Laplace Transformations and Transfer Functions

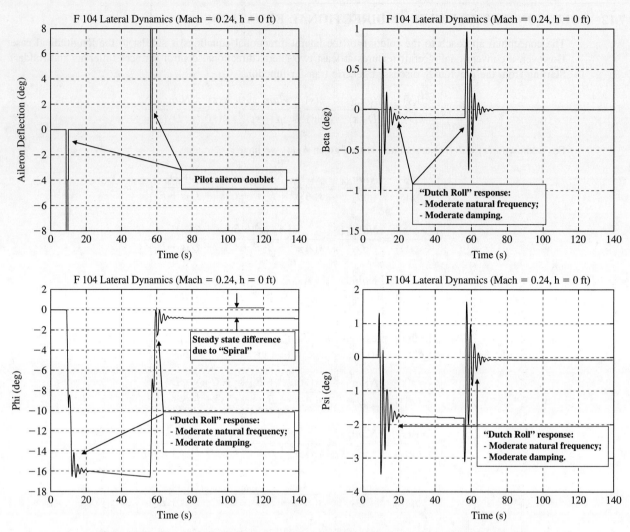

Figure 7.9 Time histories Following an Aileron Maneuver for the F-104 (approach conditions)

The roots of the lateral directional characteristic equation associated with the dutch roll, spiral, and rolling modes for the given aircraft at the given flight conditions are shown in Figure 7.11.

The preceding modeling approach can be applied without loss of generality to the more general case of multiple inputs through deflections of multiple control surfaces. Assuming, for example, the availability of asymmetric canards or asymmetric spoilers on the wing, the lateral directional responses will be given by

$$\beta(s) = \frac{\beta(s)}{\delta_A(s)}\delta_A(s) + \frac{\beta(s)}{\delta_R(s)}\delta_R(s) + \frac{\beta(s)}{\delta_{SPOILERS}(s)}\delta_{SPOILERS} + \frac{\beta(s)}{\delta_{CANARDS}(s)}\delta_{CANARDS} + \ldots$$

$$\Rightarrow \beta(t) = L^{-1}[\beta(s)]$$

$$\phi(s) = \frac{\phi(s)}{\delta_A(s)}\delta_A(s) + \frac{\phi(s)}{\delta_R(s)}\delta_R(s) + \frac{\phi(s)}{\delta_{SPOILERS}(s)}\delta_{SPOILERS} + \frac{\phi(s)}{\delta_{CANARDS}(s)}\delta_{CANARDS} + \ldots$$

$$\Rightarrow \phi(t) = L^{-1}[\phi(s)]$$

$$\psi(s) = \frac{\psi(s)}{\delta_A(s)}\delta_A(s) + \frac{\psi(s)}{\delta_R(s)}\delta_R(s) + \frac{\psi(s)}{\delta_{SPOILERS}(s)}\delta_{SPOILERS} + \frac{\psi(s)}{\delta_{CANARDS}(s)}\delta_{CANARDS} + \ldots$$

$$\Rightarrow \psi(t) = L^{-1}[\psi(s)]$$

7.12 Solution of the Lateral Directional Equations 381

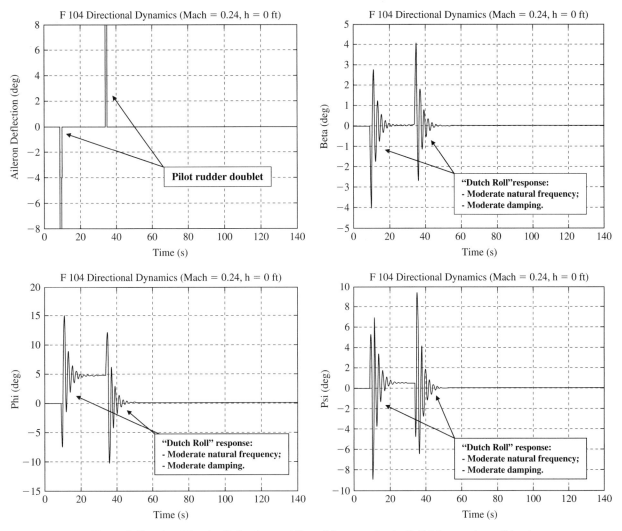

Figure 7.10 Time Histories Following an Aileron Maneuver for the F-104 (approach conditions)

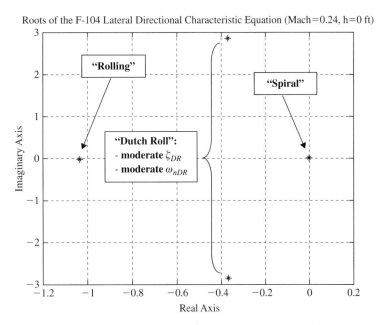

Figure 7.11 Roots of the Longitudinal Characteristic Equation for the F-104 (approach conditions)

7.13 ROLLING APPROXIMATION

As discussed in the previous sections the lateral directional dynamics features three dynamic modes: the dutch roll, the spiral, and the rolling modes. From a dynamic analysis it can be deduced that both the dutch roll and the rolling are essentially 3D modes and simultaneously involve the variables $(\beta(t), \phi(t), \psi(t))$. This is due to the fact that the rolling moment is typically coupled with the yawing moment. However, the rolling mode, which is associated with pure aileron deflection, can be considered a pure first order dynamic model. An analysis of this mode leads to important and useful results.

Consider the previously introduced transfer functions equations associated with aileron deflections.

$$\begin{bmatrix} (sV_{P_1} - Y_\beta) & -(sY_p + g) & s(V_{P_1} - Y_r) \\ -L_\beta & s(s - L_p) & -s(sI_1 + L_r) \\ -N_\beta & -s(sI_2 + N_p) & s(s - N_r) \end{bmatrix} \begin{Bmatrix} \dfrac{\beta(s)}{\delta_A(s)} \\ \dfrac{\phi(s)}{\delta_A(s)} \\ \dfrac{\psi(s)}{\delta_A(s)} \end{Bmatrix} = \begin{Bmatrix} Y_{\delta_A} \\ L_{\delta_A} \\ N_{\delta_A} \end{Bmatrix}$$

An aircraft subjected to an ailerons-only maneuver will experience negligible perturbations in the variables $\beta(t), \psi(t)$. Therefore, we have a pure rolling motion with $\psi(t) = \beta(t) \approx 0 \Rightarrow \psi(s) = \beta(s) = 0$. Thus, the preceding equations will become

$$\begin{bmatrix} \cancel{(sV_{P_1} - Y_\beta)} & -(sY_p + g) & \cancel{s(V_{P_1} - Y_r)} \\ \cancel{-L_\beta} & s(s - L_p) & \cancel{-s(sI_1 + L_r)} \\ \cancel{-N_\beta} & \cancel{-s(sI_2 + N_p)} & \cancel{s(s - N_r)} \end{bmatrix} \begin{Bmatrix} \cancel{\dfrac{\beta(s)}{\delta_A(s)}} \\ \dfrac{\phi(s)}{\delta_A(s)} \\ \cancel{\dfrac{\psi(s)}{\delta_A(s)}} \end{Bmatrix} = \begin{Bmatrix} \cancel{Y_{\delta_A}} \\ L_{\delta_A} \\ \cancel{N_{\delta_A}} \end{Bmatrix}$$

leading to

$$(s^2 - L_P s)\frac{\phi(s)}{\delta_A(s)} = L_{\delta_A} \Rightarrow \frac{\phi(s)}{\delta_A(s)} = \frac{L_{\delta_A}}{s(s - L_P)}$$

The goal is to derive an expression for the roll angular rate $p(t)$ from the preceding relationship. This could be accomplished either using the path $(\phi(s) \Rightarrow \phi(t) \Rightarrow p(t))$ or the path $(\phi(s) \Rightarrow p(s) \Rightarrow p(t))$. Using the first path, assuming a constant step aileron input $\delta_A(t) = \delta_A^*$ such that $\delta_A(s) = \dfrac{\delta_A^*}{s}$, we have

$$\phi(s) = \frac{L_{\delta_A}\delta_A(s)}{s(s - L_P)} = \frac{L_{\delta_A}\delta_A^*}{s^2(s - L_P)}$$

Following the application of the PFE method, we have

$$\phi(s) = \frac{L_{\delta_A}\delta_A^*}{s^2(s - L_P)} = \frac{k_{11}}{s} + \frac{k_{12}}{s^2} + \frac{k_2}{(s - L_p)}$$

where: $k_{11} = \dfrac{d}{ds}\left(\dfrac{s^2 L_{\delta_A}\delta_A^*}{s^2(s - L_P)}\right)\bigg|_{s=0} = \dfrac{d}{ds}\left(\dfrac{L_{\delta_A}\delta_A^*}{(s - L_P)}\right)\bigg|_{s=0} = -\dfrac{L_{\delta_A}\delta_A^*}{L_P^2}$

7.13 Rolling Approximation

$$k_{12} = s^2 \frac{L_{\delta_A} \delta_A^*}{s^2(s - L_P)}\bigg|_{s=0} = -\frac{L_{\delta_A} \delta_A^*}{L_P}$$

$$k_2 = (s - L_p)\frac{L_{\delta_A} \delta_A^*}{s^2(s - L_P)}\bigg|_{s=L_P} = \frac{L_{\delta_A} \delta_A^*}{s^2}\bigg|_{s=L_P} = \frac{L_{\delta_A} \delta_A^*}{L_P^2}$$

Using the Inverse Laplace Transform (ILT) table from Appendix A we have

$$\phi(t) = -\frac{L_{\delta_A} \delta_A^*}{L_P^2} - \frac{L_{\delta_A} \delta_A^*}{L_P}t + \frac{L_{\delta_A} \delta_A^*}{L_P^2}e^{L_P t}$$

$$= -\frac{L_{\delta_A} \delta_A^*}{L_P}t + \frac{L_{\delta_A} \delta_A^*}{L_P^2}\left(e^{L_P t} - 1\right)$$

Since $p(t) = \dot{\phi}(t)$ we have

$$p(t) = -\frac{L_{\delta_A} \delta_A^*}{L_P} + \frac{L_{\delta_A} \delta_A^*}{L_P^2}L_P e^{L_P t} = -\frac{L_{\delta_A} \delta_A^*}{L_P}\left(1 - e^{L_P t}\right)$$

At this point the attention focuses on the exponential term $e^{L_P t}$. Since $c_{l_p} < 0$, we have that $L_P = \frac{c_{l_p}\bar{q}Sb}{I_{xx}}\frac{b}{2V_{P_1}} < 0$. Therefore, $\lim_{t \to \infty} e^{L_P t} = 0$, leading to

$$\lim_{t \to \infty} p(t) = p_{SS} = -\frac{L_{\delta_A} \delta_A^*}{L_P}$$

For a first order system, the time constant can be introduced as *the time required by the system to reach 63 percent of its steady-state response.*[6] For a generic first order system with transfer function $G(s) = a/(s + a)$ we have $y(t) = 1 - e^{-at}$.

Since $\lim_{t \to \infty} y(t) = y_{SS} = 1$, $y(t)$ reaches $63\% y_{SS} = 0.63$ when $t = T = \frac{1}{a}$ such that

$$y(T) = 1 - e^{-a\frac{1}{a}} = 1 - e^{-1} = 0.63$$

Using the previous first order system property, the roll time constant $T_R(\text{sec})$ can be introduced. Following the previous approach we have

$$p(T_R) = -\frac{L_{\delta_A} \delta_A^*}{L_P}\left(1 - e^{L_P T_R}\right) = -\frac{L_{\delta_A} \delta_A^*}{L_P}\left(1 - e^{L_P - \frac{1}{L_P}}\right)$$

$$= -\frac{L_{\delta_A} \delta_A^*}{L_P}\left(1 - e^{-1}\right) = -\frac{L_{\delta_A} \delta_A^*}{L_P}\left(1 - \frac{1}{e}\right) = p_{SS}\left(1 - \frac{1}{e}\right) = 0.63\, p_{SS}$$

Therefore, *the rolling time constant $T_R(\text{sec})$ is the time required for the aircraft to reach 63 percent of its steady-state roll rate.*

$$T_R = -\frac{1}{L_P} = -\frac{1}{\frac{c_{l_p}\bar{q}Sb}{I_{xx}}\frac{b}{2V_{P_1}}} = \frac{I_{xx} 2V_{P_1}}{c_{l_p}\bar{q}Sb^2}$$

leading to $T_R = -\frac{I_{xx} 2V_{P_1}}{c_{l_p}\bar{q}Sb^2} = -\frac{I_{xx} 2V_{P_1}}{c_{l_p}\frac{1}{2}\rho V_{P_1}^2}Sb^2 = -\frac{4I_{xx}}{c_{l_p}\rho V_{P_1} S b^2}$. Recall that c_{l_p} is negative.

The following qualitative conclusions can be drawn for T_R:

- With all the parameters being constant, an increase in V_{P_1} leads to lower T_R values;
- With all the parameters being constant, a decrease in I_{xx} (for example, due to the consumption of the fuel in the wing tank or to the drop of payloads under the wing) leads to lower T_R values;
- With all the parameters being constant, a decrease in altitude (increase in air density ρ) leads to slightly lower T_R values.

As previously mentioned, the rolling time constant T_R is a very important parameter for fighter aircraft. Fast rolling capabilities are critical for fighter aircraft in combat situations. Chapter 10 will provide specific details on the rolling requirements for all different classes of aircraft. Figure 7.12 shows the difference between the rolling characteristics of a fighter aircraft (McDonnell Douglas F-4) and a widebody jetliner (Boeing B747-200).

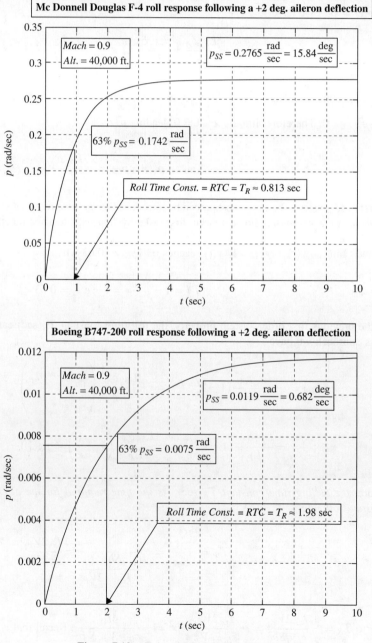

Figure 7.12 F-4 and B747 Rolling Responses

7.14 SUMMARY OF LATERAL DIRECTIONAL EQUATIONS

The derivation of the lateral directional equations is summarized in Figure 7.13.

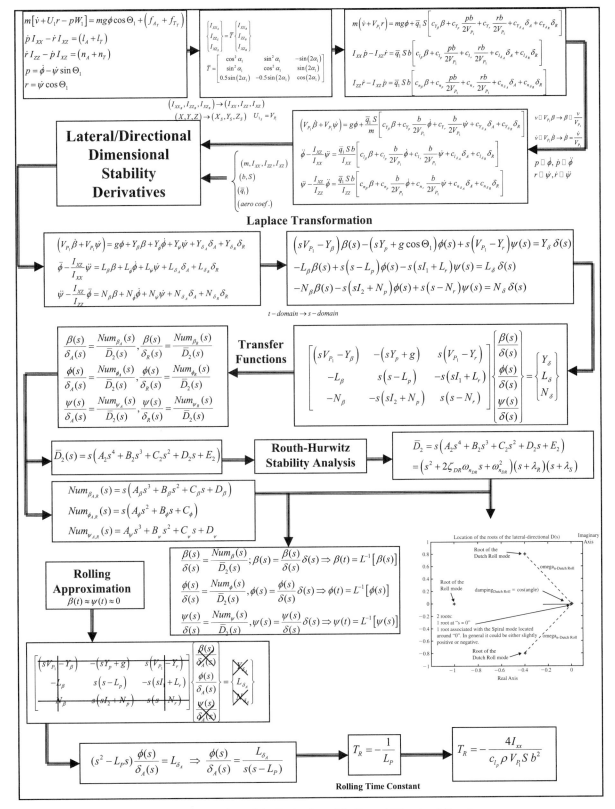

Figure 7.13 Overview of the Derivation of the Lateral Directional Equations

7.15 SENSITIVITY ANALYSIS FOR THE AIRCRAFT DYNAMICS

The previous section has shown the typical responses associated with the longitudinal and lateral directional dynamics of the aircraft. It is clear that the behavior of the responses depends on the location of the roots of the longitudinal and lateral directional characteristic equations ($\overline{D}_1(s), \overline{D}_2(s)$), which, ultimately, translates into the values of the previously introduced coefficients:

$$\zeta_{SP}, \omega_{n_{SP}}, \zeta_{Ph}, \omega_{n_{SP}} \quad \text{for the longitudinal dynamics}$$
$$\zeta_{DR}, \omega_{n_{SP}}, T_R, T_S \quad \text{for the lateral}-\text{directional dynamics}$$

Thus, the aircraft handling qualities (to be discussed in Chapter 10) depend on a set of satisfactory values for the preceding coefficients within ranges specified in the FAR 23 and FAR 25 (for commercial aircraft and general aviation) and the MIL-F-8785C (for military aircraft).

In the event that an initial design does not provide satisfactory values for the parameters within the entire operational flight envelope of the aircraft, there are three conceptual approaches toward the goal of meeting the specifications for these parameters.

The first approach is based on the introduction of a specific class of flight control systems, known as stability augmentation systems (SASs). These systems are designed with the goal of modifying the aircraft response in closed loop toward improving key parameters, such as $\zeta_{SP}, \zeta_{Ph}, \zeta_{DR}, T_S, T_R$. The resulting aircraft response will then exhibit closed-loop desirable values for the preceding coefficients as opposed to the open-loop unsatisfactory or marginal values. The design of SASs is beyond the scope of this book. Refer to[7,13,16] on this topic.

The second approach is based on a partial aerodynamic redesign of the aircraft consisting in varying key specific parameters, leading to desirable values of the associated handling qualities. A word of caution should be used while introducing this approach. In fact, while there are several aerodynamic degrees of freedom for enhancing the aircraft dynamic response to meet the specifications in the handling qualities, it should be clear that some key design components (for example, aerodynamic design of the wing) need to be fixed in the early phases of the design and should not be the target of multiple design iterations. Additionally, it should be emphasized that modifications introduced for improving one specific handling quality parameter typically lead to deterioration in other handling quality parameters.

A third approach is actually a hybrid approach combining the two previous approaches: a combination of (partial) aerodynamic redesign for some handling qualities parameters along with the introduction of SAS to improve other handling qualities parameters.

This section focuses on introducing specific tools for the implementation of the second approach. Particularly, a designer is required to have a detailed understanding of how the handling qualities are affected by

- Flight conditions;
- Aircraft geometry;
- Aircraft mass and inertial properties;
- Dimensionless aerodynamic stability and control derivatives.

The formulation of specific relationships to follow the variations of specific handling qualities parameters (for example, $\zeta_{SP}, \zeta_{DR}, T_R, \ldots$) is known in general as stability sensitivity analysis (SSA). A conceptual block diagram to outline the meaning of the SSA is provided in Figure 7.14.

From a mathematical point of view the problem can be set up as a typical gradient of a multivariable function problem. For a completely accurate SSA, the designer is required to evaluate the location of the roots of the entire longitudinal and lateral directional characteristic equations as individual aerodynamic coefficients, inertial characteristics, and geometric data are varied. However, as outlined in a previous section, the short period dynamics can be modeled using the short period approximation, which is known to provide very accurate results. Mathematically, this implies that the reduced order characteristic equation ($\overline{D}_{1_{SP}}(s)$) associated with this approximation has the same roots obtained from the solution of the full longitudinal characteristic equation ($\overline{D}_1(s)$). This greatly simplifies the sensitivity analysis associated with the short period parameters. Unfortunately, the sensitivity analysis for the phugoid parameters requires the solution of the full $\overline{D}_1(s)$.

A somewhat different scenario occurs for the sensitivity analysis of the lateral directional parameters. The rolling approximation previously introduced is known to provide very accurate results using the first order model approximation for the rolling dynamics; however, the key rolling coefficients also greatly affect the dutch roll parameters. Therefore, the sensitivity analysis for the lateral directional parameters $\zeta_{DR}, \omega_{n_{SP}}, T_R, T_S$ requires the solution of the full $\overline{D}_2(s)$.

7.15 Sensitivity Analysis for the Aircraft Dynamics

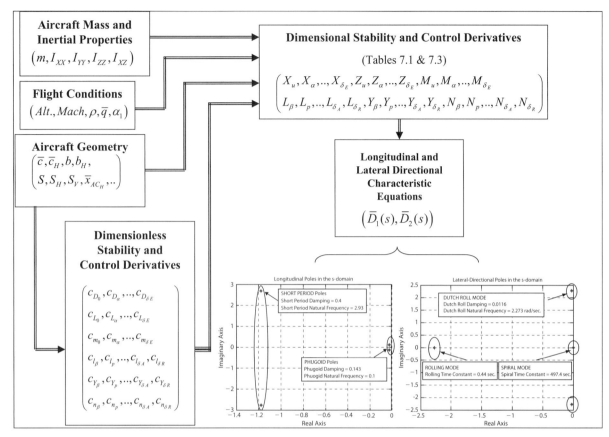

Figure 7.14 Conceptual Block Diagram of the Stability Sensitivity Analysis

A word of caution is necessary about the overall sensitivity analysis process. The availability of advanced codes—such as AAA (Advanced Aircraft Analysis) by DARCorporation[17] and Simgen by Bihrle Applied Research[18]—provide access to extensive and detailed SSA; however, these advanced tools should be used carefully to avoid the risk of losing the aerodynamic and geometric implications of the SSA process.

A typical risk is to perform the SSA with respect to a single aerodynamic coefficient without understanding that a change on a specific stability or control derivative due to a variation of a specific parameter is also likely to imply changes to other stability and control derivatives. Therefore, it is recommended, whenever possible, to obtain approximate closed-form relationships for the dimensionless stability derivatives prior to performing a sensitivity analysis study. The importance of this specific issue is clearly shown in the following sections.

7.15.1 Short Period Sensitivity Analysis

As demonstrated in Section 7.6, the use of the short period approximation has shown to provide essentially the same results achieved through the solution of the full longitudinal characteristic equations. The (realistic) assumptions used in the short period approximation are summarized:

$$Z_{\dot{\alpha}} \ll V_{P_1} \rightarrow Z_{\dot{\alpha}} \approx 0, \quad Z_q \ll V_{P_1} \rightarrow Z_q \approx 0, \quad \sin\theta_1 \approx \theta_1 \approx 0, \quad M_{T_\alpha} \approx 0$$

These assumptions yield the reduced order short period characteristic equation:

$$\overline{D}_{SP}(s) = s^2 - \left(M_{\dot{\alpha}} + M_q + \frac{Z_\alpha}{V_{P_1}}\right)s + \left(\frac{Z_\alpha M_q}{V_{P_1}} - M_\alpha\right) = s^2 + 2\zeta_{SP}\,\omega_{nSP}\,s + \omega_{nSP}^2$$

leading to

$$\omega_{nSP} = \sqrt{\left(\frac{Z_\alpha M_q}{V_{P_1}} - M_\alpha\right)}, \quad \zeta_{SP} = -\frac{\left(M_{\dot\alpha} + M_q + \frac{Z_\alpha}{V_{P_1}}\right)}{2\omega_{nSP}}$$

Thus, the values of the short period handling quality parameters (ω_{nSP}, ζ_{SP}) depend on the values of the following limited set of dimensional stability derivatives:

$$M_{\dot\alpha} = \frac{c_{m\dot\alpha}\,\bar{q}_1\,S\,\bar{c}^2}{2\,I_{YY}\,V_{P_1}}, \quad M_q = \frac{c_{mq}\,\bar{q}_1\,S\,\bar{c}^2}{2\,I_{YY}\,V_{P_1}}, \quad M_\alpha = \frac{c_{m\alpha}\,\bar{q}_1\,S\,\bar{c}}{I_{YY}}, \quad Z_\alpha = \frac{-\bar{q}_1\,S(c_{L\alpha} + c_{D_1})}{m}$$

In addition to specific dimensionless stability derivatives, each of the preceding dimensional stability derivatives is a function of the following sets of parameters:

Set #1: aircraft geometry ($S, \bar{c}$)
Set #2: flight conditions ($\bar{q}_1, V_{P_1}$)
Set #3: inertial characteristics (m, I_{YY})

Therefore, we have

$$\begin{cases} M_{\dot\alpha} = M_{\dot\alpha}(\text{Set \#1, Set \#2, Set \#3}, c_{m_{\dot\alpha}}) \\ M_q = M_q(\text{Set \#1, Set \#2, Set \#3}, c_{mq}) \\ M_\alpha = M_\alpha(\text{Set \#1, Set \#2, Set \#3}, c_{m\alpha}) \\ Z_\alpha = Z_\alpha(\text{Set \#1, Set \#2, Set \#3}, c_{L\alpha}, c_{D_1}) \end{cases}$$

From the modeling process outlined in Chapter 3, the following empirical relationships were derived for the involved dimensionless stability derivatives:

$$c_{m_{\dot\alpha}} \approx -2\,c_{L_{\alpha_H}}\eta_H \frac{S_H}{S}(\bar{x}_{AC_H} - \bar{x}_{CG})^2 \frac{d\varepsilon}{d\alpha}$$

$$c_{mq} \approx -2.2\,c_{L_{\alpha_H}}\eta_H \frac{S_H}{S}(\bar{x}_{AC_H} - \bar{x}_{CG})^2$$

$$c_{L\alpha} = c_{L_{\alpha_W}} + c_{L_{\alpha_H}}\eta_H \frac{S_H}{S}\left(1 - \frac{d\varepsilon}{d\alpha}\right)$$

$$c_{m\alpha} = c_{L\alpha}(\bar{x}_{CG} - \bar{x}_{AC}) = c_{L\alpha}\,\text{SM}$$
$$= c_{L_{\alpha_W}}(\bar{x}_{CG} - \bar{x}_{AC_{WB}}) - c_{L_{\alpha_H}}\eta_H \frac{S_H}{S}\left(1 - \frac{d\varepsilon}{d\alpha}\right)(\bar{x}_{AC_H} - \bar{x}_{CG})$$

Note that the value of c_{D_1} (which is included in Z_α) depends on the longitudinal trimming condition α_1 (since $c_{D_1} = c_{T_{X_1}} \approx c_{D_0} + c_{D_\alpha}\alpha_1$). However, numerically we have the condition $c_{D_1} \ll c_{L\alpha}$.

Given the preceding modeling, the short period handling quality parameters (ω_{nSP}, ζ_{SP}) are functions of a very large set of parameters. Some of these parameters (such as $S, \bar{c}, m$) are considered fixed quantities early in the design; some other parameters (such as $\bar{q}_1, V_{P_1}$) are known to change nominally within the aircraft flight envelope. Thus, the short period sensitivity analysis focuses on the effects associated with the variation of the following parameters:

- S_H/S (horizontal tail/wing surface ratio) affecting $c_{L\alpha}, c_{m\alpha}, c_{mq}, c_{m_{\dot\alpha}}$;
- $\bar{x}_{CG}$ (longitudinal location of center of gravity) affecting $c_{m\alpha}, c_{mq}, c_{m_{\dot\alpha}}$;
- $\bar{x}_{AC_H}$ (longitudinal location of center of gravity) affecting $c_{m\alpha}, c_{mq}, c_{m_{\dot\alpha}}$;
- $c_{L_{\alpha_{WB}}}$ (wing lift-curve slow) affecting $c_{L\alpha}, c_{m\alpha}$ directly and $c_{m_{\dot\alpha}}$ indirectly through the downwash effect;
- I_{YY} (moment of inertia around the Y-axis) affecting directly $M_{\dot\alpha}, M_q, M_\alpha$.

The geometric parameters $S_H/S, \bar{x}_{CG}, \bar{x}_{AC_H}$ are shown in Figure 7.15.

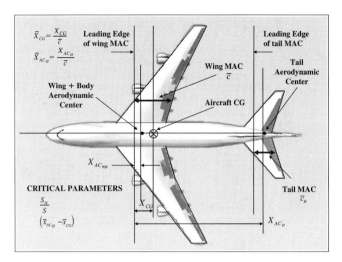

Figure 7.15 Aircraft Geometric Parameters for Longitudinal Sensitivity Analysis

7.15.1.1 Short Period Sensitivity Analysis: Variations of S_H/S

Given the relationships for $c_{L\alpha}, c_{m\alpha}, c_{mq}, c_{m\dot\alpha}$, for a given wing design with known $c_{L\alpha_{WB}}$, for a given (wing + horizontal tail) system with known quantities for the parameters $c_{L\alpha_H}, \eta_H, (\bar{x}_{AC_H} - \bar{x}_{CG}), d\varepsilon/d\alpha$, it is possible to derive closed-form relationships for $c_{L\alpha}, c_{m\alpha}, c_{mq}, c_{m\dot\alpha}$ as a function of the parameter S_H/S. Therefore, the following gradients can be evaluated:

$$\frac{\Delta c_{m\dot\alpha}}{\Delta\left(\frac{S_H}{S}\right)}, \frac{\Delta c_{mq}}{\Delta\left(\frac{S_H}{S}\right)}, \frac{\Delta c_{m\alpha}}{\Delta\left(\frac{S_H}{S}\right)}, \frac{\Delta c_{L\alpha}}{\Delta\left(\frac{S_H}{S}\right)}$$

leading to

$$\Delta c_{m\dot\alpha}\Big|_{\frac{S_H}{S}} = \frac{\Delta c_{m\dot\alpha}}{\Delta\left(\frac{S_H}{S}\right)} \Delta\left(\frac{S_H}{S}\right) \Rightarrow \Delta M_{\dot\alpha}\Big|_{\frac{S_H}{S}}$$

$$\Delta c_{mq}\Big|_{\frac{S_H}{S}} = \frac{\Delta c_{mq}}{\Delta\left(\frac{S_H}{S}\right)} \Delta\left(\frac{S_H}{S}\right) \Rightarrow \Delta M_q\Big|_{\frac{S_H}{S}}$$

$$\Delta c_{m\alpha}\Big|_{\frac{S_H}{S}} = \frac{\Delta c_{m\alpha}}{\Delta\left(\frac{S_H}{S}\right)} \Delta\left(\frac{S_H}{S}\right) \Rightarrow \Delta M_\alpha\Big|_{\frac{S_H}{S}}$$

$$\Delta c_{L\alpha}\Big|_{\frac{S_H}{S}} = \frac{\Delta c_{L\alpha}}{\Delta\left(\frac{S_H}{S}\right)} \Delta\left(\frac{S_H}{S}\right) \Rightarrow \Delta Z_\alpha\Big|_{\frac{S_H}{S}}$$

Using these gradients, the variations of the parameters $(\omega_{nSP}, \zeta_{SP})$ can then be evaluated. A numerical example is shown in Figure 7.16 for the Learjet 24 aircraft (Appendix B, Aircraft 6) at high cruise and maximum weight conditions.

The trends in the figure show that the short period damping coefficient ζ_{SP} is very sensitive to variations in the size of the horizontal tail, as expected given the direct relationship between the tail size and the aerodynamic coefficients $c_{L\alpha}, c_{m\alpha}, c_{mq}, c_{m\dot\alpha}$. On the other side, the short period natural frequency ω_{nSP} is only mildly affected by variation of the tail size since this parameter is mainly only function of M_α and the $\Delta M_\alpha\big|_{\frac{S_H}{S}}$ gradient is fairly small.

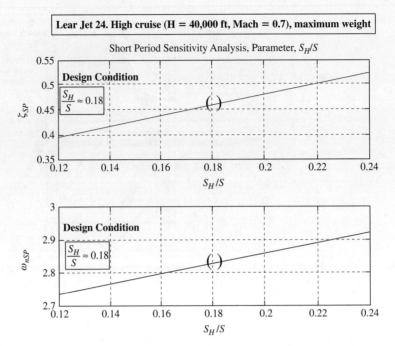

Figure 7.16 Variations of $(\omega_{nSP}, \zeta_{SP})$ Following Variations of S_H/S

7.15.1.2 Short Period Sensitivity Analysis: Variations of $\bar{x}_{CG}$

The longitudinal location of the center of gravity has major implications on the longitudinal dynamic stability. The primary aerodynamic coefficient involved is $c_{m_\alpha} = c_{L_\alpha}(\bar{x}_{CG} - \bar{x}_{AC})$, with direct implications on M_α and ω_{nSP}. A decrease in $\bar{x}_{CG}$ (associated with a shift of the center of gravity toward the nose) leads to higher static margin, higher c_{m_α}, higher M_α, and, consequently, higher ω_{nSP}. The opposite trend occurs for a center of gravity shift toward the tail. As shown in these relationships, $\bar{x}_{CG}$ also affects the values of the aerodynamic coefficients c_{m_q}, $c_{m_{\dot{\alpha}}}$. A forward shift of the center of gravity leads to higher values of c_{m_q}, $c_{m_{\dot{\alpha}}}$ and, consequently, to higher values for $M_{\dot{\alpha}}$, M_q and higher values for the numerator of ζ_{SP} as shown:

$$\zeta_{SP} = -\frac{\left(M_{\dot{\alpha}} + M_q + \dfrac{Z_\alpha}{V_{P_1}}\right)}{2\omega_{nSP}}$$

However, the increase in the numerator is offset by a more significant increase in the denominator, leading to lower values of ζ_{SP} with forward center of gravity locations. This trend can be quantified through the calculations of the following gradients:

$$\frac{\Delta c_{m_{\dot{\alpha}}}}{\Delta \bar{x}_{CG}}, \frac{\Delta c_{m_q}}{\Delta \bar{x}_{CG}}, \frac{\Delta c_{m_\alpha}}{\Delta \bar{x}_{CG}}$$

leading to

$$\Delta c_{m_{\dot{\alpha}}}\big|_{\bar{x}_{CG}} = \frac{\Delta c_{m_{\dot{\alpha}}}}{\Delta \bar{x}_{CG}} \Delta \bar{x}_{CG} \Rightarrow \Delta M_{\dot{\alpha}}\big|_{\bar{x}_{CG}}$$

$$\Delta c_{m_q}\big|_{\bar{x}_{CG}} = \frac{\Delta c_{m_q}}{\Delta \bar{x}_{CG}} \Delta \bar{x}_{CG} \Rightarrow \Delta M_q\big|_{\bar{x}_{CG}}$$

$$\Delta c_{m_\alpha}\big|_{\bar{x}_{CG}} = \frac{\Delta c_{m_\alpha}}{\Delta \bar{x}_{CG}} \Delta \bar{x}_{CG} \Rightarrow \Delta M_\alpha\big|_{\bar{x}_{CG}}$$

Using these gradients, the variations of the parameters $(\omega_{nSP}, \zeta_{SP})$ can then be evaluated. A numerical example is shown in Figure 7.17 for the Learjet 24 aircraft at high cruise and maximum weight conditions.

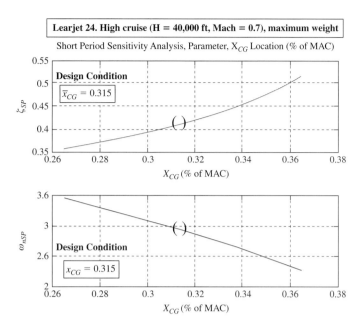

Figure 7.17 Variations of $(\omega_{nSP}, \zeta_{SP})$ Following Variations of $\bar{x}_{CG}$

As expected, the trends shown confirm that the short period natural frequency ω_{nSP} is very sensitive to shifts in the longitudinal location of the center of gravity. In general, it can be said that the gradient $\Delta M_\alpha|_{\bar{x}_{CG}}$ is more significant than the gradients $\Delta M_q|_{\bar{x}_{CG}}, \Delta M_{\dot\alpha}|_{\bar{x}_{CG}}$. Note that the analysis assumes a center of gravity in front of the aircraft aerodynamic center for stability purposes. The analysis loses significance for center of gravity locations approaching or exceeding the aerodynamic center because the overall dynamic stability will be lost.

7.15.1.3 Short Period Sensitivity Analysis: Variations of $\bar{x}_{AC_H}$

One of most effective ways of modifying the short period handling qualities is to change the location of the horizontal tail, which changes the moment arm of the aerodynamic force generated at the tail and, ultimately, affects the pitching moment contributions. From an analysis of the relationships for $c_{L_\alpha}, c_{m_\alpha}, c_{m_q}, c_{m_{\dot\alpha}}$, it is clear that variations in the values of $\bar{x}_{AC_H}$ directly affect c_{m_α}, c_{m_q}, indirectly affect c_{m_α} (through variations of $\bar{x}_{AC}$), and c_{L_α} (through variations of the downwash effect $d\varepsilon/d\alpha$), as described in Chapter 3. It is intuitive that an increase in the value of $\bar{x}_{AC_H}$ leads to substantial increase in the value of ζ_{SP}, while ω_{nSP} is affected to a much lower extent.

From a mathematical point of view it is easier to express this dependency in terms of the distance $(\bar{x}_{AC_H} - \bar{x}_{CG})$ rather than the location $\bar{x}_{AC_H}$. Therefore, the functionalities can be quantified through the calculation of the following gradients:

$$\frac{\Delta c_{m_{\dot\alpha}}}{\Delta(\bar{x}_{AC_H} - \bar{x}_{CG})}, \frac{\Delta c_{m_q}}{\Delta(\bar{x}_{AC_H} - \bar{x}_{CG})}, \frac{\Delta c_{m_\alpha}}{\Delta(\bar{x}_{AC_H} - \bar{x}_{CG})}, \frac{\Delta c_{L_\alpha}}{\Delta(\bar{x}_{AC_H} - \bar{x}_{CG})}$$

leading to

$$\Delta c_{m_{\dot\alpha}}|_{(\bar{x}_{AC_H} - \bar{x}_{CG})} = \frac{\Delta c_{m_{\dot\alpha}}}{\Delta(\bar{x}_{AC_H} - \bar{x}_{CG})} \Delta(\bar{x}_{AC_H} - \bar{x}_{CG}) \Rightarrow \Delta M_{\dot\alpha}|_{(\bar{x}_{AC_H} - \bar{x}_{CG})}$$

$$\Delta c_{m_q}|_{(\bar{x}_{AC_H} - \bar{x}_{CG})} = \frac{\Delta c_{m_q}}{\Delta(\bar{x}_{AC_H} - \bar{x}_{CG})} \Delta(\bar{x}_{AC_H} - \bar{x}_{CG}) \Rightarrow \Delta M_q|_{(\bar{x}_{AC_H} - \bar{x}_{CG})}$$

$$\Delta c_{m_\alpha}|_{(\bar{x}_{AC_H} - \bar{x}_{CG})} = \frac{\Delta c_{m_\alpha}}{\Delta(\bar{x}_{AC_H} - \bar{x}_{CG})} \Delta(\bar{x}_{AC_H} - \bar{x}_{CG}) \Rightarrow \Delta M_\alpha|_{(\bar{x}_{AC_H} - \bar{x}_{CG})}$$

$$\Delta c_{L_\alpha}|_{(\bar{x}_{AC_H} - \bar{x}_{CG})} = \frac{\Delta c_{L_\alpha}}{\Delta(\bar{x}_{AC_H} - \bar{x}_{CG})} \Delta(\bar{x}_{AC_H} - \bar{x}_{CG}) \Rightarrow \Delta Z_\alpha|_{(\bar{x}_{AC_H} - \bar{x}_{CG})}$$

Using these gradients, the variations of the parameters (ω_{nSP}, ζ_{SP}) can then be evaluated. A numerical example is shown in Figure 7.18 for the Learjet 24 aircraft at high cruise and maximum weight conditions.

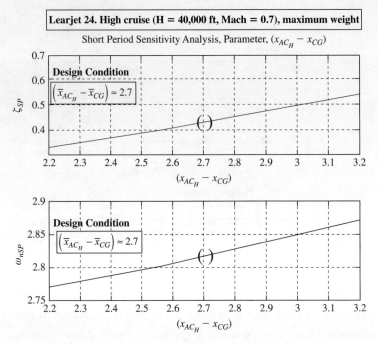

Figure 7.18 Variations of (ω_{nSP}, ζ_{SP}) Following Variations of ($x_{AC_H} - \bar{x}_{CG}$)

The trends shown confirm a very high sensitivity of (ζ_{SP}) with variations in the location of the horizontal tail while (ω_{nSP}) is only marginally affected.

7.15.1.4 Short Period Sensitivity Analysis. Variations of $c_{L_{\alpha W}}$

Although the wing lift curve slope is a key player in the overall longitudinal aerodynamic, variations in the wing lift curve slope do not significantly affect the short period parameters (ω_{nSP}, ζ_{SP}). This is clear following a brief analysis of the preceding relationships for $c_{L_\alpha}, c_{m_\alpha}, c_{m_q}, c_{m_{\dot\alpha}}$. Variations in the value of $c_{L_{\alpha W}}$—associated with a change in the wing design—directly affect c_{L_α} (through $c_{L_{\alpha W}}$ itself and through changes in the downwash effect $d\varepsilon/d\alpha$, as described in Chapter 2), and c_{m_α}. A mild variation is also expected for $c_{m_{\dot\alpha}}$ through the downwash effect.

The weak functionalities can be quantified through the calculation of the following gradients:

$$\frac{\Delta c_{m_{\dot\alpha}}}{\Delta c_{L_{\alpha W}}}, \frac{\Delta c_{m_q}}{\Delta c_{L_{\alpha W}}}, \frac{\Delta c_{m_\alpha}}{\Delta c_{L_{\alpha W}}}, \frac{\Delta c_{L_\alpha}}{\Delta c_{L_{\alpha W}}}$$

leading to

$$\Delta c_{m_{\dot\alpha}}\big|_{c_{L_{\alpha W}}} = \frac{\Delta c_{m_{\dot\alpha}}}{\Delta c_{L_{\alpha W}}} \Delta c_{L_{\alpha W}} \Rightarrow \Delta M_{\dot\alpha}\big|_{c_{L_{\alpha W}}}$$

$$\Delta c_{m_q}\big|_{c_{L_{\alpha W}}} = \frac{\Delta c_{m_q}}{\Delta c_{L_{\alpha W}}} \Delta c_{L_{\alpha W}} \Rightarrow \Delta M_q\big|_{c_{L_{\alpha W}}}$$

$$\Delta c_{m_\alpha}\big|_{c_{L_{\alpha W}}} = \frac{\Delta c_{m_\alpha}}{\Delta c_{L_{\alpha W}}} \Delta c_{L_{\alpha W}} \Rightarrow \Delta M_\alpha\big|_{c_{L_{\alpha W}}}$$

$$\Delta c_{L_\alpha}\big|_{c_{L_{\alpha W}}} = \frac{\Delta c_{L_\alpha}}{\Delta c_{L_{\alpha W}}} \Delta c_{L_{\alpha W}} \Rightarrow \Delta Z_\alpha\big|_{c_{L_{\alpha W}}}$$

Using these gradients, the variations of the parameters (ω_{nSP}, ζ_{SP}) can then be evaluated. Once again, a numerical example is shown in Figure 7.19 below for the Learjet 24 aircraft at high cruise and maximum weight conditions.

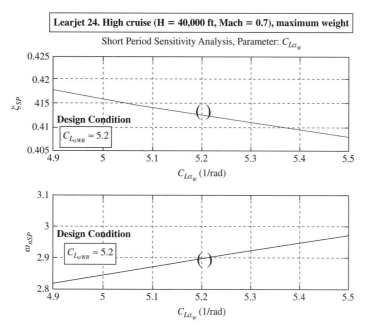

Figure 7.19 Variations of $(\omega_{nSP}, \zeta_{SP})$ Following Variations of $c_{L_{\alpha W}}$.

The plots shown confirm the limited correlation between $c_{L_{\alpha W}}$ and the short period handling qualities.

7.15.1.5 Short Period Sensitivity Analysis. Variations of I_{YY}

Design variations in the moment of inertia I_{YY} do not affect the values of the aerodynamic coefficients $c_{L_\alpha}, c_{m_\alpha}, c_{m_q}, c_{m_{\dot\alpha}}$ but directly affect the values of the dimensional stability derivatives $M_{\dot\alpha}, M_q, M_\alpha$. Particularly, variations in I_{YY} directly affect M_α, which is primarily related to the parameter ω_{nSP}. The relationship between I_{YY} and ζ_{SP} is instead weaker, since the numerators in $M_{\dot\alpha}, M_q$ have substantially larger numerical values.

A numerical example is shown in Figure 7.20 for the Learjet 24 aircraft at high cruise and maximum weight conditions.

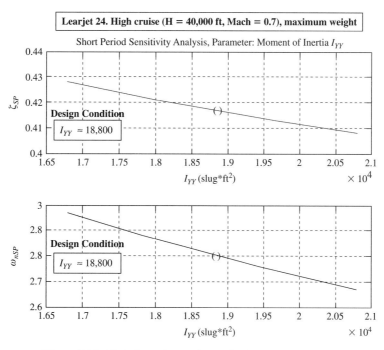

Figure 7.20 Variations of ω_{nSP}, ζ_{SP} Following Variations of I_{YY}

The plots clearly show the direct relationship between changes in the longitudinal moment of inertia and the short period natural frequency.

Note that nominal in-flight variations of I_{YY} are associated with (substantial) amounts of fuel consumption typically associated with long oversea flights, which, on the other side, do not imply substantial shifts in the location of the center of gravity. Other non-nominal and undesirable variations are associated with fuel sloshes or sudden shifts of cargo items within the aircraft cargo bay, which are also associated with undesirable shifts in the location of the center of gravity.

7.15.2 Phugoid Sensitivity Analysis

The sensitivity analysis for the phugoid dynamic characteristics is conceptually similar yet fairly different with respect to the previous short period analysis. In general, when comparing the functionalities of the two longitudinal modes, the phugoid parameters are less dependent on the aircraft geometry and more dependent on a few key aerodynamic coefficients.

A well-known dynamic problem associated with the phugoid is the tuck, discussed in Section 7.3 and Chapter 6, Section 6.3. From an aerodynamic point of view this problem is associated with a sudden shift of the aircraft aerodynamic center associated with a small variation in Mach during the transonic crisis. The involved aerodynamic derivative is the pitching moment coefficient c_{m_u}. From a stability point of view this issue implies the degeneration of the phugoid mode from a second order system to a pair of first order systems, leading eventually to one unstable first order mode as the sign for c_{m_u} reverses from positive to negative. The numerical value of this derivative affects the C_1, D_1, E_1 coefficients of the longitudinal characteristic equation $\overline{D}_1(s)$.

A numerical example is shown in Figure 7.21 for the Learjet 24 aircraft at high cruise and maximum weight conditions.

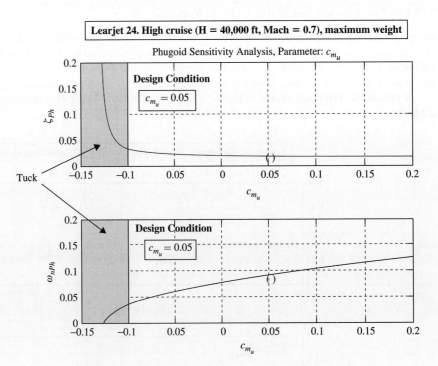

Figure 7.21 Variations of ω_{nPh}, ζ_{Ph} Following Variations of c_{m_u}

The occurrence of the tuck essentially voids the meaning of the $(\omega_{nPh}, \zeta_{Ph})$ for $c_{m_u} < -0.1$. Figure 7.22 provides a zoomed version of the previous plot.

7.15 Sensitivity Analysis for the Aircraft Dynamics

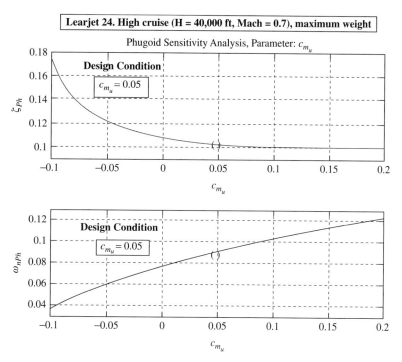

Figure 7.22 Variations of ω_{nPh}, ζ_{Ph} Following Variations of c_{m_u} (zoomed)

The plots show that an increase in the value of c_{m_u} (from negative values to positive values) leads to a substantial decrease of ζ_{Ph}, which levels off for $c_{m_u} \geq 0.05$ while ω_{nPh} steadily increases as c_{m_u} increases.

The degeneration of the phugoid mode with the transformation of the associated second order system to two first order systems (one of which eventually becoming unstable) is shown in Figure 7.23, displaying the roots of the characteristic equation $\overline{D}_1(s)$ as c_{m_u} decreases from 0.2 to -0.13.

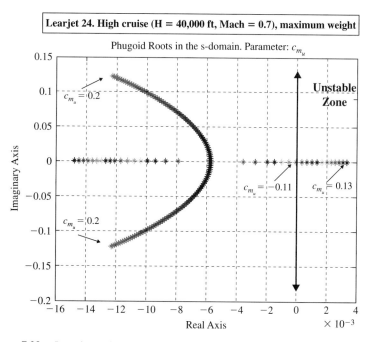

Figure 7.23 Locations of the Phugoid Roots on the S-domain for $c_{m_u} \in [-0.13, 0.2]$

On the other side, the wing lift curve and the aircraft drag curve slope aerodynamic coefficients ($c_{L_{\alpha W}}$, c_{D_α}) (which are two other important longitudinal aerodynamic derivatives) have virtually no effect on the values of (ω_{nPh}, ζ_{Ph}), as shown in Figures 7.24 and 7.25. Note that the variations in the downwash effect associated with the change in the value of $c_{L_{\alpha W}}$ need to be properly considered as well as the variations of c_{D_1} associated with the variations of c_{D_α}.

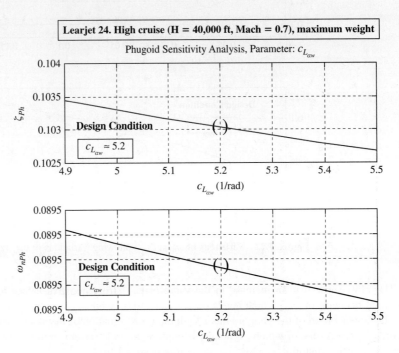

Figure 7.24 Variations of (ω_{nPh}, ζ_{Ph}) Following Variations of $c_{L_{\alpha W}}$

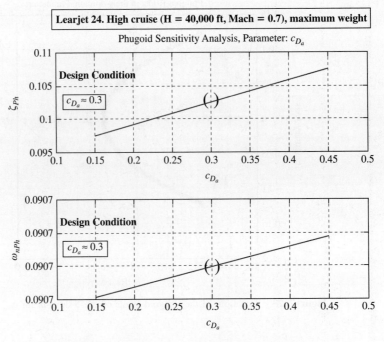

Figure 7.25 Variations of (ω_{nPh}, ζ_{Ph}) Following Variations of c_{D_α}

An interesting discussion is generated by the phugoid sensitivity analysis associated with variations in the total lift coefficient c_{L_1}. In fact, it is virtually impossible and aerodynamically unrealistic to conceive that c_{L_1} can be individually modified. A somewhat more realistic analysis is given by the variations of c_{L_1} (through, for example, variations of the trimming conditions) while holding constant the ratio c_{L_1}/c_{D_1}. The numerical results associated with this analysis are shown in Figure 7.26 considering a constant ratio of $c_{L_1}/c_{D_1} = 12$. Note that the ratio c_{L_1}/c_{D_1} substantially decreases during the aircraft approach to landing.

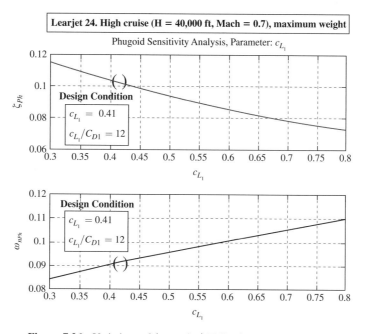

Figure 7.26 Variations of $(\omega_{nPh}, \zeta_{Ph})$ Following Variations of c_{L_1}

The phugoid sensitivity analysis is concluded with the analysis of the effect of the variations of the center of gravity location and the pitching moment of inertia $(\bar{x}_{CG}, I_{YY})$. These two parameters have shown a strong relationship with the short period handling qualities coefficients $(\omega_{nSP}, \zeta_{SP})$; on the contrary, $(\bar{x}_{CG}, I_{YY})$ have very little influence on the numerical values of $(\omega_{nPh}, \zeta_{Ph})$, as shown in Figures 7.27 and 7.28.

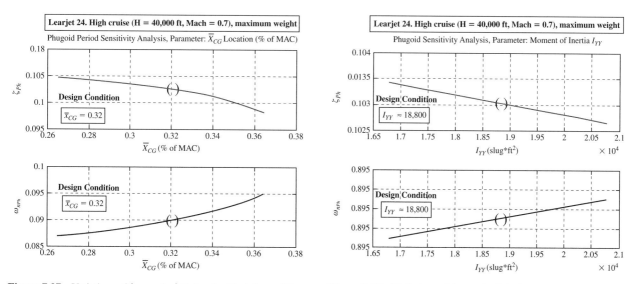

Figure 7.27 Variations of $(\omega_{nPh}, \zeta_{Ph})$ Following Variations of $\bar{x}_{CG}$

Figure 7.28 Variations of $(\omega_{nPh}, \zeta_{Ph})$ Following Variations of I_{YY}

7.15.3 Sensitivity Analysis for the Lateral Directional Parameters

The sensitivity analysis for the lateral directional dynamic modes (dutch roll, rolling, and spiral) is conceptually similar to the previous sensitivity analysis for the short period and phugoid longitudinal modes but practically very different. This is due to two fundamental reasons.

The first reason is that for the longitudinal analysis there is a substantial aerodynamic separation between the short period and phugoid modes; in other words, these two modes are associated with different aerodynamic coefficients. In terms of the location of the roots of the longitudinal characteristic equation, this separation is due to the fact that the phugoid second order system is dominant with respect to the short period second order system. For the lateral directional analysis, there is a strong aerodynamic correlation between the dutch roll, rolling, and spiral modes (with the strong independency of the rolling mode). A few critical coefficients, such as $c_{l_\beta}, c_{n_\beta}, c_{l_p}, c_{n_r}$, affect all these modes.

The second reason is related to modeling issues. For the longitudinal sensitivity analysis, it is possible to use a few critical relationships of the key aerodynamic coefficients associated with the short period approximation as functions of the key aircraft geometric parameters $(S_H/S, \bar{x}_{CG}, \bar{x}_{AC_H})$. Unfortunately, it is impossible to derive a similar set of relationships for the key lateral directional aerodynamic coefficients as functions of the key lateral directional geometric parameters. This makes a closed-form approach to the lateral directional sensitivity analysis virtually impossible.

The key lateral directional geometric parameters $(S_V/S, X_{VS}/b, Z_{VS}/b)$ are shown in Figure 7.29.

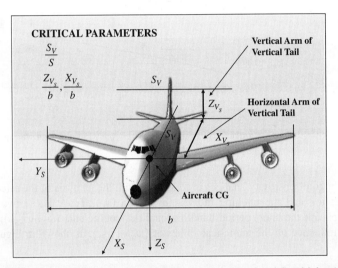

Figure 7.29 Aircraft Geometric Parameters for Lateral Directional Sensitivity Analysis

As a consequence of the issues discussed, the lateral directional sensitivity analysis can only be set up as the analysis of the variation of the lateral directional dynamic parameters $(\zeta_{DR}, \omega_{n_{DR}}, T_R, T_S)$ following variations of the values for individual aerodynamic coefficients and the moments of inertia (I_{xx}, I_{zz}). A word of caution is necessary for this type of analysis. Although mathematically it is possible to vary one coefficient at the time—for example, c_{l_β}—and evaluate its effects on $(\zeta_{DR}, \omega_{n_{DR}}, T_R, T_S)$, it should be clear that whatever parameter leads to variations in the value of c_{l_β} will also induce non-negligible variations in the values of other critical coefficients, for example, $c_{n_\beta}, c_{l_p}, c_{n_r}$ and others. Therefore, great care should be exercised in using an approach based on the variation of a single specific aerodynamic coefficient.

Given the previously described boundary conditions, the key lateral directional handling quality parameters $(\zeta_{DR}, \omega_{n_{DR}}, T_R, T_S)$ are functions of the following aerodynamic coefficients [which are related to the geometric coefficients $(S_V/S, X_{VS}/b, Z_{VS}/b)$] and inertial parameters:

$$c_{l_\beta}, c_{n_\beta}, c_{l_p}, c_{n_r}, c_{y_\beta}$$
$$I_{xx}, I_{zz}$$

Using the aerodynamic data of the same aircraft in Section 7.15 (at Mach = 0.7, Altitude = 40,000 ft., W = 13,000 lbs), Figure 7.30 shows the variations of $(\zeta_{DR}, \omega_{n_{DR}}, T_R, T_S)$ for numerical variations of c_{l_β}, the aircraft dihedral effect modeled in Chapter 4.

7.15 Sensitivity Analysis for the Aircraft Dynamics 399

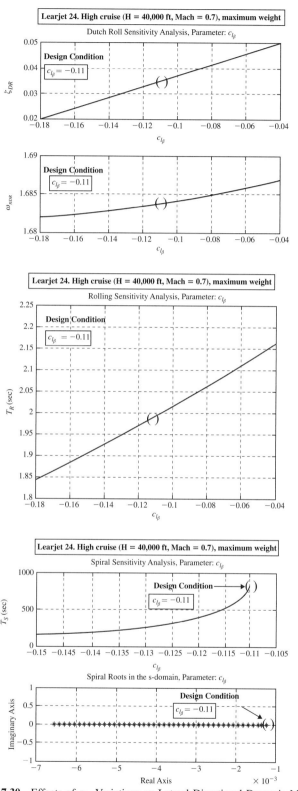

Figure 7.30 Effects of c_{l_β} Variations on Lateral Directional Dynamic Modes

Variations of c_{l_β} greatly affect the values of the dutch roll damping ζ_{DR} and the spiral time constant T_S while having very little effect on the dutch roll natural frequency $\omega_{n_{DR}}$ and the rolling time constant T_R. In particular, a negative increase in the value of c_{l_β}—that is, an increase in the aircraft dihedral effect—leads to a desirable improvement in the spiral performance—that is, a decrease in the value of T_S—at the expense of an undesirable decrease of the important dutch roll damping coefficient ζ_{DR}.

An equally critical role is played by the important aerodynamic coefficient c_{n_β}, whose effect is shown in Figure 7.31. Variations in the numerical value of c_{n_β} affect both dutch roll damping coefficient ζ_{DR} and natural frequency $\omega_{n_{DR}}$. Additionally, c_{n_β} has a marginal influence on the rolling time constant T_R and a critical influence on the spiral performance, to the level that large values of c_{n_β} can lead to an unstable spiral mode. Typically, the final design

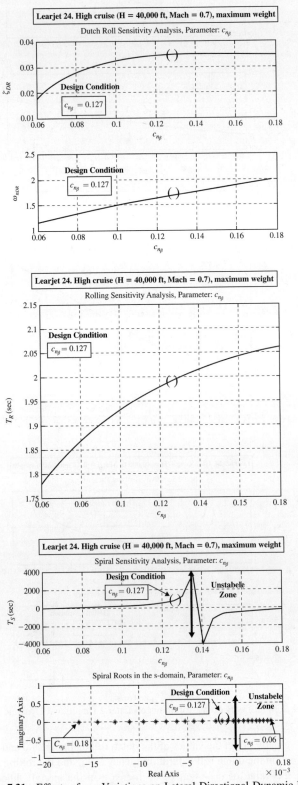

Figure 7.31 Effects of c_{n_β} Variations on Lateral Directional Dynamic Modes

results in a trade-off configuration where c_{n_β} is high enough for an acceptable value of the dutch roll damping coefficient ζ_{DR} and low enough for ensuring at least a stable spiral mode (although with large values of the time constant T_S). However, as mentioned previously, the design for an acceptable c_{n_β} value cannot be decoupled from the design for the dihedral effect c_{l_β} because these two important aerodynamic coefficients have opposite effects on the dutch roll and the spiral. An interesting trend is revealed in Figure 7.32, which shows the migration of the imaginary roots associated with the dutch roll as c_{n_β} is varied from higher to lower values. The decrease in the damping coefficient ζ_{DR} is clearly

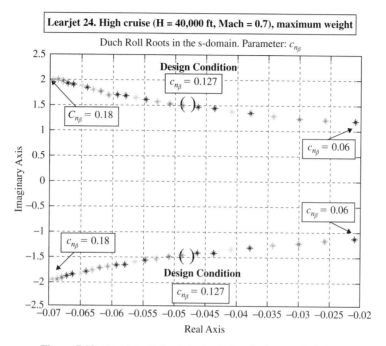

Figure 7.32 Dutch Roll Roots in the S-domain for c_{n_β} Variations

visible, leading to an unstable dutch roll as the values of c_{n_β} approach zero and eventually become negative. The analysis of the location of the dutch roll roots becomes an important tool for the flight control systems engineer designing control laws for closed-loop stability augmentation systems[7,13,16].

As shown in Figure 7.33, the aerodynamic coefficient c_{n_r} plays a role very similar to the role played by c_{n_β}. Recall that both coefficients are very closely related to the size of the vertical tail (S_V/S) and its horizontal distance with respect to the aircraft center of gravity (X_{VS}/b). Particularly, Figure 7.33 shows that variations of c_{n_r} strongly affect the dutch roll damping coefficient ζ_{DR} and have virtually no influence on the natural frequency $\omega_{n_{DR}}$; however, just like c_{n_β}, c_{n_r} is strongly related to the value of the spiral time constant T_S. Just as for c_{n_β}, the final design results in a configuration in which c_{n_r} is low enough (in terms of absolute value) for an acceptable value of the dutch roll damping coefficient ζ_{DR} and high enough for ensuring at least a stable spiral mode (although with large values of the time constant T_S).

A very specific role is played by the aerodynamic coefficient c_{l_p}. As shown in Section 7.13, this parameter is the only aerodynamic coefficient affecting the rolling time constant T_R. In fact, the 1 degree-of-freedom approximation for the modeling of the rolling mode provided the relationship

$$T_R = -\frac{1}{L_p} = -\frac{1}{\left(\dfrac{c_{l_p}\bar{q}\,S\,b}{I_{xx}}\dfrac{b}{2V_{p_1}}\right)} = -\frac{2\,I_{xx}\,V_{p_1}}{c_{l_p}\bar{q}\,S\,b^2}$$

As shown in Figure 7.34, variations of c_{l_p} have virtually no effect on the dutch roll parameters; however, they do substantially affect the spiral time constant T_S. A typical design configuration always represents a trade-off between desirable rolling performance, in terms of the low values of the time constant T_R, and a stable spiral mode.

Finally, Figure 7.35 shows the effect of the aerodynamic coefficient c_{y_β} on the lateral directional dynamic modes. Among all the considered aerodynamic coefficients c_{y_β} is the one providing the least amount of influence on the numerical values for $(\zeta_{DR}, \omega_{n_{DR}}, T_R, T_S)$. Particularly, Figure 3.35 shows a mild functionality on the dutch roll damping coefficient ζ_{DR} while having virtually no effect on the others. The aerodynamic origin of this relationship is because the value of c_{y_β} is a strong function of the geometric parameter (S_V/S), which has a major influence on the value of the dutch roll damping coefficient ζ_{DR}, as already shown in the sensitivity analysis for c_{n_β} and c_{n_r}.

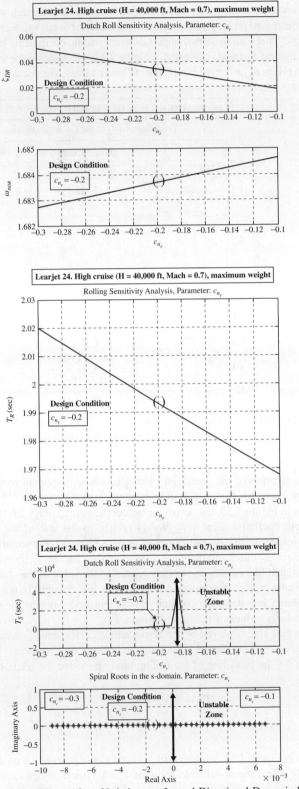

Figure 7.33 Effects of c_{n_r} Variations on Lateral Directional Dynamic Modes

7.15 Sensitivity Analysis for the Aircraft Dynamics 403

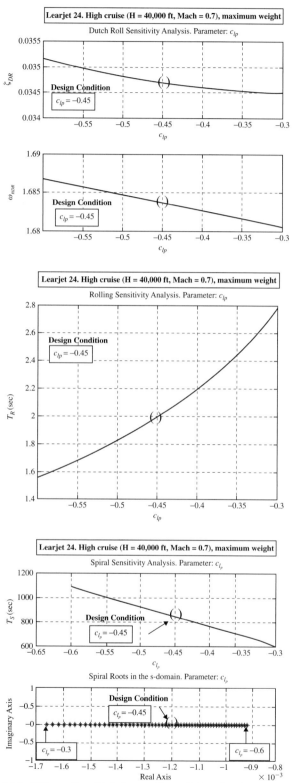

Figure 7.34 Effects of c_{l_p} Variations on Lateral Directional Dynamic Modes

404 Chapter 7 Solution of the Aircraft Equations of Motion Based on Laplace Transformations and Transfer Functions

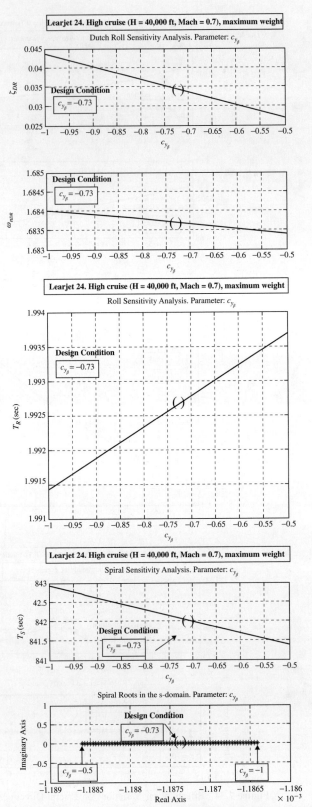

Figure 7.35 Effects of c_{y_β} Variations on Lateral Directional Dynamic Modes

Finally, Figures 7.36 and 7.37 show the functionalities between the lateral directional moments of inertia I_{xx}, I_{zz} and the lateral directional dynamic modes. As expected, the values of I_{xx} are strongly related to the values of the rolling time constant T_R and only moderately affect the dutch roll parameters (mainly ζ_{DR}). Conversely, the values of I_{zz} mainly affect the dutch roll parameters ($\omega_{n_{DR}}$, ζ_{DR}) and have virtually no effect on the rolling and spiral time constants (T_R, T_S).

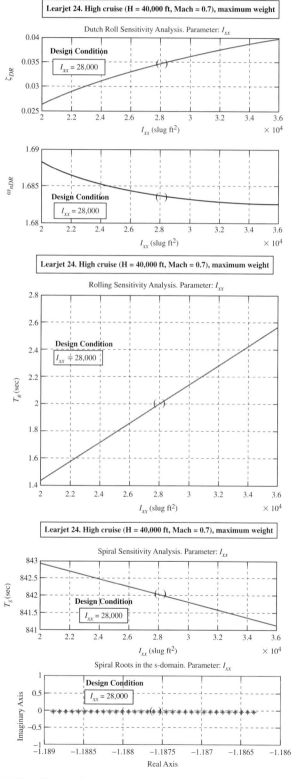

Figure 7.36 Effects of I_{XX} Variations on Lateral Directional Dynamic Modes

406 Chapter 7 Solution of the Aircraft Equations of Motion Based on Laplace Transformations and Transfer Functions

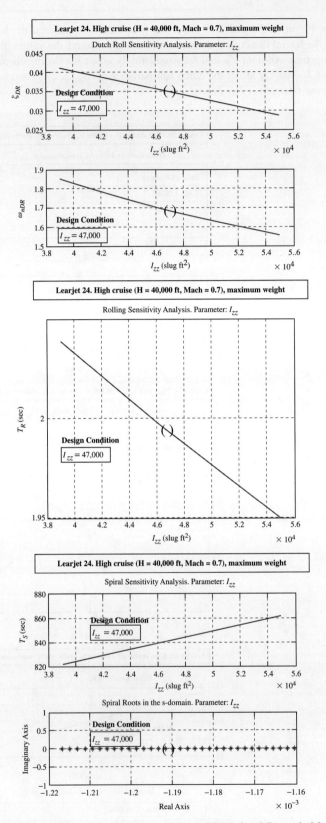

Figure 7.37 Effects of I_{ZZ} Variations on Lateral Directional Dynamic Modes

7.16 SUMMARY

This chapter has provided the students with a detailed description of the approach leading to the solution of the aircraft equations of motion along with an analysis of the aircraft dynamic characteristics. Specifically, the approach originated from the application of the Laplace transformation to the solution of the aircraft equations of motion leading to the transfer function concept. The key dynamic characteristics of the aircraft longitudinal and lateral directional dynamics have been discussed with a clear description of the short period, phugoid, dutch roll, rolling, and spiral dynamic modes. Finally, a detailed description of the stability sensitivity analysis- an important phase of the aircraft design- has been provided.

REFERENCES

1. Lanchester, F. W. *Aerodonetics*. Constable and Co. Ltd, London, 1908.
2. Bryan, G. H. *Stability in Aviation*. Macmillan and Co., London, 1911.
3. Roskam, J. *Airplane Flight Dynamics and Automatic Flight Controls – Part I*. Design, Analysis, and Research Corporation, Lawrence, KS, 1995.
4. Cook, M. *Flight Dynamics Principles* (2nd ed.). Elsevier Aerospace Engineering Series, 2007.
5. Calcara, M. *Elementi di Dinamica del Velivolo*, Collana di Aeronautica, CUEN, Napoli, 1981.
6. Nise, N. *Control Systems Engineering* (5th ed.). John Wiley & Sons, 2007.
7. Stevens, B. L. and Lewis, F. L. *"Aircraft Control and Simulation."* John Wiley & Sons, 2003.
8. Schmidt, L.V. *Introduction to Aircraft Flight Dynamics*. AIAA Education Series, 1998.
9. Stengel, R. *Flight Dynamics*, Princeton University Press, 2004.
10. Hull, D. G. *Fundamentals of Airplane Flight Mechanics*, Springer, 2007.
11. Boiffier, J. L. *The Dynamics of Flight. The Equations*, John Wiley & Sons, 1998.
12. Yechout, T. R., Morris, S. L., Bossert, D. E., Hallgren, W. F. *Introduction to Aircraft Flight Mechanics*. AIAA (Education Series), 2003.
13. Pamadi, B. N. *Performance, Stability, Dynamics, and Control of Airplanes* (2nd ed.). AIAA (Education Series), 2004.
14. Philipps, W. F. *Mechanics of Flight*. John Wiley & Sons, 2004.
15. Ving, N. X. *Flight Mechanics of High-Performance Aircraft*. Cambridge, Aerospace Series, 1993.
16. Roskam, J. *Airplane Flight Dynamics and Automatic Flight Controls—Part II*. Design, Analysis, and Research Corporation, Lawrence, KS, 1995.
17. Advanced Aircraft Analysis, developed by Design, Analysis, and Research Corporation (Web site: www.darcorp.com), Lawrence, KS.
18. SimGen, developed by Bihrle Applied Research Inc. (Web site: www.bihrle.com), Hampton, VA.

STUDENT SAMPLE PROBLEMS

Student Sample Problem 7.1

Consider the McDonnell Douglas F-4 aircraft at subsonic cruise flight conditions (Appendix B, Aircraft 10). Using the provided aircraft data, perform the analysis of the aircraft longitudinal dynamic stability at the given flight condition using the Routh–Hurwitz stability criteria.

Solution of Student Sample Problem 7.1

The general expression for the longitudinal characteristic equation is given by

$$\overline{D}_1(s) = A_1 s^4 + B_1 s^3 + C_1 s^2 + D_1 s + E_1$$

with the coefficients given by

$$A_1 = \left(V_{P_1} - Z_{\dot{\alpha}}\right)$$

$$B_1 = -(V_{P_1} - Z_{\dot{\alpha}})(X_u + X_{T_u} + M_q) - Z_\alpha - M_{\dot{\alpha}}(Z_q + V_{P_1})$$

$$C_1 = (X_u + X_{T_u})[M_q(V_{P_1} - Z_{\dot{\alpha}}) + Z_\alpha + M_{\dot{\alpha}}(V_{P_1} + Z_q)]$$
$$+ M_q Z_\alpha - Z_u X_\alpha + M_{\dot{\alpha}} g \sin\Theta_1 - (M_\alpha + M_{T_\alpha})(V_{P_1} + Z_q)$$

$$D_1 = g \sin\Theta_1[(M_\alpha + M_{T_\alpha}) - M_{\dot{\alpha}}(X_u + X_{T_u})] + g \cos\Theta_1[M_{\dot{\alpha}} Z_u + (M_u + M_{T_u})(V_{P_1} - Z_{\dot{\alpha}})]$$
$$- X_\alpha(M_u + M_{T_u})(Z_q + V_{P_1}) + Z_u X_\alpha M_q + (X_u + X_{T_u})[(M_\alpha + M_{T_\alpha})(Z_q + V_{P_1}) - M_q Z_\alpha]$$

$$E_1 = g \cos\Theta_1[Z_u(M_\alpha + M_{T_\alpha}) - Z_\alpha(M_u + M_{T_u})] + g \sin\Theta_1[(M_u + M_{T_u})X_\alpha - (X_u + X_{T_u})(M_\alpha + M_{T_\alpha})]$$

where the expressions for the longitudinal dimensional stability derivatives are given in Table 7.1.

For the given aircraft at the given flight conditions, the numerical expression for the longitudinal characteristic equation is given by

$$\overline{D}_1(s) = A_1 s^4 + B_1 s^3 + C_1 s^2 + D_1 s + E_1 = 877 s^4 + 1109 s^3 + 7097 s^2 - 4.966 s - 11.42$$

The Routh–Hurwitz array associated with the above CE is given by

$$\begin{array}{c|ccc} s^4 & 877 & 7097 & -11.42 \\ s^3 & 1109 & -4.966 & - \\ s^2 & k_{11} & k_{12} & - \\ s & k_{21} & - & - \\ s^0 & -11.42 & - & - \end{array}$$

where

$$k_{11} = \frac{B_1 C_1 - A_1 D_1}{B_1} = 7100.93,$$

$$k_{12} = \frac{B_1 E_1}{B_1} = E_1 = -11.42,$$

$$k_{21} = \frac{D_1 k_{11} - B_1 k_{12}}{k_{11}} = \frac{D_1\left(\dfrac{B_1 C_1 - A_1 D_1}{B_1}\right) - B_1 E_1}{\dfrac{B_1 C_1 - A_1 D_1}{B_1}}$$

$$= \frac{D_1(B_1 C_1 - A_1 D_1) - B_1^2 E_1}{B_1 C_1 - A_1 D_1} = -3.18$$

The analysis of the first column of the Routh–Hurwitz array provides

$$\begin{array}{c|ccc} s^4 & 877 & 7097 & -11.42 \\ s^3 & 1109 & -4.966 & - \\ s^2 & 7100.93 & k_{12} & - \\ s & -3.18 & - & - \\ s^0 & -11.42 & - & - \end{array}$$

According to the Routh–Hurwitz stability criteria, there is *one* sign change in the first column of the Routh–Hurwitz array, which implies *one* unstable root associated with the characteristic equation. Using MATLAB®, the roots of $\overline{D}_1(s)$ are given by

$$-0.5678 + 2.6351\,i,\ -0.5678 - 2.6351\,i,\ 0.0403,\ -0.0399$$

These results confirm a degenerated phugoid mode with an unstable root. This dynamic instability is associated with the tuck problem; it is essentially due to the drastic change in the location of the aircraft aerodynamic center around the transonic region.

Student Sample Problem 7.2

Consider again the McDonnell Douglas F-4 aircraft at subsonic cruise flight conditions (Appendix B, Aircraft 10). Using the provided aircraft data, perform the analysis of the aircraft lateral directional dynamic stability at the given flight condition using the Routh–Hurwitz stability criteria.

Solution of Student Sample Problem 7.2

The general expression for the lateral directional characteristic equation is given by

$$\overline{D}_2(s) = s(A_2 s^4 + B_2 s^3 + C_2 s^2 + D_2 s + E_2)$$

with the coefficients given by

$$A_2 = V_{P_1}(1 - I_1 I_2)$$

$$B_2 = -Y_\beta(1 - I_1 I_2) - V_{P_1}(L_p + N_r + I_1 N_p + I_2 L_r)$$

$$C_2 = V_{P_1}(L_p N_r - N_p L_r) + Y_\beta(L_p + N_r + I_1 N_p + I_2 L_r) - Y_p(L_\beta + I_1 N_\beta)$$
$$+ V_{P_1}(L_\beta I_2 + N_\beta) - Y_r(L_\beta I_2 + N_\beta)$$

$$D_2 = -Y_\beta(L_p N_r - N_p L_r) + Y_p(L_\beta N_r - L_r N_\beta) - g(L_\beta + I_1 N_\beta)$$
$$+ V_{P_1}(L_\beta N_p - L_p N_\beta) - Y_r(L_\beta N_p - L_p N_\beta)$$

$$E_2 = g(L_\beta N_r - L_r N_\beta)$$

with $I_1 = \dfrac{I_{XZ}}{I_{XX}}, I_2 = \dfrac{I_{XZ}}{I_{ZZ}}$

where the relationships for the lateral directional dimensional stability derivatives are given in Table 7.3. For the given aircraft at the given flight conditions, the numerical expression for the longitudinal characteristic equation is given by

$$\overline{D}_2(s) = s(A_2 s^4 + B_2 s^3 + C_2 s^2 + D_2 s + E_2) = s(1{,}732 s^4 + 1{,}835 s^3 + 10{,}870 s^2 + 8{,}219 s + 23.43)$$

The Routh–Hurwitz array associated with the preceding CE is given by

s^4	1.732	10,870	23.43
s^3	1,835	8,219	–
s^2	k_{11}	k_{12}	–
s	k_{21}	–	–
s^0	23.43	–	–

where

$$k_{11} = \frac{B_2 C_2 - A_2 D_2}{B_2} = 3112.34,$$

$$k_{12} = \frac{B_2 E_2}{B_2} = E_2 = 23.43,$$

$$k_{21} = \frac{D_2 k_{11} - B_2 k_{12}}{k_{11}} = \frac{D_2\left(\dfrac{B_2 C_2 - A_2 D_2}{B_2}\right) - B_2 E_2}{\dfrac{B_2 C_2 - A_2 D_2}{B_2}} = 8{,}205.19$$

The analysis of the first column of the Routh–Hurwitz array provides

$$\begin{array}{c|ccc} s^4 & 1.732 & 10{,}870 & 23.43 \\ s^3 & 1{,}835 & 8{,}219 & - \\ s^2 & 3{,}112.34 & k_{12} & - \\ s & 8{,}205.19 & - & - \\ s^0 & 23.43 & - & - \end{array}$$

According to the Routh–Hurwitz stability criteria, the lateral directional dynamics are stable since there are NO sign changes in the first column of the Routh–Hurwitz array.

Using MATLAB®, excluding the trivial root at $s = 0$, the roots of $\overline{D}_2(s)$ are given by

$$-0.1381 + 2.4573i \quad -0.1381 - 2.4573i, \quad -0.7804, \; -0.0029$$

confirming stable lateral directional dynamics with suitable roots associated with the dutch roll, rolling, and spiral dynamic modes.

Student Sample Problem 7.3

Consider the Cessna T37 aircraft at cruise flight conditions (Appendix B, Aircraft 2). From the available data first find the full longitudinal characteristic equations, the associated roots, and evaluate the short period and phugoid characteristics. Next, find the characteristic equations associated with the short period and phugoid approximations. For both dynamic modes evaluate the error differences in terms of natural frequency and damping between the full and the approximated characteristic equations.

Solution of Student Sample Problem 7.3

The full longitudinal characteristic equation for the Cessna T37 aircraft at cruise flight conditions is given by

$$\overline{D}_1(s) = 457s^4 + 2{,}107s^3 + 9{,}917s^2 + 117.3s + 88.23$$

The associated roots are given by

$$-2.3003 + 4.0440i \quad -2.3003 - 4.0440i, \quad -0.0050 + 0.0943i \quad -0.0050 - 0.0943i$$

leading to the following dynamic characteristics:

$$\omega_{n_{SP}} = 4.652 \frac{\text{rad}}{\text{sec}}, \quad \zeta_{SP} = 0.494$$

$$\omega_{n_{Ph}} = 0.0944 \frac{\text{rad}}{\text{sec}}, \quad \zeta_{Ph} = 0.0529$$

These values can be considered to be the true dynamic longitudinal characteristics for the aircraft at the given flight condition. Next, consider the relationships for the natural frequency and the damping associated with the short period approximation:

$$\omega_{nSP_{APPROX}} = \sqrt{\left(\frac{Z_\alpha M_q}{V_{P_1}} - M_\alpha\right)}$$

$$\zeta_{SP_{APPROX}} = -\frac{\left(M_{\dot{\alpha}} + M_q + \dfrac{Z_\alpha}{V_{P_1}}\right)}{2\sqrt{\left(\dfrac{Z_\alpha M_q}{V_{P_1}} - M_\alpha\right)}}$$

Using the aircraft data in Appendix B in the dimensional stability derivatives in Table 7.1, we have the following results:

$$\omega_{nSP_{APPROX}} = \sqrt{\left(\frac{(-437.4)(-2.477)}{456} - (-19.4)\right)} = \sqrt{2.376 + 19.4} = 4.66 \frac{\text{rad}}{\text{sec}}$$

$$\zeta_{SP_{APPROX}} = -\frac{\left((-1.1553) + (-2.48) + \frac{(-437.41)}{(456)}\right)}{2 \; 4.66} = \frac{4.594}{9.32} = 0.4929$$

with the following percentage errors:

$$e|_{\omega_{nSP}} = \frac{|\omega_{nSP_{APPROX}} - \omega_{nSP}|}{\omega_{nSP}}\% = \frac{|4.66 - 4.652|}{4.652}\% = 0.17\%$$

$$e|_{\zeta_{nSP}} = \frac{|\zeta_{SP_{APPROX}} - \zeta_{SP}|}{\zeta_{SP}}\% = \frac{|0.4929 - 0.494|}{0.494}\% = 0.22\%$$

These results confirm the accuracy associated with the short period approximation.

Consider now the relationships provided by the phugoid approximation:

$$\omega_{nPh} = \frac{g}{V_{P_1}}\sqrt{2}$$

$$\zeta_{Ph} = \frac{(c_{D_u} - c_{T_{X_u}})}{2\sqrt{2} \; c_{L_1}}$$

Using the aircraft data in Appendix B, we have the following results:

$$\omega_{nPh_{APPROX}} = \frac{32.17}{456}\sqrt{2} = 0.099 \frac{\text{rad}}{\text{sec}}$$

$$\zeta_{Ph_{APPROX}} = \frac{(0 - (-0.07))}{2\sqrt{2} \; 0.378} = \frac{0.07}{1.069} = 0.065$$

with the following percentage errors:

$$e|_{\omega_{nPh}} = \frac{|\omega_{nPh_{APPROX}} - \omega_{nPh}|}{\omega_{nPh}}\% = \frac{|0.099 - 0.0944|}{0.0944}\% = 4.9\%$$

$$e|_{\zeta_{nPh}} = \frac{|\zeta_{Ph_{APPROX}} - \zeta_{Ph}|}{\zeta_{Ph}}\% = \frac{|0.065 - 0.0529|}{0.0529}\% = 22.9\%$$

These results confirm that, unlike the short period approximation, the phugoid approximation leads to results with a substantial amount of error.

Student Sample Problem 7.4

Perform a complete MATLAB®-based analysis of the linearized longitudinal dynamics for the Learjet 24 aircraft at low weight, cruise conditions (Appendix B, Aircraft 6)

Solution of Student Sample Problem 7.4

```
% Airplane type: Learjet 24
% Flight Condition: Cruise Low Weight at altitude of 40,000 ft

% Reference Geometry
S = 230;    % (ft^2)
```

```
cbar = 7.0; % (ft)
b = 34.0; %((ft)
xcg_bar=0.32 %x_cg dimensionless location along MAC)

% Flight condition data
U1 = 677; %(ft/s)
M = 0.7; % Mach number
alpha1 = 1.5/57.3; % (rad)
theta1 = alpha1;
q1 = 134.6; % (lbs/ft^2)
g = 32.2;   % (ft/s^2)
rad2deg=57.3;
deg2rad=1/57.3;

% Mass and inertial data
W = 9000; % (lbs)
m = (W/g); % (slugs)
IxxB = 6000; % (slug*ft^2)
IyyB = 17800; % (slug*ft^2)
IzzB = 25000; % (slug*ft^2)
IxzB = 1400; % (slug*ft^2)

% Steady State Coefficients
CL1 = 0.28;
CD1 = 0.0279;
CTx1 = 0.0279;
Cm1 = 0;
CmT1 = 0;

% Longitudinal Stability Derivatives
% (dimensionless - along stability Axes)
CD0 = 0.0216;
CDu = 0.104;
CDalpha = 0.22;
CTxu = -0.07;
CL0 = 0.13;
CLu = 0.28;
CLalpha = 5.84;
CLalphadot = 2.2;
CLq = 4.7;
Cm0 = 0.050;
Cmu = 0.070;
Cmalpha = -0.64;
Cmalphadot = -6.7;
Cmq = -15.5;
CmTu = -0.003;
CmTalpha = 0;

% Longitudinal Control Derivatives
%(dimensionless - along stability Axes)
CDdeltaE = 0;
CLdeltaE = 0.46;
CmdeltaE = -1.24;
CDih = 0;
CLih = 0.94;
Cmih = -2.5;

% Longitudinal Dimensional Stability Derivatives
Xu = ((-q1)*S*(CDu+(2*CD1)))/(m*U1); % (ft/sec^2)/(ft/sec)
XTu = ((q1*S*(CTxu+(2*CTx1)))/(m*U1));
Xalpha =((-q1)*S*(CDalpha - CL1))/m;  %(ft/sec^2)/rad
Zu = ((-q1)*S*(CLu+(2*CL1)))/(m*U1); % (ft/sec^2)/(ft/sec)
Zalpha = ((-q1)*S*(CLalpha + CD1))/m;  %(ft/sec^2)/rad
Zalphadot = -(q1*S*cbar*CLalphadot)/(2*m*U1); % (ft/sec^2)/(rad/sec)
Zq = -(q1*S*cbar*CLq)/(2*m*U1); % (ft/sec^2)/(rad/sec)
Mu = (q1*S*cbar*(Cmu+(2*Cm1)))/(IyyB*U1); %(rad/sec^2)/(ft/sec)
```

```
MTu = (q1*S*cbar*(CmTu+(2*CmT1)))/(IyyB*U1);  %(rad/sec^2)/(ft/sec)
Malpha = (q1*S*cbar*Cmalpha)/IyyB;  % (rad/sec^2)/rad
MTalpha = (q1*S*cbar*CmTalpha)/IyyB; % (rad/sec^2)/rad
Malphadot = (q1*S*cbar^2*Cmalphadot)/(2*IyyB*U1);  % (rad/sec^2)/(rad/sec)
Mq = (q1*S*cbar^2*Cmq)/(2*IyyB*U1);  % (rad/sec^2)/(rad/sec)

% Longitudinal Dimensional Control Derivatives
XdeltaE = -(q1*S*CDdeltaE)/m;
ZdeltaE = -(q1*S*CLdeltaE)/m; % (ft/sec^2)/rad
MdeltaE = (q1*S*cbar*CmdeltaE)/IyyB; % (rad/sec^2)/rad

% Coefficients of the NUM(s) of u-Transfer Function
Au = XdeltaE*(U1-Zalphadot);Bu = -XdeltaE*(((U1-Zalphadot)*Mq)+Zalpha+(Malphadot*(U1+Zq))
+(ZdeltaE*Xalpha));
Cu = (XdeltaE*((Mq*Zalpha)+(Malphadot*g*sin(theta1))-((Malpha+MTalpha)*(U1+Zq))))+(ZdeltaE*
((-Malphadot*g*cos(theta1))-(Xalpha*Mq)))+(MdeltaE*((Xalpha*(U1+Zq))
-((U1-Zalphadot)*g*cos(theta1))));Du = (XdeltaE*(Malpha+MTalpha)*g*sin(theta1))
-(ZdeltaE*Malpha*g*cos(theta1))+(MdeltaE*((Zalpha*g*cos(theta1))
-(Xalpha*g*sin(theta1))));Nu = [Au Bu Cu Du];

% Coefficients of the NUM(s) of alpha-Transfer Function
Aalpha = ZdeltaE;
Balpha = (XdeltaE*Zu)+(ZdeltaE*(-Mq-(Xu+XTu)))+(MdeltaE*(U1+Zq));
Calpha = (XdeltaE*(((U1+Zq)*(Mu+MTu))-(Mq*Zu)))
+(ZdeltaE*Mq*(Xu+XTu))+(MdeltaE*((-g*sin(theta1))-((U1+Zq)*(Xu+XTu))));
Dalpha = -(XdeltaE*(Mu+MTu)*g*sin(theta1))+(ZdeltaE*(Mu+MTu)*g*cos(theta1))
+(MdeltaE*(((Xu+XTu)*g*sin(theta1))-(Zu*g*cos(theta1))));
Nalpha = [Aalpha Balpha Calpha Dalpha];

% Coefficients of the NUM(s) of theta-Transfer Function
Atheta = (ZdeltaE*Malphadot)+(MdeltaE*(U1-Zalphadot));
Btheta = (XdeltaE*((Zu*Malphadot)+((U1-Zalphadot)*(Mu+MTu))))+(ZdeltaE*
((Malpha+MTalpha)-(Malphadot*(Xu+XTu))))+(MdeltaE*(-Zalpha-((U1-Zalphadot)
*(Xu+XTu))));Ctheta = (XdeltaE*(((Malpha+MTu)*Zu)-(Zalpha*(Mu+MTu))))
+(ZdeltaE*((-(Malpha+MTu)*(Xu+XTu))+(Xalpha*(Mu+MTu))))
+(MdeltaE*((Zalpha*(Xu+XTu))-(Xalpha*Zu)));Ntheta = [Atheta Btheta Ctheta];

% Coefficients of the Longitudinal Characteristic Equation (DEN(s))
A1 = U1-Zalphadot;
B1 = -(U1-Zalphadot)*(Xu+XTu+Mq)-Zalpha-(Malphadot*(U1+Zq));
C1 = Xu+(XTu*(Mq*(U1-Zalphadot)+Zalpha+(Malphadot*(U1+Zq))))
+(Mq*Zalpha)-(Zu*Xalpha)+(Malphadot*g*sin(theta1))-((Malpha+MTalpha)*(U1+Zq));
D1 = g*sin(theta1)*(Malpha+MTalpha-(Malphadot*(Xu+XTu)))
+(g*cos(theta1)*((Zu*Malphadot)+((Mu+MTu)*(U1-Zalphadot))))+((Mu+MTu)*(-Xalpha*(U1+Zq)))
+(Zu*Xalpha*Mq)+((Xu+XTu)*(((Malpha+MTalpha)*(U1+Zq))-(Mq*Zalpha)));
E1 = (g*cos(theta1)*(((Malpha+MTalpha)*Zu)-(Zalpha*(Mu+MTu))))
+(g*sin(theta1)*(((Mu+MTu)*Xalpha)-((Xu+XTu)*(Malpha+MTalpha))));
Dbar1 = [A1 B1 C1 D1 E1];

% Check of Dynamic Stability via Routh-Hurwitz Stability Criteria
Routh = D1*((B1*C1)-(A1*D1))-((B1^2)*E1);
if A1<=0
printf('Longitudinal dynamic stability NOT satisfied!');
elseif B1<=0
printf('Longitudinal dynamic stability NOT satisfied!');
elseif C1<=0
printf('Longitudinal dynamic stability NOT satisfied!');
elseif D1<=0
fprintf('Longitudinal dynamic stability NOT satisfied!');
elseif E1<=0
printf('Longitudinal dynamic stability NOT satisfied!');
elseif Routh<=0
printf('Longitudinal dynamic stability NOT satisfied!');
end

% Time column vector for simulation (12001 time steps)
t = [0:0.025:300]';
```

```
% Small library of pilot elevator maneuvers
Library = menu('SAMPLES OF ELEVATOR INPUTS', 'Single Doublet Impulse', 'Multiple Doublet Impul-
ses', 'Single Doublet', 'Multiple Doublets')
% Single Doublet Impulse
if Library == 1
  clf;
  echo off;
  for i = 1:200
    de(i,1)=0.0/57.3;
  end;
  for i = 201:205
    de(i,1)=-4/57.3;
  end;
  for i = 206:1200
    de(i,1)=0.0/57.3;
  end;
  for i = 1201:1205
    de(i,1)=4/57.3;
  end;
  for i = 1206:12001
    de(i,1)=0.0/57.3;
  end;
  echo on;
end;
% Multiple Doublets Impulse
if Library == 2
  clf;
  echo off;
  for i = 1:200
    de(i,1)=0.0/57.3;
  end;
  for i = 201:215
    de(i,1)=-4/57.3;
  end;
  for i = 216:400
    de(i,1)=0.0/57.3;
  end;
  for i = 401:415
    de(i,1)=4/57.3;
  end;
  for i = 416:800
    de(i,1)=0.0/57.3;
  end;
  for i = 801:815
    de(i,1)=-2/57.3;
  end;
  for i = 816:1000
    de(i,1)=0.0/57.3;
  end;
  for i = 1001:1015
    de(i,1)=2/57.3;
  end;
  for i = 1016:12001
    de(i,1)=0.0/57.3;
  end;
  echo on;
end;
% Single Doublet
if Library == 3
  clf;
  echo off;
  for i = 1:200
    de(i,1)=0.0/57.3;
  end;
  for i = 201:400
    de(i,1)=-2/57.3;
```

```
        end;
      for i = 401:1200
         de(i,1)=0.0/57.3;
      end;
      for i = 1201:1400
         de(i,1)=2/57.3;
      end;
      for i = 1401:12001
         de(i,1)=0.0/57.3;
      end;
      echo on;
end;
% Multiple Doublets
if Library == 4
   clf;
   echo off;
   for i = 1:200
      de(i,1)=0.0/57.3;
   end;
   for i = 201:300
      de(i,1)=-2/57.3;
   end;
   for i = 301:1600
      de(i,1)=0.0/57.3;
   end;
   for i = 1601:1700
      de(i,1)=2/57.3;
   end;
   for i = 1701:3000
      de(i,1)=0.0/57.3;
   end;
   for i = 3001:3100
      de(i,1)=-1/57.3;
   end;
   for i = 3101:4400
      de(i,1)=0.0/57.3;
   end;
   for i = 4401:4500
      de(i,1)=1/57.3;
   end;
   for i = 4501:12001
      de(i,1)=0.0/57.3;
   end;
   echo on;
end;

% Simulation
sys_1=tf(Nalpha, Dbar1);
alpha=lsim(sys_1, de,t);
sys_2=tf(Nu, Dbar1);
u=lsim(sys_2,de,t);
sys_3=tf(Ntheta,Dbar1);
theta=lsim(sys_3,de,t);

% Plot Results
ans = menu('PLOTS', 'Alpha','U','Theta','Elevator','All');
if ans == 1
   close all;
   alpha_deg=alpha*rad2deg;
   plot(t,alpha_deg,'r');
   title('Small Perturbation Alpha vs.Time');
   xlabel('Time (s)');
   ylabel('Alpha (deg)');
   grid on;
end;
if ans == 2
```

```
        close all;
        plot(t,u,'g');
        title('Small Perturbation Velocity vs.Time');
        xlabel('Time (s)');
        ylabel('Velocity(ft/s)');
        grid on;
    end;
    if ans == 3
        close all;
        theta_deg=theta*rad2deg;
        plot(t,theta_deg,'b');
        title('Small Perturbation Theta vs.Time');
        xlabel('Time (s)');
        ylabel('Theta (deg)');
        grid on;
    end;
    if ans == 4
        close all;
        de_deg=de*rad2deg;
        plot(t,de_deg,'m');
        title('Elevator Deflection vs.Time');
        xlabel('Time (s)');
        ylabel('Elevator Deflection (deg)');
        grid on;
    end;
    if ans == 5
        close all;
        alpha_deg=alpha*rad2deg;
        theta_deg=theta*rad2deg;
        de_deg=de*rad2deg;
        subplot(221),plot(t,alpha_deg,'r');
        title('Alpha vs.Time');
        xlabel('Time (s)');
        ylabel('Alpha (deg)');
        grid on;
        subplot(222),plot(t,u,'g');
        title('Velocity vs.Time');
        xlabel('Time (s)');
        ylabel('Velocity(ft/s)');
        grid on;
        subplot(223),plot(t,theta_deg,'b');
        title('Theta vs.Time');
        xlabel('Time (s)');
        ylabel('Theta (deg)');
        grid on;
        subplot(224),plot(t,de_deg,'m');
        title('Elevator Deflection vs.Time');
        xlabel('Time (s)');
        ylabel('Elevator Deflection (deg)');
        grid on;
    end;

    % plot the poles on the s-domain
    poles = roots(Dbar1);
    figure,
    plot(poles, '*');
    grid on
    title('Longitudinal Poles in the s-domain');
    xlabel('Real Axis');
    ylabel('Imaginary Axis');
    omega_sp=sqrt(abs(poles(1,1)^2));
    omega_ph=sqrt(abs(poles(3,1)^2));
    damp_sp=abs(real(poles(1,1)))/omega_sp;
    damp_ph=abs(real(poles(3,1)))/omega_ph;
```

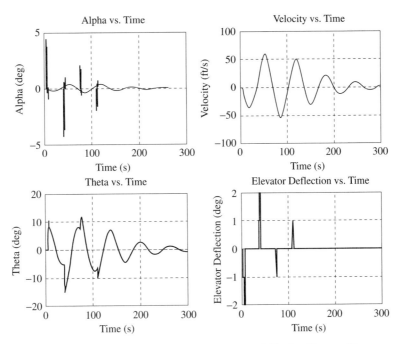

Figure SSP7.4.1 Learjet 24: Longitudinal Responses Following Elevator Maneuver

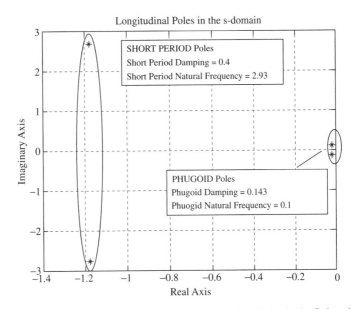

Figure SSP7.4.2 Learjet 24: Location of Longitudinal Poles in the S-domain

Most of the preceding code performs the calculations of the dimensional stability and control derivatives (Table 7.1) starting from the dimensionless counterpart. Another section of the codes prepares a small library of different elevator maneuvers. Only a few lines of the code perform the actual simulation, followed by the organization of the simulation plots for displaying purposes.

Following the modeling described in the preceding sections, the code is self-explanatory. However, refer to the following key MATLAB® commands, which allow complex simulations to be executed with limited programming efforts.

The first command is the command '*tf*'. This command allows defining a dynamic system as a transfer function with a given *Num(s)* and a given *Den(s)*. For example, the command line

$$\text{sys\_1} = \text{tf}(\text{Nalpha}, \text{Dbar1});$$

allows us to define a dynamic system sys_1 as a transfer function defined by the numerator $Num_\alpha(s) = A_\alpha s^3 + B_\alpha s^2 + C_\alpha s + D_\alpha$ and the denominator (characteristic equation) $\overline{D}_1 = A_1 s^4 + B_1 s^3 + C_1 s^2 + D_1 s + E_1 = (s^2 + 2\zeta_{SP}\omega_{n_{SP}} s + \omega_{n_{SP}}^2)(s^2 + 2\zeta_{PH}\omega_{n_{PH}} s + \omega_{n_{PH}}^2)$.

The second command is the command '*lsim*'. This powerful command is a one-step command which allows solving for the partial fraction expansion (PFE) and the inverse laplace transform (ILT). Refer to Appendix A for more details about the PFE method and the ILT process. The outcome of the '*lsim*' command is essentially the output for a given input over a specified time interval. For example, the command line

$$\text{alpha} = \text{lsim}(\text{sys\_1}, \text{de}, \text{t});$$

calculates the α (angle of attack) time histories over the specified time interval *t* for the specified elevator maneuver (within the library of available elevator maneuvers) using the transfer function defined by $Num_\alpha(s)$ and $\overline{D}_1(s)$ and identified by 'sys_1'.

In terms of the dynamic characteristics of the Learjet 24 aircraft, the analysis of the location of the poles of the longitudinal characteristic equation $\overline{D}_1(s)$ shows

$$\zeta_{SP} = 0.4, \quad \omega_{n_{SP}} = 2.93$$
$$\zeta_{Ph} = 0.143, \quad \omega_{n_{Ph}} = 0.1$$

These values reveal well behaved longitudinal dynamics with desirable damping for both short period and phugoid. Refer to Chapter 10 for a detailed understanding of the suitable ranges for these important parameters as well as a detailed explanation of the overall handling qualities for the aircraft system.

Using the modeling described in the previous MATLAB® code, the simulation of the aircraft longitudinal dynamics can be also performed within Simulink®, as shown in Figure SSP7.4.3, providing identical time histories.

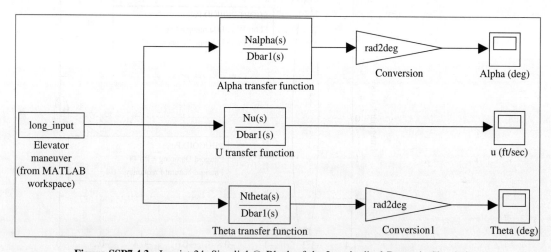

Figure SSP7.4.3 Learjet 24: Simulink® Block of the Longitudinal Dynamic Simulation

Student Sample Problem 7.5

Perform a complete MATLAB®-based analysis of the linearized lateral directional dynamics for the Learjet 24 aircraft at low weight, cruise conditions (Appendix B, Aircraft 6).

Solution of Student Sample Problem 7.5

```
% Airplane type: Learjet 24
% Flight Condition: Cruise Low Weight at altitude of 40,000 ft

% Reference Geometry
S = 230;    % (ft^2)
```

```
cbar = 7.0; % (ft)
b = 34.0; %((ft)
xcg_bar=0.32 %x_cg dimensionless location along MAC)

% Flight condition data
U1 = 677; %(ft/s)
M = 0.7; % Mach number
alpha1 = 1.5/57.3; % (rad)
theta1 = alpha1;
q1 = 134.6; % (lbs/ft^2)
g =32.2;  % (ft/s^2)
rad2deg=57.3;
deg2rad=1/57.3;

% Mass and inertial data
W = 9000; % (lbs)
m = W/g; % (slugs)
IxxB = 6000; % (slug*ft^2)
IyyB = 17800; % (slug*ft^2)
IzzB = 25000; % (slug*ft^2)
IxzB = 1400; % (slug*ft^2)
% Lateral-Directional Stability Derivatives
% (dimensionless - along stability axes)
Clbeta = -0.100;
Clp = -0.450;
Clr = 0.140;
Cybeta = -0.730;
Cyp = 0;
Cyr = 0.400;
Cnbeta = 0.124;
CnTbeta = 0;
Cnp = -0.022;
Cnr = -0.200;

% Lateral-Directional Control Derivatives
% (dimensionless - along stability axes)
Cldeltaa = 0.178;
Cldeltar = 0.021;
Cydeltaa = 0;
Cydeltar = 0.140;
Cndeltaa = -0.020;
Cndeltar = -0.074;

% Transformation of products and moment of inertia from body to stability axis
A = [(cos(alpha1))^2, (sin(alpha1))^2, -sin(2*alpha1);
    (sin(alpha1))^2, (cos(alpha1))^2, sin(2*alpha1);
    (0.5*sin(2*alpha1)), (-0.5*sin(2*alpha1)), cos(2*alpha1)];
 ib = [IxxB, IzzB, IxzB]';
 c = A*ib;
% Transformation to Stability axis
 Ixx = c(1,1);
 Izz = c(2,1);
 Ixz = c(3,1);
 Abar1 = (Ixz/Ixx);
 Bbar1 = (Ixz/Izz);

% Lateral-Directional Dimensional Stability Derivatives
Ybeta = (q1*S*Cybeta)/m;
Yp = (q1*S*b*Cyp)/(2*m*U1);
Yr = (q1*S*b*Cyr)/(2*m*U1);
Lbeta = (q1*S*b*Clbeta)/Ixx;
Lp = (q1*S*(b^2)*Clp)/(2*Ixx*U1);
Lr = (q1*S*(b^2)*Clr)/(2*Ixx*U1);
Nbeta = (q1*S*b*Cnbeta)/Izz;
NTbeta = (q1*S*b*CnTbeta)/Izz;
Np = (q1*S*(b^2)*Cnp)/(2*Izz*U1);
```

```
Nr = (q1*S*(b^2)*Cnr)/(2*Izz*U1);

% Lateral-Directional Dimensional Control Derivatives
Ydeltaa = (q1*S*Cydeltaa)/m;
Ydeltar = (q1*S*Cydeltar)/m;
Ldeltaa = (q1*S*b*Cldeltaa)/Ixx;
Ldeltar = (q1*S*b*Cldeltar)/Ixx;
Ndeltaa = (q1*S*b*Cndeltaa)/Izz;
Ndeltar = (q1*S*b*Cndeltar)/Izz;

% Coefficients of the NUM(s) of beta-Transfer Function
% (Aileron Input)
AbetaI = Ydeltaa*(1-(Abar1*Bbar1));
BbetaI = -Ydeltaa*(Nr+Lp+(Abar1*Np)+(Bbar1*Lr))+(Yp*(Ldeltaa+(Ndeltaa*Abar1)))+(Yr*((Ldeltaa*Bbar1)+Ndeltaa))-((U1*(Ldeltaa*Bbar1)+Ndeltaa));
CbetaI = Ydeltaa*((Lp*Nr)-(Np*Lr))+(Yp*((Ndeltaa*Lr)-(Ldeltaa*Nr)))+(g*cos(theta1)*(Ldeltaa+(Ndeltaa*Abar1)))+(Yr*((Ldeltaa*Np)-(Ndeltaa*Lp)))-(U1*((Ldeltaa*Np)-(Ndeltaa*Lp)));
DbetaI = g*cos(theta1)*((Ndeltaa*Lr)-(Ldeltaa*Nr));
NbetaI = [AbetaI BbetaI CbetaI DbetaI]

% Coefficients of the NUM(s) of beta-Transfer Function
% (Rudder Input)
AbetaII = Ydeltar*(1-(Abar1*Bbar1));
BbetaII  =  -Ydeltar*(Nr+Lp+(Abar1*Np)+(Bbar1*Lr))+(Yp*(Ldeltar+(Ndeltar*Abar1)))+(Yr*((Ldeltar*Bbar1)+Ndeltar))-((U1*(Ldeltar*Bbar1)+Ndeltar));
CbetaII = Ydeltar*((Lp*Nr)-(Np*Lr))+(Yp*((Ndeltar*Lr)-(Ldeltar*Nr)))+(g*cos(theta1)*(Ldeltar+(Ndeltar*Abar1)))+(Yr*((Ldeltar*Np)-(Ndeltar*Lp)))-(U1*((Ldeltar*Np)-(Ndeltar*Lp)));
DbetaII = g*cos(theta1)*((Ndeltar*Lr)-(Ldeltar*Nr));
NbetaII = [AbetaII BbetaII CbetaII DbetaII]

% Coefficients of the NUM(s) of phi-Transfer Function
% (Aileron Input)
AphiI = U1*(Ldeltaa+(Ndeltaa*Abar1));
BphiI  =  U1*((Ndeltaa*Lr)-(Ldeltaa*Nr))-(Ybeta*(Ldeltaa+(Ndeltaa*Abar1)))+(Ydeltaa*(Lbeta+(Nbeta*Abar1)+(NTbeta*Abar1)));
CphiI = -Ybeta*((Ndeltaa*Lr)-(Ldeltaa*Nr))+(Ydeltaa*((Lr*Nbeta)+(Lr*NTbeta)-(Nr*Lbeta)))+((U1-Yr)*((Nbeta*Ldeltaa)+(NTbeta*Ldeltaa)-(Lbeta*Ndeltaa)));
NphiI = [AphiI BphiI CphiI]
% Coefficients of the NUM(s) of phi-Transfer Function
% (Rudder Input)
AphiII = U1*(Ldeltar+(Ndeltar*Abar1));
BphiII = U1*((Ndeltar*Lr)-(Ldeltar*Nr))-(Ybeta*(Ldeltar+(Ndeltar*Abar1)))+(Ydeltar*(Lbeta+(Nbeta*Abar1)+(NTbeta*Abar1)));
CphiII = -Ybeta*((Ndeltar*Lr)-(Ldeltar*Nr))+(Ydeltar*((Lr*Nbeta)+(Lr*NTbeta)-(Nr*Lbeta)))+((U1-Yr)*((Nbeta*Ldeltar)+(NTbeta*Ldeltar)-(Lbeta*Ndeltar)));
NphiII = [AphiII BphiII CphiII]

% Coefficients of the NUM(s) of psi-Transfer Function
% (Aileron Input)
ApsiI = U1*(Ndeltaa+(Ldeltaa*Bbar1));
BpsiI = U1*((Ldeltaa*Np)-(Ndeltaa*Lp))-(Ybeta*(Ndeltaa+(Ldeltaa*Bbar1)))+(Ydeltaa*((Lbeta*Bbar1)+Nbeta+NTbeta));
CpsiI = -Ybeta*((Ldeltaa*Np)-(Ndeltaa*Lp))+(Yp*((Nbeta*Ldeltaa)+(NTbeta*Ldeltaa)-(Lbeta*Ndeltaa)))+(Ydeltaa*((Lbeta*Np)-(Nbeta*Lp)-(NTbeta*Lp)));
DpsiI = g*cos(theta1)*((Nbeta*Ldeltaa)+(NTbeta*Ldeltaa)-(Lbeta*Ndeltaa));
NpsiI = [ApsiI BpsiI CpsiI DpsiI]

% Coefficients of the NUM(s) of psi-Transfer Function
% (Rudder Input)
ApsiII = U1*(Ndeltar+(Ldeltar*Bbar1));BpsiII = U1*((Ldeltar*Np)-(Ndeltar*Lp))-(Ybeta*(Ndeltar+(Ldeltar*Bbar1)))+(Ydeltar*((Lbeta*Bbar1)+Nbeta+NTbeta));
CpsiII = -Ybeta*((Ldeltar*Np)-(Ndeltar*Lp))
+(Yp*((Nbeta*Ldeltar)+(NTbeta*Ldeltar)-(Lbeta*Ndeltar)))
+(Ydeltar*((Lbeta*Np)-(Nbeta*Lp)-(NTbeta*Lp)));
DpsiII = g*cos(theta1)*((Nbeta*Ldeltar)+(NTbeta*Ldeltar)-(Lbeta*Ndeltar));
NpsiII = [ApsiII BpsiII CpsiII DpsiII]
```

```
% Coefficients for the Lateral-Directional Characteristic Equation
A2 = U1*(1-(Abar1*Bbar1));
B2 = -Ybeta*(1-(Abar1*Bbar1))-(U1*(Lp+Nr+(Abar1*Np)+(Bbar1*Lr)));
C2  =  U1*((Lp*Nr)-(Lr*Np))+(Ybeta*(Nr+Lp+(Abar1*Np)+(Bbar1*Lr)))-(Yp*(Lbeta+(Nbeta*Abar1)
+(NTbeta*Abar1)))+(U1*((Lbeta*Bbar1)+Nbeta+NTbeta))-(Yr*((Lbeta*Bbar1)+Nbeta+NTbeta));D2 =
-Ybeta*((Lp*Nr)-(Lr*Np))+(Yp*((Lbeta*Nr)-(Nbeta*Lr)-(NTbeta*Lr)))-(g*cos(theta1)*(Lbeta
+(Nbeta*Abar1)
+(NTbeta*Abar1)))+(U1*((Lbeta*Np)-(Nbeta*Lp)-(NTbeta*Lp)))-(Yr*((Lbeta*Np)-(Nbeta*Lp)-
(NTbeta*Lp)));E2 = g*cos(theta1)*((Lbeta*Nr)-(Nbeta*Lr)-(NTbeta*Lr));Dbar2 = [A2 B2 C2 D2 E2]

% Check of Dynamic Stability via Routh-Hurwitz Stability Criteria
Routh = D2*((B2*C2)-(A2*D2))-((B2^2)*E2);
if A2<=0
printf('Lateral-directional dynamic stability NOT satisfied!')
elseif B2<=0
printf('Lateral-directional dynamic stability NOT satisfied!')
elseif C2<=0
printf('Lateral-directional dynamic stability NOT satisfied!')
elseif D2<=0
printf('Lateral-directional dynamic stability NOT satisfied!')
elseif E2<=0
printf('Lateral-directional dynamic stability NOT satisfied!')
elseif Routh<=0
printf('Lateral-directional dynamic stability NOT satisfied!')
end

% time column vector for simulation (12001 time steps)
t = [0:0.025:300]';
% Small library of pilot ailerons/rudder maneuvers
Library = menu('SAMPLES OF AILERONS/RUDDER INPUTS', 'Single Doublet Impulse
Aileron', 'Multiple Doublet Impulses Aileron', 'Single Doublet Aileron',
'Multiple Doublets Aileron','Single Doublet Impulse Rudder', 'Multiple Doublet
Impulses Rudder', 'Single Doublet Rudder', 'Multiple Doublets Rudder')
% Single Impulse Aileron
if Library == 1
  for i = 1:200
    da(i,1)=0.0/57.3;
  end;
  for i = 201:205
    da(i,1)=-8/57.3;
  end;
  for i = 206:3200
    da(i,1)=0.0/57.3;
  end;
  for i = 3201:3205
    da(i,1)=5/57.3;
  end;
  for i = 3206:12001
    da(i,1)=0.0/57.3;
  end;
end;
% Multiple Impulse Aileron
if Library == 2
  echo off;
  for i = 1:200
    da(i,1)=0.0/57.3;
  end;
  for i = 201:205
    da(i,1)=-5/57.3;
  end;
  for i = 206:900
    da(i,1)=0.0/57.3;
  end;
  for i = 901:905
    da(i,1)=5/57.3;
  end;
```

```
        for i = 906:1800
            da(i,1)=0.0/57.3;
        end;
        for i = 1801:1805
            da(i,1)=-2/57.3;
        end;
        for i = 1806:3000
            da(i,1)=0.0/57.3;
        end;
        for i = 3001:3005
            da(i,1)=2/57.3;
        end;
        for i = 3006:12001
            da(i,1)=0.0/57.3;
        end;
    end;
    % Single Step Aileron
    if Library == 3
        for i = 1:380
            da(i,1)=0.0/57.3;
        end;
        for i = 381:400
            da(i,1)=-4/57.3;
        end;
        for i = 401:3280
            da(i,1)=0.0/57.3;
        end;
        for i = 3281:3300
            da(i,1)=4/57.3;
        end;
        for i = 3301:12001
            da(i,1)=0.0/57.3;
        end;
    end;
    % Multiple Steps Aileron
    if Library == 4
        for i = 1:380
            da(i,1)=0.0/57.3;
        end;
        for i = 381:400
            da(i,1)=-4/57.3;
        end;
        for i = 401:1680
            da(i,1)=0.0/57.3;
        end;
        for i = 1681:1700
            da(i,1)=4/57.3;
        end;
        for i = 1701:3080
            da(i,1)=0.0/57.3;
        end;
        for i = 3081:3100
            da(i,1)=-2/57.3;
        end;
        for i = 3101:4080
            da(i,1)=0.0/57.3;
        end;
        for i = 4081:4100
            da(i,1)=2/57.3;
        end;
        for i = 4101:12001
            da(i,1)=0.0/57.3;
        end;
    end;
    % Single Impulse Doublet Rudder
    if Library == 5
```

```
    for i = 1:200
        dr(i,1)=0.0/57.3;
    end;
    for i = 201:205
        dr(i,1)=-4/57.3;
    end;
    for i = 206:1200
        dr(i,1)=0.0/57.3;
    end;
    for i = 1201:1205
        dr(i,1)=4/57.3;
    end;
    for i = 1206:12001
        dr(i,1)=0.0/57.3;
    end;
end;
% Multiple Impulse Doublet Rudder
if Library == 6
    for i = 1:200
        dr(i,1)=0.0/57.3;
    end;
    for i = 201:215
        dr(i,1)=-4/57.3;
    end;
    for i = 216:400
        dr(i,1)=0.0/57.3;
    end;
    for i = 401:415
        dr(i,1)=4/57.3;
    end;
    for i = 416:800
        dr(i,1)=0.0/57.3;
    end;
    for i = 801:815
        dr(i,1)=-4/57.3;
    end;
    for i = 816:1000
        dr(i,1)=0.0/57.3;
    end;
    for i = 1001:1015
        dr(i,1)=4/57.3;
    end;
    for i = 1016:12001
        dr(i,1)=0.0/57.3;
    end;
end;
% Single Doublet Rudder
if Library == 7
    for i = 1:360
        dr(i,1)=0.0/57.3;
    end;
    for i = 361:400
        dr(i,1)=-5/57.3;
    end;
    for i = 401:1360
        dr(i,1)=0.0/57.3;
    end;
    for i = 1361:1400
        dr(i,1)=5/57.3;
    end;
    for i = 1401:12001
        dr(i,1)=0.0/57.3;
    end;
end;
% Multiple Doublets Rudder
if Library == 8
```

```
       for i = 1:300
         dr(i,1)=0.0/57.3;
       end;
       for i = 361:400
         dr(i,1)=-5/57.3;
       end;
       for i = 401:760
         dr(i,1)=0.0/57.3;
       end;
       for i = 761:800
         dr(i,1)=5/57.3;
       end;
       for i = 801:1160
         dr(i,1)=0.0/57.3;
       end;
       for i = 1161:1200
         dr(i,1)=-2.5/57.3;
       end;
       for i = 1201:1860
         dr(i,1)=0.0/57.3;
       end;
       for i = 1861:1900
         dr(i,1)=2.5/57.3;
       end;
       for i = 1901:12001
         dr(i,1)=0.0/57.3;
       end;
    end;

    % simulation
    if Library <= 4
    sys_1=tf(NbetaI, Dbar2);
    betaI=lsim(sys_1, da, t);
    sys_2=tf(NphiI, Dbar2);
    phiI=lsim(sys_2,da,t);
    sys_3=tf(NpsiI,Dbar2);
    psiI=lsim(sys_3,da,t);
    end;
    if Library > 4
    sys_4=tf(NbetaII, Dbar2);
    betaII=lsim(sys_4,dr,t);
    sys_5=tf(NphiII, Dbar2);
    phiII=lsim(sys_5,dr,t);
    sys_6=tf(NpsiII, Dbar2);
    psiII=lsim(sys_6,dr,t);
    end;

    % Plot results
    if Library <= 4
    ans_a = menu('RESPONSES TO AILERON MANEUVER', 'Beta','Phi','Psi','Aileron','All')
    if ans_a == 1
      close all;
      betaI_deg=betaI*rad2deg;
      plot(t,betaI_deg,'r');
      title('Beta vs. Time for Aileron Maneuver');
      xlabel('Time (s)');
      ylabel('Beta (deg)');
      grid on;
    end;
    if ans_a == 2
      close all;
      phiI_deg=phiI*rad2deg;
      plot(t,phiI_deg,'k');
      title('Phi vs. Time for Aileron Maneuver');
      xlabel('Time (s)');
      ylabel('Phi (deg)');
```

```
    grid on;
end;
if ans_a == 3
  close all;
  psiI_deg=psiI*rad2deg;
  plot(t,psiI_deg,'b');
  title('Psi vs. Time for Aileron Maneuver');
  xlabel('Time (s)');
  ylabel('Psi (deg)');
  grid on;
end;
if ans_a == 4
  close all;
  da_deg=da*rad2deg;
  plot(t,da_deg,'m');
  title('Aileron Deflection vs. Time');
  xlabel('Time (s)');
  ylabel('Aileron Deflection (deg)');
  grid on;
end;
if ans_a == 5
  close all;
  betaI_deg=betaI*rad2deg;
  phiI_deg=phiI*rad2deg;
  psiI_deg=psiI*rad2deg;
  da_deg=da*rad2deg;
  subplot(221),plot(t,betaI_deg,'r');
  title('Beta vs. Time for Aileron Maneuver');
  xlabel('Time (s)');
  ylabel('Beta (deg)');
  grid on;
  subplot(222),plot(t,phiI_deg,'k');
  title('Phi vs. Time for Aileron Maneuver');
  xlabel('Time (s)');
  ylabel('Phi (deg)');
  grid on;
  subplot(223),plot(t,psiI_deg,'b');
  title('Psi vs. Time for Aileron Maneuver');
  xlabel('Time (s)');
  ylabel('Psi (deg)');
  grid on;
  subplot(224),plot(t,da_deg,'m');
  title('Aileron Deflection vs. Time');
  xlabel('Time (s)');
  ylabel('Aileron Deflection (deg)');
  grid on;
end;
  end;
  if Library >4
    ans_r = menu('RESPONSES TO RUDDER MANEUVER','Beta','Phi','Psi','Rudder','All')
if ans_r == 1
  close all;
  betaII_deg=betaII*rad2deg;
  plot(t,betaII_deg,'r');
  title('Beta vs. Time for Rudder Manuever');
  xlabel('Time (s)');
  ylabel('Beta (deg)');
  grid on;
end;
if ans_r == 2
  close all;
  phiII_deg=phiII*rad2deg;
  plot(t,phiII_deg,'k');
  title('Phi vs. Time for Rudder Maneuver');
  xlabel('Time (s)');
  ylabel('Phi (deg)');
```

```
        grid on;
      end;
      if ans_r == 3
        close all;
        psiII_deg=psiII*rad2deg;
        plot(t,psiII_deg,'b');
        title('Psi vs. Time for Rudder Maneuver');
        xlabel('Time (s)');
        ylabel('Psi (deg)');
        grid on;
      end;
      if ans_r == 4
        close all;
        dr_deg=dr*rad2deg;
        plot(t,dr_deg,'m');
        title('Rudder Deflection vs. Time');
        xlabel('Time (s)');
        ylabel('Rudder Deflection (deg)');
        grid on;
      end;
      if ans_r == 5
        close all;
        betaII_deg=betaII*rad2deg;
        phiII_deg=phiII*rad2deg;
        psiII_deg=psiII*rad2deg;
        dr_deg=dr*rad2deg;
        subplot(221),plot(t,betaII_deg,'r');
        title('Beta vs. Time for Rudder Maneuver');
        xlabel('Time (s)');
        ylabel('Beta (deg)');
        grid on;
        subplot(222),plot(t,phiII_deg,'k');
        title('Phi vs. Time for Rudder Maneuver');
        xlabel('Time (s)');
        ylabel('Phi (deg)');
        grid on;
        subplot(223),plot(t,psiII_deg,'b');
        title('Psi vs. Time for Rudder Maneuver');
        xlabel('Time (s)');
        ylabel('Psi (deg)');
        grid on;
        subplot(224),plot(t,dr_deg,'m');
        title('Rudder Deflection vs. Time');
        xlabel('Time (s)');
        ylabel('Rudder Deflection (deg)');
        grid on;
      end;
        end;
% plot the poles on the s-domain
poles = roots(Dbar2)
figure,
plot(poles, '*');
grid on
title('Lateral-Directional Poles in the s-domain');
xlabel('Real Axis');
ylabel('Imaginary Axis');
omega_dr=sqrt(abs(poles(2,1)^2));
damp_dr=abs(real(poles(2,1)))/omega_dr;
t_rolling=-1/(poles(1,1));
t_spiral=-1/(poles(4,1));
```

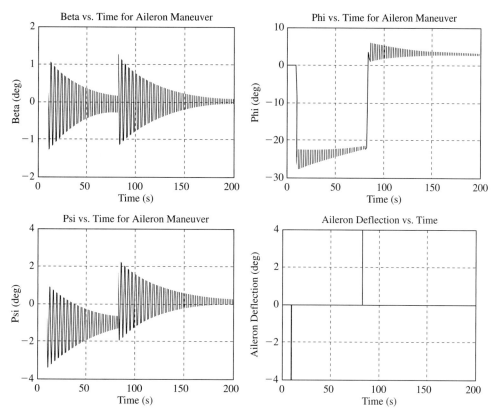

Figure SSP7.5.1 Learjet 24: Lateral Directional Responses Following Aileron Maneuver

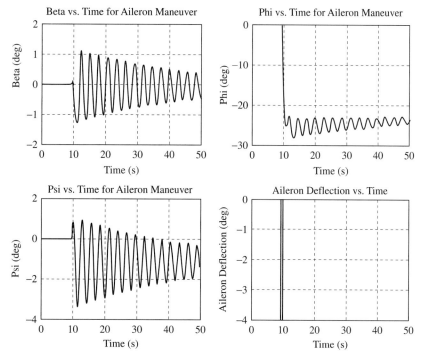

Figure SSP7.5.2 Learjet 24: Lateral Directional Responses Following Aileron Maneuver (ZOOMED)

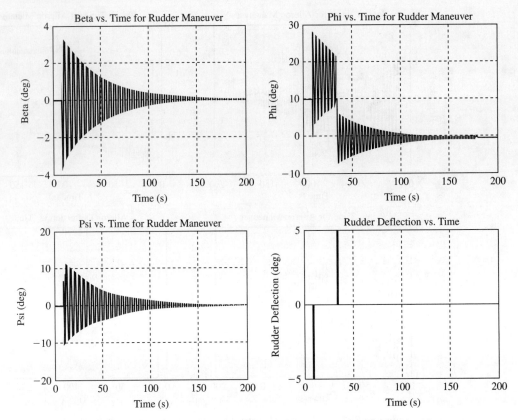

Figure SSP7.5.3 Learjet 24: Lateral Directional Responses Following Rudder Maneuver

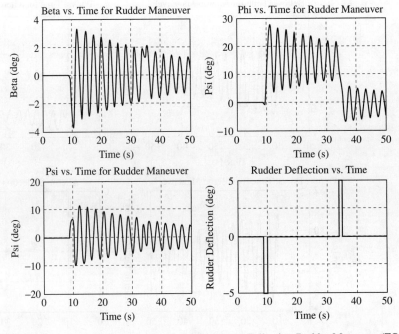

Figure SSP7.5.4 Learjet 24: Lateral Directional Responses Following Rudder Maneuver (ZOOMED)

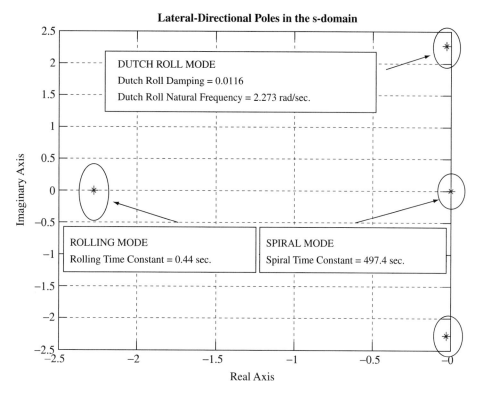

Figure SSP7.5.5 Learjet 24: Location of Lateral Directional Poles in the S-domain

The location of the poles of the lateral directional characteristic equation $\overline{D}_2(s)$ is shown below

$$\zeta_{DR} = 0.0116, \quad \omega_{n_{SP}} = 2.273$$
$$T_R = 0.44 \text{ sec}$$
$$T_S = 497.4 \text{ sec}$$

These values indicate a marginal overall quality for the lateral directional dynamics with a low damping for the dutch roll mode along with a marginally stable spiral mode. The roll time constant indicates fairly good rolling performance, which is typical for this class of aircraft. Particularly, the value of the dutch roll damping is lower than the acceptable range for this class of aircraft; this problem can be addressed with the introduction of a stability augmentation closed-loop system. The design of this class of systems is beyond the purpose of this text book; descriptions for these systems can be found in.[7,13,16] Essentially, these systems allow achieving desirable values for the above preceding parameters through the use of a feedback control scheme.

Refer to Chapter 10 for an understanding of the suitable ranges for these important parameters as well as a detailed explanation of the overall handling qualities for the aircraft system.

Using the modeling described in the previous MATLAB® code, the simulation of the aircraft lateral directional dynamics can be also performed within Simulink®, as shown in Figure 7.5.6, providing identical time histories.

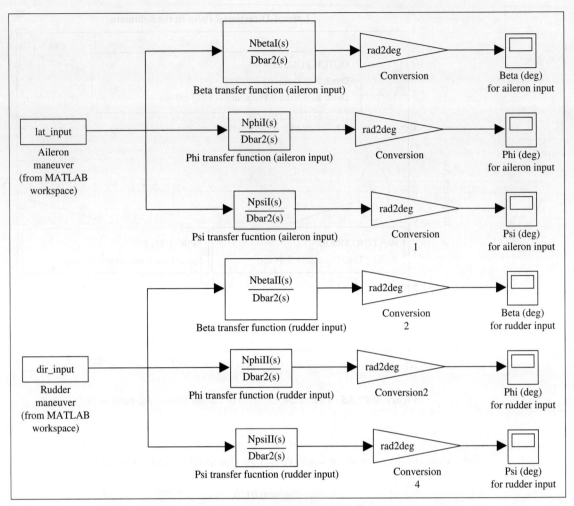

Figure SSP7.5.6 Learjet 24: Simulink® Block of the Lateral Directional Simulation

PROBLEMS

Problem 7.1

Consider the SIAI Marchetti S211 aircraft at cruise 1 flight conditions (Appendix B, Aircraft 8). From the available data first find the full longitudinal characteristic equations along with the associated roots, and evaluate the short period and phugoid characteristics. Next, find the characteristic equations associated with the short period and phugoid approximations. For both dynamic modes evaluate the error differences in terms of natural frequency and damping between the full and the approximated characteristic equations.

Problem 7.2

Consider the Mac Donnell Douglas F-4 aircraft at subsonic cruise flight conditions (Appendix B, Aircraft 10). From the available data first find the full longitudinal characteristic equations along with the associated roots, and evaluate the short period and phugoid characteristics. Next, find the characteristic equations associated with the short period and phugoid approximations. For both dynamic modes evaluate the error differences in terms of natural frequency and damping between the full and the approximated characteristic equations.

Problem 7.3

Consider the Beech 99 aircraft at high cruise flight conditions (Appendix B, Aircraft 3). From the available data first find the full longitudinal characteristic equations along with the associated roots, and evaluate the short period and phugoid characteristics. Next, find the characteristic equations associated with the short period and phugoid approximations. For both dynamic modes evaluate the error differences in terms of natural frequency and damping between the full and the approximated characteristic equations.

Problem 7.4

Consider an aircraft static model mounted on a scale in a wind tunnel. The interface between the scale and the model is such that the model is only able to rotate around its $Y_S = Y$ stability axis. Since the model is not able to perform a climb or a descent, for all practical purposes we have $\alpha = \theta, \dot{\alpha} = \dot{\theta}$. Therefore, the single degree of freedom equation describing the dynamic of the system is given by

$$\ddot{\theta} = M_\alpha \theta + (M_{\dot{\alpha}} + M_q)\dot{\theta} + M_{\delta_E} \delta_E$$

Derive relationships for the short period natural frequency and damping from the given equation and compare them with the same relationships derived using the short period approximation.

Problem 7.5

Consider again the Mc Donnell Douglas F-4 aircraft at subsonic cruise flight conditions (Appendix B, Aircraft 10). From the available dimensionless aerodynamic derivatives, geometric data, inertial data, and flight conditions at subsonic cruise conditions, develop a MATLAB® code performing a complete analysis of the aircraft lateral directional dynamics. Clearly identify the roots and the characteristics associated with the dutch roll, the rolling, and the spiral modes. Plot the lateral directional responses of the aircraft following short doublets on ailerons and rudder. Be careful in selecting the magnitude of the maneuvers such that the aircraft response does not go over the limitations associated with the small perturbation assumptions.

Problem 7.6

Consider the Cessna T37 aircraft at cruise flight conditions (Appendix B, Aircraft 2). Using the empirical modeling for the key longitudinal stability and control derivatives described in Chapter 3, along with the relationships for the dimensional stability derivatives in Table 7.1, perform a short period sensitivity analysis using the available dimensionless aerodynamic derivatives, geometric data, inertial data, and flight conditions at cruise conditions. Specifically, focus on the dependencies of the short period characteristics with the following parameters:

$$\frac{S_H}{S}, \bar{x}_{CG}, I_{YY}$$

Problem 7.7

Consider again the Cessna T37 aircraft at cruise flight conditions (Appendix B, Aircraft 2). Using the relationships for the dimensional stability derivatives in Table 7.3, perform a lateral directional sensitivity analysis using the available dimensionless aerodynamic derivatives, geometric data, inertial data, and flight conditions at cruise conditions. Specifically, focus on the dependencies of the dutch roll, rolling, and spiral characteristics with the following parameters:

$$c_{l_\beta}, c_{n_\beta}, c_{l_p}, c_{n_r}, c_{y_\beta}$$
$$I_{xx}, I_{zz}$$

Chapter 8

State Variable Modeling of the Aircraft Dynamics

TABLE OF CONTENTS

8.1 Introduction
8.2 Introduction to State Variables for Nonlinear Systems
8.3 Introduction to State Variables for Linear/Linearized Systems
8.4 State Variable Modeling of the Longitudinal Dynamics
8.5 State Variable Modeling of the Lateral Directional Dynamics
8.6 Augmentation of the Aircraft State Variable Modeling
    8.6.1 Modeling of the Altitude (h)
    8.6.2 Modeling of the Flight Path Angle (γ)
    8.6.3 Modeling of the Engine Dynamics
    8.6.4 Modeling of the Actuator Dynamics
    8.6.5 Modeling of the Atmospheric Turbulence
8.7 Summary of State Variable Modeling of the Aircraft Dynamics
8.8 Summary
    References
    Student Sample Problems
    Problems

8.1 INTRODUCTION

The previous chapter has described the well-known Laplace transformation (LT)-based approach for the solution of the aircraft dynamic equations. The approach features the important concept of transfer function (TF) expressing the relationship between an input and an output variable in the s-domain. With some minor modifications, this modeling tool was derived from the work by Lanchester and Bryan[1,2] in the early 1900s.

The key advantage of this approach is its relative simplicity, assuming that the LT-based solution of the equations is automated through MATLAB®-based software. Additionally, the approach allows us to gain valuable insights and a detailed understanding of the aircraft dynamics through the analysis of the characteristic equations leading to the analysis of the different longitudinal and lateral directional dynamic modes.[3]

However, the approach of TF-based modeling has a key drawback; in fact, its applicability is oriented toward simple single-input–single-output (SISO) systems. The extension of the TF-based modeling to multi-input–multi-output (MIMO) systems is possible but typically not practical from a computational point of view. In fact, while the TF-based solution of the aircraft dynamics outlined in Chapter 7 allows simple analysis of specific aircraft responses, the aircraft is really a MIMO system.

An alternative approach allowing a more efficient modeling of the aircraft MIMO system is an appealing alternative. Specifically, the analysis of the dynamics of MIMO systems can be performed using a different modeling approach based on the introduction of state variables.[4,5] Thus, a detailed state variable-based modeling of the aircraft system is the objective of this chapter.

8.2 INTRODUCTION TO STATE VARIABLES FOR NONLINEAR SYSTEMS

Appendix A, Section 5 provides a review of first and second order SISO linear dynamic systems modeled through the general approach of transfer functions (TF). TFs are typically used to model SISO linear systems described by linear constant coefficients differential equations (DEs).

Consider a generic n-th order dynamic system with multiple (m) inputs and instrumented with sensors providing measurements for (l) measurable outputs. In general terms, the system can be described by the following system of differential equations, known as the generic state variable model:

$$\begin{cases} \dot{x}_1(t) = f_1(x_1(t), x_2(t), x_3(t), \ldots, x_n(t), u_1(t), u_2(t), u_3(t), \ldots, u_m(t)) \\ \dot{x}_2(t) = f_2(x_1(t), x_2(t), x_3(t), \ldots, x_n(t), u_1(t), u_2(t), u_3(t), \ldots, u_m(t)) \\ \quad \ldots \\ \dot{x}_n(t) = f_n(x_1(t), x_2(t), x_3(t), \ldots, x_n(t), u_1(t), u_2(t), u_3(t), \ldots, u_m(t)) \end{cases}$$

$$\begin{cases} y_1(t) = g_1(x_1(t), x_2(t), x_3(t), \ldots, x_n(t), u_1(t), u_2(t), u_3(t), \ldots, u_m(t)) \\ y_2(t) = g_2(x_1(t), x_2(t), x_3(t), \ldots, x_n(t), u_1(t), u_2(t), u_3(t), \ldots, u_m(t)) \\ \quad \ldots \\ y_l(t) = g_l(x_1(t), x_2(t), x_3(t), \ldots, x_n(t), u_1(t), u_2(t), u_3(t), \ldots, u_m(t)). \end{cases}$$

where the first system is known as the state equations and the second set is known as the output equations. Specifically, we have

$(x_1(t), x_2(t), \ldots, x_n(t))$, defined as the state variables

$(u_1(t), u_2(t), \ldots, u_m(t))$, defined as the system inputs

$(y_1(t), y_2(t), \ldots, y_l(t))$, defined as the system (measurable) outputs

DEFINITION. The state variables (SV) of a system are defined as a minimal set of variables $(x_1(t), x_2(t), x_3(t), \ldots, x_n(t))$ such that the knowledge of these variables at any initial time t_0 *in addition to* information on the input excitation $(u_1(t), u_2(t), u_3(t), \ldots, u_m(t))$ applied at any time $t > t_0$ are sufficient to determine the state of the system at any time $t > t_0$.

NOTE: The state variables are not necessarily measurable outputs of the system.

NOTE: The selection of a set of state variables might not be unique.

NOTE: n state variables are necessary to model a n-th order dynamic system.

8.3 INTRODUCTION TO STATE VARIABLES FOR LINEAR/LINEARIZED SYSTEMS

For nonlinear dynamic systems with multiple (m) inputs and multiple (l) outputs (MIMO), the functions $(f_1(x_1(t), \ldots, x_n(t), u_1(t), \ldots, u_m(t)), \ldots, f_n(x_1(t), \ldots, x_n(t), u_1(t), \ldots, u_m(t)))$ and $(g_1(x_1(t), \ldots, x_n(t), u_1(t), \ldots, u_m(t)), \ldots, g_l(x_1(t), \ldots, x_n(t), u_1(t), \ldots, u_m(t)))$ are nonlinear. However, if the system is linear—or if it can be linearized around a set of operating initial conditions—the preceding state variable model reduces itself to the form:

$$\begin{Bmatrix} \dot{x}_1(t) \\ \dot{x}_2(t) \\ \ldots \\ \dot{x}_n(t) \end{Bmatrix} = A_{nxn} \begin{Bmatrix} x_1(t) \\ x_2(t) \\ \ldots \\ x_n(t) \end{Bmatrix} + B_{nxm} \begin{Bmatrix} u_1(t) \\ u_2(t) \\ \ldots \\ u_m(t) \end{Bmatrix}$$

$$\begin{Bmatrix} y_1(t) \\ y_2(t) \\ \ldots \\ y_l(t) \end{Bmatrix} = C_{lxn} \begin{Bmatrix} x_1(t) \\ x_2(t) \\ \ldots \\ x_n(t) \end{Bmatrix} + D_{lxm} \begin{Bmatrix} u_1(t) \\ u_2(t) \\ \ldots \\ u_m(t) \end{Bmatrix}$$

where

$[A]_{nxn}$ is known as the state matrix

$[B]_{nxm}$ is known as the control matrix

$[C]_{lxn}$ is known as the observation matrix

$[D]_{lxm}$ is known as the input Observation matrix

In a more compacted matrix notation, we have

$$\{\dot{\bar{x}}\}_{n\times 1} = \overline{\overline{A}}_{n\times n}\{\bar{x}\}_{n\times 1} + \overline{\overline{B}}_{n\times m}\{\bar{u}\}_{m\times 1}$$
$$\{\bar{y}\}_{l\times 1} = \overline{\overline{C}}_{l\times n}\{\bar{x}\}_{n\times 1} + \overline{\overline{D}}_{l\times m}\{\bar{u}\}_{m\times 1}$$

The preceding linear system featuring m inputs and l outputs could also be modeled with a matrix of transfer functions. Particularly, using the superimposition of effects associated with the linearity of the system, the Laplace transform of the generic i-th output of the system is given by

$$Y_i(s) = \left(\frac{Y_i(s)}{U_1(s)}\right) \cdot U_1(s) + \left(\frac{Y_i(s)}{U_2(s)}\right) \cdot U_2(s) + \cdots \left(\frac{Y_i(s)}{U_m(s)}\right) \cdot U_m(s)$$
$$= G_{i1}(s) \cdot U_1(s) + G_{i2}(s) \cdot U_2(s) + \cdots G_{im}(s) \cdot U_m(s)$$

Considering all the l measureable outputs subjected to m inputs we have

$$\begin{cases} Y_1(s) = G_{11}(s) \cdot U_1(s) + G_{12}(s) \cdot U_2(s) + \cdots G_{1m}(s) \cdot U_m(s) \\ Y_2(s) = G_{21}(s) \cdot U_1(s) + G_{22}(s) \cdot U_2(s) + \cdots G_{2m}(s) \cdot U_m(s) \\ \quad\quad\quad\quad \ldots\ldots \\ Y_i(s) = G_{i1}(s) \cdot U_1(s) + G_{i2}(s) \cdot U_2(s) + \cdots G_{im}(s) \cdot U_m(s) \\ \quad\quad\quad\quad \ldots\ldots \\ Y_l(s) = G_{l1}(s) \cdot U_1(s) + G_{l2}(s) \cdot U_2(s) + \cdots G_{lm}(s) \cdot U_m(s) \end{cases}$$

Therefore, we can introduce

$$\{\overline{Y}(s)\}_{l\times 1} = \begin{Bmatrix} Y_1(s) \\ Y_2(s) \\ \ldots \\ Y_i(s) \\ \ldots \\ Y_l(s) \end{Bmatrix}_{l\times 1}, \quad \{\overline{U}(s)\}_{m\times 1} = \begin{Bmatrix} U_1(s) \\ U_2(s) \\ \ldots \\ U_j(s) \\ \ldots \\ U_m(s) \end{Bmatrix}_{m\times 1}$$

such that the following matrix of transfer functions can be defined:

$$\frac{\{\overline{Y}(s)\}_{l\times 1}}{\{\overline{U}(s)\}_{m\times 1}} = \left[\overline{\overline{G}}(s)\right]_{l\times m} = \begin{bmatrix} G_{11}(s) & G_{12}(s) & \ldots & G_{1j}(s) & \ldots & G_{1m}(s) \\ G_{21}(s) & G_{22}(s) & \ldots & G_{2j}(s) & \ldots & G_{2m}(s) \\ \ldots & \ldots & \ldots & \ldots & \ldots & \ldots \\ G_{i1}(s) & G_{i2}(s) & \ldots & G_{ij}(s) & \ldots & G_{im}(s) \\ \ldots & \ldots & \ldots & \ldots & \ldots & \ldots \\ G_{l1}(s) & G_{l2}(s) & \ldots & G_{lj}(s) & \ldots & G_{lm}(s) \end{bmatrix}_{l\times m}$$

where all the transfer functions in the matrix shown share the same characteristic equation.

Based on this modeling, it is possible to derive a closed-form relationship between the state variable model of the system and its matrix of transfer functions. Consider again the state variable model (where the dimensions have been omitted for clarity purposes):

$$\{\dot{\bar{x}}\} = \overline{\overline{A}}\{\bar{x}\} + \overline{\overline{B}}\{\bar{u}\}$$
$$\{\bar{y}\} = \overline{\overline{C}}\{\bar{x}\} + \overline{\overline{D}}\{\bar{u}\}$$

Assuming zero initial conditions, the application of the Laplace transformation to the shown matrix expression leads to

$$s\overline{X}(s) \cancel{-\bar{x}(0)} = \overline{\overline{A}}\,\overline{X}(s) + \overline{\overline{B}}\,\overline{U}(s)$$
$$\overline{Y}(s) = \overline{\overline{C}}\,\overline{X}(s) + \overline{\overline{D}}\,\overline{U}(s)$$

Solving for $\overline{X}(s)$ from the first relationship above leads to

$$s\overline{X}(s) - \overline{\overline{A}}\,\overline{X}(s) = \overline{\overline{B}}\,\overline{U}(s) \quad \rightarrow \quad \overline{X}(s) = [s\overline{\overline{I}} - \overline{\overline{A}}]^{-1}\,\overline{\overline{B}}\,\overline{U}(s)$$

therefore

$$\overline{Y}(s) = \overline{\overline{C}}\left[s\overline{\overline{I}} - \overline{\overline{A}}\right]^{-1}\overline{\overline{B}}\,\overline{U}(s) + \overline{\overline{D}}\,\overline{U}(s)$$
$$= \left\{\overline{\overline{C}}\left[s\overline{\overline{I}} - \overline{\overline{A}}\right]^{-1}\overline{\overline{B}} + \overline{\overline{D}}\right\}\overline{U}(s)$$

where $\overline{\overline{I}}$ is the $(n \times n)$ identity matrix, leading to

$$\boxed{\frac{\overline{Y}(s)}{\overline{U}(s)} = \left[\overline{\overline{G}}(s)\right] = \left\{\overline{\overline{C}}\left[s\overline{\overline{I}} - \overline{\overline{A}}\right]^{-1}\overline{\overline{B}} + \overline{\overline{D}}\right\}}$$

This derivation also shows an important property of the relationship between the matrix of the transfer functions and the state variable model of a linear system, that is the *roots of the characteristic equation of a system are also the eigenvalues of the state matrix A*.

The general relationships between the differential equations (describing the system dynamics) and the different modeling tools (transfer functions and state variable model) are shown in the conceptual block diagram in Figure 8.1.

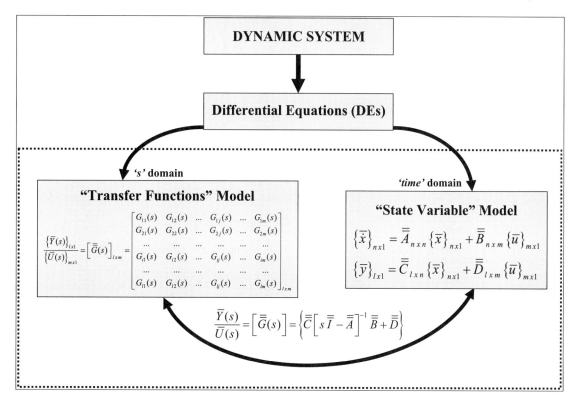

Figure 8.1 Relationship between Differential Equations and TF and SV Models

8.4 STATE VARIABLE MODELING OF THE LONGITUDINAL DYNAMICS

The drawbacks of the TF-based modeling have been discussed in Section 8.1. Sections 8.4 and 8.5 will describe the alternative state variable (SV) modeling.[6,7]

Although the TF-based modeling has allowed a detailed understanding of the key stability issues in the aircraft dynamics, the advantages of SV modeling for describing the linearized aircraft dynamics are very clear. Assuming only four conventional control surfaces (elevators, stabilators, ailerons, and rudder) and a minimum of ten dynamic variables, a TF-based model requires the evaluation of forty transfer functions, which implies the calculation of forty different numerators and two denominators (that is, the longitudinal and the lateral directional characteristic equations). Additional transfer functions will be required for modeling the thrust control. The modeling has to be repeated at each different flight condition within the aircraft operating envelope, making the application of the TF-based modeling computationally demanding. On the other side, a single SV model at each flight condition provides the same solutions.

As for the TF-based modeling, the longitudinal dynamics will be first discussed, followed by the lateral directional dynamics.

The starting point for the development of the SV model of the longitudinal dynamics is the set of longitudinal equations in Chapter 7:[6,7]

$$\dot{u} = (X_u + X_{T_u})u + X_\alpha \alpha - g\cos\Theta_1 \theta + X_{\delta_E}\delta_E$$
$$V_{P_1}\dot{\alpha} = Z_u u + Z_\alpha \alpha + Z_{\dot{\alpha}}\dot{\alpha} - g\sin\Theta_1 \theta + (Z_q + V_{P_1})\dot{\theta} + Z_{\delta_E}\delta_E$$
$$\ddot{\theta} = (M_u + M_{T_u})u + (M_\alpha + M_{T_\alpha})\alpha + M_{\dot{\alpha}}\dot{\alpha} + M_q\dot{\theta} + M_{\delta_E}\delta_E$$

Note that the elevators are here assumed to be the only longitudinal control surface. The modeling can be extended to include the presence of the stabilators as well as additional longitudinal (symmetric) control surfaces, shown in Figure 8.2.

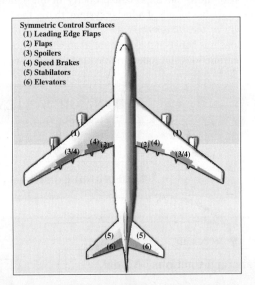

Figure 8.2 Longitudinal Control Surfaces

Using the relationships $(q = \dot{\theta}, \quad \dot{q} = \ddot{\theta})$, the equations can be rewritten as

$$\dot{u} = (X_u + X_{T_u})u + X_\alpha \alpha - g\cos\Theta_1 \theta + X_{\delta_E}\delta_E$$
$$(V_{P_1} - Z_{\dot{\alpha}})\dot{\alpha} = Z_u u + Z_\alpha \alpha - g\sin\Theta_1 \theta + (Z_q + V_{P_1})q + Z_{\delta_E}\delta_E$$
$$\dot{q} = (M_u + M_{T_u})u + (M_\alpha + M_{T_\alpha})\alpha + M_{\dot{\alpha}}\dot{\alpha} + M_q q + M_{\delta_E}\delta_E$$
$$\dot{\theta} = q$$

A close analysis of these equations reveals that the second equation in $(\dot{\alpha})$ is embedded into the third equation in (q). Solving the second equation and inserting it in the third equation leads to

$$\dot{u} = (X_u + X_{T_u})u + X_\alpha \alpha - g\cos\Theta_1 \theta + X_{\delta_E}\delta_E$$
$$\dot{\alpha} = \frac{Z_u}{(V_{P_1} - Z_{\dot{\alpha}})}u + \frac{Z_\alpha}{(V_{P_1} - Z_{\dot{\alpha}})}\alpha - \frac{g\sin\Theta_1}{(V_{P_1} - Z_{\dot{\alpha}})}\theta + \frac{(Z_q + V_{P_1})}{(V_{P_1} - Z_{\dot{\alpha}})}q + \frac{Z_{\delta_E}}{(V_{P_1} - Z_{\dot{\alpha}})}\delta_E$$
$$\dot{q} = M_{\dot{\alpha}}\left[\frac{Z_u}{(V_{P_1} - Z_{\dot{\alpha}})}u + \frac{Z_\alpha}{(V_{P_1} - Z_{\dot{\alpha}})}\alpha - \frac{g\sin\Theta_1}{(V_{P_1} - Z_{\dot{\alpha}})}\theta + \frac{(Z_q + V_{P_1})}{(V_{P_1} - Z_{\dot{\alpha}})}q + \frac{Z_{\delta_E}}{(V_{P_1} - Z_{\dot{\alpha}})}\delta_E\right]$$
$$+ (M_u + M_{T_u})u + (M_\alpha + M_{T_\alpha})\alpha + {} + M_q q + M_{\delta_E}\delta_E$$
$$\dot{\theta} = q$$

8.4 State Variable Modeling of the Longitudinal Dynamics 437

Neglecting the contribution associated with the propulsive dimensional derivatives (M_{T_u}, M_{T_α}), the preceding equations can be rearranged as in the following:

$$\dot{u} = (X_u + X_{T_u})u + X_\alpha \alpha - g\cos\Theta_1 \theta + X_{\delta_E}\delta_E$$

$$\dot{\alpha} = \frac{Z_u}{(V_{P_1} - Z_{\dot{\alpha}})}u + \frac{Z_\alpha}{(V_{P_1} - Z_{\dot{\alpha}})}\alpha - \frac{g\sin\Theta_1}{(V_{P_1} - Z_{\dot{\alpha}})}\theta + \frac{(Z_q + V_{P_1})}{(V_{P_1} - Z_{\dot{\alpha}})}q + \frac{Z_{\delta_E}}{(V_{P_1} - Z_{\dot{\alpha}})}\delta_E$$

$$\dot{q} = \left[M_{\dot{\alpha}}\left(\frac{Z_u}{(V_{P_1} - Z_{\dot{\alpha}})}\right) + M_u\right]u + \left[M_{\dot{\alpha}}\left(\frac{Z_\alpha}{(V_{P_1} - Z_{\dot{\alpha}})}\right) + M_\alpha\right]\alpha + \left[M_{\dot{\alpha}}\left(-\frac{g\sin\Theta_1}{(V_{P_1} - Z_{\dot{\alpha}})}\right)\right]\theta$$

$$+ \left[M_{\dot{\alpha}}\left(\frac{(Z_q + V_{P_1})}{(V_{P_1} - Z_{\dot{\alpha}})}\right) + M_q\right]q + \left[M_{\dot{\alpha}}\left(\frac{Z_{\delta_E}}{(V_{P_1} - Z_{\dot{\alpha}})}\right) + M_{\delta_E}\right]\delta_E$$

$$\dot{\theta} = q$$

The objective is to reduce the longitudinal equations to the SV model:

$$\dot{x}_{Long} = A_{Long} x_{Long} + B_{Long} u_{Long}$$
$$y_{Long} = C_{Long} x_{Long} + D_{Long} u_{Long}$$

(x_{Long}) represents the set of longitudinal state variables and is given by

$$x_{Long} = \begin{Bmatrix} u \\ \alpha \\ q \\ \theta \end{Bmatrix}$$

(u_{Long}) is known as the longitudinal input column. If only (δ_E) is considered as a longitudinal control surface (u_{Long}) reduces itself to a scalar $(u_{Long} = \{\delta_E\})$. Therefore, the longitudinal state equations take on the form

$$\begin{Bmatrix} \dot{u} \\ \dot{\alpha} \\ \dot{q} \\ \dot{\theta} \end{Bmatrix} = A_{Long} \begin{Bmatrix} u \\ \alpha \\ q \\ \theta \end{Bmatrix} + B_{Long}\{\delta_E\}$$

$$= \begin{bmatrix} X'_u & X'_\alpha & X'_q & X'_\theta \\ Z'_u & Z'_\alpha & Z'_q & Z'_\theta \\ M'_u & M'_\alpha & M'_q & M'_\theta \\ 0 & 0 & 1 & 0 \end{bmatrix} \begin{Bmatrix} u \\ \alpha \\ q \\ \theta \end{Bmatrix} + \begin{bmatrix} X'_{\delta_E} \\ Z'_{\delta_E} \\ M'_{\delta_E} \\ 0 \end{bmatrix} \{\delta_E\}$$

where $[A_{Long}]$ is known as the longitudinal state matrix and $[B_{Long}]$ is known as the longitudinal input matrix (in this case a column matrix since $(u_{Long} = \{\delta_E\})$ is a scalar). The coefficients of $[A_{Long}]$ and $[B_{Long}]$ are given by

$$X'_u = (X_u + X_{T_u}), \quad X'_\alpha = X_\alpha, \quad X'_\theta = -g\cos\Theta_1, \quad X'_q = 0, \quad X'_{\delta_E} = X_{\delta_E}$$

$$Z'_u = \frac{Z_u}{(V_{P_1} - Z'_{\dot{\alpha}})}, \quad Z'_\alpha = \frac{Z_\alpha}{(V_{P_1} - Z'_{\dot{\alpha}})}, \quad Z'_q = \frac{(Z_q + V_{P_1})}{(V_{P_1} - Z'_{\dot{\alpha}})},$$

$$Z'_\theta = -\frac{g\sin\Theta_1}{(V_{P_1} - Z'_{\dot{\alpha}})}, \quad Z'_{\delta_E} = \frac{Z_{\delta_E}}{(V_{P_1} - Z'_{\dot{\alpha}})}$$

$$M'_u = M_{\dot{\alpha}}Z'_u + M_u, \quad M'_\alpha = M_{\dot{\alpha}}Z'_\alpha + M_\alpha$$

$$M'_\theta = M_{\dot{\alpha}}Z'_\theta, \quad M'_q = M_{\dot{\alpha}}Z'_q + M_q$$

$$M'_{\delta_E} = M_{\dot{\alpha}}Z'_{\delta_E} + M_{\delta_E}$$

As discussed in Section 8.3, the eigenvalues of $[A_{Long}]$ are coincident with the roots of the longitudinal characteristic equations (associated with the short period and the phugoid modes).

The level of complexity associated with the derivation of the longitudinal output equations ($y_{Long} = C_{Long} x_{Long} + D_{Long} u_{Long}$) is dependent on the selection of the elements of the longitudinal 'output vector' (y_{Long}). This vector represents the aircraft longitudinal parameters which are measured by the on-board sensors. The availability of these measurements is critical for the design of flight control laws. In general (y_{Long}) contains the entire (x_{Long}) along with a number of additional parameters (for example, linear accelerations (a_X, a_Z), climb angle (γ), altitude (h), and others). Without loss of generality, (y_{Long}) is here assumed to include only the longitudinal vertical acceleration (a_Z) in addition to the longitudinal state variable column (x_{Long}). Therefore

$$y_{Long} = \begin{Bmatrix} a_Z \\ x_{Long} \end{Bmatrix} = \begin{Bmatrix} a_Z \\ u \\ \alpha \\ q \\ \theta \end{Bmatrix}$$

Next, the modeling for a_Z is required. A general relationship for a_Z is given by

$$a_Z = \frac{\sum F_Z}{m}$$

Since all the aerodynamic forces have been modeled within the stability axes X_S, Y_S, Z_S, an initial issue to be resolved is that a_Z has to be expressed with respect to the stability axes Z_S. Therefore, the third equation of the small perturbation Conservation of Linear Momentum Equations (CLMEs) expressed in the body axes

$$m[\dot{w} - U_1 q] = -mg\,\theta \sin\Theta_1 + (f_{A_Z} + f_{T_Z})$$

takes on the following form along the stability axes:

$$m[\dot{w} - V_{P_1} q] = -mg\,\gamma \sin\Theta_1 + (f_{A_Z} + f_{T_Z})$$

Since $\dot{w} = V_{P_1}\dot{\alpha}, \gamma = \theta - \alpha$, we have

$$m[V_{P_1}\dot{\alpha} - V_{P_1} q] = -mg \sin\Theta_1\,\theta + mg \sin\Theta_1 \alpha + (f_{A_Z} + f_{T_Z})$$

Thus, the aerodynamic and thrust vertical forces along the stability axes can be expressed using

$$(f_{A_Z} + f_{T_Z}) = m[V_{P_1}\dot{\alpha} - V_{P_1} q] + mg \sin\Theta_1\,\theta - mg \sin\Theta_1 \alpha$$

Therefore, an expression for a_Z is given by

$$a_Z = \frac{(f_{A_Z} + f_{T_Z})}{m} = \frac{m[V_{P_1}\dot{\alpha} - V_{P_1} q] + mg \sin\Theta_1 \theta - mg \sin\Theta_1 \alpha}{m}$$

$$= [V_{P_1}\dot{\alpha} - V_{P_1} q] + g \sin\Theta_1 \theta - g \sin\Theta_1 \alpha$$

From the previous modeling of the state variables, an expression for $\dot{\alpha}$ is given by

$$\dot{\alpha} = Z'_u u + Z'_\alpha \alpha + Z'_q q + Z'_\theta \theta + Z'_{\delta_E} \delta_E$$

Therefore

$$a_Z = [V_{P_1}(Z'_u u + Z'_\alpha \alpha + Z'_q q + Z'_\theta \theta + Z'_{\delta_E} \delta_E) - V_{P_1} q] + g \sin\Theta_1 \theta - g \sin\Theta_1 \alpha$$

8.4 State Variable Modeling of the Longitudinal Dynamics

Grouping the terms with the same variables we have

$$a_Z = (V_{P_1}Z'_u)u + (V_{P_1}Z'_\alpha - g\sin\Theta_1)\alpha + [V_{P_1}(Z'_q - 1)]q + (V_{P_1}Z'_\theta + g\sin\Theta_1)\theta + (V_{P_1}Z'_{\delta_E})\delta_E$$

Rearranging in a matrix format, the longitudinal output equations take on the form:

$$\begin{Bmatrix} a_z \\ u \\ \alpha \\ q \\ \theta \end{Bmatrix} = C_{Long} \begin{Bmatrix} u \\ \alpha \\ q \\ \theta \end{Bmatrix} + D_{Long}\{\delta_E\}$$

$$= \begin{bmatrix} Z''_\alpha & Z''_u & Z''_q & Z''_\theta \\ 1 & 0 & 0 & 0 \\ 0 & 1 & 0 & 0 \\ 0 & 0 & 1 & 0 \\ 0 & 0 & 0 & 1 \end{bmatrix} \begin{Bmatrix} u \\ \alpha \\ q \\ \theta \end{Bmatrix} + \begin{bmatrix} Z''_{\delta_E} \\ 0 \\ 0 \\ 0 \\ 0 \end{bmatrix} \{\delta_E\}$$

where $[C_{Long}]$ is known as the longitudinal observation matrix and $[D_{Long}]$ is known as the longitudinal output control matrix (in this case a column matrix, since $(u_{Long} = \{\delta_E\})$ is a scalar). The coefficients of $[C_{Long}]$ and $[D_{Long}]$ are given by

$$Z''_u = Z'_u V_{P_1}, \quad Z''_\alpha = Z'_\alpha V_{P_1} - g\sin\Theta_1, \quad Z''_q = (Z'_q - 1)V_{P_1}$$
$$Z''_\theta = Z'_\theta V_{P_1} + g\sin\Theta_1, \quad Z''_{\delta_E} = Z'_{\delta_E} V_{P_1}$$

As stated previously, in the more general case where the modeling of additional longitudinal inputs is required (u_{Long}) is a column and $[B_{Long}]$ and $[D_{Long}]$ become matrices. For example, consider the general case in which the longitudinal dynamic is controlled through deflections of canard surfaces, stabilators, and throttle settings (for the engine thrust) in addition to deflection of the elevators. In this case the longitudinal inputs are given by

$$\{u_{Long}\} = \begin{Bmatrix} \delta_E \\ i_H \\ \delta_{Canards} \\ \delta_{Throttle} \end{Bmatrix} = \begin{Bmatrix} \delta_E \\ i_H \\ \delta_C \\ \delta_T \end{Bmatrix}$$

The longitudinal state equations will therefore be given by

$$\begin{Bmatrix} \dot u \\ \dot\alpha \\ \dot q \\ \dot\theta \end{Bmatrix} = A_{Long} \begin{Bmatrix} u \\ \alpha \\ q \\ \theta \end{Bmatrix} + B_{Long} \begin{Bmatrix} \delta_E \\ i_H \\ \delta_C \\ \delta_T \end{Bmatrix}$$

$$= \begin{bmatrix} X'_u & X'_\alpha & X'_q & X'_\theta \\ Z'_u & Z'_\alpha & Z'_q & Z'_\theta \\ M'_u & M'_\alpha & M'_q & M'_\theta \\ 0 & 0 & 1 & 0 \end{bmatrix} \begin{Bmatrix} u \\ \alpha \\ q \\ \theta \end{Bmatrix} + \begin{bmatrix} X'_{\delta_E} & X'_{i_H} & X'_{\delta_C} & X'_{\delta_T} \\ Z'_{\delta_E} & Z'_{i_H} & Z'_{\delta_C} & Z'_{\delta_T} \\ M'_{\delta_E} & M'_{i_H} & M'_{\delta_C} & M'_{\delta_T} \\ 0 & 0 & 0 & 0 \end{bmatrix} \begin{Bmatrix} \delta_E \\ i_H \\ \delta_C \\ \delta_T \end{Bmatrix}$$

where the additional coefficients in the $[B_{Long}]$ matrix are given by

$$X'_{i_H} = X_{i_H}, \quad X'_{\delta_C} = X_{\delta_C}, \quad X'_{\delta_T} = X_{\delta_T}$$

$$Z'_{i_H} = \frac{Z_{i_H}}{(V_{P_1} - Z'_{\dot\alpha})}, \quad Z'_{\delta_C} = \frac{Z_{\delta_C}}{(V_{P_1} - Z'_{\dot\alpha})}, \quad Z'_{\delta_T} = \frac{Z_{\delta_T}}{(V_{P_1} - Z'_{\dot\alpha})}$$

$$M'_{i_H} = M_{\dot\alpha} Z'_{i_H} + M_{i_H}, \quad M'_{\delta_C} = M_{\dot\alpha} Z'_{\delta_C} + M_{\delta_C}, \quad M'_{\delta_T} = M_{\dot\alpha} Z'_{\delta_T} + M_{\delta_T}$$

The longitudinal dimensional derivatives $(X_{i_H}, X_{\delta_C}, X_{\delta_T}, Z_{i_H}, Z_{\delta_C}, Z_{\delta_T}, M_{i_H}, M_{\delta_C}, M_{\delta_T})$ are associated with the dimensionless aerodynamic derivatives $(c_{D_{iH}}, c_{D_{\delta C}}, c_{L_{iH}}, c_{L_{\delta C}}, c_{m_{iH}}, c_{m_{\delta C}})$ and propulsion derivatives $(c_{X_{\delta T}}, c_{Z_{\delta T}}, c_{m_{\delta T}})$. Similarly, the longitudinal output equations will be given by

$$\begin{Bmatrix} a_z \\ \alpha \\ u \\ q \\ \theta \end{Bmatrix} = C_{Long} \begin{Bmatrix} \alpha \\ u \\ q \\ \theta \end{Bmatrix} + D_{Long} \begin{Bmatrix} \delta_E \\ i_H \\ \delta_C \\ \delta_T \end{Bmatrix}$$

$$= \begin{bmatrix} Z''_\alpha & Z''_u & Z''_q & Z''_\theta \\ 1 & 0 & 0 & 0 \\ 0 & 1 & 0 & 0 \\ 0 & 0 & 1 & 0 \\ 0 & 0 & 0 & 1 \end{bmatrix} \begin{Bmatrix} \alpha \\ u \\ q \\ \theta \end{Bmatrix} + \begin{bmatrix} Z''_{\delta E} & Z''_{iH} & Z''_{\delta C} & Z''_{\delta C} \\ 0 & 0 & 0 & 0 \\ 0 & 0 & 0 & 0 \\ 0 & 0 & 0 & 0 \\ 0 & 0 & 0 & 0 \end{bmatrix} \begin{Bmatrix} \delta_E \\ i_H \\ \delta_C \\ \delta_T \end{Bmatrix}$$

where the additional coefficients in the $[D_{Long}]$ matrix are given by

$$Z''_{i_H} = Z'_{i_H} V_{P_1},\ Z''_{\delta_C} = Z'_{\delta_C} V_{P_1},\ Z''_{\delta_T} = Z'_{\delta_T} V_{P_1}$$

8.5 STATE VARIABLE MODELING OF THE LATERAL DIRECTIONAL DYNAMICS

The state variable modeling of the aircraft lateral directional dynamics is performed with a similar procedure. The starting point for this modeling is given by the set of equations in Chapter 7:[6,7]

$$(V_{P_1}\dot\beta + V_{P_1}\dot\psi) = g\cos\Theta_1 \phi + Y_\beta \beta + Y_{\dot\phi}\dot\phi + Y_{\dot\psi}\dot\psi + Y_{\delta_A}\delta_A + Y_{\delta_R}\delta_R$$

$$\ddot\phi - \frac{I_{XZ}}{I_{XX}}\ddot\psi = L_\beta \beta + L_{\dot\phi}\dot\phi + L_{\dot\psi}\dot\psi + L_{\delta_A}\delta_A + L_{\delta_R}\delta_R$$

$$\ddot\psi - \frac{I_{XZ}}{I_{ZZ}}\ddot\phi = N_\beta \beta + N_{\dot\phi}\dot\phi + N_{\dot\psi}\dot\psi + N_{\delta_A}\delta_A + N_{\delta_R}\delta_R$$

where $Y_{\dot\phi} = Y_p$, $Y_{\dot\psi} = Y_r$, $L_{\dot\phi} = L_p, L_{\dot\psi} = L_r$, $N_{\dot\phi} = N_p$, $N_{\dot\psi} = N_r$.
Next, using the relationships $(r \approx \dot\psi, \dot r \approx \ddot\psi, p \approx \dot\phi, \dot p \approx \ddot\phi)$, the equations can be rearranged and rewritten as

$$(V_{P_1}\dot\beta) = Y_\beta \beta + Y_p p + (Y_r - V_{P_1})r + g\cos\Theta_1 \phi + Y_{\delta_A}\delta_A + Y_{\delta_R}\delta_R$$

$$\dot p - \frac{I_{XZ}}{I_{XX}}\dot r = L_\beta \beta + L_p p + L_r r + L_{\delta_A}\delta_A + L_{\delta_R}\delta_R$$

$$\dot r - \frac{I_{XZ}}{I_{ZZ}}\dot p = N_\beta \beta + N_p p + N_r r + N_{\delta_A}\delta_A + N_{\delta_R}\delta_R$$

The second and third equations are coupled and need to be solved separately. The solution of the second equation in terms of $(\dot p)$ provides:

$$\dot p = L_\beta \beta + L_p p + L_r r + L_{\delta_A}\delta_A + L_{\delta_R}\delta_R + \frac{I_{XZ}}{I_{XX}}\dot r$$

The substitution of the $(\dot p)$ equation in the $(\dot r)$ equation leads to

$$\dot r - \frac{I_{XZ}}{I_{ZZ}}\left[L_\beta \beta + L_p p + L_r r + L_{\delta_A}\delta_A + L_{\delta_R}\delta_R + \frac{I_{XZ}}{I_{XX}}\dot r\right]$$

$$= N_\beta \beta + N_p p + N_r r + N_{\delta_A}\delta_A + N_{\delta_R}\delta_R$$

8.5 State Variable Modeling of the Lateral Directional Dynamics

Using $I_1 = I_{XZ}/I_{XX}$, $I_2 = I_{XZ}/I_{ZZ}$, the $(\dot{r})$ equation can be simplified to

$$\dot{r} - I_1 I_2 \dot{r} = I_2 \left[L_\beta \beta + L_p p + L_r r + L_{\delta_A} \delta_A + L_{\delta_R} \delta_R \right]$$
$$+ N_\beta \beta + N_p p + N_r r + N_{\delta_A} \delta_A + N_{\delta_R} \delta_R$$

$$\dot{r} - I_1 I_2 \dot{r} = (I_2 L_\beta + N_\beta)\beta + (I_2 L_p + N_p)p + (I_2 L_r + N_r)r$$
$$+ (I_2 L_{\delta_A} + N_{\delta_A})\delta_A + (I_2 L_{\delta_R} + N_{\delta_R})\delta_R$$

and finally

$$\dot{r} = \frac{(I_2 L_\beta + N_\beta)}{(1 - I_1 I_2)}\beta + \frac{(I_2 L_p + N_p)}{(1 - I_1 I_2)}p + \frac{(I_2 L_r + N_r)}{(1 - I_1 I_2)}r$$
$$+ \frac{(I_2 L_{\delta_A} + N_{\delta_A})}{(1 - I_1 I_2)}\delta_A + \frac{(I_2 L_{\delta_R} + N_{\delta_R})}{(1 - I_1 I_2)}\delta_R$$

The insertion of the above $(\dot{r})$ equation into the $(\dot{p})$ equation leads to

$$\dot{p} = L_\beta \beta + L_p p + L_r r + L_{\delta_A} \delta_A + L_{\delta_R} \delta_R$$
$$+ I_1 \left[\frac{(I_2 L_\beta + N_\beta)}{(1 - I_1 I_2)}\beta + \frac{(I_2 L_p + N_p)}{(1 - I_1 I_2)}p + \frac{(I_2 L_r + N_r)}{(1 - I_1 I_2)}r + \frac{(I_2 L_{\delta_A} + N_{\delta_A})}{(1 - I_1 I_2)}\delta_A + \frac{(I_2 L_{\delta_R} + N_{\delta_R})}{(1 - I_1 I_2)}\delta_R \right]$$

Rearranging the terms in the equation will lead to

$$\dot{p} = \left(L_\beta + I_1 \frac{(I_2 L_\beta + N_\beta)}{(1 - I_1 I_2)} \right)\beta + \left(L_p + I_1 \frac{(I_2 L_p + N_p)}{(1 - I_1 I_2)} \right)p + \left(L_r + I_1 \frac{(I_2 L_r + N_r)}{(1 - I_1 I_2)} \right)r$$
$$+ \left(L_{\delta_A} + I_1 \frac{(I_2 L_{\delta_A} + N_{\delta_A})}{(1 - I_1 I_2)} \right)\delta_A + \left(L_{\delta_R} + I_1 \frac{(I_2 L_{\delta_R} + N_{\delta_R})}{(1 - I_1 I_2)} \right)\delta_R$$

Following simplifications the final expression for the $(\dot{p})$ equation is given by

$$\dot{p} = \frac{(L_\beta + I_1 N_\beta)}{(1 - I_1 I_2)}\beta + \frac{(L_p + I_1 N_p)}{(1 - I_1 I_2)}p + \frac{(L_r + I_1 N_r)}{(1 - I_1 I_2)}r + \frac{(L_{\delta_A} + I_1 N_{\delta_A})}{(1 - I_1 I_2)}\delta_A + \frac{(L_{\delta_R} + I_1 N_{\delta_R})}{(1 - I_1 I_2)}\delta_R$$

Thus, the final equations are given by

$$\dot{\beta} = \frac{Y_\beta}{V_{P_1}}\beta + \frac{Y_p}{V_{P_1}}p + \frac{(Y_r - V_{P_1})}{V_{P_1}}r + \frac{g \cos \Theta_1}{V_{P_1}}\phi + \frac{Y_{\delta_A}}{V_{P_1}}\delta_A + \frac{Y_{\delta_R}}{V_{P_1}}\delta_R$$

$$\dot{p} = \frac{(L_\beta + I_1 N_\beta)}{(1 - I_1 I_2)}\beta + \frac{(L_p + I_1 N_p)}{(1 - I_1 I_2)}p + \frac{(L_r + I_1 N_r)}{(1 - I_1 I_2)}r + \frac{(L_{\delta_A} + I_1 N_{\delta_A})}{(1 - I_1 I_2)}\delta_A + \frac{(L_{\delta_R} + I_1 N_{\delta_R})}{(1 - I_1 I_2)}\delta_R$$

$$\dot{r} = \frac{(I_2 L_\beta + N_\beta)}{(1 - I_1 I_2)}\beta + \frac{(I_2 L_p + N_p)}{(1 - I_1 I_2)}p + \frac{(I_2 L_r + N_r)}{(1 - I_1 I_2)}r + \frac{(I_2 L_{\delta_A} + N_{\delta_A})}{(1 - I_1 I_2)}\delta_A + \frac{(I_2 L_{\delta_R} + N_{\delta_R})}{(1 - I_1 I_2)}\delta_R$$

along with the kinematic relationship $\dot{\phi} = p + \tan \Theta_1 \, r$.

As in the previous case of the longitudinal equations, the objective is to reduce the preceding lateral directional equations to the state variable format using the general relationships:

$$\dot{x}_{Lat\ Dir} = A_{Lat\ Dir}\, x_{Lat\ Dir} + B_{Lat\ Dir}\, u_{Lat\ Dir}$$

$$y_{Lat\ Dir} = C_{Lat\ Dir}\, x_{Lat\ Dir} + D_{Lat\ Dir}\, u_{Lat\ Dir}$$

where the first equation is known as the lateral directional state equation and the second equation is known as the lateral directional output equation. The lateral directional state variable vector ($x_{Lat\ Dir}$) and input vector ($u_{Lat\ Dir}$) are given by

$$x_{Lat\ Dir} = \begin{Bmatrix} \beta \\ p \\ r \\ \phi \end{Bmatrix}, \quad u_{Lat\ Dir} = \begin{Bmatrix} \delta_A \\ \delta_R \end{Bmatrix}$$

Therefore, the lateral directional equations using state variable modeling are given by

$$\begin{Bmatrix} \dot{\beta} \\ \dot{p} \\ \dot{r} \\ \dot{\phi} \end{Bmatrix} = A_{Lat\ Dir} \begin{Bmatrix} \beta \\ p \\ r \\ \phi \end{Bmatrix} + B_{Lat\ Dir} \begin{Bmatrix} \delta_A \\ \delta_R \end{Bmatrix}$$

$$= \begin{bmatrix} Y'_\beta & Y'_p & Y'_r & Y'_\phi \\ L'_\beta & L'_p & L'_r & 0 \\ N'_\beta & N'_p & N'_r & 0 \\ 0 & 1 & \tan\Theta_1 & 0 \end{bmatrix} \begin{Bmatrix} \beta \\ p \\ r \\ \phi \end{Bmatrix} + \begin{bmatrix} Y'_{\delta_A} & Y'_{\delta_R} \\ L'_{\delta_A} & L'_{\delta_R} \\ N'_{\delta_A} & N'_{\delta_R} \\ 0 & 0 \end{bmatrix} \begin{Bmatrix} \delta_A \\ \delta_R \end{Bmatrix}$$

where $[A_{Lat\ Dir}]$ and $[B_{Lat\ Dir}]$ are known as the lateral directional state matrix and input matrix, respectively. The coefficients of $[A_{Lat\ Dir}]$ and $[B_{Lat\ Dir}]$ are given by

$$Y'_\beta = \frac{Y_\beta}{V_{P_1}}, \quad Y'_p = \frac{Y_p}{V_{P_1}}, \quad Y'_r = \frac{(Y_r - V_{P_1})}{V_{P_1}}, \quad Y'_\phi = \frac{g\cos\Theta_1}{V_{P_1}}, \quad Y'_{\delta_A} = \frac{Y_{\delta_A}}{V_{P_1}}, \quad Y'_{\delta_R} = \frac{Y_{\delta_R}}{V_{P_1}}$$

$$L'_\beta = \frac{(L_\beta + I_1 N_\beta)}{(1 - I_1 I_2)}, \quad L'_p = \frac{(L_p + I_1 N_p)}{(1 - I_1 I_2)}, \quad L'_r = \frac{(L_r + I_1 N_r)}{(1 - I_1 I_2)}$$

$$L'_{\delta_A} = \frac{(L_{\delta_A} + I_1 N_{\delta_A})}{(1 - I_1 I_2)}, \quad L'_{\delta_R} = \frac{(L_{\delta_R} + I_1 N_{\delta_R})}{(1 - I_1 I_2)}$$

$$N'_\beta = \frac{(I_2 L_\beta + N_\beta)}{(1 - I_1 I_2)}, \quad N'_p = \frac{(I_2 L_p + N_p)}{(1 - I_1 I_2)}, \quad N'_r = \frac{(I_2 L_r + N_r)}{(1 - I_1 I_2)}$$

$$N'_{\delta_A} = \frac{(I_2 L_{\delta_A} + N_{\delta_A})}{(1 - I_1 I_2)}, \quad N'_{\delta_R} = \frac{(I_2 L_{\delta_R} + N_{\delta_R})}{(1 - I_1 I_2)}$$

As for its longitudinal equivalent, a very important property of the matrix $[A_{Lat\ Dir}]$ is that its eigenvalues are coincident with the roots of the lateral directional characteristic equations (associated with the dutch roll, rolling, and spiral modes).

As for the longitudinal case, the level of complexity associated with the derivation of the lateral directional output equations ($y_{Lat\ Dir} = C_{Lat\ Dir}\, x_{Lat\ Dir} + D_{Lat\ Dir}\, u_{Lat\ Dir}$) is dependent on the selection of the elements of the lateral directional output column ($y_{Lat\ Dir}$). This column contains the aircraft lateral directional parameters which are measured by the on-board sensors and available for the design of flight control laws. In general ($y_{Lat\ Dir}$) contains ($x_{Lat\ Dir}$) and a number of additional parameters (for example, lateral linear accelerations (a_Y), heading angle (ψ), and others). Without loss of generality, ($y_{Lat\ Dir}$) is here assumed to include the lateral acceleration (a_Y) in addition to the lateral directional state variable vector ($x_{Lat\ Dir}$). Therefore

$$y_{Lat\,Dir} = \begin{Bmatrix} a_Y \\ x_{Lat\,Dir} \end{Bmatrix} = \begin{Bmatrix} a_Y \\ \beta \\ p \\ r \\ \phi \end{Bmatrix}$$

The modeling for the lateral acceleration (a_Y) is conceptually similar to the modeling of the vertical acceleration (a_Z) but somewhat simpler since the transformation from the body axes to the stability axes is not necessary (since $Y = Y_S$). Therefore, considering the small perturbation second equation of the Conservation of Linear Momentum Equations (CLMEs):

$$m[\dot{v} + V_{P_1} r] = mg \cos\Theta_1 \phi + (f_{A_Y} + f_{T_Y})$$

Using $\dot{v} = V_{P_1} \dot{\beta}$ we have

$$m[V_{P_1} \dot{\beta} + V_{P_1} r] = mg \cos\Theta_1 \phi + (f_{A_Y} + f_{T_Y})$$

leading to $(f_{A_Y} + f_{T_Y}) = m[V_{P_1} \dot{\beta} + V_{P_1} r] - mg \cos\Theta_1 \phi$.

Next, a relationship for (a_Y) is derived using

$$a_Y = \frac{(f_{A_Y} + f_{T_Y})}{m} = [V_{P_1} \dot{\beta} + V_{P_1} r] - g\cos\Theta_1 \phi$$

Since $\dot{\beta} = Y'_\beta \beta + Y'_p p + Y'_r r + Y'_\phi \phi + Y'_{\delta_A} \delta_A + Y'_{\delta_R} \delta_R$ we have

$$a_Y = [V_{P_1}(Y'_\beta \beta + Y'_p p + Y'_r r + Y'_\phi \phi + Y'_{\delta_A} \delta_A + Y'_{\delta_R} \delta_R) + V_{P_1} r] - g\cos\Theta_1 \phi$$

Grouping the terms with the same variables we have

$$a_Y = (V_{P_1} Y'_\beta)\beta + (V_{P_1} Y'_p)p + [V_{P_1}(Y'_r + 1)]r + [V_{P_1}(Y'_\phi - g\cos\Theta_1)]\phi + (V_{P_1} Y'_{\delta_A})\delta_A + (V_{P_1} Y'_{\delta_A})\delta_R$$

Rearranging in a matrix format, the lateral directional output equations take on the form

$$\begin{Bmatrix} a_Y \\ \beta \\ p \\ r \\ \phi \end{Bmatrix} = C_{Lat\,Dir} \begin{Bmatrix} \beta \\ p \\ r \\ \phi \end{Bmatrix} + D_{Lat\,Dir} \begin{Bmatrix} \delta_A \\ \delta_R \end{Bmatrix}$$

$$= \begin{bmatrix} Y''_\beta & Y''_p & Y''_r & Y''_\phi \\ 1 & 0 & 0 & 0 \\ 0 & 1 & 0 & 0 \\ 0 & 0 & 1 & 0 \\ 0 & 0 & 0 & 1 \end{bmatrix} \begin{Bmatrix} \beta \\ p \\ r \\ \phi \end{Bmatrix} + \begin{bmatrix} Y''_{\delta_A} & Y''_{\delta_R} \\ 0 & 0 \\ 0 & 0 \\ 0 & 0 \\ 0 & 0 \end{bmatrix} \begin{Bmatrix} \delta_A \\ \delta_R \end{Bmatrix}$$

where $[C_{Lat\,Dir}]$ and $[D_{Lat\,Dir}]$ are the lateral directional observation matrix and lateral directional output control matrix, respectively. The coefficients of $[C_{Lat\,Dir}]$ and $[D_{Lat\,Dir}]$ are given by

$$Y''_\beta = Y'_\beta V_{P_1},\; Y''_p = Y'_p V_{P_1},\; Y''_r = V_{P_1}(Y'_r + 1),\; Y''_\phi = Y'_\phi V_{P_1} - g\cos\Theta_1$$
$$Y''_{\delta_A} = Y'_{\delta_A} V_{P_1},\; Y''_{\delta_R} = Y'_{\delta_R} V_{P_1}$$

If additional lateral directional (asymmetric) control surfaces are introduced, these equations need to be appropriately modified. For example, consider the availability of asymmetric spoilers and flaperons, in addition to the conventional ailerons and rudder, as shown in Figure 8.3.

444 Chapter 8 State Variable Modeling of the Aircraft Dynamics

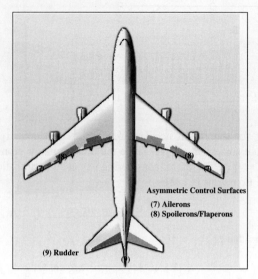

Figure 8.3 Lateral-Directional Control Surfaces

In this case the lateral directional inputs are given by

$$\{u_{Lat\ Dir}\} = \begin{Bmatrix} \delta_A \\ \delta_R \\ \delta_{Spoilers} \\ \delta_{Flaperons} \end{Bmatrix} = \begin{Bmatrix} \delta_A \\ \delta_R \\ \delta_S \\ \delta_F \end{Bmatrix}$$

The lateral directional state equations will be given by

$$\begin{Bmatrix} \dot{\beta} \\ \dot{p} \\ \dot{r} \\ \dot{\phi} \end{Bmatrix} = A_{Lat\ Dir} \begin{Bmatrix} \beta \\ p \\ r \\ \phi \end{Bmatrix} + B_{Lat\ Dir} \begin{Bmatrix} \delta_A \\ \delta_R \\ \delta_S \\ \delta_F \end{Bmatrix}$$

$$= \begin{bmatrix} Y'_\beta & Y'_p & Y'_r & Y'_\phi \\ L'_\beta & L'_p & L'_r & 0 \\ N'_\beta & N'_p & N'_r & 0 \\ 0 & 1 & tg\Theta_1 & 0 \end{bmatrix} \begin{Bmatrix} \beta \\ p \\ r \\ \phi \end{Bmatrix} + \begin{bmatrix} Y'_{\delta A} & Y'_{\delta R} & Y'_{\delta S} & Y'_{\delta F} \\ L'_{\delta A} & L'_{\delta R} & L'_{\delta S} & L'_{\delta F} \\ N'_{\delta A} & N'_{\delta R} & N'_{\delta S} & N'_{\delta F} \\ 0 & 0 & 0 & 0 \end{bmatrix} \begin{Bmatrix} \delta_A \\ \delta_R \\ \delta_S \\ \delta_F \end{Bmatrix}$$

where the additional coefficients in the $[B_{Lat\ Dir}]$ matrix are given by

$$Y'_{\delta_S} = \frac{Y_{\delta_S}}{V_{P_1}}, \quad Y'_{\delta_F} = \frac{Y_{\delta_F}}{V_{P_1}}$$

$$L'_{\delta_S} = \frac{(L_{\delta_S} + I_1 N_{\delta_S})}{(1 - I_1 I_2)}, \quad L'_{\delta_F} = \frac{(L_{\delta_F} + I_1 N_{\delta_F})}{(1 - I_1 I_2)}$$

$$N'_{\delta_S} = \frac{(I_2 L_{\delta_S} + N_{\delta_S})}{(1 - I_1 I_2)}, \quad N'_{\delta_F} = \frac{(I_2 L_{\delta_F} + N_{\delta_F})}{(1 - I_1 I_2)}$$

The new lateral directional dimensional derivatives $(Y_{\delta_S}, Y_{\delta_F}, L_{\delta_S}, L_{\delta_F}, N_{\delta_S}, N_{\delta_F})$ are associated with the dimensionless aerodynamic derivatives $(c_{Y_{\delta S}}, c_{Y_{\delta F}}, c_{l_{\delta S}}, c_{l_{\delta F}}, c_{n_{\delta S}}, c_{n_{\delta F}})$. Similarly, the lateral directional output equations will be given by

$$\begin{Bmatrix} a_Y \\ \beta \\ p \\ r \\ \phi \end{Bmatrix} = C_{Lat\,Dir} \begin{Bmatrix} \beta \\ p \\ r \\ \phi \end{Bmatrix} + D_{Lat\,Dir} \begin{Bmatrix} \delta_A \\ \delta_R \\ \delta_S \\ \delta_F \end{Bmatrix}$$

$$= \begin{bmatrix} Y''_\beta & Y''_p & Y''_r & Y''_\phi \\ 1 & 0 & 0 & 0 \\ 0 & 1 & 0 & 0 \\ 0 & 0 & 1 & 0 \\ 0 & 0 & 0 & 1 \end{bmatrix} \begin{Bmatrix} \beta \\ p \\ r \\ \phi \end{Bmatrix} + \begin{bmatrix} Y''_{\delta A} & Y''_{\delta R} & Y''_{\delta S} & Y''_{\delta F} \\ 0 & 0 & 0 & 0 \\ 0 & 0 & 0 & 0 \\ 0 & 0 & 0 & 0 \\ 0 & 0 & 0 & 0 \end{bmatrix} \begin{Bmatrix} \delta_A \\ \delta_R \\ \delta_S \\ \delta_F \end{Bmatrix}$$

where the additional coefficients in the $[D_{Lat\,Dir}]$ matrix are given by

$$Y''_{\delta_S} = Y'_{\delta_S} V_{P_1}, \quad Y''_{\delta_F} = Y'_{\delta_F} V_{P_1}$$

8.6 AUGMENTATION OF THE AIRCRAFT STATE VARIABLE MODELING

The previously introduced state variable modeling is a flexible tool which allows the modeling of additional dynamic variables of interest for simulation purposes as well as additional inputs acting on the system dynamics. The resulting system will be an augmented model with respect to the formulations outlined in Sections 8.4 and 8.5.

8.6.1 Modeling of the Altitude (h)

The need for modeling the altitude for simulation purposes is quite obvious. Recalling the flight path equations (FPEs) in a matrix format from Chapter 1, we have

$$\begin{Bmatrix} \dot{X}' \\ \dot{Y}' \\ \dot{Z}' \end{Bmatrix} = \begin{bmatrix} \cos\Psi\cos\Theta & -\sin\Psi\cos\Phi + \cos\Psi\sin\Theta\sin\Phi & \sin\Psi\sin\Phi + \cos\Psi\sin\Theta\cos\Phi \\ \sin\Psi\cos\Theta & \cos\Psi\cos\Phi + \sin\Psi\sin\Theta\sin\Phi & -\sin\Phi\cos\Psi + \sin\Psi\sin\Theta\cos\Phi \\ -\sin\Theta & \cos\Theta\sin\Phi & \cos\Theta\cos\Phi \end{bmatrix} \begin{Bmatrix} U \\ V \\ Z \end{Bmatrix}$$

where $\{\dot{X}'\ \dot{Y}'\ \dot{Z}'\}$, $\{U\ V\ Z\}$ are the velocity components along the Earth-based inertial frame and the body frame, respectively. Since (Z') is positive pointing toward the center of the Earth, we can state

$$\dot{h} = -\dot{Z}' = U\sin\Theta - V\cos\Theta\sin\Phi - W\cos\Theta\cos\Phi$$

Using the small perturbation assumptions ($\sin x \approx x$, $\cos x \approx 1$), starting from a steady-state condition with small (ϕ) and negligible (v), we have

$$\dot{h} = -\dot{Z}' \approx V_{P_1}\theta - w$$

Using the relationship

$$\tan(\alpha) \approx \alpha \approx \frac{w}{V_{P_1}} \quad \rightarrow \quad w \approx V_{P_1}\alpha$$

the following expression for the height can be derived:

$$\dot{h} = -\dot{Z}' \approx V_{P_1}\theta - V_{P_1}\alpha = V_{P_1}(\theta - \alpha)$$

The resulting augmented longitudinal state variable equations are given by

$$\begin{Bmatrix} \dot{u} \\ \dot{\alpha} \\ \dot{q} \\ \dot{\theta} \\ \dot{h} \end{Bmatrix} = \begin{bmatrix} X'_u & X'_\alpha & X'_q & X'_\theta & 0 \\ Z'_u & Z'_\alpha & Z'_q & Z'_\theta & 0 \\ M'_u & M'_\alpha & M'_q & M'_\theta & 0 \\ 0 & 0 & 1 & 0 & 0 \\ 0 & -V_{P_1} & 0 & V_{P_1} & 0 \end{bmatrix} \begin{Bmatrix} u \\ \alpha \\ q \\ \theta \\ h \end{Bmatrix} + \begin{bmatrix} X'_{\delta E} \\ Z'_{\delta E} \\ M'_{\delta E} \\ 0 \\ 0 \end{bmatrix} \{\delta_E\}$$

8.6.2 Modeling of the Flight Path Angle (γ)

An important parameter for simulation purposes is the flight path angle γ, also known as climb angle (when positive) or descent angle (when negative). The flight path is simply given by

$$\gamma = \theta - \alpha$$

The relationship is algebraic and it does not involve a derivative; therefore, the flight path angle can simply be added to the outputs leading to the following longitudinal output equations:

$$\begin{Bmatrix} a_z \\ u \\ \alpha \\ q \\ \theta \\ \gamma \end{Bmatrix} = \begin{bmatrix} Z''_\alpha & Z''_u & Z''_q & Z''_\theta \\ 1 & 0 & 0 & 0 \\ 0 & 1 & 0 & 0 \\ 0 & 0 & 1 & 0 \\ 0 & 0 & 0 & 1 \\ 0 & -1 & 0 & 1 \end{bmatrix} \begin{Bmatrix} u \\ \alpha \\ q \\ \theta \end{Bmatrix} + \begin{bmatrix} Z''_{\delta E} \\ 0 \\ 0 \\ 0 \\ 0 \\ 0 \end{bmatrix} \{\delta_E\}$$

8.6.3 Modeling of the Engine Dynamics

For a detailed modeling of jet-powered aircraft, it might be needed to include the engine dynamics in terms of variation of thrust following a deflection of the throttle. For most cases, the engine response can be modeled using the following transfer function:

$$\frac{T(s)}{\delta_T(s)} = \frac{a}{(1 + s\, b)}$$

where T is the thrust, δ_T is the throttle deflection, and (a, b) are essentially constants whose values depend on the characteristics of the propulsion system.

Rearranging the above relationship we have

$$T(s)(1 + sb) = a\, \delta_T(s) \rightarrow sbT(s) = a\, \delta_T(s) - T(s) \rightarrow sT(s) = \frac{1}{b}(a\, \delta_T(s) - T(s))$$

Taking the inverse Laplace transform (ILT) we have

$$\dot{T}(t) = \frac{1}{b}(a\, \delta_T(t) - T(t))$$

Thus, the following augmented longitudinal state variable model can be introduced featuring T as an additional state variable and δ_T as an additional input:

$$\begin{Bmatrix} \dot{u} \\ \dot{\alpha} \\ \dot{q} \\ \dot{\theta} \\ \dot{T} \end{Bmatrix} = \begin{bmatrix} X'_u & X'_\alpha & X'_q & X'_\theta & 0 \\ Z'_u & Z'_\alpha & Z'_q & Z'_\theta & 0 \\ M'_u & M'_\alpha & M'_q & M'_\theta & 0 \\ 0 & 0 & 1 & 0 & 0 \\ 0 & 0 & 0 & 0 & -\frac{1}{b} \end{bmatrix} \begin{Bmatrix} u \\ \alpha \\ q \\ \theta \\ T \end{Bmatrix} + \begin{bmatrix} X'_{\delta E} & 0 \\ Z'_{\delta E} & 0 \\ M'_{\delta E} & 0 \\ 0 & 0 \\ 0 & \frac{a}{b} \end{bmatrix} \begin{Bmatrix} \delta_E \\ \delta_T \end{Bmatrix}$$

8.6.4 Modeling of the Actuator Dynamics

The modeling introduced so far has not included the dynamics of the actuators, which ultimately deflect the control surface. In general, actuators exhibit first order responses that can be modeled using the following transfer function:

$$\frac{\delta_X(s)}{\delta_{PILOT\ COMMAND_X}(s)} = \frac{\delta_X(s)}{\delta_{PC_X}(s)} = \frac{c_X}{(s + c_X)}$$

where $x = (E, A, R, \ldots)$ indicates elevators, ailerons, rudder, or any other control surface, PC_X indicates the pilot commanded deflection for the specific control surface, and c_X is a function of the characteristics of the hardware and, ultimately, a function of the size of the control surface. In general, low values of c_X are used for the modeling of the dynamics of the actuators for large control surfaces with low deployment speed (such as elevators, stabilators, or rudders); conversely, higher values of c_X are used for the modeling of the dynamics of the actuators for small control surfaces with high deployment speed (such as ailerons). Rearranging the above relationship we have

$$\delta_X(s)(s + c_X) = c_X \, \delta_{PC_X}(s) \to s \, \delta_X(s) = c_X \, \delta_{PC_X}(s) - c_X \, \delta_X(s)$$

Taking the inverse Laplace transform (ILT) we have

$$\dot{\delta}_X(t) = c \, \delta_{PC_X}(t) - c \delta_X(t)$$

For example, for elevators, rudder, and ailerons we have

$$\dot{\delta}_E(t) = c_E \, \delta_{PC_E}(t) - c_E \, \delta_E(t)$$
$$\dot{\delta}_A(t) = c_A \, \delta_{PC_A}(t) - c_A \, \delta_A(t)$$
$$\dot{\delta}_R(t) = c_R \, \delta_{PC_R}(t) - c_R \, \delta_R(t)$$

Thus, the state variable model can be augmented by introducing $\delta_X(t)$ as an additional state variable and $\delta_{PC_X}(t)$ as input replacing the previous $\delta_X(t)$. For example, for the longitudinal state variable model we have

$$\begin{Bmatrix} \dot{u} \\ \dot{\alpha} \\ \dot{q} \\ \dot{\theta} \\ \dot{\delta}_E \end{Bmatrix} = \begin{bmatrix} X'_u & X'_\alpha & X'_q & X'_\theta & 0 \\ Z'_u & Z'_\alpha & Z'_q & Z'_\theta & 0 \\ M'_u & M'_\alpha & M'_q & M'_\theta & 0 \\ 0 & 0 & 1 & 0 & 0 \\ 0 & 0 & 0 & 0 & -c_E \end{bmatrix} \begin{Bmatrix} u \\ \alpha \\ q \\ \theta \\ \delta_E \end{Bmatrix} + \begin{bmatrix} X'_{\delta E} & 0 \\ Z'_{\delta E} & 0 \\ M'_{\delta E} & 0 \\ 0 & 0 \\ 0 & c_E \end{bmatrix} \begin{Bmatrix} \delta_E \\ \delta_{PC_E} \end{Bmatrix}$$

Similarly, the augmented lateral directional state variable model is given by

$$\begin{Bmatrix} \dot{\beta} \\ \dot{p} \\ \dot{r} \\ \dot{\phi} \\ \dot{\delta}_A \\ \dot{\delta}_R \end{Bmatrix} = \begin{bmatrix} Y'_\beta & Y'_p & Y'_r & Y'_\phi & 0 & 0 \\ L'_\beta & L'_p & L'_p & 0 & 0 & 0 \\ N'_\beta & N'_p & N'_r & 0 & 0 & 0 \\ 0 & 1 & tg(\Theta_1) & 0 & 0 & 0 \\ 0 & 0 & 0 & 0 & -c_A & 0 \\ 0 & 0 & 0 & 0 & 0 & -c_R \end{bmatrix} \begin{Bmatrix} \beta \\ p \\ r \\ \phi \\ \delta_A \\ \delta_R \end{Bmatrix} + \begin{bmatrix} Y'_{\delta A} & Y'_{\delta R} & 0 & 0 \\ L'_{\delta A} & L'_{\delta R} & 0 & 0 \\ N'_{\delta A} & N'_{\delta R} & 0 & 0 \\ 0 & 0 & 0 & 0 \\ 0 & 0 & c_A & 0 \\ 0 & 0 & 0 & c_R \end{bmatrix} \begin{Bmatrix} \delta_A \\ \delta_R \\ \delta_{PC_A} \\ \delta_{PC_R} \end{Bmatrix}$$

8.6.5 Modeling of the Atmospheric Turbulence

The detailed modeling of the atmospheric turbulence acting on the aircraft requires the availability of a turbulence model featuring statistical parameters of the turbulence itself. There are two widely known atmospheric turbulence models, the Von Karman model and the Dryden model. Refer to[8] for an understanding of the statistical characteristics of these two models. In general terms, atmospheric turbulence can be considered as additional states acting on the aircraft dynamics as perturbations on either the linear components of the velocity, indicated as $\{u_G \; v_G \; w_G\}$, or on the angle of attack and sideslip angle, indicated as $\{\alpha_G \; \beta_G\}$, where the subscript G stands for gust.

8.7 SUMMARY OF STATE VARIABLE MODELING OF THE AIRCRAFT DYNAMICS

Using the modeling in Sections 8.4 and 8.5 and invoking the decoupling between the longitudinal and the lateral directional dynamics, the state variable equations of the entire aircraft dynamic are given by

$$\begin{Bmatrix} \dot{u} \\ \dot{\alpha} \\ \dot{q} \\ \dot{\theta} \\ \dot{\beta} \\ \dot{p} \\ \dot{r} \\ \dot{\phi} \end{Bmatrix} = \begin{bmatrix} A_{Long} & 0 \\ 0 & A_{Lat\,Dir} \end{bmatrix} \begin{Bmatrix} u \\ \alpha \\ q \\ \theta \\ \beta \\ p \\ r \\ \phi \end{Bmatrix} + \begin{bmatrix} B_{Long} & 0 \\ 0 & B_{Lat\,Dir} \end{bmatrix} \begin{Bmatrix} \delta_E \\ \delta_A \\ \delta_R \end{Bmatrix}$$

Therefore, using the expressions for A_{Long}, $A_{Lat\,Dir}$, B_{Long}, $B_{Lat\,Dir}$ the state equations for the (longitudinal + lateral directional) aircraft dynamics are given by

$$\begin{Bmatrix} \dot{u} \\ \dot{\alpha} \\ \dot{q} \\ \dot{\theta} \\ \dot{\beta} \\ \dot{p} \\ \dot{r} \\ \dot{\phi} \end{Bmatrix} = \begin{bmatrix} X'_u & X'_\alpha & X'_q & X'_\theta & & & & \\ Z'_u & Z'_\alpha & Z'_q & Z'_\theta & & & 0 & \\ M'_u & M'_\alpha & M'_q & M'_\theta & & & & \\ 0 & 0 & 1 & 0 & & & & \\ & & & & Y'_\beta & Y'_p & Y'_r & Y'_\phi \\ & & 0 & & L'_\beta & L'_p & L'_r & 0 \\ & & & & N'_\beta & N'_p & N'_r & 0 \\ & & & & 0 & 1 & tg\Theta_1 & 0 \end{bmatrix} \begin{Bmatrix} u \\ \alpha \\ q \\ \theta \\ \beta \\ p \\ r \\ \phi \end{Bmatrix} + \begin{bmatrix} X'_{\delta E} & & \\ Z'_{\delta E} & & 0 \\ M'_{\delta E} & & \\ 0 & & \\ & Y'_{\delta A} & Y'_{\delta R} \\ 0 & L'_{\delta A} & L'_{\delta R} \\ & N'_{\delta A} & N'_{\delta R} \\ & 0 & 0 \end{bmatrix} \begin{Bmatrix} \delta_E \\ \delta_A \\ \delta_R \end{Bmatrix}$$

Similarly, using the relationships derived in Sections 8.4 and 8.5, the output equations of the entire aircraft dynamics are given by

$$\begin{Bmatrix} a_z \\ \alpha \\ u \\ q \\ \theta \\ a_Y \\ \beta \\ p \\ r \\ \phi \end{Bmatrix} = \begin{bmatrix} C_{Long} & 0 \\ 0 & C_{Lat\,Dir} \end{bmatrix} \begin{Bmatrix} \alpha \\ u \\ q \\ \theta \\ \beta \\ p \\ r \\ \phi \end{Bmatrix} + \begin{bmatrix} D_{Long} & 0 \\ 0 & D_{Lat\,Dir} \end{bmatrix} \begin{Bmatrix} \delta_E \\ \delta_A \\ \delta_R \end{Bmatrix}$$

Using the expressions for C_{Long}, $C_{Lat\,Dir}$, D_{Long}, $D_{Lat\,Dir}$ the state equations for the (longitudinal + lateral directional) aircraft dynamics are given by

$$\begin{Bmatrix} a_z \\ \alpha \\ u \\ q \\ \theta \\ a_Y \\ \beta \\ p \\ r \\ \phi \end{Bmatrix} = \begin{bmatrix} Z''_\alpha & Z''_u & Z''_q & Z''_\theta & & & & \\ 1 & 0 & 0 & 0 & & & & \\ 0 & 1 & 0 & 0 & & & 0 & \\ 0 & 0 & 1 & 0 & & & & \\ 0 & 0 & 0 & 1 & & & & \\ & & & & Y''_\beta & Y''_p & Y''_r & Y''_\phi \\ & & & & 1 & 0 & 0 & 0 \\ & & 0 & & 0 & 1 & 0 & 0 \\ & & & & 0 & 0 & 1 & 0 \\ & & & & 0 & 0 & 0 & 1 \end{bmatrix} \begin{Bmatrix} \alpha \\ u \\ q \\ \theta \\ \beta \\ p \\ r \\ \phi \end{Bmatrix} + \begin{bmatrix} Z''_{\delta E} & & \\ 0 & & \\ 0 & & 0 \\ 0 & & \\ 0 & & \\ & Y''_{\delta A} & Y''_{\delta R} \\ & 0 & 0 \\ 0 & 0 & 0 \\ & 0 & 0 \\ & 0 & 0 \end{bmatrix} \begin{Bmatrix} \delta_E \\ \delta_A \\ \delta_R \end{Bmatrix}$$

The entire modeling process leading to the preceding total aircraft state variable model is shown in Figure 8.4. As shown in Section 8.6 the total model can be augmented by introducing additional states and inputs to include additional parameters.

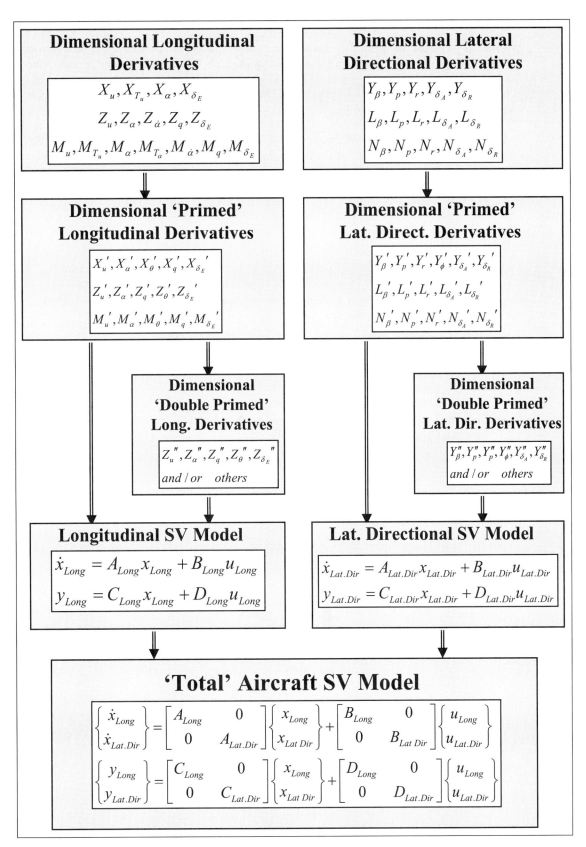

Figure 8.4 Conceptual Block Diagram of the Aircraft State Variable Model

8.8 SUMMARY

While the Laplace transformation-based TF approach described in Chapter 7 has provided a valuable understanding of the overall aircraft dynamic characteristic, TF-based modeling is computationally not very efficient for the modeling of a true multi-input–multi-output (MIMO) dynamic system such as an aircraft. Therefore, the specific objective of this chapter was to introduce the state variable modeling as an alternative modeling to the TF-based modeling.

Following the general definitions of state variables for linear and nonlinear systems, the application of the state variables to the modeling of the aircraft dynamics has been shown, with details leading first to the separate formulation of the longitudinal and the lateral directional state variable models followed by the merging of the two models into a single aircraft state variable model. Furthermore, the augmentation of the state variable model to include additional dynamic variable and additional inputs simulation purposes has been discussed.

The next chapter will focus on introducing a number of simulation tools.

REFERENCES

1. Lanchester, F. W. *Aerodonetics*. Constable and Co. Ltd, London, 1908.
2. Bryan, G. H. *Stability in Aviation*. Macmillan and Co., London, 1911.
3. Roskam, J. *Airplane Flight Dynamics and Automatic Flight Controls – Part I*. Design, Analysis, and Research Corporation, Lawrence, KS, 1995.
4. Nise, N. *Control Systems Engineering* (5th ed.). John Wiley & Sons, 2007.
5. Ogata, K. *Modern Control Engineering* (4th ed.). Prentice Hall, 2005.
6. McLean, D. *Automatic Flight Control Systems*. Prentice Hall International, 1990.
7. Roskam, J. *Airplane Flight Dynamics and Automatic Flight Controls – Part II*. Design, Analysis, and Research Corporation, Lawrence, KS, 1995.
8. Roskam, J. *Airplane Flight Dynamics and Automatic Flight Controls – Part II*. Design, Analysis, and Research Corporation, Lawrence, KS, 1979.

STUDENT SAMPLE PROBLEMS

Student Sample Problem 8.1

Consider the well-known (mass-damper-spring) mechanical system shown in Figure SSP8.1.1.

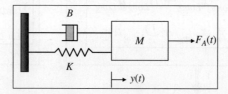

Figure SSP8.1.1 Mass-Spring-Damper System

The system is subjected to the excitation (input) of the applied force $F_A(t)$ applied to the mass M. The measurable output of the system is the displacement $y(t)$ of the mass M. Find the state variable model of the system.

Solution of Student Sample Problem 8.1

The free body diagram (FBD) of the system is shown in Figure SSP8.1.2:

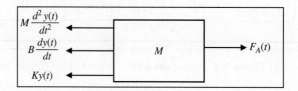

Figure SSP8.1.2 Free Body Diagram on the System in Figure SSP8.1.1

From the FBD the equation describing the system dynamics is derived:

$$M\frac{d^2y(t)}{dt^2} + B\frac{dy(t)}{dt} + Ky(t) = F_A(t)$$

The above system is a second order system with one input and one measurable output. Thus, $(n = 2, m = 1, l = 1)$. Two state variables are necessary to describe the system dynamics. A suitable set of state variables is given by

$$x_1(t) = y(t)$$
$$x_2(t) = \frac{dy(t)}{dt} = \dot{y}(t)$$

Using these state variables and the preceding differential equation, the state variable model is given by

$$\dot{x}_1(t) = x_2(t) = \frac{dy(t)}{dt} = \dot{y}(t)$$

$$\dot{x}_2(t) = \frac{d^2y(t)}{dt^2} = \ddot{y}(t) = -\frac{1}{M}\left[F_A(t) - B\frac{dy(t)}{dt} - Ky(t)\right]$$

Since $F_A(t)$ is the input $u(t)$ of the system, using the above state variables, we have

$$\dot{x}_1(t) = x_2(t)$$
$$\dot{x}_2(t) = -\frac{1}{M}[u(t) - Bx_2(t) - Kx_1(t)]$$

along with the output equation: $y(t) = x_1(t)$.

Therefore, the state variable model in a matrix format is given by

$$\begin{Bmatrix} \dot{x}_1(t) \\ \dot{x}_2(t) \end{Bmatrix} = \begin{bmatrix} 0 & 1 \\ -\frac{K}{M} & -\frac{B}{M} \end{bmatrix} \begin{Bmatrix} x_1(t) \\ x_2(t) \end{Bmatrix} + \begin{bmatrix} 0 \\ \frac{1}{M} \end{bmatrix} u(t)$$

$$y(t) = \begin{bmatrix} 1 & 0 \end{bmatrix} \begin{Bmatrix} x_1(t) \\ x_2(t) \end{Bmatrix} + [0]\, u(t)$$

Student Sample Problem 8.2

Consider the double (mass-damper-spring) mechanical system shown in Figure SSP8.2.1.

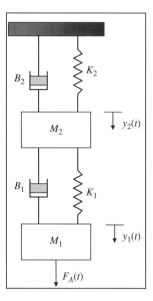

Figure SSP8.2.1 Double (Mass-Spring-Damper) System

The system has two masses, two springs, and two dampers along with one applied force $F_A(t)$ applied to M_1 and two measurable outputs, that is the displacements of M_1 and M_2.

The system is subjected to the excitation (input) of the applied force $F_A(t)$ applied to the mass M_1. The measurable outputs of the system are the displacements $y_1(t), y_2(t)$ of the masses M_1, M_2.

- Find the state variable model of the system;
- Assuming generic values for the parameters $M_1, M_2, B_1, B_2, K_1, K_2$ (with the condition that the system is stable), implement the state variable model in MATLAB®;
- Plot $y_1(t), y_2(t)$, assuming a generic $F_A(t) = 1$.

Solution of Student Sample Problem 8.2

Note the presence of a relative motion between M_1 and M_2 associated with the spring and the damper between the two masses. Following an analysis of the different inertial, friction, and elastic forces in the system, the FBD are given by the following (Figure SSP8.2.2):

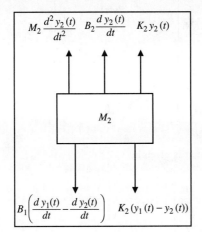
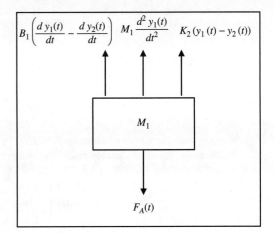

Figure SSP8.2.2 Free Body Diagrams on the System in Figure SSP8.2.1

From the FBDs the equations of the system are given by the following:
For the mass M_1:

$$M_1 \frac{d^2 y_1(t)}{dt^2} + B_1 \left(\frac{d y_1(t)}{dt} - \frac{d y_2(t)}{dt} \right) + K_1(y_1(t) - y_2(t)) = F_A(t)$$

For the mass M_2:

$$M_2 \frac{d^2 y_2(t)}{dt^2} + B_2 \frac{d y_2(t)}{dt} + K_2 \, y_2(t)$$

$$- B_1 \left(\frac{d y_1(t)}{dt} - \frac{d y_2(t)}{dt} \right) - K_1 \, (y_1(t) - y_2(t)) = 0$$

The system is therefore a fourth order system with one input and two measurable outputs ($n = 4, m = 1, l = 2$). A suitable selection of state variables is given by

$$x_1(t) = y_1(t)$$
$$x_2(t) = \frac{d y_1(t)}{dt} = \dot{y}_1(t)$$
$$x_3(t) = y_2(t)$$
$$x_4(t) = \frac{d y_2(t)}{dt} = \dot{y}_2(t)$$

Rewriting the preceding equations using the state variables and replacing $F_A(t)$ with $u(t)$ leads to the following:

$$\dot{x}_1(t) = x_2(t)$$
$$\dot{x}_2(t) = \frac{1}{M_1}[u(t) - B_1(x_2(t) - x_4(t)) - K_1(x_1(t) - x_3(t))]$$
$$\dot{x}_3(t) = x_4(t)$$
$$\dot{x}_4(t) = \frac{1}{M_2}[B_1(x_2(t) - x_4(t)) + K_1(x_1(t) - x_3(t)) - B_2 x_4(t) - K_2 x_3(t)]$$

along with the output equations: $y_1(t) = x_1(t)$, $y_2(t) = x_3(t)$.

Thus, the state variable model in a matrix format is given by

$$\begin{Bmatrix} \dot{x}_1(t) \\ \dot{x}_2(t) \\ \dot{x}_3(t) \\ \dot{x}_4(t) \end{Bmatrix} = \begin{bmatrix} 0 & 1 & 0 & 0 \\ -\frac{K_1}{M_1} & -\frac{B_1}{M_1} & \frac{K_1}{M_1} & \frac{B_1}{M_1} \\ 0 & 0 & 0 & 1 \\ \frac{K_1}{M_2} & \frac{B_1}{M_2} & -\frac{(K_1+K_2)}{M_2} & -\frac{(B_1+B_2)}{M_2} \end{bmatrix} \begin{Bmatrix} x_1(t) \\ x_2(t) \\ x_3(t) \\ x_4(t) \end{Bmatrix} + \begin{bmatrix} 0 \\ \frac{1}{M_1} \\ 0 \\ 0 \end{bmatrix} \{u(t)\}$$

$$\begin{Bmatrix} y_1(t) \\ y_2(t) \end{Bmatrix} = \begin{bmatrix} 1 & 0 & 0 & 0 \\ 0 & 0 & 1 & 0 \end{bmatrix} \begin{Bmatrix} x_1(t) \\ x_2(t) \\ x_3(t) \\ x_4(t) \end{Bmatrix}$$

A sample MATLAB® script file for the modeling of this system is included here, assuming generic values for the parameters $M_1, M_2, B_1, B_2, K_1, K_2$ (with the condition that the system is stable).

```
%
% Chapter VIII
% Student Sample Problem 8.2
%
m1=5; m2=10;
b1=2; b2=4;
k1=10; k2=20;
a=[0 1 0 0
   -(k1/m1) -(b1/m1) (k1/m1) (b1/m1)
   0 0 0 1
   (k1/m2) (b1/m2) -((k1+k2)/m2) -((b1+b2)/m2)];
b=[0 1/m1 0 0]';
c=[1 0 0 0
   0 0 1 0];
d=[0 0]';
sys=ss(a,b,c,d);
t=[0:0.01:20];
size_t=max(size(t));
for i=1:size_t;
  u(i,1)=1;
end;
y=lsim(sys,u,t);
y1=y(:,1);
y2=y(:,2);
subplot(211); grid on; plot(t,y1,'b');xlabel('Time (sec)');ylabel('y1');
subplot(212); grid on; plot(t,y2,'k');xlabel('Time (sec)');ylabel('y2');
```

Refer to the MATLAB® command '*ss*'. This command, conceptually similar to the '*tf*' command discussed in Chapter 7, implements the state variable modeling once the matrices of the state variable model have been formulated. The time simulation of the system is then performed through the command '*lsim*', also discussed in Chapter 7, following the definition of the input to the system.

The response of the system is shown in Figure SSP8.2.3:

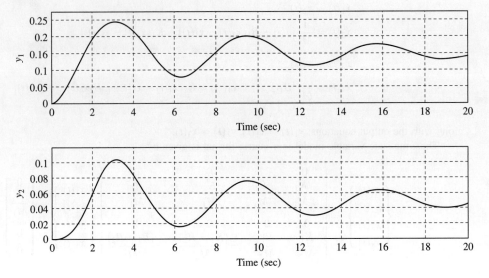

Figure SSP8.2.3 Response of the System in Figure SSP8.2.1

Student Sample Problem 8.3

Consider a linear dynamic system described by the following set of DEs with constant coefficients:

$$\begin{cases} \ddot{y}_1(t) + 4\dot{y}_1(t) - 3y_2(t) = u_1(t) \\ \dot{y}_2(t) + \dot{y}_1(t) + y_1(t) + 2y_2(t) = u_2(t) \end{cases}$$

The system is subjected to two inputs $(u_1(t), u_2(t))$ and has two measurable outputs $(y_1(t), y_2(t))$. Therefore, a TF-based modeling of the system will require the evaluation of a (2×2) matrix of transfer functions.

Find the matrix of transfer functions under the following conditions:

- Derivation from the DEs;
- Derivation from the state variable model.

Solution of Student Sample Problem 8.3
Derivation of Transfer Function (TF) Matrix Directly from the Des

Following the application of the Laplace transformation to the above system we have:

$$\begin{cases} s^2 Y_1(s) + 4s Y_1(s) - 3Y_2(s) = U_1(s) \\ sY_2(s) + sY_1(s) + Y_1(s) + 2Y_2(s) = U_2(s) \end{cases}$$

leading to the following matrix format:

$$\begin{bmatrix} (s^2 + 4s) & -3 \\ (s+1) & (s+2) \end{bmatrix} \begin{Bmatrix} Y_1(s) \\ Y_2(s) \end{Bmatrix} = \begin{Bmatrix} U_1(s) \\ U_2(s) \end{Bmatrix}$$

By invoking the superimposition of effect property of linear systems, we can solve the above system first considering $u_2(t) = 0 \to U_2(s) = 0$ and, next, considering $u_1(t) = 0 \to U_1(s) = 0$. The goal is to find all the $G_{ij}(s)$ transfer functions in the relationship:

$$Y_1(s) = G_{11}(s) \cdot U_1(s) + G_{12}(s) \cdot U_2(s)$$
$$Y_2(s) = G_{21}(s) \cdot U_1(s) + G_{22}(s) \cdot U_2(s)$$

Therefore, consider the system:

$$\begin{bmatrix} (s^2 + 4s) & -3 \\ (s+1) & (s+2) \end{bmatrix} \begin{Bmatrix} Y_1(s) \\ Y_2(s) \end{Bmatrix} = \begin{Bmatrix} U_1(s) \\ 0 \end{Bmatrix}$$

Using Cramer's rule we have:

$$Y_1(s) = \frac{\begin{vmatrix} U_1(s) & -3 \\ 0 & (s+2) \end{vmatrix}}{\begin{vmatrix} (s^2+4s) & -3 \\ (s+1) & (s+2) \end{vmatrix}} = \frac{(s+2) \cdot U_1(s)}{(s^2+4s)(s+2) + 3(s+1)}$$

$$= \frac{(s+2) \cdot U_1(s)}{s^3 + 6s^2 + 11s + 1} \quad \rightarrow \quad \frac{Y_1(s)}{U_1(s)} = G_{11}(s) = \frac{(s+2)}{s^3 + 6s^2 + 11s + 1}$$

$$Y_2(s) = \frac{\begin{vmatrix} (s^2+4s) & U_1(s) \\ (s+1) & 0 \end{vmatrix}}{\begin{vmatrix} (s^2+4s) & -3 \\ (s+1) & (s+2) \end{vmatrix}} = \frac{-(s+1) \cdot U_1(s)}{(s^2+4s)(s+2) + 3(s+1)}$$

$$= \frac{-(s+1) \cdot U_1(s)}{s^3 + 6s^2 + 11s + 1} \quad \rightarrow \quad \frac{Y_2(s)}{U_1(s)} = G_{21}(s) = \frac{-(s+1)}{s^3 + 6s^2 + 11s + 1}$$

Next, consider the system:

$$\begin{bmatrix} (s^2+4s) & -3 \\ (s+1) & (s+2) \end{bmatrix} \begin{Bmatrix} Y_1(s) \\ Y_2(s) \end{Bmatrix} = \begin{Bmatrix} 0 \\ U_2(s) \end{Bmatrix}$$

Using Cramer's rule again we have

$$Y_1(s) = \frac{\begin{vmatrix} 0 & -3 \\ U_2(s) & (s+2) \end{vmatrix}}{\begin{vmatrix} (s^2+4s) & -3 \\ (s+1) & (s+2) \end{vmatrix}} = \frac{3 \cdot U_2(s)}{(s^2+4s)(s+2) + 3(s+1)}$$

$$= \frac{3 \cdot U_2(s)}{s^3 + 6s^2 + 11s + 1} \quad \rightarrow \quad \frac{Y_1(s)}{U_2(s)} = G_{12}(s) = \frac{3}{s^3 + 6s^2 + 11s + 1}$$

$$Y_2(s) = \frac{\begin{vmatrix} (s^2+4s) & 0 \\ (s+1) & U_2(s) \end{vmatrix}}{\begin{vmatrix} (s^2+4s) & -3 \\ (s+1) & (s+2) \end{vmatrix}} = \frac{(s^2+4s) \cdot U_2(s)}{(s^2+4s)(s+2) + 3(s+1)}$$

$$= \frac{(s^2+4s) \cdot U_2(s)}{s^3 + 6s^2 + 11s + 1} \quad \rightarrow \quad \frac{Y_2(s)}{U_2(s)} = G_{22}(s) = \frac{(s^2+4s)}{s^3 + 6s^2 + 11s + 1}$$

Thus the matrix of transfer functions is given by

$$\overline{\overline{G}}(s) = \begin{bmatrix} G_{11}(s) & G_{12}(s) \\ G_{21}(s) & G_{22}(s) \end{bmatrix} = \begin{bmatrix} \dfrac{(s+2)}{s^3 + 6s^2 + 11s + 1} & \dfrac{3}{s^3 + 6s^2 + 11s + 1} \\ \dfrac{-(s+1)}{s^3 + 6s^2 + 11s + 1} & \dfrac{(s^2+4s)}{s^3 + 6s^2 + 11s + 1} \end{bmatrix}$$

Derivation of Transfer Function (TF) Matrix through the SV Model

Starting from

$$\begin{cases} \ddot{y}_1(t) + 4\dot{y}_1(t) - 3y_2(t) = u_1(t) \\ \dot{y}_2(t) + \dot{y}_1(t) + y_1(t) + 2y_2(t) = u_2(t) \end{cases}$$

select the following set of SVs:

$$x_1(t) = y_1(t)$$
$$x_2(t) = \dot{y}_1(t)$$
$$x_3(t) = y_2(t)$$

where $y_1(t), y_2(t)$ are also outputs of the system, in addition of being SVs.

Chapter 8 State Variable Modeling of the Aircraft Dynamics

Using the above DEs, the selected SVs lead to the following differentions:

$$\dot{x}_1(t) = \dot{y}_1(t) = x_2(t)$$
$$\dot{x}_2(t) = \ddot{y}_1(t) = -4x_2(t) + 3x_3(t) + u_1(t)$$
$$\dot{x}_3(t) = \dot{y}_2(t) = -x_2(t) - x_1(t) - 2x_3(t) + u_2(t)$$

Reducing the above relationships to a matrix format leads to the SV model of the system:

$$\begin{Bmatrix} \dot{x}_1(t) \\ \dot{x}_2(t) \\ \dot{x}_3(t) \end{Bmatrix} = \begin{bmatrix} 0 & 1 & 0 \\ 0 & -4 & 3 \\ -1 & -1 & -2 \end{bmatrix} \begin{Bmatrix} x_1(t) \\ x_2(t) \\ x_3(t) \end{Bmatrix} + \begin{bmatrix} 0 & 0 \\ 1 & 0 \\ 0 & 1 \end{bmatrix} \begin{Bmatrix} u_1(t) \\ u_2(t) \end{Bmatrix}$$

$$\begin{Bmatrix} y_1(t) \\ y_2(t) \end{Bmatrix} = \begin{bmatrix} 1 & 0 & 0 \\ 0 & 0 & 1 \end{bmatrix} \begin{Bmatrix} x_1(t) \\ x_2(t) \\ x_3(t) \end{Bmatrix} + \begin{bmatrix} 0 & 0 \\ 0 & 0 \end{bmatrix} \begin{Bmatrix} u_1(t) \\ u_2(t) \end{Bmatrix}$$

Once the matrices of the SV model are available, the matrix of the transfer functions $\overline{\overline{G}}(s)$ can be derived using the relationship:

$$\frac{\overline{Y}(s)}{\overline{U}(s)} = \left[\overline{\overline{G}}(s)\right] = \left\{\overline{\overline{C}} \left[s\overline{\overline{I}} - \overline{\overline{A}}\right]^{-1} \overline{\overline{B}} + \overline{\overline{D}}\right\}$$

Note that in this specific case, the $\overline{\overline{D}}$ matrix is a zero matrix.

A key step is the evaluation of $\left[s\overline{\overline{I}} - \overline{\overline{A}}\right]^{-1}$. Starting from

$$\left[s\overline{\overline{I}} - \overline{\overline{A}}\right] = \begin{bmatrix} s & 0 & 0 \\ 0 & s & 0 \\ 0 & 0 & s \end{bmatrix} - \begin{bmatrix} 0 & 1 & 0 \\ 0 & -4 & 3 \\ -1 & -1 & -2 \end{bmatrix} = \begin{bmatrix} s & -1 & 0 \\ 0 & s+4 & -3 \\ 1 & 1 & s+2 \end{bmatrix}$$

Using the adjoint/transpose approach for the evaluation of the inverse of the matrix (described in Appendix A.2), we have

$$\left[s\overline{\overline{I}} - \overline{\overline{A}}\right]^{-1} = \frac{adj\left[s\overline{\overline{I}} - \overline{\overline{A}}\right]^T}{\left|s\overline{\overline{I}} - \overline{\overline{A}}\right|}$$

First, the determinant in the denominator is given by

$$\left|s\overline{\overline{I}} - \overline{\overline{A}}\right| = s[(s+4)(s+2) + 3] + 1 \cdot 3$$
$$= s(s^2 + 6s + 11) + 3 = s^3 + 6s^2 + 11s + 3$$

Next, $adj\left[s\overline{\overline{I}} - \overline{\overline{A}}\right]^T$ is found as

$$adj\left[s\overline{\overline{I}} - \overline{\overline{A}}\right] = \begin{bmatrix} (s^2 + 6s + 11) & -3 & -(s+4) \\ (s+2) & s(s+2) & -(s+1) \\ 3 & 3s & s(s+4) \end{bmatrix}$$

$$\rightarrow adj\left[s\overline{\overline{I}} - \overline{\overline{A}}\right]^T = \begin{bmatrix} (s^2 + 6s + 11) & (s+2) & 3 \\ -3 & s(s+2) & 3s \\ -(s+4) & -(s+1) & s(s+4) \end{bmatrix}$$

leading to $\left[s\overline{\overline{I}} - \overline{\overline{A}}\right]^{-1} = \dfrac{\begin{bmatrix} (s^2 + 6s + 11) & (s+2) & 3 \\ -3 & s(s+2) & 3s \\ -(s+4) & -(s+1) & s(s+4) \end{bmatrix}}{s^3 + 6s^2 + 11s + 3}$

Next, $\overline{\overline{C}}\left[s\overline{\overline{I}} - \overline{\overline{A}}\right]^{-1}$ is found as

$$\overline{\overline{C}}\left[s\overline{\overline{I}} - \overline{\overline{A}}\right]^{-1} = \begin{bmatrix} 1 & 0 & 0 \\ 0 & 0 & 1 \end{bmatrix} \frac{\begin{bmatrix} (s^2 + 6s + 11) & (s+2) & 3 \\ -3 & s(s+2) & 3s \\ -(s+4) & -(s+1) & s(s+4) \end{bmatrix}}{s^3 + 6s^2 + 11s + 3}$$

$$= \frac{\begin{bmatrix} (s^2 + 6s + 11) & (s+2) & 3 \\ -(s+4) & -(s+1) & s(s+4) \end{bmatrix}}{s^3 + 6s^2 + 11s + 3}$$

Next, $\overline{\overline{C}}\left[s\overline{\overline{I}} - \overline{\overline{A}}\right]^{-1} \overline{\overline{B}}$ is found using

$$\overline{\overline{C}}\left[s\overline{\overline{I}} - \overline{\overline{A}}\right]^{-1} \overline{\overline{B}} = \frac{\begin{bmatrix} (s^2 + 6s + 11) & (s+2) & 3 \\ -(s+4) & -(s+1) & s(s+4) \end{bmatrix}}{s^3 + 6s^2 + 11s + 3} \cdot \begin{bmatrix} 0 & 0 \\ 1 & 0 \\ 0 & 1 \end{bmatrix}$$

$$= \frac{\begin{bmatrix} (s+2) & 3 \\ -(s+1) & s(s+4) \end{bmatrix}}{s^3 + 6s^2 + 11s + 3}$$

leading to the conclusion

$$\overline{\overline{G}}(s) = \overline{\overline{C}}\left[s\overline{\overline{I}} - \overline{\overline{A}}\right]^{-1} \overline{\overline{B}} = \begin{bmatrix} G_{11}(s) & G_{12}(s) \\ G_{21}(s) & G_{22}(s) \end{bmatrix}$$

$$= \begin{bmatrix} \dfrac{(s+2)}{s^3 + 6s^2 + 11s + 1} & \dfrac{3}{s^3 + 6s^2 + 11s + 1} \\ \dfrac{-(s+1)}{s^3 + 6s^2 + 11s + 1} & \dfrac{(s^2 + 4s)}{s^3 + 6s^2 + 11s + 1} \end{bmatrix}$$

Student Sample Problem 8.4

Consider the Simulink®-based MIMO system shown in Figure SSP8.4.1.

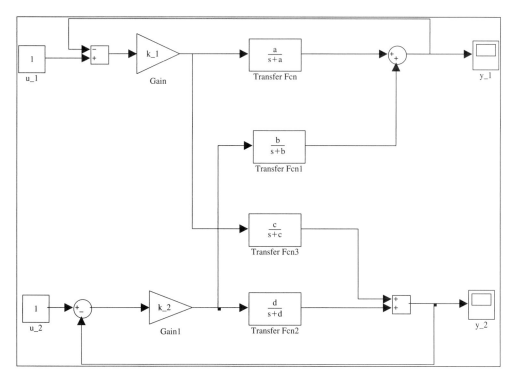

Figure SSP8.4.1 Simulink®-based MIMO System

- Find the state variable model associated with the preceding system;
- Assuming generic values for the parameters a, b, c, d, K_1, K_2, implement the state variable model in MATLAB®;
- Plot $y_1(t), y_2(t)$, assuming generic step inputs $u_1(t) = 1, u_2(t) = 1$;
- Validate the modeling in Simulink®.

Solution of Student Sample Problem 8.4

An analysis of the preceding block diagram reveals that the system has two inputs and two measurable outputs. Additionally, there are four first order components—each of them being a generic first order system (GFOS), shown in Appendix A.5—leading to a fourth order system. A suitable selection of state variables is represented by four state variables, each of them being the output of a GFOS. Additionally, recall that these state variables are defined in the s-domain, since the system is a continuous-time system. The selected state variables are shown in Figure SSP8.4.2.

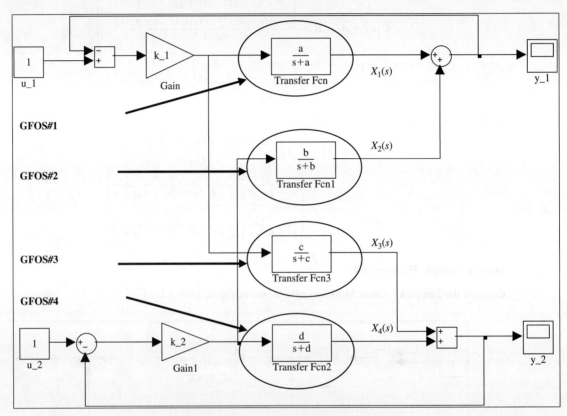

Figure SSP8.4.2 Simulink®-based MIMO System

For the given system the objective is to derive the state variable model:

$$\begin{Bmatrix} \dot{x}_1(t) \\ \dot{x}_2(t) \\ \ldots \\ \dot{x}_n(t) \end{Bmatrix} = A_{nxn} \begin{Bmatrix} x_1(t) \\ x_2(t) \\ \ldots \\ x_n(t) \end{Bmatrix} + B_{nxm} \begin{Bmatrix} u_1(t) \\ u_2(t) \\ \ldots \\ u_m(t) \end{Bmatrix}$$

$$\begin{Bmatrix} y_1(t) \\ y_2(t) \\ \ldots \\ y_l(t) \end{Bmatrix} = C_{lxn} \begin{Bmatrix} x_1(t) \\ x_2(t) \\ \ldots \\ x_n(t) \end{Bmatrix} + D_{lxm} \begin{Bmatrix} u_1(t) \\ u_2(t) \\ \ldots \\ u_m(t) \end{Bmatrix}$$

In this case we have $(n = 4, m = 2, l = 2)$.

Note that from the block diagram the output relationships are given by

$$Y_1(s) = (X_1(s) + X_2(s))$$
$$Y_2(s) = (X_3(s) + X_4(s))$$

The derivation of the state variable model starts from the analysis of the block diagram. In particular, the signal $X_1(s)$ is given by

$$X_1(s) = \frac{a}{(s+a)} K_1(U_1(s) - Y_1(s)) = \frac{a}{(s+a)} K_1(U_1(s) - X_1(s) - X_2(s))$$

Solving for $(s\,X_1(s))$ leads to

$$(s+a)X_1(s) = aK_1(U_1(s) - X_1(s) - X_2(s))$$
$$sX_1(s) = -aX_1(s) + aK_1(U_1(s) - X_1(s) - X_2(s))$$

Using inverse Laplace transformation $(ILT(s\,X_1(s)) = \dot{x}_1(t),\ ILT(X_1(s)) = x_1(t))$, the preceding relationship can be reduced to the time domain using

$$\boxed{\begin{aligned}\dot{x}_1(t) &= -a\,x_1(t) - aK_1\,x_1(t) - aK_1\,x_2(t) + aK_1 u_1(t) \\ &= -a(1+K_1)x_1(t) - aK_1\,x_2(t) + aK_1 u_1(t)\end{aligned}}$$

Similarly, the signal $X_2(s)$ is given by

$$X_2(s) = \frac{b}{(s+b)} K_2(U_2(s) - Y_2(s)) = \frac{b}{(s+b)} K_2(U_2(s) - X_3(s) - X_4(s))$$

Solving for $(s\,X_2(s))$ gives

$$(s+b)X_2(s) = bK_2(U_2(s) - X_3(s) - X_4(s))$$
$$sX_2(s) = -bX_2(s) + bK_2(U_2(s) - X_3(s) - X_4(s))$$

Using inverse Laplace transformation $(ILT(sX_2(s)) = \dot{x}_2(t),\ ILT(X_2(s)) = x_2(t))$, the preceding relationship can be reduced to the time domain using

$$\boxed{\dot{x}_2(t) = -b\,x_2(t) + bK_2(u_2(t) - x_3(t) - x_4(t))}$$

The signal $X_3(s)$ is given by

$$X_3(s) = \frac{c}{(s+c)} K_1(U_1(s) - Y_1(s)) = \frac{c}{(s+c)} K_1(U_1(s) - X_1(s) - X_2(s))$$

Solving for $(s\,X_3(s))$ leads to

$$(s+c)X_3(s) = c\,K_1(U_1(s) - X_1(s) - X_2(s))$$
$$s\,X_3(s) = -c\,X_3(s) + c\,K_1(U_1(s) - X_1(s) - X_2(s))$$

Using inverse Laplace transformation $(ILT(s\,X_3(s)) = \dot{x}_3(t), ILT(X_3(s)) = x_3(t))$, the preceding relationship can be reduced to the time domain using

$$\boxed{\dot{x}_3(t) = -c\,x_3(t) + cK_1(u_1(t) - x_1(t) - x_2(t))}$$

Finally, the signal $X_4(s)$ is given by

$$X_4(s) = \frac{d}{(s+d)} K_2(U_2(s) - Y_2(s)) = \frac{d}{(s+d)} K_2(U_2(s) - X_3(s) - X_4(s))$$

Solving for $(s\,X_4(s))$ gives

$$(s+d)X_4(s) = dK_2(U_2(s) - X_3(s) - X_4(s))$$
$$sX_4(s) = -dX_4(s) + dK_2(U_2(s) - X_3(s) - X_4(s))$$

Using inverse Laplace transformation $(ILT(s\,X_4(s)) = \dot{x}_4(t),\ ILT(X_4(s)) = x_4(t))$, the preceding relationship can be reduced to the time domain using

$$\boxed{\begin{aligned}\dot{x}_4(t) &= -dx_4(t) + dK_2(u_2(t) - x_3(t) - x_4(t)) \\ &= -dK_2\,x_3(t) - d(1+K_2)x_4(t) + dK_2\,u_2(t)\end{aligned}}$$

Rearranging the previous state variable equations in a matrix format we have

$$\begin{Bmatrix}\dot{x}_1(t)\\\dot{x}_2(t)\\\dot{x}_3(t)\\\dot{x}_4(t)\end{Bmatrix} = \begin{bmatrix}-a(1+K_1) & -aK_1 & 0 & 0\\ 0 & -b & -bK_2 & -bK_2\\ -cK_1 & -cK_1 & -c & 0\\ 0 & 0 & -dK_2 & -d(1+K_2)\end{bmatrix}\begin{Bmatrix}x_1(t)\\x_2(t)\\x_3(t)\\x_4(t)\end{Bmatrix} + \begin{bmatrix}aK_1 & 0\\ 0 & bK_2\\ cK_1 & 0\\ 0 & dK_2\end{bmatrix}\begin{Bmatrix}u_1(t)\\u_2(t)\end{Bmatrix}$$

$$\begin{Bmatrix}y_1(t)\\y_2(t)\end{Bmatrix} = \begin{bmatrix}1 & 1 & 0 & 0\\ 0 & 0 & 1 & 1\end{bmatrix}\begin{Bmatrix}x_1(t)\\x_2(t)\\x_3(t)\\x_4(t)\end{Bmatrix} + \begin{bmatrix}0 & 0\\ 0 & 0\end{bmatrix}\begin{Bmatrix}u_1(t)\\u_2(t)\end{Bmatrix}$$

A sample MATLAB® script file for the modeling of the preceding system is included here, assuming generic values for the parameters a, b, c, d, K_1, K_2 (with the condition that the system is stable).

```
%
% Chapter VIII
% Student Sample Problem 8.4
%
a=2; b=4; c=6; d=8;
k1=5; k2=6;
aa=[ -a*(1+k1) -a*k1 0 0
    0 -b -b*k2 -b*k2
    -c*k1 -c*k1 -c 0
    0 0 -d*k2 -d*(1+k2)];
bb=[a*k1 0 c*k1 0
    0 b*k2 0 d*k2]';
cc=[1 1 0 0
    0 0 1 1];
dd=[0 0
    0 0];
sys=ss(aa,bb,cc,dd);
t=[0:0.01:10];
size_t=max(size(t));
for i=1:size_t;
  u1(i)=1;
  u2(i)=1;
end;
u1=u1'; u2=u2';
u=[u1 u2];
y=lsim(sys,u,t);
y1=y(:,1);
y2=y(:,2);
subplot(211); grid on; plot(t,y1,'b');xlabel('Time (sec)');ylabel('y1');
subplot(212); grid on; plot(t,y2,'k');xlabel('Time (sec)');ylabel('y2');
```

Refer to the MATLAB® commands '*ss*' and '*lsim*'. The response of the system to generic unit inputs is shown in Figure SSP8.4.3:

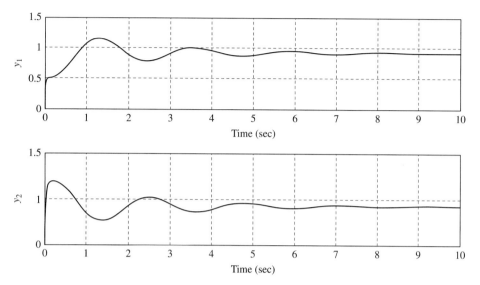

Figure SSP8.4.3 Response of the System in Figure SSP8.4.1

The validation of the preceding modeling can be performed in Simulink® by comparing the outputs of the state variable model with the outputs of the original system, as shown in Figure SSP8.4.4.

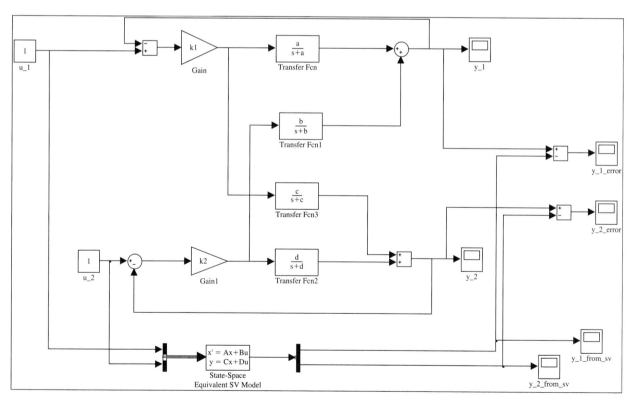

Figure SSP8.4.4 Simulink® Validation of the State Variable Model

The process will show negligible values of the differences between the outputs of the state variable model and the outputs of the original system, shown as y_{1Error} and y_{2Error}.

Student Sample Problem 8.5

Consider the SIAI Marchetti S211 aircraft at cruise condition 2 (Appendix B, Aircraft 8).

- Derive the state variable model for the longitudinal dynamics.
- Implement the state variable model in MATLAB®.
- Using a generic longitudinal maneuver, plot $y_{Long} = \{a_Z\ u\ \alpha\ q\ \theta\}$.

Solution of Student Sample Problem 8.5

Using the outlined procedure, for the selected aircraft at the selected flight condition we have

Longitudinal Dimensional Stability Derivatives

x_u =	−0.023564	(ft/sec^2)/(ft/sec)
x_Tu =	0.0011782	(ft/sec^2)/(ft/sec)
x_alpha =	8.8074	(ft/sec^2)/(rad)
x_de =	0	(ft/sec^2)/(rad)
Z_u =	−0.14139	(ft/sec^2)/(ft/sec)
Z_alpha =	−760.3304	(ft/sec^2)/(rad)
Z_alpha_dot =	−2.6722	(ft/sec^2)/(rad/sec)
Z_q =	−6.3624	(ft/sec^2)/(rad/sec)
Z_de =	−48.1657	(ft/sec^2)/(rad)
M_u =	0	(rad/sec^2)/(ft/sec)
M_Tu =	0	(rad/sec^2)/(ft/sec)
M_alpha =	−4.6157	(rad/sec^2)/(rad)
M_T_alpha =	0	(rad/sec^2)/(rad)
M_alpha_dot =	−0.85359	(rad/sec^2)/(rad/sec)
M_q =	−1.5738	(rad/sec^2)/(rad/sec)
M_de =	−15.7703	(rad/sec^2)/(rad)

Primed Longitudinal Dimensional Stability Derivatives

Alpha-Dot Coefficients	
z_alpha_prime =	−1.296
z_u_prime =	−0.000241
z_q_prime =	0.9846
z_de_prime =	−0.0821
z_theta_prime =	0

U-Dot Coefficients	
x_alpha_prime =	41.0074
x_u_prime =	−0.023564
x_q_prime =	0
x_theta_prime =	−32.2
x_de_prime =	0

Q-Dot Coefficients	
m_alpha_prime =	−3.5094
m_u_prime =	0.00020571
m_q_prime =	−2.4142
m_theta_prime =	0
m_de_prime =	−15.7002

Double Primed Longitudinal Dimensional Stability Derivatives

a_z Coefficients	
z_alpha_pprime =	−756.8672
z_u_pprime =	−0.14074
z_q_pprime =	−8.9935
z_theta_pprime =	0
z_de_pprime =	−47.9463

A_Long

−1.2960	−0.0002	0.9846	0
41.0074	−0.0236	0	−32.2000
−3.5094	0.0002	−2.4142	0
0	0	1	0

B_Long

−0.0821
0
−15.7002
0

C_Long

−756.8672	−0.1407	−8.9935	0
1	0	0	0
0	1	0	0
0	0	1	0
0	0	0	1

D_Long

−47.9463
0
0
0
0

A sample MATLAB® script file for the modeling of the preceding system is included here, assuming a generic longitudinal maneuver.

```
%
% Chapter VIII
% Student Sample Problem 8.5
%
a_long=[-1.2960 -0.0002 0.9846 0
  41.0074 -0.0236 0 -32.2000
  -3.5094 0.0002 -2.4142 0
     0 0 1 0];
b_long=[-0.0821 0 -15.7002 0]';
c_long=[-756.8672 -0.1407 -8.9935 0
     1 0 0 0
     0 1 0 0
     0 0 1 0
     0 0 0 1];
d_long=[-47.9463 0 0 0 0]';
sys=ss(a_long,b_long,c_long,d_long);
t=[0:0.01:120];
size_t=max(size(t));
for i=1:size_t;
   de(i)=0;
end;
for i=200:300;
   de(i)=-3/57.3;
end;
for i=700:800;
   de(i)=3/57.3;
end;
y=lsim(sys,de,t);
az=y(:,1)*(1/32.17)+1;
alpha=y(:,2)*57.3;
u=y(:,3);
q=y(:,4)*57.3;
theta=y(:,5)*57.3;
de=de*57.3;
figure(1);
subplot(311); grid on; plot(t,az,'k');
xlabel('Time (sec)');ylabel('a_{z} (g)');
subplot(312); grid on; plot(t,alpha,'k');
xlabel('Time (sec)');ylabel('alpha (deg)');
subplot(313); grid on; plot(t,u,'k');
xlabel('Time (sec)');ylabel('u (ft/sec)');
figure(2);
subplot(311); grid on; plot(t,q,'k');
xlabel('Time (sec)');ylabel('q (deg/sec)');
subplot(312); grid on; plot(t,theta,'k');
xlabel('Time (sec)');ylabel('theta (deg)');
subplot(313); grid on; plot(t,de,'k');
xlabel('Time (sec)');ylabel('de (deg)');
```

Refer to the MATLAB® commands 'ss' and 'lsim'. The outputs of the model for a generic longitudinal maneuver are shown in Figure SSP8.5.1.

The preceding plots show a very well damped short period mode and a very low damped phugoid mode, as expected for this size and class of aircraft.

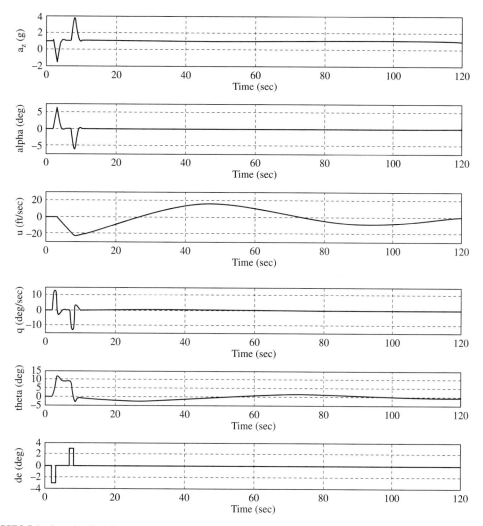

Figure SSP8.5.1 Longitudinal Outputs of the SIAI S-211 Aircraft at Flight Condition #2 for a Generic Longitudinal Maneuver

Student Sample Problem 8.6

Consider the SIAI Marchetti S211 aircraft at cruise condition 2 (Appendix B, Aircraft 8).

- Derive the state variable model for the lateral directional dynamics;
- Implement the state variable model in MATLAB®;
- Using a generic rudder maneuver, plot $y_{Lat\ Dir} = \{a_Y\ \beta\ p\ r\ \phi\}$.

Solution of Student Sample Problem 8.6

For the selected aircraft at the selected flight condition we have the following:

Lateral Directional Dimensional Stability Derivatives

Y_beta =	−137.6164	(ft/sec^2)/(rad)
Y_p =	−0.37185	(ft/sec^2)/(rad/sec)
Y_r =	1.9212	(ft/sec^2)/(rad/sec)
Y_da =	0	(ft/sec^2)/(rad)
Y_dr =	3.8533	(ft/sec^2)/(rad)
L_beta =	−62.2247	(rad/sec^2)/(rad)
L_p =	−4.9676	(rad/sec^2)/(rad/sec)
L_r =	3.9486	(rad/sec^2)/(rad/sec)
L_da =	56.5679	(rad/sec^2)/(rad)
L_dr =	28.284	(rad/sec^2)/(rad)
N_beta =	14.6839	(rad/sec^2)/(rad)
N_T_beta =	0	(rad/sec^2)/(rad)
N_p =	0.1556	(rad/sec^2)/(rad/sec)
N_r =	−0.50568	(rad/sec^2)/(rad/sec)
N_da =	−0.43188	(rad/sec^2)/(rad)
N_dr =	−10.3651	(rad/sec^2)/(rad)

Primed Lateral Dimensional Stability Derivatives

R-Dot Coefficients	
n_beta_prime =	13.1748
n_p_prime =	0.030895
n_r_prime =	−0.40815
n_phi_prime =	0
n_da_prime =	0.99368
n_dr_prime =	−9.6945

P-Dot Coefficients	
l_beta_prime =	−60.0566
l_p_prime =	−4.9625
l_r_prime =	3.8814
l_phi_prime =	0
l_da_prime =	56.7314
l_dr_prime =	26.6886

Beta-Dot Coefficients	
y_beta_prime =	−0.2356
y_p_prime =	−0.0006
y_r_prime =	−0.99671
y_phi_prime =	0.0551
y_da_prime =	0
y_dr_prime =	0.006598

Double Primed Lateral Dimensional Stability Derivatives

a_y Coefficients	
y_beta_pprime =	−137.6164
y_p_pprime =	−0.37185
y_r_pprime =	1.9212
y_phi_pprime =	0
y_da_pprime =	0
y_dr_pprime =	3.8533

A_Lat_Dir

−0.2356	−0.0006	−0.9967	0.0551
−60.0566	−4.9625	3.8814	0
13.1748	0.0309	−0.4081	0
0	1	0.0157	0

B_Lat_Dir

0	0.0066
56.7314	26.6886
0.9937	−9.6945
0	0

C_Lat_Dir

−137.6164	−0.3718	1.9212	0
1	0	0	0
0	1	0	0
0	0	1	0
0	0	0	1

D_Lat_Dir

0	3.8533
0	0
0	0
0	0
0	0

A sample MATLAB® script file for the modeling of the preceding system is included here, assuming a generic longitudinal maneuver.

```
%
% Chapter VIII
% Student Sample Problem 8.6
%
a_latdir=[ -0.2356 -0.0006 -0.9967 0.0551
 -60.0566 -4.9625 3.8814 0
  13.1748 0.0309 -0.4081 0
      0 1 0.0157 0];
b_latdir=[0 56.7314 0.9937 0
  0.0066 26.6886 -9.6945 0]';
c_latdir=[-137.6164 -0.3718 1.9212 0
      1 0 0 0
      0 1 0 0
      0 0 1 0
      0 0 0 1];
d_latdir=[0 0 0 0 0
  3.8533 0 0 0 0]';
sys=ss(a_latdir,b_latdir,c_latdir,d_latdir);
t=[0:0.01:60];
size_t=max(size(t));
for i=1:size_t;
  dr(i,1)=0;
  da(i,1)=0;
end;
for i=200:300;
  dr(i,1)=-6/57.3;
end;
```

```
for i=2200:2300;
   dr(i,1)=6/57.3;
end;
u=[da,dr];
y=lsim(sys,u,t);
ay=y(:,1)*(1/32.17);
beta=y(:,2)*57.3;
p=y(:,3)*57.3;
r=y(:,4)*57.3;
phi=y(:,5)*57.3;
dr=dr*57.3;
figure(1);
subplot(311); grid on; plot(t,ay,'k');
xlabel('Time (sec)');ylabel('a_{y} (g)');
subplot(312); grid on; plot(t,beta,'k');
xlabel('Time (sec)');ylabel('beta (deg)');
subplot(313); grid on; plot(t,p,'k');
xlabel('Time (sec)');ylabel('p (deg/sec)');
figure(2);
subplot(311); grid on; plot(t,r,'k');
xlabel('Time (sec)');ylabel('r (deg/sec)');
subplot(312); grid on; plot(t,phi,'k');
xlabel('Time (sec)');ylabel('phi (deg)');
subplot(313); grid on; plot(t,dr,'k');
xlabel('Time (sec)');ylabel('dr (deg)');
```

Refer to the MATLAB® commands '*ss*' and '*lsim*'. The outputs of the model for a generic rudder maneuver are shown in Figure SSP8.6.1.

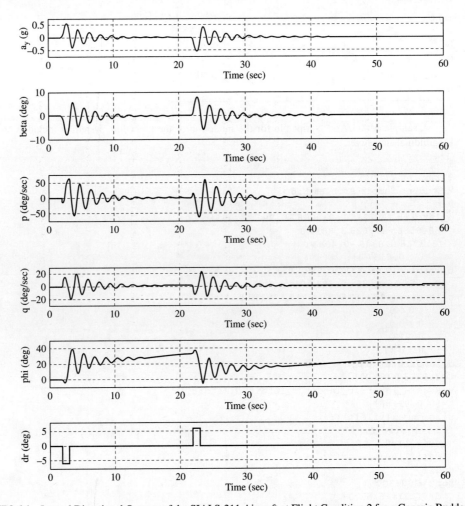

Figure SSP8.6.1 Lateral Directional Outputs of the SIAI S-211 Aircraft at Flight Condition 2 for a Generic Rudder Maneuver

A similar set of outputs can be produced for a generic aileron maneuver. The preceding plots show a relatively low damped dutch roll mode, as expected for this size and class of aircraft.

Student Sample Problem 8.7

Consider the longitudinal and the lateral directional SV models for the SIAI Marchetti S211 aircraft in SSP 8.5 and SSP 8.6. Provide the complete SV model.

Solution of Student Sample Problem 8.7

The general form of the complete aircraft SV model is given by

$$\begin{Bmatrix} \dot{u} \\ \dot{\alpha} \\ \dot{q} \\ \dot{\theta} \\ \dot{\beta} \\ \dot{p} \\ \dot{r} \\ \dot{\phi} \end{Bmatrix} = \begin{bmatrix} A_{Long} & 0 \\ 0 & A_{Lat\,Dir} \end{bmatrix} \begin{Bmatrix} u \\ \alpha \\ q \\ \theta \\ \beta \\ p \\ r \\ \phi \end{Bmatrix} + \begin{bmatrix} B_{Long} & 0 \\ 0 & B_{Lat\,Dir} \end{bmatrix} \begin{Bmatrix} \delta_E \\ \delta_A \\ \delta_R \end{Bmatrix}$$

$$\begin{Bmatrix} a_z \\ \alpha \\ u \\ q \\ \theta \\ a_Y \\ \beta \\ p \\ r \\ \phi \end{Bmatrix} = \begin{bmatrix} C_{Long} & 0 \\ 0 & C_{Lat\,Dir} \end{bmatrix} \begin{Bmatrix} \alpha \\ u \\ q \\ \theta \\ \beta \\ p \\ r \\ \phi \end{Bmatrix} + \begin{bmatrix} D_{Long} & 0 \\ 0 & D_{Lat\,Dir} \end{bmatrix} \begin{Bmatrix} \delta_E \\ \delta_A \\ \delta_R \end{Bmatrix}$$

For the given aircraft at the given flight condition we have

A Matrix

−1.2960	−0.0002	0.9846	0	0	0	0	0
41.0074	−0.0236	0	−32.2000	0	0	0	0
−3.5094	0.0002	−2.4142	0	0	0	0	0
0	0	1	0	0	0	0	0
0	0	0	0	−0.235	−0.0006	−0.997	0.0551
0	0	0	0	−60.05	−4.963	3.881	0
0	0	0	0	13.175	0.0309	−0.4081	0
0	0	0	0	0	1	0.0157	0

B Matrix

−0.0821	0	0
0	0	0
−15.7002	0	0
0	0	0
0	0	0.0066
0	56.7314	26.6886
0	0.9937	−9.6945
0	0	0

C Matrix

−756.87	−0.14	−8.99	0	0	0	0	0
1	0	0	0	0	0	0	0
0	1	0	0	0	0	0	0
0	0	1	0	0	0	0	0
0	0	0	1	0	0	0	0
0	0	0	0	−137.62	−0.372	1.92	0
0	0	0	0	1	0	0	0
0	0	0	0	0	1	0	0
0	0	0	0	0	0	1	0
0	0	0	0	0	0	0	1

D Matrix

−47.9463	0	0
0	0	0
0	0	0
0	0	0
0	0	0
0	0	3.8533
0	0	0
0	0	0
0	0	0
0	0	0

PROBLEMS

Problem 8.1

Consider a linear dynamic system described by the following DE with constant coefficients:

$$\dddot{y}(t) + 5\ddot{y}(t) + 7\dot{y}(t) + 3y(t) = u(t)$$

Find the state variable model associated with this system. Verify that the eigenvalues of the state matrix A are coincident with the roots of the characteristic equation of the preceding system.

Problem 8.2

Consider a linear dynamic system described by the following set of DEs with constant coefficients:

$$\begin{cases} \dot{y}_1(t) + 5y_2(t) = u_1(t) \\ \dot{y}_2(t) - 3y_1(t) + 2y_2(t) = u_2(t) \end{cases}$$

Find the state variable model of the above system.

The system is subjected to two inputs $(u_1(t), u_2(t))$ and has two measurable outputs $(y_1(t), y_2(t))$. Therefore, a TF-based modeling of the system will require the evaluation of a (2×2) matrix of transfer functions. Find the matrix of transfer functions under the following conditions:

- Derivation *directly* from the DEs;
- Derivation from the state variable model.

Problem 8.3

Consider a linear dynamic system described by the following set of DEs with constant coefficients:

$$\begin{cases} \dot{y}_1(t) + 7y_2(t) = 3\, u_1(t) \\ \ddot{y}_2(t) + 3\dot{y}_2(t) + 2y_1(t) + 5y_2(t) = 4\, u_2(t) \end{cases}$$

Find the state variable model of the above system.

The system is subjected to two inputs $(u_1(t), u_2(t))$ and has two measurable outputs $(y_1(t), y_2(t))$. TF-based modeling of the system will require the evaluation of a (2×2) matrix of transfer functions. Find the matrix of transfer functions under the following conditions:

- Derivation *directly* from the DEs;
- Derivation from the state variable model.

Problem 8.4

Consider the Learjet 24 aircraft at approach (Appendix B, Aircraft 6). Using the data provided in Appendix B, derive an "extended version" of longitudinal SV model (shown here) featuring one additional input (i_H) and one additional output (a_x).

$$\begin{Bmatrix} \dot{u} \\ \dot{\alpha} \\ \dot{q} \\ \dot{\theta} \end{Bmatrix} = A_{Long} \begin{Bmatrix} u \\ \alpha \\ q \\ \theta \end{Bmatrix} + B_{Long} \begin{Bmatrix} \delta_E \\ i_H \end{Bmatrix}$$

$$\begin{Bmatrix} a_z \\ a_x \\ u \\ \alpha \\ q \\ \theta \end{Bmatrix} = C_{Long} \begin{Bmatrix} u \\ \alpha \\ q \\ \theta \end{Bmatrix} + D_{Long} \begin{Bmatrix} \delta_E \\ i_H \end{Bmatrix}$$

Verify that the eigenvalues of the state matrix A_{Long} are coincident with the roots of the longitudinal characteristic equation provided in Appendix B (Aircraft 6).

Problem 8.5

Consider the Learjet 24 aircraft at approach (Appendix B, Aircraft 6). Using the data provided in Appendix B, derive *only* the state equations of the longitudinal SV model shown here:

$$\begin{Bmatrix} \dot{u} \\ \dot{\alpha} \\ \dot{q} \\ \dot{\theta} \end{Bmatrix} = A_{Long} \begin{Bmatrix} u \\ \alpha \\ q \\ \theta \end{Bmatrix} + B_{Long} \begin{Bmatrix} \delta_E \\ i_H \end{Bmatrix}$$

Next, consider as baseline the nominal values for the following dimensionless stability derivatives:

$$c_{L_\alpha}, c_{m_\alpha}, c_{m_{\dot{\alpha}}}, c_{m_q}, c_{m_u}$$

Starting from the nominal value, for *each* of the preceding stability derivatives build a column vector where the value of the selected derivative (among the eight derivatives shown) varies from -20 percent to $+20$ percent with respect to the nominal value with increments of 1 percent.

For each stability derivative plot the migration of the eigenvalues of A_{Long} in the s-domain. Highlight the conditions where loss of dynamic stability occurs.

Plot $\omega_{n_{SP}}, \zeta_{SP}, \omega_{n_{Ph}}, \zeta_{Ph}$ versus each stability derivative.

Problem 8.6

Consider the McDonnell Douglas F-4 aircraft at subsonic conditions (Appendix B, Aircraft 10). Using the data provided in Appendix B, derive *only* the state equations of the longitudinal SV model shown:

$$\begin{Bmatrix} \dot{u} \\ \dot{\alpha} \\ \dot{q} \\ \dot{\theta} \end{Bmatrix} = A_{Long} \begin{Bmatrix} u \\ \alpha \\ q \\ \theta \end{Bmatrix} + B_{Long} \begin{Bmatrix} \delta_E \\ i_H \end{Bmatrix}$$

Next, consider as baseline the nominal values for the following dimensionless stability derivatives:

$$c_{L_\alpha}, c_{m_\alpha}, c_{m_{\dot\alpha}}, c_{m_q}, c_{m_u}$$

Starting from the nominal value, for *each* of the preceding stability derivatives build a column vector where the value of the selected derivative (among the eight derivatives shown) varies from -20 percent to $+20$ percent with respect to the nominal value with increments of 1 percent.

For each stability derivative plot the migration of the eigenvalues of A_{Long} in the s-domain. Highlight the conditions where loss of dynamic stability occurs.

Plot $\omega_{n_{SP}}, \zeta_{SP}, \omega_{n_{Ph}}, \zeta_{Ph}$ versus each stability derivative.

Problem 8.7

Consider the Boeing B747-200 aircraft at high-altitude cruise conditions (Appendix B, Aircraft 7). Using the data provided in Appendix B, derive only the state equations of the longitudinal SV model shown:

$$\begin{Bmatrix} \dot u \\ \dot\alpha \\ \dot q \\ \dot\theta \end{Bmatrix} = A_{Long} \begin{Bmatrix} u \\ \alpha \\ q \\ \theta \end{Bmatrix} + B_{Long} \begin{Bmatrix} \delta_E \\ i_H \end{Bmatrix}$$

Next, consider as baseline the nominal values for the following dimensionless stability derivatives:

$$c_{L_\alpha}, c_{m_\alpha}, c_{m_{\dot\alpha}}, c_{m_q}, c_{m_u}$$

Starting from the nominal value, for each of the preceding stability derivatives build a column vector where the value of the selected derivative (among the eight derivatives shown) varies from -20 percent to $+20$ percent with respect to the nominal value with increments of 1 percent.

For each stability derivative plot the migration of the eigenvalues of A_{Long} in the s-domain. Highlight the conditions where loss of dynamic stability occurs.

Plot $\omega_{n_{SP}}, \zeta_{SP}, \omega_{n_{Ph}}, \zeta_{Ph}$ versus each stability derivative.

Problem 8.8

Consider the Learjet 24 aircraft at approach (Appendix B, Aircraft 6). Using the data provided in Appendix B, derive an extended version of lateral directional SV model (shown here).

$$\begin{Bmatrix} \dot\beta \\ \dot p \\ \dot r \\ \dot\phi \end{Bmatrix} = A_{Lat\,Dir} \begin{Bmatrix} \beta \\ p \\ r \\ \phi \end{Bmatrix} + B_{Lat\,Dir} \begin{Bmatrix} \delta_A \\ \delta_R \end{Bmatrix}$$

$$\begin{Bmatrix} a_Y \\ \beta \\ p \\ r \\ \phi \end{Bmatrix} = C_{Lat\,Dir} \begin{Bmatrix} \beta \\ p \\ r \\ \phi \end{Bmatrix} + D_{Lat\,Dir} \begin{Bmatrix} \delta_A \\ \delta_R \end{Bmatrix}$$

Verify that the eigenvalues of the state matrix $A_{Lat\,Dir}$ are coincident with the roots of the lateral directional characteristic equation provided in Appendix B (Aircraft 6).

Problem 8.9

Consider the Learjet 24 aircraft at approach (Appendix B, Aircraft 6). Using the data provided in Appendix B, derive only the state equations of the lateral directional SV model shown:

$$\begin{Bmatrix} \dot{\beta} \\ \dot{p} \\ \dot{r} \\ \dot{\phi} \end{Bmatrix} = A_{Lat\,Dir} \begin{Bmatrix} \beta \\ p \\ r \\ \phi \end{Bmatrix} + B_{Lat\,Dir} \begin{Bmatrix} \delta_A \\ \delta_R \end{Bmatrix}$$

Next, consider as baseline the nominal values for the following dimensionless stability derivatives:

$$c_{l_\beta}, c_{n_\beta}, c_{l_p}, c_{n_r}$$

Starting from the nominal value, for each of the preceding stability derivatives build a column vector where the value of the selected derivative (among the eight derivatives shown) varies from -20 percent to $+20$ percent with respect to the nominal value with increments of 1 percent.

For each stability derivative plot the migration of the eigenvalues of $A_{Lat.Dir}$ in the s-domain. Highlight the conditions where loss of dynamic stability occurs.

Plot $\omega_{n_{DR}}, \zeta_{DR}, T_R, T_S$ versus each stability derivative.

Problem 8.10

Consider the McDonnell Douglas F-4 aircraft at subsonic conditions (Appendix B, Aircraft 10). Using the data provided in Appendix B, derive only the state equations of the lateral directional SV model shown:

$$\begin{Bmatrix} \dot{\beta} \\ \dot{p} \\ \dot{r} \\ \dot{\phi} \end{Bmatrix} = A_{Lat\,Dir} \begin{Bmatrix} \beta \\ p \\ r \\ \phi \end{Bmatrix} + B_{Lat\,Dir} \begin{Bmatrix} \delta_A \\ \delta_R \end{Bmatrix}$$

Next, consider as baseline the nominal values for the following dimensionless stability derivatives:

$$c_{l_\beta}, c_{n_\beta}, c_{l_p}, c_{n_r}$$

Starting from the nominal value, for each of the preceding stability derivatives build a column vector where the value of the selected derivative (among the eight derivatives shown) varies from -20 percent to $+20$ percent with respect to the nominal value with increments of 1 percent.

For each stability derivative plot the migration of the eigenvalues of $A_{Lat.Dir}$ in the s-domain. Highlight the conditions where loss of dynamic stability occurs.

Plot $\omega_{n_{DR}}, \zeta_{DR}, T_R, T_S$ versus each stability derivative.

Problem 8.11

Consider the Boeing B747-200 aircraft at high-altitude cruise conditions (Appendix B, Aircraft 7). Using the data provided in Appendix B, derive *only* the state equations of the lateral directional SV model shown:

$$\begin{Bmatrix} \dot{\beta} \\ \dot{p} \\ \dot{r} \\ \dot{\phi} \end{Bmatrix} = A_{Lat\,Dir} \begin{Bmatrix} \beta \\ p \\ r \\ \phi \end{Bmatrix} + B_{Lat\,Dir} \begin{Bmatrix} \delta_A \\ \delta_R \end{Bmatrix}$$

Next, consider as baseline the nominal values for the following dimensionless stability derivatives:

$$c_{l_\beta}, c_{n_\beta}, c_{l_p}, c_{n_r}$$

Starting from the nominal value, for each of the preceding stability derivatives build a column vector where the value of the selected derivative (among the eight derivatives shown) varies from -20 percent to $+20$ percent with respect to the nominal value with increments of 1 percent.

For each stability derivative plot the migration of the eigenvalues of $A_{Lat.Dir}$ in the s-domain. Highlight the conditions where loss of dynamic stability occurs.

Plot $\omega_{n_{DR}}, \zeta_{DR}, T_R, T_S$ versus each stability derivative.

Problem 8.12

Consider the McDonnell Douglas F-4 aircraft at subsonic conditions (Appendix B, Aircraft 10). Note that the aircraft features the stabilators (i_H) for longitudinal control. Using the data provided in Appendix B, derive the complete SV model (shown here) as outlined in Section 8.6.

$$\begin{Bmatrix} \dot{u} \\ \dot{\alpha} \\ \dot{q} \\ \dot{\theta} \\ \dot{\beta} \\ \dot{p} \\ \dot{r} \\ \dot{\phi} \end{Bmatrix} = \begin{bmatrix} A_{Long} & 0 \\ 0 & A_{Lat\,Dir} \end{bmatrix} \begin{Bmatrix} u \\ \alpha \\ q \\ \theta \\ \beta \\ p \\ r \\ \phi \end{Bmatrix} + \begin{bmatrix} B_{Long} & 0 \\ 0 & B_{Lat\,Dir} \end{bmatrix} \begin{Bmatrix} i_H \\ \delta_A \\ \delta_R \end{Bmatrix}$$

$$\begin{Bmatrix} a_z \\ \alpha \\ u \\ q \\ \theta \\ a_Y \\ \beta \\ p \\ r \\ \phi \end{Bmatrix} = \begin{bmatrix} C_{Long} & 0 \\ 0 & C_{Lat\,Dir} \end{bmatrix} \begin{Bmatrix} \alpha \\ u \\ q \\ \theta \\ \beta \\ p \\ r \\ \phi \end{Bmatrix} + \begin{bmatrix} D_{Long} & 0 \\ 0 & D_{Lat\,Dir} \end{bmatrix} \begin{Bmatrix} i_H \\ \delta_A \\ \delta_R \end{Bmatrix}$$

Verify that the eigenvalues of the state matrices A_{Long} and $A_{Lat.Dir}$ are coincident with the roots of the longitudinal and lateral directional characteristic equations provided in Appendix B (Aircraft 6).

Next, implement the preceding model in Matlab and plot the aircraft dynamic responses following a simple double doublet maneuver on each control surface starting from the initial conditions provided in Appendix B. Use small magnitudes for these maneuvers (max. +/− 5 deg.). Also, the angular deflections need to be converted in *rad* (to be consistent with the provided aerodynamic data).

Problem 8.13

Consider the Boeing B747-200 aircraft at high-altitude cruise conditions (Appendix B, Aircraft 7). Note that the aircraft features both stabilators (i_H) and elevators (δ_E) or longitudinal control. Derive the extended SV) model (shown here (featuring one additional longitudinal input (i_H) and one additional longitudinal output (a_x).

$$\begin{Bmatrix} \dot{u} \\ \dot{\alpha} \\ \dot{q} \\ \dot{\theta} \\ \dot{\beta} \\ \dot{p} \\ \dot{r} \\ \dot{\phi} \end{Bmatrix} = \begin{bmatrix} A_{Long} & 0 \\ 0 & A_{Lat\,Dir} \end{bmatrix} \begin{Bmatrix} u \\ \alpha \\ q \\ \theta \\ \beta \\ p \\ r \\ \phi \end{Bmatrix} + \begin{bmatrix} B_{Long} & 0 \\ 0 & B_{Lat\,Dir} \end{bmatrix} \begin{Bmatrix} i_H \\ \delta_E \\ \delta_A \\ \delta_R \end{Bmatrix}$$

$$\begin{Bmatrix} a_z \\ a_x \\ \alpha \\ u \\ q \\ \theta \\ a_Y \\ \beta \\ p \\ r \\ \phi \end{Bmatrix} = \begin{bmatrix} C_{Long} & 0 \\ 0 & C_{Lat\,Dir} \end{bmatrix} \begin{Bmatrix} \alpha \\ u \\ q \\ \theta \\ \beta \\ p \\ r \\ \phi \end{Bmatrix} + \begin{bmatrix} D_{Long} & 0 \\ 0 & D_{Lat\,Dir} \end{bmatrix} \begin{Bmatrix} i_H \\ \delta_E \\ \delta_A \\ \delta_R \end{Bmatrix}$$

Verify that the eigenvalues of the state matrices A_{Long} and $A_{Lat.Dir}$ are coincident with the roots of the longitudinal and lateral directional characteristic equations provided in Appendix B (Aircraft 7).

Next, implement the preceding model in Simulink and plot the aircraft dynamic responses following a simple double doublet maneuver on each control surface starting from the initial conditions provided in Appendix B. Use small magnitudes for these maneuvers (max. +/− 5 deg.). Also, the angular deflections need to be converted in *rad* (to be consistent with the provided aerodynamic data).

Problem 8.14

Consider the Learjet 24 aircraft at approach (Appendix B, Aircraft 6) performing the following maneuver (see Figure P8.14.1)

$$i_H = \begin{cases} 0 & 0 \leq t < 100 \\ 1° & 100 \leq t < 500 \\ 0 & 500 \leq t < 900 \\ -1° & 900 \leq t < 1300 \\ 0 & 1300 \leq t \leq 1700 \end{cases}$$

where the time (t) is given in seconds. Using the extended version of longitudinal SV model plot the aircraft dynamic responses (u, α, q, θ).

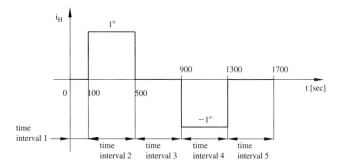

Figure P8.14.1 Longitudinal Maneuver for the Learjet 24

- Find and plot the components of the aircraft velocity expressed in the Earth fixed frame;
- Find and plot the aircraft flight path. Assume that the initial altitude is 300 ft and $\gamma_1 = 0$, i.e $\Theta_1 = \alpha_1$;
- Find the flight path angle (γ) when the aircraft reach steady-state at each one of the five time intervals defined by the maneuver. Use the following two approaches:

    Approach 1: Using the data from the dynamic responses and the fact that $\gamma = \theta - \alpha$

    Approach 2: Calculating the slope of the aircraft flight path;
- Complete the following table: Table 8.14.1.

Table 8.14.1 Longitudinal Configurations for the Learjet 24

Time Interval	Altitude Trend	Steady-State Aircraft Pitch Angle	Steady-State Aircraft Angle of Attack	Steady-State Path Angle	Steady-state Horizontal Velocity	Sign of Steady-state W
1	Constant	5°	5°	0°	170 ft/sec	Zero
2						
3	Constant	5°	5°	0°	170 ft/sec	Zero
4						
5	Constant	5°	5°	0°	170 ft/sec	Zero

Chapter 9

Introduction to Modern Flight Simulation Codes

TABLE OF CONTENTS

- 9.1 Introduction
- 9.2 Introduction to the Flight Dynamics & Control (FDC) Toolbox
  - 9.2.1 Equations of Motion within the FDC Simulation Environment
  - 9.2.2 FDC Modeling of Beaver Aerodynamic Forces and Moments
  - 9.2.3 Alternative Approach for FDC Modeling of Aerodynamic Forces and Moments
  - 9.2.4 Case Study #1: FDC Modeling of Look-Up Tables Based Aerodynamic Coefficients
  - 9.2.5 FDC Modeling of the Gravity Force
  - 9.2.6 FDC Modeling of the Atmospheric Turbulence Force
  - 9.2.7 FDC Modeling of the Beaver Propulsive Forces and Moments
  - 9.2.8 Case Study #2: FDC Modeling of Propulsive Forces and Moments
  - 9.2.9 Auxiliary FDC Blocks
  - 9.2.10 Additional FDC Blocks
- 9.3 Introduction to the Aerospace Blockset by Mathworks
  - 9.3.1 General Organization of the Aerospace Blockset
  - 9.3.2 Introduction to the Environment Library
  - 9.3.3 Introduction to the Flight Parameters Library
  - 9.3.4 Introduction to the Equations of Motion Library
  - 9.3.5 Introduction to the Aerodynamics Library
  - 9.3.6 Introduction to the Propulsion Library
  - 9.3.7 Introduction to the Utilities Library
  - 9.3.8 Introduction to the Mass Properties Library
  - 9.3.9 Introduction to the Actuators Library
  - 9.3.10 Introduction to the GNC and Animation Libraries
- 9.4 Introduction to AIRLIB
  - 9.4.1 AIRLIB's Strucure
  - 9.4.2 Generic Aircraft Model: Continuous-time Block
  - 9.4.3 Generic Aircraft Model: Discrete-time Block
  - 9.4.4 Collection of Aircraft Models
  - 9.4.5 Alternative Model Implementation
  - 9.4.6 Additional Tools within AIRLIB: The Function '*air3m*'
  - 9.4.7 Additional Tools within AIRLIB: The Function '*ab2dv*'
- 9.5 Summary
- References
- Student Sample Problems

9.1 INTRODUCTION

Chapters 7 and 8 have described modeling approaches (the transfer function—based modeling and the state variable—based modeling) which lead ultimately to the solution of the small perturbations equations of motion. Furthermore, a number of MATLAB® sample codes have been presented, featuring specific and very useful MATLAB® commands which automate the solutions of the aircraft equations of motion as opposed to manual solutions of the preceding equations. Nevertheless, the given modeling approaches represent the modern-day automated versions of the very same approaches introduced in 1910.[1,2]

The transfer function modeling approach based on the introduction of Laplace transformations has shown the great advantage of allowing a detailed understanding of the aircraft dynamic characteristics through the analysis of the longitudinal and lateral directional dynamic modes. In fact, the analysis of the characteristic equations of the longitudinal and lateral directional dynamics has naturally led to the introduction of the longitudinal dynamic modes (short period and phugoid) and lateral directional dynamic modes (dutch roll, spiral, and rolling). Of particular interest are the short period and rolling approximations, which allow very helpful insights in the aircraft dynamic behavior and correlate the aerodynamic modeling with the aircraft handling qualities.

As demonstrated in Chapter 8, the main limitation of the transfer function approach is that it is suitable for single-input–single-output (SISO) systems. The analysis of both longitudinal and lateral directional dynamics of the aircraft require approximately ten inputs (including all the longitudinal and lateral control surfaces, throttle settings, landing gears, flaps, and so on) and approximately fifteen outputs (including linear accelerations, angular velocities, Euler angles, aerodynamic angles, altitude, and so on) leading to the need of evaluating approximately 150 transfer functions. The modeling needs to be repeated at each different flight condition within the aircraft operating envelope, which makes the application of the transfer function concept practically unfeasible. This specific drawback of the transfer function approach could be overcome by the use of state variable models, which provide great flexibility for modeling the multi-input—multi-output aircraft dynamics.

Both transfer function and state variable modeling approaches provide solutions of the aircraft equations of motion, which have been linearized following the introduction of the small perturbation assumptions. Specifically, both methods allow for the solution of the following equations, outlined in Chapter 1:

$$\begin{aligned}
&m[\dot{u} + Q_1 w + qW_1 - R_1 v - rV_1] = -mg\theta \cos\Theta_1 + (f_{A_X} + f_{T_X}) \\
&m[\dot{v} + U_1 r + uR_1 - P_1 w - pW_1] = -mg\theta \sin\Phi_1 \sin\Theta_1 + mg\phi \cos\Phi_1 \cos\Theta_1 + (f_{A_Y} + f_{T_Y}) \\
&m[\dot{w} + P_1 v + pV_1 - Q_1 u - U_1 q] = -mg\theta \cos\Phi_1 \sin\Theta_1 - mg\phi \sin\Phi_1 \cos\Theta_1 + (f_{A_Z} + f_{T_Z}) \\
&\dot{p}I_{XX} - \dot{r}I_{XZ} - (P_1 q + Q_1 p)I_{XZ} + (R_1 q + Q_1 r)(I_{ZZ} - I_{YY}) = (l_A + l_T) \\
&\dot{q}I_{YY} + (P_1 r + pR_1)(I_{XX} - I_{ZZ}) + (2P_1 p - 2R_1 r)I_{XZ} = (m_A + m_T) \\
&\dot{r}I_{ZZ} - \dot{p}I_{XZ} + (P_1 q + pQ_1)(I_{YY} - I_{XX}) + (Q_1 r + R_1 q)I_{XZ} = (n_A + n_T)
\end{aligned}$$

Small Perturbation Conservation of Linear and Angular Momentum equations (CLMEs and CAMEs) embedding the Gravity Equations (GEs)

Nevertheless, the development of a more advanced flight simulation environment (used for either training or research purposes) requires the solution of the full nonlinear and coupled equations of motion given by

$$\begin{aligned}
&m(\dot{U} + QW - RV) = mg_X + (F_{A_X} + F_{T_X}) \\
&m(\dot{V} + UR - PW) = mg_Y + (F_{A_Y} + F_{T_Y}) \\
&m(\dot{W} + PV - QU) = mg_Z + (F_{A_Z} + F_{T_Z}) \\
&\dot{P}I_{XX} - \dot{R}I_{XZ} - PQI_{XZ} + RQ(I_{ZZ} - I_{YY}) = L_A + L_T \\
&\dot{Q}I_{YY} + PR(I_{XX} - I_{ZZ}) + (P^2 - R^2)I_{XZ} = M_A + M_T \\
&\dot{R}I_{ZZ} - \dot{P}I_{XZ} + PQ(I_{YY} - I_{XX}) + QRI_{XZ} = N_A + N_T
\end{aligned}$$

Conservation of Linear and Angular Momentum equations (CLMEs and CAMEs)

$$\begin{Bmatrix} \dot{\Phi} \\ \dot{\Theta} \\ \dot{\Psi} \end{Bmatrix} = \begin{bmatrix} 1 & \sin(\Phi)\tan(\Theta) & \cos(\Phi)\tan(\Theta) \\ 0 & \cos(\Phi) & -\sin(\Phi) \\ 0 & \sin(\Phi)\sec(\Theta) & \cos(\Phi)\sec(\Theta) \end{bmatrix} \begin{Bmatrix} P \\ Q \\ R \end{Bmatrix}$$

Kinematic Equations (KEs)

478 Chapter 9 Introduction to Modern Flight Simulation Codes

$$g_X = -g \sin \Theta$$
$$g_Y = g \cos \Theta \sin \Phi$$
$$g_Z = g \cos \Theta \cos \Phi$$

Gravity Equations (KEs)

Furthermore, the nonlinear nature of the preceding equations could be further complicated if one wishes to analyze the aircraft dynamic response at high angles of attack, where the aerodynamic behavior is also nonlinear. An additional level of complexity can be introduced if one wishes to analyze the time dependencies of the inertial and mass characteristics of the aircraft by modeling fuel consumption, in-flight refueling, dropping of payloads, and so on.

A generic block diagram showing the conceptual integration of all the preceding equations is shown in Figure 9.1 (from Chapter 1).

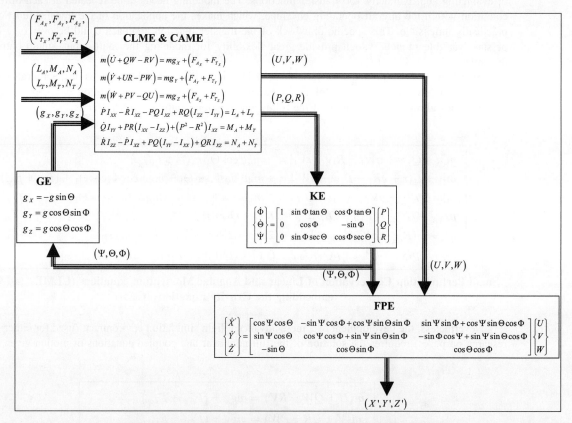

Figure 9.1 Integration of the CLMEs, CAMEs, KEs, FPEs, and GEs

A generic flight simulation environment needs—at least—the following generic components:

- an input device to generate commands to the aircraft system (for example, a joystick);
- a simulation environment (where the dynamics of a specific aircraft are modeled and the aircraft equations of motion are solved);
- a graphic environment (including both cockpit and outside graphics).

The goal of this chapter is to introduce undergraduate students to simulation environments where the full nonlinear and coupled aircraft equations of motion are solved as driven by the different forces and moments acting on the aircraft.

The availability of fairly inexpensive computational power and the de facto adoption of the software package MATLAB® and its toolbox Simulink®[3] as the universal modeling and simulation environment by the scientific and academic community have allowed the spreading of the development of powerful aircraft simulation codes. Therefore, advanced simulation codes (which only a few years ago could only be implemented on high-performance computers) can now be run on desktops and laptops with real-time performance using numerical methods for the integrations of the equations.

The chapter starts with a review of FDC, a powerful and freely available simulation package based on MATLAB®/Simulink®[3] developed by Marc Rauw.[4] Next, the Aerospace Blockset toolbox, produced by The Mathworks,[3] is introduced with a review of its key features. Finally, AIRLIB, a freely available library of mathematical models, is reviewed. AIRLIB has been developed by Dr. Giampiero Campa at the department of Mechanical and Aerospace Engineering at West Virginia University. The models within AIRLIB can be interfaced within either FDC or Aerospace Blockset.

9.2 INTRODUCTION TO THE FLIGHT DYNAMICS & CONTROL (FDC) TOOLBOX

In recent years the Flight Dynamics & Control (FDC) software package (free software available through the web site *www.dutchroll.com*[4]) has gained a widespread popularity within the academic and research community due to its extreme flexibility and remarkable capabilities.

FDC is a MATLAB® and Simulink®-based software developed specifically for the design and analysis of aircraft dynamics and control systems. The toolbox has been implemented around a general nonlinear aircraft model developed using a modular design approach. The toolbox also includes several MATLAB® routines for extracting steady-state flight conditions and determining linearized models for user-specified trim conditions, along with Simulink® models of atmospheric disturbances, and several help-utilities. Although of no specific interest for the purposes of this textbook, FDC also provides several examples of autopilot systems for the analysis of the closed-loop dynamics.

FDC features a detailed mathematical model for a general aviation aircraft; however, as mentioned above, it can be easily reconfigured for the mathematical model of any aircraft at the desired level of accuracy with linear or nonlinear aerodynamic modeling.

The following sections provide descriptions of the main components of the FDC environment and demonstrate key features enabling an undergraduate student to develop a simple but very realistic simulation environment. Refer to[4] for a detailed and informative discussion of all the issues relative to the development of a flight simulation environment, as well as a detailed description of all FDC features and capabilities.

9.2.1 Equations of Motion within the FDC Simulation Environment

The main level of the FDC environment is shown in Figure 9.2.

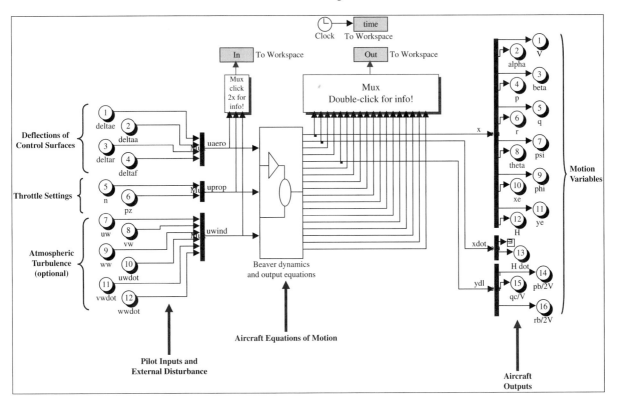

Figure 9.2 FDC First Level

480 Chapter 9 Introduction to Modern Flight Simulation Codes

The twelve inputs of the aircraft system are divided into the following categories:

- Aerodynamic control surfaces = $[\delta_{ELEVATORS}, \delta_{AILERONS}, \delta_{RUDDER}, \delta_{FLAPS}]$ (rad);
- Throttle settings = [engine speed (RPM), manifold pressure (Hg)];
- Atmospheric turbulence = $[\;U_W, V_W, W_W(\text{m/sec}), \dot{U}_W, \dot{V}_W, \dot{W}_W(\text{m/sec}^2)]$ along body axes.

The twenty-seven outputs of the aircraft system are divided into the following categories:

12 aircraft states = $[Vel\;\;\alpha\;\;\beta\;\;p\;\;q\;\;r\;\;\theta\;\;\phi\;\;x_{Earth}\;\;y_{Earth}\;\;Alt]$;
12 derivatives of the aircraft states = $[\dot{V}el\;\;\dot\alpha\;\;\dot\beta\;\;\dot p\;\;\dot q\;\;\dot r\;\;\dot\theta\;\;\dot\phi\;\;\dot x_{Earth}\;\;\dot y_{Earth}\;\;\dot{Alt}]$;
3 dimensionless components of the angular velocity along the body axes = $[pb/2\;Vel,\;q\bar{c}/2\;Vel,\;rb/2\;Vel]$.

During the simulation, a total of eighty-nine variables in addition to the twelve inputs are saved in the MATLAB® workspace.

The first level of the aircraft equations of motion is shown in Figure 9.3.

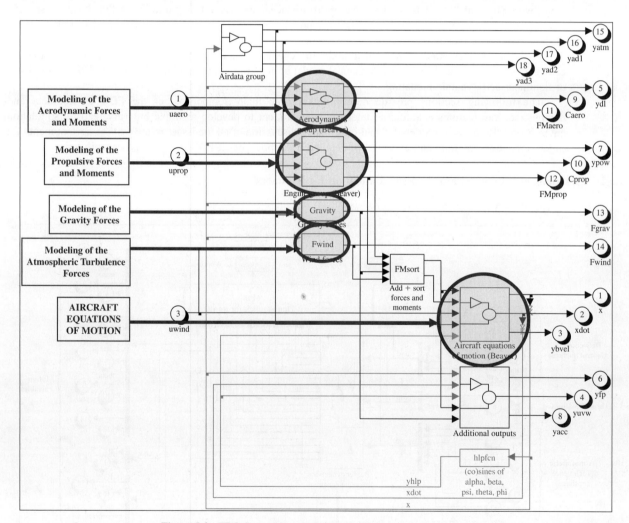

Figure 9.3 FDC Second Level of the Aircraft Equations of Motion

The preceding block includes the blocks modeling all the forces and moments driving the aircraft dynamics. The feeding of all the different forces and moments along with the loop of the integration of the equations is shown in the third level FDC block of the aircraft equations of motions in Figure 9.4.

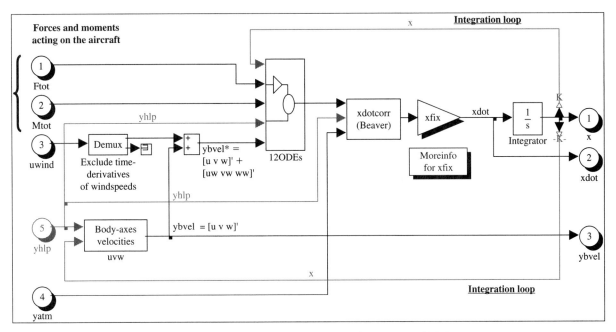

Figure 9.4 FDC Third Level of the Aircraft Equations of Motion

The preceding block includes twelve of the fifteen equations introduced in Chapter 1, as shown in the fourth level FDC block in Figure 9.5.

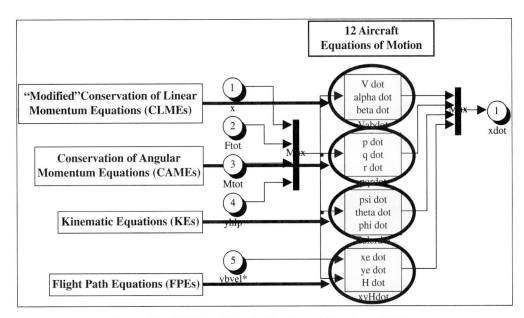

Figure 9.5 FDC Fourth Level of the Aircraft Equations of Motion

The first set of equations is the 'modified' version of the Conservation of Linear Momentum equations (CLMEs) introduced in Chapter 1.

$$m(\dot{U} + QW - RV) = mg_X + (F_{A_X} + F_{T_X})$$
$$m(\dot{V} + UR - PW) = mg_Y + (F_{A_Y} + F_{T_Y})$$
$$m(\dot{W} + PV - QU) = mg_Z + (F_{A_Z} + F_{T_Z})$$

The modification involves the second and third equations, which in FDC are written in terms of the aerodynamic angles (α, β) in lieu of the linear velocities (W, V) using the transformation of variables:

$$\tan(\alpha) \approx \sin(\alpha) \approx \alpha \approx \frac{W}{Vel}$$

$$\tan(\beta) \approx \sin(\beta) \approx \beta \approx \frac{V}{Vel}$$

Additionally, the preceding equations are modified to include the components of the forces associated with the atmospheric turbulence, assumed to be caused by random perturbations in the components of the linear velocity along the body axes.

The second set of equations is the Conservation of Angular Momentum Equations (CAMEs):

$$\dot{P}I_{XX} - \dot{R}I_{XZ} - PQI_{XZ} + RQ(I_{ZZ} - I_{YY}) = L_A + L_T$$
$$\dot{Q}I_{YY} + PR(I_{XX} - I_{ZZ}) + (P^2 - R^2)I_{XZ} = M_A + M_T$$
$$\dot{R}I_{ZZ} - \dot{P}I_{XZ} + PQ(I_{YY} - I_{XX}) + QRI_{XZ} = N_A + N_T$$

The third set of equations is the Kinematic Equations (KEs):

$$\dot{\Phi} = P + \sin(\Phi)\tan(\Theta)Q + \cos(\Phi)\tan(\Theta)R$$
$$\dot{\Theta} = \cos(\Phi)Q - \sin(\Phi)R$$
$$\dot{\Psi} = \sin(\Phi)\sec(\Theta)Q + \cos(\Phi)\sec(\Theta)R$$

Finally, the fourth set of equations is the Flight Path Equations (FPEs):

$$\begin{Bmatrix} \dot{X}' \\ \dot{Y}' \\ \dot{Z}' \end{Bmatrix} = \begin{Bmatrix} U_1 \\ V_1 \\ W_1 \end{Bmatrix} = \begin{bmatrix} \cos\Psi\cos\Theta & -\sin\Psi\cos\Phi + \cos\Psi\sin\Theta\sin\Phi & \sin\Psi\sin\Phi + \cos\Psi\sin\Theta\cos\Phi \\ \sin\Psi\cos\Theta & \cos\Psi\cos\Phi + \sin\Psi\sin\Theta\sin\Phi & -\sin\Phi\cos\Psi + \sin\Psi\sin\Theta\cos\Phi \\ -\sin\Theta & \cos\Theta\sin\Phi & \cos\Theta\cos\Phi \end{bmatrix} \begin{Bmatrix} U \\ V \\ W \end{Bmatrix}$$

The inputs to the CLMEs and CAMEs are the aerodynamic and thrust forces and moments blended with the gravity forces and forces associated with the atmospheric turbulence, as shown in Figure 9.6.

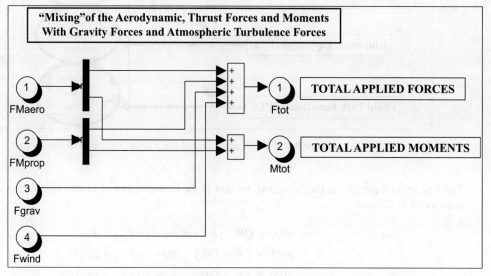

Figure 9.6 Mixing of the Different Forces and Moments Acting on the Aircraft

The modeling of the aerodynamic and thrust forces and moments along with the modeling of the gravity and atmospheric forces will be briefly reviewed in the next sections.

9.2.2 FDC Modeling of Beaver Aerodynamic Forces and Moments

The high-level modeling of the aerodynamic forces and moments is shown in Figure 9.7.

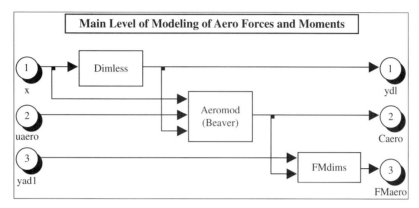

Figure 9.7 High-Level Modeling of Aerodynamic Forces and Moments

The *Dimless* block calculates the dimensionless components of the angular velocities $[pb/2\ Vel,\ q\bar{c}/Vel,\ rb/2\ Vel]$. The core of the aerodynamic modeling is performed in the block shown in Figure 9.8.

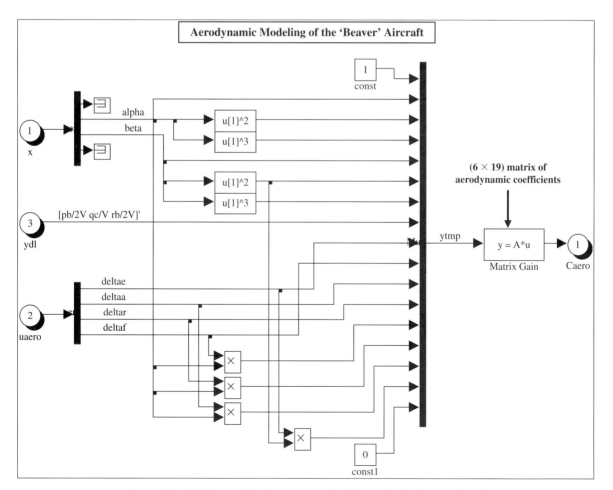

Figure 9.8 Modeling of the Aerodynamic Coefficients for the Beaver Aircraft

FDC's flexibility in accommodating the aerodynamic modeling to any desired level of accuracy resides in the capabilities of modifying the general block diagram in Figure 9.8. In case of the default Beaver aircraft model, the aerodynamic coefficients are assumed to be functions of the following variables:[5]

$$\left(\alpha, \alpha^2, \alpha^3, \beta, \beta^2, \beta^3, \frac{pb}{2Vel}, \frac{q\bar{c}}{Vel}, \frac{rb}{2Vel}, \delta_E, \delta_A, \delta_R, \delta_{FLAPS}, \alpha\delta_{FLAPS}, \alpha\delta_R, \alpha\delta_A, \delta_E\beta^2\right)$$

The preceding variables are expanded to include a 1 and a 0 for the purpose of modeling the aerodynamic bias coefficients and for the purpose of adding a placeholder for an additional variable. The resulting set of variables is given by

$$\left(1, \alpha, \alpha^2, \alpha^3, \beta, \beta^2, \beta^3, \frac{pb}{2Vel}, \frac{q\bar{c}}{Vel}, \frac{rb}{2Vel}, \delta_E, \delta_A, \delta_R, \delta_{FLAPS}, \alpha\delta_{FLAPS}, \alpha\delta_R, \alpha\delta_A, \delta_E\beta^2, 0\right)$$

The preceding aerodynamic model of the Beaver aircraft expresses the coefficients of the aerodynamic forces and moments as nonlinear functions of a combination of a subset of the state variables and the aerodynamic control inputs.[5] The model also includes longitudinal-lateral cross coupling effects and unsteady aerodynamics. Specifically, using a first order Taylor expansion, we have the following relationships for each of the aerodynamic force and moment coefficients:[5]

$$c_{X_A} = c_{X_0} + c_{X_\alpha}\alpha + c_{X_{\alpha^2}}\alpha^2 + c_{X_{\alpha^3}}\alpha^3 + c_{X_q}\frac{q\bar{c}}{Vel} + c_{X_{\delta_R}}\delta_R + c_{X_{\delta_{FLAPS}}}\delta_{FLAPS} + c_{X_{\alpha\delta_{FLAPS}}}\alpha\delta_{FLAPS}$$

$$c_{Y_A} = c_{Y_0} + c_{Y_\beta}\beta + c_{Y_p}\frac{pb}{2Vel} + c_{Y_r}\frac{rb}{2Vel} + c_{Y_{\delta_A}}\delta_A + c_{Y_{\delta_R}}\delta_R + c_{Y_{\alpha\delta_R}}\alpha\delta_R$$

$$c_{Z_A} = c_{Z_0} + c_{Z_\alpha}\alpha + c_{Z_{\alpha^3}}\alpha^3 + c_{Z_q}\frac{q\bar{c}}{Vel} + c_{Z_{\delta_E}}\delta_E + c_{Z_{\delta_{FLAPS}}}\delta_{FLAPS} + c_{Z_{\alpha\delta_{FLAPS}}}\alpha\delta_{FLAPS} + c_{Z_{\delta_E\beta^2}}\delta_E\beta^2$$

$$c_{l_A} = c_{l_0} + c_{l_\beta}\beta + c_{l_p}\frac{pb}{2Vel} + c_{l_r}\frac{rb}{2Vel} + c_{l_{\delta_A}}\delta_A + c_{l_{\delta_R}}\delta_R + c_{l_{\alpha\delta_A}}\alpha\delta_A$$

$$c_{m_A} = c_{m_0} + c_{m_\alpha}\alpha + c_{m_{\alpha^2}}\alpha^2 + c_{m_{\beta^2}}\beta^2 + c_{m_q}\frac{q\bar{c}}{Vel} + c_{m_r}\frac{rb}{2Vel} + c_{m_{\delta_E}}\delta_E + c_{m_{\delta_{FLAPS}}}\delta_{FLAPS}$$

$$c_{n_A} = c_{n_0} + c_{n_\beta}\beta + c_{n_{\beta^3}}\beta^3 + c_{n_p}\frac{pb}{2Vel} + c_{n_q}\frac{q\bar{c}}{Vel} + c_{n_r}\frac{rb}{2Vel} + c_{n_{\delta_A}}\delta_A + c_{n_{\delta_R}}\delta_R$$

Thus, the entire modeling of the Beaver aerodynamic forces and moments is introduced in the Simulink® environment by loading the *(6 × 19)* matrix of aerodynamic coefficients relative to the breakdown in the coefficients shown previously. This set of coefficients is only relative to a specific flight condition. Also note that according to the modeling, several of those coefficients are zero.

Once the preceding coefficients $\left(c_{X_A}, c_{Y_A}, c_{Z_A}, c_{l_A}, c_{m_A}, c_{n_A}\right)$ have been modeled, they are converted into actual forces and moments using

$$F_{X_A} = c_{X_A}\bar{q}S, \quad F_{Y_A} = c_{Y_A}\bar{q}S, \quad F_{Z_A} = c_{Z_A}\bar{q}S$$

$$L_A = c_{l_A}\bar{q}Sb, \quad M_A = c_{m_A}\bar{q}S\bar{c}, \quad N_A = c_{n_A}\bar{q}Sb$$

through the block shown in Figure 9.9:

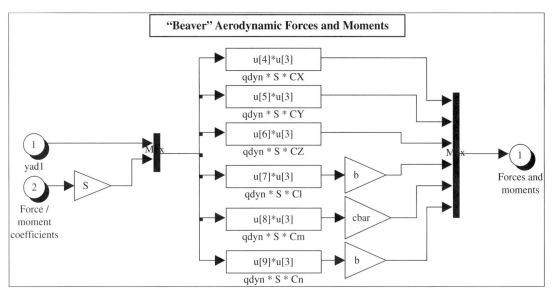

Figure 9.9 Modeling of the Aerodynamic Forces and Moments for the Beaver Aircraft

9.2.3 Alternative Approach for FDC Modeling of Aerodynamic Forces and Moments

It should be underlined that a key difference with the modeling of the aerodynamic forces and moments outlined in Chapter 3 and 4 is that within FDC the aerodynamic forces are modeled along the body axes $X\ Y\ Z$ rather than the stability axes $X_S\ Y_S\ Z_S$ (here also referred as wind axes $X_W\ Y_W\ Z_W$). Since we have the condition $Y = Y_S$, all the other coefficients will be unchanged, except the coefficients of the aerodynamic forces along X and Z. Specifically, for low values of the angle of attack α, we have

$$c_X = -c_D \cos(\alpha) + c_L \sin(\alpha) \approx -c_D + c_L \alpha$$
$$c_Z = -c_L \cos(\alpha) + c_D \sin(\alpha) \approx -c_L - c_D \alpha$$

Therefore, in a general case, the aerodynamic modeling developed along the stability axes can be adapted to the solution of the equations of the motion along the body axes $X\ Y\ Z$ using the preceding relationships, in lieu of the solution along the stability axes outlined in Chapter 3.

Regardless of whether the aerodynamic modeling is performed along the stability or the body axes, this example of the Beaver aircraft represents the aerodynamic modeling at only one specific point in its flight envelope. This implies that the modeling becomes invalid if the user deviates substantially from the assigned flight condition following, for example, a maneuver involving a large change in airspeed and altitude. If accurate modeling of the aerodynamic forces and moments is required, it is necessary to introduce multidimensional look-up tables for the modeling of the aerodynamic coefficients. For this task Simulink® currently provides a variety of tools, as shown in Figure 9.10.

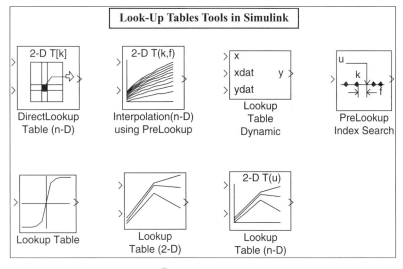

Figure 9.10 Simulink® Tools for Aerodynamic Look-Up Tables

Refer to a "tour" of the different look-up table tools for a detailed evaluation of the capabilities of each of those tools. Of particular interest are the 1D, 2D, and n-D look-up tables, which are extensively used for the modeling of the aerodynamic coefficients as function of key independent variables. An example of 1D modeling has been provided in the case study in Chapters 3 and 4 for the modeling of the aerodynamic coefficients of the Cessna CJ3 aircraft.

In the general case, an aerodynamic coefficient can be a function of one or more variables. For example, it is well known from Chapter 2 that

$$c_{L_\alpha} = f(Mach)$$

It is also known that outside a certain range of linear behavior c_{L_α} is also a nonlinear function of α. Therefore, we would have

$$c_{L_\alpha} = f(\alpha, Mach)$$

In general, other aerodynamic factors can also affect the values of c_{L_α}, such as the deflections of flaps and the deployment of landing gears. Additionally, c_{L_α} can be a strong nonlinear function of α. Therefore, a generic nonlinear relationship for c_{L_α} could be expressed through an n-D look-up table (LUT) with

$$c_{L_\alpha} = f(\alpha, \alpha^2, \ldots, Mach, \delta_{FLAPS}, \delta_{LANDING\ GEARS}, \ldots)$$

The level of complexity of the LUT is in general a function of the following factors:

- Required modeling accuracy, depending on the intended use of the simulation environments (for example, simulator for training purposes versus recreational simulator);
- Aircraft geometry (conventional architecture with limited number of aerodynamic control surfaces versus unconventional architecture with a large number of aerodynamic control surfaces);
- Flight envelope (subsonic aircraft versus supersonic aircraft);
- Range of inertial characteristics (modeling of fuel consumption, wide changes in center of gravity locations, and so on);
- Complexity of the flight control laws.

It should be pointed out that in flight simulators developed for high-performance military and civilian aircraft, the modeling equations, the format used for the development of these LUTs, and the data in the LUTs are proprietary information. Finally, it should be emphasized that accurate documentation of the large volume of data leading to the development of the LUTs is absolutely required. In fact, typical flight simulation codes undergo several refinements through the aircraft operational life, as more detailed aerodynamic data or refinements become available, as well as modifications to the design of the aircraft are introduced starting from the initial design configuration.

The next section provides an interesting case study focused on the modeling of the aerodynamic coefficients for a high-performance military aircraft.

9.2.4 Case Study #1: FDC Modeling of Look-Up Tables Based Aerodynamic Coefficients

In 1990 a Fortran-based simulation code was distributed to the academic community for the 1991 AIAA GNC Student Design Control Challenge.[6] This code was developed by NASA researchers for a generic high-performance military aircraft; the modeling of the aerodynamic coefficients included in this code was based on a number of LUTs. The code also featured a detailed engine model and first and second order dynamic response for the actuators of each of the control surfaces.

Obviously, the original Fortran code was not suitable for interfacing with a Simulink®-based simulation environment. Therefore, a massive effort was conducted by researchers at West Virginia University to develop Simulink®-based 1D, 2D, and n-D LUTs for the modeling of each of the aerodynamic coefficients.[7] As discussed in the previous section, the aerodynamic modeling is often customized to the specific aircraft. This aircraft in particular featured speed brakes and a stabilator capable of both symmetric and differential deflections. The resulting modeling was quite peculiar and was based on the following relationships for the coefficients of the aerodynamic forces expressed along the stability axes $X_S\ Y_S\ Z_S$:[7-11]

$$c_D = 1.02 \cdot c_{D_{Wing+Body+Tail}} + \Delta c_{D_{Alt}} + \Delta c_{D_{NOZZLE}} \cdot Sw_{NOZ} + \Delta c_{D_{SB}} \cdot Sw_{SB45}$$

$$c_Y = c_{Y_{Wing+Body+Tail}} + c_{Y_{\delta A}} \cdot \delta_A + c_{Y_{\delta DT}} \cdot \delta_{DT} - \Delta c_{Y_{\delta R}} \cdot E_{Y\delta R}$$

$$c_L = 0.95 \cdot c_{L_{Wing+Body+Tail}} + \Delta c_{L_{nz}} \cdot n_z + Sw_{SB45} \cdot \left[\Delta c_{L_0} + \Delta c_{L_{\alpha 1}} \cdot \alpha_1 + \Delta c_{L_{\alpha 2}} \cdot (\alpha_2 - 10°) \right]$$

where

Alt = aircraft altitude (ft)

δ_{DT} = differential tail deflection

δ_{HT} = symmetric tail deflection

n_Z = vertical load factor

Sw_{NOZZLE}, Sw_{SB45} are weighting coefficients relative to the actuation of the engine nozzle and the deployment of the speed brakes

$E_{Y\delta_R}$ = rudder lateral effectiveness coefficient

Similarly, the relationships for the coefficients of the aerodynamic moments expressed along the stability axes $X_S Y_S Z_S$ are given by

$$c_l = c_{l_{Wing+Body+Tail}} + c_{l_{\delta A}} \cdot \delta_A + c_{l_{\delta DT}} \cdot \delta_{DT} - \Delta c_{l\delta R} \cdot E_{l\delta R} + c_{l_P} \cdot \hat{p} + c_{l_R} \cdot \hat{r} + \Delta c_{l_{SB}} \cdot Sw_{SB45}$$

$$c_m = c_{m_{Wing+Body+Tail}} + \Delta c_{m_{nz}} \cdot n_z + c_{m_q} \cdot \hat{q} + c_{m_{\dot{\alpha}}} \cdot \hat{\dot{\alpha}} + c_{L_{Wing+Body+Tail}} \cdot \hat{L}$$
$$+ Sw_{SB45} \cdot \left(\Delta c_{m_0} + c_{L_{Wing+Body+Diff.Tail}} \cdot \hat{L}_{SB} \right)$$

$$c_n = c_{n_{Wing+Body+Tail}} + c_{n_{\delta A}} \cdot \delta_A + c_{n_{\delta DT}} \cdot \delta_{DT} + c_{n_{\delta R}} \cdot \delta_R \cdot E_{n\delta R} + c_{n_P} \cdot \hat{p} + c_{n_R} \cdot \hat{r} + \Delta c_{n_{SB}} \cdot Sw_{SB45}$$

where

$E_{l\delta_R}, E_{n\delta_R}$ are rudder rolling and yawing effectiveness coefficients

$\hat{p} = \dfrac{pb}{2V}, \hat{r} = \dfrac{rb}{2V}$ are the dimensionless roll and yaw angular velocities

$\hat{\dot{\alpha}} = \dfrac{\dot{\alpha}\bar{c}}{2V}, \hat{q} = \dfrac{q\bar{c}}{2V}$ are the dimensionless α-rate of change and pitch angular velocity

$\hat{L}$ = dimensionless distance between the application point of the (Body + Wing + Tail) lift and the aircraft center of gravity

$\hat{L}_{SB}$ = dimensionless distance between the application point of the speed brake lift and the aircraft center of gravity

Next, a more detailed modeling is provided for each of the longitudinal aerodynamic force and moment coefficients, followed by the lateral directional aerodynamic force and moment coefficients.

Figure 9.11 shows the modeling of the lift coefficient c_L.

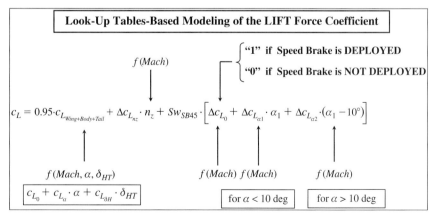

Figure 9.11 Modeling for c_L.

A typical look-up table used for the modeling of the coefficient $c_{L_{Wing+Body+Tail}}$ is shown in Figure 9.12 for a 5° symmetric tail deflection. Similar tables are used for different deflections of the symmetric stabilator.

488 Chapter 9 Introduction to Modern Flight Simulation Codes

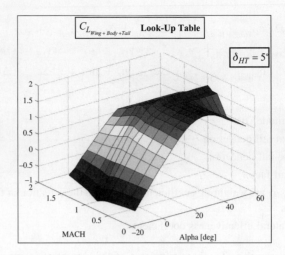

Figure 9.12 Sample Look-Up Table for $c_{L_{Wing+Body+Tail}}$

The other 1D look-up tables are shown in Figure 9.13.

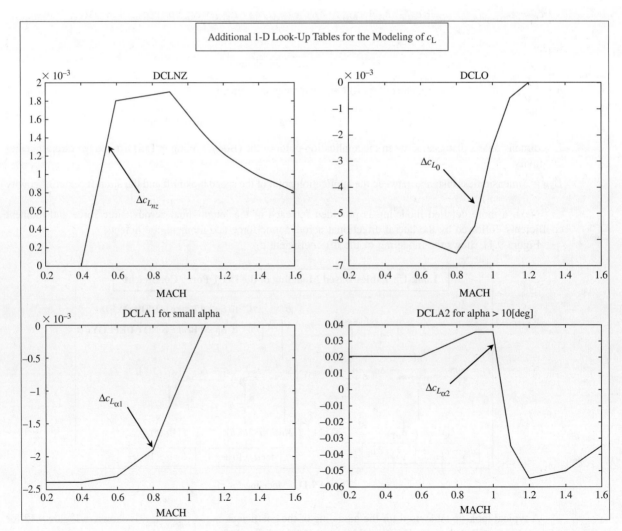

Figure 9.13 1D Look-Up Tables for c_L

Figure 9.14 shows the modeling of the drag coefficient c_D.

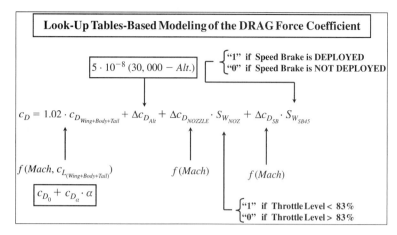

Figure 9.14 Modeling for c_D

The 2D look-up table used for the modeling of $c_{D_{Wing+Body+Tail}}$ is shown in Figure 9.15.

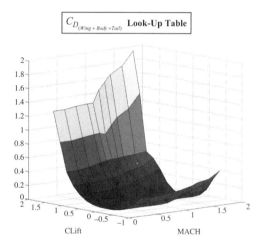

Figure 9.15 2D Look-Up Table for $c_{D\ Wing+Body+Tail}$

Similarly, Figure 9.16 shows the modeling of the pitching moment coefficient c_m.

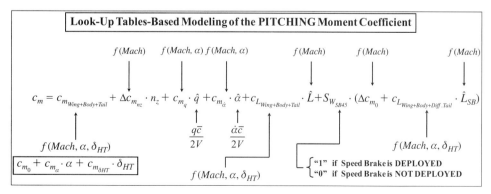

Figure 9.16 Modeling for c_m

A sample 3D look-up table for the modeling of $c_{m_{Wing+Body+Tail}}$ and the 2D look-up tables for the modeling of $c_{m_{\dot{\alpha}}}$, c_{m_q} are shown in Figures 9.17 and 9.18, respectively.

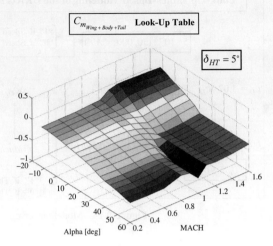

Figure 9.17 Sample Look-Up Table for $c_{m_{Wing+Body+Tail}}$

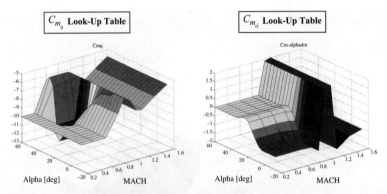

Figure 9.18 Look-Up Tables for $c_{m_{\dot{\alpha}}}$, c_{m_q}

In terms of the modeling of the lateral directional force and moments, Figures 9.19, 9.20, and 9.21 show the breakdown for the side force, rolling moment, and yawing moment coefficients, respectively.

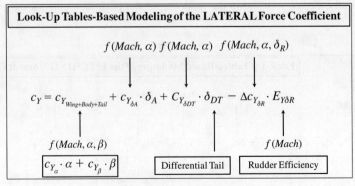

Figure 9.19 Modeling for c_Y

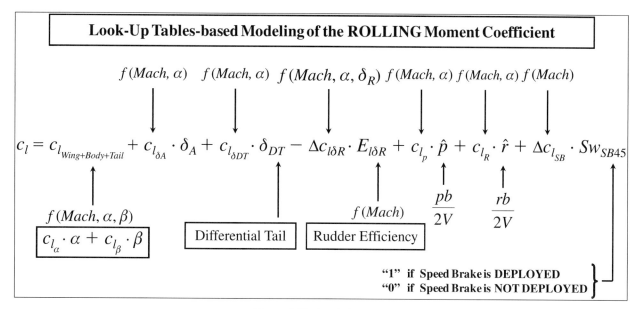

Figure 9.20 Modeling for c_l

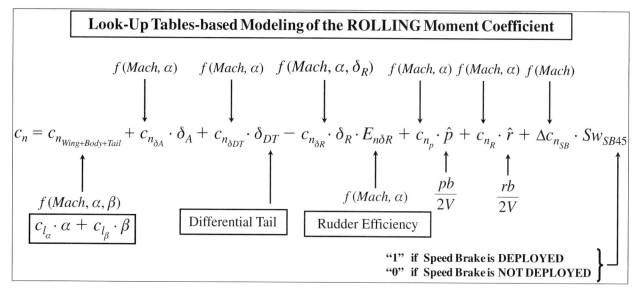

Figure 9.21 Modeling for c_n

The complete (longitudinal + lateral directional) aerodynamic modeling has been coded using the 1D, 2D, and 3D Simulink® look-up table tools and reduced to the format in Figure 9.22, which is equivalent to the modeling in Figure 9.3 for the Beaver aircraft in the previous section. It should be emphasized that the aerodynamic coefficients are first converted from the stability axes to the body axes, prior to the calculation of the actual aerodynamic forces and moments, as previously discussed.

492 Chapter 9 Introduction to Modern Flight Simulation Codes

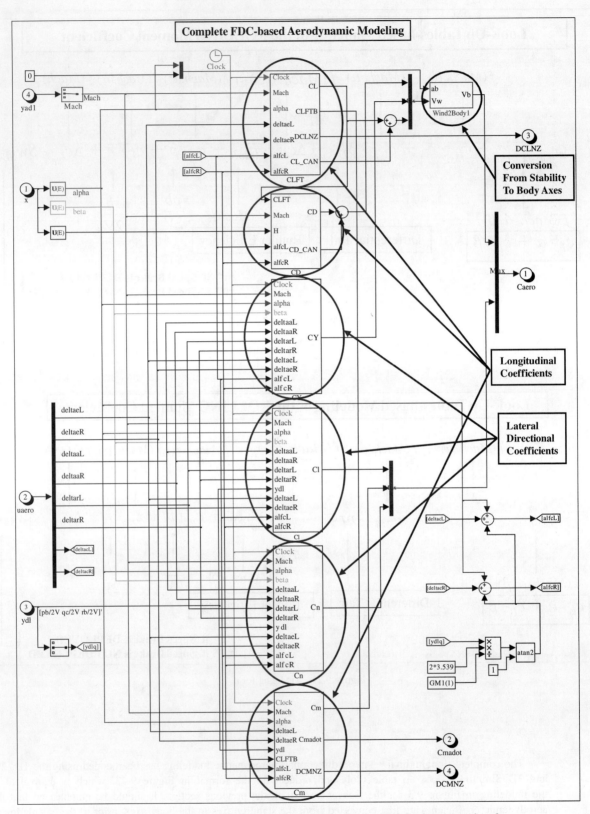

Figure 9.22 Modeling of the Aerodynamic Coefficients for a High-Performance Aircraft

9.2.5 FDC Modeling of the Gravity Force

As discussed in Chapter 1, the components of the gravity force along the body axes are calculated using the following equations:

$$F_{G_X} = mg_X = -mg \sin \Theta$$
$$F_{G_Y} = mg_Y = mg \cos \Theta \sin \Phi$$
$$F_{G_Z} = mg_Z = mg \cos \Theta \cos \Phi$$

Note that it is common to find the gravity equations embedded directly in the Conservation of Linear Momentum equations.

FDC provides a direct and straightforward path of the gravity force components through the block shown in Figure 9.23.

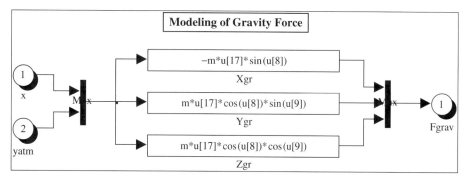

Figure 9.23 FDC Modeling of the Gravity Force along the Body Axes

9.2.6 FDC Modeling of the Atmospheric Turbulence Force

Atmospheric turbulences are essentially external disturbances acting on the aircraft dynamics. As described in Chapter 8, they are random in nature, and, therefore, a statistical modeling of their characteristics is required. There are two main statistical models used for the modeling of the atmospheric turbulence. The first model is the Dryden turbulence model, and the second is the Von Karman turbulence model.[12] Rauw[4] provides a detailed description of the Dryden turbulence model and its statistical description. An analysis of the statistical characterization of the Dryden turbulence is beyond the level of this textbook. However, it is critical to understand how the atmospheric turbulence acts as an input to the aircraft dynamics. Essentially, the atmospheric turbulence is modeled within FDC as random components of the linear velocities along the body axes and are indicated using

$$U_W, V_W, W_W$$

where the subscript w indicates *wind*. Therefore, the force associated with the atmospheric turbulence enters the aircraft dynamics through the Conservation of Linear Momentum equations (CLMEs)

$$F_{W_X} = m(\dot{U}_W + QW_W - RV_W)$$
$$F_{W_Y} = m(\dot{V}_W + U_W R - PW_W)$$
$$F_{W_Z} = m(\dot{W}_W + PV_W - QU_W)$$

This modeling is shown in Figure 9.24, where the body axes components of the angular velocity (P, Q, R) are found from the Conservation of Angular Momentum equations (CAMEs).

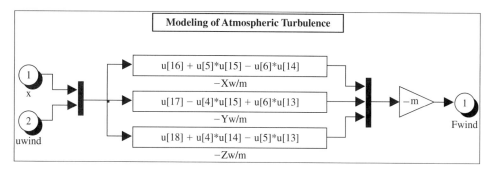

Figure 9.24 FDC Modeling of the Atmospheric Turbulence along the Body Axes

Ultimately, the force associated with the atmospheric turbulence is modeled as an additional applied force acting on the RHS of the CLMEs along with the gravity, aerodynamic force, and thrust forces, as shown:

$$m(\dot{U} + QW - RV) = F_{G_X} + F_{W_X} + F_{A_X} + F_{P_X}$$
$$m(\dot{V} + UR - PW) = F_{G_Y} + F_{W_Y} + F_{A_Y} + F_{P_Y}$$
$$m(\dot{W} + PV - QU) = F_{G_Z} + F_{W_Z} + F_{A_Z} + F_{P_Z}$$

9.2.7 FDC Modeling of the Beaver Propulsive Forces and Moments

FDC features a simple modeling of the propulsive forces and moments. The approach is suitable for the modeling of propeller-based propulsion systems for aircraft with a limited flight envelope, such as the Beaver aircraft. The high-level modeling of the propulsive forces and moments is shown in Figure 9.25.

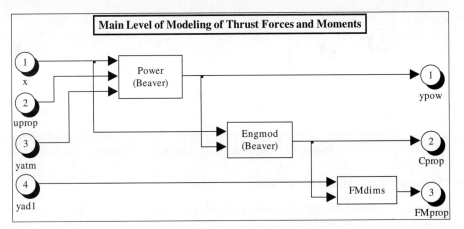

Figure 9.25 FDC Modeling of Thrust Forces and Moments

The modeling is based on the determination of the nondimensional pressure increase across the propeller and the propeller power in ($N\ m/sec$),[4,5] as shown in the block in Figure 9.26.

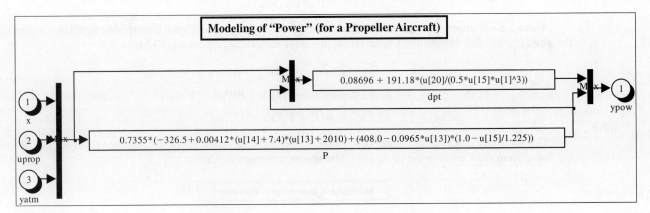

Figure 9.26 FDC Modeling of Beaver Engine Power

The inputs to this block are the state vector (x), the engine *RPM*, and manifold pressure, along with the parameters of the standard atmospheric models.

Next, the nondimensional pressure increase across the propeller (*dpt*) and the power ($N\ m/sec$) are one of the inputs of the block evaluating the propulsive force and moment coefficients, shown in Figure 9.27.

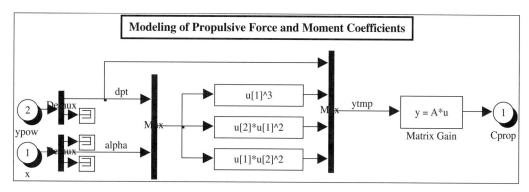

Figure 9.27 FDC Modeling of Beaver Propulsive Force and Moment Coefficients

Specifically, using the modeling procedure outlined in,[5] the propulsive force and moment coefficients for a propeller-based thrust system are given by:

$$c_{X_P} = c_{X_{P_{dpt}}} dpt + c_{X_{P_{\alpha dpt^2}}} (\alpha dpt^2)$$
$$c_{Y_P} = 0$$
$$c_{Z_P} = c_{Z_{P_{dpt}}} dpt$$
$$c_{l_P} = c_{l_{P_{\alpha^2 dpt}}} (\alpha^2 dpt)$$
$$c_{m_P} = c_{m_{P_{dpt}}} dpt$$
$$c_{n_P} = c_{n_{P_{dpt^3}}} dpt^3$$

Next, the preceding propulsive coefficients are used to calculate the associated propulsive forces and moments using the conventional modeling procedure

$$F_{P_X} = c_{X_P} \bar{q} S, \quad F_{P_Y} = c_{Y_P} \bar{q} S, \quad F_{P_Z} = c_{Z_P} \bar{q} S$$
$$L_P = c_{l_P} \bar{q} S b, \quad M_P = c_{m_P} \bar{q} S \bar{c}, \quad N_P = c_{n_P} \bar{q} S b$$

also shown in Figure 9.28 below.

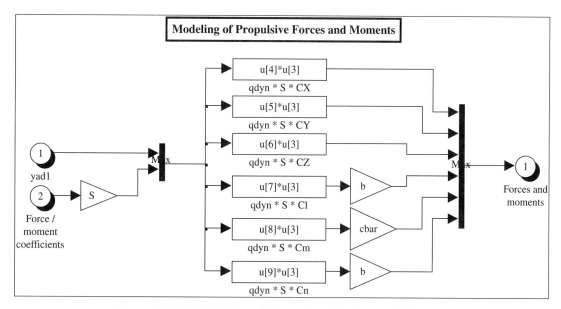

Figure 9.28 FDC Modeling of Beaver Propulsive Forces and Moments

Note that the subscript P (for Propulsive) in the preceding coefficients replaces for all practical purposes the subscript T (for Thrust) used in Chapters 1 and 5 to indicate the forces and moments associated with the propulsion system.

The modeling described is a simple example of modeling of the propulsive forces. Different tools can be used for the task of modeling the forces and moments associated with more complex propulsion systems, such as turbo-prop and jet aircraft. In these cases, the engine performance can be modeled using 1D, 2D, or 3D look-up tables, as shown in the next section.

9.2.8 Case Study #2: FDC Modeling of Propulsive Forces and Moments

One of FDC's trademarks is its flexibility in accommodating different modeling with different levels of accuracy and complexity. The goal of this section is to show an example of the modeling of the propulsive forces and moments for a generic jet-type high-performance aircraft.[6,7]

For this type of propulsion system, the engine thrust can be modeled using 3D tables as functions of Mach and the angle of attack. Typically, there are three main thrust levels, that is, *idle*, *military power*, and *maximum power*; there are also three levels of fuel flow associated with these conditions.

The general modeling for the thrust from the jet engine can be obtained using the block shown in Figure 9.29.

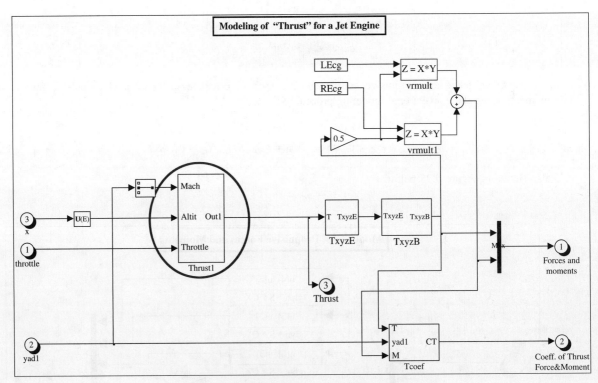

Figure 9.29 FDC Modeling of Thrust for a Jet Engine

Unlike the modeling in Figures 9.27 and 9.28, the modeling in Figure 9.29 calculates directly the propulsive forces and moments (expressed in the body axes) in addition to the propulsive dimensionless coefficients. The main block in the figure features the 3D look-up tables where the thrust force from the propulsion system is evaluated as a function of Mach and the angle of attack for different levels, as shown in Figure 9.30. The block includes conversions from the MKS unit system, which is used in FDC.

9.2 Introduction to the Flight Dynamics & Control (FDC) Toolbox 497

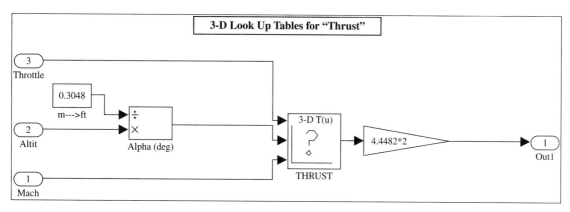

Figure 9.30 3D Look-Up Tables for Thrust

The look-up tables are shown in Figure 9.31.

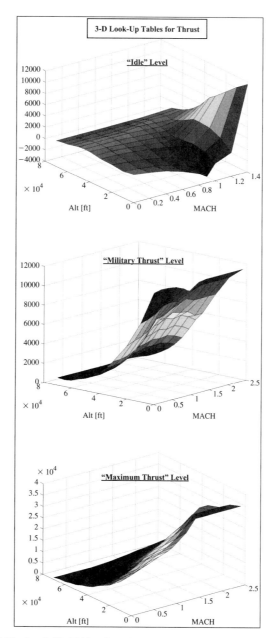

Figure 9.31 Look-Up Tables for a Jet-Engine for Different Levels of Thrust

498 Chapter 9 Introduction to Modern Flight Simulation Codes

Finally, a detailed modeling of the propulsion system should also include the change in the aircraft inertial characteristics associated with the fuel consumption, assuming that the location of the center of gravity of the different fuel tanks and their distances from the aircraft center of gravity are known geometric parameters. For this purpose, the fuel flow needs to be properly modeled. A typical set of fuel flow look-up tables for different throttle levels is shown in Figure 9.32.

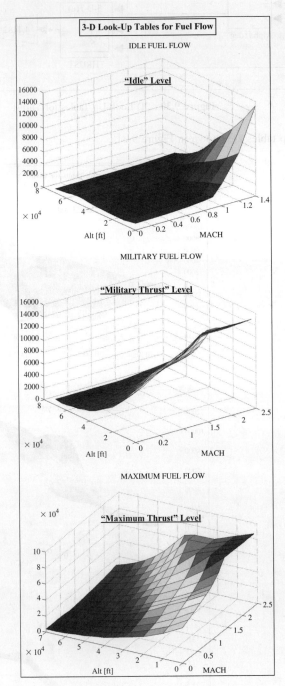

Figure 9.32 Fuel Flow Look-Up Tables for a Jet-Engine for Different Levels of Thrust

9.2.9 Auxiliary FDC Blocks

FDC also provides two additional blocks of utilities and features, which feed inputs to all the different blocks discussed in the previous sections.

The first block of utilities is the Airdata Group block, shown in Figure 9.33.

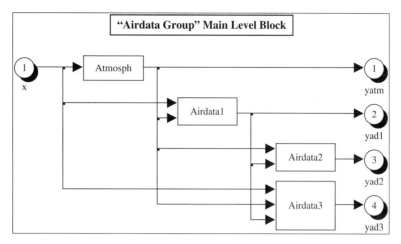

Figure 9.33 FDC Airdata Group Main Level Block

Within the Airdata Group block, the first block is given by the Atmospheric block, shown in Figure 9.34.

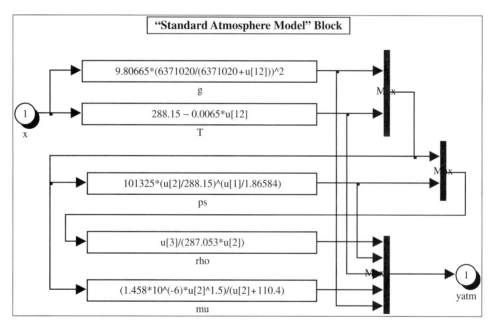

Figure 9.34 FDC Standard Atmospheric Model Block

This block features the conventional calculations of the parameters of the standard atmosphere model (SAM).[13] Specifically, FDC features the ICAO SAM. The outputs of the SAM block are:

- gravity acceleration g (m sec^{-2});
- air temperature T (°K);
- air pressure p_S (N m^{-2});
- air density ρ (Kg m^{-3});
- coefficient of dynamic viscosity μ.

The outputs of the SAM block are used throughout FDC for the solution of the aircraft equations of motion outlined in the previous sections.

The second block within the Airdata Group is known as the Air Data Block 1 and is shown in Figure 9.35.

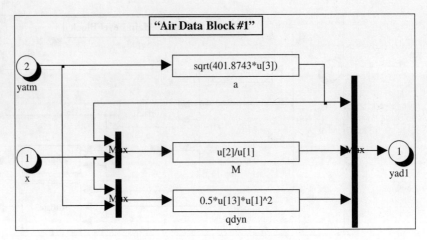

Figure 9.35 FDC Air Data Block 1

The outputs of this block are:

- speed of sound a (m sec^{-1});
- Mach number M;
- dynamic pressure $q_{dyn} = \bar{q}$ (N m^{-2}).

These parameters are also used throughout FDC for the solutions of the aircraft equations of motion.

The third block within the Airdata Group is known as the Air Data Block 2 and is shown in Figure 9.36.

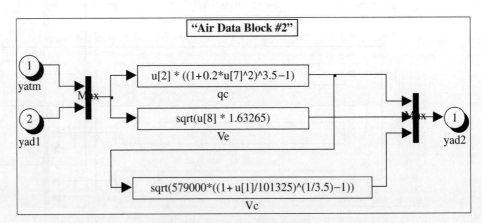

Figure 9.36 FDC Air Data Block 2

The outputs of this block are:

- impact pressure q_c (N m^{-2});
- equivalent airspeed V_e (m sec^{-1});
- calibrated airspeed V_c (m sec^{-1}).

These parameters are calculated for reference purposes only and are not used for the solution of the aircraft equations of motion.

Finally, the fourth block within the Airdata Group is known as the Air Data Block 3 and is shown in Figure 9.37.

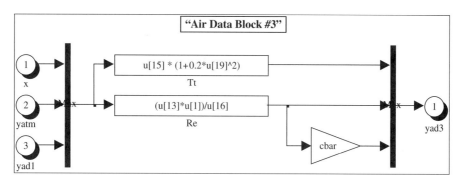

Figure 9.37 FDC Air Data Block 3

The outputs of this block are

- total temperature T_t (°K);
- Reynold number per unit length Re (m^{-1}).

As for the Air Data Block 2, the preceding parameters are not used for the solution of the aircraft equations of motion.

The second block of utilities is the Additional Outputs block, shown in Figure 9.38.

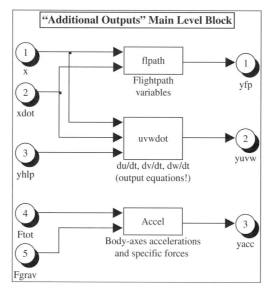

Figure 9.38 FDC Additional Outputs Main Level Block

None of the blocks in the Additional Outputs main block calculates parameters used in the solutions of the aircraft equations of motion. However, the parameters in these blocks are very useful simulation parameters.

The first block within the Additional Outputs main block is the Flight Path block, shown in Figure 9.39.

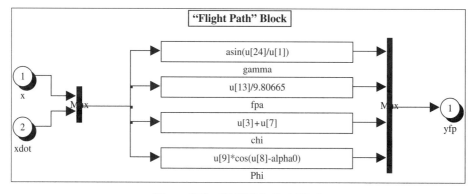

Figure 9.39 FDC Flight Path Block

The outputs of this block are:

- flight path angle γ(rad);
- flight path acceleration (in units of g) *fpa*;
- azimuth angle χ(rad);
- bank angle ϕ(rad).

The second block within the Additional Outputs main block is the Time Derivative block, shown in Figure 9.40.

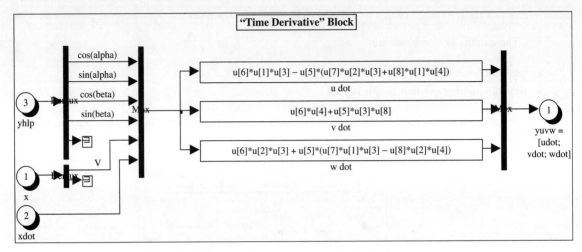

Figure 9.40 FDC Time Derivative Block

The outputs of this block are

- rate of change in the airspeed along the x body axis $\dot{u}$(m sec^{-2});
- rate of change in the airspeed along the y body axis $\dot{v}$(m sec^{-2});
- rate of change in the airspeed along the z body axis $\dot{w}$(m sec^{-2}).

Note that these parameters are not part of the state variables and, as stated before, are not used for the solution of the equations of motion.

Finally, the third block within the Additional Outputs main block is the Acceleration block, shown in Figure 9.41.

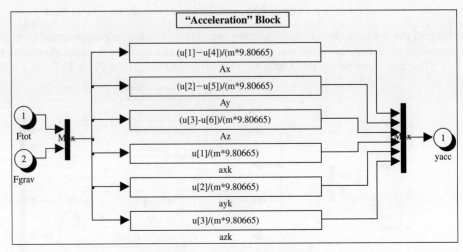

Figure 9.41 FDC Acceleration Block

The outputs of this block are essentially the virtual outputs of accelerometers along the body axes if the accelerometers were ideally placed at the aircraft center of gravity.[4] These accelerations are given by the kinematic body accelerations minus the components of the gravity acceleration and are calculated using:

- kinematic acceleration along the x body axis in units of g: $A_X = (F_X - F_{G_X})/W$;
- kinematic acceleration along the y body axis in units of g: $A_Y = (F_Y - F_{G_Y})/W$;
- kinematic acceleration along the z body axis in units of g: $A_Z = (F_Z - F_{G_Z})/W$.

9.2.10 Additional FDC Blocks

Interested readers are strongly recommended to browse all the other FDC features and capabilities. The available features include a library of radio-navigation tools, such as the Instrumental Landing System (ILS) and the Very high frequency Omnidirectional Radio range (VOR) system.[4,14–17] Another important feature is a library of wind and atmospheric turbulence models. A detailed description of these libraries requires technical knowledge at graduate level. Finally, FDC also features some simple but very clear examples of designs of specific autopilot systems, which again require technical knowledge at the graduate level.

9.3 INTRODUCTION TO THE AEROSPACE BLOCKSET BY MATHWORKS

Another powerful modeling tool for the development of aircraft simulation environments is the Aerospace Blockset toolbox, produced and distributed by The Mathworks.

Aerospace Blockset (AB) presents a collection of features and blocks for the solution of the aircraft equations of motion as well as the modeling of all different forces and moments acting on the aircraft. Refer to.[18] An important feature is its simple integration with graphic packages for a complete visualization of the system dynamics. Specifically, AB can be interfaced with FlightGear,[19] an open source software package available through a GNU General Public License, and Virtual Reality Toolbox (VRT), a graphic simulation package developed by The Mathworks.[20] The purpose of the sections following is to highlight the key features of this powerful tool.

9.3.1 General Organization of the Aerospace Blockset

The Aerospace Blockset (AB) package is organized in a set of ten libraries, as shown in Figure 9.42 (in alphabetical order).

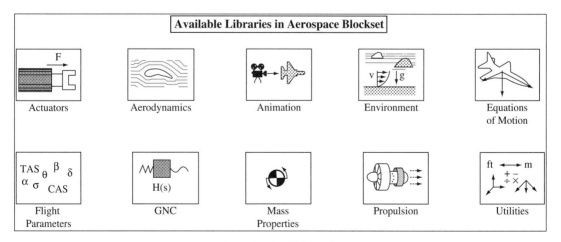

Figure 9.42 AB Libraries

Each of these libraries contains a number of sub libraries with different and specific tools. The next sections will summarize the main features of these libraries.

9.3.2 Introduction to the Environment Library

The Environment library, shown in Figure 9.43. contains the following sublibraries:

- Atmosphere sub library;
- Gravity sub library;
- Wind sub library.

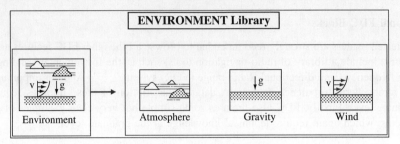

Figure 9.43 AB Environment Library

The Atmosphere sub library is shown in Figure 9.44.

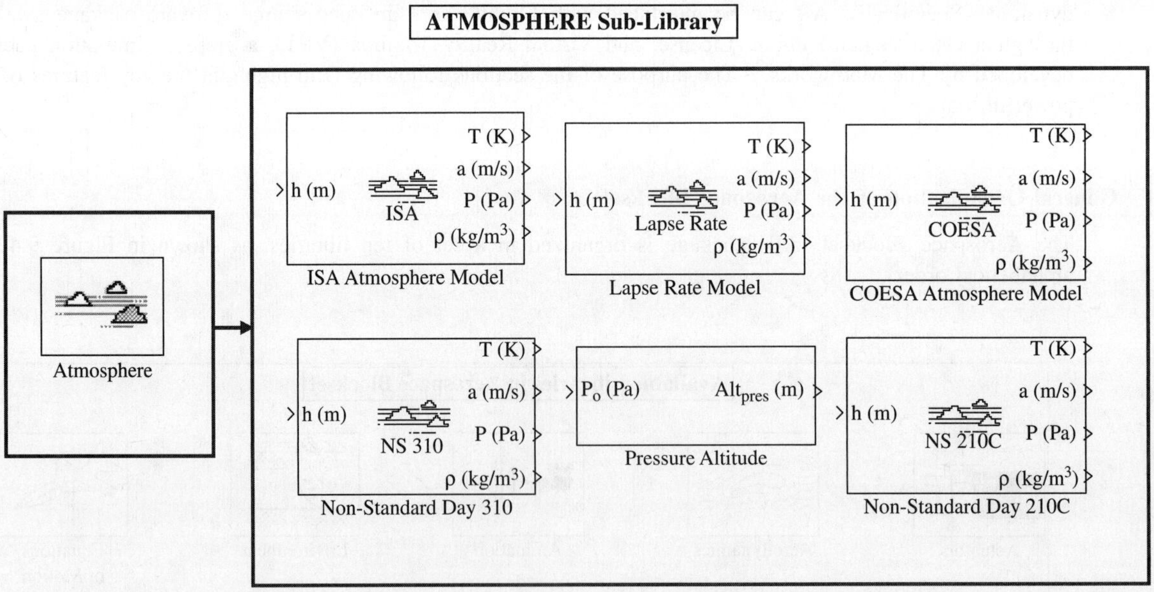

Figure 9.44 AB Atmosphere Sublibrary

The Atmosphere sub library contains several atmospheric models, including the standard ISA and COESA atmospheric models, as well as atmospheric models for nonstandard atmospheric conditions. It also includes a pressure altitude relationship.

The Gravity sub library, shown in Figure 9.45, contains two blocks. The first block provides the value for the gravity acceleration at any point on the Earth. The second block provides a complete set of information of the magnetic field at any point on the Earth.

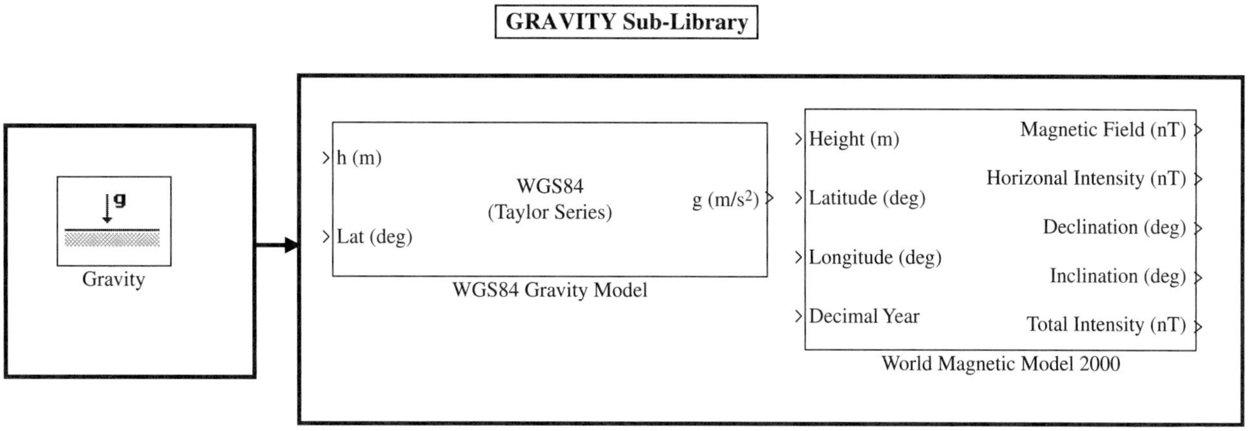

9.3 Introduction to the Aerospace Blockset by Mathworks

Figure 9.45 AB Gravity Sublibrary

Finally, the Wind sub library, shown in Figure 9.46, contains a number of blocks for the modeling of atmospheric turbulence, including the Dryden and Von Karman turbulence models (in continuous and discrete time)[12] as well as wind shear models.

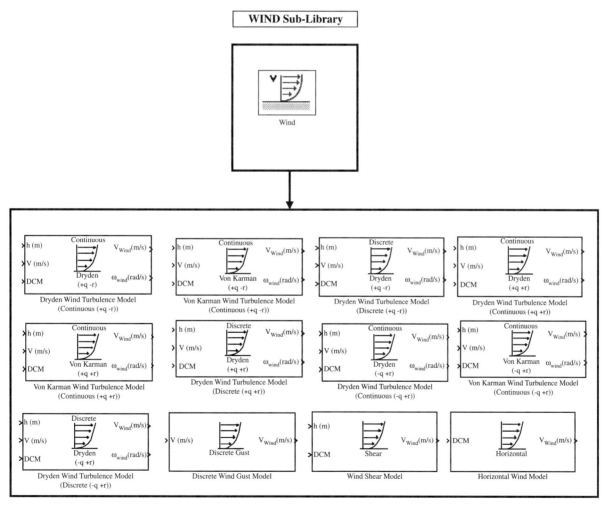

Figure 9.46 AB Wind Sublibrary

9.3.3 Introduction to the Flight Parameters Library

The Flight Parameter library, shown in Figure 9.47, contains a number of utility blocks for the calculation of several flight parameters, including the aerodynamic angles, airspeed components along the wind or stability axes, dynamic pressure, Mach number, calibrated air speed (CAS), as well as the density, pressure, and temperature ratios according to the standard atmospheric model.

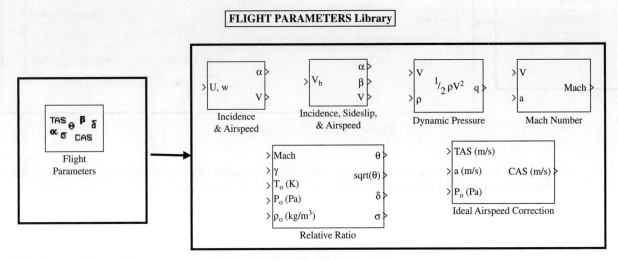

Figure 9.47 AB Flight Parameters Library

9.3.4 Introduction to the Equations of Motion Library

The Equations of Motion library, shown in Figure 9.48, contains two sub libraries for the simulation of three and six degree of motion dynamics, respectively.

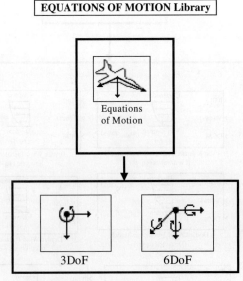

Figure 9.48 AB Equations of Motion Library

The Three Degree of Motion sub library, shown in Figure 9.49, contains three blocks for the simulation of three DOF dynamics with constant mass, varying mass, and varying inertial properties.

The Six Degree of Motion sub library, shown in Figure 9.50, contains six blocks for the simulation of six DOF dynamics with constant mass, varying mass, and varying inertial properties. These blocks are essentially the extended version of the blocks from the Three DOF sub library; however, they are formulated for the orientation of the aircraft expressed using either the conventional Euler angles or the more advanced quaternion-based orientation system.[21]

9.3 Introduction to the Aerospace Blockset by Mathworks 507

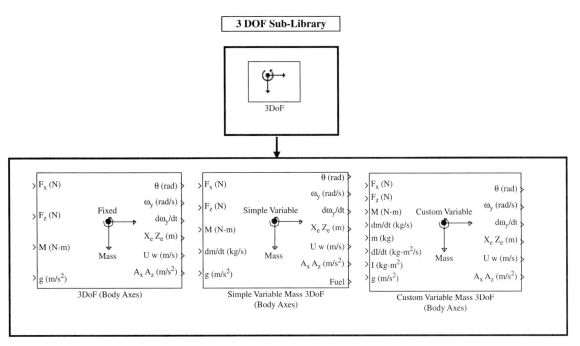

Figure 9.49 AB "Three DOF Sublibrary"

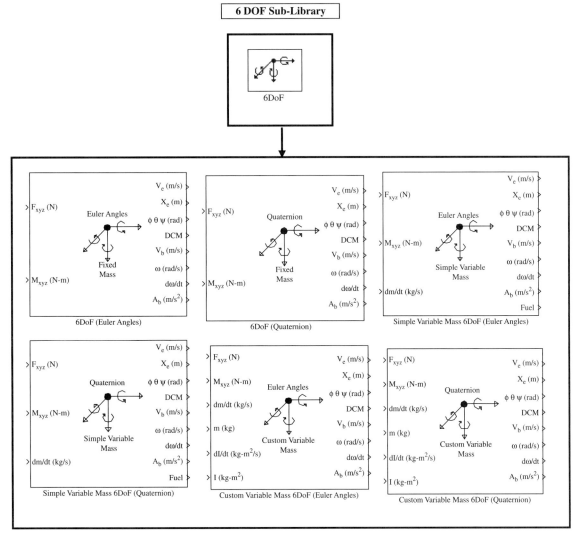

Figure 9.50 AB Six DOF Sublibrary

9.3.5 Introduction to the Aerodynamics Library

The Aerodynamics library, shown in Figure 9.51, is a single block library, providing the modeling of the aerodynamic forces and moments along the body axes using the relationships:

$$F_{X_A} = c_{X_A}\overline{q}S, \quad F_{Y_A} = c_{Y_A}\overline{q}S, \quad F_{Z_A} = c_{Z_A}\overline{q}S$$
$$L_A = c_{l_A}\overline{q}Sb, \quad M_A = c_{m_A}\overline{q}S\overline{c}, \quad N_A = c_{n_A}\overline{q}Sb$$

Each of the aerodynamic coefficients can be entered as a single-value (from a single point linearized aerodynamic modeling) or it can be modeled through the use of sets of customized 2D or 3D look-up tables, as shown in the previous sections.

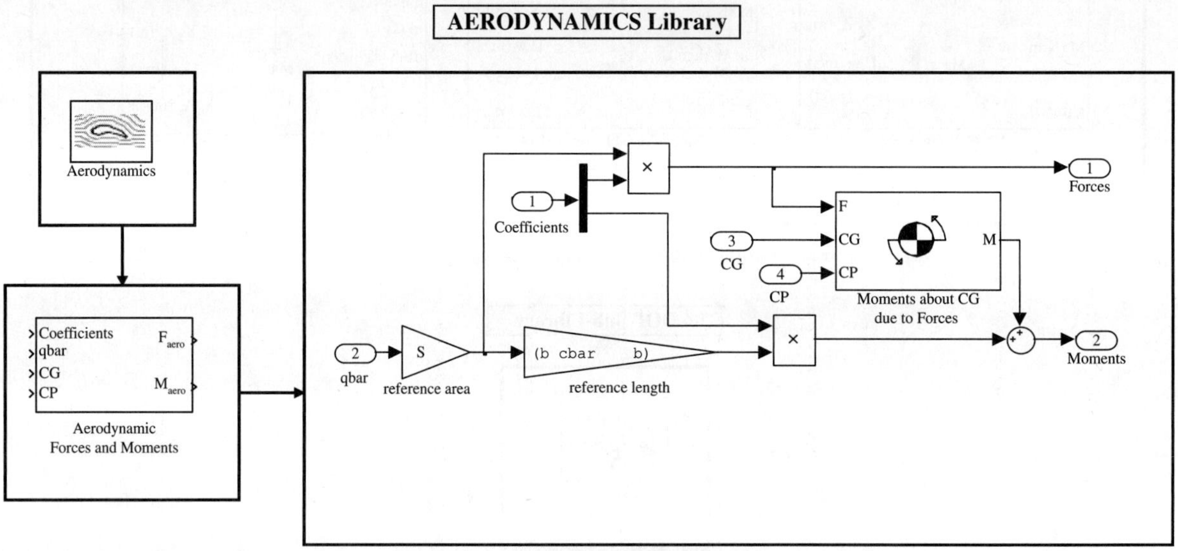

Figure 9.51 AB Aerodynamics Library

9.3.6 Introduction to the Propulsion Library

Like the previous library, the Propulsion library, shown in Figure 9.52, is a single block library featuring only the modeling of a turbofan engine system. No other propulsion system is included.

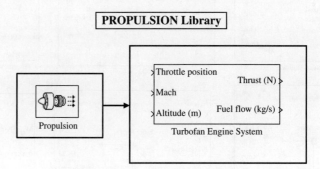

Figure 9.52 AB Propulsion Library

The Turbofan Engine System block, shown in Figure 9.53, features look-up tables for the engine response as a function of Mach as well as the calculations for the engine thrust as function of the altitude. Additionally, the block calculates the fuel consumption rate, which can be used for the calculation of the varying mass in the appropriate Equations of Motion blocks.

9.3 Introduction to the Aerospace Blockset by Mathworks   509

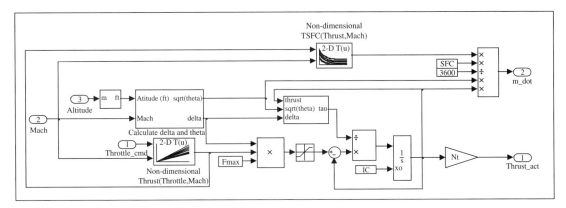

Figure 9.53 AB Turbofan Engine System Block

9.3.7 Introduction to the Utilities Library

An important library is the Utilities library, shown in Figure 9.54, which includes the following sub libraries:

- Axes transformation sub library;
- Unit conversions sub library;
- Math operations sub library.

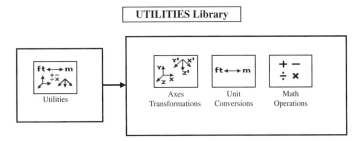

Figure 9.54 AB Utilities Library

The Axes Transformation sub library includes a large number of blocks allowing transformations from/to a number of reference systems. The most common blocks (for transformations to/from Euler angles-based, quaternion-based, and cosine matrix-based systems) are shown in Figure 9.55.

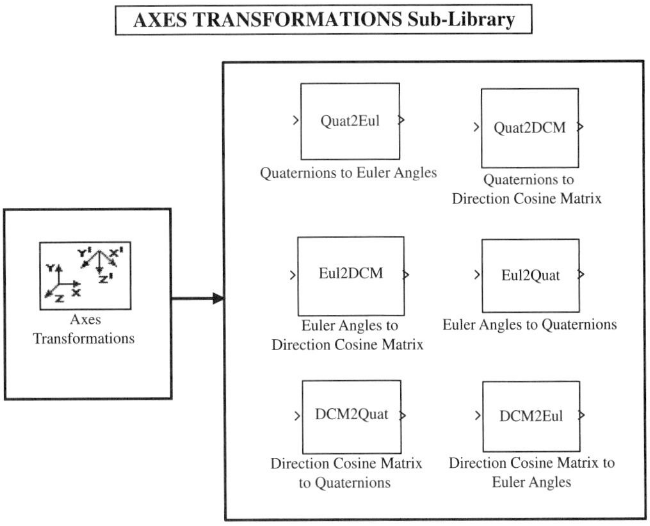

Figure 9.55 AB Axes Transformation Sublibrary

The Unit Conversion sub library includes a large number of blocks allowing transformations from/to different units for a number of parameters, as shown in Figure 9.56.

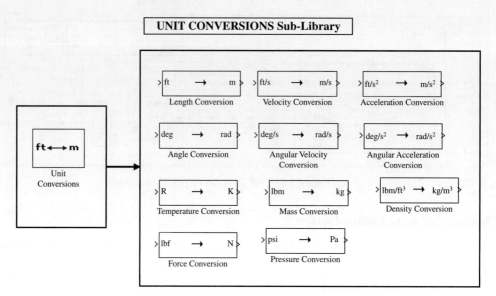

Figure 9.56 AB Unit Convesions Sublibrary

Finally, the Math Operations sub library includes a number of blocks for common matrix and trigonometric operations, as shown in Figure 9.57.

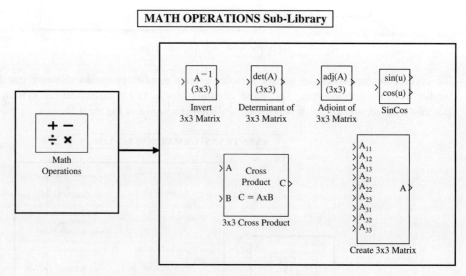

Figure 9.57 AB Math Operations Sublibrary

9.3.8 Introduction to the Mass Properties Library

The Mass Properties library, shown in Figure 9.58, features four blocks for the evaluation of the aircraft center of gravity, as well as the aircraft inertial characteristics, in terms of moments and products of inertia.

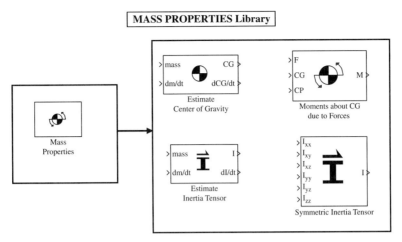

Figure 9.58 AB Mass Properties Library

9.3.9 Introduction to the Actuators Library

The Actuators library features two blocks for the modeling of aircraft actuators with second order response without and with saturation, as shown in Figure 9.59. These blocks can accurately model the response of hydraulic or electro-hydraulic actuators installed in most commercial and military aircraft, excluding general aviation aircraft.[12]

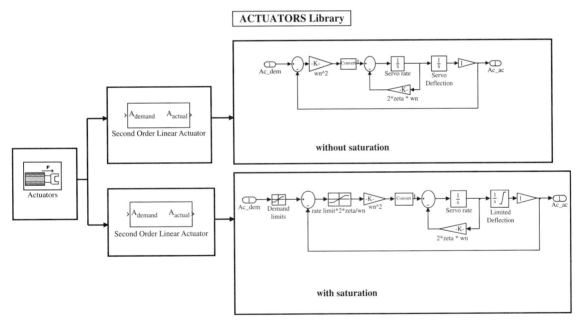

Figure 9.59 AB Actuators Library

9.3.10 Introduction to the GNC and Animation Libraries

Finally, AB features two additional libraries for advanced applications.

The GNC library includes a variety of schemes for the modeling of closed-loop schemes for aircraft flight control systems (including different autopilot functions on longitudinal and lateral directional channels, and different classes of stability augmentation systems) as well as different schemes for aircraft navigation purposes. Refer to graduate level courses in flight control systems for a detailed understanding of the blocks in this library.[12, 22, 23]

The Animation library features a number of blocks for visualizing flight paths and trajectories, as well as for interfacing with a graphic package for the visualization of the aircraft dynamics.

9.4 INTRODUCTION TO AIRLIB

AIRLIB is a freely available library of aircraft models to be used with Simulink® 3.1 (MATLAB® 5.3) or later versions.[24] It is essentially a library of different aircraft mathematical models suitable for applications within simulation at open-loop and closed-loop conditions. The key components of the AIRLIB are discussed next.

9.4.1 AIRLIB's Strucure

The library is based on two blocks. The first block implements a general nonlinear *continuous time* aircraft model; the second block represents a *discrete-time* version of the same model. Both blocks are based on the FDC toolbox[4] introduced in Section 9.2. The main window of the library is shown in Figure 9.60.

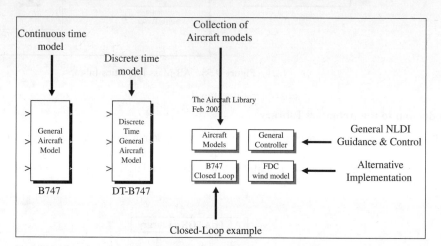

Figure 9.60 AIRLIB Main Window

The four buttons on the right-hand side lead, respectively, a collection of aircraft models, a general nonlinear dynamics inversion (NLDI) controller, a closed-loop simulation example featuring the mathematical model of a Boeing 747-200, and a simulation featuring an alternative (but equivalent) implementation developed for verification purposes. The discussion of the features of the last three buttons is beyond the technical purpose of this textbook.

9.4.2 Generic Aircraft Model: Continuous-time Block

This block implements a general nonlinear aircraft model with linearized aerodynamics.
The three inputs to the block are:

- A *(6 × 1)* vector containing wind velocity and acceleration in body frame expressed in m/s and m/s², respectively ($[u_w \ v_w \ w_w \ \dot{u}_w \ \dot{v}_w \ \dot{w}_w]^T$);
- A *(6 × 1)* vector containing external forces and moments in body frame expressed in Newtons (including the thrust force generated by the engines);
- A *(4 × 1)* vector containing elevators, ailerons, rudder, and stabilators deflections in radians (that is $\delta_E = (\delta_{E\_Right} + \delta_{E\_Left})/2$, $\delta_A = (\delta_{A\_Right} - \delta_{A\_Left})/2$, $\delta_R = (\delta_{R\_Right} + \delta_{R\_Left})/2$, $i_H = (i_{H\_Right} + i_{H\_Left})/2$), where the positive deflection for the horizontal surfaces is assumed to be trailing edge down (TED), and the positive rudder(s) deflection is assumed to be trailing edge left (TEL).

The state vector of the block is:

- $[V, \alpha, \beta, p, q, r, \phi, \theta, \phi, x, y, H]^T$, where the velocity is in m/s, the angles in radians, the angular rates in radians/sec and the flat-Earth positions and altitude (with respect to the sea level) in meters.

The three outputs of the block are:

- State vector;
- Time derivative of the state vector;

- *(12 × 1)* vector containing:
  - time derivatives of the velocity in body axis ($\dot{u}, \dot{v}, \dot{w}$ in m/s^2);
  - Dimensionless components of the angular velocity ($pb/2V$, $q\bar{c}/2V$, $rb/2V$);
  - Three components of the *measured* acceleration in body frame (that is $[Ax, Ay, Az] = (F - F_{gravity})/m$);
  - Three components of the *total* acceleration in body frame (that is $[a_x, a_y, a_z] = F/m$).

The equations of motions are written assuming that the Earth is flat, non-rotating, and that the aircraft is rigid, with constant mass m and thrust force along its x body axis:

$$\dot{V} = \frac{1}{m}\left(-\frac{1}{2}\rho S V^2 C_D \cos\beta + \frac{1}{2}\rho S V^2 C_Y \sin\beta + T\cos\alpha\cos\beta\right)$$

$$- g(\sin\theta\cos\alpha\cos\beta - \cos\theta\sin\phi\sin\beta - \cos\theta\cos\phi\sin\alpha\cos\beta)$$

$$\dot{\alpha} = \frac{1}{mV\cos\beta}\left[-\frac{1}{2}\rho S V^2 C_L - T\sin\alpha + mg(\cos\theta\cos\phi\cos\alpha + \sin\theta\sin\alpha)\right]$$

$$+ q - (p\cos\alpha + r\sin\alpha)\tan\beta$$

$$\dot{\beta} = \frac{1}{mV}\left[\frac{1}{2}\rho S V^2 C_D \sin\beta + \frac{1}{2}\rho S V^2 C_Y \cos\beta - T\cos\alpha\sin\beta + mg(\sin\theta\cos\alpha\sin\beta\right.$$

$$\left. + \cos\theta\sin\phi\cos\beta - \cos\theta\cos\phi\sin\alpha\sin\beta)\right] + p\sin\alpha - r\cos\alpha$$

$$\begin{bmatrix}\dot{p}\\ \dot{q}\\ \dot{r}\end{bmatrix} = M_1\begin{bmatrix}p^2\\ q^2\\ r^2\end{bmatrix} + M_2\begin{bmatrix}qr\\ pr\\ pq\end{bmatrix} + \bar{q}SM_0\begin{bmatrix}bC_l\\ \bar{c}C_m\\ bC_n\end{bmatrix}$$

$$\begin{bmatrix}\dot{\theta}\\ \dot{\psi}\\ \dot{\phi}\end{bmatrix} = \begin{bmatrix}q\cos\phi - r\sin\phi\\ q\sin\phi\sec\theta + r\cos\phi\sec\theta\\ p + q\sin\phi\tan\theta + r\cos\phi\tan\theta\end{bmatrix}$$

$$\dot{x} = V[\cos\beta\cos\alpha\cos\theta\cos\psi + \sin\beta(\sin\phi\sin\theta\cos\psi - \cos\phi\sin\psi)$$

$$+ \cos\beta\sin\alpha(\cos\phi\sin\theta\cos\psi + \sin\phi\sin\psi)]$$

$$\dot{y} = V[\cos\beta\cos\alpha\cos\theta\sin\psi + \sin\beta(\sin\phi\sin\theta\sin\psi + \cos\phi\cos\psi)$$

$$+ \cos\beta\sin\alpha(\cos\phi\sin\theta\sin\psi - \sin\phi\cos\psi)]$$

$$\dot{H} = V(\cos\beta\cos\alpha\sin\theta - \sin\beta\sin\phi\cos\theta - \cos\beta\sin\alpha\cos\phi\cos\theta)$$

with ρ being the air density, S the wing surface, and T the thrust force along the x body axis. The matrices M_0, M_1, M_2, contain a combination of inertial coefficients:

$$M_0 = \frac{1}{\det(I)}\begin{bmatrix}I_{yy}I_{zz} - J_{yz}J_{yz} & J_{xy}I_{zz} + J_{yz}J_{xz} & J_{xy}J_{yz} + I_{yy}J_{xz}\\ J_{xy}I_{zz} + J_{yz}J_{xz} & I_{xx}I_{zz} - J_{xz}J_{xz} & J_{yz}I_{xx} + J_{xy}J_{xz}\\ J_{xy}J_{yz} + I_{yy}J_{xz} & J_{yz}I_{xx} + J_{xy}J_{xz} & I_{xx}I_{yy} - J_{xy}J_{xy}\end{bmatrix}$$

$$M_1 = M_0\begin{bmatrix}0 & J_{yz} & -J_{yz}\\ -J_{xz} & 0 & J_{xz}\\ J_{xy} & -J_{xy} & 0\end{bmatrix}$$

$$M_2 = M_0\begin{bmatrix}I_{yy} - I_{zz} & -J_{xy} & J_{xz}\\ J_{xy} & I_{zz} - I_{xx} & -J_{yz}\\ -J_{xz} & J_{yz} & I_{xx} - I_{yy}\end{bmatrix}$$

where I is the inertia matrix of the aircraft:

$$I = \begin{bmatrix} I_x & -J_{xy} & -J_{xz} \\ -J_{xy} & I_y & -J_{yz} \\ -J_{xz} & -J_{yz} & I_z \end{bmatrix}$$

The variables $c_D, c_Y, c_L, c_l, c_m, c_n$ in the linear and angular velocity equations are the aerodynamic coefficients that model the aerodynamic forces and moments on the aircraft and are given by

$$c_D = c_{D0} + c_{D\alpha}\alpha + c_{Dq}\frac{\bar{c}}{2V}q + c_{D\delta_E}\delta_E + c_{Di_H}i_H$$

$$c_L = c_{L0} + c_{L\alpha}\alpha + c_{Lq}\frac{\bar{c}}{2V}q + c_{L\delta_E}\delta_E + c_{Li_H}i_H$$

$$c_m = c_{m0} + c_{m\alpha}\alpha + c_{mq}\frac{\bar{c}}{2V}q + c_{m\delta_E}\delta_E + c_{mi_H}i_H$$

$$c_Y = c_{Y0} + c_{Y\beta}\beta + c_{Yp}\frac{b}{2V}p + c_{Yr}\frac{b}{2V}r + c_{Y\delta_A}\delta_A + c_{Y\delta_R}\delta_R$$

$$c_l = c_{l0} + c_{l\beta}\beta + c_{lp}\frac{b}{2V}p + c_{lr}\frac{b}{2V}r + c_{l\delta_A}\delta_A + c_{l\delta_R}\delta_R$$

$$c_n = c_{n0} + c_{n\beta}\beta + c_{np}\frac{b}{2V}p + c_{nr}\frac{b}{2V}r + c_{n\delta_A}\delta_A + c_{n\delta_R}\delta_R$$

Each aircraft is entirely specified by the geometric, inertial, and aerodynamic parameters typed on the mask (which is accessible by double-clicking the block), as shown in Figure 9.61.

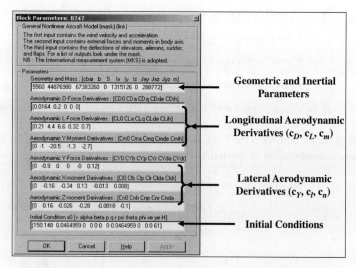

Figure 9.61 Mask of the General Nonlinear Aircraft Model Block

The default parameters are relative to a Boeing 747-200 aircraft in a straight and level flight at a constant speed of 150.148 m/s and at an altitude of 61 m.[25]

The implementation of the continuous time block is based on the FDC toolbox discussed in Section 9.2. In fact, looking under the mask of the general aircraft models leads to the same structure that is used for the implementation of the Beaver aircraft in FDC, as shown in Figure 9.62.

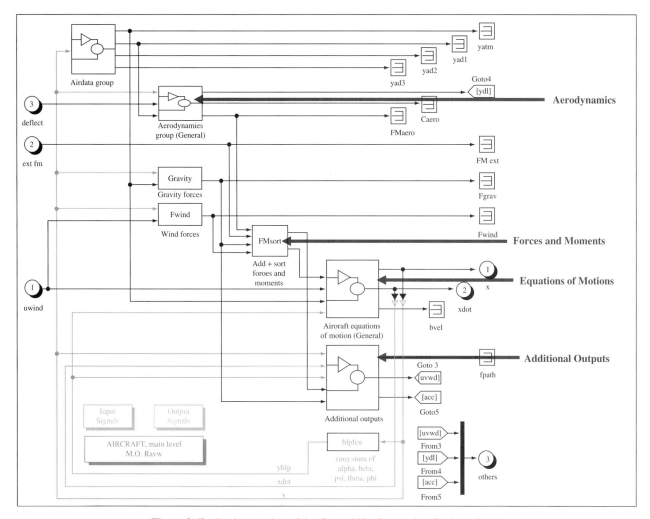

Figure 9.62 Implementation of the General Nonlinear Aircraft Model Block

There are, however, the following differences with respect to FDC:

- The general aircraft model is not restricted to the De Havilland Beaver aircraft; in fact, any aircraft can be simulated by simply changing the geometric, inertial, and aerodynamic parameters in the mask. Thirteen different aircraft models are provided as examples;
- Within AIRLIB, the stability-axis drag and lift coefficients c_D and c_L are used to represent the longitudinal aerodynamic forces, while in the FDC implementation of the Beaver aircraft the body axis coefficients c_X and c_Z are used;
- Within AIRLIB, the pitch rate q is normalized as $q\bar{c}/2V$, instead of $q\bar{c}/V$, as used in FDC, to be consistent with most of the technical literature. This means that, for the same aircraft, the values of the aerodynamic coefficients depending on q such as c_{Dq}, c_{Lq}, and c_{mq} in AIRLIB must be doubled with respect to the corresponding values for FDC.

9.4.3 Generic Aircraft Model: Discrete-time Block

The only difference between the continuous time and discrete time blocks is that in the aircraft equation of motion subsystem, the continuous time integrator is replaced with a discrete time integrator (with a sampling time selected as the last element of the first vector in the mask), as shown in Figure 9.63.

This is equivalent to replacing the continuous time derivatives in the equation of motion with their Euler discrete-time (backward) approximations. This typically results in a faster (although slightly less accurate) simulation, and can be useful whenever the block is part of a scheme that must be used to generate executable code, since discrete-time integration algorithms are typically more suitable for real-time implementations.

516 Chapter 9 Introduction to Modern Flight Simulation Codes

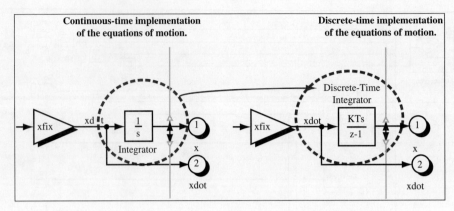

Figure 9.63 Implementation of the Discrete-Time Aircraft Model Block

9.4.4 Collection of Aircraft Models

From the main AIRLIB Simulink® scheme a total of thirteen different aircraft models can be accessed, as shown in Figure 9.64.

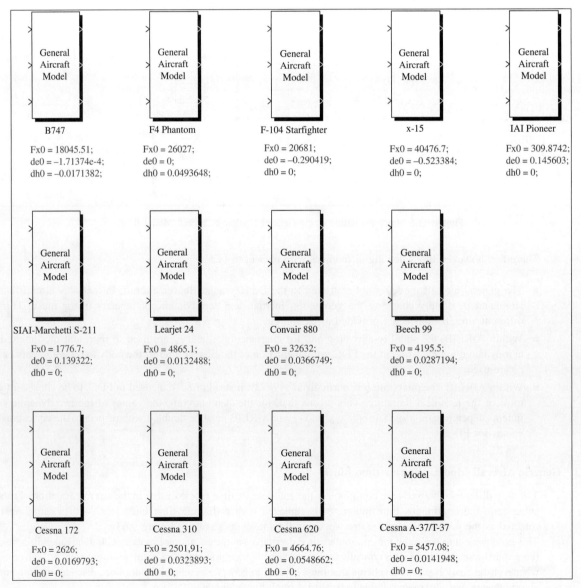

Figure 9.64 Available Aircraft Models

Specifically, the aircraft models are

- Boeing B747-200;[25]
- McDonnell Douglas F-4;[25]
- Lockheed F-104;[25]
- NASA X15 (experimental aircraft);
- Generic large UAV;
- SIAI-Marchetti S211;[25]
- Learjet 24;[25]
- Convair 880;
- Beech 99;[25]
- Cessna 182;[25]
- Cessna 310;[25]
- Cessna 620;[25]
- Cessna T-37.[25]

Each of these blocks is a link to the continuous-time general aircraft model with different parameters in the mask. The aerodynamic parameters are calculated for a given steady and level flight condition, described by the initial condition vector in the mask. Forces and deflections relative to those conditions are also specified for each aircraft.

Each block is ready to be used and can be just dragged and dropped in any Simulink® scheme. Breaking the link between any of this block and the AIRLIB library (by right-clicking on the block and choosing *"Link Options"-> "Disable Link"* first, and *"Link Options"-> "Break Link"* then) allows the model to be executed without having the AIRLIB library in the MATLAB® path.

9.4.5 Alternative Model Implementation

From the main AIRLIB Simulink® scheme one can access a comparison between the standard AIRLIB McDonnell Douglas F-4 aircraft model and an equivalent model in which the internal blocks have been rearranged in a modular fashion, similar to the way blocks are arranged within aircraft modeled with the Aerospace Blockset toolbox.

As shown in Figure 9.65, there is a general block (which is equivalent to the six DOF block in the Aerospace Blockset toolbox) that implements a six DOF rigid object subject to initial conditions and external forces and moments. The aerodynamic forces and moments are then calculated as functions of inputs, states, and wind conditions and added as inputs to the six DOF block.

The execution of the scheme reveals that there are no differences between the standard implementation in the Aerospace Blockset toolbox and the implementation within AIRLIB. As further verification, replacing the six DOF block with the one shipped with the Aerospace Toolbox also does not lead to any difference.

9.4.6 Additional Tools within AIRLIB: The Function '*air3m*'

The MATLAB® function '*air3m*' can be used to find trim conditions for a given aircraft model. The function can be invoked using

$$[X0, U0] = \text{'air3m'}(model\_name, V, H, G)$$

where '*model_name*' is the name of the Simulink® scheme containing the aircraft model. The first argument, V, is the desired aircraft speed, the second argument, H, is the desired altitude, and the third argument, G, is gamma, the desired flight path vertical angle. The resulting function outputs $X0$ and $U0$ are the state and input vectors corresponding to the desired speed, altitude and flight path. If V, H, and G are vectors, then trim conditions are found for each combination of them, and the resulting 5D matrices $X0$ and $U0$ are read to be used with the ready to be used with the interpolate matrix block from the Aerospace Blockset. Typing *"help air3m"* from the MATLAB® command line yields additional help, examples, and details.

518 Chapter 9 Introduction to Modern Flight Simulation Codes

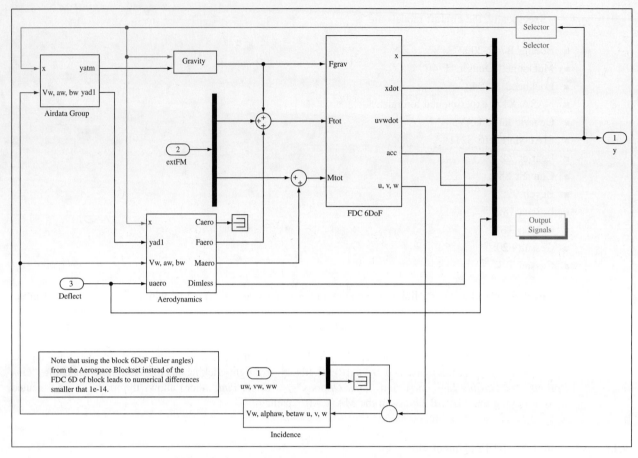

Figure 9.65 Alternative Implementation of the Aircraft Model

9.4.7 Additional Tools within AIRLIB: The Function 'ab2dv'

The MATLAB® function '*ab2dv*' can be used to recover the aircraft aerodynamic derivatives given the longitudinal and lateral linear model matrices, the inertial data and the trim conditions. The function can be invoked using:

$$[CD, CL, CY, Cl, Cm, Cn] = \text{ab2dv}(Alg, Blg, Alt, Blt, GM, deltaE\_0, alpha\_0, V0, h0, T0);$$

For the longitudinal matrices *Alg* and *Blg* (of dimensions *(3 × 3)* and *(3 × 1)*, respectively) the states are V (m/s), α(rad), and q (rad/s), while the input is δ_E(rad). For the lateral matrices *Alt* and *Blt* (of dimensions *(3 × 3)* and *(3 × 2)*, respectively) the states are β(rad), p(rad/s), and r(rad/s), while the inputs are δ_A(rad) and δ_R(rad). The *(10 × 1)* vector GM is the vector containing *cbar*, b, S, I_x, I_z, J_{xy}, J_{xz}, J_{yz} and the aircraft mass. The last five parameters are δ_E(rad), α(rad), V(m/s), H(m), and thrust force (N), relative to the trim condition corresponding to the given lateral and longitudinal linear matrices.

9.5 SUMMARY

This chapter has introduced powerful, realistic, but yet simple tools commonly available for developing detailed simulation environments for the aircraft dynamics. These tools allow the evaluation of the aircraft dynamic response with different levels of complexity for the modeling of the aircraft dynamic, aerodynamic and propulsive characteristics. Additionally, these codes can be interfaced with graphic packages for a complete 3D visualization.

REFERENCES

1. Lanchester, F. W. *Aerodonetics*. Constable and Co. Ltd, London, 1908.
2. Bryan, G. H. *Stability in Aviation*. Macmillan and Co., London, 1911.

3. *www.mathworks.com*.
4. Rauw, M. FDC 1.2—A SIMULINK Toolbox for Flight Dynamics and Control Analysis. User Manual, *www.dutchroll.com*, May 2001.
5. Tjee, R.T.H., Mulder, J. A. *Stability and Control Derivatives of the De Havilland DHC-2 'Beaver' Aircraft*. Report LR-556, Delft University of Technology, Faculty of Aerospace Engineering, Delft, The Netherlands, 1988.
6. Brumbaugh, R.W. "An Aircraft Model for the AIAA Controls Design Challenge. NASA CR186019, December 1991.
7. Perhinschi, M. G., Campa G., Napolitano, M. R., Lando, M., Massotti, L., Fravolini, M.L. "Modeling and Simulation of a Fault-Tolerant Flight Control System. *International Journal of Modeling & Simulation*, Volume 26, Number 1, 2006, pp. 1–10.
8. Perhinschi M. G., Lando M., Massotti L., Campa G., Napolitano, M. R., Fravolini M. L. "On-Line Parameter Estimation Issues for the NASA IFCS F-15 Fault Tolerant Systems." Proceeding of American Control Conference, Anchorage, AK, 2002, pg. 191–196.
9. Perhinschi, M. G., Napolitano, M. R., Campa G., Fravolini, M. L. A Simulation Environment for Testing and Research of Neurally Augmented Fault Tolerant Control Laws Based on Non-Linear Dynamic Inversion. Proceedings of the 2004 AIAA Modeling and Simulation Technology (MST) conference, Providence, RI, August 2004.
10. Perhinschi, M. G., Napolitano, M. R., Campa G., Fravolini, M. L., "Integration of Fault Tolerant System for Sensor and Actuator Failures within the WVU NASA F-15 Simulator", Proceedings of the AIAA 2003 Guidance Navigation and Control (GNC) Conference, AIAA Paper 2003-5644, Austin, TX, August 2003.
11. Perhinschi, M. G., Napolitano, M. R., Campa, G. "A Simulation Environment for Design and Testing of Aircraft Adaptive Fault-Tolerant Control Systems." *Journal of Aircraft Engineering and Aerospace Technology*, Volume 80, Issue 6, 2008, pp. 620–632.
12. Roskam, J. *Airplane Flight Dynamics and Automatic Flight Controls – Part II*. Design, Analysis, and Research Corporation, Lawrence, Kansas, 1995.
13. Ruijgrok, G.J.J. *Elements of Airplane Performance*. Delft University Press, Delft, The Netherlands, 1990.
14. Anon. International Standards and Recommended Practices. Annex 10, Volume I, Part I: Equipment and Systems, ICAO, Montreal, Canada, 1968.
15. Abbink, F. J. "Vliegtuiginstrumentate I/II." Lecture Notes D-34 (in Dutch), Delft University of Technology, Faculty of Aerospace Engineering, Delft, 1984.
16. Bauss, W. *Radio Navigation Systems for Aviation and Maritime Use*. AGARDograph 63, Pergamon Press, UK, 1963.
17. Kendal, B. *Manual of Avionics* (2nd ed.). BSP Professional Books, UK, 1987.
18. Aerospace Blockset. *User's Guide*. The Mathworks.
19. *www.flightgear.org*.
20. Virtual Reality Toolbox. *User's Guide*. The Mathworks.
21. Phillips, W. F. *Mechanics of Flight*. John Wiley & Sons, 2004.
22. Pamadi, B. N. *Performance, Stability, Dynamics, and Control of Airplanes* (2nd ed.). AIAA (Education Series), 2004.
23. Stevens, B. L., and Lewis F. L. *Aircraft Control and Simulation*. John Wiley & Sons, 2003.
24. http://www.mathworks.com/matlabcentral/fileexchange/3019-airlib
25. Roskam J. *Airplane Flight Dynamics and Automatic Flight Controls – Part I*. Design, Analysis, and Research Corporation, Lawrence, KS, 1995.

STUDENT SAMPLE PROBLEMS

Student Sample Problem 9.1

Using the aerodynamic model of the default Beaver DC-2 aircraft provided with the FDC simulation environment, perform a qualitative sensitivity analysis for the following longitudinal parameters:

$$\omega_{nSP}, \zeta_{SP} \quad \text{(short period mode)}$$
$$\omega_{nPh}, \zeta_{Ph} \quad \text{(phugoid mode)}$$

Solution of Student Sample Problem 9.1

The aerodynamic modeling of the Beaver aircraft is contained in the *(6 × 19)* AM matrix described in Section 9.2.2 and shown in Figure SSP9.1.1.[5]

$$AM = \begin{bmatrix} C_{X_0} & C_{Y_0} & C_{Z_0} & C_{l_0} & C_{m_0} & C_{n_0} \\ C_{X_\alpha} & 0 & C_{Z_\alpha} & 0 & C_{m_\alpha} & 0 \\ C_{X_{\alpha^2}} & 0 & 0 & 0 & C_{m_{\alpha^2}} & 0 \\ C_{X_{\alpha^3}} & 0 & C_{Z_{\alpha^3}} & 0 & 0 & 0 \\ 0 & C_{Y_\beta} & 0 & C_{l_\beta} & 0 & C_{n_\beta} \\ 0 & 0 & 0 & 0 & C_{m_{\beta^2}} & 0 \\ 0 & 0 & 0 & 0 & 0 & C_{n_{\beta^3}} \\ 0 & C_{Y_P} & 0 & C_{l_p} & 0 & C_{n_p} \\ C_{X_q} & 0 & C_{Z_P} & 0 & C_{m_q} & C_{n_q} \\ 0 & C_{Y_r} & 0 & C_{l_r} & C_{m_r} & C_{n_r} \\ 0 & 0 & C_{Z_{\delta_e}} & 0 & C_{m_{\delta_e}} & 0 \\ C_{X_{\delta_l}} & 0 & C_{Z_{\delta_l}} & 0 & C_{m_{\delta_l}} & 0 \\ 0 & C_{Y_{\delta_a}} & 0 & C_{l_{\delta_a}} & 0 & C_{n_{\delta_a}} \\ C_{X_{\delta_r}} & C_{Y_{\delta_r}} & 0 & C_{l_{\delta_r}} & 0 & C_{n_{\delta_r}} \\ C_{X_{\alpha\delta_l}} & 0 & C_{Z_{\alpha\delta_l}} & 0 & 0 & 0 \\ 0 & C_{Y_{\delta_{ra}}} & 0 & 0 & 0 & 0 \\ 0 & 0 & 0 & C_{l_{\delta_a\alpha}} & 0 & 0 \\ 0 & 0 & C_{Z_{\delta_e\beta^2}} & 0 & 0 & 0 \\ 0 & C_{Y_\beta} & 0 & 0 & 0 & 0 \end{bmatrix}^T \begin{matrix} 1 \\ 2 \\ 3 \\ 4 \\ 5 \\ 6 \\ 7 \\ 8 \\ 9 \\ 10 \\ 11 \\ 12 \\ 13 \\ 14 \\ 15 \\ 16 \\ 17 \\ 18 \\ 19 \end{matrix}$$

Table D.1 : Coefficients of the aerodynamic model of the 'Beaver'

Figure SSP9.1.1 AM Matrix within FDC [5]

The matrix allows the modeling of six aerodynamic coefficients as functions of the following nineteen variables:

$$\left(1, \alpha, \alpha^2, \alpha^3, \beta, \beta^2, \beta^3, \frac{pb}{2Vel}, \frac{q\bar{c}}{Vel}, \frac{rb}{2Vel}, \delta_E, \delta_A, \delta_R, \delta_{FLAPS}, \alpha\delta_{FLAPS}, \alpha\delta_R, \alpha\delta_A, \delta_E\beta^2, 0\right)$$

where the 1 allows the modeling of the aerodynamic biases $\left(c_{X_0}, c_{Y_0}, c_{Z_0}, c_{l_0}, c_{m_0}, c_{n_0}\right)$.

According to the sensitivity analysis outlined in Chapter 7, for a given flight condition and a given inertial configuration of the aircraft, the functionalities of the short period and phugoid modes with the specific aerodynamic coefficients are listed in Table SSP9.1.1.

Table SSP9.1.1 Functionalities between the Longitudinal Modes and the Aerodynamic Coefficients

Longitudinal Parameter	Aerodynamic Stability Derivative
ω_{SP}	$f\left(c_{m_\alpha}, c_{m_q}, c_{Z_\alpha}\right)$
ζ_{SP}	$f\left(c_{m_q}, c_{Z_\alpha}, c_{m_\alpha}, c_{m_{\dot\alpha}}\right)$
ω_{Ph}	$f\left(c_{X_\alpha}, c_{Z_\alpha}, c_{Z_1}\right)$
ζ_{Ph}	$f\left(c_{X_\alpha}, c_{Z_\alpha}, c_{Z_1}\right)$

Note that the functionality of $\left(c_{m_u}\right)$ with the phugoid is not here considered since the Beaver aircraft operates at very low subsonic airspeed.

Using the default values of the coefficients in the AM matrix as the baseline configuration, the values of the aerodynamic coefficients in Table SSP9.1.1 have been modified. The resulting analysis leads to the qualitative conclusions summarized in Table SSP9.1.2.

Table SSP9.1.2 Results of the Longitudinal Sensitivity Analysis for the Beaver Aircraft

Dynamic Parameter	$(\uparrow, \approx, \downarrow)$	Aerodynamic Stability Derivative	Absolute Value($\uparrow$)
$\omega_{n_{SP}}$	$\uparrow$	c_{m_α}	$\uparrow$
$\omega_{n_{SP}}$	$\approx$	c_{m_q}	$\uparrow$
$\omega_{n_{SP}}$	$\approx$	c_{Z_α}	$\uparrow$
ζ_{SP}	$\approx \uparrow$	c_{Z_α}	$\uparrow$
ζ_{SP}	$\uparrow$	$c_{m_{\dot\alpha}}$	$\uparrow$
ζ_{SP}	$\uparrow$	c_{m_q}	$\uparrow$
ζ_{SP}	$\downarrow$	c_{m_α}	$\uparrow$
$\omega_{n_{Ph}}$	$\approx$	c_{Z_α}	$\uparrow$
$\omega_{n_{Ph}}$	$\approx$	c_{X_α}	$\uparrow$
$\omega_{n_{Ph}}$	$\uparrow$	c_{Z_1}	$\uparrow$
ζ_{Ph}	$\approx$	c_{Z_α}	$\uparrow$
ζ_{Ph}	$\approx \uparrow$	c_{X_α}	$\uparrow$
ζ_{Ph}	$\downarrow$	c_{Z_1}	$\uparrow$

Student Sample Problem 9.2

Using the aerodynamic model of the default Beaver DC-2 aircraft provided with the FDC simulation environment, perform a qualitative sensitivity analysis for the following lateral-directional parameters:

$$\omega_{nDR}, \zeta_{DR} \quad \text{(DUTCH ROLL mode)}$$
$$T_R \quad \text{(ROLLING mode)}$$
$$T_S \quad \text{(SPIRAL mode)}$$

Solution of Student Sample Problem 9.2

Again, the aerodynamic modeling of the Beaver aircraft is contained in the (6×19) AM matrix described in Section 9.2.2 and shown in Figure SSP9.2.1.[5]

$$AM = \begin{bmatrix}
C_{X_0} & C_{Y_0} & C_{Z_0} & C_{l_0} & C_{m_0} & C_{n_0} \\
C_{X_\alpha} & 0 & C_{Z_\alpha} & 0 & C_{m_\alpha} & 0 \\
C_{X_{\alpha^2}} & 0 & 0 & 0 & C_{m_{\alpha^2}} & 0 \\
C_{X_{\alpha^3}} & 0 & C_{Z_{\alpha^3}} & 0 & 0 & 0 \\
0 & C_{Y_\beta} & 0 & C_{l_\beta} & 0 & C_{n_\beta} \\
0 & 0 & 0 & 0 & C_{m_{\beta^2}} & 0 \\
0 & 0 & 0 & 0 & 0 & C_{n_{\beta^3}} \\
0 & C_{Y_p} & 0 & C_{l_p} & 0 & C_{n_p} \\
C_{X_q} & 0 & C_{Z_p} & 0 & C_{m_q} & C_{n_q} \\
0 & C_{Y_r} & 0 & C_{l_r} & C_{m_r} & C_{n_r} \\
0 & 0 & C_{Z_{\delta_e}} & 0 & C_{m_{\delta_e}} & 0 \\
C_{X_{\delta_t}} & 0 & C_{Z_{\delta_t}} & 0 & C_{m_{\delta_t}} & 0 \\
0 & C_{Y_{\delta_a}} & 0 & C_{l_{\delta_a}} & 0 & C_{n_{\delta_a}} \\
C_{X_{\delta_r}} & C_{Y_{\delta_r}} & 0 & C_{l_{\delta_r}} & 0 & C_{n_{\delta_r}} \\
C_{X_{\alpha\delta_t}} & 0 & C_{Z_{\alpha\delta_t}} & 0 & 0 & 0 \\
0 & C_{Y_{\delta_{r\alpha}}} & 0 & 0 & 0 & 0 \\
0 & 0 & 0 & C_{l_{\delta_a\alpha}} & 0 & 0 \\
0 & 0 & C_{Z_{\delta_e\beta^2}} & 0 & 0 & 0 \\
0 & C_{Y_\beta} & 0 & 0 & 0 & 0
\end{bmatrix}^T \begin{matrix} 1 \\ 2 \\ 3 \\ 4 \\ 5 \\ 6 \\ 7 \\ 8 \\ 9 \\ 10 \\ 11 \\ 12 \\ 13 \\ 14 \\ 15 \\ 16 \\ 17 \\ 18 \\ 19 \end{matrix}$$

Table D.1 : Coefficients of the aerodynamic model of the 'Beaver'

Figure SSP9.2.1 AM Matrix within FDC[5]

According to the Sensitivity Analysis outlined in Chapter 7, for a given flight condition and a given inertial configuration of the aircraft, the functionalities of the dutch roll, spiral, and rolling modes with the specific aerodynamic coefficients are listed in Table SSP9.2.1.

Table SSP9.2.1 Functionalities between the Lat. Dir. Modes and the Aerodynamic Coefficients

Lateral Directional Parameter	Aerodynamic Stability Derivative
$\omega_{n_{DR}}$	$f\left(c_{l_\beta}, c_{n_\beta}\right)$
ζ_{DR}	$f\left(c_{n_\beta}, c_{l_\beta}, c_{n_r}, c_{Y_\beta}\right)$
T_{Spiral}	$f\left(c_{l_\beta}, c_{l_r}, c_{n_r}, c_{n_\beta}\right)$
T_{Roll}	$f\left(c_{l_p}, c_{l_\beta}\right)$

Using the default values of the coefficients in the AM matrix as the baseline configuration, the values of the aerodynamic coefficients in Table SSP9.2.1 have been modified. The resulting analysis leads to the qualitative conclusions summarized in Table SSP9.2.2.

Table SSP9.2.2 Results of the Lateral Directional Sensitivity Analysis for the Beaver Aircraft

Dynamic Parameter	$(\uparrow, \approx, \downarrow)$	Aerodynamic Stability Derivative	Absolute Value ($\uparrow$)
$\omega_{n_{DR}}$	$\approx$	c_{l_β}	$\uparrow$
$\omega_{n_{DR}}$	$\uparrow$	c_{n_β}	$\uparrow$
ζ_{DR}	$\downarrow$	c_{l_β}	$\uparrow$
ζ_{DR}	$\uparrow$	c_{n_β}	$\uparrow$
ζ_{DR}	$\uparrow$	c_{n_r}	$\uparrow$
ζ_{DR}	$\uparrow$	c_{Y_β}	$\uparrow$
T_{Spiral}	$\downarrow$	c_{l_β}	$\uparrow$
T_{Spiral}	No trend (stability issues)	c_{n_β}	$\uparrow$
T_{Spiral}	No trend (stability issues)	c_{n_r}	$\uparrow$
T_{Roll}	$\downarrow \approx$	c_{l_β}	$\uparrow$
T_{Roll}	$\downarrow$	c_{l_p}	$\uparrow$

# Chapter 10

Pilot Ratings and Aircraft Handling Qualities

TABLE OF CONTENTS

10.1 Introduction
10.2 Aircraft Flight Envelope
10.3 Levels of Aircraft Flying Qualities: Cooper-Harper Pilot Rating
    10.3.1 Aircraft Control Authority
    10.3.2 Pilot Workload
    10.3.3 Pilot Compensation
    10.3.4 Levels of Flying Qualities
10.4 Classes of Aircraft
10.5 Classification of Aircraft Maneuvers and Mission Profile
10.6 Flying Quality Requirements for the Longitudinal Dynamics
    10.6.1 Longitudinal Control Forces
    10.6.2 Requirements for the Damping for the Phugoid Mode
    10.6.3 Requirements for the Short Period Mode
10.7 Flying Quality Requirements for the Lateral Directional Dynamics
    10.7.1 Lateral Directional Control Forces
    10.7.2 Requirements for the Dutch Roll Mode
    10.7.3 Requirements for the Spiral Mode
    10.7.4 Requirements for the Rolling Mode
    10.7.5 Requirements for the Roll Control Effectiveness
    10.7.6 Additional Requirements for Steady Sideslips
10.8 Summary
    References

10.1 INTRODUCTION

The goal of this chapter is to provide a description of the requirements on the characteristics of the aircraft dynamics and on the physical effort required by the pilots for maneuvering different aircraft. These requirements are mandated by the competent regulating agencies; in the United States the regulating agencies are the Federal Aviation Administration (FAA) for civilian and commercial aircraft and the Department of Defense (DOD) for military aircraft.

Following a successful design and an initial flight testing, a new aircraft is required to undergo a flight testing phase leading to the airworthiness certification, where the competent agency verifies that every single aspect of the aircraft dynamics, for every single operating flight condition at every possible maneuver, falls within specified ranges of values for a workload of the pilot within specified bounds. While recent advances in flight simulation technology allow the validation of most of these requirements prior to the certification phase, ultimately flight testing provides the final verification leading to the release of the airworthiness certificates for the given aircraft.

Within this context, we will refer to the MIL-F-8785C[1] as the main source of requirements for the military aircraft, with some requirements also reported in the MIL-STD-1797A[2] and MIL-F-9490D.[3] Also, we will refer to the Federal Aviation Regulation 23 (FAR 23)[4] and the Joint Aviation Requirements for Very Light Airplanes (JAR-VLA)[5] as the source of requirements for civilian aircraft with a take-off weight less than 5,000 lbs, whereas the Federal Aviation Regulation 25 (FAR 25) provides the requirements for aircraft with a take-off weight more than 5,000 lbs.[4]

The chapter includes the following sections. An initial section describes the concept of aircraft flight envelope. Next, the concept of levels of flying qualities is introduced, followed by the introduction of the concepts of classes of aircraft and categories of maneuvers. Additionally, an important assessment tool, known as the pilot Cooper-Harper rating scale, is discussed. Next, the key aspects of the longitudinal flying quality requirements, in terms of both pilot physical workload and characteristics of the aircraft response, are introduced, and followed by a similar review of the lateral directional flying quality requirements. A final section concludes and summarizes the chapter.

10.2 AIRCRAFT FLIGHT ENVELOPE

Depending on its propulsion system and its dynamic and aerodynamic characteristics, each aircraft operates in a specific range of airspeed (in terms of Mach number) and altitude. All the possible airspeed and altitude conditions of an aircraft are visualized in the flight envelope.[6] Every aircraft performs the specific mission for which it has been designed within its flight envelope. As part of the flight testing program leading to issuing the airworthiness certificates, each new aircraft has to be designed to exhibit acceptable flying qualities and dynamic characteristics not only at its typical operating conditions but at every single point in its flight envelope. A later section will provide a list of all classes of aircraft. Without loss of generality, Figures 10.1, 10.2, and 10.3 provide an approximated flight envelope for three different types of aircraft (general aviation and regional transport, jetliner, and military fighter).

The difference in the size of the flight envelope is evident between different classes of aircraft. Even more dramatically evident is the difference between the approximated flight envelope of the fastest World War I aircraft (1915) and the USAF Lockheed SR-71 Black Bird aircraft (1965), shown in Figure 10.4.

The enormous difference between the two flight envelopes is by itself a tribute to the amazing progress by mankind in aviation in a very short amount of time (approximately only fifty years!). It should be pointed out that this rate of progress has not been matched in any other form of transportation.

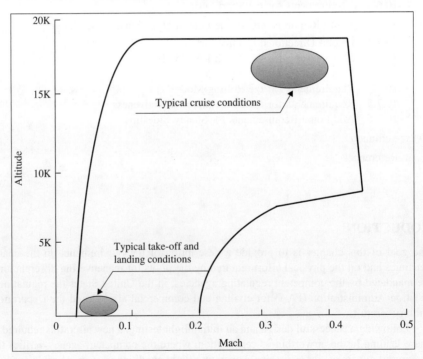

Figure 10.1 Typical Flight Envelope for a General Aviation and Regional Commuter Aircraft

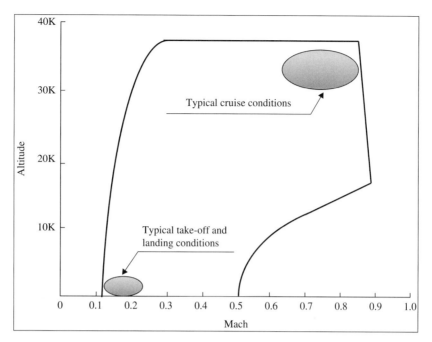

Figure 10.2 Typical Flight Envelope for a Commercial Jetliner

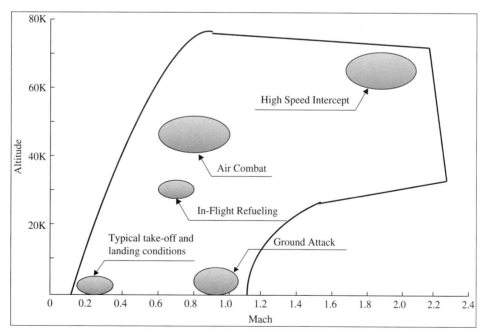

Figure 10.3 Typical Flight Envelope for a Fighter Aircraft

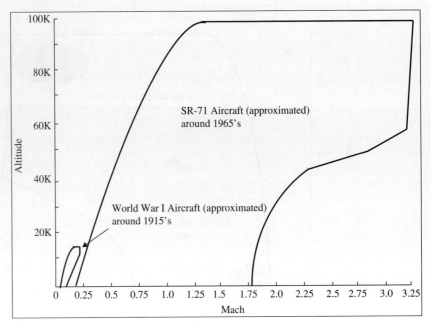

Figure 10.4 Flight Envelopes for World War I Fighter and SR-71 Aircraft

10.3 LEVELS OF AIRCRAFT FLYING QUALITIES: COOPER-HARPER PILOT RATING

Within this context, desirable flying qualities are defined as a combination of characteristics both in terms of piloting the aircraft as well as in terms of the aircraft response in itself. Specifically, this includes the analysis of three factors: control authority, pilot workload, and pilot compensation. These factors are individually discussed next, leading to a general discussion of the flying qualities.[6]

10.3.1 Aircraft Control Authority

Control authority is here defined as the capability by the pilot of generating appropriate aerodynamic and thrust forces and moments. For a successful certification, an aircraft must demonstrate adequate control authority for the following tasks:

- To maintain steady-state, rectilinear flight (for example: cruise conditions);
- To maintain steady-state maneuvering (for example: circular holding pattern);
- To take-off and to climb safely;
- To conduct an approach and to land safely.

These characteristics must be present with all the installed thrust performing nominally but also with a propulsive failure involving the loss of up to 50 percent of the installed thrust (as discussed in Chapter 6). The engine-out condition is also known in the literature as asymmetric power condition.

10.3.2 Pilot Workload

Within this context pilot workload is defined as the physical effort by the pilot while at the controls at the aircraft, that is, while he/she applies physical force(s) through his/her arms and his/her feet to the control commands in the cockpit of the aircraft. Clearly, the amount of pilot workload necessary for the different maneuvers within the flight envelope is an important parameter for the airworthiness of the aircraft. The following parameters related to the pilot workload are subjected to regulation:

- Maximum required stick force;
- Stick force per g;
- Stick force-speed gradient;
- Maximum aileron wheel force.

Particularly, specific ranges are provided for each of these parameters,

A discussion about the pilot workload naturally leads to a specific classification of the flight control systems commonly available on aircraft. Particularly, flight control systems are mainly divided in two classes, reversible and irreversible flight control systems. Samples of both systems are shown in Figure 10.5 and Figure 10.6.

A reversible flight control system is a system where the pilot feels the external dynamic pressure associated with a given flight condition. Therefore, through the physical hardware connecting the control commands and the aerodynamic control surfaces (either on the horizontal/vertical tail or on the wing), the pilot has a quantitative feedback of the external dynamic pressure; in turn, based on this feedback, the pilot would know about the physical force he/she needs to apply on the control commands to achieve the desired deflection of the specific aerodynamic control surfaces.

The top sketch in Figure 10.5 shows a conventional reversible flight control system (RFCS) typically implemented in general aviation or commuter aircraft featuring an all-mechanical linkage between the control commands in the aircraft cockpit and the aerodynamic control surfaces. The bottom sketch in the figure shows a similar system, with the only difference being the presence of a servo-tab mechanism deflecting in the opposite direction of the intended direction for the control surface. Essentially, the purpose of the tab mechanism is to decrease the workload required by the pilot.

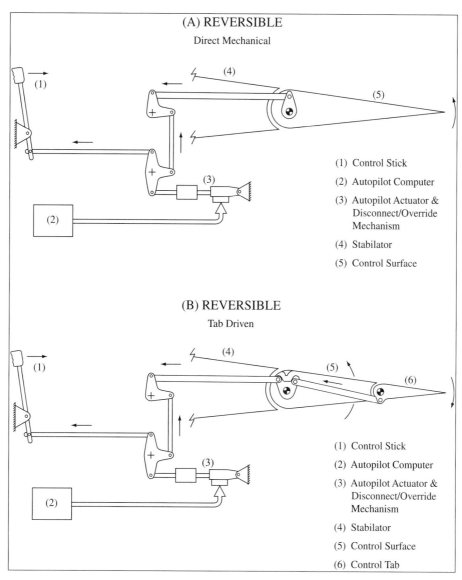

Figure 10.5 Samples of Reversible Flight Control Systems

528   Chapter 10   Pilot Ratings and Aircraft Handling Qualities

Conversely, an irreversible flight control system (IFCS) is a system where the pilot does not feel the external dynamic pressure due to the use of hydraulic boosted actuators for the deflection of the control surface; therefore, for all practical purposes the control commands act like a joy stick with very minimum physical effort by the pilot. This class of flight control systems is today used for high-performance military aircraft as well as on all modern jetliners. While the advantages of IFCS for the reduction of the pilot's workload are clear, a potential drawback of IFCS is a false sense of control authority by the pilot. In fact, due to a lack of feedback on the external dynamic pressure, the pilot can be lead to enter high g flight conditions, which could be unsafe for the aircraft structural integrity as well as his/her own physical tolerance to these conditions. Typically, this problem is solved through the introduction in the control commands of devices providing artificial resistant elastic forces proportional to the external dynamic pressure; this mechanism, known as artificial $\bar{q}$-feel compensation, practically acts as a reminder to the pilot of the external dynamic pressure and it has been effective in limiting the g excursions in maneuvered flight for high-performance military aircraft.

The top sketch of Figure 10.6 shows an IFCS with a mechanical linkage between the control command at the aircraft cockpit and the hydraulically boosted actuator at the control surface. The bottom sketch of the figure shows instead an IFCS with fly-by-wire technology where the mechanical linkages have been replaced by an electric cable and the control signal from the cockpit to the hydraulic actuator is an electric voltage.

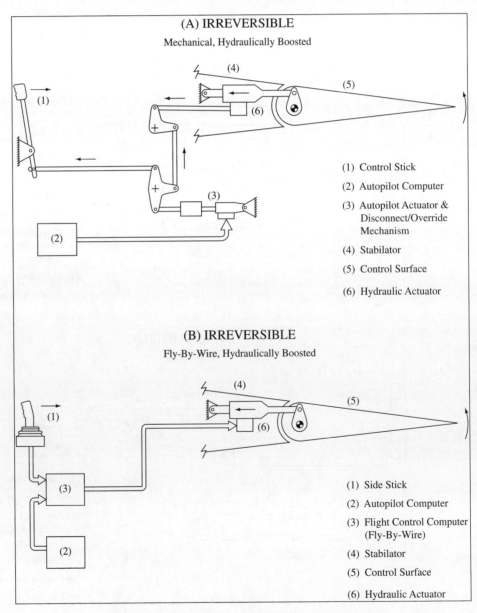

Figure 10.6 Samples of Irreversible Flight Control Systems

Fly-by-wire (FBW) technology was introduced in the 1970s along with digital flight computers. Virtually all high-performance military aircraft and commercial jetliners since the 1980s feature this technology. Today, the FBW technology has been transitioned to the fly-by-fiber (FBF) technology, where electric cables are being replaced by optic fibers.

It should be emphasized that within this context the handling qualities requirements in terms of pilot workload are provided for reversible flight control systems, where the pilot workload is proportional to the external dynamic pressure.

In addition to the difference between reversible and irreversible flight control systems, an important design issue to be considered is that several modern aircraft, both military and civilian, are designed to satisfy the requirements for handling qualities only through the use of different types of stability augmentation systems (SAS). Therefore, for all practical purposes, these aircraft exhibit an appropriate dynamic behavior only at closed-loop conditions; in fact, a large number of modern aircraft are actually dynamically unstable at open-loop conditions. It is clear that for these aircraft, failures to the components of the flight control systems (including computers, sensors, actuators, and so on) might have very important effects on the flying qualities, up to the point of compromising the overall dynamic stability. Therefore, it is critical to incorporate in the regulations a detailed analysis of all the failure scenarios leading to a relationship between acceptable flying qualities and the probability of the occurrence of certain types of failures. This has led historically to a quadruple physical redundancy of some of the key components of the flight control systems (including computers and sensors).

10.3.3 Pilot Compensation

An additional parameter is the pilot compensation, which is defined as the mental effort by the pilot while at the controls of the aircraft. An aircraft might require a substantial pilot compensation if its response following a pilot command is either too slow or too fast. In fact, if the aircraft reacts too fast or too slow to an input maneuver by the pilot, then the pilot is required to compensate through adjusting his/her own gain or by leading (or anticipating) the aircraft response. Clearly, pilot compensation is another important parameter affecting the overall flight qualities of an aircraft.

10.3.4 Levels of Flying Qualities

Considering the parameters defined above—control authority, pilot workload, pilot compensation—for any operating condition within the flight envelope and for every category of maneuvers there are three levels of flying qualities:

Level I: Satisfactory and desirable for the given maneuver at the given flight conditions.

Level II: Adequate and sufficient for the given maneuver at the given flight conditions but with an increase in the workload and compensation by the pilot, leading to a mild deterioration of the mission's effectiveness.

Level III: Safe control and guidance of the aircraft but with a large increase in the workload and compensation by the pilot leading to a substantial drop in the mission's effectiveness.

It should be emphasized that an aircraft for which the manufactures seeks the airworthiness certification must exhibit Level I flying qualities for all the nominal operating conditions. Level II and Level III are allowed only if resulting from failures to components of the flight control system, with the failure probability shown in Table 10.1.

Table 10.1 Flying Quality Levels vs. Probability of Failure Occurrence

Probability of Failure Occurrence Within Operational Flight Envelope	Military Aircraft	Civilian Equivalent
Level II (following failure)	$< 10^{-2}$ per flight	$< 10^{-4}$ per flight
Level III (following failure)	$< 10^{-4}$ per flight	$< 10^{-6}$ per flight

Furthermore, within each level of flying qualities there are three sublevels. Specifically, we have

- Sublevel #1, #2, and #3 for Level I;
- Sublevel #4, #5, and #6 for Level II;
- Sublevel #7, #8, and #9 for Level III.

An additional sublevel #10 indicates flying qualities leading to potentially catastrophic and unrecoverable conditions.

The aircraft flying qualities are assessed in the flight testing certification phase using a rating chart known as the Cooper-Harper pilot rating scale (CHPRS). CHPRS can be defined as a systematic diagram with specific decisions associated with the pilot thresholds for levels and sublevels of flying qualities. It should be emphasized that while CHPRS is a fixed and very detailed tool, the rating by two pilots relative to the same aircraft at the same flight conditions performing the same maneuver might be slightly different. The CHPRS is shown in Figure 10.7.[6]

The final goal of the regulating agencies is to ensure flying qualities associated with desirable dynamic characteristics in the aircraft response, along with a reasonable pilot workload, minimum pilot compensation, adequate mission effectiveness, and high flight safety for all the classes of aircraft and categories of maneuvers. The classification of the aircraft and the characterization of the different maneuvers of an aircraft will be described in the following sections.

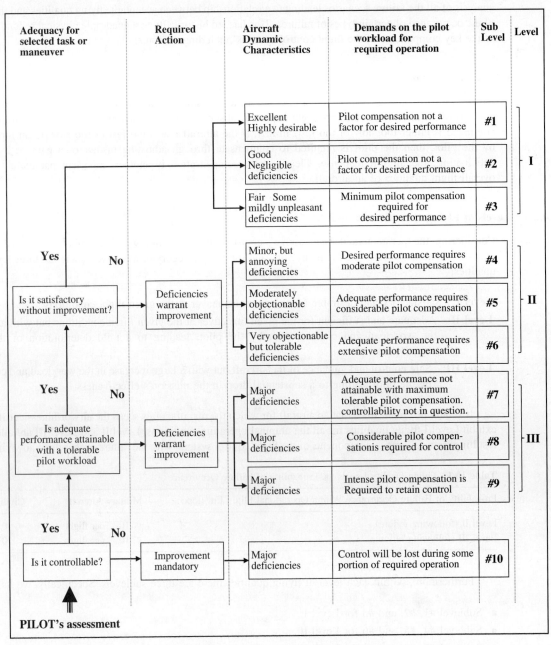

Figure 10.7 Cooper-Harper Pilot Rating Scale

10.4 CLASSES OF AIRCRAFT

Table 10.2 shows all the different classes of military aircraft, according to the MIL-F-8785C[1], along with examples of civilian equivalent aircraft.

Table 10.2 Classes of Military Aircraft along with Examples of Civilian Equivalent[1]

Classes of Military Aircraft (according to MIL-F-8785C)	Examples	Civilian Equivalent
Class I: Small & Light Airplanes.		
■ Light utility;	■ Cessna T-41;	■ Cessna 210;
■ Primary trainer;	■ Beech T-34C.	■ Piper Tomahawk.
■ Light observation and/or reconnaissance.		
Class II: Medium weight, low-to-medium maneuverability airplanes.		
■ Heavy utility/search and rescue;	■ Fairchild C-119;	■ Boeing 727;
■ Light or medium transport/cargo/tanker;	■ Grumman E-2C;	■ Boeing 737;
■ Early warning/electronic counter-measures/ airborne command, control or communications relay;	■ Boeing E-3A;	■ McDD DC-9;
	■ Lockheed S-3A;	■ McDD MD-80;
■ Anti-submarine;	■ Lockheed C-130;	■ Airbus A 320.
■ Assault transport;	■ Fairchild A-10;	
■ Reconnaissance;	■ Aeritalia G222;	
■ Tactical Bomber;	■ Douglas B-60;	
■ Heavy Attack;	■ Grumman A-6;	
■ Trainer for Class II aircraft.	■ Beech T-1A.	
Class III: Large, heavy, low-to-medium maneuverability airplanes.		
■ Heavy transport/cargo/tanker;	■ McDD C-17;	■ McDD MD-11;
■ Heavy bomber;	■ Boeing B-52H;	■ Boeing 747;
■ Patrol/early warning/electronic counter-measures/ airborne command, control or communications relay;	■ Lockheed P-3;	■ Boeing 777;
	■ Lockheed C-5;	■ Airbus 330;
■ Trainer for Class III.	■ Boeing E-3D;	■ Airbus 340;
	■ Boeing KC-135.	■ Airbus 380.
Class IV: High maneuverability airplanes.		
■ Fighter/interceptor;	■ Lockheed F-22;	■ Pitts Special.
■ Attack;	■ McDD F-4;	
■ Tactical reconnaissance;	■ McDD F-15;	
■ Observation;	■ Lockheed SR-71;	
■ Trainer for Class IV aircraft.	■ Northrop T-38.	

In the following sections, it will be clear how the requirements on the aircraft dynamic characteristics will depend on the specific class of the aircraft.

10.5 CLASSIFICATION OF AIRCRAFT MANEUVERS AND MISSION PROFILE

Table 10.3 shows all the different categories of maneuvers for military aircraft, according to the MIL-F-8785C.

A mission profile is defined as a typical sequence of maneuvers (from Categories A, B, and C) for a given aircraft of a given class.

For example, a mission profile for a Class IV fighter interceptor aircraft includes a take-off (Category C), followed by a climb and a cruise (Category B), followed eventually by air-to-air combat, weapon launch, in-flight refueling (Category A), leading to a descent (Category B), approach and landing (Category C).

Table 10.3 Categories of Aircraft Maneuvers[1]

"Non-Terminal" Maneuvers
Category A: Non-terminal maneuvers requiring rapid execution, precision tracking, and/or precise flight path control.
- Air-to-air combat;
- Ground attack;
- Weapon delivery/launch;
- Aerial recovery;
- Reconnaissance;
- In-flight refueling (as receiver aircraft);
- Terrain following;
- Anti-submarine search;
- Close formation flying.

Category B: Non-terminal maneuvers requiring slow and/or gradual maneuvers, moderately accurate flight path control without extreme precision tracking.
- Climb;
- Cruise;
- Loiter;
- In-flight refueling (as tanker);
- Descent;
- Emergency descent;
- Emergency deceleration;
- Aerial delivery.

"Terminal" Maneuvers
Category C: Terminal maneuvers requiring fast or gradual maneuvers and very accurate flight path control.
- Takeoff;
- Carrier and/or catapult takeoff;
- Approach;
- Wave-off/go-around;
- Landing.

Similarly, a mission profile for a Class III military cargo includes a take-off (Category C), followed by a climb, a cruise, and a descent (Category B), leading to an approach and landing (Category C).

10.6 FLYING QUALITY REQUIREMENTS FOR THE LONGITUDINAL DYNAMICS

The complete set of requirements for the longitudinal dynamics of the aircraft (considering all different classes of aircraft for all the category maneuvers, with reversible and irreversible flight control systems, and with or without stability augmentation systems) is very complex and voluminous. As discussed in Section 10.3, the requirements include appropriate ranges for the pilot workload, pilot compensation, as well as the ranges for the parameters of the aircraft response itself.

A detailed recollection of the entire regulations for flying qualities of the longitudinal dynamics is clearly beyond the scope of this book. Interested readers and professionals are referred to[1-5] for the complete list of requirements for both military and civilian aircraft.

The purpose of this section is to provide only a brief overview of some of the key regulations for the longitudinal dynamics. Specifically, the main requirements of the longitudinal control forces for reversible flight control systems will be reviewed; additionally, the allowed ranges of the damping and natural frequency coefficients for the two short period and phugoid longitudinal modes described in Chapter 7—ζ_{SP}, $\omega_{n_{SP}}$, ζ_{Ph}, $\omega_{n_{Ph}}$— will be presented. Additional important longitudinal flying qualities parameters such as the flight path stability and the control anticipation factor are discussed and reviewed in.[1-3,6]

10.6.1 Longitudinal Control Forces

The requirements for longitudinal control forces are here defined for aircraft with reversible flight control systems with stick or wheel control commands in the cockpit. Specifically, the longitudinal control forces are defined for the following conditions:

1. Control forces for steady-state flight condition
2. Control forces for maneuvered flight
3. Control forces for take-off and landing
4. Control forces during dives

Each of these conditions will be discussed separately.

10.6.1.1 Longitudinal Control Forces for Steady-State Flight

The requirements for the longitudinal control forces for steady-state flight include the following:

- Deployment of landing gears;
- Deployment of flaps;
- Deployment of speed brakes;
- Changing power settings;
- One engine-out condition;
- Two engines-out condition (for aircraft with four engines).

The pilot workload for transitioning to steady-state flight conditions to any of the above conditions must satisfy the requirements listed in Table 10.4 .

Table 10.4 Maximum Allowable Longitudinal Control Forces for Steady-State Flight

Force (lbs)	VLA	FAR 23	FAR 25	MIL-F-8785C
For **temporary** force application:				
■ Stick Controller;	45.0	60.0	No requirement	No requirement
■ Wheel Controller.	56.2	75.0	75.0	
For **prolonged** application:				
■ Stick and Wheel Controller.	4.5	10.0	10.0	No requirement

In addition to these requirements, in terms of changing the airspeed starting from a steady-state trim condition, a push/pull force must be applied to the stick or the wheel controller for increasing/decreasing the airspeed, respectively. Additionally, the longitudinal flying qualities requirements in[1-5] also demand a specific behavior for return-to-trim speed, expressed as a percentage of the trim airspeed.

10.6.1.2 Longitudinal Control Forces for Maneuvered Flight

The requirements for the longitudinal control forces for maneuvered flight are given in terms of a specific parameters, that is the $\partial F_S / \partial n$ (lbs/g), technically defined as the control force versus load factor gradient and also known as stick force per g.

In turn, the maximum values for the gradient $\partial F_S / \partial n$ (lbs/g) are provided either in terms of the maximum value of vertical acceleration n_{Limit} or in terms of the parameter n/α (g/rad), known as the g/α ratio. An approximate relationship for n/α is given by

$$n/\alpha \approx \frac{\overline{q}_1 c_{L_\alpha}}{W/S}$$

The maximum values for the gradient $\partial F_S/\partial n$ (lbs/g) for military and civilian aircraft are shown in Table 10.5 and Table 10.6. The differentiation between the military and civilian aircraft is quite clear, with the civilian requirements being substantially more stringent than the military requirements. It should also be pointed out that the requirements for civilian aircraft do not differentiate between different levels of flying qualities, whereas the requirement for military aircraft become progressively more tolerant from Level I to Level III.

Table 10.5 Longitudinal Control Force Limits in Maneuvering Flight (Civilian Aircraft)[4-6]

	Civilian Requirements	
VLA	FAR-23	FAR-25
$\partial F_S/\partial n > \dfrac{15.7}{n_{Limit}}$	For wheel controllers: $\partial F_S/\partial n > \dfrac{W_{TO}/140}{n_{Limit}}$ and $\dfrac{15}{n_{Limit}}$ but not more than $\dfrac{35}{n_{Limit}}$ For stick controllers: $\partial F_S/\partial n > W/140$	No requirement: The use MIL-F-8785C is recommended.

Table 10.6 Longitudinal Control Force Limits in Maneuvering Flight (Military Aircraft)[1-3,6]

Military Requirements MIL-F-8785C	Minimum allowable Gradient $\partial F_S/\partial n$ (lbs/g)	Maximum Allowable Gradient $\partial F_S/\partial n$ (lbs/g)
Stick Controller		
Level I	higher of: $\dfrac{21}{(n_{Limit}-1)}$ and 3.0	$[240/(n/\alpha)]$ but not more than 28.0 nor less than $\dfrac{56}{(n_{Limit}-1)}$*
Level II	higher of: $\dfrac{18}{(n_{Limit}-1)}$ and 3.0	$[360/(n/\alpha)]$ but not more than 42.5 nor less than $\dfrac{85}{(n_{Limit}-1)}$*
Level III	higher of: $\dfrac{12}{(n_{Limit}-1)}$ and 2.0	56
*For $n_{Limit} < 3.0$, $\partial F_S/\partial n = 28$ for Level I, and $\partial F_S/\partial n = 42.5$ for Level II		
Wheel Controller		
Level I	higher of: $\dfrac{35}{(n_{Limit}-1)}$ and 6.0	$[500/(n/\alpha)]$ but not more than 120.0 nor less than $\dfrac{120}{(n_{Limit}-1)}$*
Level II	higher of: $\dfrac{30}{(n_{Limit}-1)}$ and 6.0	$[775/(n/\alpha)]$ but not more than 182.0 nor less than $\dfrac{182}{(n_{Limit}-1)}$*
Level III	5.0	240.0

10.6.1.3 Longitudinal Control Forces for Take-Off and Landing

The requirements for the longitudinal control forces for take-off and landing are conceptually different for civilian and military aircraft. In fact, for civilian aircraft the allowed control forces for take-off and landing are considered the same as the control forces for changing configuration starting from a steady-state flight condition. Therefore, the ranges are the same as those provide in Table 10.4 (Section 10.6.1.1).

For military aircraft the maximum allowed forces for take-off and landing are dependent on the class of aircraft and are shown in Table 10.7, where L and C indicate land-based and carrier-based aircraft, respectively.

10.6 Flying Quality Requirements for the Longitudinal Dynamics

Table 10.7 Maximum Allowable Longitudinal Control Forces in Take-off and Landing (Military Aircraft)[1-3,6]

MIL-F-8785C Airplane Class	Takeoff Pull (lbs)	Takeoff Push (lbs)	Airplane Class	Landing Pull only
Nose-wheel and bicycle-gear airplanes			All gear configurations	
Classes I, IV-C	20.0	10.0	Classes I, II-C	35.0
Classes II-C, IV-L	30.0	10.0	Classes II-L	50.0
Classes II-L, III	50.0	20.0		
Tail-wheel airplanes				
Classes I, II-C, IV	20.0	10.0		
Classes II-L, III	35.0	15.0		

10.6.1.4 Longitudinal Control Forces for Dives

There are no requirements for the longitudinal control forces for dives for civilian aircraft. For military aircraft the maximum allowed longitudinal control forces for dives are shown in Table 10.8.

Table 10.8 Maximum Allowable Longitudinal Control Forces in Dives (Military Aircraft)[1-3,6]

Class of Aircraft	Dives Pull (lbs)	Push (lbs)
Stick controller. Class I, II, III, and IV (C and L)	10.0	50.0
Wheel controller. Class I, II, III, and IV (C and L)	75.0	10.0

10.6.2 Requirements for the Damping for the Phugoid Mode

The characteristics of the phugoid dynamic mode have been extensively described in Chapter 7. In terms of regulating the damping of the phugoid mode, there is a substantial difference between the military[1-3,6] and civilian requirements,[4-6] as shown in Table 10.9.

Table 10.9 Requirements for the Phugoid Damping[6]

MIL-F-8785C	VAL, FAR 23 and FAR 25
Level I $\zeta_{Ph} \geq 0.04$	Level I No requirement
Level II $\zeta_{Ph} \geq 0$	Level II No requirement
Level III $T_{2_{Ph}} \geq 55$ sec.	Level III No requirement

For Level III within the military specifications the parameter $T_{2_{Ph}}$ implies that the oscillations of the phugoid mode are allowed to double their amplitudes as long as the required time is greater than 55 sec. The parameter $T_{2_{Ph}}$ is therefore defined as time-to-double-phugoid-magnitude and can be found using the following equation:

$$M_{Ph} e^{-\zeta_{Ph} \omega_{n_{Ph}} (t_1 + T_{2_{Ph}})} = 2 M_{Ph} e^{-\zeta_{Ph} \omega_{n_{Ph}} t_1}$$

where M_{Ph} indicates the magnitude of the phugoid oscillation. Using the anti-log of the preceding relationship we will have $T_{2_{Ph}} = \dfrac{\ln(2)}{-\zeta_{Ph} \omega_{n_{Ph}}}$.

10.6.3 Requirements for the Short Period Mode

The requirements for the short period mode are provided in terms of both natural frequency $\omega_{n_{SP}}$ and damping ζ_{SP}. As shown in the sensitivity analysis in Chapter 7 the natural frequency $\omega_{n_{SP}}$ is directly related to the aircraft stability through the relationship:

$$\omega_{nSP} = \sqrt{\left(\frac{Z_\alpha M_q}{V_{P_1}} - M_\alpha\right)} \approx \sqrt{-M_\alpha} = \sqrt{-\left(\frac{c_{m_\alpha}\overline{q}S\overline{c}}{I_{YY}}\right)}$$

Thus, at a given flight condition with given inertial properties, $\omega_{n_{SP}}$ is directly a function of the aerodynamic coefficient c_{m_α}; since $c_{m_\alpha} = c_{L_\alpha}(\overline{x}_{CG} - \overline{x}_{AC}) = c_{L_\alpha} \cdot SM$, this ultimately implies that the value of $\omega_{n_{SP}}$ depends on the aircraft static margin SM. The requirements on the value of $\omega_{n_{SP}}$ for different classes of aircraft (I, II, III, and IV, C and L) at different levels of flying qualities are quite complicated; furthermore, they are also dependent on the specific categories of maneuvers (A, B, and C). Refer to[6] for a review of this specific requirement.

The requirements on the value of ζ_{SP} are more immediate; for military aircraft they are summarized in Table 10.10.

Table 10.10 Requirements for the Short Period Damping (Military Aircraft)[1–3,6]

Category A & C Maneuvers	Category B Maneuvers
Level I $\zeta_{SP} > 0.35$	Level I $\zeta_{SP} > 0.35$
Level II $\zeta_{SP} > 0.25$	Level II $\zeta_{SP} > 0.2$
Level III $\zeta_{SP} > 0.15$	Level III $\zeta_{SP} > 0.15$

Similar but somewhat more stringent requirements (higher values for ζ_{SP}) are defined for civilian aircraft, as described in the FAR 23, FAR 25, and VLA.[4–6]

10.7 FLYING QUALITY REQUIREMENTS FOR THE LATERAL DIRECTIONAL DYNAMICS

The complete set of requirements for the aircraft lateral directional dynamics (considering all different classes of aircraft for all the category maneuvers, reversible and irreversible flight control systems, and with or without stability augmentation systems) are even more complex and voluminous than the longitudinal counterpart. This is expected, since a discussion on the lateral directional dynamics requires at least two aircraft control surfaces, rudder (typically deflected through control commands exerted on the pedals) and ailerons (deflected using either a stick or a wheel). Additionally, the lateral directional dynamics involves three dynamic modes (dutch roll, spiral, and rolling) in lieu of two (phugoid and short period). Finally, the lateral directional requirements include appropriate ranges for the pilot workload, pilot compensation, as well as the ranges for the parameters of the aircraft response itself.

A detailed description of the entire regulations for flying qualities of the lateral directional dynamics is beyond the scope of this textbook. Interested readers and professionals are referred to[1–5] for the complete list of requirements for both military and civilian aircraft.

The purpose of this section is to provide a brief overview of some of the regulations for the lateral directional dynamics for reversible flight control systems. First, the requirements on the control forces for the lateral directional control surfaces will be briefly reviewed, followed by a review on the requirements for the parameters of the dutch roll, spiral, and rolling dynamic modes, along with parameters related to the effectiveness of the aircraft in terms of maneuvering.

10.7.1 Lateral Directional Control Forces

The requirements for lateral directional control forces are here defined for aircraft with reversible flight control systems with stick or wheel control commands in the cockpit. Specifically, the allowed ranges of lateral directional control forces are defined for the following conditions:

1. Control forces for rolling maneuvers
2. Control forces for holding a specific direction (heading) with asymmetric loading configurations
3. Control forces for holding a specific direction (heading) and bank angle with an engine-out condition

Each of these conditions will be discussed separately.

10.7.1.1 Control Forces for Rolling Maneuvers

For military aircraft, according to the MIL-F-8785C, the maximum control forces allowed for the aircraft rolling are shown in Table 10.11.

Table 10.11 Maximum Forces for Roll Control for Military Aircraft[1-3,6]

Level	Airplane Class	Category of Maneuvers	Maximum Allowable Stick Force (lbs)	Maximum Allowable Wheel Force (lbs)
Level I	I, II-C, IV	A, B	20.0	40.0
		C	20.0	20.0
	II-L, III	A, B	25.0	50.0
		C	25.0	25.0
Level II	I, II-C, IV	A, B	30.0	60.0
		C	20.0	20.0
	II-L, III	A, B	30.0	60.0
		C	30.0	30.0
Level III	All	All	35.0	70.0

A similar table does not exist for civilian aircraft. The only criteria used for civilian aircraft is the classification of the roll control forces between temporary and prolonged applications, as shown in Table 10.12. Note that the table applies to both roll and directional control (since it specifies the maximum forces at the rim of the wheel for the ailerons and at the pedals for the rudder).

Table 10.12 Maximum Forces for Roll and Directional Control for Civilian Aircraft[4-6]

Force (lbs)	VLA	FAR 23	FAR 25
For TEMPORARY force application:			
■ Rudder Pedal;	90.0	150.0	150.0
■ Wheel Controller.	45.0	60.0	60.0
For PROLONGED force application:			
■ Rudder Pedal;	22.5	20.0	20.0
■ Wheel Controller.	3.5	5.0	5.0

10.7.1.2 Control Forces for Holding Heading with Asymmetric Loading

For military aircraft, according to the MIL-F-8785C, the maximum control forces allowed on the pedals for the directional control of an aircraft with asymmetric loading are by

- < 100 lbs. for Level I and Level II flying qualities;
- < 180 lbs. for Level III flying qualities.

In the absence of an equivalent regulation for civilian aircraft, the maximum forces for directional control of a civilian aircraft with asymmetric loading are given in Table 10.12.

10.7.1.3 Control Forces for Holding Heading and Bank Angle with an Engine-Out

For civilian aircraft the maximum allowed forces for the rolling and directional control of an aircraft are shown in Table 10.12. In the absence of aircraft specific requirements for military aircraft, the only requirement is that the force to be exerted on the pedals for engine-out at take-off conditions should be < 180 lbs, which is consistent with the condition in Section 10.7.1.2.

10.7.2 Requirements for the Dutch Roll Mode

For civilian aircraft, the requirements for the Dutch roll mode are fairly unrestrictive, as shown in Table 10.13. Note there are no requirements on the Dutch roll natural frequency $\omega_{n_{DR}}$.

Table 10.13 Dutch Roll Requirements (Civilian Aircraft)[4–6]

FAR 23, VLA	$\zeta_{DR} > 0.052$
FAR 25	$\zeta_{DR} > 0$

For military aircraft, according to the MIL-F-8785C, the requirements for the Dutch roll mode in terms of ζ_{DR} and $\omega_{n_{DR}}$ are shown in Table 10.14.

Table 10.14 Dutch Roll Damping and Natural Frequency Requirements (Military Aircraft)[1–3,6]

Level	Category of Maneuvers	Class of Aircraft	Minimum ζ_{DR}*	Minimum $\zeta_{DR}\,\omega_{n_{DR}}$*	Minimum $\omega_{n_{DR}}$*
Level I	A (Combat and Ground Attack)	IV	0.4	—	1.0
	A (All Others)	I, IV	0.19	0.35	1.0
		II, III	0.19	0.35	0.4
	B	All	0.08	0.15	0.4
	C	I, II-C, IV	0.08	0.15	1.0
		II-L, III	0.08	0.10	0.4
Level II	All	All	0.05	0.05	0.4
Level III	All	All	0	—	0.4

*In case of conflicting regulations, the ruling requirement is the one providing the largest value of ζ_{DR}.

Particularly interesting in this table are the requirements for Level I, Class IV aircraft when performing combat and ground attack maneuvers (belonging to Category A); in fact, in this specific case, a fairly large value of ζ_{DR} for the need of the aircraft to be a stable platform for precision pointing in duel combat missions or for the purpose of precise delivery of weapons.

A special distinction is given by Class III aircraft (large, heavy, low-to-medium maneuverability aircraft), which could be exempted from the requirements and eventually follow the civilian equivalent requirements.

An additional requirement is for high maneuverability military aircraft. For these aircraft, a specific parameter of interest is given by $|\Phi/\beta|$, that is, the magnitude of the bank angle / sideslip angle ratio. If the following condition is satisfied

$$\left|\frac{\Phi}{\beta}\right| > \frac{20}{\omega_{n_{DR}}^2}$$

then the additional requirements shown in Table 10.15 apply.

Table 10.15 Additional Dutch Roll Requirements (Military Aircraft)[1–3,6]

Level I	$\Delta(\zeta_{DR}\,\omega_{n_{DR}}) = 0.014\left[(\omega_{n_{DR}}^2)\left|\frac{\Phi}{\beta}\right| - 20\right]$
Level II	$\Delta(\zeta_{DR}\,\omega_{n_{DR}}) = 0.009\left[(\omega_{n_{DR}}^2)\left|\frac{\Phi}{\beta}\right| - 20\right]$
Level III	$\Delta(\zeta_{DR}\,\omega_{n_{DR}}) = 0.005\left[(\omega_{n_{DR}}^2)\left|\frac{\Phi}{\beta}\right| - 20\right]$

10.7.3 Requirements for the Spiral Mode

Since the spiral mode is easily controlled by the pilot (or by the autopilot system), there are no requirements for the spiral mode for civilian aircraft. The requirements for military aircraft for the spiral mode are shown in Table 10.16.

Table 10.16 Requirements for the Spiral Mode (Military Aircraft)[1-3,6]

Category of Maneuvers	Level I	Level II	Level III
A , C	$T_{2_S} > 12$ sec	$T_{2_S} > 8$ sec	$T_{2_S} > 4$ sec
B	$T_{2_S} > 20$ sec	$T_{2_S} > 12$ sec	$T_{2_S} > 12$ sec

The parameter T_{2_S} is defined as the time required for doubling the amplitude of the spiral. Although the spiral is characterized by a first order behavior, the parameter T_{2_S} is conceptually similar to the previously introduced parameter $T_{2_{Ph}}$ for the phugoid mode.

10.7.4 Requirements for the Rolling Mode

The requirements for the rolling mode are very important for fighter aircraft. It should be emphasized that fast rolling capability is a key asset for a pilot engaged in an air combat. Therefore, the rolling mode requirements for Class IV aircraft are fairly demanding. The requirements for military aircraft for the rolling mode are shown in Table 10.17 below.

Table 10.17 Requirements for the Rolling Mode (Military Aircraft)[1-3,6]

Category of Maneuvers	Class of Aircraft	Level I	Level II	Level III
A	I, IV	$T_R < 1$ sec	$T_R < 1.4$ sec	$T_R < 10$ sec *
	II, III	$T_R < 1.4$ sec	$T_R < 3$ sec	—
B	ALL	$T_R < 1.4$ sec	$T_R < 3$ sec	$T_R < 10$ sec
C	I, II-C, IV	$T_R < 1$ sec	$T_R < 1.4$ sec	$T_R < 10$ sec *
	II-L, III	$T_R < 1.4$ sec	$T_R < 3$ sec	—
	* for Class IV only			

Recall that the parameter T_R is defined as the *time required for an aircraft to reach 63% of its steady state roll rate* p_{SS}. There are no specific requirements for civilian aircraft; however, it is customary for the design of a civilian aircraft for a given class to follow the military requirements in the preceding table.

10.7.5 Requirements for the Roll Control Effectiveness

Along with the previous T_R, another important lateral handling quality parameter is the roll control effectivess (RCE). Although the value of this parameter is strongly related to T_R, RCE is conceptually different. The following "floating" definition is introduced: RCE_x *is defined as the time required for the pilot to go—through a full deflection at the cockpit—from a wing level condition* $(\phi = 0°)$ *to* $(\phi = x°)$.

The requirements for military aircraft for the RCE are shown in Table 10.18 for Class I and II, Table 10.19 for Class III, and Table 10.20 for Class IV.

Table 10.18 Requirements for the Roll Control Effectiveness – Class I and II (Military Aircraft)[1]

Class of Aircraft	Level	Category of Maneuvers					
		A		B		C	
		$RCE_{60°}$	$RCE_{45°}$	$RCE_{60°}$	$RCE_{45°}$	$RCE_{60°}$	$RCE_{45°}$
I	I	1.3 sec	–	1.7 sec	–	1.3 sec	–
I	II	1.7 sec	–	2.5 sec	–	1.8 sec	–
I	III	2.6 sec	–	3.4 sec	–	2.6 sec	–
II-L	I	–	1.4 sec	–	1.9 sec	1.8 sec	–
II-L	II	–	1.9 sec	–	2.8 sec	2.5 sec	–
II-L	III	–	2.8 sec	–	3.8 sec	3.6 sec	–
II-C	I	–	1.4 sec	–	1.9 sec	–	1.0 sec
II-C	II	–	1.9 sec	–	2.8 sec	–	1.5 sec
II-C	III	–	2.8 sec	–	3.8 sec	–	2.0 sec

Table 10.19 Requirements for the Roll Control Effectiveness – Class III (Military Aircraft)[1]

Class III		Category of Maneuvers		
		A	B	C
Level	Range of Speed	$RCE_{30°}$	$RCE_{30°}$	$RCE_{30°}$
I	Low	1.8 sec	2.3 sec	2.5 sec
	Medium	1.5 sec	2.0 sec	2.5 sec
	High	2.0 sec	2.3 sec	2.5 sec
II	Low	2.4 sec	3.9 sec	4.0 sec
	Medium	2.0 sec	3.3 sec	4.0 sec
	High	2.5 sec	3.9 sec	4.0 sec
III	All	3.0 sec	5.0 sec	6.0 sec

Table 10.20 Requirements for the Roll Control Effectiveness – Class IV (Military Aircraft)[1]

Class IV		Category of Maneuvers				
		A		B		C
Level	Range of Speed	$RCE_{30°}$	$RCE_{50°}$	$RCE_{90°}$	$RCE_{90°}$	$RCE_{30°}$
I	Very Low	1.1 sec	–	–	2.0 sec	1.1 sec
	Low	1.1 sec	–	–	1.7 sec	1.1 sec
	Medium	–	–	1.3 sec	1.7 sec	1.1 sec
	High	–	1.1 sec	–	1.7 sec	1.1 sec
II	Very Low	1.6 sec	–	–	2.8 sec	1.3 sec
	Low	1.5 sec	–	–	2.5 sec	1.3 sec
	Medium	–	–	1.7 sec	2.5 sec	1.3 sec
	High	–	1.3 sec	–	2.5 sec	1.3 sec
III	Very Low	2.6 sec	–	–	3.7 sec	2.0 sec
	Low	2.0 sec	–	–	3.4 sec	2.0 sec
	Medium	–	–	2.6 sec	3.4 sec	2.0 sec
	High	–	2.6 sec	–	3.4 sec	2.0 sec

where the ranges of speed are defined as in the following:

- *Very low speed* refers to airspeeds around the aircraft stall speed;
- *Low speed* refers to airspeeds around the aircraft take-off and approach speed;
- *Medium speed* refers to airspeeds around 70 percent of the maximum aircraft speed;
- *High speed* refers to airspeeds above 70 percent of the maximum aircraft speed.

The requirements for the RCE for civilian aircraft are in general less demanding; they are shown in Table 10.21. Note the introduction of an additional version of the RCE parameter, known as $RCE_{-30°↔30°}$ and defined as the *time required for the pilot to go—through a full deflection at the cockpit—from $\phi = -30°$ to $\phi = 30°$*

Table 10.21 Requirements for the Roll Control Effectiveness (Civilian Aircraft)[4-6]

Flight Condition	Speed	Weight (lbs.)	VLA $RCE_{-30°↔30°}$	FAR 23 $RCE_{-30°↔30°}$	FAR 25 $RCE_{-30°↔30°}$
Take-off	1.2 V_{Stall}	W ≤ 6,000 lbs.	5 sec.	5 sec.	No requirement
		W > 6,000 lbs.	Not applicable	$t = \dfrac{W + 500}{1,300}$	No requirement
Landing	1.3 $V_{Stall_{PA}}$	W ≤ 6,000 lbs.	4 sec.	4 sec.	No requirement
		W > 6,000 lbs.	Not applicable	$t = \dfrac{W + 2,800}{2,200}$	No requirement

10.7.6 Additional Requirements for Steady Sideslips

In addition to the preceding requirements, there are a number of trivial requirements for both civilian and military aircraft at all flight conditions for all levels of flying qualities and for all categories of maneuvers. These requirements are listed next.

Additional Requirement #1: "A right rudder pedal force must produce a negative sideslip angle; vice versa, a left rudder pedal force must produce a positive sideslip angle." This requirement is trivially satisfied if the static stability criteria $c_{n_\beta} > 0$ is satisfied.

Additional Requirement #2: "During a left rudder maneuver an increase in right (positive) wing bank angle must be present with an increase in right (positive) sideslip angle; vice versa, during a right rudder maneuver an increase in left (negative) wing bank angle must be present with an increase in left (negative) sideslip angle." This requirement is trivially satisfied if the static stability criteria $c_{Y_\beta} < 0$ is satisfied.

Additional Requirement #3: "During a rudder maneuver a left (negative) rolling control deflection must be present with a left (negative) sideslip angle; vice versa, a right (positive) rolling control deflection must be present with a right (positive) sideslip angle." This requirement is trivially satisfied if the static stability criteria $c_{l_\beta} < 0$ is satisfied.

10.8 SUMMARY

The chapter has introduced the key aspects for the analysis of the aircraft handling qualities. First, the concepts of aircraft flight envelope, levels of handling qualities, classes of aircraft, and category of maneuvers have been introduced, along with a description of the Cooper-Harper pilot rating scale. Next, the key longitudinal flying quality requirements have been introduced and briefly discussed, followed by a similar discussion of the key lateral directional flying quality requirements; whenever possible, the differences between the requirements for military and civilian aircraft have been highlighted. A complete and detailed review of all flying and handling quality requirements (for both civilian and military aircraft) is clearly beyond the scope of a textbook; therefore, interested readers are referred to[1-5] for an exhaustive analysis of the regulations necessary for the release of the airworthiness certification for new aircraft.

REFERENCES

1. MIL-F-8785C, "Military Specification Flying Qualities of Piloted Airplanes." U.S. Air Force Flight Dynamic Laboratory, Wright Patterson Air Force Base (WPAFB), Dayton, OH, 1980.
2. MIL-STD-1797A, "Flying Qualities of Piloted Aircraft." U.S. Air Force Flight Dynamic Laboratory, Wright Patterson Air Force Base (WPAFB), Dayton, OH, 1990.

3. MIL-F-9490D, "Military Specification Flight Control Systems—General Specification for Installation and Test of Piloted Aircraft." Air Force Flight Dynamic Laboratory, Wright Patterson Air Force Base (WPAFB), Dayton, OH, 1975.
4. Code of Federal Regulations (CFR), Title 14, Part 1 to Part 59 (including FAR 25 and FAR 35). U.S. Government Printing Office, Superintendent of Documents, Mail Stop SSOP, Washington, DC, 1992.
5. Joint Aviation Requirements for Very Light Airplanes (JAR-VLA). Civil Aviation Authority, Printing and Publication Services, Cheltenham, United Kingdom, 1992.
6. Roskam, J. *Airplane Flight Dynamics and Automatic Flight Controls—Part I*. Design, Analysis, and Research Corporation, Lawrence, KS, 1995.

Appendix A

Review of Useful Topics

TABLE OF CONTENTS

- A.1 Review of Vector Operations
- A.2 Review of Matrix Operations
- A.3 Review of Center of Gravity and Inertial Properties
- A.4 Review of Application of Laplace Transform to Linear Constant Coefficients Differential Equations
- A.5 Review of First and Second Order Systems
- A.6 Review of Standard Atmospheric Model

544   Appendix A   Review of Useful Topics

A.1 REVIEW OF VECTOR OPERATIONS

Vectors are defined in the n-th space (R^n), also known as the Euclidean space. For our purposes we will consider vectors defined in the R^2 or R^3, that is 2D or 3D vectors:

2D vector: $\overline{A} = A_x \overline{i} + A_y \overline{j}$
3D vector: $\overline{A} = A_x \overline{i} + A_y \overline{j} + A_z \overline{k}$

Consider the following generic scalars k_1, k_2 and the following generic vectors:

$$\overline{A} = A_x\overline{i} + A_y\overline{j} + A_z\overline{k}, \ \overline{B} = B_x\overline{i} + B_y\overline{j} + B_z\overline{k}, \ \overline{C} = C_x\overline{i} + C_y\overline{j} + C_z\overline{k}$$

The following vectorial operations are defined for the **sum/difference of vectors** and/or **multiplication of vectors with scalars**:

$$\overline{A} + \overline{B} = \overline{B} + \overline{A}$$

$$\overline{A} + (\overline{B} + \overline{C}) = (\overline{A} + \overline{B}) + \overline{C}$$

$$\overline{A} + \overline{0} = \overline{A} \quad \text{where} \quad \overline{0} = 0\overline{i} + 0\overline{j} + 0\overline{k}$$

$$\overline{A} + (-\overline{A}) = \overline{0}$$

$$k_1(k_2\overline{A}) = (k_1 k_2)\overline{A}$$

$$k_1(\overline{A} + \overline{B}) = k_1\overline{A} + k_1\overline{B}$$

$$(k_1 + k_2)\overline{A} = k_1\overline{A} + k_2\overline{A}$$

Next, the **scalar product between vectors** is defined as:

$$\overline{A} \cdot \overline{B} = (A_x\overline{i} + A_y\overline{j} + A_z\overline{k}) \cdot (B_x\overline{i} + B_y\overline{j} + B_z\overline{k}) = A_xB_x + A_yB_y + A_zB_z = scalar$$

with the following relationships between the unity vectors $\overline{i}, \overline{j}, \overline{k}$:

$$\overline{i} \cdot \overline{i} = 1, \ \overline{j} \cdot \overline{j} = 1, \ \overline{k} \cdot \overline{k} = 1 \quad \overline{i} \cdot \overline{j} = \overline{i} \cdot \overline{k} = \overline{j} \cdot \overline{k} = 0$$

The vectorial scalar product, also known as Euclidean product, has the following properties:

$$\overline{A} \cdot \overline{B} = \overline{B} \cdot \overline{A}$$

$$(\overline{A} + \overline{B}) \cdot \overline{C} = \overline{A} \cdot \overline{C} + \overline{B} \cdot \overline{C}$$

$$(k_1\overline{A}) \cdot \overline{B} = k_1(\overline{A} \cdot \overline{B})$$

$$\overline{A} \cdot \overline{A} > 0 \quad \overline{A} \cdot \overline{A} = 0 \text{ only when } \overline{A} = \overline{0}$$

Example A.1.1

Given: $\overline{A} = 5\overline{i} - 3\overline{j} + 7\overline{k}, \ \overline{B} = 2\overline{i} - 4\overline{j} + 3\overline{k}$

$$\overline{A} \cdot \overline{B} = 10 + 12 + 21 = 43$$

The **magnitude of a vector** $\overline{A}$ (also known as the norm) is defined as:

$$\|\overline{A}\| = (\overline{A} \cdot \overline{A})^{1/2}$$

Example A.1.2

Given:

$$\overline{A} = 5\overline{i} - 3\overline{j} + 7\overline{k}, \quad \|\overline{A}\| = \left(\overline{A} \cdot \overline{A}\right)^{1/2} = \left(5^2 + 3^2 + 7^2\right)^{1/2} = (83)^{1/2} \approx 9.11$$

Next, the **cross product between vectors** is defined as:

$$\overline{A} \times \overline{B} = \begin{vmatrix} \overline{i} & \overline{j} & \overline{k} \\ A_x & A_y & A_z \\ B_x & B_y & B_z \end{vmatrix} = \overline{i}(A_y B_z - A_z B_y) + \overline{j}(A_z B_x - A_x B_z) + \overline{k}(A_x B_y - A_y B_x)$$

Thus, a vectorial cross product is essentially the determinant of a matrix. Therefore, the following relationships from linear algebra apply:

$$\overline{A} \times \overline{A} = \overline{0}$$

If $\overline{B} = k_1 \overline{A}$ then $\overline{A} \times \overline{B} = \overline{0}$

Example A.1.3

Given: $\overline{A} = 3\overline{i} - 5\overline{j} + 4\overline{k}, \overline{B} = 2\overline{i} + 4\overline{j} - 5\overline{k}$

$$\overline{A} \times \overline{B} = \begin{vmatrix} \overline{i} & \overline{j} & \overline{k} \\ 3 & -5 & 4 \\ 2 & 4 & -5 \end{vmatrix} = \overline{i}(25 - 16) + \overline{j}(8 + 15) + \overline{k}(12 + 10) = 9\overline{i} + 23\overline{j} + 22\overline{k}$$

Note that $\overline{A} \times scalar$ is not defined.
The vectorial cross product has also the property that $\overline{A} \times \overline{B} = -(\overline{B} \times \overline{A})$

Example A.1.4

Given: $\overline{A} = 3\overline{i} - 5\overline{j} + 4\overline{k}, \overline{B} = 2\overline{i} + 4\overline{j} - 5\overline{k}$

$$\overline{B} \times \overline{A} = \begin{vmatrix} \overline{i} & \overline{j} & \overline{k} \\ 2 & 4 & -5 \\ 3 & -5 & 4 \end{vmatrix} = \overline{i}(-25 + 16) + \overline{j}(-8 - 15) + \overline{k}(-12 - 10)$$

$$= -9\overline{i} - 23\overline{j} - 22\overline{k} = -(\overline{A} \times \overline{B})$$

Next, the **vectorial double cross product** is defined as $\overline{A} \times (\overline{B} \times \overline{C})$

A double cross product is a chain of two cross products; thus, it can be found using $\overline{A} \times (\overline{B} \times \overline{C}) = \overline{A} \times \overline{D}$ where $\overline{D} = (\overline{B} \times \overline{C})$. However, it might be easier to find it using the **BAC/CAB** rule following:

$$\overline{A} \times (\overline{B} \times \overline{C}) = \overline{B}(\overline{A} \cdot \overline{C}) - \overline{C}(\overline{A} \cdot \overline{B})$$

Example A.1.5

Given: $\overline{A} = 3\overline{i} - 5\overline{j} + 2\overline{k}, \overline{B} = 4\overline{i} - 3\overline{j} + 3\overline{k}, \overline{C} = 3\overline{i} + 5\overline{j} - 3\overline{k}$

Method 1

First, define: $\overline{D} = (\overline{B} \times \overline{C}) = \begin{vmatrix} \overline{i} & \overline{j} & \overline{k} \\ 4 & -3 & 3 \\ 3 & -5 & -3 \end{vmatrix} = -6\overline{i} + 21\overline{j} + 29\overline{k}$

Next, find $\overline{A} \times (\overline{B} \times \overline{C}) = \overline{A} \times \overline{D} = \begin{vmatrix} \overline{i} & \overline{j} & \overline{k} \\ 3 & -5 & 2 \\ -6 & 21 & 29 \end{vmatrix} = -187\overline{i} - 99\overline{j} + 33\overline{k}$

Method 2

$$\overline{A} \times (\overline{B} \times \overline{C}) = \overline{B}(\overline{A} \cdot \overline{C}) - \overline{C}(\overline{A} \cdot \overline{B}) = (4\bar{i} - 3\bar{j} + 3\bar{k})(9 - 25 - 6) - (3\bar{i} + 5\bar{j} - 3\bar{k})(12 + 15 + 6)$$

$$= -22(4\bar{i} - 3\bar{j} + 3\bar{k}) - 33(3\bar{i} + 5\bar{j} - 3\bar{k}) = -187\bar{i} - 99\bar{j} + 33\bar{k}$$

For dynamic systems vectors are functions of time. The previously defined operations apply also to this type of vectors.

Example A.1.6

Given: $\overline{A} = 3t\bar{i} + 5t^2\bar{j} - 6t^3\bar{k}$, $\overline{B} = 3t^2\bar{i} - 4t^3\bar{j} + 2t^2\bar{k}$

$$\overline{A} \cdot \overline{B} = (3t \cdot 3t^2) + (5t^2 \cdot -4t^3) + (-6t^3 \cdot 2t^2) = 9t^3 - 20t^5 - 12t^5 = 9t^3 - 32t^5$$

$$\overline{A} \times \overline{B} = \begin{vmatrix} \bar{i} & \bar{j} & \bar{k} \\ 3t & 5t^2 & -6t^3 \\ 3t^2 & -4t^3 & 2t^2 \end{vmatrix} = \bar{i}(10t^4 - 24t^6) + \bar{j}(-18t^5 - 6t^3) + \bar{k}(-27t^4)$$

The time derivative of a vector is defined as:

$$\dot{\overline{A}} = \frac{d\overline{A}}{dt} = \frac{dA_X}{dt}\bar{i} + \frac{dA_Y}{dt}\bar{j} + \frac{dA_Z}{dt}\bar{k} = \dot{A}_X\bar{i} + \dot{A}_Y\bar{j} + \dot{A}_Z\bar{k}$$

Example A.1.7

Given:

$$\overline{A} = 3t\bar{i} + 5t^2\bar{j} - 6t^3\bar{k}, \quad \dot{\overline{A}} = \frac{d\overline{A}}{dt} = \frac{dA_X}{dt}\bar{i} + \frac{dA_Y}{dt}\bar{j} + \frac{dA_Z}{dt}\bar{k} = 3\bar{i} + 10t\bar{j} - 18t^2\bar{k}$$

An important operation is given by $\frac{d}{dt}(\overline{A} \times \overline{B})$. The time derivative of a cross product can be found using any of the two methods outlined in the example below.

Example A.1.8

Given $\overline{C} = \overline{A} \times \overline{B} = \begin{vmatrix} \bar{i} & \bar{j} & \bar{k} \\ 3t & 5t^2 & -6t^3 \\ 3t^2 & -4t^3 & 2t^2 \end{vmatrix} = \bar{i}(10t^4 - 24t^6) + \bar{j}(-18t^5 - 6t^3) + \bar{k}(-27t^4)$

Method 1

$$\frac{d\overline{C}}{dt} = \frac{d}{dt}(\overline{A} \times \overline{B}) = \frac{d}{dt}[\bar{i}(10t^4 - 24t^6) + \bar{j}(-18t^5 - 6t^3) + \bar{k}(-27t^4)]$$

$$= (40t^3 - 144t^5)\bar{i} + (-90t^4 - 18t^2)\bar{j} + (-108t^3)\bar{k}$$

Method 2

$$\frac{d\overline{C}}{dt} = \frac{d}{dt}\left(\overline{A} \times \overline{B}\right) = \left(\dot{\overline{A}} \times \overline{B}\right) + \left(\overline{A} \times \dot{\overline{B}}\right)$$

$$\overline{A} = 3t\overline{i} + 5t^2\overline{j} - 6t^3\overline{k} \rightarrow \dot{\overline{A}} = 3\overline{i} + 10t\overline{j} - 18t^2\overline{k}$$

$$\overline{B} = 3t^2\overline{i} - 4t^3\overline{j} + 2t^2\overline{k} \rightarrow \dot{\overline{B}} = 6t\overline{i} - 12t^2\overline{j} + 4t\overline{k}$$

$$\left(\dot{\overline{A}} \times \overline{B}\right) = \begin{vmatrix} \overline{i} & \overline{j} & \overline{k} \\ 3 & 10t & -18t^2 \\ 3t^2 & -4t^3 & 2t^2 \end{vmatrix} = \left(20t^3 - 72t^5\right)\overline{i} + \left(-54t^4 - 6t^2\right)\overline{j} + \left(-42t^3\right)\overline{k}$$

$$\left(\overline{A} \times \dot{\overline{B}}\right) = \begin{vmatrix} \overline{i} & \overline{j} & \overline{k} \\ 3t & 5t^2 & -6t^3 \\ 6t & -12t^2 & 4t \end{vmatrix} = \left(20t^3 - 72t^5\right)\overline{i} + \left(-36t^4 - 12t^2\right)\overline{j} + \left(-66t^3\right)\overline{k}$$

$$\frac{d\overline{C}}{dt} = \left(40t^3 - 144t^5\right)\overline{i} + \left(-90t^4 - 18t^2\right)\overline{j} + \left(-108t^3\right)\overline{k}$$

A.2 REVIEW OF MATRIX OPERATIONS

A matrix is an array of numerical coefficients organized in rows and columns. A generic matrix A is indicated with $\overline{\overline{A}}$:

$$\overline{\overline{A}} = \begin{bmatrix} a_{11} & a_{12} & \ldots & a_{1m} \\ a_{21} & a_{22} & \ldots & a_{2m} \\ \ldots & \ldots & \ldots & \ldots \\ a_{n1} & a_{n2} & \ldots & a_{nm} \end{bmatrix}$$ 'a_{ij}' is the generic coefficient with $i = 1, \ldots, n \quad j = 1, \ldots, m$

Matrices are often indicated with a subscript specifying the number of rows (n) and the number of columns (m). For example $\overline{\overline{A}}_{4 \times 6}$ indicates a matrix with 4 rows and 6 columns. A matrix is **rectangular** when $n \neq m$; a matrix is **square** when $n = m$. The diagonal connecting the coefficients $a_{11}, a_{22}, \ldots, a_{nn}$ is called the **main diagonal** of the matrix whereas the diagonal connecting the coefficients $a_{n1}, a_{n-1,2}, \ldots, a_{1,m}$ is called the **secondary diagonal**. A special square matrix is the **Identity matrix**, shown here.

$$\overline{\overline{I}}_{n \times n} = \begin{bmatrix} 1 & 0 & \ldots & 0 \\ 0 & 1 & \ldots & 0 \\ \ldots & \ldots & \ldots & \ldots \\ 0 & 0 & \ldots & 1 \end{bmatrix}$$ where $a_{11} = a_{22} = \cdots = a_{nn} = 1, \quad a_{ij} = 0$

A square matrix is said to be **symmetric** if $a_{ij} = a_{ji}$ for $i \neq j, i = 1, \ldots, n; \; j = 1, \ldots, m$

By definition, the Identity matrix $\overline{\overline{I}}_{n \times n}$ is a symmetric matrix.

Example A.2.1

$$\overline{\overline{A}} = \begin{bmatrix} 2 & -3 & 5 & 2 \\ -3 & 3 & -2 & 1 \\ 5 & -2 & 4 & 7 \\ 2 & 1 & 7 & 8 \end{bmatrix}$$ is a symmetric matrix.

$$\overline{\overline{B}} = \begin{bmatrix} 2 & 1 & 3 \\ -1 & 4 & -5 \\ 3 & 4 & 6 \end{bmatrix}$$ is NOT a symmetric matrix (asymmetric matrix).

If $\overline{\overline{A}}$ and $\overline{\overline{B}}$ are two matrices with the same order (that is, they have the same number of rows and columns), the sum of the matrices $\overline{\overline{A}}$ and $\overline{\overline{B}}$ is defined as:

$$\overline{\overline{A}} + \overline{\overline{B}} = \overline{\overline{C}} \text{ with } c_{ij} = a_{ij} + b_{ij} \quad i = 1, \ldots, n; \; j = 1, \ldots, m$$

Similarly, we can define:

$$\overline{\overline{A}} - \overline{\overline{B}} = \overline{\overline{D}} \text{ with } d_{ij} = a_{ij} - b_{ij} \quad i = 1, \ldots, n; \; j = 1, \ldots, m$$

Example A.2.2

$$\overline{\overline{A}} = \begin{bmatrix} 1 & 2 & 3 \\ 4 & 5 & 6 \\ 7 & 8 & 9 \end{bmatrix}, \overline{\overline{B}} = \begin{bmatrix} 1 & 3 & 4 \\ 2 & 5 & 1 \\ 3 & 2 & 6 \end{bmatrix} \to \overline{\overline{C}} = \begin{bmatrix} 1+1 & 2+3 & 3+4 \\ 4+2 & 5+5 & 6+1 \\ 7+3 & 8+2 & 9+6 \end{bmatrix} = \begin{bmatrix} 2 & 5 & 7 \\ 6 & 10 & 7 \\ 10 & 10 & 15 \end{bmatrix}$$

Given a scalar k_1 and a matrix $\overline{\overline{A}}$ we can define the product $\overline{\overline{C}} = k_1 \overline{\overline{A}}$ with $c_{ij} = k_1 a_{ij}, i = 1, \ldots, n; \; j = 1, \ldots, m$.

If $\overline{\overline{A}}$ and $\overline{\overline{B}}$ are two matrices such that the number of columns of $\overline{\overline{A}}$ is equal to the number of rows of $\overline{\overline{B}}$ then we can define the product of the two matrices $\overline{\overline{A}}$ and $\overline{\overline{B}}$. The corresponding matrix $\overline{\overline{C}}$ will be given by:

$$\overline{\overline{A}}_{n_1 \times m_1} \; x \; \overline{\overline{B}}_{m_1 \times l_1} = \overline{\overline{C}}_{n_1 \times l_1}$$

Example A.2.3

$$\overline{\overline{A}} = \begin{bmatrix} 1 & 2 & 3 \\ 4 & 5 & 6 \end{bmatrix}, \; \overline{\overline{B}} = \begin{bmatrix} 1 & 2 \\ 3 & 4 \\ 5 & 6 \end{bmatrix}, \; \overline{\overline{A}}_{2x3} \; x \; \overline{\overline{B}}_{3x2} = \overline{\overline{C}}_{2x2}$$

$$\overline{\overline{C}} = \begin{bmatrix} (1x1 + 2x3 + 3x5) & (1x2 + 2x4 + 3x6) \\ (4x1 + 5x3 + 6x5) & (4x2 + 5x4 + 6x6) \end{bmatrix} = \begin{bmatrix} 22 & 28 \\ 49 & 64 \end{bmatrix}$$

Given a square matrix $\overline{\overline{A}}_{nxn}$ we define the **transpose** of $\overline{\overline{A}}_{nxn}$ the matrix $\overline{\overline{A}}^T_{nxn}$ such that:

$$\overline{\overline{A}} = \begin{bmatrix} a_{11} & a_{12} & a_{13} & a_{14} \\ a_{21} & a_{22} & a_{23} & a_{24} \\ a_{31} & a_{32} & a_{33} & a_{34} \\ a_{41} & a_{42} & a_{43} & a_{44} \end{bmatrix} \rightarrow \overline{\overline{A}}^T = \begin{bmatrix} a_{11} & a_{21} & a_{31} & a_{41} \\ a_{12} & a_{22} & a_{32} & a_{42} \\ a_{13} & a_{23} & a_{33} & a_{43} \\ a_{14} & a_{24} & a_{34} & a_{44} \end{bmatrix}$$

In other words, the coefficients of $\overline{\overline{A}}^T$ are $a^T_{ij} = a_{ji}$ for $i \neq j, i = 1, \ldots, n; \; j = 1, \ldots, m$.
The transpose of a matrix has the following properties:

$$\left(\overline{\overline{A}}^T\right)^T = \overline{\overline{A}}$$

$$\left(\overline{\overline{A}} + \overline{\overline{B}}\right)^T = \overline{\overline{A}}^T + \overline{\overline{B}}^T$$

$$\left(\overline{\overline{A}} \; \overline{\overline{B}}\right)^T = \overline{\overline{B}}^T \; \overline{\overline{A}}^T$$

Example A.2.4

$$\overline{\overline{A}} = \begin{bmatrix} 1 & 5 \\ 2 & 4 \\ 3 & 7 \end{bmatrix}, \; \overline{\overline{A}}^T = \begin{bmatrix} 1 & 2 & 3 \\ 5 & 4 & 7 \end{bmatrix}$$

Assuming that the sizes of the matrices are compatible, the following matrix operations and matrix-scalar operations are possible:

$$\overline{\overline{A}} + \overline{\overline{B}} = \overline{\overline{B}} + \overline{\overline{A}}$$

$$\overline{\overline{A}} + \left(\overline{\overline{B}} + \overline{\overline{C}}\right) = \left(\overline{\overline{A}} + \overline{\overline{B}}\right) + \overline{\overline{C}}$$

$$\overline{\overline{A}} \cdot \left(\overline{\overline{B}} \cdot \overline{\overline{C}}\right) = \left(\overline{\overline{A}} \cdot \overline{\overline{B}}\right) \cdot \overline{\overline{C}}$$

$$\overline{\overline{A}} \cdot \left(\overline{\overline{B}} + \overline{\overline{C}}\right) = \overline{\overline{A}} \cdot \overline{\overline{B}} + \overline{\overline{A}} \cdot \overline{\overline{C}}$$

$$k_1 \left(\overline{\overline{A}} + \overline{\overline{B}}\right) = k_1 \overline{\overline{A}} + k_1 \overline{\overline{B}}$$

$$(k_1 + k_2)\overline{\overline{A}} = k_1 \overline{\overline{A}} + k_2 \overline{\overline{A}}$$

$$k_1 \left(k_2 \overline{\overline{A}}\right) = k_1 k_2 \overline{\overline{A}}$$

Next, the concept of matrix rank, inverse, and determinant are briefly reviewed. First, the concept of linearly dependence (or linearly independence) is introduced.

A matrix is said to be **Linearly Independent** by row (or by column) if each of the rows (or columns) is NOT a multiple of any of the other rows (or columns). If the opposite is true then we have linear dependence by row (or by column).

Example A.2.5

$$\bar{\bar{A}} = \begin{bmatrix} 1 & 2 & 3 & 5 \\ 2 & 6 & -3 & 1 \\ 4 & 7 & 8 & -5 \\ 2 & 3 & -4 & 9 \end{bmatrix}, \quad \bar{\bar{B}} = \begin{bmatrix} 1 & 3 & 2 \\ 5 & -3 & 7 \\ 2 & 6 & 4 \end{bmatrix}$$

Upon visual inspection, for matrix $\bar{\bar{A}}$ we have:

- linear independence for the rows;
- linear independence for the columns.

Upon visual inspection, for matrix $\bar{\bar{B}}$ we have:

- linear dependence for the rows (the third row is a multiple of the first row);
- linear independence for the columns.

Consider now a generic square matrix $\bar{\bar{A}}$ with n rows and n columns. If we have linear independence for the rows and for the columns, the matrix $\bar{\bar{A}}$ is said to have **Full Rank** with $rank(\bar{\bar{A}}) = n$.

The definition of rank can also be extended to rectangular matrices with n and m columns. In fact, for a generic rectangular matrix $\bar{\bar{B}}$ we can define:

- rank by rows as the number of linearly independent rows;
- rank by columns as the number of linearly independent columns.

Example A.2.6

$$\bar{\bar{A}} = \begin{bmatrix} 2 & 5 & 3 \\ 1 & 2 & 4 \\ 3 & -7 & 8 \end{bmatrix}, \quad \bar{\bar{B}} = \begin{bmatrix} 2 & 3 & 4 & 5 \\ 1 & 2 & 5 & 7 \\ 3 & -3 & 1 & 2 \end{bmatrix}, \quad \bar{\bar{C}} = \begin{bmatrix} 2 & 4 & 3 \\ 1 & 2 & 5 \\ -3 & -6 & -7 \end{bmatrix}$$

Upon visual inspection, for the square matrix $\bar{\bar{A}}$ we have:

- rank by rows = 3;
- rank by columns = 3.

Therefore matrix $\bar{\bar{A}}$ has full rank, with rank = 3.

Upon visual inspection, for the rectangular $\bar{\bar{B}}$ we have:

- rank by rows = 3;
- rank by columns = 4.

Upon visual inspection, for the square matrix $\bar{\bar{C}}$ we have:

- rank by rows = 3;
- rank by columns = 2 (since the second column is a multiple of the firstcolumn).

Therefore matrix $\overline{\overline{C}}$ is not fully ranked, with rank = 2.

A property of the Identity matrix is that it is full ranked.

$$\overline{\overline{I}}_{n \times n} = \begin{bmatrix} 1 & 0 & \ldots & 0 \\ 0 & 1 & \ldots & 0 \\ \ldots & \ldots & \ldots & \ldots \\ 0 & 0 & \ldots & 1 \end{bmatrix}, \quad rank\left(\overline{\overline{I}}_{n \times n}\right) = n$$

Next, the concept of **Inverse** of a square matrix is introduced. The inverse of a rectangular matrix (often referred to as pseudo-inverse) is not here reviewed.

By definition, if $\overline{\overline{A}}$ is a square matrix and if a matrix $\overline{\overline{B}}$ can be found such that $\overline{\overline{A}}\,\overline{\overline{B}} = \overline{\overline{B}}\,\overline{\overline{A}} = \overline{\overline{I}}_{n \times n}$, then $\overline{\overline{A}}$ is said to be *invertible* and $\overline{\overline{B}}$ is called the inverse of $\overline{\overline{A}}$ and indicated as $\overline{\overline{B}} = \overline{\overline{A}}^{-1}$.

The inverse of a matrix has the following properties:

- An invertible matrix has ONLY one inverse;
- Necessary and sufficient condition for the inverse of a square matrix to exist is that the matrix has full rank;
- If $\overline{\overline{A}}$ and $\overline{\overline{B}}$ are invertible square matrices then.

$$\overline{\overline{A}}\,\overline{\overline{B}} = \overline{\overline{B}}\,\overline{\overline{A}} = \overline{\overline{C}} \quad \text{where the matrix } \overline{\overline{C}} \text{ is also invertible.}$$
$$\left(\overline{\overline{A}}\,\overline{\overline{B}}\right)^{-1} = \overline{\overline{B}}^{-1}\,\overline{\overline{A}}^{-1}$$
$$\left(\overline{\overline{A}}^{-1}\right)^{-1} = \overline{\overline{A}}$$

The availability of the inverse of a matrix is extremely important for solving systems of linear algebraic equations reduced to the form:

$$\overline{\overline{A}}\,\overline{x} = \overline{b}$$

where
$\overline{x}$ is the vector of the unknowns
$\overline{y}$ is the vector of the known terms
$\overline{\overline{A}}$ is the system matrix

Next, two methods will be introduced to find the inverse of a matrix:

- Adjoint-based method (best suitable for square matrices with n up to 3);
- Inverse-by-row operations method (suitable for square matrices with $n > 3$).

Adjoint-based method

Consider generic *(2 × 2)* and *(3 × 3)* square matrices. The **Determinant** is defined as a scalar calculated as:

$$\overline{\overline{A}} = \begin{bmatrix} a_{11} & a_{12} \\ a_{21} & a_{22} \end{bmatrix}, \quad \det\left(\overline{\overline{A}}\right) = a_{11}a_{22} - a_{21}a_{12}$$

$$\overline{\overline{A}} = \begin{bmatrix} a_{11} & a_{12} & a_{13} \\ a_{21} & a_{22} & a_{23} \\ a_{31} & a_{32} & a_{33} \end{bmatrix},$$

$$\det\left(\overline{\overline{A}}\right) = a_{11}(a_{22}a_{33} - a_{32}a_{23}) + a_{12}(a_{31}a_{23} - a_{21}a_{33}) + a_{13}(a_{21}a_{32} - a_{31}a_{22})$$

$- OR -$

$$\det\left(\overline{\overline{A}}\right) = a_{11}(a_{22}a_{33} - a_{32}a_{23}) + a_{21}(a_{32}a_{13} - a_{12}a_{33}) + a_{31}(a_{12}a_{23} - a_{22}a_{13})$$

Example A.2.7

$$\overline{\overline{A}} = \begin{bmatrix} 1 & 3 \\ 5 & 6 \end{bmatrix}, \quad \det(\overline{\overline{A}}) = 1 \cdot 6 - 3 \cdot 5 = 6 - 15 = -9$$

$$\overline{\overline{A}} = \begin{bmatrix} 1 & 2 & 3 \\ -4 & 5 & 6 \\ 7 & -8 & 9 \end{bmatrix},$$

$$\det(\overline{\overline{A}}) = 1((5 \cdot 9) - (-8 \cdot 6)) + 2((6 \cdot 7) - (-4 \cdot 9)) + 3((-4 \cdot -8) - (5 \cdot 7))$$
$$= 1(45 + 48) + 2(42 + 36) + 3(32 - 35) = 93 + 156 - 9 = 240$$

Next, we introduce the concept of **Co-Factor** for the coefficient of a matrix.

Consider a generic matrix:

$$\overline{\overline{A}} = \begin{bmatrix} a_{11} & a_{12} & a_{13} \\ a_{21} & a_{22} & a_{23} \\ a_{31} & a_{32} & a_{33} \end{bmatrix}$$

Consider the coefficient a_{11}. The co-factor of a_{11} is defined as the coefficient c_{11} such that $c_{11} = (a_{22}a_{33} - a_{32}a_{23})$. The calculation for c_{11} can be visualized in the following:

$$\overline{\overline{A}} = \begin{bmatrix} \cancel{a_{11}} & \cancel{a_{12}} & \cancel{a_{13}} \\ \cancel{a_{21}} & a_{22} & a_{23} \\ \cancel{a_{31}} & a_{32} & a_{33} \end{bmatrix},$$

$c_{11} = $ *product coefficients main diagonal − product coefficients sec .diagonal*

$$= (a_{22}a_{33} - a_{32}a_{23})$$

The other coefficients are found using:

$$c_{12} = (a_{31}a_{23} - a_{21}a_{33})$$
$$c_{13} = (a_{21}a_{32} - a_{31}a_{22})$$
$$c_{21} = (a_{32}a_{13} - a_{12}a_{33})$$
$$c_{22} = (a_{11}a_{33} - a_{31}a_{13})$$
$$c_{23} = (a_{31}a_{12} - a_{11}a_{32})$$
$$c_{31} = (a_{12}a_{23} - a_{22}a_{13})$$
$$c_{32} = (a_{21}a_{13} - a_{11}a_{23})$$
$$c_{33} = (a_{11}a_{22} - a_{21}a_{12})$$

A simple trick to find the co-factors of the coefficients of a matrix is to change the order of the main and secondary diagonal depending on the + or − associated with each of the coefficients. Specifically, for the coefficients associated with a + use (product coefficients main diagonal − product coefficients secondary diagonal), and for the coefficients associated with a − use (product coefficients secondary diagonal − product coefficients main diagonal), as shown next.

$$\overline{\overline{A}} = \begin{bmatrix} + & - & + \\ - & + & - \\ + & - & + \end{bmatrix}$$

Once the individual co-factors of the coefficients are evaluated, the **Co-Factor Matrix** is defined as:

$$cof\left(\overline{\overline{A}}\right) = \begin{bmatrix} c_{11} & c_{12} & c_{13} \\ c_{21} & c_{22} & c_{23} \\ c_{31} & c_{32} & c_{33} \end{bmatrix}$$

Next, the **Adjoint Matrix** is defined as:

$$adj\left(\overline{\overline{A}}\right) = \left(cof\left(\overline{\overline{A}}\right)\right)^T$$

Next, the Inverse Matrix can be found using:

$$\overline{\overline{A}}^{-1} = inv\left(\overline{\overline{A}}\right) = \frac{adj\left(\overline{\overline{A}}\right)}{det\left(\overline{\overline{A}}\right)}$$

It should be emphasized that this method is not suitable for any matrix higher than (3×3).

Example A.2.8

$$\overline{\overline{A}} = \begin{bmatrix} 3 & 2 & -1 \\ 1 & 6 & 3 \\ 2 & -4 & 0 \end{bmatrix} = \begin{bmatrix} a_{11} & a_{12} & a_{13} \\ a_{21} & a_{22} & a_{23} \\ a_{31} & a_{32} & a_{33} \end{bmatrix}$$

From visual inspection, the matrix has full rank (rank = 3).

$c_{11} = (a_{22}a_{33} - a_{32}a_{23}) = 12, \quad c_{12} = (a_{31}a_{23} - a_{21}a_{33}) = 6,$

$c_{13} = (a_{21}a_{32} - a_{31}a_{22}) = -16$

$c_{21} = (a_{32}a_{13} - a_{12}a_{33}) = 4, \quad c_{22} = (a_{11}a_{33} - a_{31}a_{13}) = 2, \quad c_{23} = (a_{31}a_{12} - a_{11}a_{32}) = 16$

$c_{31} = (a_{12}a_{23} - a_{22}a_{13}) = 12, \quad c_{32} = (a_{21}a_{13} - a_{11}a_{23}) = -10, \quad c_{33} = (a_{11}a_{22} - a_{21}a_{12}) = 16$

$$cof\left(\overline{\overline{A}}\right) = \begin{bmatrix} c_{11} & c_{12} & c_{13} \\ c_{21} & c_{22} & c_{23} \\ c_{31} & c_{32} & c_{33} \end{bmatrix} = \begin{bmatrix} 12 & 6 & -16 \\ 4 & 2 & 16 \\ 12 & -10 & 16 \end{bmatrix}$$

$$adj\left(\overline{\overline{A}}\right) = \left(cof\left(\overline{\overline{A}}\right)\right)^T = cof\left(\overline{\overline{A}}\right) = \begin{bmatrix} 12 & 4 & 12 \\ 6 & 2 & -10 \\ -16 & 16 & 16 \end{bmatrix}$$

$det\left(\overline{\overline{A}}\right) = a_{11}c_{11} + a_{12}c_{12} + a_{13}c_{13} = 3 \cdot 12 + 2 \cdot 6 - 1 \cdot (-16)$

$\quad = 36 + 12 + 16 = 64$

$$\overline{\overline{A}}^{-1} = inv\left(\overline{\overline{A}}\right) = \frac{adj\left(\overline{\overline{A}}\right)}{det\left(\overline{\overline{A}}\right)} = adj\left(\overline{\overline{A}}\right)$$

$$= \frac{\begin{bmatrix} 12 & 4 & 12 \\ 6 & 2 & -10 \\ -16 & 16 & 16 \end{bmatrix}}{64} = \begin{bmatrix} 12/64 & 4/64 & 12/64 \\ 6/64 & 2/64 & -10/64 \\ -16/64 & 16/64 & 16/64 \end{bmatrix}$$

$$= \begin{bmatrix} 0.1875 & 0.0625 & 0.1875 \\ 0.0937 & 0.0312 & -0.1562 \\ -0.25 & 0.25 & 0.25 \end{bmatrix}$$

Inverse-by-row operations method

This can be considered as a general method to find the inverse of matrices with order larger than a *(3 × 3)*. Given the complexity of the symbols, this method will be illustrated directly through a numerical example.

Example A.2.9

$$\bar{\bar{A}} = \begin{bmatrix} 1 & 2 & 3 \\ 2 & 5 & 3 \\ 1 & 0 & 8 \end{bmatrix}$$

From visual inspection, the matrix has full rank (rank = 3).

The starting point is to add an identity matrix $\bar{\bar{I}}_{3 \times 3}$ to the right of $\bar{\bar{A}}$. The objective is to manipulate numerically so that at the end of the process the identity matrix $\bar{\bar{I}}_{3 \times 3}$ will show up on the left.

$$\begin{bmatrix} 1 & 2 & 3 & | & 1 & 0 & 0 \\ 2 & 5 & 3 & | & 0 & 1 & 0 \\ 1 & 0 & 8 & | & 0 & 0 & 1 \end{bmatrix}$$

Next, add the product of (-2 multiplied the first row) to the second row and add the product of (-1 multiplied the first row) to the third row.

$$\begin{bmatrix} 1 & 2 & 3 & | & 1 & 0 & 0 \\ 0 & 1 & -3 & | & -2 & 1 & 0 \\ 0 & -2 & 5 & | & -1 & 0 & 1 \end{bmatrix}$$

Next, add the product of (2 multiplied the second row) to the third row.

$$\begin{bmatrix} 1 & 2 & 3 & | & 1 & 0 & 0 \\ 0 & 1 & -3 & | & -2 & 1 & 0 \\ 0 & 0 & -1 & | & -5 & 2 & 1 \end{bmatrix}$$

Next, multiply the third row by (-1).

$$\begin{bmatrix} 1 & 2 & 3 & | & 1 & 0 & 0 \\ 0 & 1 & -3 & | & -2 & 1 & 0 \\ 0 & 0 & 1 & | & 5 & -2 & -1 \end{bmatrix}$$

Next, add the product of (3 multiplied by the third3 row) to the second row and add the product of (-3 multiplied by the third row) to the first row.

$$\begin{bmatrix} 1 & 2 & 0 & | & -14 & 6 & 3 \\ 0 & 1 & 0 & | & 13 & -5 & -3 \\ 0 & 0 & 1 & | & 5 & -2 & -1 \end{bmatrix}$$

Finally, add the product of (-2 multiplied by the second row) to the first row.

$$\begin{bmatrix} 1 & 0 & 0 & | & -40 & 16 & 9 \\ 0 & 1 & 0 & | & 13 & -5 & -3 \\ 0 & 0 & 1 & | & 5 & -2 & -1 \end{bmatrix}$$

Thus

$$\overline{\overline{A}} = \begin{bmatrix} 1 & 2 & 3 \\ 2 & 5 & 3 \\ 1 & 0 & 8 \end{bmatrix} \Rightarrow \overline{\overline{A}}^{-1} = \begin{bmatrix} -40 & 16 & 9 \\ 13 & -5 & -3 \\ 5 & -2 & -1 \end{bmatrix}$$

To complete the example the results from the Inverse-by-row operations method are compared with the results from the Adjoint-based method.

$$\overline{\overline{A}} = \begin{bmatrix} 1 & 2 & 3 \\ 2 & 5 & 3 \\ 1 & 0 & 8 \end{bmatrix} = \begin{bmatrix} a_{11} & a_{12} & a_{13} \\ a_{21} & a_{22} & a_{23} \\ a_{31} & a_{32} & a_{33} \end{bmatrix}$$

From visual inspection, the matrix has full rank (rank = 3).

$$c_{11} = (a_{22}a_{33} - a_{32}a_{23}) = 40, \quad c_{12} = (a_{31}a_{23} - a_{21}a_{33}) = -13,$$
$$c_{13} = (a_{21}a_{32} - a_{31}a_{22}) = -5$$
$$c_{21} = (a_{32}a_{13} - a_{12}a_{33}) = -16, \quad c_{22} = (a_{11}a_{33} - a_{31}a_{13}) = 5, \quad c_{23} = (a_{31}a_{12} - a_{11}a_{32}) = 2$$
$$c_{31} = (a_{12}a_{23} - a_{22}a_{13}) = -9, \quad c_{32} = (a_{21}a_{13} - a_{11}a_{23}) = 3, \quad c_{33} = (a_{11}a_{22} - a_{21}a_{12}) = 1$$

$$cof(\overline{\overline{A}}) = \begin{bmatrix} c_{11} & c_{12} & c_{13} \\ c_{21} & c_{22} & c_{23} \\ c_{31} & c_{32} & c_{33} \end{bmatrix} = \begin{bmatrix} 40 & -13 & -5 \\ -16 & 5 & 2 \\ -9 & 3 & 1 \end{bmatrix}$$

$$adj(\overline{\overline{A}}) = \left(cof(\overline{\overline{A}})\right)T = cof(\overline{\overline{A}}) = \begin{bmatrix} 40 & -16 & -9 \\ -13 & 5 & 3 \\ -5 & 2 & 1 \end{bmatrix}$$

$$\det(\overline{\overline{A}}) = a_{11}c_{11} + a_{12}c_{12} + a_{13}c_{13} = 1 \cdot 40 + (2 \cdot -13) + 3 \cdot (-5)$$
$$= 40 - 26 - 15 = -1$$

$$\overline{\overline{A}}^{-1} = inv(\overline{\overline{A}}) = \frac{adj(\overline{\overline{A}})}{\det(\overline{\overline{A}})} = adj(\overline{\overline{A}}) = \frac{\begin{bmatrix} 40 & -16 & -9 \\ -13 & 5 & 3 \\ -5 & 2 & 1 \end{bmatrix}}{-1} = \begin{bmatrix} -40 & 16 & 9 \\ 13 & -5 & -3 \\ 5 & -2 & -1 \end{bmatrix}$$

The inverse of a matrix has immediate applications for the **Solution of System of Linear Algebraic Equations**.

Example A.2.10

Consider the following system of equations:

$$\begin{cases} x_1 + 2x_2 + 3x_3 = 5 \\ 2x_1 + 5x_2 + 3x_3 = 3 \\ x_1 + (0)x_2 + 8x_3 = 17 \end{cases}$$

In a matrix/vector format we have:

$$\begin{bmatrix} 1 & 2 & 3 \\ 2 & 5 & 3 \\ 1 & 0 & 8 \end{bmatrix} \begin{Bmatrix} x_1 \\ x_2 \\ x_3 \end{Bmatrix} = \begin{Bmatrix} 5 \\ 3 \\ 17 \end{Bmatrix}$$

In a general matrix format we have $\overline{\overline{A}}\ \overline{x} = \overline{b} \Rightarrow \overline{x} = \overline{\overline{A}}^{-1}\ \overline{b}$.

The matrix $\overline{\overline{A}}$ is provided in the previous example:

$$\overline{\overline{A}} = \begin{bmatrix} 1 & 2 & 3 \\ 2 & 5 & 3 \\ 1 & 0 & 8 \end{bmatrix}$$

Therefore:

$$\overline{x} = \begin{Bmatrix} x_1 \\ x_2 \\ x_3 \end{Bmatrix} = \begin{bmatrix} 1 & 2 & 3 \\ 2 & 5 & 3 \\ 1 & 0 & 8 \end{bmatrix}^{-1} \begin{Bmatrix} 5 \\ 3 \\ 17 \end{Bmatrix} = \begin{bmatrix} -40 & 16 & 9 \\ 13 & -5 & -3 \\ 5 & -2 & -1 \end{bmatrix} \begin{Bmatrix} 5 \\ 3 \\ 17 \end{Bmatrix}$$

$$\overline{x} = \begin{Bmatrix} x_1 \\ x_2 \\ x_3 \end{Bmatrix} = \begin{Bmatrix} -40 \cdot 5 + 16 \cdot 3 + 9 \cdot 17 \\ 13 \cdot 5 - 5 \cdot 3 - 3 \cdot 17 \\ 5 \cdot 5 - 2 \cdot 3 - 1 \cdot 17 \end{Bmatrix} = \begin{Bmatrix} 1 \\ -1 \\ 2 \end{Bmatrix}$$

An alternative method for solving systems of linear algebraic equations is **Cramer's rule**. However, this method is not suitable for systems with more than three equations.

Example A.2.11

Consider again the following system of equations:

$$\begin{cases} x_1 + 2x_2 + 3x_3 = 5 \\ 2x_1 + 5x_2 + 3x_3 = 3 \\ x_1 + (0)x_2 + 8x_3 = 17 \end{cases}$$

In a matrix/vector format we have:

$$\begin{bmatrix} 1 & 2 & 3 \\ 2 & 5 & 3 \\ 1 & 0 & 8 \end{bmatrix} \begin{Bmatrix} x_1 \\ x_2 \\ x_3 \end{Bmatrix} = \begin{Bmatrix} 5 \\ 3 \\ 17 \end{Bmatrix}$$

In a general matrix format we have $\overline{\overline{A}}\,\overline{x} = \overline{b}$.

Define $\overline{\overline{A}}_1$ the matrix $\overline{\overline{A}}$ where the first column is replaced by the known vector $\overline{b}$; next, define $\overline{\overline{A}}_2$ the matrix $\overline{\overline{A}}$ where the second column is replaced by the known vector $\overline{b}$; next, define $\overline{\overline{A}}_3$ the matrix $\overline{\overline{A}}$ where the third column is replaced by the known vector $\overline{b}$.

According to Cramer's rule the solution of the system of equations will be given by:

$$x_1 = \frac{\det\left[\overline{\overline{A}}_1\right]}{\det\left[\overline{\overline{A}}\right]}, \quad x_2 = \frac{\det\left[\overline{\overline{A}}_2\right]}{\det\left[\overline{\overline{A}}\right]}, \quad x_3 = \frac{\det\left[\overline{\overline{A}}_3\right]}{\det\left[\overline{\overline{A}}\right]}$$

For the given system we have:

$$\overline{\overline{A}}_1 = \begin{bmatrix} 5 & 2 & 3 \\ 3 & 5 & 3 \\ 17 & 0 & 8 \end{bmatrix}, \quad \overline{\overline{A}}_2 = \begin{bmatrix} 1 & 5 & 3 \\ 2 & 3 & 3 \\ 1 & 17 & 8 \end{bmatrix}, \quad \overline{\overline{A}}_3 = \begin{bmatrix} 1 & 2 & 5 \\ 2 & 5 & 3 \\ 1 & 0 & 17 \end{bmatrix}$$

leading to:

$$\det\left(\overline{\overline{A}}_1\right) = 5 \cdot 40 + 2 \cdot (51 - 24) + 3 \cdot (-85) = -1$$

$$\det\left(\overline{\overline{A}}_2\right) = 1 \cdot (24 - 51) + 5 \cdot (3 - 16) + 3 \cdot (34 - 3) = -27 - 65 + 93 = 1$$

$$\det\left(\overline{\overline{A}}_3\right) = 1 \cdot (85) + 2 \cdot (3 - 34) + 5 \cdot (-5) = 85 - 62 - 25 = -2$$

$$\det\left(\overline{\overline{A}}\right) = 1 \cdot 40 + (2 \cdot -13) + 3 \cdot (-5) = 40 - 26 - 15 = -1$$

Finally we will have:

$$x_1 = \frac{\det\left[\overline{\overline{A}}_1\right]}{\det\left[\overline{\overline{A}}\right]} = \frac{-1}{-1} = 1$$

$$x_2 = \frac{\det\left[\overline{\overline{A}}_2\right]}{\det\left[\overline{\overline{A}}\right]} = \frac{1}{-1} = -1$$

$$x_3 = \frac{\det\left[\overline{\overline{A}}_3\right]}{\det\left[\overline{\overline{A}}\right]} = \frac{-2}{-1} = 2$$

NOTE: Cramer's rule is used extensively for the determination of the transfer functions for the aircraft dynamics in Chapter 7.

A.3 REVIEW OF CENTER OF GRAVITY AND INERTIAL PROPERTIES

The concepts to be reviewed in this section are relative to a continuous mass and a discrete system of masses. Consider first a continuous 3D mass (nicknamed potatoid) where ρ is the density.

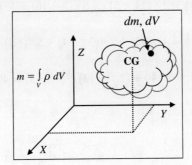

Figure A.3.1 Generic Continuous 3D Mass

For the system shown in Figure A.3.1, we can define the **Center of Gravity for a continuous system** as the point with the following coordinates:

$$x_{CG} = \frac{\int_V x \cdot \rho dV}{\int_V \rho dV}, \quad y_{CG} = \frac{\int_V y \cdot \rho dV}{\int_V \rho dV}, \quad z_{CG} = \frac{\int_V z \cdot \rho dV}{\int_V \rho dV}$$

Consider now a discrete system of masses as shown in Figure A.3.2.

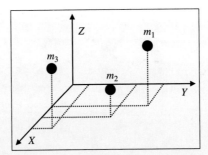

Figure A.3.2 Generic System of Discrete 3D Masses

with: m_1 located at coordinates (x_1, y_1, z_1)
m_2 located at coordinates (x_2, y_2, z_2)
m_3 located at coordinates (x_3, y_3, z_3)

For the system in Figure A.3.2, we can define the **Center of Gravity for a discrete system** as the point with the following coordinates:

$$x_{CG} = \frac{\sum_{i=1}^{3} m_i x_i}{\sum_{i=1}^{3} m_i}, \quad y_{CG} = \frac{\sum_{i=1}^{3} m_i y_i}{\sum_{i=1}^{3} m_i}, \quad z_{CG} = \frac{\sum_{i=1}^{3} m_i z_i}{\sum_{i=1}^{3} m_i}$$

Example A.3.1

Given the following system of three discrete masses:

$$m_1 = 10 \text{ lbs.}, \quad (x_1, y_1, z_1) = (3 \text{ ft}, -2 \text{ ft}, 4 \text{ ft})$$
$$m_2 = 15 \text{ lbs.}, \quad (x_2, y_2, z_2) = (2 \text{ ft}, -3 \text{ ft}, 5 \text{ ft})$$
$$m_3 = 20 \text{ lbs.}, \quad (x_3, y_3, z_3) = (-5 \text{ ft}, 3 \text{ ft}, 2 \text{ ft})$$

The center of gravity for this system will be located at:

$$x_{CG} = \frac{\sum_{i=1}^{3} m_i x_i}{\sum_{i=1}^{3} m_i} = \frac{[(10 \cdot 3) + (15 \cdot 2) + (20 \cdot -5)]}{(10 + 15 + 20)} \text{ ft} = \frac{-40}{45} \text{ ft} = -0.889 \text{ ft}$$

$$y_{CG} = \frac{\sum_{i=1}^{3} m_i y_i}{\sum_{i=1}^{3} m_i} = \frac{[(10 \cdot -2) + (15 \cdot -3) + (20 \cdot 3)]}{(10 + 15 + 20)} \text{ ft} = \frac{-5}{45} \text{ ft} = -0.111 \text{ ft}$$

$$z_{CG} = \frac{\sum_{i=1}^{3} m_i z_i}{\sum_{i=1}^{3} m_i} = \frac{[(10 \cdot 4) + (15 \cdot 5) + (20 \cdot 2)]}{(10 + 15 + 20)} \text{ ft} = \frac{155}{45} \text{ ft} = 3.444 \text{ ft}$$

Consider again the previous continuous 3D mass shown in Figure A.3.3:

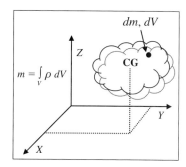

Figure A.3.3 Generic Continuous 3D Mass

Next, we can define the **Moments and Products of Inertia for a continuous system** as:

Moments of Inertia

$$I_{XX} = \int_V \rho(y^2 + z^2)dV, \quad I_{YY} = \int_V \rho(x^2 + z^2)dV, \quad I_{ZZ} = \int_V \rho(x^2 + y^2)dV$$

Products of Inertia

$$I_{XY} = I_{YX} = \int_V \rho xy \, dV, \quad I_{XZ} = I_{ZX} = \int_V \rho xz \, dV, \quad I_{YZ} = I_{ZY} = \int_V \rho yz \, dV$$

Similarly, consider again the previous discrete system of masses, shown in Figure A.3.4:

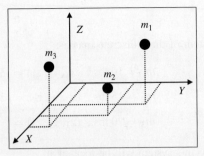

Figure A.3.4 Generic System of Discrete 3D Masses

The **Moments and Products of Inertia for a discrete system** with n masses are defined as:

$$I_{XX} = \sum_{i=1}^{n} m_i(y_i^2 + z_i^2), \quad I_{YY} = \sum_{i=1}^{n} m_i(x_i^2 + z_i^2), \quad I_{ZZ} = \sum_{i=1}^{n} m_i(x_i^2 + y_i^2)$$

$$I_{XY} = I_{YX} = \sum_{i=1}^{n} m_i(x_i y_i), \quad I_{XZ} = I_{ZX} = \sum_{i=1}^{n} m_i(x_i z_i), \quad I_{YZ} = I_{ZY} = \sum_{i=1}^{n} m_i(y_i z_i)$$

Properties of Moments and Products of Inertia

- The moments of inertia are always positive. They can only be ZERO if the mass is located at the origin of the reference axes (X, Y, Z).;
- The products of inertia can be positive, negative, or ZERO. If a product of inertia is ZERO, the mass has a plane of symmetry. For example, the (X, Z) plane is a plane of symmetry for any aircraft. Therefore, for any aircraft we have:

$$I_{XY} = I_{YX} = 0, I_{YZ} = I_{ZY} = 0, I_{XZ} = I_{ZX} \neq 0;$$

- The dimension for moments and products of inertia are $[M] \cdot [L^2]$.

Example A.3.2

Consider the discrete system of masses of Example A.3.1.

$$m_1 = 10 \text{ lbs.}, \quad (x_1, y_1, z_1) = (3 \text{ ft}, -2 \text{ ft}, 4 \text{ ft})$$
$$m_2 = 15 \text{ lbs.}, \quad (x_2, y_2, z_2) = (2 \text{ ft}, -3 \text{ ft}, 5 \text{ ft})$$
$$m_3 = 20 \text{ lbs.}, \quad (x_3, y_3, z_3) = (-5 \text{ ft}, 3 \text{ ft}, 2 \text{ ft})$$

The moments and products of inertia are given by:

$$I_{XX} = \sum_{i=1}^{n} m_i(y_i^2 + z_i^2) = 10(4+16) + 15(9+25) + 20(9+4) = 970 \; (\text{lb} \cdot \text{ft}^2)$$

$$I_{YY} = \sum_{i=1}^{n} m_i(x_i^2 + z_i^2) = 10(9+16) + 15(4+25) + 20(25+4) = 1265 \; (\text{lb} \cdot \text{ft}^2)$$

$$I_{ZZ} = \sum_{i=1}^{n} m_i(x_i^2 + y_i^2) = 10(9+4) + 15(4+9) + 20(25+9) = 1005 \; (\text{lb} \cdot \text{ft}^2)$$

$$I_{XY} = I_{YX} = \sum_{i=1}^{n} m_i(x_i y_i) = 10(3)(-2) + 15(2)(-3) + 20(-5)(3) = 450 \; (\text{lb} \cdot \text{ft}^2)$$

$$I_{XZ} = I_{ZX} = \sum_{i=1}^{n} m_i(x_i z_i) = 10(3)(4) + 15(2)(5) + 20(-5)(2) = 70 \; (\text{lb} \cdot \text{ft}^2)$$

$$I_{YZ} = I_{ZY} = \sum_{i=1}^{n} m_i(y_i z_i) = 10(-2)(4) + 15(-3)(5) + 20(3)(2) = 185 \; (\text{lb} \cdot \text{ft}^2)$$

A.3 Review of Center of Gravity and Inertial Properties

For both continuous and discrete systems, of specific interest are the moments and products of inertia with respect to a reference frame located at the center of gravity of the system (X_{CG}, Y_{CG}, Z_{CG}). These parameters can be found using:

$$I_{XX_{CG}} = \sum_{i=1}^{n} m_i \left((y_i - y_{CG})^2 + (z_i - z_{CG})^2 \right)$$

$$I_{YY_{CG}} = \sum_{i=1}^{n} m_i \left((x_i - x_{CG})^2 + (z_i - z_{CG})^2 \right)$$

$$I_{ZZ_{CG}} = \sum_{i=1}^{n} m_i \left((x_i - x_{CG})^2 + (y_i - y_{CG})^2 \right)$$

$$I_{XY_{CG}} = I_{YX_{CG}} = \sum_{i=1}^{n} m_i ((x_i - x_{CG})(y_i - y_{CG}))$$

$$I_{XZ_{CG}} = I_{ZX_{CG}} = \sum_{i=1}^{n} m_i ((x_i - x_{CG})(z_i - z_{CG}))$$

$$I_{YZ_{CG}} = I_{ZY_{CG}} = \sum_{i=1}^{n} m_i ((y_i - y_{CG})(z_i - z_{CG}))$$

Similar expressions can be introduced for a continuous system.

Example A.3.3

Consider, again, the discrete system of masses of Example A.3.1.

$$m_1 = 10 \text{ lbs.}, \quad (x_1, y_1, z_1) = (3 \text{ ft}, -2 \text{ ft}, 4 \text{ ft})$$

$$m_2 = 15 \text{ lbs.}, \quad (x_2, y_2, z_2) = (2 \text{ ft}, -3 \text{ ft}, 5 \text{ ft})$$

$$m_3 = 20 \text{ lbs.}, \quad (x_3, y_3, z_3) = (-5 \text{ ft}, 3 \text{ ft}, 2 \text{ ft})$$

$$x_{CG} = \frac{\sum_{i=1}^{3} m_i x_i}{\sum_{i=1}^{3} m_i} = -0.889 \text{ ft}, \quad y_{CG} = \frac{\sum_{i=1}^{3} m_i y_i}{\sum_{i=1}^{3} m_i} = -0.111 \text{ ft}, \quad z_{CG} = \frac{\sum_{i=1}^{3} m_i z_i}{\sum_{i=1}^{3} m_i} = 3.444 \text{ ft}$$

$$I_{XX_{CG}} = \sum_{i=1}^{n} m_i \left((y_i - y_{CG})^2 + (z_i - z_{CG})^2 \right) = 10(-2 + 0.11)^2 (4 - 3.44)^2$$
$$+ 15(-3 + 0.11)^2 (5 - 3.44)^2 + 20(3 + 0.11)^2 (2 - 3.44)^2 = 435.5 \quad (\text{lb} \cdot \text{ft}^2)$$

$$I_{YY_{CG}} = \sum_{i=1}^{n} m_i \left((x_i - x_{CG})^2 + (z_i - z_{CG})^2 \right) = 10(3 + 0.88)^2 (4 - 3.44)^2$$
$$+ 15(2 + 0.88)^2 (5 - 3.44)^2 + 20(-5 + 0.88)^2 (2 - 3.44)^2 = 695.5 \quad (\text{lb} \cdot \text{ft}^2)$$

$$I_{ZZ_{CG}} = \sum_{i=1}^{n} m_i \left((x_i - x_{CG})^2 + (y_i - y_{CG})^2 \right) = 10(3 + 0.88)^2 (-2 + 0.11)^2$$
$$+ 15(2 + 0.88)^2 (-3 + 0.11)^2 + 20(-5 + 0.88)^2 (3 + 0.11)^2 = 968.95 \quad (\text{lb} \cdot \text{ft}^2)$$

Parallel Axis Theorem

Assuming the availability of $\left(I_{XX_{CG}}, I_{YY_{CG}}, I_{ZZ_{CG}} \right)$ the moments of inertia with respect to *any* reference frame (X, Y, Z) can be calculated using:

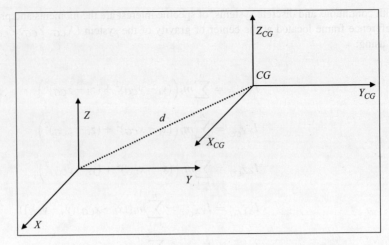

Figure A.3.5 Generic Reference Frame vs. CG Reference Frame

$$I_{XX} = I_{XX_{CG}} + Md_X^2$$
$$I_{YY} = I_{YY_{CG}} + Md_Y^2$$
$$I_{ZZ} = I_{ZZ_{CG}} + Md_Z^2$$

where:

$$M = \sum_{i=1}^{n} m_i$$

$$d_X^2 = (y_{CG}^2 + z_{CG}^2),\ d_Y^2 = (x_{CG}^2 + z_{CG}^2),\ d_Z^2 = (x_{CG}^2 + y_{CG}^2)$$

Example A.3.4 as shown in Figure A.3.5.

Consider, once again, the discrete system of masses of Example A.3.1:

$$m_1 = 10 \text{ lbs.}, \quad (x_1, y_1, z_1) = (3 \text{ ft}, -2 \text{ ft}, 4 \text{ ft})$$
$$m_2 = 15 \text{ lbs.}, \quad (x_2, y_2, z_2) = (2 \text{ ft}, -3 \text{ ft}, 5 \text{ ft})$$
$$m_3 = 20 \text{ lbs.}, \quad (x_3, y_3, z_3) = (-5 \text{ ft}, 3 \text{ ft}, 2 \text{ ft})$$

$$x_{CG} = \frac{\sum_{i=1}^{3} m_i x_i}{\sum_{i=1}^{3} m_i} = -0.889 \text{ ft}, \quad y_{CG} = \frac{\sum_{i=1}^{3} m_i y_i}{\sum_{i=1}^{3} m_i} = -0.111 \text{ ft}, \quad z_{CG} = \frac{\sum_{i=1}^{3} m_i z_i}{\sum_{i=1}^{3} m_i} = 3.444 \text{ ft}$$

From Example A.3.2:

$$I_{XX} = \sum_{i=1}^{n} m_i (y_i^2 + z_i^2) = 970 \ (\text{lb} \cdot \text{ft}^2)$$

$$I_{YY} = \sum_{i=1}^{n} m_i (x_i^2 + z_i^2) = 1265 \ (\text{lb} \cdot \text{ft}^2)$$

$$I_{ZZ} = \sum_{i=1}^{n} m_i (x_i^2 + y_i^2) = 1005 \ (\text{lb} \cdot \text{ft}^2)$$

From Example A.3.3:

$$I_{XX_{CG}} = \sum_{i=1}^{n} m_i\left((y_i - y_{CG})^2 + (z_i - z_{CG})^2\right) = 435.5 \; (\text{lb} \cdot \text{ft}^2)$$

$$I_{YY_{CG}} = \sum_{i=1}^{n} m_i\left((x_i - x_{CG})^2 + (z_i - z_{CG})^2\right) = 695.5 \; (\text{lb} \cdot \text{ft}^2)$$

$$I_{ZZ_{CG}} = \sum_{i=1}^{n} m_i\left((x_i - x_{CG})^2 + (y_i - y_{CG})^2\right) = 968.95 \; (\text{lb} \cdot \text{ft}^2)$$

Using the Parallel Axis Theorem to check the results we have:

$$I_{XX} = I_{XX_{CG}} + Md_X^2 = 435.5 + 45(0.11^2 + 3.44^2) = 968.56 \; (\text{lb} \cdot \text{ft}^2)$$

$$I_{YY} = I_{YY_{CG}} + Md_Y^2 = 695 + 45(0.88^2 + 3.44^2) = 1262.86 \; (\text{lb} \cdot \text{ft}^2)$$

$$I_{ZZ} = I_{ZZ_{CG}} + Md_Z^2 = 968.95 + (0.88^2 + 0.11^2) = 1004.35 \; (\text{lb} \cdot \text{ft}^2)$$

The two sets of results are off by a small negligible difference due to round-off errors.

A.4 REVIEW OF APPLICATION OF LAPLACE TRANSFORM TO LINEAR CONSTANT COEFFICIENTS DIFFERENTIAL EQUATIONS

One of the most widely used approaches for the solution of linear, constant coefficients differential equations is the application of Laplace transformation. Given a generic function $y(t)$, its Laplace transformation is defined as:

$$Y(s) = L[y(t)] = \int_0^\infty y(t) e^{-st} dt$$

The inverse Laplace transformation (ILT) is instead defined as:

$$L^{-1}[Y(s)] = y(t)$$

Prior to describing the properties of the LT and the ILT, it is important to recall the fundamental property which makes the LT an appealing tool for the solution of linear constant coefficients differential equations (DE). The LT is a transformation allowing to migrate from the *time*-domain to the *s*-domain. In the *s*-domain, a differential equation becomes an algebraic equation. Furthermore, this algebraic equation can be divided into a number of smaller terms for which a solution is known to exist. This breakdown process can be performed with a number of methods. In this review, the LT is coupled with the partial fraction expansion (PFE) method. Once these small terms are identified and solved, the ILT will bring the process back to the time-domain with the final solution $y(t)$. The process is summarized in Figure A.4.1.

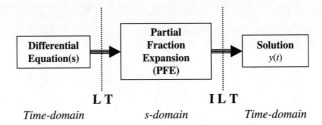

Figure A.4.1 Solution of a DE or a Set of DEs Using Laplace Transformation

The following is a list of the most important properties of the Laplace transformation (LT) operator.

Property 1 The LT operator is a *linear operator*. Therefore, given the generic scalars a, b and the functions $y_1(t)$, $y_2(t)$, the following applies:

$$L[ay_1(t) + by_2(t)] = L[ay_1(t)] + L[by_2(t)]$$
$$= aL[y_1(t)] + bL[y_2(t)] = aY_1(s) + bY_2(s)$$

Property 2 Given a differentiable function $y(t)$ and its *n-th* order derivative $y^n(t) = \frac{d^n y(t)}{dt^n}$:

$$L[y^n(t)] = s^n Y(s) - s^{n-1} y(0) - s^{n-2} y'(0) - s^{n-3} y''(0) - \cdots - y^{n-1}(0)$$

Thus, the LT of a *n-th* order derivative requires the following n initial conditions:

$$y(0), \ y'(0), \ y''(0), \ldots, y^{n-1}(0)$$

The previous general relationship takes on the following simpler forms for lower order derivatives:

A.4 Review of Application of Laplace Transform to Linear Constant Coefficients Differential Equations

$$L[y'(t)] = sY(s) - y(0) \quad \text{for a first order derivative}$$

$$L[y''(t)] = s^2 Y(s) - sy(0) - y'(0) \quad \text{for a second order derivative}$$

$$L[y'''(t)] = s^3 Y(s) - s^2 y(0) - sy'(0) - y''(0) \quad \text{for a third order derivative}$$

Property 3 Given a generic function $y(t)$ and its integral $\int_0^t y(t)dt$:

$$L\left[\int_0^t y(t)dt\right] = \frac{Y(s)}{s}$$

Property 4 Given a generic function $y(t)$:

$$y(0) = \lim_{t \to 0} y(t) = \lim_{s \to \infty} sY(s)$$

This property is known as the Initial Value Theorem (IVT).

Property 5 Given a function $y(t)$ with the property of being bounded:

$$y_{SS} = \lim_{t \to \infty} y(t) = \lim_{s \to 0} sY(s)$$

This property is known as the Final Value Theorem (FVT).

The following Table A.4.1 is a list of commonly used Laplace transformations (LT) and inverse Laplace transformations (ILT). Note the impulse or spike function, also known as the Dirac function $\delta(t)$. This function has the following property:

$$\delta(t) = \begin{cases} \infty & t = 0 \\ 0 & t > 0 \end{cases}$$

Table A.4.1 LT and ILT for Common Functions

$y(t)$	$y(s)$
$\delta(t)$	1
1	$\dfrac{1}{s}$
c	$\dfrac{c}{s}$
t	$\dfrac{1}{s^2}$
$\dfrac{t^2}{2}$	$\dfrac{1}{s^3}$
$\dfrac{t^{n-1}}{(n-1)!}$	$\dfrac{1}{s^n}$
$e^{\pm at}$	$\dfrac{1}{(s \mp a)}$
$\dfrac{t^{n-1} e^{\pm at}}{(n-1)!}$	$\dfrac{1}{(s \mp a)^n}$
$\dfrac{\sin(at)}{a}$	$\dfrac{1}{(s^2 + a^2)}$
$\cos(at)$	$\dfrac{s}{(s^2 + a^2)}$

As previously described, following the application of the LT, the goal is to derive an expression for $Y(s)$.

Example A.4.1

Given the generic second order differential equation:

$$M y''(t) + f_V y'(t) + k y(t) = a$$

with initial conditions $c_1 = y(0)$, $c_2 = y'(0)$.

The application of the LT to the individual terms in the preceding equation will lead to:

$$L[My''(t)] = ML[y''(t)] = M(s^2 Y(s) - s y(0) - y'(0))$$

$$= M(s^2 Y(s) - c_1 s - c_2)$$

$$L[f_V y'(t)] = f_V L[y'(t)] = f_V (s Y(s) - y(0)) = f_V (s Y(s) - c_1)$$

$$L[k y(t)] = k L[y(t)] = k Y(s)$$

$$L[a] = \frac{a}{s}$$

Grouping the terms we will have:

$$M(s^2 Y(s) - c_1 s - c_2) + f_V(s Y(s) - c_1) + k Y(s) = \frac{a}{s}$$

By keeping all the terms associated with $Y(s)$ on the left-hand side and transferring all the terms associated with the initial conditions on the right-hand side we will have:

$$M s^2 Y(s) + f_V s Y(s) + k Y(s) = \frac{a}{s} + M c_1 s + M c_2 + f_V c_1$$

The rationalization of the right-hand side further leads to:

$$Y(s)(M s^2 + f_V s + k) = \frac{M c_1 s^2 + (M c_2 + f_V c_1) s + a}{s}$$

providing the final expression for $Y(s)$:

$$Y(s) = \frac{M c_1 s^2 + (M c_2 + f_V c_1) s + a}{s \ (M s^2 + f_V s + k)} = \frac{Num(s)}{Den(s)}$$

A few comments about the nature of the $Y(s)$ relationship are given next.

The values of the coefficients of the numerator polynomial $Num(s)$ depend on the values of the through the parameters on the right-hand side of the DE along with the value of the parameter a associated with the forcing term in the DE. The values of the coefficients of the denominator polynomial $Den(s)$ are instead functions of the parameters on the left-hand side of the DE along with the LT of the forcing term. In particular, the order of $Den(s)$ is always equal to the sum of the order of the DE plus whatever order is associated with the LT of the forcing terms in the DE. In this case the $Den(s)$ will be a third order polynomial.

Thus, following the application of the LT the following general expression is derived for $Y(s)$:

$$Y(s) = \frac{Num(s)}{Den(s)} \quad \text{where} \ \ order(Num(s)) = m < n = order(Den(s))$$

When using the partial fraction expansion (PFE) method the three following scenarios can occur depending on the nature of the roots of the polynomial $Den(s)$:

Case 1: Real distinct roots (poles)

Case 2: Real roots (poles) with multiplicity

Case 3: Complex conjugate roots (poles)

A.4 Review of Application of Laplace Transform to Linear Constant Coefficients Differential Equations

Case #1: Real distinct roots (poles) Given:

$$Y(s) = \frac{Num(s)}{Den(s)}$$

If the n-th order polynomial $Den(s)$ has real distinct roots (poles) we will have:

$$Y(s) = \frac{Num(s)}{Den(s)} = \frac{Num(s)}{(s+s_1)(s+s_2)\cdots(s+s_n)}$$

The application of the PFE method will lead to:

$$Y(s) = \frac{Num(s)}{Den(s)} = \frac{k_1}{(s+s_1)} + \frac{k_2}{(s+s_2)} + \cdots \frac{k_n}{(s+s_n)}$$

where the coefficients $k_1, k_2, \ldots, k_n$ will be evaluated using the relationships:

$$k_1 = [(s+s_1)Y(s)]|_{s=-s_1} = \left[\cancel{(s+s_1)}\frac{Num(s)}{\cancel{(s+s_1)}(s+s_2)\cdots(s+s_n)}\right]\bigg|_{s=-s_1}$$

$$= \frac{Num(-s_1)}{(-s_1+s_2)\cdots(-s_1+s_n)}$$

$$k_2 = [(s+s_2)Y(s)]|_{s=-s_2} = \left[\cancel{(s+s_2)}\frac{Num(s)}{(s+s_1)\cancel{(s+s_2)}\cdots(s+s_n)}\right]\bigg|_{s=-s_2}$$

$$= \frac{Num(-s_2)}{(-s_2+s_1)\cdots(-s_2+s_n)}$$

$$k_n = [(s+s_n)Y(s)]|_{s=-s_n} = \left[\cancel{(s+s_n)}\frac{Num(s)}{(s+s_1)(s+s_2)\cdots\cancel{(s+s_n)}}\right]\bigg|_{s=-s_n}$$

$$= \frac{Num(-s_n)}{(-s_n+s_1)(-s_n+s_2)\cdots}$$

Example A.4.2

Given:

$$Y(s) = \frac{(5s+3)}{s(s+1)(s+2)(s+3)}$$

The application of the PFE method leads to:

$$Y(s) = \frac{k_1}{s} + \frac{k_2}{(s+1)} + \frac{k_3}{(s+2)} + \frac{k_4}{(s+3)}$$

The coefficients k_1, k_2, k_3, k_4 are evaluated using:

$$k_1 = [s\,Y(s)]|_{s=0} = \left[\cancel{s}\frac{(5s+3)}{\cancel{s}(s+1)(s+2)(s+3)}\right]\bigg|_{s=0}$$

$$= \frac{Num(0)}{(0+1)(0+2)(0+3)} = \frac{3}{6} = 0.5$$

$$k_2 = [(s+1)\,Y(s)]|_{s=-1} = \left[\cancel{(s+1)}\frac{(5s+3)}{s\cancel{(s+1)}(s+2)(s+3)}\right]\bigg|_{s=-1}$$

$$= \frac{Num(-1)}{-1\cdot(-1+2)(-1+3)} = \frac{-2}{-2} = 1$$

$$k_3 = [(s+2)\,Y(s)]\big|_{s=-2} = \left[\cancel{(s+2)}\,\frac{(5s+3)}{s(s+1)\cancel{(s+2)}(s+3)}\right]\bigg|_{s=-2}$$

$$= \frac{Num(-2)}{-2\cdot(-2+1)(-2+3)} = \frac{-7}{2} = -3.5$$

$$k_4 = [(s+3)\,Y(s)]\big|_{s=-3} = \left[\cancel{(s+3)}\,\frac{(5s+3)}{s(s+1)(s+2)\cancel{(s+3)}}\right]\bigg|_{s=-3}$$

$$= \frac{Num(-3)}{-3\cdot(-3+1)(-3+2)} = \frac{-12}{-6} = 2$$

Therefore:

$$Y(s) = \frac{Num(s)}{Den(s)} = \frac{k_1}{s} + \frac{k_2}{(s+1)} + \frac{k_3}{(s+2)} + \frac{k_4}{(s+3)}$$

$$= 0.5\,\frac{1}{s} + 1\,\frac{1}{(s+1)} - 3.5\,\frac{1}{(s+2)} + 2\,\frac{1}{(s+3)}$$

The application of the inverse Laplace transforms (ILTs) using Table A.4.1 leads to:

$$\boxed{y(t) = 0.5 + 1\cdot e^{-t} - 3.5\cdot e^{-2t} + 2\cdot e^{-3t}}$$

Case #2: Real roots (poles) with multiplicity
Given:

$$Y(s) = \frac{Num(s)}{Den(s)}$$

The n-th order polynomial $Den(s)$ has n real roots (poles); however, at least one of these roots (poles) might be repeated with multiplicity r, where r is an integer. In this case we have:

$$Y(s) = \frac{Num(s)}{Den(s)} = \frac{Num(s)}{(s+s_1)(s+s_2)\ldots(s+s_i)^r\ldots(s+s_{n-r})}$$

The application of the PFE method leads to:

$$Y(s) = \frac{Num(s)}{Den(s)} = \left[\frac{k_1}{(s+s_1)} + \frac{k_2}{(s+s_2)} + \cdots + \frac{k_{n-r}}{(s+s_{n-r})}\right]$$

$$+ \left[\frac{k_{i1}}{(s+s_i)} + \frac{k_{i2}}{(s+s_i)^2} + \cdots + \frac{k_{ir}}{(s+s_i)^r}\right]$$

Therefore, there are two sets of fractions. The first set (involving $k_1, k_2, \ldots, k_{n-r}$) is associated with the real distinct roots (located at: $-s_1, -s_2, \ldots, -s_{n-r}$). The second set (involving $k_{i1}, k_{i2}, \ldots, k_{ir}$) is associated with the real root with multiplicity r (located at $-s_i$).

The coefficients $k_1, k_2, \ldots, k_{n-r}$ associated with the first set can be evaluated using the same approach introduced for the case of distinct real roots (poles). Therefore:

$$k_1 = [(s+s_1)Y(s)]\big|_{s=-s_1} = \left[\cancel{(s+s_1)}\,\frac{Num(s)}{\cancel{(s+s_1)}(s+s_2)\ldots(s+s_i)^r\ldots(s+s_{n-r})}\right]\bigg|_{s=-s_1}$$

$$= \frac{Num(-s_1)}{(-s_1+s_2)\ldots(-s_1+s_i)^r\ldots(-s_1+s_{n-r})}$$

A.4 Review of Application of Laplace Transform to Linear Constant Coefficients Differential Equations 569

$$k_2 = [(s+s_2)Y(s)]|_{s=-s_2} = \left[\cancel{(s+s_2)} \frac{Num(s)}{(s+s_1)\cancel{(s+s_2)}\dots(s+s_i)^r\dots(s+s_{n-r})} \right]\Bigg|_{s=-s_2}$$

$$= \frac{Num(-s_2)}{(-s_2+s_1)\dots(-s_2+s_i)^r\dots(-s_2+s_{n-r})}$$

$$k_{n-r} = [(s+s_{n-r})Y(s)]|_{s=-s_{n-r}}$$

$$= \left[\cancel{(s+s_{n-r})} \frac{Num(s)}{(s+s_1)(s+s_2)\dots(s+s_i)^r\dots\cancel{(s+s_{n-r})}} \right]\Bigg|_{s=-s_{n-r}}$$

$$= \frac{Num(-s_{n-r})}{(-s_n+s_1)(-s_n+s_2)\dots(s+s_i)^r\dots}$$

The coefficients $k_{i1}, k_{i2}, \dots, k_{ir}$ associated with the second set can be instead evaluated, in descending order, using:

$$k_{ir} = [(s+s_i)^r Y(s)]|_{s=-s_i} = \left[\cancel{(s+s_i)^r} \frac{Num(s)}{(s+s_1)(s+s_2)\dots\cancel{(s+s_i)^r}\dots(s+s_{n-r})} \right]\Bigg|_{s=-s_i}$$

$$= \frac{Num(-s_1)}{(-s_1+s_2)\dots(-s_1+s_i)^r\dots(-s_1+s_{n-r})}$$

$$k_{i(r-1)} = \frac{1}{1!} \frac{d}{ds}[(s+s_i)^r Y(s)]|_{s=-s_i}$$

$$k_{i(r-2)} = \frac{1}{2!} \frac{d^2}{ds^2}[(s+s_i)^r Y(s)]|_{s=-s_i}$$

$$k_{i1} = \frac{1}{(r-1)!} \frac{d^{r-1}}{ds^{r-1}}[(s+s_i)^r Y(s)]|_{s=-s_i}$$

Example A.4.3

Given:

$$Y(s) = \frac{1}{s(s+1)^3(s+2)}$$

The application of the PFE method leads to:

$$Y(s) = \left[\frac{k_1}{s} + \frac{k_3}{(s+2)} \right] + \left[\frac{k_{21}}{(s+1)} + \frac{k_{22}}{(s+1)^2} + \frac{k_{23}}{(s+1)^3} \right]$$

The coefficients k_1, k_3 are evaluated using:

$$k_1 = [s\, Y(s)]|_{s=0} = \left[\cancel{s} \frac{1}{\cancel{s}(s+1)^3(s+2)} \right]\Bigg|_{s=0} = \frac{1}{(0+1)^3(0+2)} = \frac{1}{2} = 0.5$$

$$k_3 = [(s+2)\, Y(s)]|_{s=-2} = \left[\cancel{(s+2)} \frac{1}{s\,(s+1)^3\cancel{(s+2)}} \right]\Bigg|_{s=-2} = \frac{1}{-2(-2+1)^3} = \frac{1}{2} = 0.5$$

The coefficients k_{21}, k_{22}, k_{23} are instead evaluated using:

$$k_{23} = [(s+1)^3 Y(s)]|_{s=-1} = \left[\cancel{(s+1)^3} \frac{1}{s\,\cancel{(s+1)^3}(s+2)} \right]\Bigg|_{s=-1}$$

$$= \frac{1}{-1(-1+2)} = \frac{1}{-1} = -1$$

$$k_{22} = \frac{d}{ds}[(s+1)^3 Y(s)]|_{s=-1} = \frac{d}{ds}\left[\cancel{(s+1)^3} \frac{1}{s\cancel{(s+1)^3}(s+2)}\right]\Bigg|_{s=-1}$$

$$= \frac{d}{ds}\left[\frac{1}{s(s+2)}\right]\Bigg|_{s=-1} = \left[\frac{-1(2s+2)}{[s(s+2)]^2}\right]\Bigg|_{s=-1} = \frac{0}{1} = 0$$

$$k_{21} = \frac{1}{2!}\frac{d^2}{ds^2}[(s+1)^3 Y(s)]|_{s=-1} = \frac{1}{2!}\frac{d^2}{ds^2}\left[\frac{1}{s(s+2)}\right]\Bigg|_{s=-1} = \frac{1}{2!}\frac{d}{ds}\left[\frac{-(2s+2)}{[s(s+2)]^2}\right]\Bigg|_{s=-1}$$

$$= \frac{1}{2!}\frac{d}{ds}\left[\frac{-(2s+2)}{[s^4+4s^3+4s^2]}\right]\Bigg|_{s=-1} = \frac{1}{2!}\left[\frac{-2[s^4+4s^3+4s^2] + (2s+2)(4s^3+12s^2+8s)}{[s^4+4s^3+4s^2]^2}\right]\Bigg|_{s=-1}$$

$$= \frac{1}{2!}\left[\frac{-2[1-4+4] + \cancel{(-2+2)(-4+12-8)}}{[1-4+4]^2}\right] = \frac{1}{2!}\frac{-2}{1} = -1$$

By combining the coefficients k_{21}, k_{22}, k_{23} and k_1, k_3 in $Y(s)$:

$$Y(s) = \frac{k_1}{s} + \frac{k_3}{(s+2)} + \frac{k_{21}}{(s+1)} + \frac{k_{22}}{(s+1)^2} + \frac{k_{23}}{(s+1)^3}$$

$$= 0.5\,\frac{1}{s} + 0.5\,\frac{1}{(s+2)} - 1\frac{1}{(s+1)} + (0)\frac{1}{(s+1)^2} - 1\frac{1}{(s+1)^3}$$

The application of the Inverse Laplace Transforms (ILTs) using Table A.4.1 leads to:

$$\boxed{y(t) = 0.5 + 0.5 \cdot e^{-2t} - e^{-t} - 0.5 \cdot t^2 \cdot e^{-t}}$$

Note: Recall that

$$L^{-1}\left[\frac{1}{(s+a)^n}\right] = \frac{1}{(n-1)!}\,t^{n-1}\,e^{-at}$$

Case 3: Complex conjugate roots (poles)

Given the numerical complexity of this case (and given its relevance for the analysis of the aircraft dynamics), the application of the PFE to the case with complex conjugate roots (poles) is shown directly through an example.

Example A.4.4

Given:

$$Y(s) = \frac{2(s+1)}{s(s^2+s+2)}$$

In this case the roots (poles) are given by:

$$s_1 = 0, \quad s_2 = -\frac{1}{2} - i\frac{\sqrt{7}}{2}, \quad s_3 = s_2^* = -\frac{1}{2} + i\frac{\sqrt{7}}{2}$$

where "*" indicates the complex conjugate of an imaginary number.

The basic application of the PFE method leads to:

$$Y(s) = \frac{k_1}{s} + \frac{k_2}{\left(s + \frac{1}{2} + i\frac{\sqrt{7}}{2}\right)} + \frac{k_2^*}{\left(s + \frac{1}{2} - i\frac{\sqrt{7}}{2}\right)}$$

Note that since $s_3 = s_2^*$ (conjugate of s_2), we have that $k_3 = k_2^*$ (conjugate of k_2). Therefore, using the 'basic' PFE method we have:

$$k_1 = [sY(s)]|_{s=0} = \left[\cancel{s}\frac{2(s+1)}{\cancel{s}(s^2+s+2)}\right]\bigg|_{s=0} = \frac{2}{2} = 1$$

$$k_2 = \left[\left(s+\frac{1}{2}+i\frac{\sqrt{7}}{2}\right)Y(s)\right]\bigg|_{s=-\frac{1}{2}-i\frac{\sqrt{7}}{2}}$$

$$= \left[\cancel{\left(s+\frac{1}{2}+i\frac{\sqrt{7}}{2}\right)}\frac{2(s+1)}{s\cancel{\left(s+\frac{1}{2}+i\frac{\sqrt{7}}{2}\right)}\left(s+\frac{1}{2}-i\frac{\sqrt{7}}{2}\right)}\right]\bigg|_{s=-\frac{1}{2}-i\frac{\sqrt{7}}{2}}$$

$$= \frac{2\left(-\frac{1}{2}-i\frac{\sqrt{7}}{2}+1\right)}{\left(-\frac{1}{2}-i\frac{\sqrt{7}}{2}\right)\left(\cancel{-\frac{1}{2}}-i\frac{\sqrt{7}}{2}+\cancel{\frac{1}{2}}-i\frac{\sqrt{7}}{2}\right)} = \frac{2\left(\frac{1}{2}-i\frac{\sqrt{7}}{2}\right)}{\left(-\frac{1}{2}-i\frac{\sqrt{7}}{2}\right)(-i\sqrt{7})} = \frac{(1-i\sqrt{7})}{\left(i\frac{\sqrt{7}}{2}-\frac{7}{2}\right)}$$

The previous expression for k_2 needs to be rationalized with the goal of eliminating the imaginary component at the denominator. Given a generic imaginary expression:

$$x = \frac{a+ib}{c+id}$$

The rationalization of x will give:

$$x = \frac{a+ib}{c+id} \cdot \frac{c-id}{c-id} = \frac{(ac+bd)+i(bc-ad)}{c^2+d^2}$$

with the property that the denominator becomes a real number. For k_2 we have:

$$k_2 = \frac{(1-i\sqrt{7})}{\left(i\frac{\sqrt{7}}{2}-\frac{7}{2}\right)} \cdot \frac{\left(i\frac{\sqrt{7}}{2}+\frac{7}{2}\right)}{\left(i\frac{\sqrt{7}}{2}+\frac{7}{2}\right)} = \frac{\left(i\frac{\sqrt{7}}{2}+\frac{7}{2}+\frac{7}{2}-i\frac{7}{2}\sqrt{7}\right)}{\left(-\frac{7}{4}-\frac{49}{4}\right)}$$

$$= \frac{(7-i\cdot 3\sqrt{7})}{-14} = -\frac{1}{2}+i\frac{3}{14}\sqrt{7}$$

Since $k_3 = k_2^*$ we have $k_3 = k_2^* = -\frac{1}{2}-i\frac{3}{14}\sqrt{7}$.

Therefore, we have:

$$Y(s) = \frac{1}{s} + \frac{\left(-\frac{1}{2}+i\frac{3}{14}\sqrt{7}\right)}{\left(s+\frac{1}{2}+i\frac{\sqrt{7}}{2}\right)} + \frac{\left(-\frac{1}{2}-i\frac{3}{14}\sqrt{7}\right)}{\left(s+\frac{1}{2}-i\frac{\sqrt{7}}{2}\right)}$$

Treating the complex conjugate roots (poles) as ordinary roots, using the ILTs in Table A.4.1:

$$y(t) = 1 + \left(-\frac{1}{2} + i\frac{3}{14}\sqrt{7}\right)e^{-\left(\frac{1}{2} + i\frac{\sqrt{7}}{2}\right)t} + \left(-\frac{1}{2} - i\frac{3}{14}\sqrt{7}\right)e^{-\left(\frac{1}{2} - i\frac{\sqrt{7}}{2}\right)t}$$

While formally correct, the preceding expression for $y(t)$ is virtually untreatable. Next, an important step is to take a new approach in the modeling of k_2, k_3. As stated, k_2, k_3 are imaginary conjugate numbers. An alternative approach in handling imaginary conjugate numbers is given by the application of **Moivre's formula**. Conceptually this approach entails treating imaginary conjugate numbers as 2D vectors in the polar coordinates, in lieu of the more conventional Cartesian coordinates. See Figure A.4.2.

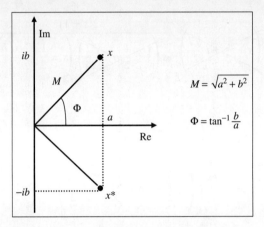

Figure A.4.2 Polar Relationships for Imaginary Roots

For example, given the complex conjugate numbers:

$$x = a + i \cdot b$$
$$x^* = a - i \cdot b$$

Therefore, we have:

$$x = a + i \cdot b = Me^{i\Phi}$$
$$x^* = a - i \cdot b = Me^{-i\Phi}$$

The application of the Moivre's formula to k_2, k_3 above leads to:

$$k_2 = -\frac{1}{2} + i\frac{3}{14}\sqrt{7} = Me^{i\Phi}$$
$$k_3 = k_2^* = -\frac{1}{2} - i\frac{3}{14}\sqrt{7} = Me^{-i\Phi}$$

with:

$$M = \sqrt{\left(-\frac{1}{2}\right)^2 + \left(\frac{3}{14}\sqrt{7}\right)^2} = \sqrt{\left(\frac{1}{4}\right) + 7 \cdot \left(\frac{3}{14}\right)^2} = 0.756$$

$$\Phi = \tan^{-1}\frac{\left(\frac{3}{14}\sqrt{7}\right)}{\left(-\frac{1}{2}\right)} = -0.848 \text{ (rad)}$$

A.4 Review of Application of Laplace Transform to Linear Constant Coefficients Differential Equations

Therefore, we have:

$$k_2 = -\frac{1}{2} + i\frac{3}{14}\sqrt{7} = Me^{i\Phi} = 0.756e^{-i\,0.848}$$

$$k_3 = k_2^* = -\frac{1}{2} - i\frac{3}{14}\sqrt{7} = Me^{-i\Phi} = 0.756\,e^{i\,0.848}$$

Thus, the previous $y(t)$ expression becomes:

$$y(t) = 1 + \left(-\frac{1}{2} + i\frac{3}{14}\sqrt{7}\right)e^{-\left(\frac{1}{2} + i\frac{\sqrt{7}}{2}\right)t} + \left(-\frac{1}{2} - i\frac{3}{14}\sqrt{7}\right)e^{-\left(\frac{1}{2} - i\frac{\sqrt{7}}{2}\right)t}$$

$$= 1 + 0.756\,e^{-i\,0.848}\,e^{-(0.5 + i\,1.323)t} + 0.756\,e^{i\,0.848}\,e^{-(0.5 - i\,1.323)t}$$

Recall that:

$$e^a \cdot e^b = e^{(a+b)}$$

Therefore, after separating the imaginary and the real terms, we have:

$$y(t) = 1 + 0.756\,e^{-i\,0.848}\,e^{-(0.5 + i\,1.323)t} + 0.756\,e^{i\,0.848}\,e^{-(0.5 - i\,1.323)t}$$

$$= 1 + 0.756\,e^{-0.5\,t}\left[e^{-i\,(0.848 + 1.323t)} + e^{i\,(0.848 + 1.323t)}\right]$$

At this point, recall the famous Euler sine and cosine formulas for a generic angle x:

$$\boxed{\begin{aligned} e^{ix} &= \cos(x) + i\sin(x) \\ e^{-ix} &= \cos(x) - i\sin(x) \end{aligned}}$$

Therefore:

$$e^{ix} + e^{-ix} = \cos(x) + \cancel{i\sin(x)} + \cos(x) - \cancel{i\sin(x)} = 2 \cdot \cos(x)$$

$$e^{ix} - e^{-ix} = \cancel{\cos(x)} + i\sin(x) - \cancel{\cos(x)} + i\sin(x) = i \cdot (2\sin(x))$$

Thus, the application of the Euler formulas leads to:

$$\left[e^{-i\,(0.848 + 1.323t)} + e^{i\,(0.848 + 1.323t)}\right] = 2 \cdot \cos(0.848 + 1.323t)$$

The final expression for $y(t)$ becomes:

$$y(t) = 1 + 0.756\,e^{-0.5t}\left[e^{-i(0.848 + 1.323t)} + e^{i\,(0.848 + 1.323t)}\right]$$

$$= 1 + 0.756\,e^{-0.5t} \cdot 2 \cdot \cos(0.848 + 1.323t)$$

$$\boxed{y(t) = 1 + 1.512 \cdot e^{-0.5t} \cdot \cos(0.848 + 1.323t)}$$

A plot of $y(t)$ is shown in Figure A.4.3.

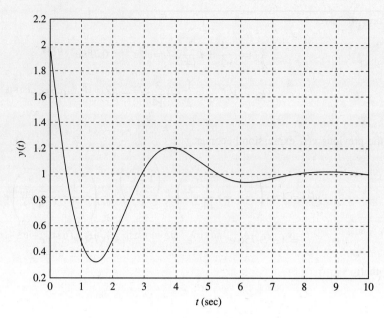

Figure A.4.3 Sample Response of a System with Imaginary Roots

The importance of Case 3 is clear for the purpose of analyzing aircraft dynamic responses, as shown in Chapter 7.

A.5 REVIEW OF FIRST AND SECOND ORDER SYSTEMS

A generic *n-th* order linear dynamic system can be considered a blending of first and second order dynamic systems. Therefore, it is important to understand the characteristics of the dynamic responses of first and second order systems for the purpose of analyzing the dynamic response of *n-th* order systems. For this purpose, the concept of generic first order system (GFOS) and generic second order system (GSOS) are introduced. The GFOS and the GSOS have the properties of having 0 as initial conditions and 1 as their steady-state response. They will be discussed next.

The GFOS is characterized by having the following transfer function:

$$GFOS(s) = \frac{Num(s)}{Den(s)} = \frac{a}{s+a}$$

Therefore, its denominator has a real root located at $s = -a$, as shown in Figure A.5.1.

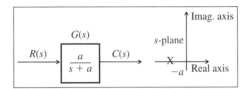

Figure A.5.1 Transfer Function of a Generic First Order System (GFOS)

Assuming a step input $u(t) = 1 \rightarrow U(s) = \frac{1}{s}$ we have:

$$Y(s) = U(s) \cdot GFOS(s) = \frac{1}{s}\left(\frac{a}{s+a}\right)$$

Following application of the PFE method, we have:

$$Y(s) = k_1 \frac{1}{s} + k_2 \left(\frac{1}{s+a}\right)$$

leading to:

$$k_1 = [sY(s)]|_{s=0} = s\frac{1}{s}\left(\frac{a}{s+a}\right)\bigg|_{s=0} = \cancel{s}\,\frac{1}{\cancel{s}}\left(\frac{a}{\cancel{s}+a}\right) = 1$$

$$k_1 = [(s+a)\,Y(s)]|_{s=-a} = (s+a)\frac{1}{s}\left(\frac{a}{s+a}\right)\bigg|_{s=-a} = \cancel{(s+a)}\frac{1}{-a}\left(\frac{a}{\cancel{s+a}}\right) = -1$$

Thus, we have:

$$Y(s) = \frac{1}{s} - 1\left(\frac{1}{s+a}\right)$$

Following the application of the ILT we have:

$$\boxed{y(t) = 1 - e^{-at}}$$

The response of a GFOS with '$a = 1$' is shown in Figure A.5.2.

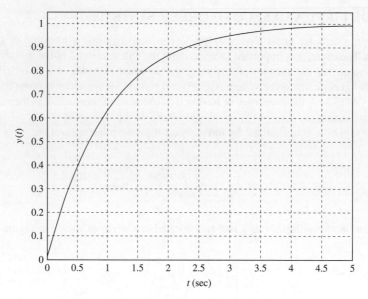

Figure A.5.2 Response of a First Order System

The following dynamic characteristics are introduced to analyze the dynamic response of 1st order systems.

- **Time Constant** (T). Time Constant is defined as the time required for a first order system to reach 63 percent of its steady-state response;
- **Rise Time** (T_R). Rise Time is defined as the time required for a first order system to go from 10 to 90 percent of its steady-state response;
- **Setting Time** (T_S). Settling Time is defined as the time required for a first order system to reach 98 percent of its steady-state response.

It can be shown that for a GFOS these parameters are given by the following relationships: $T = \frac{1}{a}$, $T_R = \frac{2.2}{a}$, $T_S = \frac{3.9}{a}$. The parameters are also shown in Figure A.5.3:

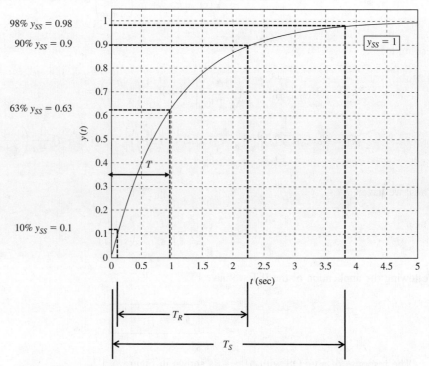

Figure A.5.3 Time Constant, Rise Time, and Settling Time for a First Order System

Based on the preceding analysis, the responses of first order and GFOS systems are function only of the parameter a, as shown in Figure A.5.4:

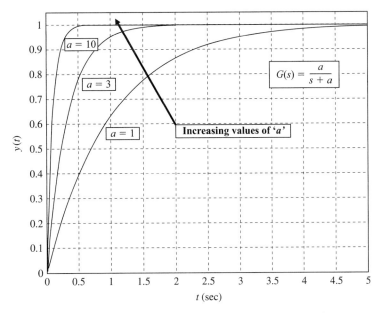

Figure A.5.4 Responses of a GFOS for Different Values of a

The GSOS is characterized by having the following transfer function:

$$GSOS(s) = \frac{Num(s)}{Den(s)} = \frac{\omega_n^2}{s^2 + 2\zeta\omega_n s + \omega_n^2}$$

with the property that the denominator has a pair of complex conjugate roots located at (see Figure A.5.5):

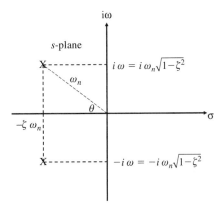

Figure A.5.5 Complex Conjugate Roots for a Second Order System

Assuming a step input $u(t) = 1 \rightarrow U(s) = \frac{1}{s}$ we have:

$$Y(s) = U(s) \cdot GSOS(s) = \frac{1}{s}\left(\frac{\omega_n^2}{s^2 + 2\zeta\omega_n s + \omega_n^2}\right)$$

Following application of the PFE method, we have:

$$Y(s) = k_1 \frac{1}{s} + k_2\left(\frac{1}{s+p_1}\right) + k_2^*\left(\frac{1}{s+p_1^*}\right)$$

where $p_2 = p_1^*$, $k_3 = k_2^*$ with $p_{1,2} = -\zeta\omega_n \pm i\omega_n\sqrt{1-\zeta^2}$.

Using the PFE method relative to the case of complex conjugate roots (Appendix A.4, Case 3), through PFE and ILT a solution is given by:

$$y(t) = 1 - \frac{1}{\sqrt{1-\zeta^2}} e^{-\zeta\omega_n t} \cos\left(\omega_n\sqrt{1-\zeta^2}\, t - \phi\right)$$

shown in Figure A.5.6, with $\phi = \tan^{-1}\left(\frac{\zeta}{\sqrt{1-\zeta^2}}\right)$.

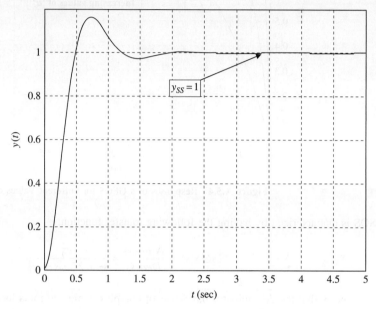

Figure A.5.6 Response of a GSOS

The following dynamic characteristics are introduced to analyze the dynamic response of second order systems.

- **Overshoot Percentage** ($OS\%$). Overshoot Percentage is defined as the percentage of the ratio between the difference of the maximum response and the steady-state response versus the steady-state response;
- **Peak Time** (T_P). Peak Time is defined as the time at which the response reaches its maximum value;
- **Rise Time** (T_R). Rise Time is defined as the time required for a second order system to go from 10 to 90 percent of its steady-state response;
- **Setting Time** (T_S). Settling Time is defined as the time required for a second order system to reach *and* stay within 2 percent of its steady-state response.

It can be shown that for a GSOS the preceding parameters are given by the following relationships:

$$OS\% = \frac{y_{MAX} - y_{SS}}{y_{SS}} = e^{-\frac{\pi\zeta}{\sqrt{1-\zeta^2}}} \Rightarrow \zeta = \sqrt{\frac{(\ln(os\%))^2}{\pi^2 + (\ln(os\%))^2}}$$

$$T_P \approx \frac{\pi}{\omega_n\sqrt{1-\zeta^2}}$$

$$T_R \approx \frac{1 + 1.1\zeta + 1.4\zeta^2}{\omega_n}$$

$$T_S \approx \frac{4}{\zeta\omega_n}$$

Note that $(OS\%)$ is the only parameter which is function only of the damping coefficient (ζ). The other parameters are functions of both damping and natural frequency (ζ, ω_n).

The previous parameters are shown in Figure A.5.7:

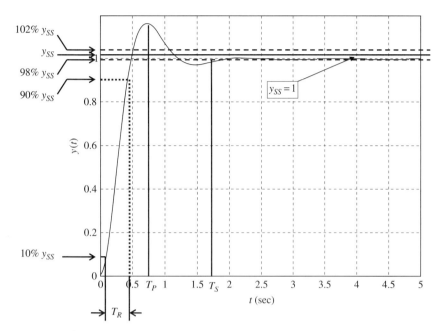

Figure A.5.7 Rise Time, Peak Time, and Settling Time for a GSOS

The influence of the damping and natural frequency (ζ, ω_n) on the response of a GSOS are shown in Figures A.5.8 and A.5.9.

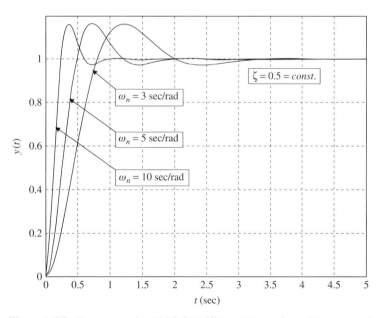

Figure A.5.8 Responses of a GFOS for Different Values of ω_n with constant ζ

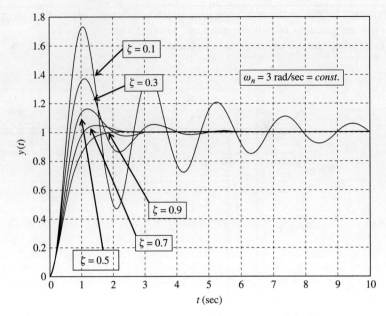

Figure A.5.9 Responses of a GFOS for Different Values of ζ with constant ω_n

A.6 REVIEW OF STANDARD ATMOSPHERIC MODEL

The performance of the aircraft depends on the conditions of the atmosphere through which it is flying. Since the atmospheric conditions of pressure, density, and temperature are constantly changing, it is desirable to define standard atmospheric conditions to be used for comparison purposes. Tests can then be performed under ambient conditions and the data reduced to predict what the test results would have been under the standard conditions. This allows data obtained under several different conditions to be directly compared.

Two standard atmosphere conditions are currently in use in the aeronautical world. The NASA (National Aeronautics and Space Administration) condition is used in the United States. The ICAN (International Commission of Air Navigation) condition is used in most European countries and some Asian countries. Both approximate the standard conditions of 40°N latitude. Both conditions are identical up to a standard altitude of 35,332 ft.; furthermore, both conditions set standard sea level conditions as:

$$\rho_0 = 0.002378 \text{ slugs/ft}^3$$
$$p_0 = 2116 \text{ lbs/ft}^2 = 760 \text{ mm } Hg = 29.921 \text{ in } Hg = 406 \text{ in } H_2O$$
$$T_0 = 59°F = 518.4°R = 288°K = 15°C$$
$$g_0 = 32.17 \text{ ft/sec}^2$$

Both NASA and ICAN conditions use the following assumptions:

Assumption 1

The Equation of State is given by:

$$P = \rho R T$$

The relationship holds true throughout the defined atmosphere, where:

P = ambient pressure (lbs/ft^2)
ρ = ambient density (slugs/ft^3)
T = ambient temperature (°R = °F + 459.4)
R = air gas constant = 1,716 ft^2/sec^2 °R

Assumption 2

A temperature lapse rate of 3.57 °F per 1,000 ft is assumed for the isothermal region. Therefore

$$T(°F) = 59° + kh$$

with $k = -3.57/1,000$.

Assumption 3

The relationship between pressure and altitude is given by the hydrostatic equation:

$$dP = -\rho g dh$$

Assumption 4

$$g = g_0 = 32.17 \text{ ft/sec}^2$$

for all the altitudes.

Using these assumptions, the derivation of the standard atmospheric model (SAM) starts from:

$$dP = -\rho g dh = -\rho g_0 dh$$

By substituting $\rho = \frac{P}{RT}$ from the Equation of State we have:

$$dP = -\rho g_0 dh = -\frac{P g_0 dh}{RT}$$

Inserting the lapse rate equation $T = T_0 + k h$, we have:

$$dP = -\frac{P g_0 dh}{RT} = -\frac{P g_0 dh}{R(T_0 + k\ h)}$$

leading to:

$$\frac{dP}{P} = -\frac{g_0}{R}\frac{dh}{(T_0 + k\ h)}$$

Next, integrate both sides from sea level standard conditions to some standard altitude h with pressure p, density ρ, and temperature T:

$$\int_{p_0}^{p}\frac{dP}{P} = \int_{0}^{h} -\frac{g_0}{R}\frac{dh}{(T_0 + k\ h)}$$

leading to:

$$\frac{P}{P_0} = \left(1 + \frac{kh}{T_0}\right)^{-\frac{g_0}{kR}}$$

Substituting the numerical values for the constants, we have the first relationship of the SAM:

$$\boxed{\delta = \frac{P}{P_0} = \left(1 + \frac{kh}{T_0}\right)^{-\frac{g_0}{kR}} = \left[1 - (6.89 \cdot 10^{-6})h\right]^{5.256}}$$

The second SAM relationship is simply a modification of the temperature lapse rate.

$$\boxed{\theta = \frac{T}{T_0} = \left(\frac{T_0 + kh}{T_0}\right) = \left(1 + \frac{kh}{T_0}\right) = \left[1 - (6.89 \cdot 10^{-6})h\right]}$$

And, finally, from the Equation of State:

$$\rho = \frac{P}{RT}, \rho_0 = \frac{P_0}{RT_0} \rightarrow \sigma = \frac{\rho}{\rho_0} = \frac{\frac{P}{RT}}{\frac{P_0}{RT_0}} = \frac{P}{P_0}\frac{T_0}{T} = \frac{\delta}{\theta}$$

leading to the third relationship of the SAM:

$$\boxed{\begin{aligned}\sigma = \frac{\rho}{\rho_0} = \frac{\delta}{\theta} &= \frac{[1-(6.89\cdot 10^{-6})h]^{5.256}}{[1-(6.89\cdot 10^{-6})h]} \\ &= [1-(6.89\cdot 10^{-6})h]^{(5.256-1)} = [1-(6.89\cdot 10^{-6})h]^{4.256}\end{aligned}}$$

A specific set of definitions is needed for defining altitudes. The SAM defined previously specifies a certain value for pressure, density, and altitude associated with each altitude. However, it would be a pure coincidence if test data were obtained under standard atmospheric conditions. To express the nonstandard conditions typically obtained, we specify the following altitude definitions.

Pressure Altitude H_P for altitude $< 35{,}332$ ft.

$$\frac{P}{P_0} = [1-(6.89\cdot 10^{-6})H_P]^{5.256} \rightarrow H_P = 1.452\cdot 10^5\left[1-\left(\frac{P}{P_0}\right)^{0.1903}\right]$$

Density Altitude H_d for altitude $< 35{,}332$ ft.

$$\sigma = \frac{\rho}{\rho_0} = [1-(6.89\cdot 10^{-6})h]^{4.256} \rightarrow H_d = 1.452\cdot 10^5\left[1-\left(\frac{\rho}{\rho_0}\right)^{0.2349}\right]$$

Temperature Altitude H_T for altitude $< 35{,}332$ ft.

$$\theta = \frac{T}{T_0} = [1-(6.89\cdot 10^{-6})h] \rightarrow H_T = 1.452\cdot 10^5\left[1-\left(\frac{T}{T_0}\right)\right]$$

Appendix B

Data for Different Aircraft

TABLE OF CONTENTS

    B.1   Introduction
    B.2   Aircraft 1—Cessna 182
    B.3   Aircraft 2—Cessna 310
    B.4   Aircraft 3—Beech 99
    B.5   Aircraft 4—Cessna T37-A
    B.6   Aircraft 5—Cessna 620
    B.7   Aircraft 6—Learjet 24
    B.8   Aircraft 7—Boeing 747-200
    B.9   Aircraft 8—SIAI Marchetti S-211
    B.10   Aircraft 9—Lockheed F-104
    B.11   Aircraft 10—McDonnell Douglas F-4
           Reference

B.1 INTRODUCTION

Appendix B provides a database for ten aircraft from different classes at different flight conditions.[1] The database includes geometry data, flight conditions, inertial characteristics (with respect to the body axes), and dimensionless aerodynamic coefficients (with respect to the stability axes). Detailed sketches and drawings for each of these aircraft are provided in Appendix C. The list of aircraft includes the following:

Aircraft 1—Cessna 182
Single piston engine, general aviation aircraft (Class I)

Aircraft 2—Cessna 310
Twin piston engine, general aviation aircraft (Class I)

Aircraft 3—Beech 99
Twin turboprop, regional commercial aviation aircraft (Class II)

Aircraft 4—Cessna T-37A
Twin jet, military training aircraft (Class II)

Aircraft 5—Cessna 620
Four piston engine, corporate aviation aircraft (Class II)

Aircraft 6—Learjet 24
Single jet, corporate aviation aircraft (Class II)

Aircraft 7—Boeing 747-200
Four jet engine, commercial jetliner (Class III)

Aircraft 8—SIAI Marchetti S-211
Single jet engine, fighter trainer (Class IV)

Aircraft 9—Lockheed F-104
Single jet engine, military fighter/interceptor aircraft (Class IV)

Aircraft 10—McDonnell Douglas F-4
Twin jet engine, military fighter/attack aircraft (Class IV)

B.2 AIRCRAFT 1—CESSNA 182

Table B1.1 Geometric Data for the Cessna 182 Aircraft

Wing surface (ft^2)	S	174
Mean Aerodynamic Chord (MAC) (ft)	$\bar{c}$	4.9
Wing Span (ft)	b	36

Table B1.2 Flight Conditions Data for the Cessna 182 Aircraft

		Climb	Cruise	Approach
Altitude (ft)	h	0	5,000	0
Mach Number	M	0.120	0.201	0.096
True Airspeed (ft/sec)	V_{P_1}	133.5	220.1	107.1
Dynamic Pressure (lbs/ft^2)	$\bar{q}$	21.2	49.6	13.6
Location of CG - % MAC	$\bar{x}_{CG}$	0.264	0.264	0.264
Steady state angle of attack (deg)	α_1	5.4	0	4

Table B1.3 Mass and Inertial Data for the Cessna 182 Aircraft

		Climb	Cruise	Approach
Mass (lbs)	m	2,650	2,650	2,650
Moment of Inertia x-axis (slug ft^2)	I_{XX_B}	948	948	948
Moment of Inertia y-axis (slug ft^2)	I_{YY_B}	1,346	1,346	1,346
Moment of Inertia z-axis (slug ft^2)	I_{ZZ_B}	1,967	1,967	1,967
Product of inertia xz-plane (slug ft^2)	I_{XZ_B}	0	0	0

Table B1.4 Longitudinal Aerodynamic Coefficients for the Cessna 182 Aircraft

	Climb	Cruise	Approach
Steady State			
c_{L_1}	0.719	0.307	1.120
c_{D_1}	0.057	0.032	0.132
c_{m_1}	0	0	0
$c_{T_{X_1}}$	0.057	0.032	0.132
$c_{m_{T_1}}$	0	0	0
Stability Derivatives			
c_{D_0}	0.027	0.027	0.0605
c_{D_u}	0	0	0
c_{D_α}	0.38	0.121	0.547
$c_{T_{X_u}}$	−0.171	−0.096	−0.396
c_{L_0}	0.307	0.307	0.807
c_{L_u}	0	0	0
c_{L_α}	4.41	4.41	4.41
$c_{L_{\dot{\alpha}}}$	1.7	1.7	1.7
c_{L_q}	3.9	3.9	3.9
c_{m_0}	0.04	0.04	0.04
c_{m_u}	0	0	0
c_{m_α}	−0.650	−0.613	−0.611
$c_{m_{\dot{\alpha}}}$	−5.57	−7.27	−5.40
c_{m_q}	−15.2	−12.4	−11.4
$c_{m_{T_u}}$	0	0	0
$c_{m_{T_\alpha}}$	0	0	0
Control Derivatives			
$c_{D_{\delta_E}}$	0	0	0
$c_{L_{\delta_E}}$	0.43	0.43	0.43
$c_{m_{\delta_E}}$	−1.369	−1.122	−1.029

Table B1.5 Longitudinal Dimensional Stability Derivatives for the Cessna 182 Aircraft

Cruise
$X_u = -0.0304$, $X_{T_u} = -0.0152$, $X_\alpha = 19.459$, $X_{\delta_E} = 0$
$Z_u = -0.2919$, $Z_\alpha = -464.71$, $Z_{\dot\alpha} = -1.98$, $Z_q = -4.542$, $Z_{\delta_E} = -44.985$
$M_u = 0$, $M_{T_u} = 0$, $M_\alpha = -19.26$, $M_{T_\alpha} = 0$, $M_{\dot\alpha} = -2.543$, $M_q = -4.337$, $M_{\delta_E} = -35.251$

Table B1.6 Longitudinal Transfer Functions for the Cessna 182 Aircraft

Cruise

$$\frac{\alpha(s)}{\delta_E(s)} = \frac{Num_\alpha(s)}{\overline{D}_1(s)} = \frac{A_\alpha s^3 + B_\alpha s^2 + C_\alpha s + D_\alpha}{A_1 s^4 + B_1 s^3 + C_1 s^2 + D_1 s + E_1}$$

$$= \frac{-44.985\, s^3 - 7794.87\, s^2 - 355.63\, s - 330.516}{222.05\, s^4 + 1985.95\, s^3 + 6262.28\, s^2 + 329.88\, s + 180.58}$$

$$\frac{u(s)}{\delta_E(s)} = \frac{Num_u(s)}{\overline{D}_1(s)} = \frac{A_u s^2 + B_u s + C_u}{A_1 s^4 + B_1 s^3 + C_1 s^2 + D_1 s + E_1}$$

$$= \frac{-875.36\, s^2 + 96137.81\, s + 498397.28}{222.05\, s^4 + 1985.95\, s^3 + 6262.28\, s^2 + 329.88\, s + 180.58}$$

$$\frac{\theta(s)}{\delta_E(s)} = \frac{Num_\theta(s)}{\overline{D}_1(s)} = \frac{A_\theta s^2 + B_\theta s + C_\theta}{A_1 s^4 + B_1 s^3 + C_1 s^2 + D_1 s + E_1}$$

$$= \frac{-7713.23\, s^2 - 15867.0\, s - 908.24}{222.05\, s^4 + 1985.95\, s^3 + 6262.28\, s^2 + 329.88\, s + 180.58}$$

$$roots(\overline{D}_1(s)) = -4.45 \pm i\, 2.825,\quad -0.022 \pm i\, 0.17$$

$$\zeta_{SP} = 0.844,\quad \omega_{nSP} = 5.27,\quad \zeta_{Ph} = 0.129,\quad \omega_{nPh} = 0.171$$

Table B1.7 Lateral Directional Aerodynamic Coefficients for the Cessna 182 Aircraft

	Climb	Cruise	Approach
Stability Derivatives			
c_{l_β}	−0.0895	−0.0923	−0.969
c_{l_p}	−0.487	−0.484	−0.494
c_{l_r}	0.1869	0.0798	0.2039
c_{Y_β}	−0.404	−0.393	−0.303
c_{Y_p}	−0.145	−0.075	−0.213
c_{Y_r}	0.267	0.214	0.201
c_{n_β}	0.0907	0.0587	0.0701
$c_{n_{T_\beta}}$	0	0	0
c_{n_p}	−0.0649	−0.0278	−0.096
c_{n_r}	−0.1199	−0.0937	−0.1151
Control Derivatives			
$c_{l_{\delta_A}}$	0.229	0.229	0.229
$c_{l_{\delta_R}}$	0.0147	0.0147	0.0147
$c_{Y_{\delta_A}}$	0	0	0
$c_{Y_{\delta_R}}$	0.187	0.187	0.187
$c_{n_{\delta_A}}$	−0.0504	−0.0216	−0.0786
$c_{n_{\delta_R}}$	−0.0805	−0.0645	−0.0604

Table B1.8 Lateral Directional Dimensional Stability Derivatives for the Cessna 182 Aircraft

Cruise

$$Y_\beta = -41.11, \ Y_P = -0.642, \ Y_r = 1.831, \ Y_{\delta_A} = 0, \ Y_{\delta_R} = 19.56$$

$$L_\beta = -30.25, \ L_P = -12.97, \ L_r = 2.14, \ L_{\delta_A} = 75.06, \ L_{\delta_R} = 4.82$$

$$N_\beta = 9.27, \ N_{T_\beta} = 0, \ N_P = -0.36, \ N_r = -1.21, \ N_{\delta_A} = -3.41, \ N_{\delta_R} = -10.19$$

Table B1.9 Lateral Directional Transfer Functions for the Cessna 182 Aircraft

Cruise

$$\frac{\beta(s)}{\delta_A(s)} = \frac{Num_{\beta A}(s)}{\overline{D}_2(s)} = \frac{s(A_{\beta A}s^3 + B_{\beta A}s^2 + C_{\beta A}s + D_{\beta A})}{s(A_2 s^4 + B_2 s^3 + C_2 s^2 + D_2 s + E_2)}$$

$$= \frac{s(0\, s^3 + 696.43\, s^2 + 17{,}900.03\, s + 2683.95)}{s(220.07\, s^4 + 3162.72\, s^3 + 6212.56\, s^2 + 30{,}261.69\, s + 539.17)}$$

$$\frac{\phi(s)}{\delta_A(s)} = \frac{Num_{\phi A}(s)}{\overline{D}_2(s)} = \frac{s(A_{\phi A}s^2 + B_{\phi A}s + C_{\phi A})}{s(A_2 s^4 + B_2 s^3 + C_2 s^2 + D_2 s + E_2)}$$

$$= \frac{s(16{,}516.80\, s^2 + 21{,}473.13\, s + 132{,}776.72)}{s(220.07\, s^4 + 3162.72\, s^3 + 6212.56\, s^2 + 30{,}261.69\, s + 539.17)}$$

$$\frac{\psi(s)}{\delta_A(s)} = \frac{Num_{\psi A}(s)}{\overline{D}_2(s)} = \frac{A_{\psi A}s^3 + B_{\psi A}s^2 + C_{\psi A}s + D_{\psi A}}{s(A_2 s^4 + B_2 s^3 + C_2 s^2 + D_2 s + E_2)}$$

$$= \frac{(-750.84\, s^3 - 15{,}813.43\, s^2 - 3{,}308.40\, s + 19{,}037.92)}{s(220.07\, s^4 + 3162.72\, s^3 + 6212.56\, s^2 + 30{,}261.69\, s + 539.17)}$$

$$\frac{\beta(s)}{\delta_R(s)} = \frac{Num_{\beta R}(s)}{\overline{D}_2(s)} = \frac{s(A_{\beta R}s^3 + B_{\beta R}s^2 + C_{\beta R}s + D_{\beta R})}{s(A_2 s^4 + B_2 s^3 + C_2 s^2 + D_2 s + E_2)}$$

$$= \frac{s(19.56\, s^3 + 2{,}497.84\, s^2 + 29{,}711.27\, s - 512.71)}{s(220.07\, s^4 + 3162.72\, s^3 + 6212.56\, s^2 + 30{,}261.69\, s + 539.17)}$$

$$\frac{\phi(s)}{\delta_R(s)} = \frac{Num_{\phi R}(s)}{\overline{D}_2(s)} = \frac{s(A_{\phi R}s^2 + B_{\phi R}s + C_{\phi R})}{s(A_2 s^4 + B_2 s^3 + C_2 s^2 + D_2 s + E_2)}$$

$$= \frac{s(1060.25\, s^2 - 3906.26\, s - 58{,}494.40)}{s(220.07\, s^4 + 3162.72\, s^3 + 6212.56\, s^2 + 30{,}261.69\, s + 539.17)}$$

$$\frac{\psi(s)}{\delta_R(s)} = \frac{Num_{\psi R}(s)}{\overline{D}_2(s)} = \frac{A_{\psi R}s^3 + B_{\psi R}s^2 + C_{\psi R}s + D_{\psi R}}{s(A_2 s^4 + B_2 s^3 + C_2 s^2 + D_2 s + E_2)}$$

$$= \frac{(-2242.09\, s^3 - 29{,}706.68\, s^2 - 2{,}770.53\, s - 8{,}464.93)}{s(220.07\, s^4 + 3162.72\, s^3 + 6212.56\, s^2 + 30{,}261.69\, s + 539.17)}$$

$$\text{roots}(\overline{D}_2(s)) = -0.6703 \pm i\, 3.1748, \ -13.013, \ -0.0179$$

$$\zeta_{DR} = 0.206, \ \omega_{nDR} = 3.245, \ T_R = 0.077, \ T_S = 55.92$$

B.3 AIRCRAFT 2—CESSNA 310

Table B2.1 Geometric Data for the Cessna 310 Aircraft

Wing Surface (ft^2)	S	175
Mean Aerodynamic Chord (MAC) (ft)	$\bar{c}$	4.79
Wing Span (ft)	b	36.9

Table B2.2 Flight Conditions Data for the Cessna 310 Aircraft

		Climb	Cruise	Approach
Altitude (ft)	h	0	8,000	0
Mach Number	M	0.160	0.288	0.124
True Airspeed (ft/sec)	V_{P_1}	179.0	312.5	137.9
Dynamic Pressure (lbs/ft^2)	$\bar{q}$	38.1	91.2	22.6
Location of CG - % MAC	$\bar{x}_{CG}$	0.33	0.33	0.33
Steady-state angle of attack (deg)	α_1	5	0	6.6

Table B2.3 Mass and Inertial Data for the Cessna 310 Aircraft

		Climb	Cruise	Approach
Mass (lbs)	m	4,600	4,600	4,600
Moment of Inertia x-axis (slug ft^2)	I_{XX_B}	8,884	8,884	8,884
Moment of Inertia y-axis (slug ft^2)	I_{YY_B}	1,939	1,939	1,939
Moment of Inertia z-axis (slug ft^2)	I_{ZZ_B}	11,001	11,001	11,001
Product of inertia xz-plane (slug ft^2)	I_{XZ_B}	0	0	0

Table B2.4 Longitudinal Aerodynamic Coefficients for the Cessna 310 Aircraft

	Climb	Cruise	Approach
Steady State			
c_{L_1}	0.690	0.288	1.163
c_{D_1}	0.054	0.031	0.171
c_{m_1}	0	0	0
$c_{T_{X_1}}$	0.054	0.031	0.171
$c_{m_{T_1}}$	0	0	0
Stability Derivatives			
c_{D_0}	0.029	0.029	0.0974
c_{D_u}	0	0	0
c_{D_α}	0.362	0.160	0.650
$c_{T_{X_u}}$	−0.162	−0.093	−0.513
c_{L_0}	0.288	0.288	0.640
c_{L_u}	0	0	0
c_{L_α}	4.58	4.58	4.58
$c_{L_{\dot\alpha}}$	4.5	5.3	4.1
c_{L_q}	8.8	9.7	8.4
c_{m_0}	0.07	0.07	0.1
c_{m_u}	0	0	0
c_{m_α}	−0.339	−0.137	−0.619
$c_{m_{\dot\alpha}}$	−14.8	−12.7	−11.4
c_{m_q}	−29.2	−26.3	−25.1
$c_{m_{T_u}}$	0	0	0
$c_{m_{T_\alpha}}$	0	0	0
Control Derivatives			
$c_{D_{\delta_E}}$	0	0	0
$c_{L_{\delta_E}}$	0.90	0.81	0.77
$c_{m_{\delta_E}}$	−2.53	−2.26	−2.16

Table B2.5 Longitudinal Dimensional Stability Derivatives for the Cessna 310 Aircraft

Climb

$$X_u = -0.0281, \; X_{T_u} = -0.0141, \; X_\alpha = 15.284, \; X_{\delta_E} = 0$$

$$Z_u = -0.3593, \; Z_\alpha = -215.94, \; Z_{\dot\alpha} = -2.806, \; Z_q = -5.487, \; Z_{\delta_E} = -41.94$$

$$M_u = 0, \; M_{T_u} = 0, \; M_\alpha = -5.58, \; M_{T_\alpha} = 0, \; M_{\dot\alpha} = -3.259, \; M_q = -6.431, \; M_{\delta_E} = -41.64$$

Table B2.6 Longitudinal Transfer Functions for the Cessna 310 Aircraft

Climb

$$\frac{\alpha(s)}{\delta_E(s)} = \frac{Num_\alpha(s)}{\overline{D}_1(s)} = \frac{A_\alpha s^3 + B_\alpha s^2 + C_\alpha s + D_\alpha}{A_1 s^4 + B_1 s^3 + C_1 s^2 + D_1 s + E_1}$$

$$= \frac{-41.939 s^3 - 7495.56 s^2 - 199.31 s - 474.58}{181.79 s^4 + 1958.13 s^3 + 2435.24 s^2 + 156.22 s + 63.59}$$

$$\frac{u(s)}{\delta_E(s)} = \frac{Num_u(s)}{\overline{D}_1(s)} = \frac{A_u s^2 + B_u s + C_u}{A_1 s^4 + B_1 s^3 + C_1 s^2 + D_1 s + E_1}$$

$$= \frac{-641.0 s^2 + 123693.04 s + 282474.70}{181.79 s^4 + 1958.13 s^3 + 2435.24 s^2 + 156.22 s + 63.59}$$

$$\frac{\theta(s)}{\delta_E(s)} = \frac{Num_\theta(s)}{\overline{D}_1(s)} = \frac{A_\theta s^2 + B_\theta s + C_\theta}{A_1 s^4 + B_1 s^3 + C_1 s^2 + D_1 s + E_1}$$

$$= \frac{-7432.72 s^2 - 9070.95 s - 598.03}{181.79 s^4 + 1958.13 s^3 + 2435.24 s^2 + 156.22 s + 63.59}$$

$$\text{roots}(\overline{D}_1(s)) = -9.3477, \; -1.3796, \; -0.022 \pm i\, 0.1632$$

$(\textit{Degenerated Short Period}) T_{SP-1} = 0.107, \; T_{SP-2} = 0.725, \; \zeta_{Ph} = 0.1338, \; \omega_{nPh} = 0.165$

Table B2.7 Lateral Directional Aerodynamic Coefficients for the Cessna 310 Aircraft

	Climb	Cruise	Approach
Stability Derivatives			
c_{l_β}	−0.0923	−0.1096	−0.0965
c_{l_p}	−0.552	−0.551	−0.566
c_{l_r}	0.1746	0.0729	0.2433
c_{Y_β}	−0.610	−0.698	−0.577
c_{Y_p}	−0.2093	−0.1410	−0.2897
c_{Y_r}	0.356	0.355	0.355
c_{n_β}	0.1552	0.1444	0.1683
$c_{n_{T_\beta}}$	0	0	0
c_{n_p}	−0.0615	−0.0257	−0.1021
c_{n_r}	−0.1561	−0.1495	−0.1947
Control Derivatives			
$c_{l_{\delta_A}}$	0.172	0.172	0.172
$c_{l_{\delta_R}}$	0.0192	0.0192	0.0192
$c_{Y_{\delta_A}}$	0	0	0
$c_{Y_{\delta_R}}$	0.230	0.230	0.230
$c_{n_{\delta_A}}$	−0.0402	−0.0168	−0.0676
$c_{n_{\delta_R}}$	−0.1152	−0.1152	−0.1152

Table B2.8 Lateral Directional Dimensional Stability Derivatives for the Cessna 310 Aircraft

Cruise

$Y_\beta = -28.425, \ Y_P = -1.005, \ Y_r = 1.710, \ Y_{\delta_A} = 0, \ Y_{\delta_R} = 10.72$

$L_\beta = -2.549, \ L_P = -1.572, \ L_r = 0.497, \ L_{\delta_A} = 4.751, \ L_{\delta_R} = 0.53$

$N_\beta = 3.473, \ N_{T_\beta} = 0, \ N_P = -0.142, \ N_r = -0.3601, \ N_{\delta_A} = -0.9, \ N_{\delta_R} = -2.578$

Table B2.9 Lateral Directional Transfer Functions for the Cessna 310 Aircraft

Cruise

$$\frac{\beta(s)}{\delta_A(s)} = \frac{Num_{\beta A}(s)}{\overline{D}_2(s)} = \frac{s(A_{\beta A}s^3 + B_{\beta A}s^2 + C_{\beta A}s + D_{\beta A})}{s(A_2 s^4 + B_2 s^3 + C_2 s^2 + D_2 s + E_2)}$$

$$= \frac{s(0\,s^3 + 168.78\,s^2 + 521.76\,s + 40.50)}{s(178.918\,s^4 + 375.15\,s^3 + 789.64\,s^2 + 1{,}134.805\,s - 25.918)}$$

$$\frac{\phi(s)}{\delta_A(s)} = \frac{Num_{\phi A}(s)}{\overline{D}_2(s)} = \frac{s(A_{\phi A}s^2 + B_{\phi A}s + C_{\phi A})}{s(A_2 s^4 + B_2 s^3 + C_2 s^2 + D_2 s + E_2)}$$

$$= \frac{s(0\,s^3 + 853.658\,s^2 + 361.74\,s + 2554.58)}{s(178.918\,s^4 + 375.15\,s^3 + 789.64\,s^2 + 1{,}134.805\,s - 25.918)}$$

$$\frac{\psi(s)}{\delta_A(s)} = \frac{Num_{\psi A}(s)}{\overline{D}_2(s)} = \frac{A_{\psi A}s^3 + B_{\psi A}s^2 + C_{\psi A}s + D_{\psi A}}{s(A_2 s^4 + B_2 s^3 + C_2 s^2 + D_2 s + E_2)}$$

$$= \frac{(-175.25\,s^3 - 401.56\,s^2 - 73.64\,s + 455.39)}{s(178.918\,s^4 + 375.15\,s^3 + 789.64\,s^2 + 1{,}134.805\,s - 25.918)}$$

$$\frac{\beta(s)}{\delta_R(s)} = \frac{Num_{\beta R}(s)}{\overline{D}_2(s)} = \frac{s(A_{\beta R}s^3 + B_{\beta R}s^2 + C_{\beta R}s + D_{\beta R})}{s(A_2 s^4 + B_2 s^3 + C_2 s^2 + D_2 s + E_2)}$$

$$= \frac{s(10.714\,s^3 + 478.77\,s^2 + 758.30\,s - 34.96)}{s(178.918\,s^4 + 375.15\,s^3 + 789.64\,s^2 + 1{,}134.805\,s - 25.918)}$$

$$\frac{\phi(s)}{\delta_R(s)} = \frac{Num_{\phi R}(s)}{\overline{D}_2(s)} = \frac{s(A_{\phi R}s^2 + B_{\phi R}s + C_{\phi R})}{s(A_2 s^4 + B_2 s^3 + C_2 s^2 + D_2 s + E_2)}$$

$$= \frac{s(104.45\,s^2 - 206.73\,s - 860.99)}{s(178.918\,s^4 + 375.15\,s^3 + 789.64\,s^2 + 1{,}134.805\,s - 25.918)}$$

$$\frac{\psi(s)}{\delta_R(s)} = \frac{Num_{\psi R}(s)}{\overline{D}_2(s)} = \frac{A_{\psi R}s^3 + B_{\psi R}s^2 + C_{\psi R}s + D_{\psi R}}{s(A_2 s^4 + B_2 s^3 + C_2 s^2 + D_2 s + E_2)}$$

$$= \frac{(-463.03\,s^3 - 774.59\,s^2 - 50.18\,s - 151.63)}{s(178.918\,s^4 + 375.15\,s^3 + 789.64\,s^2 + 1{,}134.805\,s - 25.918)}$$

$roots(\overline{D}_2(s)) = -0.2037 \pm i\,1.929, \quad -1.712, \quad +0.0225$

$\zeta_{DR} = 0.105, \quad \omega_{nDR} = 1.94, \quad T_R = 0.584, \quad T_S = -44.48 \ (Unstable \ Spiral)$

B.4 AIRCRAFT 3—BEECH 99

Table B3.1 Geometric Data for the Beech 99 Aircraft

Wing Surface (ft^2)	S	280
Mean Aerodynamic Chord (MAC) (ft)	$\bar{c}$	6.5
Wing Span (ft)	b	46

Table B3.2 Flight Conditions Data for the Beech 99 Aircraft

		Approach	Cruise (low)	Cruise (high)
Altitude (ft)	h	0	5,000	20,000
Mach Number	M	0.152	0.310	0.434
True Airspeed (ft/sec)	V_{P_1}	170	340	450
Dynamic Pressure (lbs/ft^2)	$\bar{q}$	34.2	118.3	128.2
Location of CG - % MAC	$\bar{x}_{CG}$	0.16	0.16	0.16
Steady state angle of attack (deg)	α_1	3.5	0	1.1

Table B3.3 Mass and Inertial Data for the Beech 99 Aircraft

		Approach	Cruise (low)	Cruise (high)
Mass (lbs)	m	11,000	7,000	11,000
Moment of Inertia x-axis (slug ft^2)	I_{XX_B}	15,189	10,085	15,189
Moment of Inertia y-axis (slug ft^2)	I_{YY_B}	20,250	15,148	20,250
Moment of Inertia z-axis (slug ft^2)	I_{ZZ_B}	34,141	23,046	34,141
Product of inertia xz-plane (slug ft^2)	I_{XZ_B}	4,371	1,600	4,371

Table B3.4 Longitudinal Aerodynamic Coefficients for the Beech 99 Aircraft

	Approach	Cruise (low)	Cruise (high)
Steady State			
c_{L_1}	1.15	0.211	0.306
c_{D_1}	0.162	0.0298	0.0298
c_{m_1}	0	0	0
$c_{T_{X_1}}$	0.162	0.0298	0.0298
$c_{m_{T_1}}$	0	0	0
Stability Derivatives			
c_{D_0}	0.0969	0.0270	0.0270
c_{D_u}	0	0	0
c_{D_α}	0.933	0.131	0.131
$c_{T_{X_u}}$	−0.324	−0.0596	−0.0596
c_{L_0}	0.760	0.201	0.201
c_{L_u}	0.027	0.02	0.02
c_{L_α}	6.24	5.48	5.48
$c_{L_{\dot\alpha}}$	2.7	2.5	2.5
c_{L_q}	8.1	8.1	8.1
c_{m_0}	0.10	0.05	0.05
c_{m_u}	0	0	0
c_{m_α}	−2.08	−1.89	−1.89
$c_{m_{\dot\alpha}}$	−9.1	−9.1	−9.1
c_{m_q}	−34.0	−34.0	−34.0
$c_{m_{T_u}}$	0	0	0
$c_{m_{T_\alpha}}$	0	0	0
Control Derivatives			
$c_{D_{\delta_E}}/c_{D_{i_H}}$	0/0	0/0	0/0
$c_{L_{\delta_E}}/c_{L_{i_H}}$	0.58/1.3	0.60/1.35	0.60/1.35
$c_{m_{\delta_E}}/c_{m_{i_H}}$	−1.9/−3.9	−2.0/−4.1	−2.0/−4.1

Table B3.5 Longitudinal Dimensional Stability Derivatives for the Beech 99 Aircraft

Cruise
$X_u = -0.0138$, $X_{T_u} = 0.0$, $X_\alpha = 18.27$, $X_{\delta_E} = 0$
$Z_u = -0.147$, $Z_\alpha = -575.23$, $Z_{\dot\alpha} = -1.885$, $Z_q = -6.108$, $Z_{\delta_E} = -62.64$
$M_u = 0$, $M_{T_u} = 0$, $M_\alpha = -21.79$, $M_{T_\alpha} = 0$, $M_{\dot\alpha} = -0.758$, $M_q = -2.8314$, $M_{\delta_E} = -23.06$

Table B3.6 Longitudinal Transfer Functions for the Beech 99 Aircraft

Cruise

$$\frac{\alpha(s)}{\delta_E(s)} = \frac{Num_\alpha(s)}{\overline{D}_1(s)} = \frac{A_\alpha s^3 + B_\alpha s^2 + C_\alpha s + D_\alpha}{A_1 s^4 + B_1 s^3 + C_1 s^2 + D_1 s + E_1}$$

$$= \frac{-62.641\, s^3 - 10{,}412.59\, s^2 - 129.83\, s - 107.898}{451.835\, s^4 + 2{,}197.16\, s^3 + 11{,}332.70\, s^2 + 154.027\, s + 101.963}$$

$$\frac{u(s)}{\delta_E(s)} = \frac{Num_u(s)}{\overline{D}_1(s)} = \frac{A_u s^2 + B_u s + C_u}{A_1 s^4 + B_1 s^3 + C_1 s^2 + D_1 s + E_1}$$

$$= \frac{-1{,}144.47\, s^2 + 141{,}313.74\, s + 380{,}639.61}{451.835\, s^4 + 2{,}197.16\, s^3 + 11{,}332.70\, s^2 + 154.027\, s + 101.963}$$

$$\frac{\theta(s)}{\delta_E(s)} = \frac{Num_\theta(s)}{\overline{D}_1(s)} = \frac{A_\theta s^2 + B_\theta s + C_\theta}{A_1 s^4 + B_1 s^3 + C_1 s^2 + D_1 s + E_1}$$

$$= \frac{-10{,}371.21\, s^2 - 12{,}042.51\, s - 226.33}{451.835\, s^4 + 2{,}197.16\, s^3 + 11{,}332.70\, s^2 + 154.027\, s + 101.963}$$

$$roots(\overline{D}_1(s)) = -2.4254 \pm i\, 4.374, \quad -0.0059 \pm i\, 0.0948$$

$$\zeta_{SP} = 0.485, \quad \omega_{nSP} = 5.001, \quad \zeta_{Ph} = 0.050, \quad \omega_{nPh} = 0.0625$$

Table B3.7 Lateral Directional Aerodynamic Coefficients for the Beech 99 Aircraft

	Approach	Cruise (low)	Cruise (high)
Stability Derivatives			
c_{l_β}	−0.13	−0.13	−0.13
c_{l_p}	−0.50	−0.50	−0.50
c_{l_r}	0.06	0.14	0.14
c_{Y_β}	−0.59	−0.59	−0.59
c_{Y_p}	−0.21	−0.19	−0.19
c_{Y_r}	0.39	0.39	0.39
c_{n_β}	0.120	0.080	0.080
$c_{n_{T_\beta}}$	0	0	0
c_{n_p}	−0.005	0.019	0.019
c_{n_r}	−0.204	−0.197	−0.197
Control Derivatives			
$c_{l_{\delta_A}}$	0.156	0.156	0.156
$c_{l_{\delta_R}}$	0.0087	0.0109	0.01016
$c_{Y_{\delta_A}}$	0	0	0
$c_{Y_{\delta_R}}$	0.144	0.148	0.144
$c_{n_{\delta_A}}$	−0.0012	−0.0012	−0.0012
$c_{n_{\delta_R}}$	−0.0763	−0.0772	−0.0758

Table B3.8 Lateral Directional Dimensional Stability Derivatives for the Beech 99 Aircraft

Cruise

$Y_\beta = -61.597, \ Y_P = -1.014, \ Y_r = 2.0813, \ Y_{\delta_A} = 0, \ Y_{\delta_R} = 15.0338$

$L_\beta = -14.2925, \ L_P = -2.81, \ L_r = 0.7868, \ L_{\delta_A} = 17.151, \ L_{\delta_R} = 1.1654$

$N_\beta = 3.8534, \ N_{T_\beta} = 0, \ N_P = 0.0468, \ N_r = -0.4850, \ N_{\delta_A} = -0.0578, \ N_{\delta_R} = -3.651$

Table B3.9 Lateral Directional Transfer Functions for the Beech 99 Aircraft

Cruise

$$\frac{\beta(s)}{\delta_A(s)} = \frac{Num_{\beta A}(s)}{\overline{D}_2(s)} = \frac{s(A_{\beta A}s^3 + B_{\beta A}s^2 + C_{\beta A}s + D_{\beta A})}{s(A_2 s^4 + B_2 s^3 + C_2 s^2 + D_2 s + E_2)}$$

$$= \frac{s(0\,s^3 - 888.13\,s^2 + 252.785\,s + 264.485)}{s(435.956\,s^4 + 1{,}495.337\,s^3 + 1{,}758.41\,s^2 + 5{,}051.84\,s + 124.69)}$$

$$\frac{\phi(s)}{\delta_A(s)} = \frac{Num_{\phi A}(s)}{\overline{D}_2(s)} = \frac{s(A_{\phi A}s^2 + B_{\phi A}s + C_{\phi A})}{s(A_2 s^4 + B_2 s^3 + C_2 s^2 + D_2 s + E_2)}$$

$$= \frac{s(7{,}710.172\,s^2 + 4{,}778.22\,s + 29{,}739.27)}{s(435.956\,s^4 + 1{,}495.337\,s^3 + 1{,}758.41\,s^2 + 5{,}051.84\,s + 124.69)}$$

$$\frac{\psi(s)}{\delta_A(s)} = \frac{Num_{\psi A}(s)}{\overline{D}_2(s)} = \frac{A_{\psi A}s^3 + B_{\psi A}s^2 + C_{\psi A}s + D_{\psi A}}{s(A_2 s^4 + B_2 s^3 + C_2 s^2 + D_2 s + E_2)}$$

$$= \frac{(874.80\,s^3 + 407.69\,s^2 - 26.758\,s + 2{,}086.31)}{s(435.956\,s^4 + 1{,}495.337\,s^3 + 1{,}758.41\,s^2 + 5{,}051.84\,s + 124.69)}$$

$$\frac{\beta(s)}{\delta_R(s)} = \frac{Num_{\beta R}(s)}{\overline{D}_2(s)} = \frac{s(A_{\beta R}s^3 + B_{\beta R}s^2 + C_{\beta R}s + D_{\beta R})}{s(A_2 s^4 + B_2 s^3 + C_2 s^2 + D_2 s + E_2)}$$

$$= \frac{s(14.566\,s^3 + 1{,}622.06\,s^2 + 4{,}598.91\,s - 73.76)}{s(435.956\,s^4 + 1{,}495.337\,s^3 + 1{,}758.41\,s^2 + 5{,}051.84\,s + 124.69)}$$

$$\frac{\phi(s)}{\delta_R(s)} = \frac{Num_{\phi R}(s)}{\overline{D}_2(s)} = \frac{s(A_{\phi R}s^2 + B_{\phi R}s + C_{\phi R})}{s(A_2 s^4 + B_2 s^3 + C_2 s^2 + D_2 s + E_2)}$$

$$= \frac{s(86.667\,s^2 - 1{,}225.78\,s - 21{,}560.89)}{s(435.956\,s^4 + 1{,}495.337\,s^3 + 1{,}758.41\,s^2 + 5{,}051.84\,s + 124.69)}$$

$$\frac{\psi(s)}{\delta_R(s)} = \frac{Num_{\psi R}(s)}{\overline{D}_2(s)} = \frac{A_{\psi R}s^3 + B_{\psi R}s^2 + C_{\psi R}s + D_{\psi R}}{s(A_2 s^4 + B_2 s^3 + C_2 s^2 + D_2 s + E_2)}$$

$$= \frac{(-1{,}581.606\,s^3 - 4{,}775.38\,s^2 - 427.50\,s - 1{,}524.61)}{s(435.956\,s^4 + 1{,}495.337\,s^3 + 1{,}758.41\,s^2 + 5{,}051.84\,s + 124.69)}$$

$$\text{roots}(\overline{D}_2(s)) = -0.0667 \pm i\,1.8728, \quad -3.2716, \quad -0.0249$$

$$\zeta_{DR} = 0.0356, \quad \omega_{nDR} = 1.873, \quad T_R = 0.305, \quad T_S = 40.17$$

B.5 AIRCRAFT 4—CESSNA T37-A

Table B4.1 Geometric Data for the Cessna T37-A Aircraft

Wing Surface (ft^2)	S	182
Mean Aerodynamic Chord (MAC) (ft)	$\bar{c}$	5.47
Wing Span (ft)	b	33.8

Table B4.2 Flight Conditions Data for the Cessna T37-A Aircraft

		Climb	Cruise	Approach
Altitude (ft)	h	0	30,000	0
Mach Number	M	0.313	0.459	0.143
True Airspeed (ft/sec)	V_{P_1}	349	456	160
Dynamic Pressure (lbs/ft^2)	$\bar{q}$	144.9	92.7	30.4
Location of CG - % MAC	$\bar{x}_{CG}$	0.27	0.27	0.27
Steady state angle of attack (deg)	α_1	0.7	2	4.2

Table B4.3 Mass and Inertial Data for the Cessna T37-A Aircraft

		Climb	Cruise	Approach
Mass (lbs)	m	6,360	6,360	6,360
Moment of Inertia x-axis (slug ft^2)	I_{XX_B}	7,985	7,985	7,985
Moment of Inertia y-axis (slug ft^2)	I_{YY_B}	3,326	3,326	3,326
Moment of Inertia z-axis (slug ft^2)	I_{ZZ_B}	11,183	11,183	11,183
Product of inertia xz-plane (slug ft^2)	I_{XZ_B}	0	0	0

Table B4.4 Longitudinal Aerodynamic Coefficients for the Cessna T37-A Aircraft

	Climb	Cruise	Approach
Steady State			
c_{L_1}	0.241	0.378	1.150
c_{D_1}	0.022	0.030	0.158
c_{m_1}	0	0	0
$c_{T_{X_1}}$	0.022	0.030	0.158
$c_{m_{T_1}}$	0	0	0
Stability Derivatives			
c_{D_0}	0.02	0.02	0.0689
c_{D_u}	0	0	0
c_{D_α}	0.13	0.25	0.682
$c_{T_{X_u}}$	−0.05	−0.07	−0.40
c_{L_0}	0.19	0.20	0.81
c_{L_u}	0	0	0
c_{L_α}	4.81	5.15	4.64
$c_{L_{\dot\alpha}}$	1.8	2.0	1.8
c_{L_q}	3.7	4.1	3.7
c_{m_0}	0.025	0.025	0.10
c_{m_u}	0	0	0
c_{m_α}	−0.668	−0.70	−0.631
$c_{m_{\dot\alpha}}$	−6.64	−6.95	−6.84
c_{m_q}	−14.3	−14.9	−14.0
$c_{m_{T_u}}$	0	0	0
$c_{m_{T_\alpha}}$	0	0	0
Control Derivatives			
$c_{D_{\delta_E}}$	0	0	0
$c_{L_{\delta_E}}$	0.4	0.5	0.4
$c_{m_{\delta_E}}$	−1.07	−1.12	−1.05

Table B4.5 Longitudinal Dimensional Stability Derivatives for the Cessna T37-A Aircraft

Cruise

$$X_u = -0.0111, \; X_{T_u} = -0.0019, \; X_\alpha = 10.809, \; X_{\delta_E} = 0$$

$$Z_u = -0.14, \; Z_\alpha = -437.415, \; Z_{\dot\alpha} = -1.013, \; Z_q = -2.077, \; Z_{\delta_E} = -42.222$$

$$M_u = 0, \; M_{T_u} = 0, \; M_\alpha = -19.398, \; M_{T_\alpha} = 0, \; M_{\dot\alpha} = -1.1553, \; M_q = -2.477, \; M_{\delta_E} = -31.037$$

Table B4.6 Longitudinal Transfer Functions for the Cessna T37-A Aircraft

Cruise

$$\frac{\alpha(s)}{\delta_E(s)} = \frac{Num_\alpha(s)}{\overline{D}_1(s)} = \frac{A_\alpha s^3 + B_\alpha s^2 + C_\alpha s + D_\alpha}{A_1 s^4 + B_1 s^3 + C_1 s^2 + D_1 s + E_1}$$

$$= \frac{-42.22\, s^3 - 14{,}191.75\, s^2 - 149.454\, s - 137.98}{456.96\, s^4 + 2{,}099.46\, s^3 + 9{,}914.89\, s^2 + 115.49\, s + 86.235}$$

$$\frac{u(s)}{\delta_E(s)} = \frac{Num_u(s)}{\overline{D}_1(s)} = \frac{A_u s^2 + B_u s + C_u}{A_1 s^4 + B_1 s^3 + C_1 s^2 + D_1 s + E_1}$$

$$= \frac{-456.36\, s^2 + 296{,}829.82\, s + 406{,}732.25}{456.96\, s^4 + 2{,}099.46\, s^3 + 9{,}914.89\, s^2 + 115.49\, s + 86.235}$$

$$\frac{\theta(s)}{\delta_E(s)} = \frac{Num_\theta(s)}{\overline{D}_1(s)} = \frac{A_\theta s^2 + B_\theta s + C_\theta}{A_1 s^4 + B_1 s^3 + C_1 s^2 + D_1 s + E_1}$$

$$= \frac{-7713.23\, s^2 - 15867.0\, s - 908.24}{456.96\, s^4 + 2{,}099.46\, s^3 + 9{,}914.89\, s^2 + 115.49\, s + 86.235}$$

$$\text{roots}(\overline{D}_1(s)) = -2.2923 \pm i\, 4.0483, \quad -0.0049 \pm i\, 0.0932$$

$$\zeta_{SP} = 0.493, \; \omega_{nSP} = 4.6523, \; \zeta_{Ph} = 0.0526, \; \omega_{nPh} = 0.0934$$

Table B4.7 Lateral Directional Aerodynamic Coefficients for the Cessna T37-A Aircraft

	Climb	Cruise	Approach
Stability Derivatives			
c_{l_β}	−0.0851	−0.0944	−0.0822
c_{l_p}	−0.440	−0.442	−0.458
c_{l_r}	0.0590	0.0926	0.2540
c_{Y_β}	−0.361	−0.346	−0.303
c_{Y_p}	−0.0635	−0.0827	−0.1908
c_{Y_r}	0.314	0.300	0.263
c_{n_β}	0.1052	0.1106	0.1095
$c_{n_{T_\beta}}$	0	0	0
c_{n_p}	−0.0154	−0.0243	−0.0768
c_{n_r}	−0.1433	−0.1390	−0.1613
Control Derivatives			
$c_{l_{\delta_A}}$	0.1788	0.1810	0.1788
$c_{l_{\delta_R}}$	0.015	0.015	0.015
$c_{Y_{\delta_A}}$	0	0	0
$c_{Y_{\delta_R}}$	0.2	0.2	0.2
$c_{n_{\delta_A}}$	−0.0160	−0.0254	−0.0760
$c_{n_{\delta_R}}$	−0.0365	−0.0365	−0.0365

Table B4.8 Lateral Directional Dimensional Stability Derivatives for the Cessna T37-A Aircraft

Cruise

$$Y_\beta = -29.217, \ Y_P = -0.258, \ Y_r = 0.939, \ Y_{\delta_A} = 0, \ Y_{\delta_R} = 16.889$$

$$L_\beta = -6.73, \ L_P = -1.168, \ L_r = 0.245, \ L_{\delta_A} = 12.903, \ L_{\delta_R} = 1.069$$

$$N_\beta = 5.6345, \ N_{T_\beta} = 0, \ N_P = -0.0459, \ N_r = -0.2625, \ N_{\delta_A} = -1.294, \ N_{\delta_R} = -1.859$$

Table B4.9 Lateral Directional Transfer Functions for the Cessna T37-A Aircraft

Cruise

$$\frac{\beta(s)}{\delta_A(s)} = \frac{Num_{\beta A}(s)}{\overline{D}_2(s)} = \frac{s(A_{\beta A}s^3 + B_{\beta A}s^2 + C_{\beta A}s + D_{\beta A})}{s(A_2 s^4 + B_2 s^3 + C_2 s^2 + D_2 s + E_2)}$$

$$= \frac{s(0\,s^3 + 644.017\,s^2 + 1{,}367.855\,s + 97.797)}{s(455.885\,s^4 + 682.22\,s^3 + 2{,}779.28\,s^2 + 3{,}360.827\,s + 12.35)}$$

$$\frac{\phi(s)}{\delta_A(s)} = \frac{Num_{\phi A}(s)}{\overline{D}_2(s)} = \frac{s(A_{\phi A}s^2 + B_{\phi A}s + C_{\phi A})}{s(A_2 s^4 + B_2 s^3 + C_2 s^2 + D_2 s + E_2)}$$

$$= \frac{s(5{,}891.46\,s^2 + 1{,}777.36\,s + 29{,}208.18)}{s(455.885\,s^4 + 682.22\,s^3 + 2{,}779.28\,s^2 + 3{,}360.827\,s + 12.35)}$$

$$\frac{\psi(s)}{\delta_A(s)} = \frac{Num_{\psi A}(s)}{\overline{D}_2(s)} = \frac{A_{\psi A}s^3 + B_{\psi A}s^2 + C_{\psi A}s + D_{\psi A}}{s(A_2 s^4 + B_2 s^3 + C_2 s^2 + D_2 s + E_2)}$$

$$= \frac{(-648.698\,s^3 - 1{,}000.596\,s^2 - 78.02\,s + 2{,}030.51)}{s(455.885\,s^4 + 682.22\,s^3 + 2{,}779.28\,s^2 + 3{,}360.827\,s + 12.35)}$$

$$\frac{\beta(s)}{\delta_R(s)} = \frac{Num_{\beta R}(s)}{\overline{D}_2(s)} = \frac{s(A_{\beta R}s^3 + B_{\beta R}s^2 + C_{\beta R}s + D_{\beta R})}{s(A_2 s^4 + B_2 s^3 + C_2 s^2 + D_2 s + E_2)}$$

$$= \frac{s(16.886\,s^3 + 874.84\,s^2 + 1{,}050.79\,s - 5.553)}{s(455.885\,s^4 + 682.22\,s^3 + 2{,}779.28\,s^2 + 3{,}360.827\,s + 12.35)}$$

$$\frac{\phi(s)}{\delta_R(s)} = \frac{Num_{\phi R}(s)}{\overline{D}_2(s)} = \frac{s(A_{\phi R}s^2 + B_{\phi R}s + C_{\phi R})}{s(A_2 s^4 + B_2 s^3 + C_2 s^2 + D_2 s + E_2)}$$

$$= \frac{s(499.397\,s^2 - 162.46\,s - 2{,}964.0)}{s(455.885\,s^4 + 682.22\,s^3 + 2{,}779.28\,s^2 + 3{,}360.827\,s + 12.35)}$$

$$\frac{\psi(s)}{\delta_R(s)} = \frac{Num_{\psi R}(s)}{\overline{D}_2(s)} = \frac{A_{\psi R}s^3 + B_{\psi R}s^2 + C_{\psi R}s + D_{\psi R}}{s(A_2 s^4 + B_2 s^3 + C_2 s^2 + D_2 s + E_2)}$$

$$= \frac{(-852.696\,s^3 - 970.92\,s^2 + 53.147\,s - 206.688)}{s(455.885\,s^4 + 682.22\,s^3 + 2{,}779.28\,s^2 + 3{,}360.827\,s + 12.35)}$$

$$\text{roots}(\overline{D}_2(s)) = -0.1133 \pm i\,2.4065, \quad -1.2663, \quad -0.0037$$

$$\zeta_{DR} = 0.0470, \ \omega_{nDR} = 2.4092, \ T_R = 0.790, \ T_S = 271.31$$

B.6 AIRCRAFT 5—CESSNA 620

Table B5.1 Geometric Data for the Cessna 620 Aircraft

Wing Surface (ft^2)	S	340
Mean Aerodynamic Chord (MAC) (ft)	$\bar{c}$	6.58
Wing Span (ft)	b	55.1

Table B5.2 Flight Conditions Data for the Cessna 620 Aircraft

		Climb	Cruise	Approach
Altitude (ft)	h	0	18,000	0
Mach Number	M	0.181	0.351	0.170
True Airspeed (ft/sec)	V_{P_1}	202.4	366.8	189.2
Dynamic Pressure (lbs/ft^2)	$\bar{q}$	48.7	91.1	42.6
Location of CG - % MAC	$\bar{x}_{CG}$	0.25	0.25	0.25
Steady-state angle of attack (deg)	α_1	5.0	0	6.0

Table B5.3 Mass and Inertial Data for the Cessna 620 Aircraft

		Climb	Cruise	Approach
Mass (lbs)	m	15,000	15,000	15,000
Moment of Inertia x-axis (slug ft^2)	I_{XX_B}	64,811	64,811	64,811
Moment of Inertia y-axis (slug ft^2)	I_{YY_B}	17,300	17,300	17,300
Moment of Inertia z-axis (slug ft^2)	I_{ZZ_B}	64,543	64,543	64,543
Product of inertia xz-plane (slug ft^2)	I_{XZ_B}	0	0	0

Table B5.4 Longitudinal Aerodynamic Coefficients for the Cessna 620 Aircraft

	Climb	Cruise	Approach
Steady State			
c_{L_1}	0.903	0.484	1.038
c_{D_1}	0.0750	0.0420	0.1140
c_{m_1}	0	0	0
$c_{T_{X_1}}$	0.070	0.042	0.1140
$c_{m_{T_1}}$	0	0	0
Stability Derivatives			
c_{D_0}	0.0408	0.0322	0.0628
c_{D_u}	0	0	0
c_{D_α}	0.527	0.269	0.475
$c_{T_{X_u}}$	−0.225	−0.126	−0.342
c_{L_0}	0.43	0.48	0.48
c_{L_u}	0	0	0
c_{L_α}	5.38	5.55	5.38
$c_{L_{\dot\alpha}}$	3.3	2.7	2.7
c_{L_q}	8.0	7.5	7.6
c_{m_0}	0.06	0.06	0.09
c_{m_u}	0	0	0
c_{m_α}	−1.06	−1.18	−1.0
$c_{m_{\dot\alpha}}$	−10.3	−8.17	−8.68
c_{m_q}	−24.7	−22.4	−22.8
$c_{m_{T_u}}$	0	0	0
$c_{m_{T_\alpha}}$	0	0	0
Control Derivatives			
$c_{D_{\delta_E}}$	0	0	0
$c_{L_{\delta_E}}$	0.63	0.58	0.59
$c_{m_{\delta_E}}$	−1.90	−1.73	−1.75

B.7 AIRCRAFT 6—LEARJET 24

Table B6.1 Geometric Data for the Learjet 24 Aircraft

Wing Surface (ft^2)	S	230
Mean Aerodynamic Chord (MAC) (ft)	$\bar{c}$	7.0
Wing Span (ft)	b	34.0

Table B6.2 Flight Conditions Data for the Learjet 24 Aircraft

		Approach	Cruise (Max Weight)	Cruise (Low Weight)
Altitude (ft)	h	0	40,000	40,000
Mach Number	M	0.152	0.7	0.7
True Airspeed (ft/sec)	V_{P_1}	170	677	677
Dynamic Pressure (lbs/ft^2)	$\bar{q}$	34.3	134.6	134.6
Location of CG - % MAC	$\bar{x}_{CG}$	0.32	032	0.32
Steady state angle of attack (deg)	α_1	5.0	2.7	1.5

Table B6.3 Mass and Inertial Data for the Learjet 24 Aircraft

		Approach	Cruise (Max Weight)	Cruise (Low Weight)
Mass (lbs)	m	13,000	13,000	9,000
Moment of Inertia x-axis (slug ft^2)	I_{XX_B}	28,000	28,000	6,000
Moment of Inertia y-axis (slug ft^2)	I_{YY_B}	18,800	18,800	17,800
Moment of Inertia z-axis (slug ft^2)	I_{ZZ_B}	47,000	47,000	25,000
Product of inertia xz-plane (slug ft^2)	I_{XZ_B}	1,300	1,300	1,400

Table B6.4 Longitudinal Aerodynamic Coefficients for the Learjet 24 Aircraft

	Approach	Cruise (Max Weight)	Cruise (Low Weight)
Steady State			
c_{L_1}	1.64	0.41	0.28
c_{D_1}	0.256	0.0335	0.0279
c_{m_1}	0	0	0
$c_{T_{X_1}}$	0.256	0.0335	0.0279
$c_{m_{T_1}}$	0	0	0
Stability Derivatives			
c_{D_0}	0.0431	0.0216	0.0216
c_{D_u}	0	0.104	0.104
c_{D_α}	1.06	0.30	0.22
$c_{T_{X_u}}$	-0.60	-0.07	-0.07
c_{L_0}	1.2	0.13	0.13
c_{L_u}	0.04	0.40	0.28
c_{L_α}	5.04	5.84	5.84
$c_{L_{\dot{\alpha}}}$	1.6	2.2	2.2
c_{L_q}	4.1	4.7	4.7
c_{m_0}	0.047	0.050	0.050
c_{m_u}	-0.01	0.050	0.070
c_{m_α}	-0.66	-0.64	-0.64
$c_{m_{\dot{\alpha}}}$	-5.0	-6.7	-6.7
c_{m_q}	-13.5	-15.5	-15.5
$c_{m_{T_u}}$	0.006	-0.003	-0.003
$c_{m_{T_\alpha}}$	0	0	0
Control Derivatives			
$c_{D_{\delta_E}}/c_{D_{i_H}}$	0/0	0/0	0/0
$c_{L_{\delta_E}}/c_{L_{i_H}}$	0.4/0.85	0.46/0.94	0.46/0.94
$c_{m_{\delta_E}}/c_{m_{i_H}}$	$-0.98/-2.1$	$-1.24/-2.5$	$-1.24/-2.5$

Table B6.5 Longitudinal Dimensional Stability Derivatives for the Learjet 24 Aircraft

Approach

$$X_u = -0.0589, \ X_{T_u} = -0.0101, \ X_\alpha = 11.337, \ X_{\delta_E} = 0$$

$$Z_u = -0.382, \ Z_\alpha = -103.516, \ Z_{\dot{\alpha}} = -0.644, \ Z_q = -1.65, \ Z_{\delta_E} = -7.818$$

$$M_u = -0.0002, \ M_{T_u} = 0.0001, \ M_\alpha = -1.94, \ M_{T_\alpha} = 0,$$

$$M_{\dot{\alpha}} = -0.3047, \ M_q = -0.817, \ M_{\delta_E} = -2.88$$

Table B6.6 Longitudinal Transfer Functions for the Learjet 24 Aircraft

Approach

$$\frac{\alpha(s)}{\delta_E(s)} = \frac{Num_\alpha(s)}{\overline{D}_1(s)} = \frac{A_\alpha s^3 + B_\alpha s^2 + C_\alpha s + D_\alpha}{A_1 s^4 + B_1 s^3 + C_1 s^2 + D_1 s + E_1}$$

$$= \frac{-7.8184 \ s^3 - 492.02 \ s^2 - 25.83 \ s - 34.69}{170.62 \ s^4 + 305.72 \ s^3 + 435.07 \ s^2 + 29.87 \ s + 23.14}$$

$$\frac{u(s)}{\delta_E(s)} = \frac{Num_u(s)}{\overline{D}_1(s)} = \frac{A_u s^2 + B_u s + C_u}{A_1 s^4 + B_1 s^3 + C_1 s^2 + D_1 s + E_1}$$

$$= \frac{-88.634 \ s^2 + 10,112.17 \ s + 9,166.64}{170.62 \ s^4 + 305.72 \ s^3 + 435.07 \ s^2 + 29.87 \ s + 23.14}$$

$$\frac{\theta(s)}{\delta_E(s)} = \frac{Num_\theta(s)}{\overline{D}_1(s)} = \frac{A_\theta s^2 + B_\theta s + C_\theta}{A_1 s^4 + B_1 s^3 + C_1 s^2 + D_1 s + E_1}$$

$$= \frac{-489.33 \ s^2 - 316.90 \ s - 32.001}{170.62 \ s^4 + 305.72 \ s^3 + 435.07 \ s^2 + 29.87 \ s + 23.14}$$

$$\text{roots}(\overline{D}_1(s)) = -0.881 \pm i \ 1.29, \ -0.0158 \pm i \ 0.2353$$

$$\zeta_{SP} = 0.564, \ \omega_{nSP} = 1.562, \ \zeta_{Ph} = 0.0671, \ \omega_{nPh} = 0.2358$$

Table B6.7 Lateral directional Aerodynamic Coefficients for the Learjet 24 Aircraft

	Approach	Cruise (Max Weight)	Cruise (Low Weight)
Stability Derivatives			
c_{l_β}	−0.173	−0.110	−0.100
c_{l_p}	−0.390	−0.450	−0.450
c_{l_r}	0.450	0.160	0.140
c_{Y_β}	−0.730	−0.730	−0.730
c_{Y_p}	0	0	0
c_{Y_r}	0.4	0.4	0.4
c_{n_β}	0.15	0.127	0.124
$c_{n_{T_\beta}}$	0	0	0
c_{n_p}	−0.130	−0.008	−0.022
c_{n_r}	−0.260	−0.200	−0.200
Control Derivatives			
$c_{l_{\delta_A}}$	0.149	0.178	0.178
$c_{l_{\delta_R}}$	0.014	0.019	0.021
$c_{Y_{\delta_A}}$	0	0	0
$c_{Y_{\delta_R}}$	0.140	0.140	0.140
$c_{n_{\delta_A}}$	−0.05	−0.02	−0.02
$c_{n_{\delta_R}}$	−0.074	−0.074	−0.074

Table B6.8 Lateral Directional Dimensional Stability Derivatives for the Learjet 24 Aircraft

Approach

$$Y_\beta = -14.27, \ Y_P = 0.0, \ Y_r = 0.782, \ Y_{\delta_A} = 0, \ Y_{\delta_R} = 2.736$$

$$L_\beta = -1.664, \ L_P = -0.375, \ L_r = 0.433, \ L_{\delta_A} = 1.433, \ L_{\delta_R} = 0.135$$

$$N_\beta = 0.855, \ N_{T_\beta} = 0, \ N_P = -0.0742, \ N_r = -0.148, \ N_{\delta_A} = -0.285, \ N_{\delta_R} = -0.422$$

Table B6.9 Lateral Directional Transfer Functions for the Learjet 24 Aircraft

Approach

$$\frac{\beta(s)}{\delta_A(s)} = \frac{Num_{\beta A}(s)}{\overline{D}_2(s)} = \frac{s(A_{\beta A}s^3 + B_{\beta A}s^2 + C_{\beta A}s + D_{\beta A})}{s(A_2 s^4 + B_2 s^3 + C_2 s^2 + D_2 s + E_2)}$$

$$= \frac{s(0\,s^3 + 50.152\,s^2 + 82.134\,s + 2.856)}{s(169.96\,s^4 + 103.65\,s^3 + 169.373\,s^2 + 130.123\,s - 3.96)}$$

$$\frac{\phi(s)}{\delta_A(s)} = \frac{Num_{\phi A}(s)}{\overline{D}_2(s)} = \frac{s(A_{\phi A}s^2 + B_{\phi A}s + C_{\phi A})}{s(A_2 s^4 + B_2 s^3 + C_2 s^2 + D_2 s + E_2)}$$

$$= \frac{s(244.235\,s^2 + 35.646\,s + 128.43)}{s(169.96\,s^4 + 103.65\,s^3 + 169.373\,s^2 + 130.123\,s - 3.96)}$$

$$\frac{\psi(s)}{\delta_A(s)} = \frac{Num_{\psi A}(s)}{\overline{D}_2(s)} = \frac{A_{\psi A}s^3 + B_{\psi A}s^2 + C_{\psi A}s + D_{\psi A}}{s(A_2 s^4 + B_2 s^3 + C_2 s^2 + D_2 s + E_2)}$$

$$= \frac{(-50.38\,s^3 - 40.476\,s^2 - 3.043\,s + 24.087)}{s(169.96\,s^4 + 103.65\,s^3 + 169.373\,s^2 + 130.123\,s - 3.96)}$$

$$\frac{\beta(s)}{\delta_R(s)} = \frac{Num_{\beta R}(s)}{\overline{D}_2(s)} = \frac{s(A_{\beta R}s^3 + B_{\beta R}s^2 + C_{\beta R}s + D_{\beta R})}{s(A_2 s^4 + B_2 s^3 + C_2 s^2 + D_2 s + E_2)}$$

$$= \frac{s(2.736\,s^3 + 73.027\,s^2 + 33.213\,s - 5.215)}{s(169.96\,s^4 + 103.65\,s^3 + 169.373\,s^2 + 130.123\,s - 3.96)}$$

$$\frac{\phi(s)}{\delta_R(s)} = \frac{Num_{\phi R}(s)}{\overline{D}_2(s)} = \frac{s(A_{\phi R}s^2 + B_{\phi R}s + C_{\phi R})}{s(A_2 s^4 + B_2 s^3 + C_2 s^2 + D_2 s + E_2)}$$

$$= \frac{s(23.837\,s^2 - 30.242\,s - 101.31)}{s(169.96\,s^4 + 103.65\,s^3 + 169.373\,s^2 + 130.123\,s - 3.96)}$$

$$\frac{\psi(s)}{\delta_R(s)} = \frac{Num_{\psi R}(s)}{\overline{D}_2(s)} = \frac{A_{\psi R}s^3 + B_{\psi R}s^2 + C_{\psi R}s + D_{\psi R}}{s(A_2 s^4 + B_2 s^3 + C_2 s^2 + D_2 s + E_2)}$$

$$= \frac{(-71.918\,s^3 - 32.27\,s^2 - 1.1857\,s - 18.816)}{s(169.96\,s^4 + 103.65\,s^3 + 169.373\,s^2 + 130.123\,s - 3.96)}$$

$$\text{roots}(\overline{D}_2(s)) = 0.047 \pm i\,1.0403, \quad -0.7334, \quad 0.0293$$

$$\zeta_{DR} = -0.0453 \ (Unstable \ Dutch \ Roll), \ \omega_{nDR} = 1.0413,$$

$$T_R = 1.363, \ T_S = -34.137 \ (Unstable \ Spiral)$$

B.8 AIRCRAFT 7—BOEING 747-200

Table B7.1 Geometric Data for the Boeing B747-200 Aircraft

Wing Surface (ft^2)	S	5,500
Mean Aerodynamic Chord (MAC) (ft)	$\bar{c}$	27.3
Wing Span (ft)	b	196

Table B7.2 Flight Conditions Data for the Boeing B747-200 Aircraft

		Approach	Cruise (low)	Cruise (high)
Altitude (ft)	h	0	20,000	40,000
Mach Number	M	0.198	0.65	0.90
True Airspeed (ft/sec)	V_{P_1}	221	673	871
Dynamic Pressure (lbs/ft^2)	$\bar{q}$	58	287.2	222.8
Location of CG - % MAC	$\bar{x}_{CG}$	0.25	0.25	0.25
Steady state angle of attack (deg)	α_1	8.5	2.5	2.4

Table B7.3 Mass and Inertial Data for the Boeing B747-200 Aircraft

		Approach	Cruise (low)	Cruise (high)
Mass (lbs)	m	564,000	636,636	636,636
Moment of Inertia x-axis (slug ft^2)	I_{XX_B}	13,700,000	18,200,000	18,200,000
Moment of Inertia y-axis (slug ft^2)	I_{YY_B}	30,500,000	33,100,000	33,100,000
Moment of Inertia z-axis (slug ft^2)	I_{ZZ_B}	43,100,000	49,700,000	49,700,000
Product of inertia xz-plane (slug ft^2)	I_{XZ_B}	830,000	970,000	970,000

Table B7.4 Longitudinal Aerodynamic Coefficients for the Boeing B747-200 Aircraft

	Approach	Cruise (low)	Cruise (high)
Steady State			
c_{L_1}	1.76	0.4	0.52
c_{D_1}	0.2630	0.0250	0.0450
c_{m_1}	0	0	0
$c_{T_{X_1}}$	0.2630	0.0250	0.0450
$c_{m_{T_1}}$	0	0	0
Stability Derivatives			
c_{D_0}	0.0751	0.0164	0.0305
c_{D_u}	0	0	0
c_{D_α}	1.13	0.20	0.50
$c_{T_{X_u}}$	−0.5523	−0.055	−0.950
c_{L_0}	0.92	0.21	0.29
c_{L_u}	−0.22	0.13	−0.23
c_{L_α}	5.67	4.4	5.5
$c_{L_{\dot\alpha}}$	6.7	7.0	8.0
c_{L_q}	5.65	6.6	7.8
c_{m_0}	0	0	0
c_{m_u}	0.071	0.013	−0.09
c_{m_α}	−1.45	−1.00	−1.60
$c_{m_{\dot\alpha}}$	−3.3	−4.0	−9.0
c_{m_q}	−21.4	−20.5	−25.5
$c_{m_{T_u}}$	0	0	0
$c_{m_{T_\alpha}}$	0	0	0
Control Derivatives			
$c_{D_{\delta_E}}/c_{D_{i_H}}$	0/0	0/0	0/0
$c_{L_{\delta_E}}/c_{L_{i_H}}$	0.36/0.75	0.32/0.70	0.30/0.65
$c_{m_{\delta_E}}/c_{m_{i_H}}$	−1.40/−3.0	−1.30/−2.70	1.20/−2.50

Table B7.5 Longitudinal Dimensional Stability Derivatives for the Boeing B747-200 Aircraft

Cruise (high)

$$X_u = -0.0218, \ X_{T_u} = -0.0604, \ X_\alpha = 1.2227, \ X_{\delta_E} = 0$$

$$Z_u = -0.0569, \ Z_\alpha = -339.0, \ Z_{\dot\alpha} = -7.666, \ Z_q = -7.474, \ Z_{\delta_E} = -18.341$$

$$M_u = -0.0001, \ M_{T_u} = 0.0, \ M_\alpha = -1.616, \ M_{T_\alpha} = 0,$$
$$M_{\dot\alpha} = -0.1425, \ M_q = -0.4038, \ M_{\delta_E} = -1.2124$$

Table B7.6 Longitudinal Transfer Functions for the Boeing B747-200 Aircraft

Cruise (high)

$$\frac{\alpha(s)}{\delta_E(s)} = \frac{Num_\alpha(s)}{\overline{D}_1(s)} = \frac{A_\alpha s^3 + B_\alpha s^2 + C_\alpha s + D_\alpha}{A_1 s^4 + B_1 s^3 + C_1 s^2 + D_1 s + E_1}$$

$$= \frac{-18.341 \, s^3 - 1{,}055.696 \, s^2 - 84.97 \, s - 1.995}{878.568 \, s^4 + 888.97 \, s^3 + 1{,}599.56 \, s^2 + 121.194 \, s + 1.617}$$

$$\frac{u(s)}{\delta_E(s)} = \frac{Num_u(s)}{\overline{D}_1(s)} = \frac{A_u s^2 + B_u s + C_u}{A_1 s^4 + B_1 s^3 + C_1 s^2 + D_1 s + E_1}$$

$$= \frac{-22.426 \, s^2 + 32{,}442.603 \, s + 12{,}108.424}{878.568 \, s^4 + 888.97 \, s^3 + 1{,}599.56 \, s^2 + 121.194 \, s + 1.617}$$

$$\frac{\theta(s)}{\delta_E(s)} = \frac{Num_\theta(s)}{\overline{D}_1(s)} = \frac{A_\theta s^2 + B_\theta s + C_\theta}{A_1 s^4 + B_1 s^3 + C_1 s^2 + D_1 s + E_1}$$

$$= \frac{-1{,}062.524 \, s^2 - 468.614 \, s - 31.403}{878.568 \, s^4 + 888.97 \, s^3 + 1{,}599.56 \, s^2 + 121.194 \, s + 1.617}$$

$$\text{roots}(\overline{D}_1(s)) = -0.4667 \pm i\, 1.2364, \ -0.0612, \ -0.0172$$
$$\zeta_{SP} = 0.353, \ \omega_{nSP} = 1.3215, \ T_1 = 16.34, \ T_2 = 58.05 \ (\textit{Degenerated Phugoid})$$

Table B7.7 Lateral Directional Aerodynamic Coefficients for the Boeing B747−200 Aircraft

	Approach	Cruise (low)	Cruise (high)
Stability Derivatives			
c_{l_β}	−0.281	−0.160	−0.095
c_{l_p}	−0.502	−0.340	−0.320
c_{l_r}	0.195	0.130	0.200
c_{Y_β}	−1.08	−0.90	−0.90
c_{Y_p}	0	0	0
c_{Y_r}	0	0	0
c_{n_β}	0.184	0.160	0.210
$c_{n_{T_\beta}}$	0	0	0
c_{n_p}	−0.222	−0.026	0.02
c_{n_r}	−0.360	−0.280	−0.330
Control Derivatives			
$c_{l_{\delta_A}}$	0.053	0.013	0.014
$c_{l_{\delta_R}}$	0	0.008	0.0005
$c_{Y_{\delta_A}}$	0	0	0
$c_{Y_{\delta_R}}$	0.179	0.120	0.060
$c_{n_{\delta_A}}$	0.0083	0.0018	−0.0028
$c_{n_{\delta_R}}$	−0.113	−0.10	−0.095

Table B7.8 Lateral Directional Dimensional Stability Derivatives for the Boeing B747-200 Aircraft

Cruise (high)

$$Y_\beta = -55.023, \ Y_P = 0.0, \ Y_r = 0.0, \ Y_{\delta_A} = 0, \ Y_{\delta_R} = 7.336$$

$$L_\beta = -2.114, \ L_P = -0.5054, \ L_r = 0.1932, \ L_{\delta_A} = 0.1717, \ L_{\delta_R} = 0.1057$$

$$N_\beta = 0.7725, \ N_{T_\beta} = 0, \ N_P = -0.0141, \ N_r = -0.1521, \ N_{\delta_A} = 0.0087, \ N_{\delta_R} = -0.4828$$

Table B7.9 Lateral Directional Transfer Functions for the Boeing B747-200 Aircraft

Cruise (high)

$$\frac{\beta(s)}{\delta_A(s)} = \frac{Num_{\beta A}(s)}{\overline{D}_2(s)} = \frac{s(A_{\beta A}s^3 + B_{\beta A}s^2 + C_{\beta A}s + D_{\beta A})}{s(A_2 s^4 + B_2 s^3 + C_2 s^2 + D_2 s + E_2)}$$

$$= \frac{s(0 s^3 + 6.5121 s^2 + 3.734 s + 0.883)}{s(870.783 s^4 + 628.627 s^3 + 791.372 s^2 + 438.002 s + 5.469)}$$

$$\frac{\phi(s)}{\delta_A(s)} = \frac{Num_{\phi A}(s)}{\overline{D}_2(s)} = \frac{s(A_{\phi A}s^2 + B_{\phi A}s + C_{\phi A})}{s(A_2 s^4 + B_2 s^3 + C_2 s^2 + D_2 s + E_2)}$$

$$= \frac{s(149.419 s^2 + 33.656 s + 133.07)}{s(870.783 s^4 + 628.627 s^3 + 791.372 s^2 + 438.002 s + 5.469)}$$

$$\frac{\psi(s)}{\delta_A(s)} = \frac{Num_{\psi A}(s)}{\overline{D}_2(s)} = \frac{A_{\psi A}s^3 + B_{\psi A}s^2 + C_{\psi A}s + D_{\psi A}}{s(A_2 s^4 + B_2 s^3 + C_2 s^2 + D_2 s + E_2)}$$

$$= \frac{(6.512 s^3 + 2.124 s^2 + 0.1082 s + 4.795)}{s(870.783 s^4 + 628.627 s^3 + 791.372 s^2 + 438.002 s + 5.469)}$$

$$\frac{\beta(s)}{\delta_R(s)} = \frac{Num_{\beta R}(s)}{\overline{D}_2(s)} = \frac{s(A_{\beta R}s^3 + B_{\beta R}s^2 + C_{\beta R}s + D_{\beta R})}{s(A_2 s^4 + B_2 s^3 + C_2 s^2 + D_2 s + E_2)}$$

$$= \frac{s(7.3354 s^3 + 425.974 s^2 + 218.0615 s - 2.4517)}{s(870.783 s^4 + 628.627 s^3 + 791.372 s^2 + 438.002 s + 5.469)}$$

$$\frac{\phi(s)}{\delta_R(s)} = \frac{Num_{\phi R}(s)}{\overline{D}_2(s)} = \frac{s(A_{\phi R}s^2 + B_{\phi R}s + C_{\phi R})}{s(A_2 s^4 + B_2 s^3 + C_2 s^2 + D_2 s + E_2)}$$

$$= \frac{s(100.169 s^2 - 76.546 s - 823.194)}{s(870.783 s^4 + 628.627 s^3 + 791.372 s^2 + 438.002 s + 5.469)}$$

$$\frac{\psi(s)}{\delta_R(s)} = \frac{Num_{\psi R}(s)}{\overline{D}_2(s)} = \frac{A_{\psi R}s^3 + B_{\psi R}s^2 + C_{\psi R}s + D_{\psi R}}{s(A_2 s^4 + B_2 s^3 + C_2 s^2 + D_2 s + E_2)}$$

$$= \frac{(-421.142 s^3 - 234.656 s^2 - 10.426 s - 29.807)}{s(870.783 s^4 + 628.627 s^3 + 791.372 s^2 + 438.002 s + 5.469)}$$

$$\text{roots}(\overline{D}_2(s)) = -0.0586 \pm i\, 0.9094, \quad -0.5919, \quad -0.0128$$

$$\zeta_{DR} = 0.0643, \ \omega_{nDR} = 0.9112, \ T_R = 1.689, \ T_S = 78.264$$

B.9 AIRCRAFT 8—SIAI MARCHETTI S-211

Table B8.1 Geometric Data for the SIAI Marchetti S-211 Aircraft

Wing Surface (ft^2)	S	136
Mean Aerodynamic Chord (MAC) (ft)	$\bar{c}$	5.4
Wing Span (ft)	b	26.3

Table B8.2 Flight Conditions Data for the SIAI Marchetti S-211 Aircraft

		Approach	Cruise (low)	Cruise (high)
Altitude (ft)	h	0	25,000	35,000
Mach Number	M	0.111	0.60	0.60
True Airspeed (ft/sec)	V_{P_1}	124	610	584
Dynamic Pressure (lbs/ft^2)	$\bar{q}$	18.2	198.0	125.7
Location of CG - % MAC	$\bar{x}_{CG}$	0.25	0.25	0.25
Steady-state angle of attack (deg)	α_1	8	0	0.9

Table B8.3 Mass and Inertial Data for the SIAI Marchetti S-211 Aircraft

		Approach	Cruise (low)	Cruise (high)
Mass (lbs)	m	3,500	4,000	4,000
Moment of Inertia x-axis (slug ft^2)	I_{XX_B}	750	800	800
Moment of Inertia y-axis (slug ft^2)	I_{YY_B}	4,600	4,800	4,800
Moment of Inertia z-axis (slug ft^2)	I_{ZZ_B}	5,000	5,200	5,200
Product of inertia xz-plane (slug ft^2)	I_{XZ_B}	200	200	200

Table B8.4 Longitudinal Aerodynamic Coefficients for the SIAI Marchetti S−211 Aircraft

	Approach	Cruise (low)	Cruise (high)
Steady State			
c_{L_1}	1.414	0.149	0.234
c_{D_1}	0.21	0.022	0.025
c_{m_1}	0	0	0
$c_{T_{X_1}}$	0.21	0.022	0.025
$c_{m_{T_1}}$	0	0	0
Stability Derivatives			
c_{D_0}	0.090	0.0205	0.0205
c_{D_u}	0	0.05	0.05
c_{D_α}	1.14	0.12	0.17
$c_{T_{X_u}}$	−0.45	−0.05	−0.055
c_{L_0}	0.65	0.149	0.149
c_{L_u}	0.071	0.084	0.132
c_{L_α}	5.0	5.5	5.5
$c_{L_{\dot{\alpha}}}$	3.0	4.2	4.2
c_{L_q}	9.0	10.0	10.0
c_{m_0}	−0.07	−0.08	−0.08
c_{m_u}	0	0	0
c_{m_α}	−0.60	−0.24	−0.24
$c_{m_{\dot{\alpha}}}$	−7.0	−9.6	−9.6
c_{m_q}	−15.7	−17.7	−17.7
$c_{m_{T_u}}$	0	0	0
$c_{m_{T_\alpha}}$	0	0	0
Control Derivatives			
$c_{D_{\delta_E}}/c_{D_{i_H}}$	0/0	0/0	0/0
$c_{L_{\delta_E}}/c_{L_{i_H}}$	0.39/1.0	0.38/0.99	0.35/0.99
$c_{m_{\delta_E}}/c_{m_{i_H}}$	−0.90/−2.30	−0.88/−2.30	−0.82/−2.30

Table B8.5 Longitudinal Dimensional Stability Derivatives for the SIAI Marchetti S-211 Aircraft

Approach

$$X_u = -0.078, \ X_{T_u} = -0.0055, \ X_\alpha = 6.26, \ X_{\delta_E} = 0$$

$$Z_u = -0.534, \ Z_\alpha = -119, \ Z_{\dot\alpha} = -1.492, \ Z_q = -4.48, \ Z_{\delta_E} = -8.91$$

$$M_u = 0.0, \ M_{T_u} = 0.0, \ M_\alpha = -1.75, \ M_{T_\alpha} = 0.0,$$

$$M_{\dot\alpha} = -0.445, \ M_q = -1.0, \ M_{\delta_E} = -2.625$$

Table B8.6 Longitudinal Transfer Functions for the SIAI Marchetti S-211 Aircraft

Approach

$$\frac{\alpha(s)}{\delta_E(s)} = \frac{Num_\alpha(s)}{\overline{D}_1(s)} = \frac{A_\alpha s^3 + B_\alpha s^2 + C_\alpha s + D_\alpha}{A_1 s^4 + B_1 s^3 + C_1 s^2 + D_1 s + E_1}$$

$$= \frac{-8.91 s^3 - 323.34 s^2 - 14.99 s - 43.7}{125.48 s^4 + 307.67 s^3 + 353.8 s^2 + 30.1 s + 29.13}$$

$$\frac{u(s)}{\delta_E(s)} = \frac{Num_u(s)}{\overline{D}_1(s)} = \frac{A_u s^2 + B_u s + C_u}{A_1 s^4 + B_1 s^3 + C_1 s^2 + D_1 s + E_1}$$

$$= \frac{-55.75 s^2 + 8{,}349.51 s + 9{,}529.46}{125.48 s^4 + 307.67 s^3 + 353.8 s^2 + 30.1 s + 29.13}$$

$$\frac{\theta(s)}{\delta_E(s)} = \frac{Num_\theta(s)}{\overline{D}_1(s)} = \frac{A_\theta s^2 + B_\theta s + C_\theta}{A_1 s^4 + B_1 s^3 + C_1 s^2 + D_1 s + E_1}$$

$$= \frac{-325.43 s^2 - 323.76 s - 33.38}{125.48 s^4 + 307.67 s^3 + 353.8 s^2 + 30.1 s + 29.13}$$

$$\text{roots}(\overline{D}_1(s)) = -1.23 \pm i\,1.102, \quad -0.0056 \pm i\,0.293$$

$$\zeta_{SP} = 0.742, \quad \omega_{nSP} = 1.645, \quad \zeta_{Ph} = 0.019, \quad \omega_{nPh} = 0.293$$

Table B8.7 Lateral Directional Aerodynamic Coefficients for the SIAI Marchetti S-211 Aircraft

	Approach	Cruise (low)	Cruise (high)
Stability Derivatives			
c_{l_β}	−0.140	−0.110	−0.110
c_{l_p}	−0.350	−0.390	−0.390
c_{l_r}	0.560	0.280	0.310
c_{Y_β}	−0.94	−1.0	−1.0
c_{Y_p}	−0.010	−0.140	−0.120
c_{Y_r}	0.590	0.610	0.620
c_{n_β}	0.160	0.170	0.170
$c_{n_{T_\beta}}$	0	0	0
c_{n_p}	−0.030	0.090	0.080
c_{n_r}	−0.310	−0.260	−0.260
Control Derivatives			
$c_{l_{\delta_A}}$	0.110	0.10	0.10
$c_{l_{\delta_R}}$	0.030	0.050	0.050
$c_{Y_{\delta_A}}$	0	0	0
$c_{Y_{\delta_R}}$	0.260	0.0280	0.0280
$c_{n_{\delta_A}}$	−0.030	−0.003	−0.005
$c_{n_{\delta_R}}$	−0.110	−0.120	−0.120

Table B8.8 Lateral Directional Dimensional Stability Derivatives for the Marchetti S-211 Aircraft

Approach
$Y_\beta = -21.47,\ Y_P = -0.0242,\ Y_r = 1.43,\ Y_{\delta_A} = 0,\ Y_{\delta_R} = 5.94$
$L_\beta = -11.77,\ L_P = -3.12,\ L_r = 5,\ L_{\delta_A} = 9.25,\ L_{\delta_R} = 2.52$
$N_\beta = 2.1,\ N_{T_\beta} = 0,\ N_P = -0.042,\ N_r = -0.432,\ N_{\delta_A} = -0.394,\ N_{\delta_R} = -1.445$

Table B8.9 Lateral Directional Transfer Functions for the SIAI Marchetti S-211 Aircraft

Approach

$$\frac{\beta(s)}{\delta_A(s)} = \frac{Num_{\beta A}(s)}{\overline{D}_2(s)} = \frac{s(A_{\beta A}s^3 + B_{\beta A}s^2 + C_{\beta A}s + D_{\beta A})}{s(A_2 s^4 + B_2 s^3 + C_2 s^2 + D_2 s + E_2)}$$

$$= \frac{s(0\,s^3 + 137.77\,s^2 + 499.17\,s + 64.59)}{s(119.02\,s^4 + 507.52\,s^3 + 648.91\,s^2 + 1{,}307.07\,s - 172.49)}$$

$$\frac{\phi(s)}{\delta_A(s)} = \frac{Num_{\phi A}(s)}{\overline{D}_2(s)} = \frac{s(A_{\phi A}s^2 + B_{\phi A}s + C_{\phi A})}{s(A_2 s^4 + B_2 s^3 + C_2 s^2 + D_2 s + E_2)}$$

$$= \frac{s(1{,}171.46\,s^2 + 454.20\,s + 1{,}858)}{s(119.02\,s^4 + 507.52\,s^3 + 648.91\,s^2 + 1{,}307.07\,s - 172.49)}$$

$$\frac{\psi(s)}{\delta_A(s)} = \frac{Num_{\psi A}(s)}{\overline{D}_2(s)} = \frac{A_{\psi A}s^3 + B_{\psi A}s^2 + C_{\psi A}s + D_{\psi A}}{s(A_2 s^4 + B_2 s^3 + C_2 s^2 + D_2 s + E_2)}$$

$$= \frac{(-139.61\,s^3 - 224.68\,s^2 - 35.08\,s + 471.7)}{s(119.02\,s^4 + 507.52\,s^3 + 648.91\,s^2 + 1{,}307.07\,s - 172.49)}$$

$$\frac{\beta(s)}{\delta_R(s)} = \frac{Num_{\beta R}(s)}{\overline{D}_2(s)} = \frac{s(A_{\beta R}s^3 + B_{\beta R}s^2 + C_{\beta R}s + D_{\beta R})}{s(A_2 s^4 + B_2 s^3 + C_2 s^2 + D_2 s + E_2)}$$

$$= \frac{s(5.7\,s^3 + 224.86\,s^2 + 678.92\,s - 195.26)}{s(119.02\,s^4 + 507.52\,s^3 + 648.91\,s^2 + 1{,}307.07\,s - 172.49)}$$

$$\frac{\phi(s)}{\delta_R(s)} = \frac{Num_{\phi R}(s)}{\overline{D}_2(s)} = \frac{s(A_{\phi R}s^2 + B_{\phi R}s + C_{\phi R})}{s(A_2 s^4 + B_2 s^3 + C_2 s^2 + D_2 s + E_2)}$$

$$= \frac{s(403.48\,s^2 - 766.22\,s - 1{,}534.75)}{s(119.02\,s^4 + 507.52\,s^3 + 648.91\,s^2 + 1{,}307.07\,s - 172.49)}$$

$$\frac{\psi(s)}{\delta_R(s)} = \frac{Num_{\psi R}(s)}{\overline{D}_2(s)} = \frac{A_{\psi R}s^3 + B_{\psi R}s^2 + C_{\psi R}s + D_{\psi R}}{s(A_2 s^4 + B_2 s^3 + C_2 s^2 + D_2 s + E_2)}$$

$$= \frac{(-203.96\,s^3 - 589.74\,s^2 - 56.95\,s - 373.14)}{s(119.02\,s^4 + 507.52\,s^3 + 648.91\,s^2 + 1{,}307.07\,s - 172.49)}$$

$$roots(\overline{D}_2(s)) = -0.381 \pm i\,1.757,\quad -3.626,\quad 0.1236$$

$$\zeta_{DR} = 0.212,\ \omega_{nDR} = 1.798,\ T_R = 0.276,\ T_S = -8.09$$

B.10 AIRCRAFT 9—LOCKHEED F-104

Table B9.1 Geometric Data for the Lockheed F-104 Aircraft

Wing Surface (ft²)	S	196
Mean Aerodynamic Chord (MAC) (ft)	$\bar{c}$	9.6
Wing Span (ft)	b	21.9

Table B9.2 Flight Conditions Data for the Lockheed F-104 Aircraft

		Approach	Cruise
Altitude (ft)	h	0	55,000
Mach Number	M	0.257	1.8
True Airspeed (ft/sec)	V_{P_1}	287	1,742
Dynamic Pressure (lbs/ft²)	$\bar{q}$	97.8	434.5
Location of CG - % MAC	$\bar{x}_{CG}$	0.07	0.07
Steady state angle of attack (deg)	α_1	10	2

Table B9.3 Mass and Inertial Data for the Lockheed F-104 Aircraft

		Approach	Cruise
Mass (lbs)	m	16,300	16,300
Moment of Inertia x-axis (slug ft²)	I_{XX_B}	3,600	3,600
Moment of Inertia y-axis (slug ft²)	I_{YY_B}	59,000	59,000
Moment of Inertia z-axis (slug ft²)	I_{ZZ_B}	60,000	60,000
Product of inertia xz-plane (slug ft²)	I_{XZ_B}	0	0

Table B9.4 Longitudinal Aerodynamic Coefficients for the Lockheed F-104 Aircraft

	Approach	Cruise
Steady State		
c_{L_1}	0.850	0.191
c_{D_1}	0.2634	0.0553
c_{m_1}	0	0
$c_{T_{X_1}}$	0.2634	0.0553
$c_{m_{T_1}}$	0	0
Stability Derivatives		
c_{D_0}	0.1189	0.0480
c_{D_u}	0	−0.060
c_{D_α}	0.455	0.384
$c_{T_{X_u}}$	−0.50	−0.13
c_{L_0}	0.240	0.122
c_{L_u}	0	−0.20
c_{L_α}	3.440	2.005
$c_{L_{\dot\alpha}}$	0.66	0.82
c_{L_q}	2.30	1.90
c_{m_0}	0.03	−0.028
c_{m_u}	0	0
c_{m_α}	−0.644	−1.308
$c_{m_{\dot\alpha}}$	−1.640	−2.050
c_{m_q}	−5.840	−4.830
$c_{m_{T_u}}$	0	0
$c_{m_{T_\alpha}}$	0	0
Control Derivatives		
$c_{D_{i_H}}$	0	0
$c_{L_{i_H}}$	0.684	0.523
$c_{m_{i_H}}$	−1.60	−1.31

Table B9.5 Longitudinal Dimensional Stability Derivatives for the Lockheed F-104 Aircraft

Approach

$$X_u = -0.0695, \ X_{T_u} = 0.0035, \ X_\alpha = 14.96, \ X_{\delta_E} = 0$$

$$Z_u = -0.224, \ Z_\alpha = -140.22, \ Z_{\dot\alpha} = -0.418, \ Z_q = -1.456, \ Z_{\delta_E} = -25.9$$

$$M_u = 0.0, \ M_{T_u} = 0.0, \ M_\alpha = -2.01, \ M_{T_\alpha} = 0.0,$$
$$M_{\dot\alpha} = -0.0856, \ M_q = -0.305, \ M_{\delta_E} = -5$$

Table B9.6 Longitudinal Transfer Functions for the Lockheed F-104 Aircraft

Approach

$$\frac{\alpha(s)}{\delta_E(s)} = \frac{Num_\alpha(s)}{\overline{D}_1(s)} = \frac{A_\alpha s^3 + B_\alpha s^2 + C_\alpha s + D_\alpha}{A_1 s^4 + B_1 s^3 + C_1 s^2 + D_1 s + E_1}$$

$$= \frac{-25.9 \, s^3 - 1{,}435.41 \, s^2 - 66.68 \, s - 33.65}{287.39 \, s^4 + 271.25 \, s^3 + 636.16 \, s^2 + 31.05 \, s + 13.55}$$

$$\frac{u(s)}{\delta_E(s)} = \frac{Num_u(s)}{\overline{D}_1(s)} = \frac{A_u s^2 + B_u s + C_u}{A_1 s^4 + B_1 s^3 + C_1 s^2 + D_1 s + E_1}$$

$$= \frac{-387.34 \, s^2 + 23{,}961 \, s + 20{,}955.6}{287.39 \, s^4 + 271.25 \, s^3 + 636.16 \, s^2 + 31.05 \, s + 13.55}$$

$$\frac{\theta(s)}{\delta_E(s)} = \frac{Num_\theta(s)}{\overline{D}_1(s)} = \frac{A_\theta s^2 + B_\theta s + C_\theta}{A_1 s^4 + B_1 s^3 + C_1 s^2 + D_1 s + E_1}$$

$$= \frac{-1{,}432.95 \, s^2 - 742.73 \, s - 59.51}{287.39 \, s^4 + 271.25 \, s^3 + 636.16 \, s^2 + 31.05 \, s + 13.55}$$

$$roots(\overline{D}_1(s)) = -0.45 \pm i\,1.397, \quad -0.02 \pm i\,0.1465$$
$$\zeta_{SP} = 0.307, \quad \omega_{nSP} = 1.468, \quad \zeta_{Ph} = 0.138, \quad \omega_{nPh} = 0.148$$

Table B9.7 Lateral Directional Aerodynamic Coefficients for the Lockheed F-104 Aircraft

	Approach	Cruise
Stability Derivatives		
c_{l_β}	−0.175	−0.093
c_{l_p}	−0.285	−0.272
c_{l_r}	0.265	0.154
c_{Y_β}	−1.180	−1.045
c_{Y_p}	0	0
c_{Y_r}	0	0
c_{n_β}	0.507	0.242
$c_{n_{T_\beta}}$	0	0
c_{n_p}	−0.144	−0.093
c_{n_r}	−0.753	−0.649
Control Derivatives		
$c_{l_{\delta_A}}$	0.0392	0.0173
$c_{l_{\delta_R}}$	0.0448	0.0079
$c_{Y_{\delta_A}}$	0	0
$c_{Y_{\delta_R}}$	0.329	0.087
$c_{n_{\delta_A}}$	0.0042	0.0025
$c_{n_{\delta_R}}$	−0.1645	−0.0435

Table B9.8 Lateral Directional Dimensional Stability Derivatives for the Lockheed F-104 Aircraft

Approach

$Y_\beta = -44.68, \ Y_P = 0.0, \ Y_r = 0.0, \ Y_{\delta_A} = 0, \ Y_{\delta_R} = 12.46$

$L_\beta = -13.87, \ L_P = -0.862, \ L_r = 0.8, \ L_{\delta_A} = 3.107, \ L_{\delta_R} = 3.55$

$N_\beta = 3.65, \ N_{T_\beta} = 0, \ N_P = -0.04, \ N_r = -0.21, \ N_{\delta_A} = 0.03, \ N_{\delta_R} = -1.185$

Table B9.9 Lateral Directional Transfer Functions for the Lockheed F-104 Aircraft

Approach

$$\frac{\beta(s)}{\delta_A(s)} = \frac{Num_{\beta A}(s)}{\overline{D}_2(s)} = \frac{s(A_{\beta A}s^3 + B_{\beta A}s^2 + C_{\beta A}s + D_{\beta A})}{s(A_2 s^4 + B_2 s^3 + C_2 s^2 + D_2 s + E_2)}$$

$$= \frac{s(0\,s^3 + 138.81\,s^2 + 124.5\,s + 21.149)}{s(200.58\,s^4 + 355.34\,s^3 + 1{,}817.61\,s^2 + 1{,}720.61\,s - 1.781)}$$

$$\frac{\phi(s)}{\delta_A(s)} = \frac{Num_{\phi A}(s)}{\overline{D}_2(s)} = \frac{s(A_{\phi A}s^2 + B_{\phi A}s + C_{\phi A})}{s(A_2 s^4 + B_2 s^3 + C_2 s^2 + D_2 s + E_2)}$$

$$= \frac{s(875.72\,s^2 + 327.88\,s + 3{,}407.29)}{s(200.58\,s^4 + 355.34\,s^3 + 1{,}817.61\,s^2 + 1{,}720.61\,s - 1.781)}$$

$$\frac{\psi(s)}{\delta_A(s)} = \frac{Num_{\psi A}(s)}{\overline{D}_2(s)} = \frac{A_{\psi A}s^3 + B_{\psi A}s^2 + C_{\psi A}s + D_{\psi A}}{s(A_2 s^4 + B_2 s^3 + C_2 s^2 + D_2 s + E_2)}$$

$$= \frac{(-138.8\,s^3 - 49.42\,s^2 - 4.33\,s + 372.92)}{s(200.58\,s^4 + 355.34\,s^3 + 1{,}817.61\,s^2 + 1{,}720.61\,s - 1.781)}$$

$$\frac{\beta(s)}{\delta_R(s)} = \frac{Num_{\beta R}(s)}{\overline{D}_2(s)} = \frac{s(A_{\beta R}s^3 + B_{\beta R}s^2 + C_{\beta R}s + D_{\beta R})}{s(A_2 s^4 + B_2 s^3 + C_2 s^2 + D_2 s + E_2)}$$

$$= \frac{s(8.71\,s^3 + 522.79\,s^2 + 516.97\,s - 6.8)}{s(200.58\,s^4 + 355.34\,s^3 + 1{,}817.61\,s^2 + 1{,}720.61\,s - 1.781)}$$

$$\frac{\phi(s)}{\delta_R(s)} = \frac{Num_{\phi R}(s)}{\overline{D}_2(s)} = \frac{s(A_{\phi R}s^2 + B_{\phi R}s + C_{\phi R})}{s(A_2 s^4 + B_2 s^3 + C_2 s^2 + D_2 s + E_2)}$$

$$= \frac{s(1{,}637.82\,s^2 - 62.22\,s - 1{,}004.29)}{s(200.58\,s^4 + 355.34\,s^3 + 1{,}817.61\,s^2 + 1{,}720.61\,s - 1.781)}$$

$$\frac{\psi(s)}{\delta_R(s)} = \frac{Num_{\psi R}(s)}{\overline{D}_2(s)} = \frac{A_{\psi R}s^3 + B_{\psi R}s^2 + C_{\psi R}s + D_{\psi R}}{s(A_2 s^4 + B_2 s^3 + C_2 s^2 + D_2 s + E_2)}$$

$$= \frac{(-508.72\,s^3 - 338.62\,s^2 - 5.86\,s - 109.9)}{s(200.58\,s^4 + 355.34\,s^3 + 1{,}817.61\,s^2 + 1{,}720.61\,s - 1.781)}$$

$roots(\overline{D}_2(s)) = -0.37 \pm i\,2.857, \quad -1.0346, \quad 0.001$

$\zeta_{DR} = 0.128, \ \omega_{nDR} = 2.881, \ T_R = 0.967, \ T_S = -1{,}000 \quad (\text{unstable spiral})$

B.11 AIRCRAFT 10—MCDONNELL DOUGLAS F-4

Table B10.1 Geometric Data for the McDonnell Douglas F-4 Aircraft

Wing Surface (ft^2)	S	530
Mean Aerodynamic Chord (MAC) (ft)	$\bar{c}$	16
Wing Span (ft)	b	38.7

Table B10.2 Flight Conditions Data for the McDonnell Douglas F-4 Aircraft

		Approach	Cruise (mach < 1)	Cruise (mach > 1)
Altitude (ft)	h	0	35,000	55,000
Mach Number	M	0.206	0.90	1.80
True Airspeed (ft/sec)	V_{P_1}	230	876	1,742
Dynamic Pressure (lbs/ft^2)	$\bar{q}$	62.9	283.2	434.5
Location of CG - % MAC	$\bar{x}_{CG}$	0.29	0.29	0.29
Steady-state angle of attack (deg)	α_1	11.7	2.6	3.3

Table B10.3 Mass and Inertial Data for the McDonnell Douglas F-4 Aircraft

		Approach	Cruise (mach < 1)	Cruise (mach > 1)
Mass (lbs)	m	33,200	39,000	39,000
Moment of Inertia x-axis (slug ft^2)	I_{XX_B}	23,700	25,000	25,000
Moment of Inertia y-axis (slug ft^2)	I_{YY_B}	117,500	122,200	122,200
Moment of Inertia z-axis (slug ft^2)	I_{ZZ_B}	133,700	139,800	139,800
Product of inertia xz-plane (slug ft^2)	I_{XZ_B}	1,600	2,200	2,200

Table B10.4 Longitudinal Aerodynamic Coefficients for the McDonnell Douglas F-4 Aircraft

	Approach	Cruise (mach < 1)	Cruise (mach > 1)
Steady State			
c_{L_1}	1.0	0.26	0.17
c_{D_1}	0.20	0.030	0.0480
c_{m_1}	0	0	0
$c_{T_{X_1}}$	0.20	0.030	0.0480
$c_{m_{T_1}}$	0	0	0
Stability Derivatives			
c_{D_0}	0.0269	0.0205	0.0439
c_{D_u}	0	0.027	−0.054
c_{D_α}	0.555	0.30	0.40
$c_{T_{X_u}}$	−0.45	−0.064	−0.10
c_{L_0}	0.430	0.10	0.010
c_{L_u}	0	0.270	−0.180
c_{L_α}	2.80	3.75	2.80
$c_{L_{\dot\alpha}}$	0.63	0.86	0.17
c_{L_q}	1.33	1.80	1.30
c_{m_0}	0.02	0.025	−0.025
c_{m_u}	0	−0.117	0.054
c_{m_α}	−0.098	−0.40	−0.780
$c_{m_{\dot\alpha}}$	−0.95	−1.30	−0.25
c_{m_q}	−2.0	−2.70	−2.0
$c_{m_{T_u}}$	0	0	0
$c_{m_{T_\alpha}}$	0	0	0
Control Derivatives			
$c_{D_{i_H}}$	−0.14	−0.10	−0.15
$c_{L_{i_H}}$	0.24	0.40	0.25
$c_{m_{i_H}}$	−0.322	−0.580	−0.380

Table B10.5 Longitudinal Dimensional Stability Derivatives for the McDonnell Douglas F-4 Aircraft

Cruise (mach < 1)

$$X_u = -0.0122, \ X_{T_u} = -0.0006, \ X_\alpha = -4.8986, \ X_{i_H} = 12.247$$

$$Z_u = -0.1105, \ Z_\alpha = -462.92, \ Z_{\dot\alpha} = -0.962, \ Z_q = -2.013, \ Z_{i_H} = -48.99$$

$$M_u = -0.0026, \ M_{T_u} = 0.0, \ M_\alpha = -7.86, \ M_{T_\alpha} = 0,$$
$$M_{\dot\alpha} = -0.233, \ M_q = -0.485, \ M_{i_H} = -11.397$$

Table B10.6 Longitudinal Transfer Functions for the McDonnell Douglas F-4 Aircraft

Cruise (mach < 1)

$$\frac{\alpha(s)}{i_H(s)} = \frac{Num_\alpha(s)}{\overline{D}_1(s)} = \frac{A_\alpha s^3 + B_\alpha s^2 + C_\alpha s + D_\alpha}{A_1 s^4 + B_1 s^3 + C_1 s^2 + D_1 s + E_1}$$

$$= \frac{-48.986 \, s^3 - 9{,}985.70 \, s^2 - 139.32 \, s - 35.68}{876.86 \, s^4 + 1{,}102.89 \, s^3 + 7{,}106.29 \, s^2 - 4.953 \, s - 11.154}$$

$$\frac{u(s)}{i_H(s)} = \frac{Num_u(s)}{\overline{D}_1(s)} = \frac{A_u s^2 + B_u s + C_u}{A_1 s^4 + B_1 s^3 + C_1 s^2 + D_1 s + E_1}$$

$$= \frac{10{,}738.60 \, s^2 + 13{,}610.03 \, s + 453{,}126.05}{876.86 \, s^4 + 1{,}102.89 \, s^3 + 7{,}106.29 \, s^2 - 4.953 \, s - 11.154}$$

$$\frac{\theta(s)}{i_H(s)} = \frac{Num_\theta(s)}{\overline{D}_1(s)} = \frac{A_\theta s^2 + B_\theta s + C_\theta}{A_1 s^4 + B_1 s^3 + C_1 s^2 + D_1 s + E_1}$$

$$= \frac{-9.982.46 \, s^2 - 5{,}045.91 \, s - 60.94}{876.86 \, s^4 + 1{,}102.89 \, s^3 + 7{,}106.29 \, s^2 - 4.953 \, s - 11.154}$$

$roots(\overline{D}_1(s)) = -0.6291 \pm i \, 2.7768, \quad -0.03942, \ 0.0398$

$\zeta_{SP} = 0.221, \quad \omega_{nSP} = 2.847, \quad T_1 = 25.38, \quad T_2 = -25.12$ (*Degenerated Unstable Phugoid*)

Table B10.7 Lateral Directional Aerodynamic Coefficients for the McDonnell Douglas F-4 Aircraft

	Approach	Cruise (mach < 1)	Cruise (mach > 1)
Stability Derivatives			
c_{l_β}	−0.156	−0.080	−0.025
c_{l_p}	−0.272	−0.240	−0.20
c_{l_r}	0.205	0.070	0.040
c_{Y_β}	−0.655	−0.680	−0.70
c_{Y_p}	0	0	0
c_{Y_r}	0	0	0
c_{n_β}	0.199	0.125	0.09
$c_{n_{T_\beta}}$	0	0	0
c_{n_p}	0.013	−0.036	0
c_{n_r}	−0.320	−0.270	−0.260
Control Derivatives			
$c_{l_{\delta_A}}$	0.0570	0.0420	0.0150
$c_{l_{\delta_R}}$	0.0009	0.0060	0.0030
$c_{Y_{\delta_A}}$	−0.0355	−0.0160	−0.010
$c_{Y_{\delta_R}}$	0.124	0.095	0.05
$c_{n_{\delta_A}}$	0.0041	−0.0010	−0.0009
$c_{n_{\delta_R}}$	−0.072	−0.066	−0.025

Table B10.8 Lateral Directional Dimensional Stability Derivatives for the F-4 Aircraft

Cruise (mach < 1)

$$Y_\beta = -83.277, \ Y_P = 0.0, \ Y_r = 0.0, \ Y_{\delta_A} = -1.9595, \ Y_{\delta_R} = 11.634$$

$$L_\beta = -18.559, \ L_P = -1.23, \ L_r = 0.3587, \ L_{\delta_A} = 9.743, \ L_{\delta_R} = 1.392$$

$$N_\beta = 5.195, \ N_{T_\beta} = 0, \ N_P = -0.0331, \ N_r = -0.2479, \ N_{\delta_A} = -0.0416, \ N_{\delta_R} = -2.743$$

Table B10.9 Lateral Directional Transfer Functions for the McDonnell Douglas F-4 Aircraft

Cruise (mach < 1)

$$\frac{\beta(s)}{\delta_A(s)} = \frac{Num_{\beta A}(s)}{\overline{D}_2(s)} = \frac{s(A_{\beta A}s^3 + B_{\beta A}s^2 + C_{\beta A}s + D_{\beta A})}{s(A_2 s^4 + B_2 s^3 + C_2 s^2 + D_2 s + E_2)}$$

$$= \frac{s(-1.954\, s^3 + 217.376\, s^2 + 636.116\, s + 76.306)}{s(873.63\, s^4 + 1{,}380.79\, s^3 + 5{,}301.04\, s^2 + 6{,}769.83\, s + 87.004)}$$

$$\frac{\phi(s)}{\delta_A(s)} = \frac{Num_{\phi A}(s)}{\overline{D}_2(s)} = \frac{s(A_{\phi A}s^2 + B_{\phi A}s + C_{\phi A})}{s(A_2 s^4 + B_2 s^3 + C_2 s^2 + D_2 s + E_2)}$$

$$= \frac{s(8{,}538.56\, s^2 + 2{,}951.76\, s + 43{,}861.61)}{s(873.63\, s^4 + 1{,}380.79\, s^3 + 5{,}301.04\, s^2 + 6{,}769.83\, s + 87.004)}$$

$$\frac{\psi(s)}{\delta_A(s)} = \frac{Num_{\psi A}(s)}{\overline{D}_2(s)} = \frac{A_{\psi A}s^3 + B_{\psi A}s^2 + C_{\psi A}s + D_{\psi A}}{s(A_2 s^4 + B_2 s^3 + C_2 s^2 + D_2 s + E_2)}$$

$$= \frac{(-220.28\, s^3 - 358.73\, s^2 - 44.79\, s\ + 1{,}584.53)}{s(873.63\, s^4 + 1{,}380.79\, s^3 + 5{,}301.04\, s^2 + 6{,}769.83\, s + 87.004)}$$

$$\frac{\beta(s)}{\delta_R(s)} = \frac{Num_{\beta R}(s)}{\overline{D}_2(s)} = \frac{s(A_{\beta R}s^3 + B_{\beta R}s^2 + C_{\beta R}s + D_{\beta R})}{s(A_2 s^4 + B_2 s^3 + C_2 s^2 + D_2 s + E_2)}$$

$$= \frac{s(11.60\, s^3 + 2{,}445.90\, s^2 + 3{,}053.58\, s - 20.31)}{s(873.63\, s^4 + 1{,}380.79\, s^3 + 5{,}301.04\, s^2 + 6{,}769.83\, s + 87.004)}$$

$$\frac{\phi(s)}{\delta_R(s)} = \frac{Num_{\phi R}(s)}{\overline{D}_2(s)} = \frac{s(A_{\phi R}s^2 + B_{\phi R}s + C_{\phi R})}{s(A_2 s^4 + B_2 s^3 + C_2 s^2 + D_2 s + E_2)}$$

$$= \frac{s(1{,}508.12\, s^2 - 639.43\, s - 38{,}337.14)}{s(873.63\, s^4 + 1{,}380.79\, s^3 + 5{,}301.04\, s^2 + 6{,}769.83\, s + 87.004)}$$

$$\frac{\psi(s)}{\delta_R(s)} = \frac{Num_{\psi R}(s)}{\overline{D}_2(s)} = \frac{A_{\psi R}s^3 + B_{\psi R}s^2 + C_{\psi R}s + D_{\psi R}}{s(A_2 s^4 + B_2 s^3 + C_2 s^2 + D_2 s + E_2)}$$

$$= \frac{(-2{,}428.66\, s^3 - 3{,}160.98\, s^2 - 203.3\, s - 1{,}388.38)}{s(873.63\, s^4 + 1{,}380.79\, s^3 + 5{,}301.04\, s^2 + 6{,}769.83\, s + 87.004)}$$

$$\text{roots}(\overline{D}_2(s)) = -0.1155 \pm i\, 2.393, \quad -1.337, \quad -0.013$$

$$\zeta_{DR} = 0.0482, \quad \omega_{nDR} = 2.395, \quad T_R = 0.748, \quad T_S = 76.9$$

REFERENCE

1. Roskam, J. *Airplane Flight Dynamics and Automatic Flight Controls — Part I*. Design, Analysis, and Research Corporation, Lawrence, KS, 1995.

Appendix C

Detailed Drawings for Different Aircraft

TABLE OF CONTENTS

C.1	Introduction	
C.2	Aircraft 1—Aeritalia Fiat G-91	
C.3	Aircraft 2—Beech 99	
C.4	Aircraft 3—Boeing B52	
C.5	Aircraft 4—Boeing B727-200	
C.6	Aircraft 5—Boeing B737-600	
C.7	Aircraft 6—Boeing B747-200	
C.8	Aircraft 7—Boeing B757-200	
C.9	Aircraft 8—Boeing B767-200	
C.10	Aircraft 9—Cessna Citation CJ3	
C.11	Aircraft 10—Cessna T37	
C.12	Aircraft 11—General Dynamics F-16	
C.13	Aircraft 12—Grumman F-14	
C.14	Aircraft 13—Learjet 24	
C.15	Aircraft 14—Lockheed F-104	
C.16	Aircraft 15—Lockheed F-22	
C.17	Aircraft 16—Lockheed L-1011	
C.18	Aircraft 17—McDonnell Douglas C-17	
C.19	Aircraft 18—McDonnell Douglas DC-8	
C.20	Aircraft 19.1—McDonnell Douglas DC-9 Series 10	
C.21	Aircraft 19.2—McDonnell Douglas DC-9 Series 30	
C.22	Aircraft 19.3—McDonnell Douglas DC-9 Series 40	
C.23	Aircraft 19.4—McDonnell Douglas DC-9 Series 50	
C.24	Aircraft 20—McDonnell Douglas DC-10	
C.25	Aircraft 21—McDonnell Douglas F-4	
C.26	Aircraft 22—McDonnell Douglas F-15	
C.27	Aircraft 23—Rockwell B-1	
C.28	Aircraft 24—SIAI Marchetti S211	
C.29	Aircraft 25—Supermarine Spitfire	

C.1 INTRODUCTION

Appendix C provides 3D drawings for twenty-five aircraft from different classes. Additionally, detailed geometric parameters have been derived from basic geometric information and can be used for the modeling of the aerodynamic coefficients, as discussed in Chapters 2, 3, and 4. The list of aircraft includes the following:

- Aeritalia Fiat G-91
- Beech 99
- Boeing B52
- Boeing B727-200
- Boeing B737-600
- Boeing B747-200
- Boeing B757-200
- Boeing B767-200
- Cessna CJ3
- Cessna T37
- General Dynamics F-16
- Grumman F-14
- Learjet 24
- Lockheed F-104
- Lockheed F-22
- Lockheed L-1011
- McDonnell Douglas C-17
- McDonnell Douglas DC-8
- McDonnell Douglas DC-9 Series 10, Series 30, Series 40, Series 50
- McDonnell Douglas DC-10
- McDonnell Douglas F-4
- McDonnell Douglas F-15
- Rockwell B-1
- SIAI Marchetti S211
- Supermarine Spitfire

C.2 AIRCRAFT 1—AERITALIA FIAT G-91

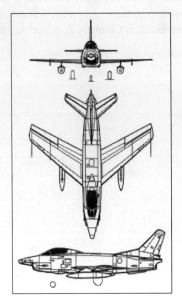

Figure C-A1.1 3D View of the Aeritalia Fiat G91 Aircraft
(Source: www.aviastar.org/index2.html)

Geometric Parameters and Typical Flight Conditions

$$Alt. = 25{,}000 \text{ ft}, \ Mach = 0.49$$
$$S = 177 \text{ ft}^2, \ b = 28.3 \text{ ft}, \ Height = 14.75 \text{ ft}, \ Length = 36.5 \text{ ft}$$
$$(Estimated) \ \bar{x}_{CG} = 0.26, \ W = 9{,}300 \text{ lbs}$$

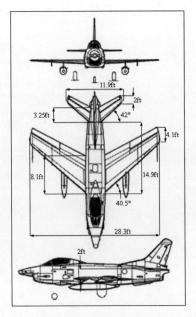

Figure C-A1.2 General Dimensions of the Aeritalia Fiat G91 Aircraft

C.2 Aircraft 1—Aeritalia Fiat G-91   **619**

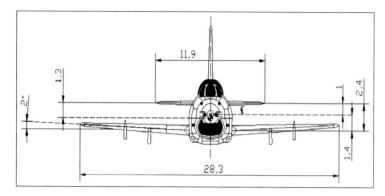

Figure C-A1.3 Front CAD View of the Aeritalia Fiat G91 Aircraft

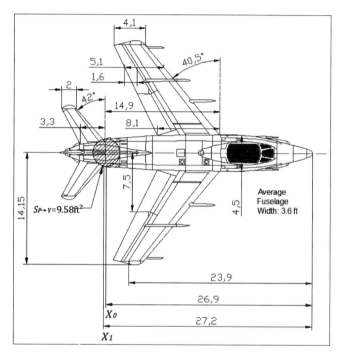

Figure C-A1.4 Top CAD View of the Aeritalia Fiat G91 Aircraft

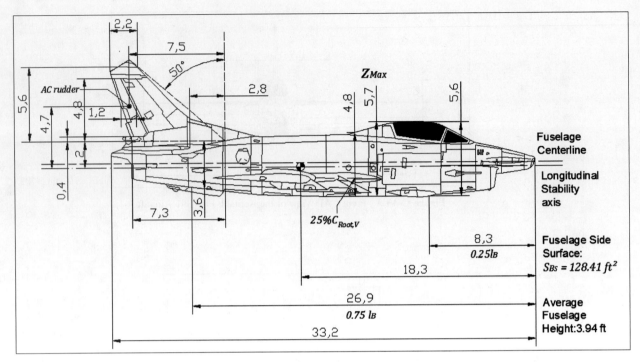

Figure C-A1.5 Side CAD View of the Aeritalia Fiat G91 Aircraft

Table C-A1.1 Geometric Parameters for the Aerodynamic Modeling of the Aeritalia G91 Aircraft

Geometric Parameters	G91 Aircraft	Geometric Parameters	G91 Aircraft
A [ft]	23.9	X_{HV} [ft]	2.8
b [ft]	28.3	X_{WH_r} [ft]	14.9
b_H [ft]	11.9	X_1 [ft]	27.2
b_V [ft]	5.6	y_{A_I} [ft]	7.5
$\bar{c}$ [ft]	6.3	y_{A_O} [ft]	14.1
$\bar{c}_{Aileron}$ [ft]	1.6	y_{R_I} [ft]	0.5
$\bar{c}_R$ [ft]	1.2	y_{R_F} [ft]	4.8
$\bar{c}_{Wing(atAileron)}$ [ft]	5.1	y_V [ft]	2
c_r [ft]	8.1	Z_H [ft]	-1
c_{r_H} [ft]	3.2	Z_{R_S} [ft]	4.7
c_{r_V} [ft]	7.3	z_1 [ft]	5.6
c_T [ft]	4.1	z_2 [ft]	3.9
c_{T_H} [ft]	2	Z_{H_S} [ft]	-1.3
c_{T_V} [ft]	2.2	z_{max} [ft]	5.7
d [ft]	4.8	Z_W [ft]	1.4
l_b [ft]	33.2	Z_{WH_r} [ft]	2.4
l_{cg} [ft]	18.3	Γ_H [deg]	0
r_1 [ft]	3.6	Γ_W [deg]	2
S [ft^2]	177	ε_H [deg] (assumed)	2
S_{B_S} [ft]	128.4	ε_W [deg] (assumed)	2
$S_{f_{AVG}}$ [ft^2]	11.1	Λ_{LE} [deg]	40.5
$S_{P \to V}$ [ft^2]	9.6	Λ_{LE_H} [deg]	42
w_{max} [ft]	4.5	Λ_{LE_V} [deg]	50
X_{AC_R} [ft]	7.5		

C.3 AIRCRAFT 2—BEECH 99

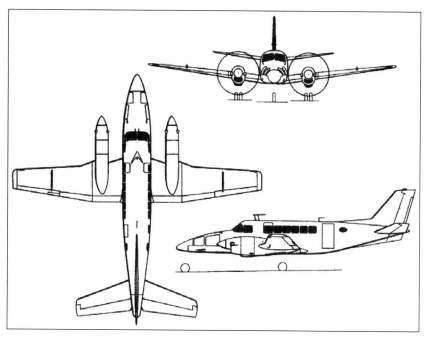

Figure C-A2.1 3D View of the Beech 99 Aircraft
(Source: www.aviastar.org/index2.html)

Geometric Parameters and Typical Flight Conditions (see Appendix B, Aircraft 3)

$$Alt. = 5{,}000 \text{ ft}, \ Mach = 0.31$$
$$V_{P_1} = 340 \text{ ft/sec}, \ \bar{q}_1 = 118.3 \text{ lbs/ft}^2, \ \alpha_1 = 0°$$
$$S = 280 \text{ ft}^2, \ \bar{c} = 6.5 \text{ ft}, \ b = 46 \text{ ft}$$
$$\bar{x}_{CG} = 0.16, \ W = 7{,}000 \text{ lb}$$

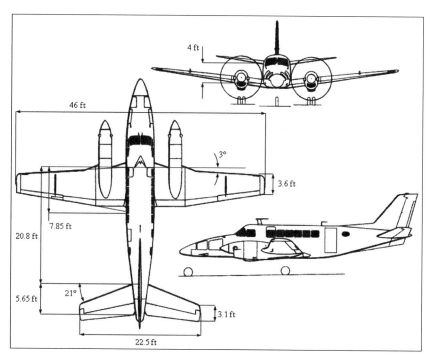

Figure C-A2.2 General Dimensions of the Beech 99 Aircraft

622 Appendix C Detailed Drawings for Different Aircraft

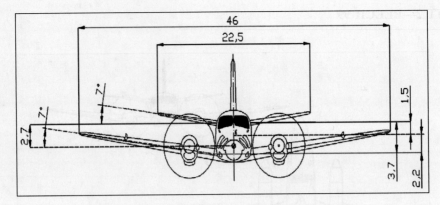

Figure C-A2.3 Front CAD View of the Beech 99 Aircraft

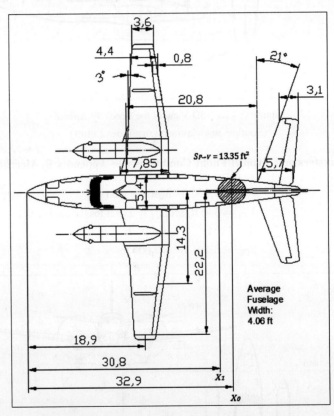

Figure C-A2.4 Top CAD View of the Beech 99 Aircraft

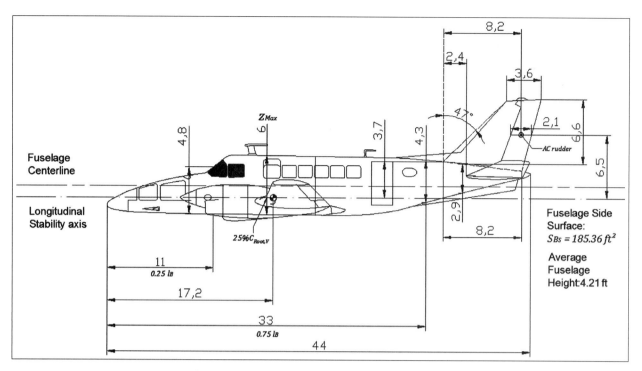

Figure C-A2.5 Side CAD View of the Beech 99 Aircraft

Table C-A2.1 Geometric Parameters for the Aerodynamic Modeling of the Beech 99 Aircraft

Geometric Parameters	B99 Aircraft	Geometric Parameters	B99 Aircraft
A [ft]	18.9	X_{HV} [ft]	2.4
b [ft]	46	X_{WH_r} [ft]	20.8
b_H [ft]	22.5	X_1 [ft]	30.8
b_V [ft]	6.8	y_{A_I} [ft]	14.3
$\bar{c}$ [ft]	6.5	y_{A_O} [ft]	22.2
$\bar{c}_{Aileron}$ [ft]	0.8	y_{R_I} [ft]	0
$\bar{c}_R$ [ft]	2.1	y_{R_F} [ft]	6.8
$\bar{c}_{Wing(atAileron)}$ [ft]	4.4	y_V [ft]	3.7
c_r [ft]	7.9	Z_H [ft]	−1.5
c_{r_H} [ft]	5.7	Z_{R_S} [ft]	6.5
c_{r_V} [ft]	8.2	z_1 [ft]	4.8
c_T [ft]	3.6	z_2 [ft]	4.3
c_{T_H} [ft]	3.1	Z_{H_S} [ft]	−2.7
c_{T_V} [ft]	3.6	z_{max} [ft]	6
d [ft]	6	Z_W [ft]	2.2
l_b [ft]	44	Z_{WH_r} [ft]	3.7
l_{cg} [ft]	17.2	Γ_H [deg]	7
r_1 [ft]	2.9	Γ_W [deg]	7
S [ft^2]	280	ε_H [deg] (assumed)	2
S_{B_S} [ft^2]	185.3	ε_W [deg] (assumed)	2
$S_{f_{AVG}}$ [ft^2]	13.4	Λ_{LE} [deg]	3
$S_{P \to V}$ [ft^2]	13.4	Λ_{LE_H} [deg]	21
w_{max} [ft]	5.4	Λ_{LE_V} [deg]	47
X_{AC_R} [ft]	8.2		

C.4 AIRCRAFT 3—BOEING B52

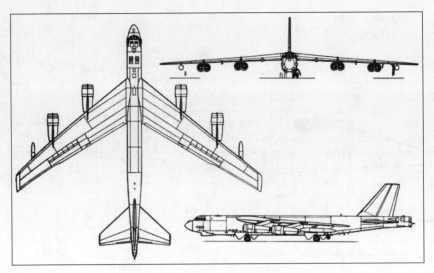

Figure C-A3.1 3D View of the Boeing B52 Aircraft
(Source: www.aviastar.org/index2.html)

Geometric Parameters and Typical Flight Conditions

$Alt. = 35,000$ ft, $Mach = 0.7$

$S = 4,000$ ft^2, $b = 185$ ft, $Length = 161$ ft, $Height = 40.7$ ft

$(Estimated)$ $\bar{x}_{CG} = 0.35$, $W = 430,000$ lbs

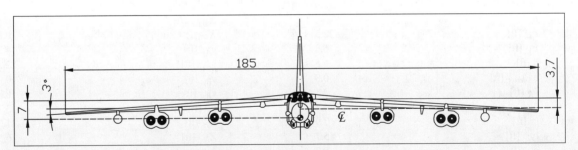

Figure C-A3.2 Front CAD View of the B-52 Aircraft

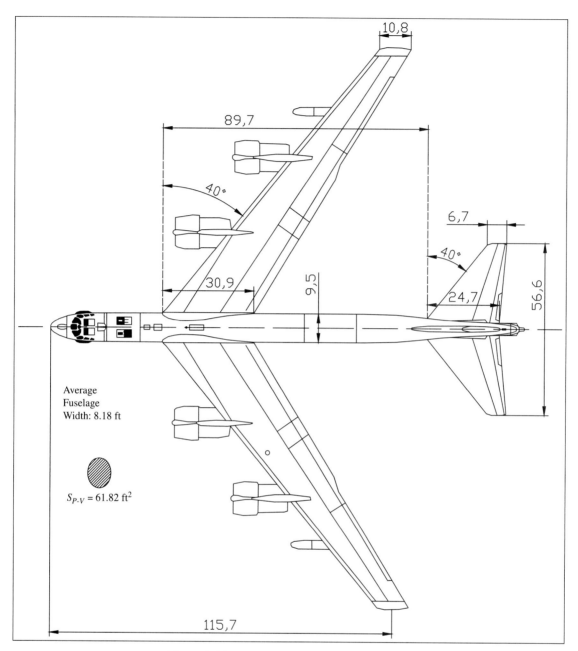

Figure C-A3.3 Top CAD View of the B-52 Aircraft

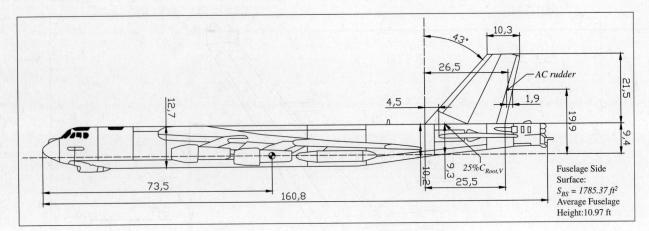

Figure C-A3.4 Side CAD View of the B52 Aircraft

Table C-A3.1 Geometric Parameters for the Aerodynamic Modeling of the B52 Aircraft

Geometric Parameters	B-52 Aircraft	Geometric Parameters	B-52 Aircraft
A [ft]	115.7	X_{WH_r} [ft]	89.7
b [ft]	185.0	y_{A_I} [ft]	0.0
b_H [ft]	56.6	y_{A_O} [ft]	0.0
b_V [ft]	21.5	y_{R_I} [ft]	0.0
$\bar{c}_R$ [ft]	1.9	y_{R_F} [ft]	21.5
c_r [ft]	30.9	y_V [ft]	9.4
c_{r_H} [ft]	24.7	Z_H [ft]	−3.7
c_{r_V} [ft]	25.5	Z_{R_S} [ft]	19.9
c_T [ft]	10.8	z_1 [ft]	12.7
c_{T_H} [ft]	6.7	z_2 [ft]	10.2
c_{T_V} [ft]	10.3	Z_{H_S} [ft]	−7.0
d [ft]	12.7	z_{max} [ft]	12.7
l_b [ft]	160.8	Z_W [ft]	−3.7
l_{cg} [ft]	73.5	Z_{WH_r} [ft]	0.0
r_1 [ft]	9.3	Γ_H [deg]	0.0
S_{B_S} [ft]	1785.4	Γ_W [deg]	−3.0
$S_{f_{AVG}}$ [ft²]	9.3	ε_H [deg] (assumed)	2.0
$S_{P \rightarrow V}$ [ft²]	61.8	ε_W [deg] (assumed)	2.0
w_{max} [ft]	9.5	Λ_{LE} [deg]	40.0
X_{AC_R} [ft]	26.5	Λ_{LE_H} [deg]	40.0
X_{HV} [ft]	4.5	Λ_{LE_V} [deg]	43.0

C.5　AIRCRAFT 4—BOEING B727-200

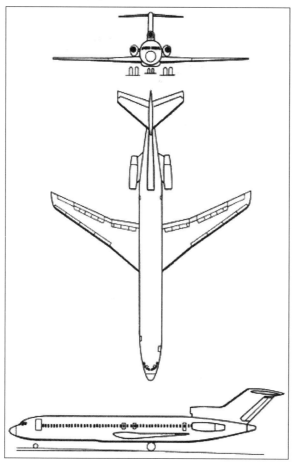

Figure C-A4.1　3D View of the Boeing B727 Aircraft
(Source: www.aviastar.org/index2.html)

Geometric Parameters and Typical Flight Conditions

$Alt. = 30{,}000$ ft, $Mach = 0.8$
$S = 1{,}650$ ft^2, $b = 108$ ft, $Length = 138$ ft, $Height = 34$ ft
$(Estimated)\ \bar{x}_{CG} = 0.25$, $W = 125{,}000$ lbs

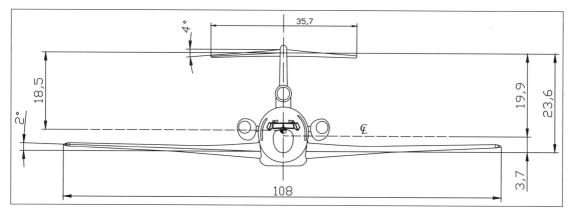

Figure C-A4.2　Front CAD View of the B727-200 Aircraft
(derived from www.boeing.com/commercial/airports/3_view.html)

628 Appendix C Detailed Drawings for Different Aircraft

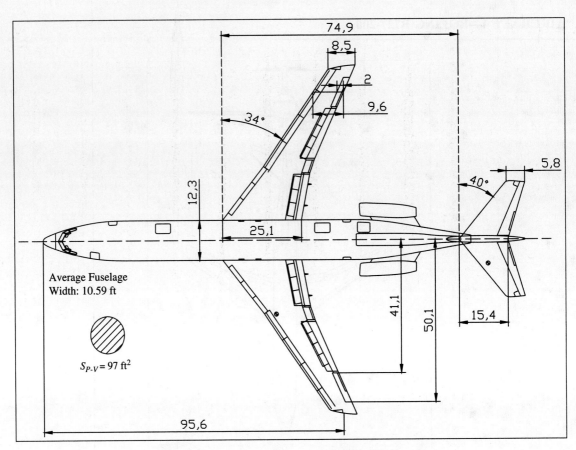

Figure C-A4.3 Top CAD View of the B727-200 Aircraft
(derived from www.boeing.com/commercial/airports/3_view.html)

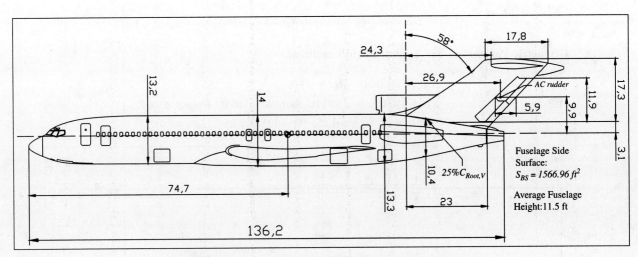

Figure C-A4.4 Side CAD View of the B727-200 Aircraft
(derived from www.boeing.com/commercial/airports/3_view.html)

Table C-A4.1 Geometric Parameters for the Aerodynamic Modeling of the B727-200 Aircraft

Geometric Parameters	B727-200 Aircraft	Geometric Parameters	B727-200 Aircraft
A [ft]	95.6	X_{HV} [ft]	24.3
b [ft]	108.0	X_{WH_r} [ft]	74.9
b_H [ft]	35.7	y_{A_I} [ft]	41.1
b_V [ft]	17.3	y_{A_O} [ft]	50.1
$\bar{c}_{Aileron}$ [ft]	2.0	y_{R_I} [ft]	0.0
$\bar{c}_R$ [ft]	5.9	y_{R_F} [ft]	11.9
$\bar{c}_{Wing(atAileron)}$ [ft]	9.6	y_V [ft]	3.1
c_r [ft]	25.1	Z_H [ft]	−19.9
c_{r_H} [ft]	15.4	Z_{R_S} [ft]	9.9
c_{r_V} [ft]	23.0	z_1 [ft]	13.2
c_T [ft]	8.5	z_2 [ft]	13.3
c_{T_H} [ft]	5.8	Z_{H_S} [ft]	−18.5
c_{T_V} [ft]	17.8	$z_{\max}$ [ft]	14.0
d [ft]	14.0	Z_W [ft]	3.7
l_b [ft]	136.2	Z_{WH_r} [ft]	23.6
l_{cg} [ft]	74.7	Γ_H [deg]	−4.0
r_1 [ft]	10.4	Γ_W [deg]	2.0
S_{B_S} [ft]	1567.0	ε_H [deg] (assumed)	2.0
$S_{f_{AVG}}$ [ft^2]	9.3	ε_W [deg] (assumed)	2.0
$S_{P \to V}$ [ft^2]	97.0	Λ_{LE} [deg]	34.0
$w_{\max}$ [ft]	12.3	Λ_{LE_H} [deg]	40.0
X_{AC_R} [ft]	26.9	Λ_{LE_V} [deg]	58.0

C.6 AIRCRAFT 5—BOEING B737-600

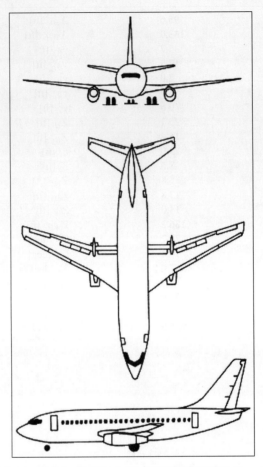

Figure C-A5.1 3D View of the Boeing B737-600 Aircraft
(Source: http://www.aviastar.org/index2.html)

Geometric Parameters and Typical Flight Conditions

$Alt. = 30{,}000$ ft, $Mach = 0.8$
$S = 980$ ft^2, $b = 112.6$ ft, $Length = 102.5$ ft, $Height = 40$ ft
$(Estimated)\ \bar{x}_{CG} = 0.25$, $W = 80{,}000$ lbs

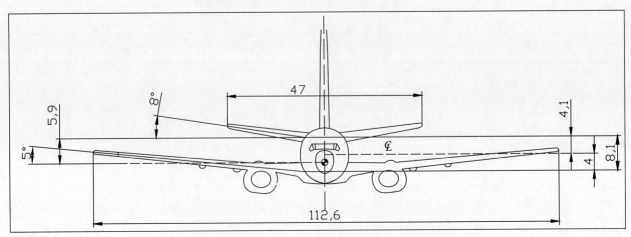

Figure C-A5.2 Front CAD View of the B737-600 Aircraft
(derived from www.boeing.com/commercial/airports/3_view.html)

C.6 Aircraft 5—Boeing B737-600

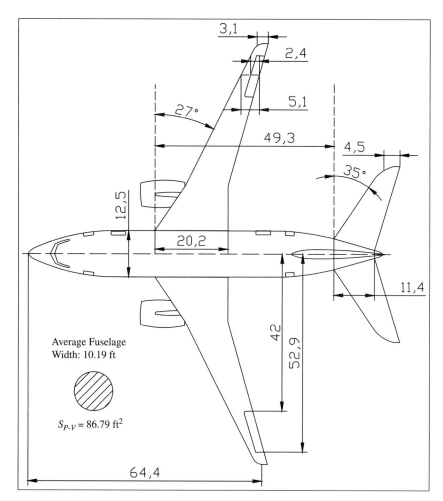

Figure C-A5.3 Top CAD View of the B737-600 Aircraft
(derived from www.boeing.com/commercial/airports/3_view.html)

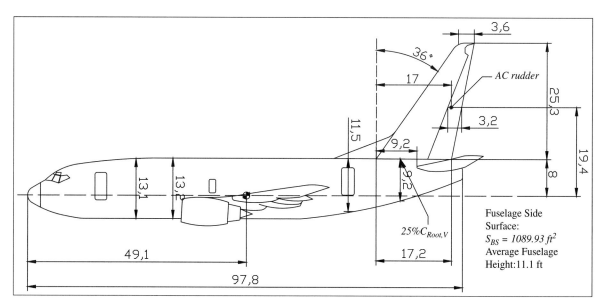

Figure C-A5.4 Side CAD View of the B737-600 Aircraft
(derived from www.boeing.com/commercial/airports/3_view.html)

Table C-A5.1 Geometric Parameters for the Aerodynamic Modeling of the B737-600 Aircraft

Geometric Parameters	B737-600 Aircraft	Geometric Parameters	B737-600 Aircraft
A [ft]	64.4	X_{HV} [ft]	9.2
b [ft]	112.6	X_{WH_r} [ft]	49.3
b_H [ft]	47.0	y_{A_I} [ft]	42.0
b_V [ft]	25.3	y_{A_O} [ft]	52.9
$\bar{c}_{Aileron}$ [ft]	2.4	y_{R_I} [ft]	0.0
$\bar{c}_R$ [ft]	3.2	y_{R_F} [ft]	25.3
$\bar{c}_{Wing(atAileron)}$ [ft]	5.1	y_V [ft]	8.0
c_r [ft]	20.2	Z_H [ft]	−4.1
c_{r_H} [ft]	11.4	Z_{R_S} [ft]	19.4
c_{r_V} [ft]	17.2	z_1 [ft]	13.1
c_T [ft]	3.1	z_2 [ft]	11.5
c_{T_H} [ft]	4.5	Z_{H_S} [ft]	−5.9
c_{T_V} [ft]	3.6	z_{max} [ft]	13.2
d [ft]	13.2	Z_W [ft]	4.0
l_b [ft]	97.8	Z_{WH_r} [ft]	8.1
l_{cg} [ft]	49.1	Γ_H [deg]	8.0
r_1 [ft]	9.2	Γ_W [deg]	5.0
S_{B_S} [ft]	1089.9	ε_H [deg] (assumed)	2.0
$S_{f_{AVG}}$ [ft^2]	9.3	ε_W [deg] (assumed)	2.0
$S_{P \to V}$ [ft^2]	86.8	Λ_{LE} [deg]	27.0
w_{max} [ft]	12.5	Λ_{LE_H} [deg]	35.0
X_{AC_R} [ft]	17.0	Λ_{LE_V} [deg]	36.0

C.7 AIRCRAFT 6—BOEING B747-200

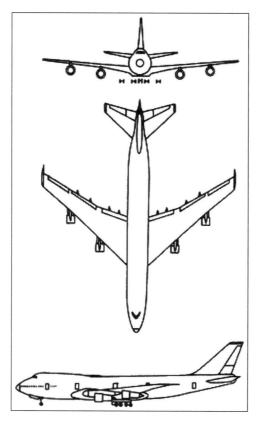

Figure C-A6.1 3D View of the Boeing B747 200 Aircraft
(Source: www.aviastar.org/index2.html)

Geometric Parameters and Typical Flight Conditions (see Appendix B, Aircraft 7)

$Alt. = 40{,}000$ ft, $Mach = 0.9$, $V_{P_1} = 871$ ft/sec, $\bar{q}_1 = 222.8$ lbs/ft^2, $\alpha_1 = 2.4°$
$S = 5{,}500$ ft^2, $\bar{c} = 27.3$ ft, $b = 196$ ft, $\bar{x}_{CG} = 0.25$, $W = 636{,}636$ lbs

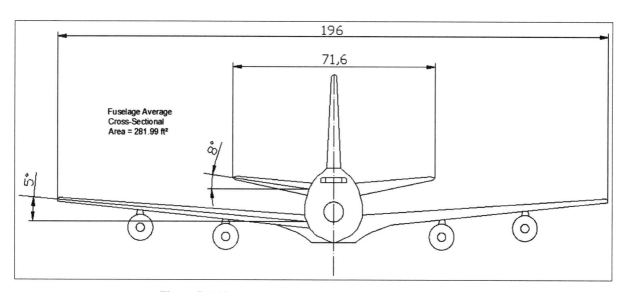

Figure C-A6.2 Front CAD View of the Boeing B747-200 Aircraft

634 Appendix C Detailed Drawings for Different Aircraft

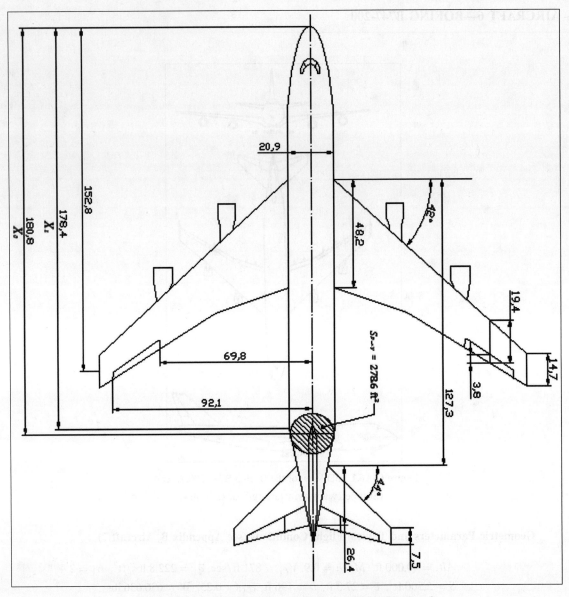

Figure C-A6.3 Top CAD View of the Boeing B747-200 Aircraft

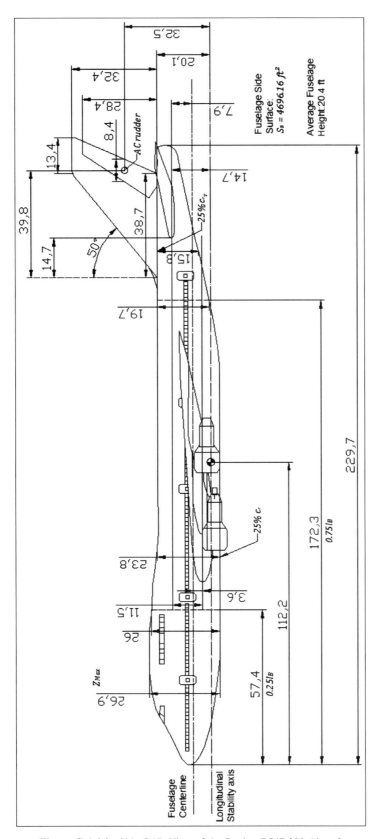

Figure C-A6.4 Side CAD View of the Boeing B747-200 Aircraft

Table C-A6.1 Geometric Parameters for the Aerodynamic Modeling of the Boeing B747-200 Aircraft

Geometric Parameters	B747 Aircraft	Geometric Parameters	B747 Aircraft
A [ft]	152.8	X_{HV} [ft]	14.7
b [ft]	196	X_{WH_r} [ft]	127.3
b_H [ft]	71.6	X_1 [ft]	178.4
b_V [ft]	32.4	y_{A_I} [ft]	69.8
$\bar{c}$ [ft]	27.3	y_{A_O} [ft]	92.1
$\bar{c}_{Aileron}$ [ft]	3.8	y_{R_I} [ft]	0
$\bar{c}_R$ [ft]	8.4	y_{R_F} [ft]	28.4
$\bar{c}_{Wing(atAileron)}$ [ft]	19.4	y_V [ft]	20.1
c_r [ft]	48.2	Z_H [ft]	−7.9
c_{r_H} [ft]	26.4	Z_{R_S} [ft]	32.5
c_{r_V} [ft]	38.7	z_1 [ft]	26
c_T [ft]	14.7	z_2 [ft]	19.7
c_{T_H} [ft]	7.5	Z_{H_S} [ft]	−14.7
c_{T_V} [ft]	13.4	z_{max} [ft]	26.9
d [ft]	23.7	Z_W [ft]	3.6
l_b [ft]	229.7	Z_{WH_r} [ft]	11.5
l_{cg} [ft]	112.2	Γ_H [deg]	8
r_1 [ft]	15.8	Γ_W [deg]	5
S [ft^2]	5500	ε_H [deg] (assumed)	2
S_{B_S} [ft]	4696.2	ε_W [deg] (assumed)	2
$S_{f_{AVG}}$ [ft^2]	282.0	Λ_{LE} [deg]	41.5
$S_{P \to V}$ [ft^2]	278.6	Λ_{LE_H} [deg]	44
w_{max} [ft]	20.9	Λ_{LE_V} [deg]	50
X_{AC_R} [ft]	39.8		

C.8 AIRCRAFT 7—BOEING B757-200

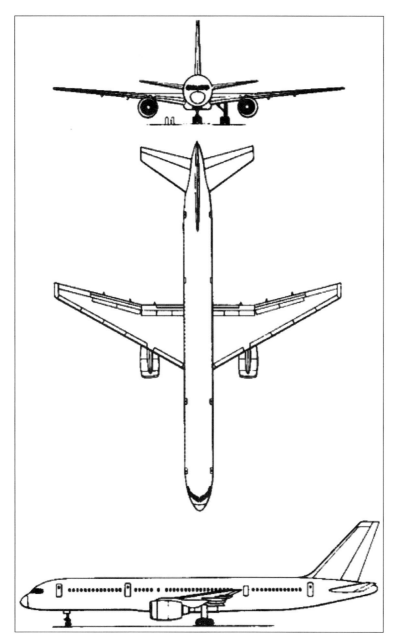

Figure C-A7.1 3D View of the Boeing B757 Aircraft
(Source: www.aviastar.org/index2.html)

Geometric Parameters and Typical Flight Conditions

$Alt. = 33{,}000$ ft, $Mach = 0.83$
$S = 1{,}951$ ft^2, $b = 124.8$ ft, $Length = 155.25$ ft, $Height = 44.5$ ft
(*Estimated*) $\bar{x}_{CG} = 0.25$, $W = 210{,}000$ lbs

638 Appendix C Detailed Drawings for Different Aircraft

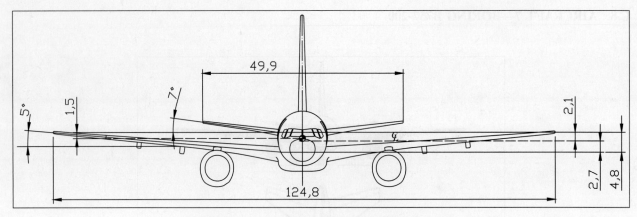

Figure C-A7.2 Front CAD View of the B757-200 Aircraft
(derived from www.boeing.com/commercial/airports/3_view.html)

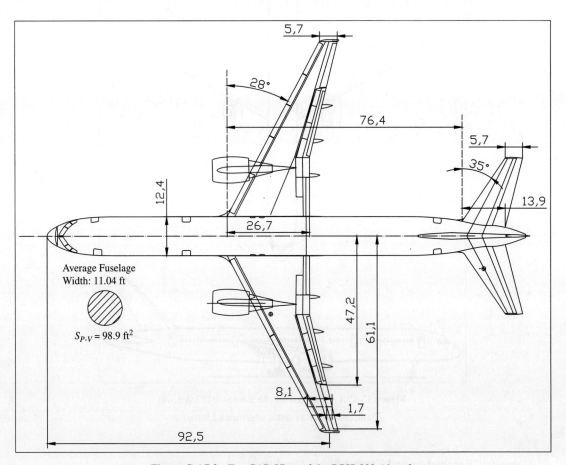

Figure C-A7.3 Top CAD View of the B757-200 Aircraft
(derived from www.boeing.com/commercial/airports/3_view.html)

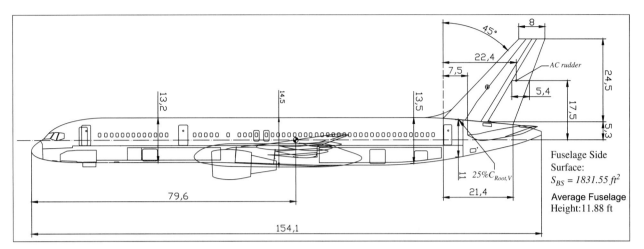

Figure C-A7.4 Side CAD View of the B757-200 Aircraft
(derived from www.boeing.com/commercial/airports/3_view.html)

Table C-A7.1 Geometric Parameters for the Aerodynamic Modeling of the B757-200 Aircraft

Geometric Parameters	B757-200 Aircraft	Geometric Parameters	B757-200 Aircraft
A [ft]	92.5	X_{HV} [ft]	7.5
b [ft]	124.8	X_{WH_r} [ft]	76.4
b_H [ft]	49.9	y_{A_I} [ft]	47.2
b_V [ft]	24.5	y_{A_O} [ft]	61.1
$\bar{c}_{Aileron}$ [ft]	1.7	y_{R_I} [ft]	0.0
$\bar{c}_R$ [ft]	5.4	y_{R_F} [ft]	24.5
$\bar{c}_{Wing(atAileron)}$ [ft]	8.1	y_V [ft]	5.3
c_r [ft]	26.7	Z_H [ft]	-2.1
c_{r_H} [ft]	13.9	Z_{R_S} [ft]	17.5
c_{r_V} [ft]	21.4	z_1 [ft]	13.2
c_T [ft]	5.7	z_2 [ft]	13.5
c_{T_H} [ft]	5.7	Z_{H_S} [ft]	-1.5
c_{T_V} [ft]	8.0	z_{max} [ft]	14.5
d [ft]	14.5	Z_W [ft]	2.7
l_b [ft]	154.1	Z_{WH_r} [ft]	4.8
l_{cg} [ft]	79.6	Γ_H [deg]	7.0
r_1 [ft]	11.0	Γ_W [deg]	5.0
S_{B_S} [ft]	1831.6	ε_H [deg] (assumed)	2.0
$S_{f_{AVG}}$ [ft^2]	9.3	ε_W [deg] (assumed)	2.0
$S_{P \to V}$ [ft^2]	98.9	Λ_{LE} [deg]	28.0
w_{max} [ft]	12.4	Λ_{LE_H} [deg]	35.0
X_{AC_R} [ft]	22.4	Λ_{LE_V} [deg]	45.0

C.9 AIRCRAFT 8—BOEING B767-200

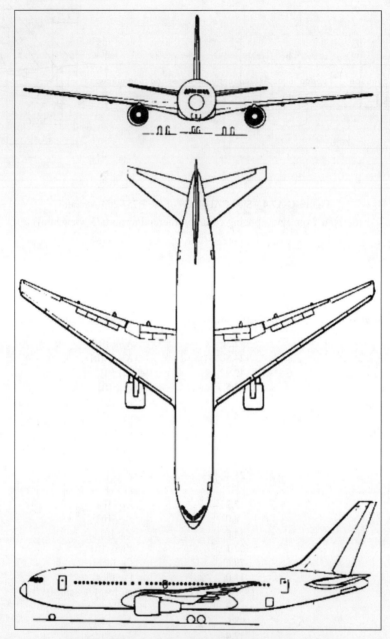

Figure C-A8.1 3D View of the Boeing B767-200 Aircraft
(Source: www.aviastar.org/index2.html)

Geometric Parameters and Typical Flight Conditions

$Alt. = 35{,}000$ ft, $Mach = 0.85$

$S = 3{,}050$ ft^2, $b = 156$ ft, $Length = 159$ ft, $Height = 52$ ft

$(Estimated)\ \bar{x}_{CG} = 0.25$, $W = 270{,}000$ lbs

C.9 Aircraft 8—Boeing B767-200 **641**

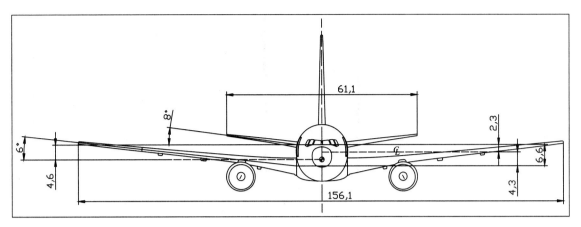

Figure C-A8.2 Front CAD View of the B767-200 Aircraft
(derived from www.boeing.com/commercial/airports/3_view.html)

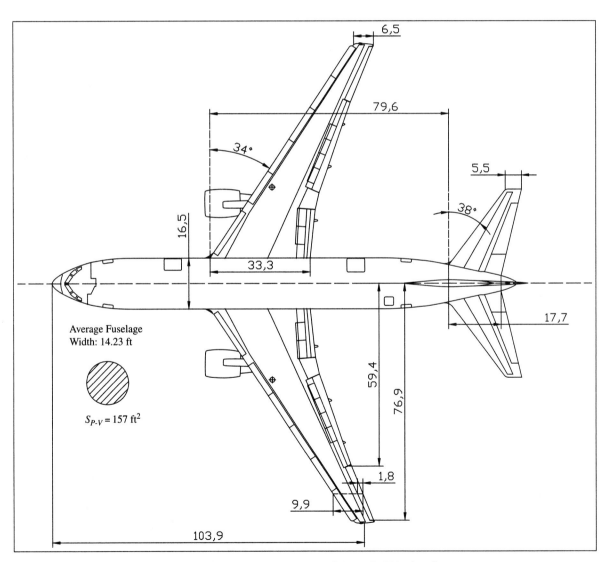

Figure C-A8.3 Top CAD View of the B767-200 Aircraft
(derived from www.boeing.com/commercial/airports/3_view.html)

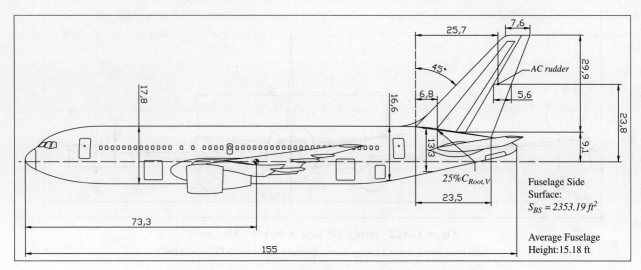

Figure C-A8.4 Side CAD View of the B767-200 Aircraft
(derived from www.boeing.com/commercial/airports/3_view.html)

Table C-A8.1 Geometric Parameters for the Aerodynamic Modeling of the B767-200 Aircraft

Geometric Parameters	B767-200 Aircraft	Geometric Parameters	B767-200 Aircraft
A [ft]	103.9	X_{HV} [ft]	6.8
b [ft]	156.1	X_{WH_r} [ft]	79.6
b_H [ft]	61.1	y_{A_I} [ft]	59.4
b_V [ft]	29.9	y_{A_O} [ft]	76.9
$\bar{c}_{Aileron}$ [ft]	1.8	y_{R_I} [ft]	0.0
$\bar{c}_R$ [ft]	5.6	y_{R_F} [ft]	29.9
$\bar{c}_{Wing(atAileron)}$ [ft]	9.9	y_V [ft]	9.1
c_r [ft]	33.3	Z_H [ft]	-2.3
c_{r_H} [ft]	17.7	Z_{R_S} [ft]	23.8
c_{r_V} [ft]	23.5	z_1 [ft]	17.8
c_T [ft]	6.5	z_2 [ft]	16.6
c_{T_H} [ft]	5.5	Z_{H_S} [ft]	-4.6
c_{T_V} [ft]	7.6	z_{max} [ft]	17.8
d [ft]	17.8	Z_W [ft]	4.3
l_b [ft]	155.0	Z_{WH_r} [ft]	6.6
l_{cg} [ft]	73.3	Γ_H [deg]	8.0
r_1 [ft]	13.3	Γ_W [deg]	6.0
S_{B_S} [ft]	2353.2	ε_H [deg] (assumed)	2.0
$S_{f_{AVG}}$ [ft^2]	9.3	ε_W [deg] (assumed)	2.0
$S_{P \to V}$ [ft^2]	157.0	Λ_{LE} [deg]	34.0
w_{max} [ft]	16.5	Λ_{LE_H} [deg]	38.0
X_{AC_R} [ft]	25.7	Λ_{LE_V} [deg]	45.0

C.10 AIRCRAFT 9—CESSNA CITATION CJ3

Geometric Parameters and Typical Flight Conditions

$Alt. = 33{,}000$ ft, $Mach = 0.7$, $V_{P_1} = 675$ ft/sec, $\alpha_1 \approx 2°$
$S = 294.1$ ft^2, $\bar{c} = 5.86$ ft, $b = 53.3$ ft, $\bar{x}_{CG} \approx 0.25$, $W \approx 13{,}000$ lbs

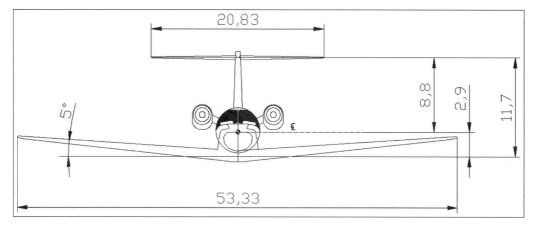

Figure C-A9.1 Front CAD View of the Cessna CJ3 Aircraft
(created by the author from the document: http://textron.vo.llnwd.net/o25/CES/cessna_aircraft_docs/citation/cj3/cj3_s&d.pdf)

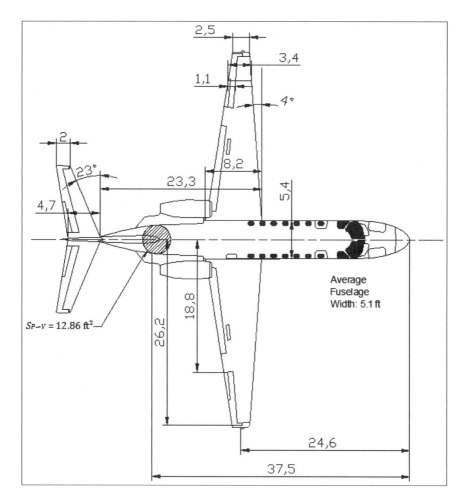

Figure C-A9.2 Top CAD View of the Cessna CJ3 Aircraft
(created by the author from the document: http://textron.vo.llnwd.net/o25/CES/cessna_aircraft_docs/citation/cj3/cj3_s&d.pdf)

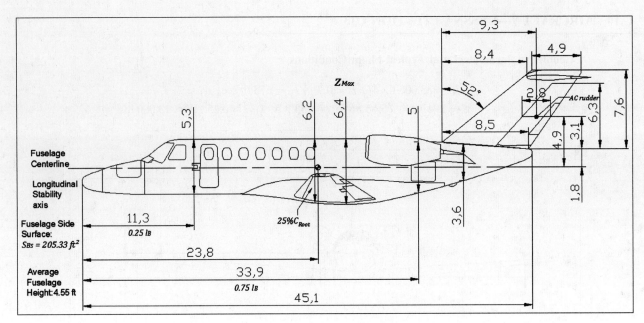

Figure C-A9.3 Lateral CAD View of the Cessna CJ3 Aircraft
(created by the author from the document: http://textron.vo.llnwd.net/o25/CES/cessna_aircraft_docs/citation/cj3/cj3_s&d.pdf)

Table C-A9.1 Geometric Parameters for the Aerodynamic Modeling of the Cessna CJ3 Aircraft

Geometric Parameters	CJ3 Aircraft	Geometric Parameters	CJ3 Aircraft
A [ft]	24.6	X_{AC_R} [ft]	9.3
b [ft]	53.3	X_{HV} [ft]	8.4
b_H [ft]	20.8	X_{WH_r} [ft]	23.3
b_V [ft]	7.6	y_{A_I} [ft]	18.8
$\bar{c}$ [ft]	5.9	y_{A_O} [ft]	26.2
$\bar{c}_{Aileron}$ [ft]	1.1	y_{R_I} [ft]	0
$\bar{c}_R$ [ft]	2.8	y_{R_F} [ft]	6.3
$\bar{c}_{Wing(atAileron)}$ [ft]	3.4	y_V [ft]	1.8
c_r [ft]	8.2	Z_H [ft]	8.8
c_{r_H} [ft]	4.7	Z_{R_S} [ft]	4.9
c_{r_V} [ft]	8.5	z_1 [ft]	5.3
c_T [ft]	2.5	z_2 [ft]	5
c_{T_H} [ft]	2	Z_{H_S} [ft]	−8.8
c_{T_V} [ft]	4.9	$z_{\max}$ [ft]	6.4
d [ft]	6.2	Z_W [ft]	2.9
l_b [ft]	45.1	Z_{WH_r} [ft]	11.7
l_{cg} [ft]	23.8	Γ_H [deg]	0
r_1 [ft]	3.6	Γ_W [deg]	5
S [ft^2]	294.1	ε_H [deg] (assumed)	2
S_{B_S} [ft]	205.3	ε_W [deg] (assumed)	2
$S_{f_{AVG}}$ [ft^2]	18.2	Λ_{LE} [deg]	4
$S_{P \to V}$ [ft^2]	12.9	Λ_{LE_H} [deg]	23
$w_{\max}$ [ft]	5.4	Λ_{LE_V} [deg]	52

C.11 AIRCRAFT 10—CESSNA T37

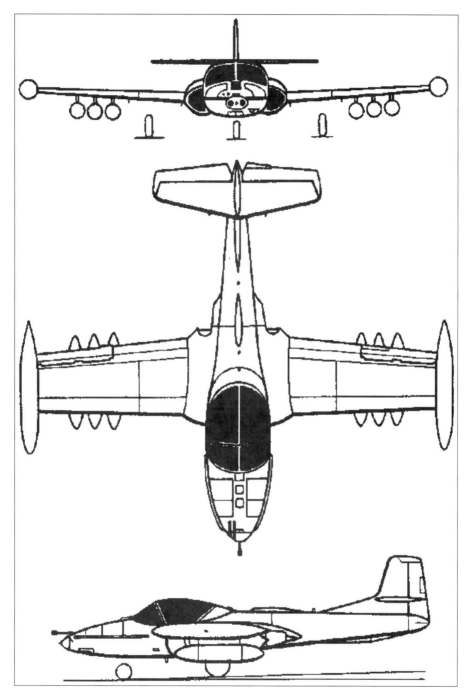

Figure C-A10.1 3D View of the Cessna T37 Aircraft
(Source: www.aviastar.org/index2.html)

Geometric Parameters and Typical Flight Conditions (see Appendix B, Aircraft 4)

$$Alt. = 30{,}000 \text{ ft}, \ Mach = 0.459$$
$$V_{P_1} = 456 \text{ ft/sec}, \bar{q}_1 = 92.7 \text{ lbs/ft}^2, \ \alpha_1 = 2°$$
$$S = 182 \text{ ft}^2, \ \bar{c} = 5.47 \text{ ft}, \ b = 33.8 \text{ ft}$$
$$\bar{x}_{CG} = 0.27, \ W = 6{,}360 \text{ lbs}$$

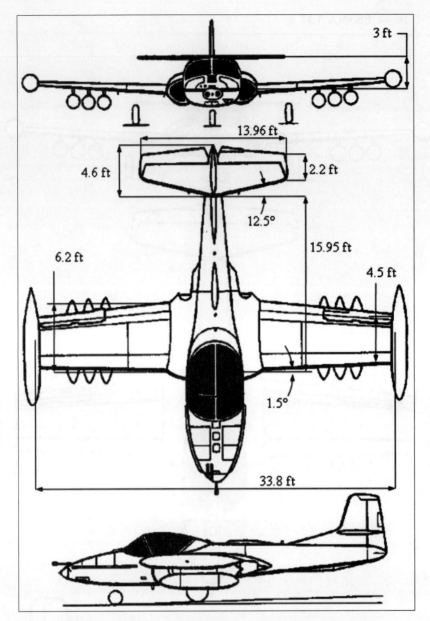

Figure C-A10.2 General Dimensions of the Cessna T-37 Aircraft

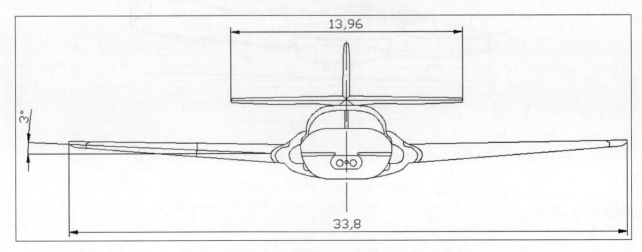

Figure C-A10.3 Front CAD View of the Cessna T-37 Aircraft

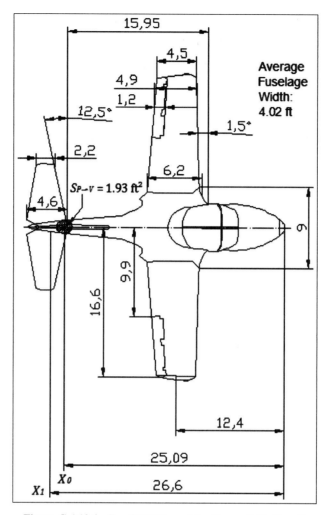

Figure C-A10.4 Top CAD View of the Cessna T-37 Aircraft

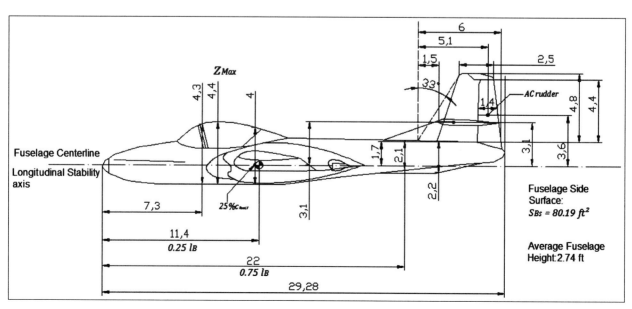

Figure C-A10.5 Side CAD View of the Cessna T-37 Aircraft

Table C-A10.1 Geometric Parameters for the Aerodynamic Modeling of the Cessna T-37 Aircraft

Geometric Parameters	T37 Aircraft	Geometric Parameters	T37 Aircraft
A [ft]	12.4	X_{HV} [ft]	1.5
b [ft]	33.8	X_{WH_r} [ft]	15.9
b_H [ft]	14.0	X_1 [ft]	26.6
b_V [ft]	4.8	y_{A_I} [ft]	9.9
$\bar{c}$ [ft]	5.5	y_{A_O} [ft]	16.6
$\bar{c}_{Aileron}$ [ft]	1.2	y_{R_I} [ft]	0
$\bar{c}_R$ [ft]	1.4	y_{R_F} [ft]	4.4
$\bar{c}_{Wing(atAileron)}$ [ft]	4.9	y_V [ft]	1.7
c_r [ft]	6.2	Z_H [ft]	-3.1
c_{r_H} [ft]	4.6	Z_{R_S} [ft]	3.6
c_{r_V} [ft]	6	z_1 [ft]	4.3
c_T [ft]	4.5	z_2 [ft]	2.1
c_{T_H} [ft]	2.2	Z_{H_S} [ft]	-3.1
c_{T_V} [ft]	2.5	z_{max} [ft]	4.4
d [ft]	4	Z_W [ft]	0
l_b [ft]	29.2	Z_{WH_r} [ft]	3
l_{cg} [ft]	11.4	Γ_H [deg]	0
r_1 [ft]	2.2	Γ_W [deg]	3
S [ft^2]	182	ε_H [deg] (assumed)	2
S_{B_S} [ft]	80.2	ε_W [deg] (assumed)	2
$S_{f_{AVG}}$ [ft^2]	8.7	Λ_{LE} [deg]	1.5
$S_{P \rightarrow V}$ [ft^2]	1.9	Λ_{LE_H} [deg]	12.5
w_{max} [ft]	9	Λ_{LE_V} [deg]	33
X_{AC_R} [ft]	5.1		

C.12 AIRCRAFT 11—GENERAL DYNAMICS F-16

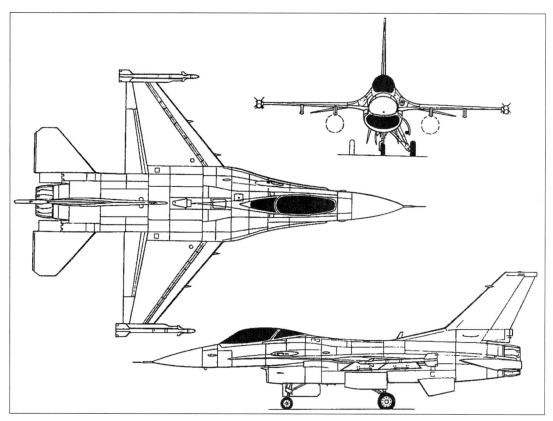

Figure C-A11.1 3D View of the General Dynamic F-16 Aircraft
(Source: www.aviastar.org/index2.html)

Geometric Parameters and Typical Flight Conditions

$Alt. = 45{,}000$ ft, $Mach = 0.85$
$S = 280$ ft^2, $b = 32.8$ ft, $Length = 49$ ft, $Height = 16.5$ ft
$(Estimated)\ \bar{x}_{CG} = 0.25$, $W = 20{,}000$ lbs

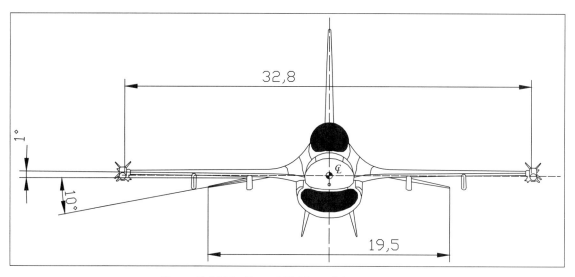

Figure C-A11.2 Front CAD View of the F-16 Aircraft

650 Appendix C Detailed Drawings for Different Aircraft

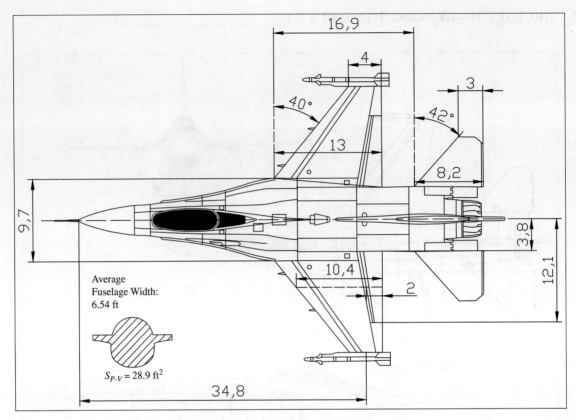

Figure C-A11.3 Top CAD View of the F-16 Aircraft

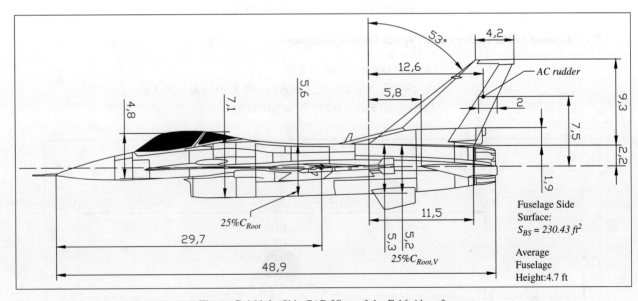

Figure C-A11.4 Side CAD View of the F-16 Aircraft

Table C-A11.1 Geometric Parameters for the Aerodynamic Modeling of the F-16 Aircraft

Geometric Parameters	F-16 Aircraft	Geometric Parameters	F-16 Aircraft
A [ft]	34.8	X_{HV} [ft]	5.8
b [ft]	32.8	X_{WH_r} [ft]	16.9
b_H [ft]	19.5	y_{A_I} [ft]	3.8
b_V [ft]	9.3	y_{A_O} [ft]	12.1
$\bar{c}_{Aileron}$ [ft]	2.0	y_{R_I} [ft]	1.9
$\bar{c}_R$ [ft]	2.0	y_{R_F} [ft]	9.3
$\bar{c}_{Wing(atAileron)}$ [ft]	10.4	y_V [ft]	2.2
c_r [ft]	13.0	Z_H [ft]	0.0
c_{r_H} [ft]	8.2	Z_{R_S} [ft]	7.5
c_{r_V} [ft]	11.5	z_1 [ft]	4.8
c_T [ft]	4.0	z_2 [ft]	5.3
c_{T_H} [ft]	3.0	Z_{H_S} [ft]	0.0
c_{T_V} [ft]	4.2	$z_{\max}$ [ft]	7.1
d [ft]	5.6	Z_W [ft]	0.0
l_b [ft]	48.9	Z_{WH_r} [ft]	0.0
l_{cg} [ft]	29.7	Γ_H [deg]	-10.0
r_1 [ft]	5.2	Γ_W [deg]	1.0
S_{B_S} [ft]	230.4	ε_H [deg] (assumed)	2.0
$S_{f_{AVG}}$ [ft^2]	9.3	ε_W [deg] (assumed)	2.0
$S_{P \to V}$ [ft^2]	28.9	Λ_{LE} [deg]	40.0
$w_{\max}$ [ft]	9.7	Λ_{LE_H} [deg]	42.0
X_{AC_R} [ft]	12.6	Λ_{LE_V} [deg]	53.0

C.13 AIRCRAFT 12—GRUMMAN F-14

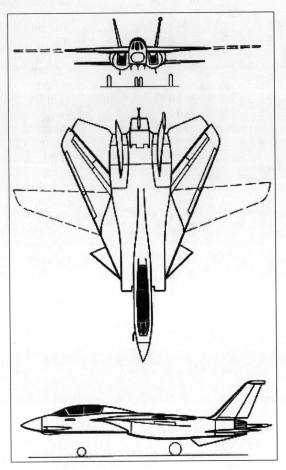

Figure C-A12.1 3D View of the Grumman F-14 Aircraft
(Source: www.aviastar.org/index2.html)

Geometric Parameters and Typical Flight Conditions

$Alt. = 35{,}000$ ft, $Mach = 0.85$

(*Estimated*) $S = 700$ ft^2, $b = 38$ ft/64 ft, $Length = 62$ ft, $Height = 16$ ft

(*Estimated*) $\bar{x}_{CG} = 0.25$, $W = 40{,}000$ lbs

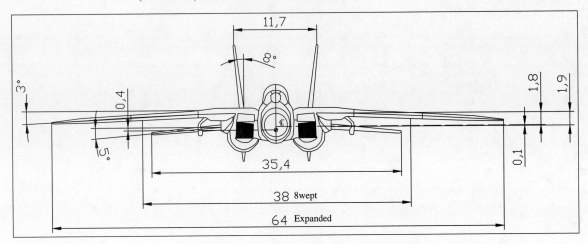

Figure C-A12.2 Front CAD View of the F-14 Aircraft

C.13 Aircraft 12—Grumman F-14

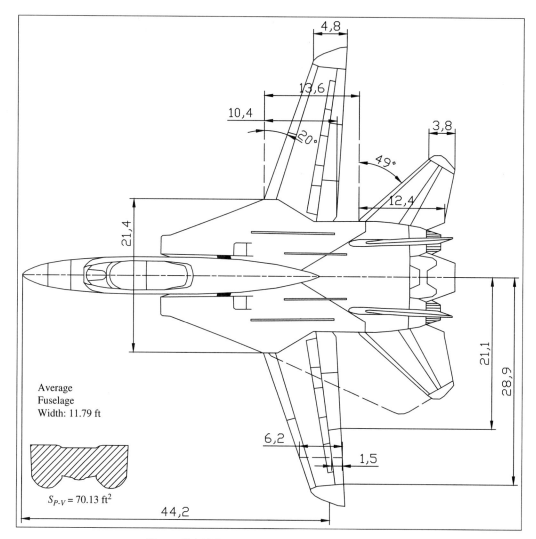

Figure C-A12.3 Top CAD View of the F-14 Aircraft

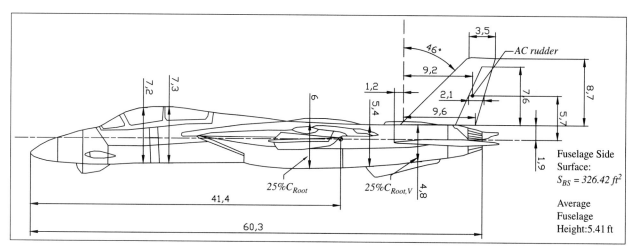

Figure C-A12.4 Side CAD View of the F-14 Aircraft

Table C-A12.1 Geometric Parameters for the Aerodynamic Modeling of the F-14 Aircraft

Geometric Parameters	F-14 Aircraft	Geometric Parameters	F-14 Aircraft
A [ft]	44.2	X_{HV} [ft]	1.2
b [ft]	64.0	X_{WH_r} [ft]	13.6
b_H [ft]	35.4	y_{A_I} [ft]	21.1
b_V [ft]	8.7	y_{A_O} [ft]	28.9
$\bar{c}_{Aileron}$ [ft]	1.5	y_{R_I} [ft]	0.0
$\bar{c}_R$ [ft]	2.1	y_{R_F} [ft]	7.6
$\bar{c}_{Wing(atAileron)}$ [ft]	6.2	y_V [ft]	1.9
c_r [ft]	10.4	Z_H [ft]	0.1
c_{r_H} [ft]	12.4	Z_{R_S} [ft]	5.7
c_{r_V} [ft]	9.6	z_1 [ft]	7.2
c_T [ft]	4.8	z_2 [ft]	5.4
c_{T_H} [ft]	3.8	Z_{H_S} [ft]	−0.4
c_{T_V} [ft]	3.5	z_{max} [ft]	7.3
d [ft]	6.0	Z_W [ft]	−1.8
l_b [ft]	60.3	Z_{WH_r} [ft]	1.9
l_{cg} [ft]	41.4	Γ_H [deg]	−5.0
r_1 [ft]	4.8	Γ_W [deg]	−3.0
S_{B_S} [ft]	326.4	ε_H [deg] (assumed)	2.0
$S_{f_{AVG}}$ [ft^2]	9.3	ε_W [deg] (assumed)	2.0
$S_{P \to V}$ [ft^2]	70.1	Λ_{LE} [deg]	20.0
w_{max} [ft]	21.4	Λ_{LE_H} [deg]	49.0
X_{AC_R} [ft]	9.2	Λ_{LE_V} [deg]	46.0

C.14 AIRCRAFT 13—LEARJET 24

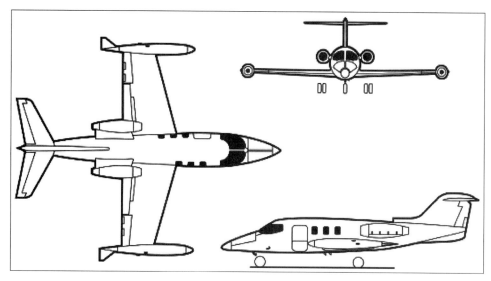

Figure C-A13.1 3D View of the Learjet 24 Aircraft
(Source: www.aerospaceweb.org)

Geometric Parameters and Typical Flight Conditions

$Alt. = 40,000$ ft, $Mach = 0.7$
($Estimated$) $S = 230$ ft^2, $b = 34$ ft, $Length = 46$ ft, $Height = 12$ ft
($Estimated$) $\bar{x}_{CG} = 0.32$, $W = 40,000$ lbs

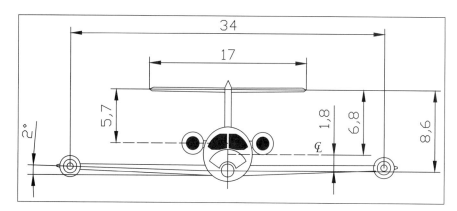

Figure C-A13.2 Front CAD View of the Learjet 24 Aircraft

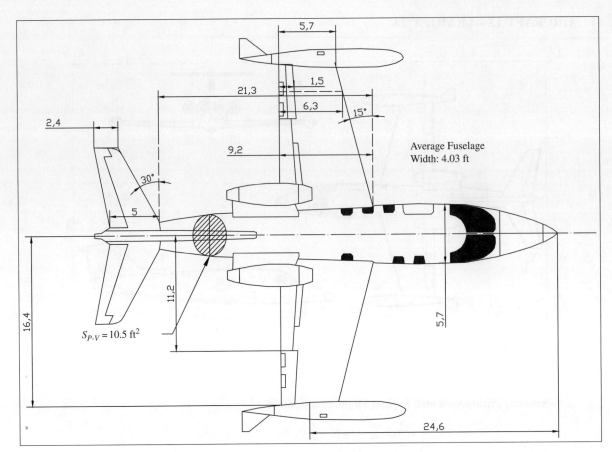

Figure C-A13.3 Top CAD View of the Learjet 24 Aircraft

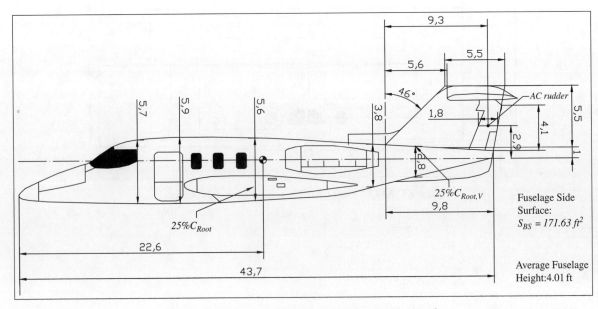

Figure C-A13.4 Side CAD View of the Learjet 24 Aircraft

Table C-A13.1 Geometric Parameters for the Aerodynamic Modeling of the Learjet 24 Aircraft

Geometric Parameters	Learjet 24 Aircraft	Geometric Parameters	Learjet 24 Aircraft
A [ft]	24.6	X_{AC_R} [ft]	9.9
b [ft]	34	X_{HV} [ft]	5.6
b_H [ft]	17	X_{WH_r} [ft]	21.3
b_V [ft]	5.5	y_{A_I} [ft]	11.2
$\bar{c}$ [ft]	7	y_{A_O} [ft]	16.4
$\bar{c}_{Aileron}$ [ft]	1.5	y_{R_I} [ft]	0
$\bar{c}_R$ [ft]	1.8	y_{R_F} [ft]	4.1
$\bar{c}_{Wing(atAileron)}$ [ft]	6.3	y_V [ft]	1
c_r [ft]	9.2	Z_H [ft]	−6.8
c_{r_H} [ft]	5	Z_{R_S} [ft]	2.9
c_{r_V} [ft]	9.8	z_1 [ft]	5.7
c_T [ft]	5.7	z_2 [ft]	3.8
c_{T_H} [ft]	2.4	Z_{H_S} [ft]	−5.7
c_{T_V} [ft]	5.5	z_{max} [ft]	5.9
d [ft]	5.6	Z_W [ft]	1.8
l_b [ft]	43.7	Z_{WH_r} [ft]	8.6
l_{cg} [ft]	22.6	Γ_H [deg]	0
r_1 [ft]	2.8	Γ_W [deg]	2
S [ft²]	230	ε_H [deg] (assumed)	2
S_{B_S} [ft]	171.6	ε_W [deg] (assumed)	2
$S_{f_{AVG}}$ [ft²]	9.3	Λ_{LE} [deg]	15
$S_{P \to V}$ [ft²]	10.5	Λ_{LE_H} [deg]	30
w_{max} [ft]	5.7	Λ_{LE_V} [deg]	46

C.15 AIRCRAFT 14—LOCKHEED F-104

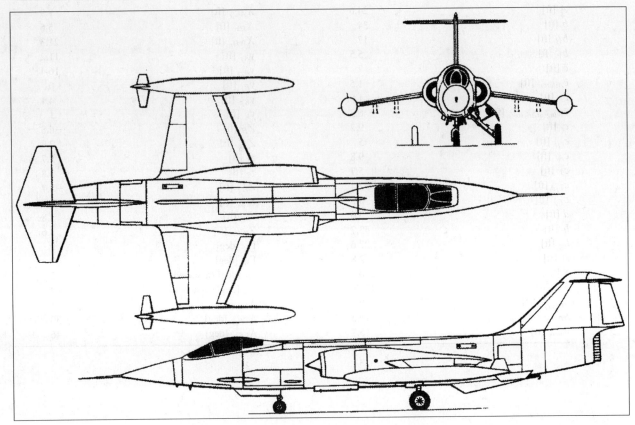

Figure C-A14.1 3D View of the Lockheed F-104 Aircraft
(Source: www.aviastar.org/index2.html)

Geometric Parameters and Typical Flight Conditions (see Appendix B, Aircraft 9)

$Alt. = 35,000$ ft, $Mach = 0.85 - 2.4$
$S = 196$ ft^2, $b = 22$ ft, $Length = 54.75$ ft, $Height = 13.5$ ft
$\bar{x}_{CG} = 0.07$, $W = 24,000$ lbs

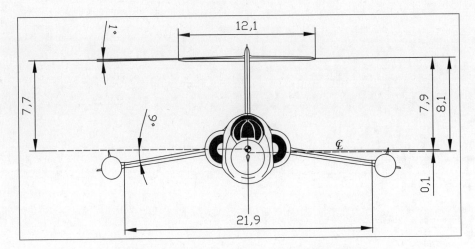

Figure C-A14.2 Front CAD View of the F-104 Aircraft

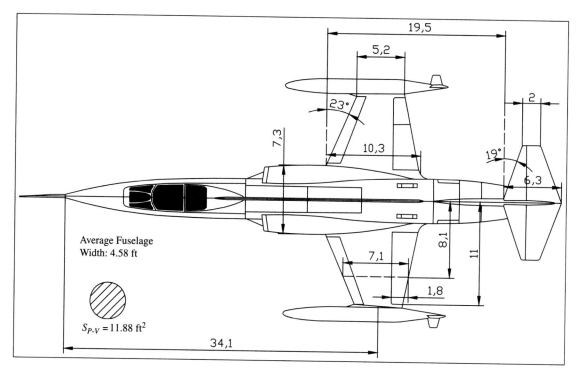

Figure C-A14.3 Top CAD View of the F-104 Aircraft

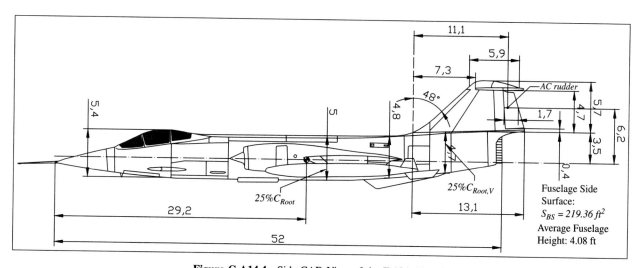

Figure C-A14.4 Side CAD View of the F-104 Aircraft

Table C-A14.1 Geometric Parameters for the Aerodynamic Modeling of the F-104 Aircraft

Geometric Parameters	F-104 Aircraft	Geometric Parameters	F-104 Aircraft
A [ft]	34.1	X_{AC_R} [ft]	11.1
b [ft]	21.9	X_{HV} [ft]	7.3
b_H [ft]	12.1	X_{WH_r} [ft]	19.5
b_V [ft]	5.7	y_{A_I} [ft]	8.1
$\bar{c}$ [ft]	9.6	y_{A_O} [ft]	11.0
$\bar{c}_{Aileron}$ [ft]	1.8	y_{R_I} [ft]	0.4
$\bar{c}_R$ [ft]	1.7	y_{R_F} [ft]	4.7
$\bar{c}_{Wing(atAileron)}$ [ft]	7.1	y_V [ft]	3.5
c_r [ft]	10.3	Z_H [ft]	−8.1
c_{r_H} [ft]	6.3	Z_{R_S} [ft]	6.2
c_{r_V} [ft]	13.1	z_1 [ft]	5.4
c_T [ft]	5.2	z_2 [ft]	4.8
c_{T_H} [ft]	2.0	Z_{H_S} [ft]	−7.7
c_{T_V} [ft]	5.9	z_{max} [ft]	5.4
d [ft]	5.0	Z_W [ft]	0.1
l_b [ft]	52.0	Z_{WH_r} [ft]	7.9
l_{cg} [ft]	29.2	Γ_H [deg]	−1.0
r_1 [ft]	4.7	Γ_W [deg]	−9.0
S [ft^2]	196.0	ε_H [deg] (assumed)	2.0
S_{B_S} [ft]	219.4	ε_W [deg] (assumed)	2.0
$S_{f_{AVG}}$ [ft^2]	9.3	Λ_{LE} [deg]	23.0
$S_{P \rightarrow V}$ [ft^2]	11.9	Λ_{LE_H} [deg]	19.0
w_{max} [ft]	7.3	Λ_{LE_V} [deg]	48.0

C.16 AIRCRAFT 15—LOCKHEED F-22

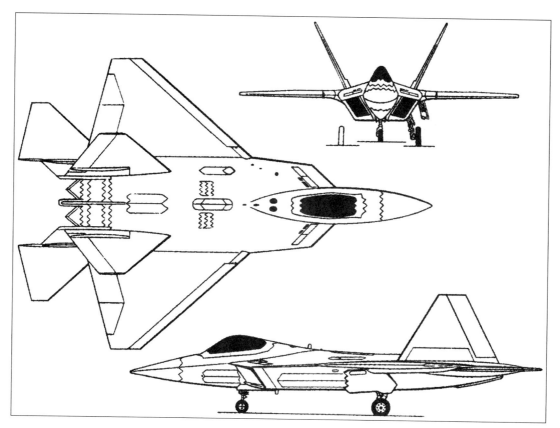

Figure C-A15.1 3D View of the Lockheed F-22 Aircraft
(Source: www.aviastar.org/index2.html)

Geometric Parameters and Typical Flight Conditions

$$Alt. = 35{,}000 \text{ ft}, \quad Mach = 0.85 - 1.8$$
$$b = 44.5 \text{ ft}, \quad Length = 62 \text{ ft}$$
$$(Estimated) \; \bar{x}_{CG} = 0.15, \quad W = 54{,}000 \text{ lbs}$$

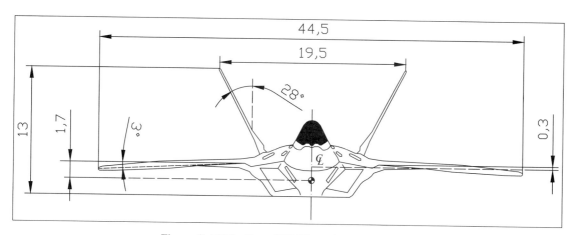

Figure C-A15.2 Front CAD View of the F-22 Aircraft

662 Appendix C Detailed Drawings for Different Aircraft

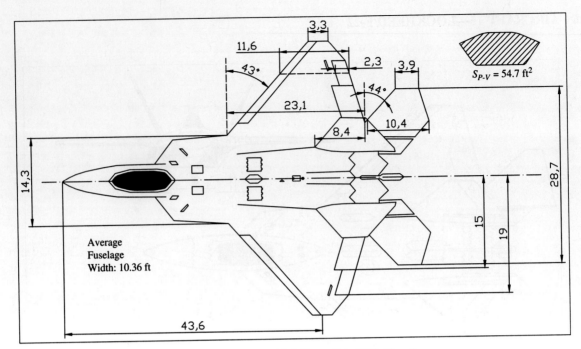

Figure C-A15.3 Top CAD View of the F-22 Aircraft

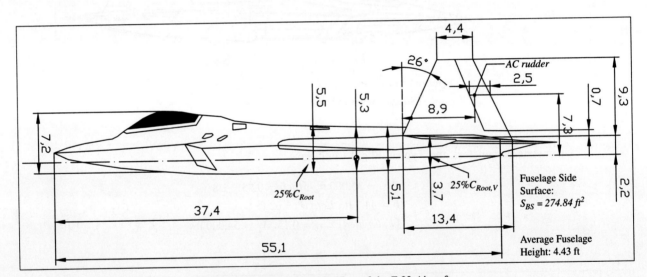

Figure C-A15.4 Side CAD View of the F-22 Aircraft

Table C-A15.1 Geometric Parameters for the Aerodynamic Modeling of the F-22 Aircraft

Geometric Parameters	F-22 Aircraft	Geometric Parameters	F-22 Aircraft
A [ft]	43.6	X_{HV} [ft]	8.4
b [ft]	44.5	X_{WH_r} [ft]	23.1
b_H [ft]	28.7	y_{A_I} [ft]	15.0
b_V [ft]	9.3	y_{A_O} [ft]	19.0
$\bar{c}_{Aileron}$ [ft]	2.3	y_{R_I} [ft]	0.7
$\bar{c}_R$ [ft]	2.5	y_{R_F} [ft]	9.3
$\bar{c}_{Wing(atAileron)}$ [ft]	11.6	y_V [ft]	2.2
c_r [ft]	23.1	Z_H [ft]	-0.3
c_{r_H} [ft]	10.4	Z_{R_S} [ft]	7.3
c_{r_V} [ft]	13.4	z_1 [ft]	7.2
c_T [ft]	3.3	z_2 [ft]	5.1
c_{T_H} [ft]	3.9	Z_{H_S} [ft]	-1.7
c_{T_V} [ft]	4.4	z_{max} [ft]	7.2
d [ft]	5.5	Z_W [ft]	-0.3
l_b [ft]	55.1	Z_{WH_r} [ft]	0.0
l_{cg} [ft]	37.4	Γ_H [deg]	0.0
r_1 [ft]	3.7	Γ_W [deg]	-3.0
S_{B_S} [ft]	274.8	ε_H [deg] (assumed)	2.0
$S_{f_{AVG}}$ [ft^2]	9.3	ε_W [deg] (assumed)	2.0
$S_{P \rightarrow V}$ [ft^2]	54.7	Λ_{LE} [deg]	43.0
w_{max} [ft]	14.3	Λ_{LE_H} [deg]	44.0
X_{AC_R} [ft]	8.9	Λ_{LE_V} [deg]	26.0

C.17 AIRCRAFT 16—LOCKHEED L-1011

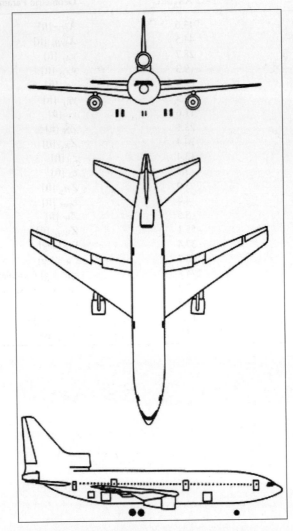

Figure C-A16.1 3D View of the Lockheed L-1011 Aircraft
(Source: www.aviastar.org/index2.html)

Geometric Parameters and Typical Flight Conditions

$Alt. = 35,000$ ft, $Mach = 0.85$
$S = 3,360$ ft^2, $b = 155.3$ ft, $Length = 177.1$ ft, $Height = 55.5$ ft
($Estimated$) $\bar{x}_{CG} = 0.25$, $W = 350,000$ lbs

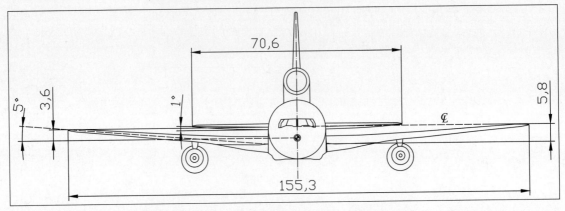

Figure C-A16.2 Front CAD View of the L-1011 Aircraft

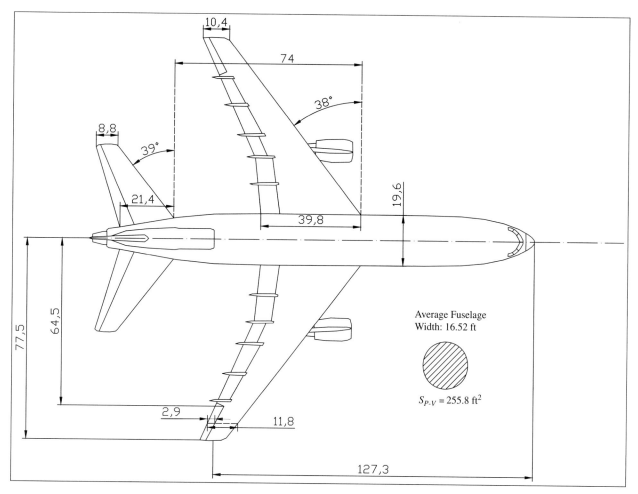

Figure C-A16.3 Top CAD View of the L-1011 Aircraft

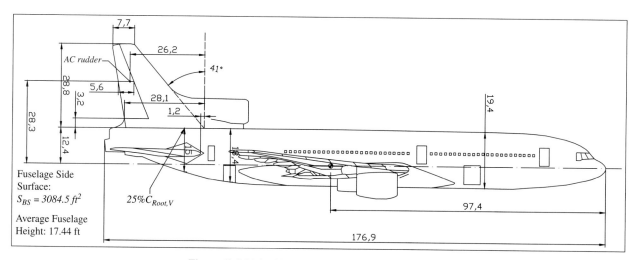

Figure C-A16.4 Side CAD View of the L-1011 Aircraft

Table C-A16.1 Geometric Parameters for the Aerodynamic Modeling of the L-1011 Aircraft

Geometric Parameters	L-1011 Aircraft	Geometric Parameters	L-1011 Aircraft
A [ft]	127.3	X_{HV} [ft]	1.2
b [ft]	155.3	X_{WH_r} [ft]	74.0
b_H [ft]	70.6	y_{A_I} [ft]	64.5
b_V [ft]	28.8	y_{A_O} [ft]	77.5
$\bar{c}_{Aileron}$ [ft]	2.9	y_{R_I} [ft]	3.2
$\bar{c}_R$ [ft]	5.6	y_{R_F} [ft]	28.8
$\bar{c}_{Wing(atAileron)}$ [ft]	11.8	y_V [ft]	12.4
c_r [ft]	39.8	Z_H [ft]	0.0
c_{r_H} [ft]	21.4	Z_{R_S} [ft]	28.3
c_{r_V} [ft]	28.1	z_1 [ft]	19.4
c_T [ft]	10.4	z_2 [ft]	18.4
c_{T_H} [ft]	8.8	Z_{H_S} [ft]	−3.6
c_{T_V} [ft]	7.7	z_{max} [ft]	19.4
d [ft]	19.4	Z_W [ft]	5.8
l_b [ft]	176.9	Z_{WH_r} [ft]	5.8
l_{cg} [ft]	97.4	Γ_H [deg]	1.0
r_1 [ft]	15.0	Γ_W [deg]	5.0
S_{B_S} [ft]	3084.5	ε_H [deg] (assumed)	2.0
$S_{f_{AVG}}$ [ft^2]	9.3	ε_W [deg] (assumed)	2.0
$S_{P \to V}$ [ft^2]	255.8	Λ_{LE} [deg]	38.0
w_{max} [ft]	19.6	Λ_{LE_H} [deg]	39.0
X_{AC_R} [ft]	26.2	Λ_{LE_V} [deg]	41.0

C.18 AIRCRAFT 17—MCDONNELL DOUGLAS C-17

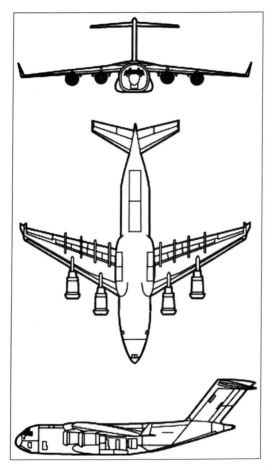

Figure C-A17.1 3D Views of the McDonnell DouglasC-17 Aircraft
(Source: www.aviastar.org/index2.html)

Geometric Parameters and Typical Flight Conditions

$$Alt. = 30{,}000 \text{ ft}, \quad Mach = 0.7$$
$$b = 170 \text{ ft}, \quad Length = 174 \text{ ft}, \quad Height = 55 \text{ ft}$$
$$(Estimated) \; \bar{x}_{CG} = 0.25, \quad W = 500{,}000 \text{ lbs}$$

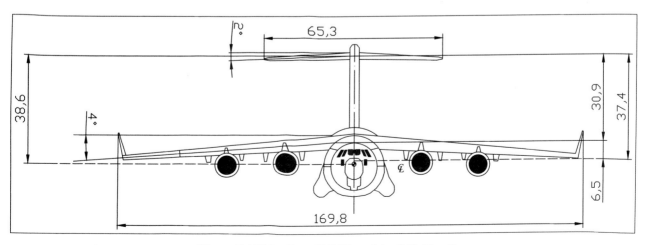

Figure C-A17.2 Front CAD View of the C-17 Aircraft

668 Appendix C Detailed Drawings for Different Aircraft

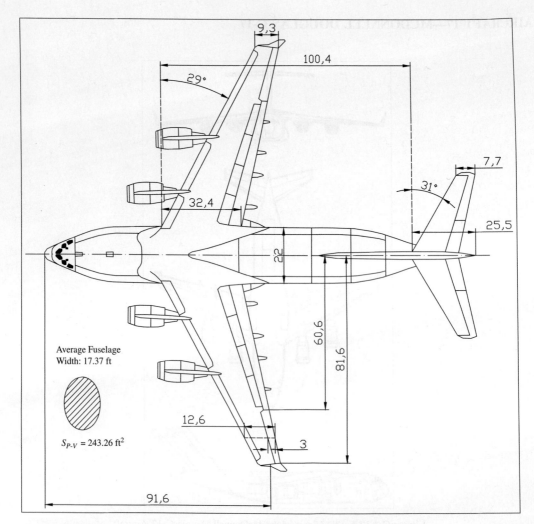

Figure C-A17.3 Top CAD View of the C-17 Aircraft

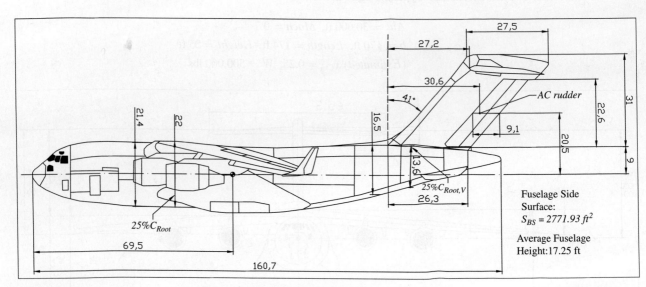

Figure C-A17.4 Side CAD View of the C-17 Aircraft

C.19 Aircraft 18—McDonnell Douglas DC-8

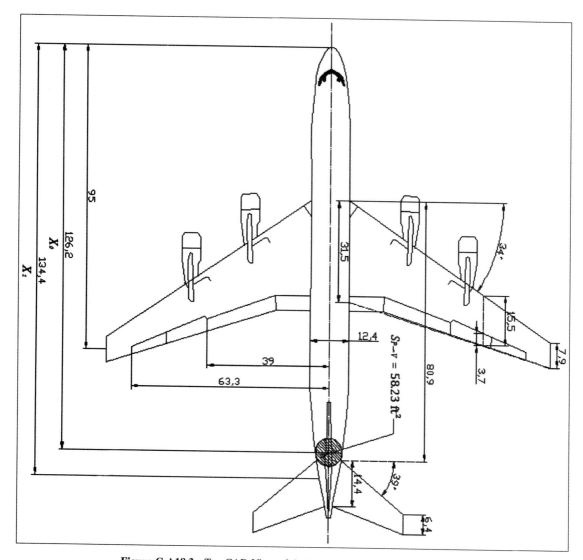

Figure C-A18.3 Top CAD View of the McDonnell Douglas DC-8 Aircraft

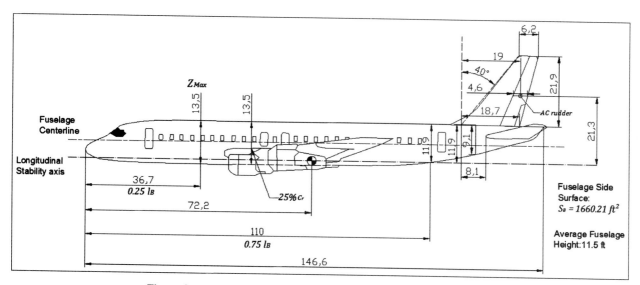

Figure C-A18.4 Side CAD View of the McDonnell Douglas DC-8 Aircraft

Table C-A18.1 Geometric Parameters for the Aerodynamic Modeling of the DC-8 Aircraft

Aircraft Dimension	DC-8 Value	Aircraft Dimension	DC-8 Value
A [ft]	95	X_{HV} [ft]	8.1
b [ft]	142.5	X_{WH_r} [ft]	80.9
b_H [ft]	47.9	X_1 [ft]	134.4
b_V [ft]	21.9	y_{A_I} [ft]	39
$\bar{c}$ [ft]	22.1	y_{A_O} [ft]	63.3
$\bar{c}_{Aileron}$ [ft]	3.7	y_{R_I} [ft]	0
$\bar{c}_R$ [ft]	4.6	y_{R_F} [ft]	21.9
$\bar{c}_{Wing(atAileron)}$ [ft]	15.5	y_V [ft]	11.9
c_r [ft]	31.5	Z_{R_S} [ft]	21.3
c_{r_H} [ft]	14.4	z_1 [ft]	13.5
c_{r_V} [ft]	18.7	z_2 [ft]	11.9
c_T [ft]	7.9	Z_H [ft]	3.1
c_{T_H} [ft]	6.4	Z_{H_S} [ft]	8.7
c_{T_V} [ft]	6.2	z_{max} [ft]	13.5
d [ft]	13.5	Z_W [ft]	4.2
l_b [ft]	146.6	Z_{WH_r} [ft]	7.3
l_{cg} [ft]	72.2	Γ_H [deg]	11
r_1 [ft]	9.1	Γ_W [deg]	6
S [ft^2]	2773	ε_H [deg]	2
S_{B_S} [ft^2]	1660.2	ε_W [deg]	2
$S_{f_{AVG}}$ [ft^2]	95.7	Λ_{LE} [deg]	34
$S_{P \to V}$ [ft^2]	58.2	Λ_{LE_H} [deg]	39
w_{max} [ft]	12.4	Λ_{LE_V} [deg]	40
X_{AC_R} [ft]	19		

C.20 AIRCRAFT 19.1—MCDONNELL DOUGLAS DC-9 SERIES 10

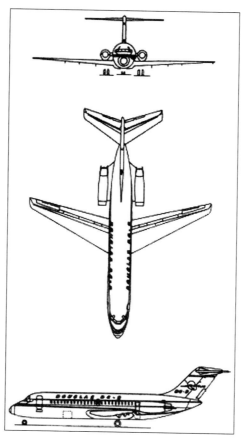

Figure C-A19.1 3D View of the McDonnell Douglas DC-9 Series 10 Aircraft (Source: www.aviastar.org/index2.html)

Geometric Parameters and Typical Flight Conditions

$Alt. = 30,000$ ft, $Mach = 0.75$
$S = 925$ ft^2, $b = 88$ ft, $Length = 104$ ft, $Height = 27$ ft,
$(Estimated)$ $\bar{x}_{CG} = 0.25$, $W = 65,000$ lbs

674 Appendix C Detailed Drawings for Different Aircraft

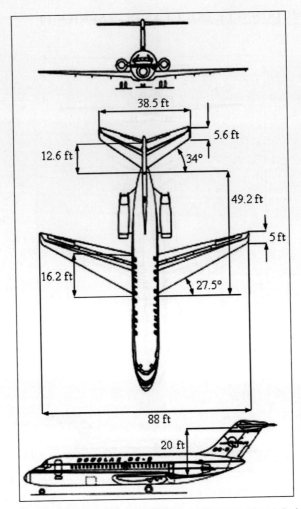

Figure C-A19.2 General Dimensions of the McDonnell Douglas DC-9 Series 10 Aircraft

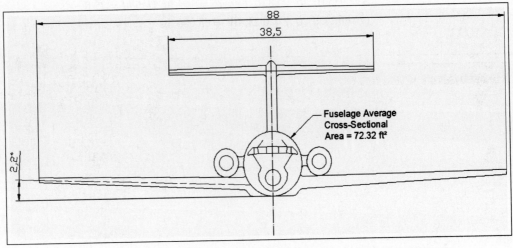

Figure C-A19.3 Front CAD View of the McDonnell Douglas DC-9 Series 10 Aircraft

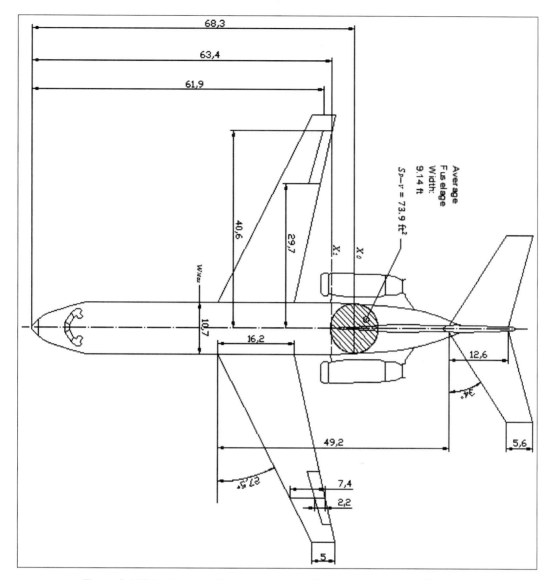

Figure C-A19.4 Top CAD View of the McDonnell Douglas DC-9 Series 10 Aircraft

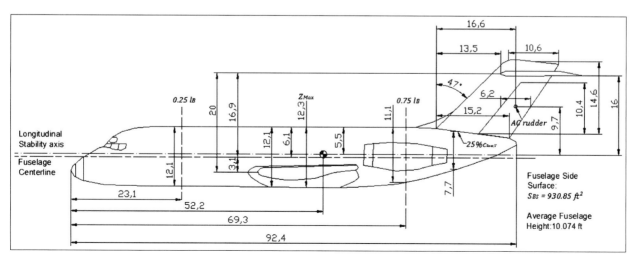

Figure C-A19.5 Side CAD View of the McDonnell Douglas DC-9 Series 10 Aircraft

Table C-A19.1 Geometric Parameters for the Aerodynamic Modeling of the DC-9 Series 10 Aircraft

Aircraft Dimension	DC-9-Series 10 Value	Aircraft Dimension	DC-9-Series 10 Value
A [ft]	60.7	X_{HV} [ft]	11.9
b [ft]	89.4	X_{WH_r} [ft]	49.5
b_H [ft]	36.9	X_1 [ft]	64
b_V [ft]	13.6	y_{A_I} [ft]	27.1
$\bar{c}$ [ft]	10.9	y_{A_O} [ft]	38
$\bar{c}_{Aileron}$ [ft]	2.2	y_{R_I} [ft]	0
$\bar{c}_R$ [ft]	6.2	y_{R_F} [ft]	10.4
$\bar{c}_{Wing(atAileron)}$ [ft]	7.4	y_V [ft]	5.6
c_r [ft]	15.6	Z_{R_S} [ft]	9
c_{r_H} [ft]	10.9	z_1 [ft]	11.5
c_{r_V} [ft]	13.7	z_2 [ft]	10.7
c_T [ft]	3.9	Z_H [ft]	17.8
c_{T_H} [ft]	4	Z_{H_S} [ft]	16
c_{T_V} [ft]	10.9	z_{max} [ft]	11.7
d [ft]	11.5	Z_W [ft]	3.1
l_b [ft]	92.1	Z_{WH_r} [ft]	20
l_{cg} [ft]	49.2	Γ_H [deg]	0
r_1 [ft]	7.7	Γ_W [deg]	2.2
S [ft^2]	928	ε_H [deg]	0
S_{B_S} [ft^2]	896.6	ε_W [deg]	2
$S_{f_{AVG}}$ [ft^2]	72.3	Λ_{LE} [deg]	28
$S_{P \rightarrow V}$ [ft^2]	89.5	Λ_{LE_H} [deg]	35
w_{max} [ft]	11	Λ_{LE_V} [deg]	45
X_{AC_R} [ft]	15		

C.21 AIRCRAFT 19.2—MCDONNELL DOUGLAS DC-9 SERIES 30

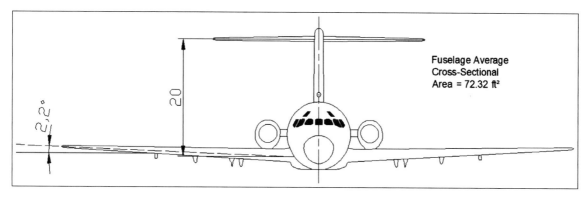

Figure C-A19.6 Front CAD View of the McDonnell Douglas DC-9 Series 30 Aircraft

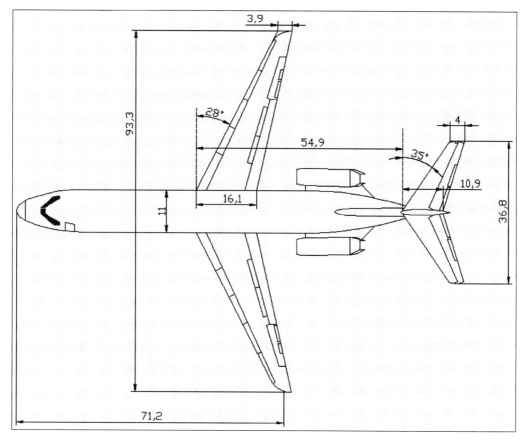

Figure C-A19.7 Top CAD View of the McDonnell Douglas DC-9 Series 30 Aircraft

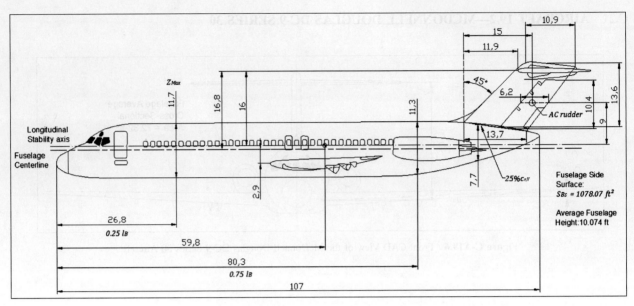

Figure C-A19.8 Side CAD View of the McDonnell Douglas DC-9 Series 30 Aircraft

C.22 AIRCRAFT 19.3—MCDONNELL DOUGLAS DC-9 SERIES 40

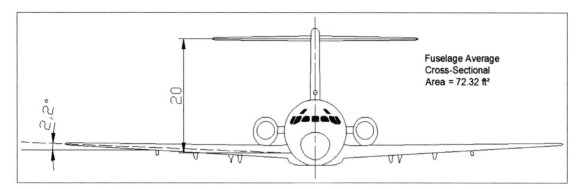

Figure C-A19.9 Front CAD View of the McDonnell Douglas DC-9 Series 40 Aircraft

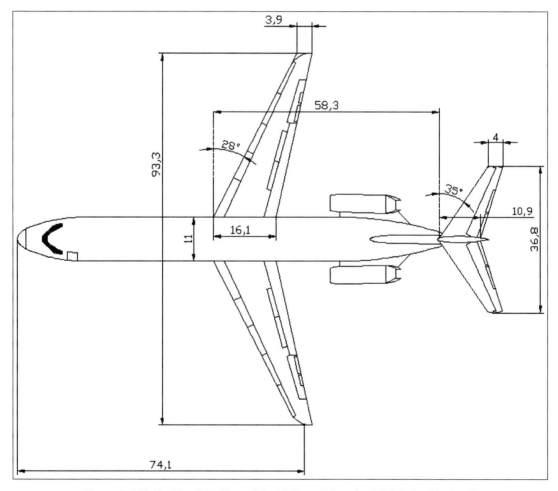

Figure C-A19.10 Top CAD View of the McDonnell Douglas DC-9 Series 40 Aircraft

680 Appendix C Detailed Drawings for Different Aircraft

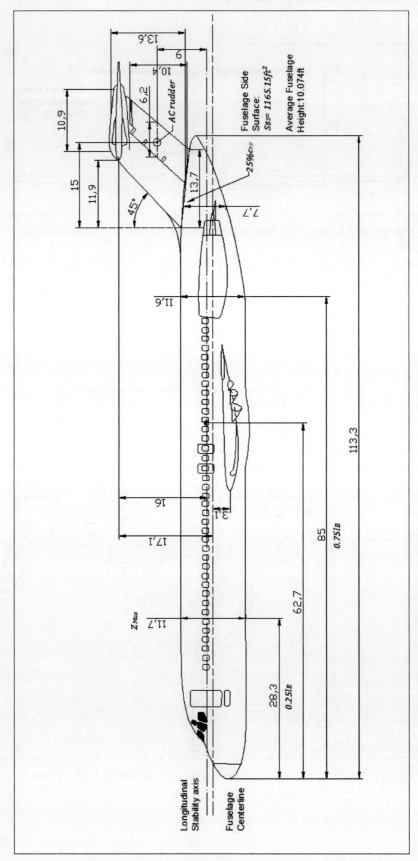

Figure C-A19.11 Side CAD View of the McDonnell Douglas DC-9 Series 40 Aircraft

C.23 AIRCRAFT 19.4—MCDONNELL DOUGLAS DC-9 SERIES 50

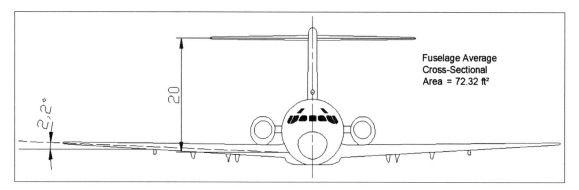

Figure C-A19.12 Front CAD View of the McDonnell Douglas DC-9 Series 50 Aircraft

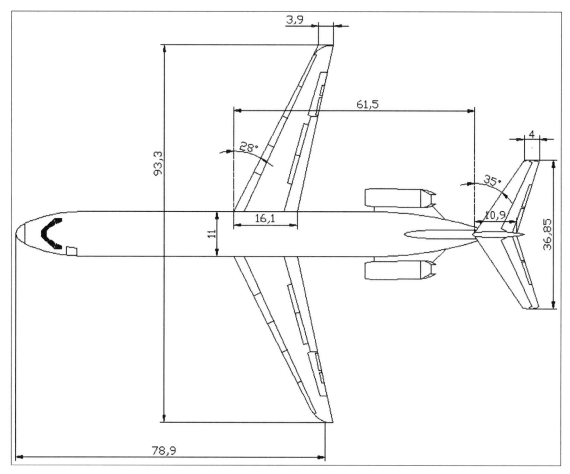

Figure C-A19.3 Top CAD View of the McDonnell Douglas DC-9 Series 50 Aircraft

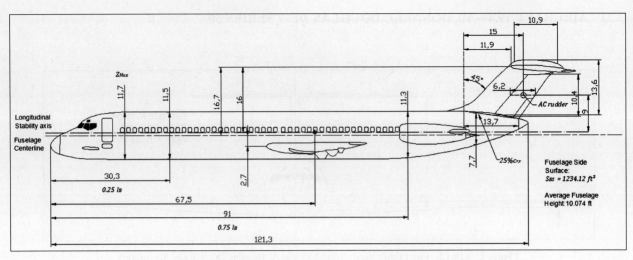

Figure C-A19.14 Side CAD View of the McDonnell Douglas DC-9 Series 50 Aircraft

C.24 AIRCRAFT 20—MCDONNELL DOUGLAS DC-10

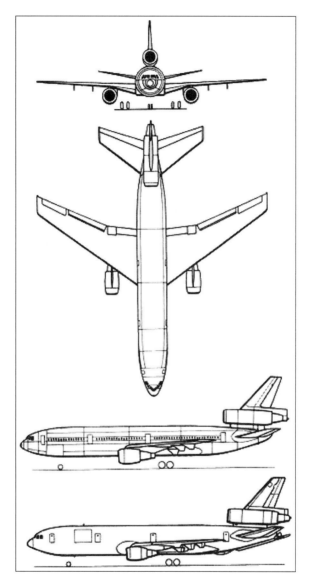

Figure C-A20.1 3D View of the McDonnell DC-10 Aircraft
(Source: www.aviastar.org/index2.html)

Geometric Parameters and Typical Flight Conditions

$Alt. = 30,000$ ft, $Mach = 0.85$
$S = 3,550$ ft^2, $b = 165.4$ ft, $Length = 181.5$ ft, $Height = 58$ ft,
($Estimated$) $\bar{x}_{CG} = 0.25$, $W = 365,000$ lbs

684 Appendix C Detailed Drawings for Different Aircraft

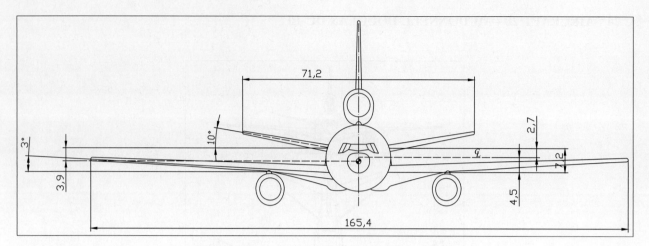

Figure C-A20.2 Front CAD View of the DC-10 Aircraft

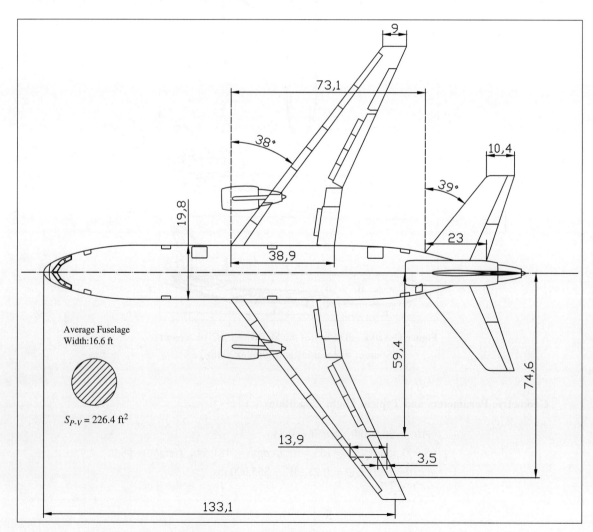

Figure C-A20.3 Top CAD View of the DC-10 Aircraft

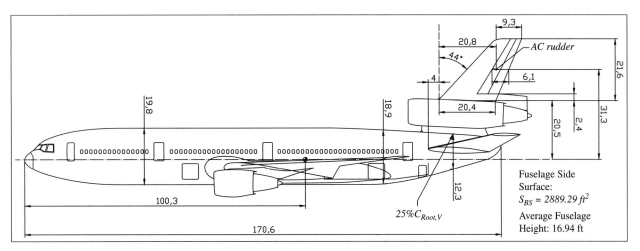

Figure C-A20.4 Side CAD View of the DC-10 Aircraft

Table C-A20.1 Geometric Parameters for the Aerodynamic Modeling of the DC-10 Aircraft

Geometric Parameters	DC-10 Aircraft	Geometric Parameters	DC-10 Aircraft
A [ft]	133.1	X_{HV} [ft]	4.0
b [ft]	165.4	X_{WH_r} [ft]	73.1
b_H [ft]	71.2	y_{A_I} [ft]	59.4
b_V [ft]	21.6	y_{A_O} [ft]	74.6
$\bar{c}_{Aileron}$ [ft]	3.5	y_{R_I} [ft]	2.4
$\bar{c}_R$ [ft]	6.1	y_{R_F} [ft]	21.6
$\bar{c}_{Wing(atAileron)}$ [ft]	13.9	y_V [ft]	31.3
c_r [ft]	38.9	Z_H [ft]	-2.7
c_{r_H} [ft]	23.0	Z_{R_S} [ft]	31.3
c_{r_V} [ft]	20.4	z_1 [ft]	19.8
c_T [ft]	9.0	z_2 [ft]	18.9
c_{T_H} [ft]	10.4	Z_{H_S} [ft]	-3.9
c_{T_V} [ft]	9.3	z_{max} [ft]	19.8
d [ft]	19.8	Z_W [ft]	4.5
l_b [ft]	170.6	Z_{WH_r} [ft]	7.2
l_{cg} [ft]	100.3	Γ_H [deg]	10.0
r_1 [ft]	12.3	Γ_W [deg]	3.0
S_{B_S} [ft]	2889.3	ε_H [deg] (assumed)	2.0
$S_{f_{AVG}}$ [ft^2]	9.3	ε_W [deg] (assumed)	2.0
$S_{P \to V}$ [ft^2]	226.4	Λ_{LE} [deg]	38.0
w_{max} [ft]	19.8	Λ_{LE_H} [deg]	39.0
X_{AC_R} [ft]	20.8	Λ_{LE_V} [deg]	44.0

686 Appendix C Detailed Drawings for Different Aircraft

C.25 AIRCRAFT 21—MCDONNELL DOUGLAS F-4

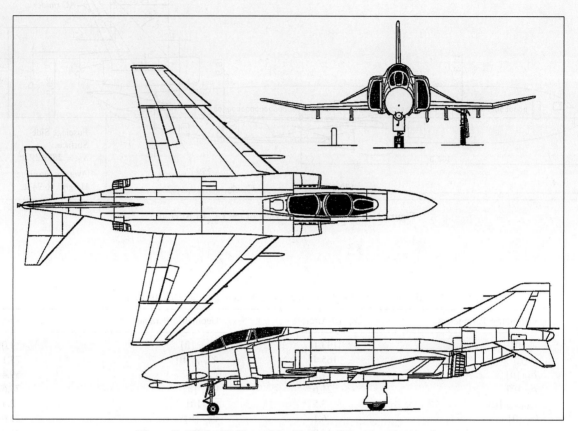

Figure C-A21.1 3D View of the McDonnell Douglas F4 Aircraft
(Source: www.aviastar.org/index2.html)

Geometric Parameters and Typical Flight Conditions (see Appendix B, Aircraft 10)

$Alt. = 35{,}000$ ft, $Mach = 0.9$, $\bar{q}_1 = 283.2$ lbs/ft^2, $\alpha_1 = 2.6°$
$S = 530$ ft^2, $b = 38.7$ ft, $\bar{c} = 16$ ft, $Length = 63.75$ ft, $Height = 16.5$ ft, $\bar{x}_{CG} = 0.29$, $W = 39{,}000$ lbs

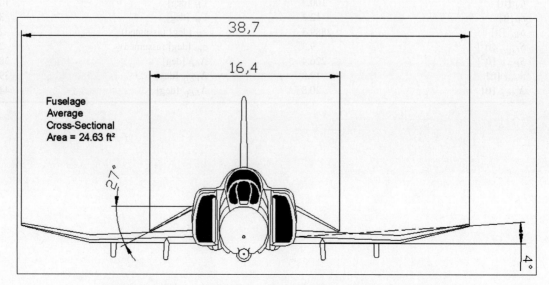

Figure C-A21.2 Front CAD View of the McDonnell Douglas F-4 Aircraft

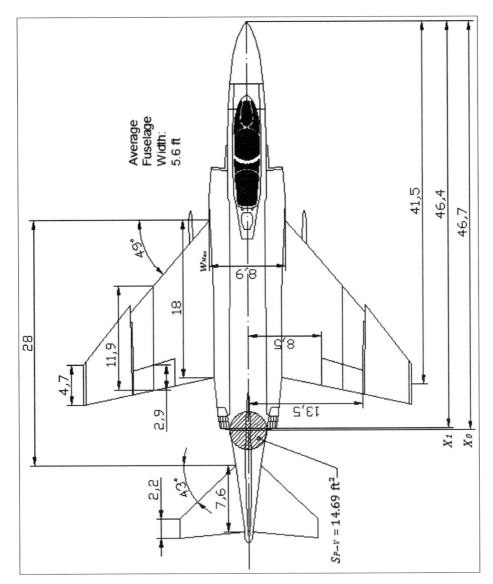

Figure C-A21.3 Top CAD View of the McDonnell Douglas F-4 Aircraft

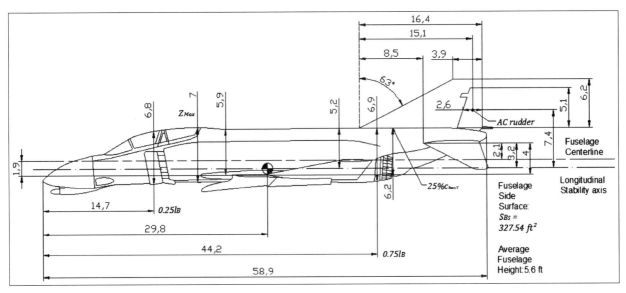

Figure C-A21.4 Side CAD View of the McDonnell Douglas F-4 Aircraft

Table C-A21.1 Geometric Parameters for the Aerodynamic Modeling of the F-4 Aircraft

Geometric Parameters	F4 Aircraft	Geometric Parameters	F4 Aircraft
A [ft]	41.5	X_{HV} [ft]	8.5
b [ft]	38.7	X_{WH_r} [ft]	28.0
b_H [ft]	16.4	X_1 [ft]	46.4
b_V [ft]	6.2	y_{A_I} [ft]	8.5
$\bar{c}$ [ft]	16.0	y_{A_O} [ft]	13.5
$\bar{c}_{Aileron}$ [ft]	2.9	y_{R_I} [ft]	0.0
$\bar{c}_R$ [ft]	2.6	y_{R_F} [ft]	5.1
$\bar{c}_{Wing(atAileron)}$ [ft]	11.9	y_V [ft]	5.2
c_r [ft]	18.0	Z_{R_S} [ft]	7.4
c_{r_H} [ft]	7.6	z_1 [ft]	6.8
c_{r_V} [ft]	16.4	z_2 [ft]	6.9
c_T [ft]	4.7	Z_H [ft]	2.1
c_{T_H} [ft]	2.2	Z_{H_S} [ft]	3.2
c_{T_V} [ft]	3.9	z_{max} [ft]	7.0
d [ft]	5.9	Z_W [ft]	1.9
l_b [ft]	58.9	Z_{WH_r} [ft]	4.0
l_{cg} [ft]	29.8	Γ_H [deg]	−27
r_1 [ft]	6.2	Γ_W [deg]	4.0
S [ft^2]	530	ε_H [deg] (assumed)	2
S_{B_S} [ft]	327.5	ε_W [deg] (assumed)	2
$S_{f_{AVG}}$ [ft^2]	24.6	Λ_{LE} [deg]	48.5
$S_{P \to V}$ [ft^2]	14.7	Λ_{LE_H} [deg]	43.0
w_{max} [ft]	8.9	Λ_{LE_V} [deg]	63
X_{AC_R} [ft]	15.1		

C.26 AIRCRAFT 22—MCDONNELL DOUGLAS F-15

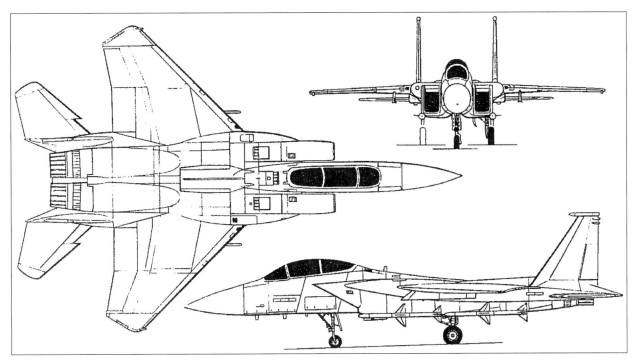

Figure C-A22.1 3D View of the McDonnell Douglas F15 Aircraft
(Source: www.aviastar.org/index2.html)

Geometric Parameters and Typical Flight Conditions

$Alt. = 37,000$ ft, $Mach = 0.9$
$S = 608$ ft^2, $b = 42.8$ ft, $Length = 63$ ft, $Height = 18.75$ ft
$(Estimated)\ \bar{x}_{CG} = 0.3$, $W = 50,000$ lbs

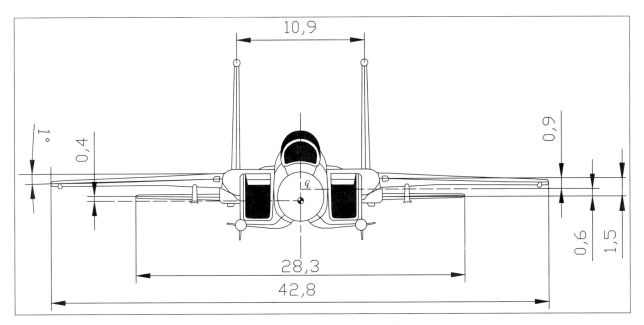

Figure C-A22.2 Front CAD View of the F-15 Aircraft

690 Appendix C Detailed Drawings for Different Aircraft

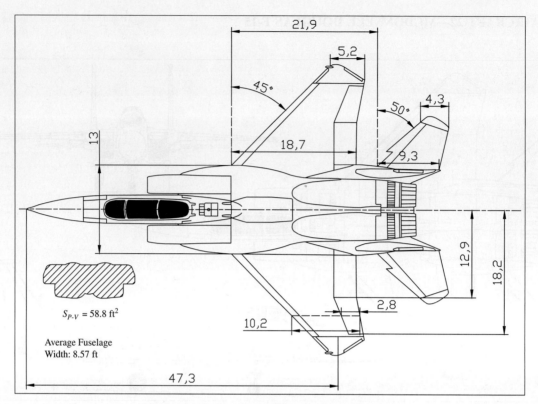

Figure C-A22.3 Top CAD View of the F-15 Aircraft

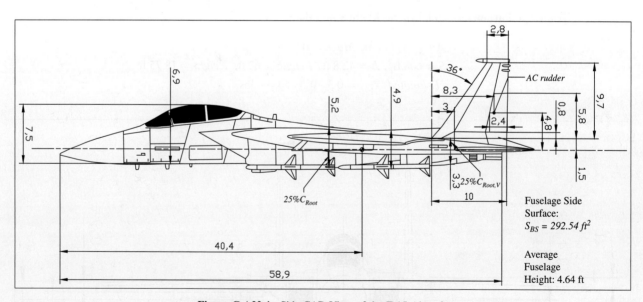

Figure C-A22.4 Side CAD View of the F-15 Aircraft

Table C-A22.1 Geometric Parameters for the Aerodynamic Modeling of the F-15 Aircraft

Geometric Parameters	F-15 Aircraft	Geometric Parameters	F-15 Aircraft
A [ft]	47.3	X_{HV} [ft]	3.0
b [ft]	42.8	X_{WH_r} [ft]	21.9
b_H [ft]	28.3	y_{A_I} [ft]	12.9
b_V [ft]	9.7	y_{A_O} [ft]	18.2
$\bar{c}_{Aileron}$ [ft]	2.8	y_{R_I} [ft]	0.8
$\bar{c}_R$ [ft]	2.4	y_{R_F} [ft]	5.8
$\bar{c}_{Wing(atAileron)}$ [ft]	10.2	y_V [ft]	1.5
c_r [ft]	18.7	Z_H [ft]	0.6
c_{r_H} [ft]	9.3	Z_{R_S} [ft]	4.8
c_{r_V} [ft]	10.0	z_1 [ft]	6.9
c_T [ft]	5.2	z_2 [ft]	4.9
c_{T_H} [ft]	4.3	Z_{H_S} [ft]	0.4
c_{T_V} [ft]	2.8	z_{max} [ft]	7.5
d [ft]	5.2	Z_W [ft]	-0.9
l_b [ft]	58.9	Z_{WH_r} [ft]	1.5
l_{cg} [ft]	40.4	Γ_H [deg]	0.0
r_1 [ft]	3.3	Γ_W [deg]	-1.0
S_{B_S} [ft]	292.5	ε_H [deg] (assumed)	2.0
$S_{f_{AVG}}$ [ft^2]	9.3	ε_W [deg] (assumed)	2.0
$S_{P \to V}$ [ft^2]	58.8	Λ_{LE} [deg]	45.0
w_{max} [ft]	13.0	Λ_{LE_H} [deg]	50.0
X_{AC_R} [ft]	8.3	Λ_{LE_V} [deg]	36.0

C.27 AIRCRAFT 23—ROCKWELL B-1

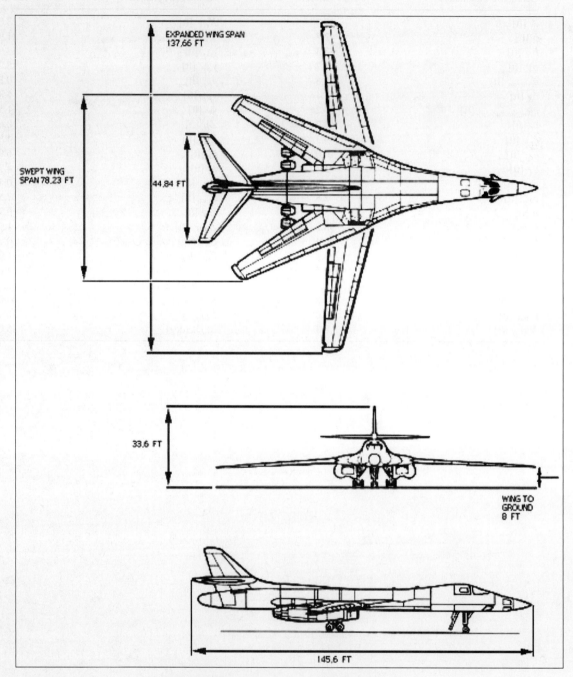

Figure C-A23.1 3D View of the Rockwell B1 Aircraft
(Source: www.aviastar.org/index2.html)

Geometric Parameters and Typical Flight Conditions

$Alt. = 33{,}000$ ft, $Mach = 0.7$
$S = 1{,}950$ ft^2, $b = 78.25$ ft/137.7 ft, $Length = 145.6$ ft, $Height = 33.6$ ft
$(estimated)\ \bar{x}_{CG} = 0.25$, $W = 440{,}000$ lbs

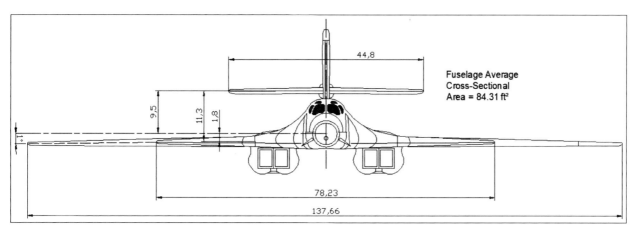

Figure C-A23.2 Front CAD View of the Rockwell B-1 Aircraft

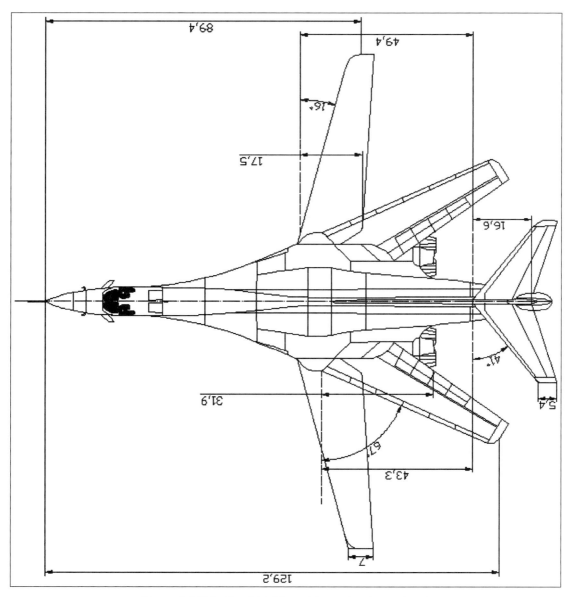

Figure C-A23.3 Top CAD View of the Rockwell B-1 Aircraft

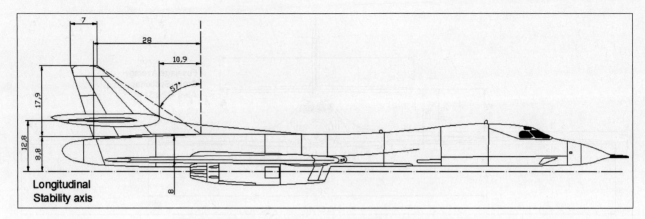

Figure C-A23.4 Side CAD View of the Rockwell B-1 Aircraft

Table C-A23.1 Geometric Parameters for the Aerodynamic Modeling of the Rockwell B-1 Aircraft

Aircraft Dimension	B1 (Extended wing) Value	B1 (Swept Wing) Value
A[ft]	89.4	129.2
b[ft]	137.7	78.2
b_H[ft]	44.8	44.8
b_V[ft]	17.9	17.9
$\bar{c}$[ft]	13.0	22.1
c_r[ft]	17.5	31.9
c_{r_H}[ft]	16.6	16.6
c_{r_V}[ft]	28	28
c_T[ft]	7	7
c_{T_H}[ft]	5.4	5.4
c_{T_V}[ft]	7	7
r_1[ft]	8	8
S[ft^2]	1950	1521.6
$S_{f_{AVG}}$[ft^2]	84.3	84.3
$S_{P \to V}$[ft^2]	13.0	22.1
X_{HV}[ft]	10.9	10.9
X_{WH_r}[ft]	49.4	43.3
y_V [ft]	8.8	8.8
Z_H [ft]	9.5	9.5
Z_{H_S}[ft]	12.8	12.8
Z_W[ft]	1.8	1.8
Z_{WH_r}[ft]	11.3	11.3
Γ_H[deg]	0	0
Γ_W [deg]	−1	−1
ε_H[deg]	2	2
ε_W[deg]	2	2
Λ_{LE}[deg]	16	67
Λ_{LE_H}[deg]	41	41
Λ_{LE_V}[deg]	57	57

C.28 AIRCRAFT 24—SIAI MARCHETTI S211

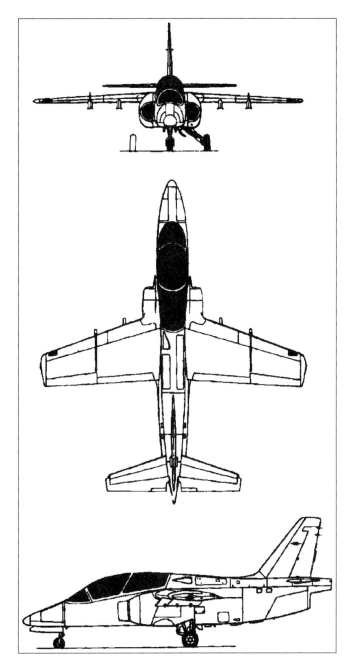

Figure C-A24.1 3D View of the SIAI Marchetti S211 Aircraft
(Source: www.aviastar.org/index2.html)

Geometric Parameters and Typical Flight Conditions (see Appendix B, Aircraft 8)

$Alt. = 25{,}000$ ft, $Mach = 0.6$
$V_{P_1} = 610$ ft/sec, $\bar{q}_1 = 198$ lbs/ft^2, $\alpha_1 = 0°$
$S = 136$ ft^2, $\bar{c} = 5.4$ ft, $b = 26.3$ ft
$\bar{x}_{CG} = 0.25$, $W = 4{,}000$ lb

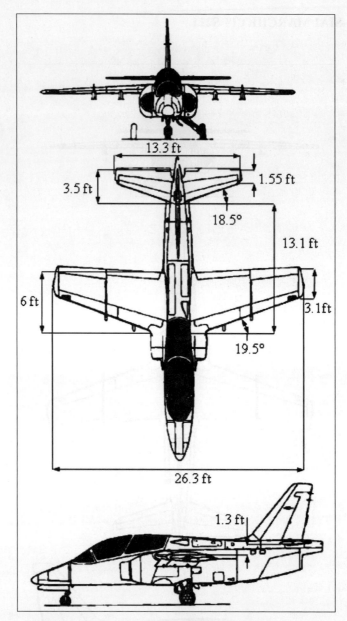

Figure C-A24.2 General Dimensions of the SIAI S-211 Aircraft

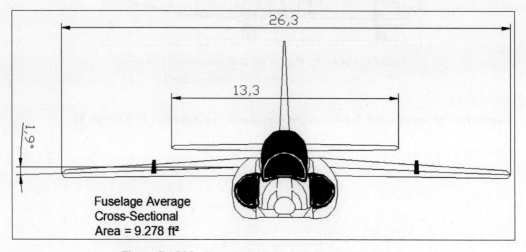

Figure C-A24.3 Front CAD View of the SIAI S-211 Aircraft

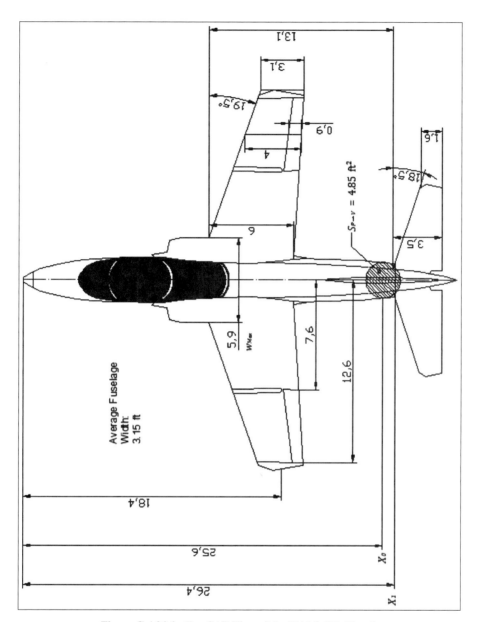

Figure C-A24.4 Top CAD View of the SIAI S-211 Aircraft

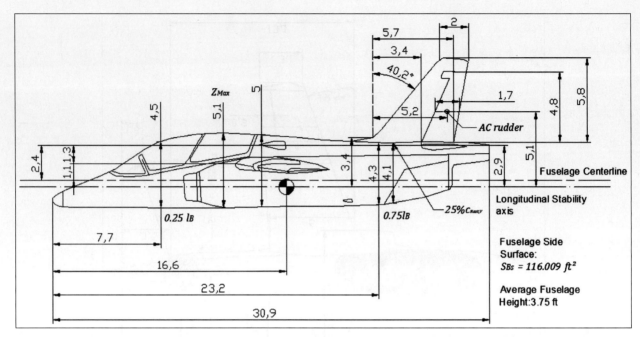

Figure C-A24.5 Side CAD View of the SIAI S-211 Aircraft

Table C-A24.1 Geometric Parameters for the Aerodynamic Modeling of the SIAI S-211 Aircraft

Geometric Parameters	S211 Aircraft	Geometric Parameters	S211 Aircraft
A[ft]	18.4	X_{HV}[ft]	3.4
b[ft]	26.3	X_{WH_r}[ft]	13.1
b_H[ft]	13.3	X_1[ft]	26.4
b_V[ft]	5.8	y_{A_I}[ft]	7.6
$\bar{c}$[ft]	5.4	y_{A_O}[ft]	12.6
$\bar{c}_{Aileron}$[ft]	0.9	y_{R_I}[ft]	0.0
$\bar{c}_R$[ft]	1.7	y_{R_F}[ft]	4.8
$\bar{c}_{Wing(atAileron)}$[ft]	4.0	y_V[ft]	3.0
c_r[ft]	6.0	Z_{R_S}[ft]	5.1
c_{r_H}[ft]	3.5	z_1[ft]	4.5
c_{r_V}[ft]	5.7	z_2[ft]	4.3
c_T[ft]	3.1	Z_H[ft]	2.4
c_{T_H}[ft]	1.55	Z_{H_S}[ft]	2.9
c_{T_V}[ft]	2.0	z_{max}[ft]	5.1
d[ft]	5	Z_W[ft]	-1.1
l_b[ft]	30.9	Z_{WH_r}[ft]	1.3
l_{cg}[ft]	16.6	Γ_H[deg]	0
r_1[ft]	4.1	Γ_W[deg]	-1.9
S[ft^2]	136	ε_H[deg]	0
S_{B_S}[ft]	116.0	ε_W[deg] (assumed)	2
$S_{f_{AVG}}$[ft^2]	9.3	Λ_{LE}[deg]	19.5
S_{PV}[ft^2]	4.9	Λ_{LE_H}[deg]	18.5
w_{max}[ft]	5.9	Λ_{LE_V}[deg]	40
X_{AC_R}[ft]	5.2		

C.29　AIRCRAFT 25—SUPERMARINE SPITFIRE

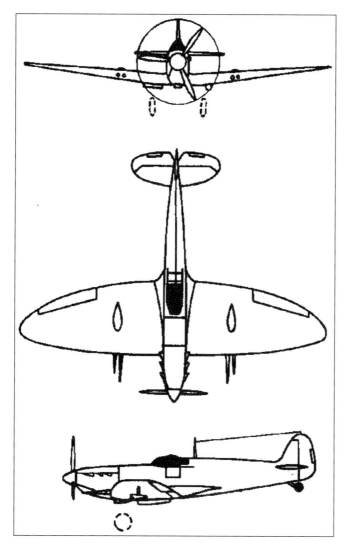

Figure C-A25.1　3D View of the Supermarine Spitfire Aircraft
(Source: www.aviastar.org/index2.html)

Geometric Parameters and Typical Flight Conditions

$$Alt. = 33{,}000 \text{ ft}, \ Mach = 0.35$$
$$S = 242 \text{ ft}^2, \ b = 37 \text{ ft}, \ Length = 30 \text{ ft}, Height = 10 \text{ ft}$$
$$(estimated) \ \bar{x}_{CG} = 0.25, \ W = 6{,}200 \text{ lbs}$$

700 Appendix C Detailed Drawings for Different Aircraft

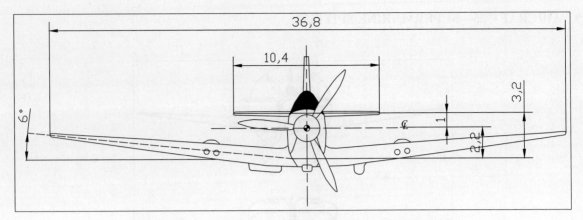

Figure C-A25.2 Front CAD View of the Supermarine Spitfire Aircraft

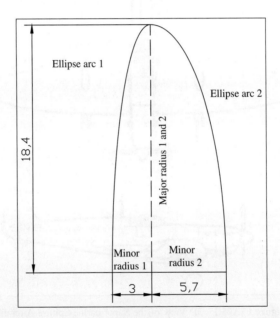

Figure C-A25.3 Geometry of the Supermarine Spitfire Aircraft Elliptical Wing (approximated)

C.29 Aircraft 25—Supermarine Spitfire 701

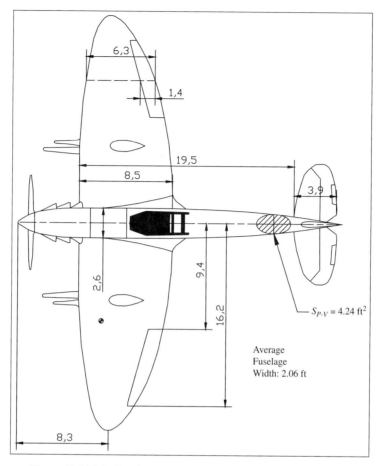

Figure C-A25.4 Top CAD View of the Supermarine Spitfire Aircraft

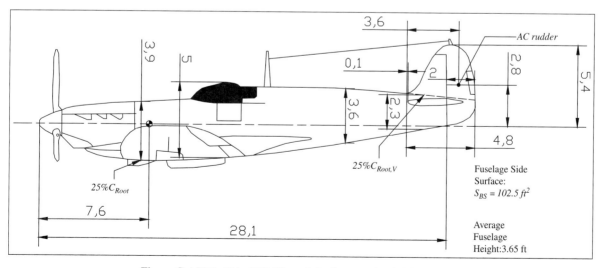

Figure C-A25.5 Side CAD View of the Supermarine Spitfire Aircraft

Table C-A25.1 Geometric Parameters for the Aerodynamic Modeling of the Spitfire Aircraft

Geometric Parameters	Spitfire Aircraft	Geometric Parameters	Spitfire Aircraft
A[ft]	8.3	X_{HV}[ft]	0.1
b[ft]	36.8	X_{WH_r}[ft]	19.5
b_H[ft]	10.4	y_{A_I}[ft]	9.4
b_V[ft]	5.4	y_{A_O}[ft]	16.2
$\bar{c}_{Aileron}$[ft]	1.4	y_{R_I}[ft]	0.0
$\bar{c}_R$[ft]	2.0	y_{R_F}[ft]	5.4
$\bar{c}_{Wing(atAileron)}$[ft]	6.3	y_V[ft]	0.0
c_r[ft]	8.5	Z_H[ft]	−1.0
c_{r_H}[ft]	3.9	Z_{R_S}[ft]	2.8
c_{r_V}[ft]	4.8	z_1[ft]	3.9
d[ft]	3.9	z_2[ft]	3.6
l_b[ft]	28.1	Z_{H_S}[ft]	−1.0
l_{cg}[ft]	7.6	z_{max}[ft]	5.0
r_1[ft]	2.3	Z_W[ft]	2.2
S[ft^2]	249.6	Z_{WH_r}[ft]	3.2
S_{B_S}[ft]	102.5	Γ_H[deg]	0.0
$S_{f_{AVG}}$[ft^2]	9.3	Γ_W [deg]	6.0
S_{PV}[ft^2]	4.2	ε_H[deg] (assumed)	2.0
w_{max}[ft]	2.6	ε_W[deg] (assumed)	2.0
X_{AC_R}[ft]	3.6		

Index

Abbott, I. H., 38
Actuator dynamics, SV modeling of, 446–447
Actuators library, 511
Advanced Aircraft Analysis (AAA), 60
Aerodynamic angles, 306
Aerodynamic center (AC), 53–57, 89–91
  for wing and wing+fuselage, 53–57
Aerodynamic forces and moments, 57–60
  frequency-domain techniques, 60
  lateral directional force and moments, 57
  longitudinal forces and moment, 57
  modeling, 57–60
  real-time PID techniques, 60
  on wing section, 37
Aerodynamic modeling, basic concepts, 37–77, *See also* Wing lift curve slope; Wing planforms; Wing sections
Aerospace blockset (AB), 503–511
  Actuators library, 511
  Aerodynamics library, 508
  Environment library, 504–505
  Equations of Motion library, 506–507
  Flight Parameters library, 506
  GNC and Animation libraries, 511
  Mass Properties library, 510–511
  by mathworks, 503–511
  organization of, 503
  Propulsion library, 508–509
  six DOF sublibrary, 507
  'three DOF sublibrary,' 506–507
  Utilities library, 509–510
'Air3m' function, 517–518
Aircraft control surfaces, 136
Aircraft equations of motion, 1–36
  Conservation of the Angular Momentum Equations (CAMEs), 6–10
  Conservation of the Linear Momentum Equations (CLMEs), 3–6
  development, 24
  Euler angles, 11–12
  Flight Path Equations (FPEs), 12–14
  Gravity Equations (GEs), 16
  Inverse Kinematic Equations (IKEs), 15
  Kinematic Equations (KEs), 14–15
  perturbed conditions, 19–22
  solution, 352–431, *See also under* Laplace transformations; Routh–Hurwitz analysis
  at steady-state conditions, 17–19
Aircraft models, collection of, 516–517
AIRLIB, 512–518
  ab2dv function, 518
  'air3m' function, 517–518
  AIRLIB Simulink® scheme, 516
  continuous-time block, 512–515
  strucure, 512
Airspeed on the wing planform, 44
Altitude, state variable (SV) modeling of, 445
Angle γ, 274
Angular velocity, 306
Anhedral angle, 142
Atmosphere sub library, 504
Atmospheric turbulence force, FDC modeling of, 493–494
  Dryden turbulence model, 493
  Von Karman turbulence model, 493
Atmospheric turbulence, SV modeling of, 447

Auxiliary FDC blocks, 498–503
  standard atmosphere model (SAM), 499
  FDC Acceleration Block, 502
  FDC Airdata Group Main Level Block, 499
  FDC Flight Path Block, 501
  FDC Time Derivative Block, 502
  outputs of SAM, 499
Average cross-sectional area along fuselage, 156

B747 aircraft
  bank angle into operational engines following engine out for, 339
  engine out condition for, 334
  rudder compensation
    vs. airspeed for engine-out condition for, 337
    vs. airspeed for two engines out condition for, 339
  thrust moment modeling
    following engine out on, 336
    following two engines out on, 338
  wing planform, 42–48
Basic aircraft performance, 268–296
  angle γ, 274
  drag, 273
  forces acting on aircraft on vertical plane, 273
  lift, 273
  thrust, 273
  weight, 273
Beaver aerodynamic forces and moments, 483–485
Beaver propulsive forces and moments, FDC modeling of, 494–496
Boeing 747
  with a nose down pitching moment, 317
  with a nose up pitching moment, 318
Brake horsepower, 269
Bryan, G. H., 1
By-pass ratio (BPR), 272
  high BPR turbofan engine, 272
  low BPR turbofan engine, 272

Calibrated air speed (CAS), 506
Caminez 447 Engine, 270
Canards (δ_C), 196
Cessna 182, 340
Cessna CJ3 aircraft, 110–126
  DRAG aerodynamic coefficients, modeling, 126–127
  longitudinal aerodynamics, modeling for, 110–119
    wing aerodynamic center, location, 117
    aircraft lift-slope coefficient, 15
    c_{L1} modeling, 110–111
    c_{Lq} vs. Mach, 118
    c_{Lu} modeling, 111
    $c_{L\alpha}$ modeling, 111–115
    $c_{L\alpha}$ vs. Mach, 115
    $c_{L\alpha}, c_{Lq}$ modeling, 116–118
    $c_{L\delta E}, c_{LiH}$ modeling, 119
    downwash effect, modeling, 114–115
    wing geometric parameters, 111
    horizontal tail geometric parameters, 111–112

  horizontal tail lift slope coefficient, 114
  LIFT aerodynamic coefficients, 110–119
  wing lift-slope coefficient, 111, 113
  wing-tail geometric parameters, 112
  pitching aerodynamic coefficients, 120–126
Cessna U206, 281
Classes of aircraft, 531
Coefficients, aerodynamic, 196–198
  canards (δ_C), 196
  flaperons, 196
  importance, 196–198
  leading-edge flaps, 196
  ruddervators, 196
  speedbrakes, 196
  Spoilerons, 196
  spoilers, 196
  trailing-edge flaps, 196
Computation fluid dynamics (CFD) analysis, 57–59
  postprocessing, 58
  preprocessing, 58
  simulation, 58
Conservation of Angular Momentum equations (CAMEs), 6–10, 352–353, 477–478, 481, 493
  'ad hoc' modification to, 10
  with gyroscopic effects, 10–11
  with rotor effects, 10–11
Conservation of the Linear Momentum Equations (CLMEs), 3–6
Continuous-time block, 512–515
  inputs to, 512
  state vector of, 512
Control surfaces, effectiveness of, 48–53
  geometry of, 52
  range for, 52
Cooper-Harper pilot rating scale (CHPRS), 526–530
  aircraft control authority, 526
  irreversible flight control system (IFCS), 528
  pilot compensation, 529
  pilot workload, 526–529
  reversible flight control system (RFCS), 527
Cramer's rule, 356

Differential thrust, 288
Discrete-time block, 515–516
Dives, longitudinal control forces for, 535
Downwash effect, 48–53
  geometric coefficients for, 51
Drag, 273
Dryden turbulence model, 493
Dutch Roll mode, 538
  lateral directional dynamic, 377–378
Dynamic aircraft stability, 306
Dynamic pressure ratios (η), 52

Earth-based reference system, 2
Engine toe-in angle, 288
Environment library, 504–505
  atmosphere sub library, 504
  gravity sub library, 504–505
  wind sub library, 505
Equations of Motion library, 505–506
Equations of motion within FDC simulation environment, 479–483

703

Euler angles, 11–12
Exposed wing surface, 42

F_{AY1} modeling, 137–147
   $c_{Y\beta}$ conceptual modeling, 138–139
   $c_{Y\beta}$ mathematical modeling, 140–147
   $c_{Y\delta A}$ modeling, 147
   $c_{Y\delta R}$ modeling, 147–149
   wing-body interference factor, 140
   wing-fuselage integration, 141
Fixed-pitch propellers, 293
Flaperons, 196
Flat-Earth assumption, 2
Flight dynamics & control (FDC) toolbox, 479–503, See also Auxiliary FDC blocks
   atmospheric turbulence force, FDC modeling of, 493–494
   auxiliary FDC blocks, 498–503
   Beaver propulsive forces and moments, FDC modeling of, 494–496
   different forces and moments, mixing of, 482
      Beaver aerodynamic forces and moments, 483–485
      Simulink® tools for, 478
   equations of motion within FDC simulation environment, 479–483
      first level, 479
      fourth level, 481
      second level, 480
      third level, 481
   gravity force, FDC modeling of, 493
   look-up tables based aerodynamic coefficients, FDC modeling of, 486–492
   thrust forces and moments, FDC modeling of, 494
Flight envelope, aircraft, 524–526
   for a Commercial Jetliner, 525
   for Fighter Aircraft, 525
   for a General Aviation and Regional Commuter Aircraft, 524
Flight Parameters library, 506
Flight path angle (γ), SV modeling, 446
Flight Path Equations (FPEs), 12–14, 482
Fly-by-wire (FBW) technology, 529
Flying qualities, 526–531, See also Cooper-Harper pilot rating scale (CHPRS)
   levels of, 529–530
   quality requirements for
      lateral directional dynamics, 536–541, See also under Lateral directional dynamic modes
      longitudinal dynamics, 532–536, See also under Longitudinal dynamics
Frequency-domain techniques, 60
Fuselage diameter, 111
Fuselage stations, 55, 121

General nonlinear aircraft model block
   implementation of, 515
   mask of, 514
Generic aircraft model, 515–516
   discrete-time block, 515–516
GNC and Animation libraries, 511
Gravity Equations (GEs), 16, 478
Gravity force, FDC modeling of, 493
Gravity sub library, 504

Handling qualities, aircraft, 523–541
Hoak, D. E., 81–82, 86, 197–198

Inertial frame, 2
'In-line' piston engine, 269
Inverse Kinematic Equations (IKEs), 15, 21
Inverse Laplace transform (ILT), 447
Inverted Kinematic Equations (IKEs), 352
Irreversible flight control system (IFCS), 528

Jet propulsion, 293

Kinematic equations (KEs), 14–15, 477

L_{A1} modeling, 149–168
   airspeed at vertical tail, 154
   $c_{l\beta}$ conceptual modeling, 150–155
   $c_{l\beta}$ mathematical modeling, 155–166
   $c_{l\delta A}$ modeling, 160–166
   $c_{l\delta R}$ modeling, 166–168
   compressibility correction factor, 157–158
   dihedral angles for wing and horizontal tail, 150
   F-4 horizontal tail anhedral angle Γ_H, 154
   McDonnell Douglas F-4 aircraft, 153
   rolling moment effectiveness, 163–165
   wing + body dihedral effect, 151
Lanchester, F. W., 1
Laplace transformations
   aircraft equations of motion solution based on, 352–431
   application to lateral directional small perturbation equations, 371–375
   dimensional lateral directional stability derivatives, 373
   application to longitudinal small perturbation equations, 353–358
Lateral directional aerodynamic forces and moments, modeling, 135–267
   aircraft control surfaces, 136
   c_D stability and control derivatives, 190
   c_L stability and control derivatives, 191
   c_l stability and control derivatives, 194
   c_m stability and control derivatives, 192
   c_n stability and control derivatives, 195
   c_Y stability and control derivatives, 193
   F_{AY1} modeling, 137–149, See also individual entry
   L_{A1} modeling, 149–168, See also individual entry
   N_{A1} modeling, 168–177, See also individual entry
   steady-state lateral directional force and moments, 136
Lateral directional dynamic modes
   control forces for
      holding heading and bank angle with an engine-out, 537
      holding heading with asymmetric loading, 537
      rolling maneuvers, 537
   Dutch roll, 377–378
   flying quality requirements for, 532–536
      Dutch Roll Mode, 538
      lateral directional control forces, 536
      rolling mode, 539
      spiral mode, 539
   rolling, 377–378
   spiral, 377–378
   state variable modeling of, 440–445
      control surfaces, 444
Lateral directional equations, solution of, 379–381
Lateral directional parameters, sensitivity analysis for, 398–406

   c_{lp} variations effect on, 403
   $c_{l\beta}$ variations effect on, 399
   c_{nr} variations effect on, 402
   $c_{n\beta}$ variations effect on, 400
   $c_{y\beta}$ variations effect on, 404
   Dutch roll roots in, 401
   geometric parameters for, 398
   I_{XX} variations effect on, 405
   I_{ZZ} variations effect on, 406
Leading-edge flaps, 196
Learjet 24, 361
   trim conditions for, 320
Least squares (LS) algorithms, 59
Library, Aerodynamics library, 508
Lift, 273
Lift Chart, 322–331
Lift distribution, wing section, 42
Linear velocity, 306
Linear/linearized systems, state variables for, 433–435
Longitudinal aerodynamic forces and moments, modeling of, 78–134, See also Cessna CJ3 aircraft
   F_{AX1} modeling, 80–82
   F_{AZ1} modeling, 83–86
   M_{A1} modeling, 87–89
   McDonnell Douglas F4 fighter, 81
Longitudinal dynamics
   flying quality requirements for, 532–536
      for dives, 535
      longitudinal control forces, 533
      for maneuvered flight, 533–534
      for Phugoid mode, 535
      for short period mode, 536
      for steady-state flight, 533
      for take-off and landing, 534–535
   modes, 360–361
      Phugoid, 360–361
      short period, 360–361
   state variable modeling of, 435–440
      control surfaces, 436
      longitudinal input column, 439
      longitudinal input matrix, 439
      longitudinal output control matrix, 440
      longitudinal state matrix, 439
Longitudinal equations, solution of, 361–363
Longitudinal small perturbation
   aerodynamic forces and moments, 91–96
   c_D stability and control derivatives, 97
   c_L stability and control derivatives, 98
   c_m stability and control derivatives, 99
   modeling of (c_{D1}, c_{L1}, c_{m1}), 93
   modeling of (c_{Du}, c_{Lu}, c_{mu}), 93–94
   equations, Laplace transformations application to, 353–358
Longitudinal time-histories for F-104, 362
Look-up tables based aerodynamic coefficients, FDC modeling of, 486–492

Mach number, 81, 83, 113, 115
Maneuvered flight, longitudinal control forces for, 533–534
Maneuvers, 531–532
   categories of, 532
   'Non-Terminal' Maneuvers, 532
   'Terminal' Maneuvers, 532
Mass Properties library, 510–511
MATLAB®, 358–369, 478
Maximum likelihood (ML) method, 59
Mean aerodynamic chord (MAC), 44
Mission profile, 531–532
Modern flight simulation codes, 476–522

MATLAB® sample codes, 476
Multi-input–multi-output (MIMO) systems, 432

N_{A1} modeling, 168–177
 $c_{n\beta}$ conceptual modeling, 169–172
 $c_{n\beta}$ mathematical modeling, 172–174
 $c_{n\delta A}$ modeling, 174–175
 lateral force acting on the vertical tail, 171
 Yawing moment contribution, 170–171
Navier-Stokes equations, 58
Nonlinear dynamics inversion (NLDI) controller, 512
Nonlinear systems, state variables for, 433
Nonstraight wing planform, 43

One-engine lnoperative (OEI), 291
Oswald efficiency factor, 49, 82, 279

Parameter identification (PID) from flight data, 57, 59
 least squares (LS) algorithms, 59
 maximum likelihood (ML) method, 59
Partial fraction expansion (PFE) method, 361, 379, 418
Perturbation aerodynamic forces and moments, 91–96, See also under Small perturbation
Perturbation conditions, aircraft equations of motion at, 19–22
 large perturbation, 20
 medium perturbation, 20
 small perturbation, 20
  equations from steady-state level flight, 22–23
  small perturbation CAMEs, 21
  small perturbation CLMEs, 21
  small perturbation IKEs, 22
Phillips, Warren F., 269
Phugoid, 360–361
 approximation, 363–366
  Phugoid mode damping, requirements for, 535
 Phugoid sensitivity analysis, 394–397
Pilot ratings and aircraft handling qualities, 523–541
 aircraft flight envelope, 524–526
 for steady sideslips, 541
Piston propeller aircraft engines, 269–270
 advantages, 270
 drawbacks, 270
 'in-line' piston engine, 269
 reciprocating engine, 269
 rotary design, 269
 V engines, 269
Polhamus formula, 46, 51, 65, 71, 86, 89, 101–102, 106, 113–114, 143, 197, 202–203, 213, 238
Power at level flight, 274–278
 aircraft drag (parasite+induced) vs. airspeed, 276
 maximum aerodynamic efficiency, 275
 minimum aerodynamic drag, 275–277
 minimum power required, 277–278
 power required vs. airspeed, 278
Power available, determination of, 282–287
 longitudinal aerodynamic and thrust forces, 283
Power required, determination, 279–282
Prandtl relationship, 274
Propulsion library, 508–509
Propulsion systems, 268–273

piston engine (propeller) aircraft engines, 269–270
Ramjet aircraft engines, 273
turbofan aircraft engines, 272
turbojet aircraft engines, 271
turboprop aircraft engines, 270–271
Propulsive forces and moments, FDC modeling of, 496–498
 3D look-up tables for thrust, 497
 jet-engine for different levels of thrust
 fuel flow look-up tables for, 498
 look-up tables for, 497

Ramjet aircraft engines, 273
Rauw, M., 479, 493
Real-time PID techniques, 60
Reciprocating engine, 269
Reference frames and assumptions, 2–3
 Earth-based reference system, 2
 flat-Earth assumption, 2
 inertial enough reference, 4
 inertial frame, 2
Reversible flight control system (RFCS), 527
Reynolds number, 81, 172–173
Rocket propulsion, 293
Roll control effectiveness (RCE), 539
Rolling approximation, 384–384
Roskam, J., 82, 86, 93, 197–198, 293–294
Rotary design engine, 269
Routh–Hurwitz analysis, 408–410
 of lateral directional stability, 376–377
 of longitudinal stability, 358–360
Ruddervators, 196

Sensitivity analysis for aircraft dynamics, 359, 386–406, See also Short period sensitivity analysis
 for lateral directional parameters, 398–406
 stability augmentation systems (SASs), 386
 stability sensitivity analysis (SSA), 386
Short period approximation, 363–366
Short period mode, 360–361, 536
Short period sensitivity analysis, 387–394
 variations of $c_{L\alpha WB}$, 392–393
 variations of I_{YY}, 393–394
 variations of S_H/S, 389–390
 variations of $\bar{x}_{AC_H}$, 391–392
 variations of $\bar{x}_{CG}$, 390–391
Short take-off and landing (STOL) aircraft, 270
Simulink®, 479, 484–485, 512, 516
Single-input–single-output (SISO) systems, 432, 477
Small perturbation lateral directional aerodynamic force and moments, 178–189
 in angular velocities, 178
 β-derivatives modeling, 180
 c_{lp} modeling, 181–183
 c_{lr} modeling, 185–187
 c_{np} modeling, 183–185
 c_{nr} modeling, 187–189
 c_{Yp} modeling, 180
 c_{Yr} modeling, 185
 geometric parameters of vertical tail, 185, 187
 in linear velocities, 178
 rolling damping parameters, 181–182
 wing twist angle, 184, 186
Small perturbation thrust forces and moments, modeling of, 291–294
 lateral-directional, 291

longitudinal, 291
Speedbrakes, 196
Spiral mode, lateral directional dynamic, 377–378
Spoilerons, 196
Spoilers, 196
Stability augmentation systems (SASs), 386
Stability axes, 79
Stability sensitivity analysis (SSA), 386
Stability, aircraft, 305–351
 dynamic stability, 306
 perturbation terms, 306
  aerodynamic angles, 306
  angular velocity, 306
  linear velocity, 306
 starting conditions for analysis of, 306
 static stability, 306–312, See also Static aircraft stability
Standard atmosphere model (SAM), 284, 499
State variable (SV) modeling of aircraft dynamics, 432–475
 augmentation of, 445–447
 actuator dynamics, modeling of, 446–447
 altitude, modeling of, 445
 atmospheric turbulence, modeling of, 447
 engine dynamics, modeling of, 446
 flight path angle (γ), modeling, 446
 of lateral directional dynamics, 440–445
 for linear/linearized systems, 433–435
 of longitudinal dynamics, 435–440
 for nonlinear systems, 433
Static aircraft stability, 306–312
 criteria for, 306–312
Steady-state conditions, aircraft equations of motion at, 18–19
 aircraft equations of motion at, 18–19
 CAMEs at, 18
 CLMEs at, 18
 KEs at, 19
Steady-state flight, longitudinal control forces for, 533
Steady-state lateral directional force and moments, 136
Steady-state level flight, small perturbation equations from, 22–23
Steady-state straight flight
 lateral directional analysis of, 332–340, See also B747 aircraft
 longitudinal analysis of, 313–322
  aircraft weight, 316
  airspeed, 316
  altitude (through the density), 316
  body and stability axes, 313
  propulsion system, 314
  stabilator+elevator horizontal tail configuration, 313
Steady-state thrust forces and moments, 288–291
 differential thrust, 288
 engine toe-in angle, 288
 lateral directional, 290–291
 longitudinal, 290–291
 thrust vectoring, 288
Straight wing planform, 43
 geometric characteristics, 44

Take-off and landing, longitudinal control forces for, 534–535
Taylor expansion, 50

Thrust forces and moments
　modeling, 268–304
　　FDC modeling of, 494
　　lateral directional, 288
　　longitudinal, 288
　　steady-state thrust forces and moments, 288–291
Thrust, 273
　vectoring, 288
Trailing-edge flaps, 196
Transfer functions, aircraft equations of motion solution based on, 352–431
Trim conditions, design for, 305–351
Trim diagram (TD), 322, 324–329
　c_L vs c_m Chart (Trim Diagram), 328–329
　for different classes of aircraft, 329–330
　for thrust axis above/below center of gravity, 329–331
Tuck, 359
Turbofan aircraft engines, 272
Turbojet aircraft engines, 271

Turboprop aircraft engines, 270–271
Two-section nonstraight wing planform, 44

Utilities library, 509–510

V engines, 269
Variable-pitch propellers, 293
Vertical Take Off Landing (VTOL) aircraft, 33
Von Doenhoff, A. E., 38
Von Karman turbulence model, 493, 505

Weathercock effect, 169
Weight, 273
Wind sub library, 504
Wind tunnel analysis, 57–58
Wind tunnel data and empirical "Build-Up" analysis, correlation, 60
Wing lift curve slope, 42–48
Wing planforms, 42–48
　airspeed on, 44
　B747 wing planform, 42

　generic straight wing planform, 43
　nonstraight wing planform, 43
　two-section nonstraight wing planform, 45
Wing sections, 37–42
　aerodynamic characteristics for, 37–42
　aerodynamic forces on, 39
　c_d vs. α curve, 41
　c_l vs. c_d curve, 40
　c_l vs. α curve, 40
　c_m vs. α curve, 41
　3D Wing as sequence of 2D wing sections, 38
　geometric parameters of, 38
　identifying parameters, 38
　lift distribution, 49
　NACA 0012 wing section, 39
　wing-tail distances, 50

Yawing moment contribution, 170